Advances in Intelligent Systems and Computing

Volume 819

Series editor

Janusz Kacprzyk, Polish Academy of Sciences, Warsaw, Poland
e-mail: kacprzyk@ibspan.waw.pl

The series "Advances in Intelligent Systems and Computing" contains publications on theory, applications, and design methods of Intelligent Systems and Intelligent Computing. Virtually all disciplines such as engineering, natural sciences, computer and information science, ICT, economics, business, e-commerce, environment, healthcare, life science are covered. The list of topics spans all the areas of modern intelligent systems and computing such as: computational intelligence, soft computing including neural networks, fuzzy systems, evolutionary computing and the fusion of these paradigms, social intelligence, ambient intelligence, computational neuroscience, artificial life, virtual worlds and society, cognitive science and systems, Perception and Vision, DNA and immune based systems, self-organizing and adaptive systems, e-Learning and teaching, human-centered and human-centric computing, recommender systems, intelligent control, robotics and mechatronics including human-machine teaming, knowledge-based paradigms, learning paradigms, machine ethics, intelligent data analysis, knowledge management, intelligent agents, intelligent decision making and support, intelligent network security, trust management, interactive entertainment, Web intelligence and multimedia.

The publications within "Advances in Intelligent Systems and Computing" are primarily proceedings of important conferences, symposia and congresses. They cover significant recent developments in the field, both of a foundational and applicable character. An important characteristic feature of the series is the short publication time and world-wide distribution. This permits a rapid and broad dissemination of research results.

More information about this series at http://www.springer.com/series/11156

Sebastiano Bagnara · Riccardo Tartaglia
Sara Albolino · Thomas Alexander
Yushi Fujita
Editors

Proceedings of the 20th Congress of the International Ergonomics Association (IEA 2018)

Volume II: Safety and Health, Slips, Trips and Falls

Springer

Editors
Sebastiano Bagnara
University of the Republic of San Marino
San Marino, San Marino

Riccardo Tartaglia
Centre for Clinical Risk Management
    and Patient Safety, Tuscany Region
Florence, Italy

Sara Albolino
Centre for Clinical Risk Management
    and Patient Safety, Tuscany Region
Florence, Italy

Thomas Alexander
Fraunhofer FKIE
Bonn, Nordrhein-Westfalen
Germany

Yushi Fujita
International Ergonomics Association
Tokyo, Japan

ISSN 2194-5357   ISSN 2194-5365 (electronic)
Advances in Intelligent Systems and Computing
ISBN 978-3-319-96088-3   ISBN 978-3-319-96089-0 (eBook)
https://doi.org/10.1007/978-3-319-96089-0

Library of Congress Control Number: 2018950646

This Springer imprint is published by the registered company Springer Nature Switzerland AG
The registered company address is: Gewerbestrasse 11, 6330 Cham, Switzerland

Preface

The Triennial Congress of the International Ergonomics Association is where and when a large community of scientists and practitioners interested in the fields of ergonomics/human factors meet to exchange research results and good practices, discuss them, raise questions about the state and the future of the community, and about the context where the community lives: the planet. The ergonomics/human factors community is concerned not only about its own conditions and perspectives, but also with those of people at large and the place we all live, as Neville Moray (Tatcher et al. 2018) taught us in a memorable address at the IEA Congress in Toronto more than twenty years, in 1994.

The Proceedings of an IEA Congress describes, then, the actual state of the art of the field of ergonomics/human factors and its context every three years.

In Florence, where the XX IEA Congress is taking place, there have been more than sixteen hundred (1643) abstract proposals from eighty countries from all the five continents. The accepted proposal has been about one thousand (1010), roughly, half from Europe and half from the other continents, being Asia the most numerous, followed by South America, North America, Oceania, and Africa. This Proceedings is indeed a very detailed and complete state of the art of human factors/ergonomics research and practice in about every place in the world.

All the accepted contributions are collected in the Congress Proceedings, distributed in ten volumes along with the themes in which ergonomics/human factors field is traditionally articulated and IEA Technical Committees are named:

I. Healthcare Ergonomics (ISBN 978-3-319-96097-5).
II. Safety and Health and Slips, Trips and Falls (ISBN 978-3-319-96088-3).
III. Musculoskeletal Disorders (ISBN 978-3-319-96082-1).
IV. Organizational Design and Management (ODAM), Professional Affairs, Forensic (ISBN 978-3-319-96079-1).
V. Human Simulation and Virtual Environments, Work with Computing Systems (WWCS), Process control (ISBN 978-3-319-96076-0).

VI. Transport Ergonomics and Human Factors (TEHF), Aerospace Human Factors and Ergonomics (ISBN 978-3-319-96073-9).
VII. Ergonomics in Design, Design for All, Activity Theories for Work Analysis and Design, Affective Design (ISBN 978-3-319-96070-8).
VIII. Ergonomics and Human Factors in Manufacturing, Agriculture, Building and Construction, Sustainable Development and Mining (ISBN 978-3-319-96067-8).
IX. Aging, Gender and Work, Anthropometry, Ergonomics for Children and Educational Environments (ISBN 978-3-319-96064-7).
X. Auditory and Vocal Ergonomics, Visual Ergonomics, Psychophysiology in Ergonomics, Ergonomics in Advanced Imaging (ISBN 978-3-319-96058-6).

Altogether, the contributions make apparent the diversities in culture and in the socioeconomic conditions the authors belong to. The notion of well-being, which the reference value for ergonomics/human factors is not monolithic, instead varies along with the cultural and societal differences each contributor share. Diversity is a necessary condition for a fruitful discussion and exchange of experiences, not to say for creativity, which is the "theme" of the congress.

In an era of profound transformation, called either digital (Zisman & Kenney, 2018) or the second machine age (Bnynjolfsson & McAfee, 2014), when the very notions of work, fatigue, and well-being are changing in depth, ergonomics/human factors need to be creative in order to meet the new, ever-encountered challenges. Not every contribution in the ten volumes of the Proceedings explicitly faces the problem: the need for creativity to be able to confront the new challenges. However, even the more traditional, classical papers are influenced by the new conditions.

The reader of whichever volume enters an atmosphere where there are not many well-established certainties, but instead an abundance of doubts and open questions: again, the conditions for creativity and innovative solutions.

We hope that, notwithstanding the titles of the volumes that mimic the IEA Technical Committees, some of them created about half a century ago, the XX Triennial IEA Congress Proceedings may bring readers into an atmosphere where doubts are more common than certainties, challenge to answer ever-heard questions is continuously present, and creative solutions can be often encountered.

Acknowledgment

A heartfelt thanks to Elena Beleffi, in charge of the organization committee. Her technical and scientific contribution to the organization of the conference was crucial to its success.

References

Brynjolfsson E., A, McAfee A. (2014) The second machine age. New York: Norton.

Tatcher A., Waterson P., Todd A., and Moray N. (2018) State of science: Ergonomics and global issues. Ergonomics, 61 (2), 197–213.

Zisman J., Kenney M. (2018) The next phase in digital revolution: Intelligent tools, platforms, growth, employment. Communications of ACM, 61 (2), 54–63.

Sebastiano Bagnara
Chair of the Scientific Committee, XX IEA Triennial World Congress
Riccardo Tartaglia
Chair XX IEA Triennial World Congress
Sara Albolino
Co-chair XX IEA Triennial World Congress

Organization

Organizing Committee

Riccardo Tartaglia (Chair IEA 2018)	Tuscany Region
Sara Albolino (Co-chair IEA 2018)	Tuscany Region
Giulio Arcangeli	University of Florence
Elena Beleffi	Tuscany Region
Tommaso Bellandi	Tuscany Region
Michele Bellani	Humanfactor[x]
Giuliano Benelli	University of Siena
Lina Bonapace	Macadamian Technologies, Canada
Sergio Bovenga	FNOMCeO
Antonio Chialastri	Alitalia
Vasco Giannotti	Fondazione Sicurezza in Sanità
Nicola Mucci	University of Florence
Enrico Occhipinti	University of Milan
Simone Pozzi	Deep Blue
Stavros Prineas	ErrorMed
Francesco Ranzani	Tuscany Region
Alessandra Rinaldi	University of Florence
Isabella Steffan	Design for all
Fabio Strambi	Etui Advisor for Ergonomics
Michela Tanzini	Tuscany Region
Giulio Toccafondi	Tuscany Region
Antonella Toffetti	CRF, Italy
Francesca Tosi	University of Florence
Andrea Vannucci	Agenzia Regionale di Sanità Toscana
Francesco Venneri	Azienda Sanitaria Centro Firenze

Scientific Committee

Sebastiano Bagnara (President of IEA2018 Scientific Committee)	University of San Marino, San Marino
Thomas Alexander (IEA STPC Chair)	Fraunhofer-FKIE, Germany
Walter Amado	Asociación de Ergonomía Argentina (ADEA), Argentina
Massimo Bergamasco	Scuola Superiore Sant'Anna di Pisa, Italy
Nancy Black	Association of Canadian Ergonomics (ACE), Canada
Guy André Boy	Human Systems Integration Working Group (INCOSE), France
Emilio Cadavid Guzmán	Sociedad Colombiana de Ergonomia (SCE), Colombia
Pascale Carayon	University of Wisconsin-Madison, USA
Daniela Colombini	EPM, Italy
Giovanni Costa	Clinica del Lavoro "L. Devoto," University of Milan, Italy
Teresa Cotrim	Associação Portuguesa de Ergonomia (APERGO), University of Lisbon, Portugal
Marco Depolo	University of Bologna, Italy
Takeshi Ebara	Japan Ergonomics Society (JES)/Nagoya City University Graduate School of Medical Sciences, Japan
Pierre Falzon	CNAM, France
Daniel Gopher	Israel Institute of Technology, Israel
Paulina Hernandez	ULAERGO, Chile/Sud America
Sue Hignett	Loughborough University, Design School, UK
Erik Hollnagel	University of Southern Denmark and Chief Consultant at the Centre for Quality Improvement, Denmark
Sergio Iavicoli	INAIL, Italy
Chiu-Siang Joe Lin	Ergonomics Society of Taiwan (EST), Taiwan
Waldemar Karwowski	University of Central Florida, USA
Peter Lachman	CEO ISQUA, UK
Javier Llaneza Álvarez	Asociación Española de Ergonomia (AEE), Spain
Francisco Octavio Lopez Millán	Sociedad de Ergonomistas de México, Mexico

Donald Norman	University of California, USA
José Orlando Gomes	Federal University of Rio de Janeiro, Brazil
Oronzo Parlangeli	University of Siena, Italy
Janusz Pokorski	Jagiellonian University, Cracovia, Poland
Gustavo Adolfo Rosal Lopez	Asociación Española de Ergonomia (AEE), Spain
John Rosecrance	State University of Colorado, USA
Davide Scotti	SAIPEM, Italy
Stefania Spada	EurErg, FCA, Italy
Helmut Strasser	University of Siegen, Germany
Gyula Szabò	Hungarian Ergonomics Society (MET), Hungary
Andrew Thatcher	University of Witwatersrand, South Africa
Andrew Todd	ERGO Africa, Rhodes University, South Africa
Francesca Tosi	Ergonomics Society of Italy (SIE); University of Florence, Italy
Charles Vincent	University of Oxford, UK
Aleksandar Zunjic	Ergonomics Society of Serbia (ESS), Serbia

Contents

Safety and Health

Learning Role-Playing Game Scenario Design for Crisis Management Training: From Pedagogical Targets to Action Incentives

Pierrick Duhamel[⊠], Sylvain Brohez, Christian Delvosalle,
Agnès Van Daele, and Sylvie Vandestrate

University of Mons, Place du Parc, 20, 7000 Mons, Belgium
pierrick.duhamel@umons.ac.be

Abstract. Emergency and crisis management requires, from operatives and decision-makers, specific knowledge that cannot be acquired through theoretical course or real-life practice only [1]. Besides, developing practical exercises adapted for agents and their needs is even more difficult when the system where they operate is complex [2]. It is therefore necessary to develop such exercises according to both rigorous and flexible methodology.

Since 2015, the Expert'Crise project, funded by the European Social Fund, has organized seven exercises mainly in hazardous chemical companies. During such exercises, trainees play their own role in their usual working place. Hence, arrangements must be made [3] to isolate trainees from real environment and establish exercise diegesis [4]. Through a trial and error experience, we developed a design methodology for crisis management Learning Role-Playing Game [5] scenario.

Scenario design starts from trainees' statement of requirements leading to pedagogical targets, chosen from an existing classification [6]. Then, because emergency sequences are often similar, we developed a framework for our scenario based on the narrative storyline of Campbell [7], describing the steps of an emergency sequence. Nevertheless, because pedagogical targets change depending on exercise, this storyline varies and includes dedicated "situation-tasks" that target competences previously identified. These situation-tasks aim to "force" trainees to do actions under special circumstances through serious game interface [8], and using its gameplay and diegesis. However, "situations" may not lead obviously to a "task", and incentives must be introduced to help trainees performing the task, unlike disturbances [9] that can also be added.

Keywords: Education and training · Safety and health · Crisis management
Seveso

1 Introduction

1.1 Background

Despite the progress of major accident prevention since the 1970's and accident such as those of Flixborough or Seveso, there are still industrial disasters. Proof, if needed, that prevention alone is not enough and safety needs emergency management when a

S. Bagnara et al. (Eds.): IEA 2018, AISC 819, pp. 3–12, 2019.
https://doi.org/10.1007/978-3-319-96089-0_1

disaster occurs. Yet emergency and crisis management requires specific competences that cannot be acquired through theoretical course or real-life practice only [10]. They must therefore be trained through practical exercises adapted to their environment, working habits, education, and professional needs. Moreover, European regulation known as "SEVESO 3 Directive" requires hazardous companies to test their emergency planning every three years, for instance with exercises.

Nevertheless, developing such exercise is a complex task that requires dedicated skills [11]. This kind of training already exists but is mainly oriented to public organization [12] or outsourced to consulting firms. Therefore, because organizing exercise in industrial plant is complicated, expensive, and comes with uncertain outcomes, companies may be reluctant to organize such exercises [13]. To test the emergency management with exercises, SHE managers need a methodology to easily set up internal exercises matching with their goals.

Launched in 2015 and funded by the European Social Fund, the Expert'Crise project develops emergency and crisis management training for hazardous industries and critical infrastructures. These training are based on a pedagogy combining theoretical courses and immersive practical exercises. Through a trial and error experience, and the literature from different fields such as pedagogy, dramaturgy, game design, and crisis management, a design methodology for crisis and emergency management Learning Role-Playing Game [5] scenario was developed.

1.2   Context of the Development of the Design Methodology

Eight walloon SEVESO companies have participated to one of the six exercises organized between 2015 and 2017. Eight other exercises are currently under preparation for 2018. Companies, their environment and the type of exercises [8] proposed vary from an exercise to another and cover a wide scope of situation.

Exercises were held on industrial sites and used rooms, tools and communication devices available to operatives and decision-makers both in their work-life and during emergency. This configuration allows immersive situation for trainees without destabilizing them [14]. In addition, it provides an emergency system test for companies in accordance to European regulations [15]. Exercises are mainly functional [8], focused on decision-maker functions [16]. The operational part is often simulated through the control of information flow entering the crisis room. Therefore, these trainings target members of the crisis management team and key persons in the warning chain.

Exercises rely on material arrangement and human organization. Human organization refers to facilitators regulating the exercise [9] near trainees (or at distance) and observers consigning what happened during the event. Experimental device is composed of cameras, microphones, projectors and speaker that allow immersing trainee in the diegesis and, on another hand, capturing multimedia flows that are lived-streamed to distant facilitators to help them adjusting the scenario and saved to complete observers' notes for later analysis.

The framework of the training offered remained the same throughout the project: one to six theoretical course modules, briefing of some or all the trainees, simulation, "hot" debriefing right after the simulation, analyse of the simulation [17] and "cold" debriefing. Nevertheless, inside this framework, each step have evolved to take previous mistakes, bias, limits and suggestions of improvement into account.

2 Methodology: From Pedagogical Target to Action Incentive

As Like games, Learning Role-Playing Games need rules, medium, context and players [18]. Rules and medium are defined by the category of game but context and players vary depending on the company. Therefore, before considering developing a scenario for the exercise, these two elements must be properly defined with the company and will allow to build the proper diegesis – i.e. what is true in the context of the simulation – for the exercise. Then the methodology is split in two main parts. The first one analyzes needs and wishes, and emergency system of the company to propose a pedagogical and organizational framework for the exercise. The second part designs exercise content inside the previously defined framework.

These two parts are not isolated from each other, and the development may step from one to the other depending on the progress of the pedagogical engineering and the reactivity of industrial contact person. In addition, the methodology is iterative so some steps are looped several times (Fig. 1).

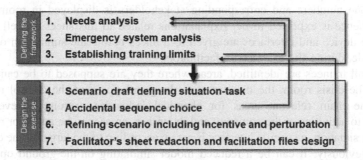

Defining the framework
1. Needs analysis
2. Emergency system analysis
3. Establishing training limits

Design the exercise
4. Scenario draft defining situation-task
5. Accidental sequence choice
6. Refining scenario including incentive and perturbation
7. Facilitator's sheet redaction and facilitation files design

Fig. 1. Structure of the emergency exercise design methodology

2.1 The Framework: Defining Needs to Specify the Scope of LRPG

After the company agreed to participate to the training, a meeting is scheduled to deal with needs and expectations of the firm. This **first** meeting aims to have a first scope of the training company expectations including target audience identification, resources that can be used and operational (e.g. evacuation or intervention exercise, warning chain test...) and/or global (e.g. improving reactivity, internal or external communication...) objectives. These objectives formulated by the company must be interpreted to fit emergency and crisis exercises pedagogical targets identified in literature [6].

The **second** step is about processing emergency plan and organization documents into a mental picture of how the emergency system works. It appears emergency plans from European Seveso companies are similar [3] with separated operational on site management and strategic management in a crisis room. This common structure is convenient and allows building mental picture from an existing framework. A second meeting with the contact person is then usually needed to clarify or verify some points. This meeting may be held with other persons of the companies to have a better description of their role and seize the difference between the prescriptive plan and what they would do during an emergency according to their experience, their working procedures and habits. Therefore, it is possible to have a representative picture of how the emergency system could be expected to work during an emergency.

Once the company's needs and its emergency system is understood, the **third** step is about establishing borders of the sub-system tested during the exercise and figuring out how this sub-system will evolve during the simulation and especially how it will interact with the defined borders. This step lead to define, considering the target audience, a second category of audience: the peripheral audience. Those persons are not directly aimed by the training but they play an important role in emergency plan and are direct interlocutor to target audience so they have to participate, if possible. Nevertheless, because they are not directly targeted, facilitators mentor them by telling them how to interact with the target audience. Therefore, they can be seen as an input/output interface for facilitation. In the same way, indirect interlocutor such as medias, political stakeholders, administrative authorities and emergency services are identified according to company's needs and facilitators simulate them during the exercise through phone, mail or other means [9]. To this end, a sheet showing exercise's phone numbers and corresponding stakeholders is displayed in rooms where target audience is expected to be. Explanations related to this sheet as well as other immersive device and interfaces are given to trainees before the simulation or, if it is not possible, during the exercise by facilitators.

Once all trainees are identified, areas where they are supposed to be can be considered. The crisis room, the disaster area, the guard post and the control room are usually the main relevant areas for chemical industries exercises. Nevertheless, according to activities of the company and its wishes, areas may be added or removed. In this last situation, a specific interface is set up according to the target of the exercises defined previously. It can be a reduced model simulating on-the-ground operations, leaded interviews with intervention leader or sub-crisis unit managed by a facilitator. These simulations aims to give to the target audience an immersive experience including the correct information flow and realistic interactions with stakeholders as represented in Fig. 2.

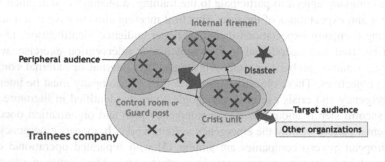

Fig. 2. System borders audiences and others organizations simulation

An analogy can be made with theatrical scenography. Indeed, this step consist in building a scenery where trainees will evolve as actors and, the same way scenography use cardboard environment to improve immersion, it may be relevant to use trick such as smoke-producing device, alert horn or other sound and light device to simulate events and strengthen realistic feeling of trainees, actors but also spectators of the simulation. Nevertheless, these tricks do not have to be fully realistic with a homogenous diegesis and can only be representative or symbolic with a heterogeneous diegesis [4], the same way an accessory – such a hat or glasses – is enough to understand a same actor plays different character.

The last border, which should be defined, is the duration of the exercise. They usually last between one and three hours for logistic reasons but longer exercise could be planned (Fig. 3).

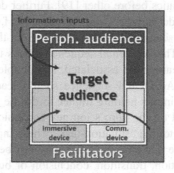

Fig. 3. Audience, simulation device and information flow

Therefore, once interfaces between trainees, environment and facilitators are defined, context and players are set and the LRPG is operable. Therefore, an exercise draft can be submitted and the scenario development initiated.

2.2 The Scenario: Designing an Interactive Story to Achieved Pedagogical Target

Once the framework is defined, scenario can be developed, it is the **fourth** step. A scenario explains the diegesis of the exercise and develops the sequence of input to the target audience, which should lead it to pedagogical targets [19]. Based on targets expressed by the company, knowledge and competences [20] are selected from an existing list [6]. At this point, situations staging competences are considered. The relevance and the ease to stage of each situations are then evaluated to choose the more adapted situation for the exercise. This first association between knowledge and situation are proto-"situation-task" [21]. A situation-task are the central part of these exercises and aims to "force" trainees to do an action (the task) under special circumstances (the situation) through serious game interface and using its specific gameplay. The task is a mean to involve trainees in a reasoning process harnessing knowledge targeted [22]. Once all knowledge, skill, and competences that will be aimed by the exercise are identified, they are grouped into pedagogic bloc associated with a situation, eventually including first information input ideas. These pedagogical blocs are integrated in a framework of the evolution of a crisis in industrial environment inspirited by the hero's journey [7] in the same way as in game design or scriptwriting methodologies [23]. The hero's journey is organized around the transition from the common world to the unnatural/uncommon world where the protagonist of the story becomes a hero. A parallel can be made with the transition from normal to crisis mode. Therefore, the different step of the hero's trip can be adapted to fit with crisis situation process as seen in Table 2.

Pedagogical blocs with prototype idea of scripting will be integrated in this framework and, based on the length of the exercise, will give a first view of how the exercise will process. The arrangement of each blocs is important because it will influence the dynamics and the stress of the exercise in addition to encourage the resolution of some problematics before other [19]. Further developments of "situation-task" is not possible before defining the "plot" of exercise, i.e. the accidental sequence.

The **fifth** step consist to select, with the contact person, the most adapted accidental sequence for the exercise. The accidental sequence means the causes of the accident, the accident, dangerous phenomenon associated, and people, environment, equipment or structure affected [24]. Accidental sequence should lead to "situation-tasks" and should be justifications for inputs helping trainee to do the wanted task.

At the **sixth** step, based on the hero's journey chronology, exercise is divided in 15 min sequences assigned to one "situation-task" where input leading to the task will be added as shown in Table 1. Some sequence may not be related with a "situation-task" especially in introduction, transition, conclusion or build-up period.

Table 1. Block chronology of the exercise

Time block	Situation-task	Pedagogical target	Incentive/Perturbation

For each time sequences, situation-tasks and pedagogical targets related are reported. In this way, the purpose of each sequence and reasons why specific task were chosen is not forgot. Then inputs leading trainees to do the task are chosen. Nevertheless, the choice of relevant and efficient inputs is not easy. Indeed, they must lead to the task to do but in a realistic and non-obvious way to keep trainees focused and in a "flow" state [26]. It is the reason why proto-situation task should integrate staging ideas in the very first step of scriptwriting. Each situation-task stages the resolution of a problem by participants and can lead to different solutions as shown in Fig. 4. The situation is usually implied by dangerous phenomenon or its consequences such as a wounded person or damaged equipment, and the trainee's reasoning process can be helped by incentive (e.g. municipal authority asking a press statement) but can also be slowed down by perturbation [21] keeping the simulation challenging for trainees.

Once inputs are chosen, the way they will be injected into the system is defined. It could be trough a call, a direct interaction with someone from the peripheral audience or the use of an immersing device such as a sound speaker playing firefighters' siren. Direct interactions between target audience and facilitation are carefully managed

Table 2. Hero's journey step and crisis management process comparison

Hero's journey step		Crisis management process [25]	
Departure	The call to adventure	First step of the warning chain	Warning
	Refusal of the call	Disbelief or minimization of the crisis. Fear to leave the normal operation mode	
	Supernatural aid	There is no supernatural aid in crisis process. Nevertheless, during exercises, facilitation can play this role	
	The crossing of the first threshold	Awareness of the gravity of the situation. Emergency plan "engagement" and reflex procedures	
Initiation	The belly of the whale	Information and action flooding. Difficulty to picture correctly the situation	Crisis management
	The road of trials	First decision-making and awareness of operational difficulties	
	The meeting with the goddess	Meeting with emergency services and information exchange	
	Woman as a temptress	Temptation to not act anymore, letting all actions to emergency services	
	Atonement with the father	Communication with authorities, medias and higher hierarchic level	
	Apotheosis	Expectation of change, improvement or the end of crisis. Domino effect if any	
Return	Refusal of the return	Expectation of recurrence, domino effect or unexpected consequences. Stay in alert	Back to the normal
	The magic flight	Last communications to authorities and media with, eventually press conference. Checklist verification	
	Rescue from without	Other stakeholders close their crisis units and emergency services leaves plant	
	The crossing of the return threshold	Report and debriefing	
	Master of the two worlds	Crisis unit closing	
	Freedom to live	End of the sequence	

because facilitator may influence target audience and decrease relevancy of observation. Nevertheless, in some situation, it is not possible to fully separate facilitation and target audience. In that case, interactions are limited – with only some fact presentation for instance – and are taken into account into further analysis.

In the end of this step, a precise timing with a 5 min meshing is established and inputs, recall, expected trainee's reaction and facilitators recommended reaction are

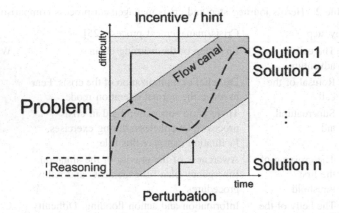

Fig. 4. Flow state diagram of a situation-task reasoning process

specified in a table as shown in Table 3. There is no need to over saturate trainees with inputs if it is not the purpose wanted. Indeed, communication between participants dispatched inside the organization will be important independently of the number of input. This precise table is then used to write the global script, which can be exposed to company's contact person for validation.

Table 3. Precise chronology and input timeline

Time block	Situation-task	Pedagogical target	Incentive/Perturbation	Facilitator sheet	Precise timing
14h00–14h15	Warning chain Crisis unit establishment	Procedures check	Imprecise alert Weak signals No feedback from operators	Intervention chief	14h00: Small event 14h00/5: Information check not successful 14h15: Warning confirmation. Major accident

Once the global script written, come the **seventh** and last step of the design methodology. Each input is integrated into the corresponding facilitator's sheet with details related to the medium and the context of injection. These sheets can be split in two groups: those dedicated to mentoring peripheral public and give it a consistent – but partial – view of what happen and how they are supposed to react, and those dedicated to distant facilitation, simulating different stakeholder and controlling immersing devices. Facilitation sheets can also include a question/answer part to help facilitator to answer to possible questions from trainees. Nevertheless, these sheets cannot anticipate all questions and reaction of trainees and facilitators should adapt the scenario to trainees' reactions.

3 Conclusion

This methodology allows to design efficiently emergency and crisis exercise for chemical industries. Nevertheless, it still needs a high level of competences to choose and arrange situations-tasks and therefore cannot be used by SHE manager. The next step of development of the methodology will consist in identifying a limited number of situation-task that can be staged in exercise to offer an user-friendly method for SHE.

References

1. Lagadec P (1991) La gestion de crise - Outils de réflexion à l'usage des décideurs. MacGraw Hill, Paris
2. Saad Noori S, Comes T, Schwarz P, Wang Y (2017) Behind the scenes of scenario-based training: understanding scenario design and requirements in high-risk and uncertain environments. In: Proceedings of the 14th ISCRAM conference, Albi, France, pp 976–987
3. Duhamel P, Brohez S, Delvosalle C, Dubois L-A, Van Daele A, Vandestrate S (2017) Le projet Expert'Crise ou la formation à la gestion de crise en milieu industriel par des exercices de mise en situation: premiers résultats. In: Récents Progrès en Génie des Procédés, vol 110
4. Morten G (2003) Interaction: the key element of larp. In: Morten G, Thorup L, Sander M (eds) When larp grows up. Projektgruppen KP03, Frederiksberg, pp 66–71
5. Mariais C, Michau F, Pernin JP (2012) A description grid to support the design of learning role-play games. Simul Gaming 43(1):23–33
6. Lapierre D (2016) Methode EVADE: Une approche intégré pour l'Evaluation et l'Aide au DEbriefing. PhD thesis, Université de Nîmes, Nîmes, France
7. Campbel J (1949) The hero with a thousand faces. New World Library, Novato
8. Tena-Chollet F, Tixier J, Dandrieux A, Slangen P (2016) Training decision-makers: existing strategies for natural and technological crisis management and specifications of an improved simulation-based tool. Saf Sci 97:144–153
9. Fréalle N, Tena-Chollet F (2017) The key role of animation in the execution of crisis management exercise. In: Proceedings of the 14th ISCRAM conference, Albi, France, pp 944–956
10. Lagadec P (2001) Les exercices de crise: pour des ruptures créatrices. In: La Lettre des Cindyniques, no 34, pp 5–9
11. Limousin P (2017) Contribution à la scénarisation pédagogique d'exercices de crise. PhD thesis, Ecole Nationale Supérieure des Mines de Saint-Etienne, Saint-Etienne, France
12. Alberta Emergency Management Agency (2012) Exercise design 100. http://www.aema. alberta.ca/documents/exercisedesign100sel63227.pdf. Accessed 09 Apr 2018
13. European Commission's Joint Research Centre and the Dutch Ministry of Social Affairs and Employment (2008) Enforcement of Seveso II: an analysis of compliance drivers and barriers in five industrial sectors: key points and conclusions
14. Cook D (2015) Flight Operations. https://www.tc.gc.ca/eng/civilaviation/publications/tp185-4-07-operations-4034.htm#instructor Accessed 09 Apr 2018

15. European Parliament and Council (2012) Directive 2012/18/EU of the European Parliament and of the Council of 4 July 2012 on the control of major-accident hazards involving dangerous substances, amending and subsequently repealing Council Directive 96/82/EC

16. Lagadec P (1995) Cellules de crise - Les conditions d'une conduite efficace. Les Editions d'Organisation, Paris

17. Vandestrate S, Dubois LA, Van Daele A (2018) Crisis management and simulation training: using activity chronicles to analyse crisis managers' behaviours. Communication presented at the 20th congress of the international ergonomics association, Florence

18. Klabbers JHG (2003) The gaming landscape: a taxonomy for classifying games and simulations. In: Proceedings of the 2003 DiGRA international conference: level up, vol 2, pp 54–68

19. Limousin P, Chapurlat V, Tixier J, Sauvagnargues S (2016) A new method and tools to scenario design for crisis management exercise. Chem Eng Trans 53:319–324

20. Winterton J, Delamare-Le Diest F, Stringfellow E (2005) Typology of knowledge, skills and competences. Cedefop, Luxembourg

21. Tena-Chollet F (2012) Elaboration d'un environnement semi-virtuel de formation à la gestion stratégique de crise basé sur la simulation multi-agents. PhD thesis, Ecole Nationale Supérieure des Mines de Saint-Etienne, France

22. Pastré P, Mayen P, Vergnaud G (2006) La didactique professionnelle. Revue française de pédagogie 154:145–198

23. Schell J (2008) The art of game design. Elsevier, Burlington

24. Debray B, Salvi O (2005) ARAMIS project: an integrated risk assessment methodology that answers the needs of various stakeholders. WIT Trans Built Environ 82:265–275

25. Dautun C (2007) Contribution à l'étude des crises de grande ampleur: connaissance et aide à ladécision pour la sécurité civile. Sciences de l'environnement. PhD thesis, Ecole Nationale Supérieure des Mines de Saint-Etienne, Saint-Etienne, Franc

26. Schmidt JA (2010) Flow in education. In: Peterson P, Baker E, McGaw B (eds) International Encyclopedia of Education, 3ième ed, Elsevier Ltd, pp 605–611

Mathematical Approach to Estimate the Peak Expiratory Flow Rate of Male Bakers in Abeokuta, Nigeria

Adekunle Ibrahim Musa[✉]

Department of Mechanical Engineering, Moshood Abiola Polytechnic,
P.M.B 2210, Abeokuta, Nigeria
kunlemusa@yahoo.com, musa.adekunle@mapoly.edu.ng

Abstract. The study presented the mathematical approach to determine the Peak expiratory flow rate of male bakers in Abeokuta, Ogun State, Nigeria with the relationship of the peak expiratory flow rate and the anthropometrical parameters. A total of One hundred and Eighty (180) individuals were investigated with ninety (90) bakers (study group) who are exposed to flour dust and ninety (90) control subjects. The entire subject both study and control group are male. Peak expiratory flow rate (PEFR) and anthropometrical parameters were measured using mini-Wright peak flow meter (PFM 20, OMRON) and Detecto PD300MDHR (Cardinal Scale manufacturing company USA) column scale respectively. PEFR measured were compared using T-test and regression analysis. A mathematical model was developed to determine the peak expiratory flow rate (PEFR) with four factors of body mass, height, age and year of exposure where applicable. The study showed that PEFR in bakers was 182.67 ± 16.34 L/min as against 287.67 ± 17.03 L/min for control group from the regression analysis. Similarly, the model revealed that baker has 182.69 L/min and 285.77 L/min for control group. The Study concluded that using the developed model will serve as a great importance to workers to determine the level of their health and subsequently prevent untimely death.

Keywords: Bakery · Flour · Dust · Workers · Peak expiratory flow rate Exposure · Asthma

1 Introduction

The degree of obstruction of airways in the lung needed to be checked as bakers have being adjudged to have highest incidence rate of occupational heart diseases (Ige and Awoyemi 2002). Different researchers have performed extensive work on the determination of peak expiratory flow rate of individuals to actually prevent the unwanted death among workers (Musa 2015; Musa et al. 2016). Baatjies et al. (2010) research established the physiological status of bakery workers like allergic conditions, respiratory problems due to the daily exposure to flour dust.

The respiratory effects of exposure to flour dust are influenced by the dose and duration of exposure (Meo 2006) and these differ from one working environment to other. Rafnsson et al. (1997) presented the peak expiratory flow rate of a healthy adult

© Springer Nature Switzerland AG 2019
S. Bagnara et al. (Eds.): IEA 2018, AISC 819, pp. 13–19, 2019.
https://doi.org/10.1007/978-3-319-96089-0_2

and non-exposed to dust between 300–600 L/min with variation of age, body weight, height and gender. Musa et al. (2016a) investigated the peak expiratory flow rate (PEFR) of female bakers with about One hundred and twenty participant. The results showed that female bakers have 158.17 ± 12.55 L/min PEFR but did not consider male bakers as participant in the research. Elebute and Femi Pearse (1971) established values of PEFR in Nigeria with 142 healthy adults participated. The study showed that the mean value of male PEFR was 582 ± 88.3 L/min and 385 ± 65.7 L/min for female respectively.

Musa et al. (2016b) also investigated the relationship between the PEFR and the anthropometric parameters to determine the model equation for the determination of female bakers PEFR. The result showed that a female baker has 158.07 L/min PEFR as against 267.96 L/min for the control group. Hence, the aim of this research is to develop a statistical model to predict the peak expiratory flow rate of male bakers and compare it with non-bakers.

2   Materials and Methods

The study was conducted in Abeokuta with One hundred and Eighty (180) participants. Peak expiratory flow rate and anthropometric parameters of the participant were measured and recorded with ninety (90) adult male bakers and ninety (90) healthy male adult selected as the control group to the bakers. The age of the participated bakers range between 21–28 years and they have involved in bakery business between 5 month and 8 years. The control group were of the same age bracket with the bakers. Both bakers and control group were assumed not to had no earlier reported systematic disease. Similarly, individual with smoking habit or suffering from any respiratory illness were exempted from the study.

The PEFR was measured with mini-Wright peak flow meter (PFM 20, OMRON) (Fig. 1). Three readings were taken from each subject in standing position and the best of the three were considered as Peak Expiratory Flow meter reading for that subject. Detecto PD300MDHR (Cardinal Scale manufacturing company, USA) column scale with digital height rod (Fig. 2) was used to measure body mass (kg) and height (cm) of the subjects simultaneously.

Questionnaire was also administered to the participant. The data collected from the questionnaire includes the detailed demographic data such as age, marital status, education level, smoking habit, duration of flour dust exposure and working experience. Figure 3 below showed the subjects performing measurement of PEFR.

Fig. 1. Mini-Wright peak flow meter

Fig. 2. Detecto (PD300DHR column scale with digital height rod)

Reference to the data collected, the relationship between the PEFR and anthropometric parameters existed and this was used to design the multi-linear regression model where PEFR remain the dependant of the anthropometric parameters.

The model follows the trends in Eqs. (1) and (2) respectively (Musa et al. 2016b).

$$Y = a + b_1 x_1 + b_2 x_2 + b_3 x_3 + \ldots + b_n x_n \tag{1}$$

$$\text{PEFR} = a + b_1(\text{Body mass}) + b_2(\text{Height}) + b_3(\text{Age}) + b_4(\text{years of exposure}) \tag{2}$$

Fig. 3. Subjects (bakers) performing measurement of PEFR

Where, a is the constant and b is the coefficient of regression. Each coefficient b represent the effect of the independent variable y. b_4 is only applicable to bakers only.

3   Results and Discussion

Table 1 below showed the descriptive statistic and the T-test (one sample test) analysis of the investigated participants.

The results obtained were expressed using mean ± standard deviation, T-test for the two groups' comparisons and regression analysis.

Table 1. Descriptive statistics/T-test (one sample test) for male subjects

	Male (study)						Male (control)				
	N	Mean	S. E mean	Std. dev	t	df	Mean	S. E mean	Std. dev	t	df
Body mass (kg)	90	61.508	0.4201	3.9851	146.422	89	65.811	0.4837	4.5885	136.065	89
Height (cm)	90	170.184	0.3953	3.7501	430.523	89	170.904	0.4538	4.3050	376.618	89
PEFR (L/min)	90	182.667	1.7223	16.3391	106.060	89	287.667	1.7951	17.0294	160.255	89
Age (yrs)	90	28.04	0.260	2.467	107.829	89	27.97	0.327	3.107	85.405	89
Yrs of expo	90	2.50	0.079	0.753	31.485	89					

Tables 2 and 3 below showed the result of multi-linear regression analysis between the peak expiratory flow rate and the anthropometric parameters from the analysis, a model to predict the peak expiratory (PEFR) of the subject was derived.

Table 2. Regression analysis between PEFR and other parameter for male bakers

Model		Unstandardized coefficients		Standardized coefficients	t	Sig.
		B	Std. error	Beta		
1	(Constant)	277.033	71.637		3.867	0.000
	Body mass	−0.480	0.477	−0.117	−1.007	0.317
	Height	−0.292	0.497	−0.067	−0.587	0.559
	Age	0.449	0.629	0.068	0.714	0.477
	Yr of expo	−11.122	2.002	−0.513	−5.555	0.000

a. Dependent Variable: PEFR

From Table 2 and the trend in Eq. (2), the model for the determination of PEFR for Male baker can be deduced and written as in Eq. (3).

Table 3. Regression analysis between PEFR and other parameter for male (control)

Model		Unstandardized coefficients		Standardized coefficients	T	Sig.
		B	Std. error	Beta		
1	(Constant)	−21.985	61.321		−0.359	0.721
	Body mass	2.035	0.359	0.548	5.669	0.000
	Height	0.953	0.347	0.241	2.748	0.007
	Age	0.461	0.525	0.084	0.877	0.383

a. Dependent Variable: PEFR

$$\text{PEFR}_{\text{Male baker}} = 277.03 - 0.48(\text{Body mass}) - 0.29(\text{height}) + 0.449(\text{Age})$$
$$- 11.12(\text{yr of exposure}) \tag{3}$$

Similarly, from Table 3 and the trend in Eq. (2), the model for the determination of PEFR for the men not exposed to dust (control study) can be deduced and written as in Eq. (4).

From Table 3 above the model for the determination of PEFR for non-dust exposed Male can be deduced as,

$$\text{PEFR}_{\text{Male baker}} = -21.99 + 2.04(\text{Body mass}) + 0.95(\text{height}) + 0.46(\text{Age}) \tag{4}$$

Equation 3 and 4 showed the regression equations where PEFR remain the dependant variable to determine the mathematical model. The applied anthropometrical parameters such as body mass, height, age and year of exposure played a major role in the determination of this model and they were considered as independent variables.

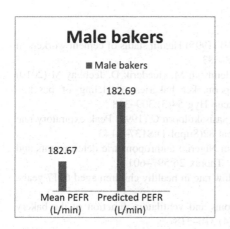

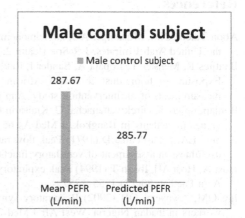

Fig. 4. Comparison of mean and predicted PEFR (L/min)

Host et al. (1994) and Verma et al. (2000) showed that an accuracy of predicted value of PEFR will be higher when age is considered along with height. Figure 4 showed the direct and the calculated measurement of PEFR for comparison. The Fig. 4 showed that the mean and predicted PEFR were closely similar (±0.02 L/min). This little difference may be due to the instrument variations and the physical characteristics of the studied individual. PEFR (L/min) predicted were based on the multi-linear regression equation and this was found very consistent when compared with other study (Benjaponpitak et al. 1999).

Musa et al. (2016b) determine the female bakers and female flour non-exposed individual model. This model was in variance with the present study but similar trend was adopted. Musa et al. (2016b) showed that the predicted PEFR for females both bakers and control study were 158.07 L/min and 267.96 L/min respectively. But the present study which were male dominant showed 182.69 L/min and 285.77 L/min respectively for bakers and the control. This result was compared with Abou Taleb et al. (1995), Rafnsson et al. (1997) and Vestbo et al. (1991) and this revealed that the bakers were not healthy and this might have affected the lung obstruction.

4   Conclusion

The results obtained in the study can be used as a standard PEFR for any male individual who are exposed or not exposed to the flour dust. These equations can also play a major role in determining the PEFR of any male individual in the absence of the hand-held devices for the measurement.

Further study is also required to develop a model that will be devoid of gender either male or female. This will allow for wide range of determining the PEFR of individual who are exposed or non-exposed to flour dust.

References

Abou Taleb ANM, Musaniger AO, Abdelmoneim RB (1995) Health status of cement workers in the United Arab Emirates. J R Soc Health 2:378–383

Baatjies R, Meijster T, Lopata A, Sander I, Raulf-Heimsoth M, Heederik D, Jeebhay M (2010) Exposure to flour dust in South African supermarket bakeries: modeling of baseline measurements of an intervention study. Ann Occup Hyg 54(3):309–318

Benjaponpitak S, Direkwattanachai C, Kraisarin C, Sasisakulporn C (1999) Peak expiratory rate values of students in Bangkok. J Med Assoc Thai 82(Suppl 1):S137–S143

Elebute EA, Femi-Pearse D (1971) Peak flow rate in Nigeria: anthropometric determinants and usefulness in assessment of ventilatory function. Thorax 26:597–601

Host A, Host AH, Ibsen T (1994) Peak expiratory flow rate in healthy children aged 6–17 years. Acta Paediatr 83(12):1255–1257

Ige OM, Awoyemi OB (2002) Respiratory symptoms and ventilatory function of the bakery workers in Ibadan Nigeria. West Afr J Med 21(4):316–318

Meo SA (2006) Dose responses of years of exposure on lung function in flour mill workers. J Occup Health 46:187–191

Musa AI (2015) Peak expiratory flow rate model for female bakers in Abeokuta, Ogun State, Southwest Nigeria. In: 2nd international conference on applied sciences and technology (ICAST 2015), 28–30 October 2015 at Technical University (formerly Kumasi Polytechnic), Kumasi, Ghana. www.Icast.kpoly.edu.gh

Musa AI, Ishola AA, Adeyemi HO (2016a) Investigation of peak expiratory flow rate of female bakers in Abeokuta, Ogun State, Nigeria. J Sci Eng Res 3(3):67–72 http://www.jsaer.com/

Musa AI, Adeyemi HO, Odunlami SA (2016b) Modelling the peak expiratory flow rate for female bakers in Abeokuta, Nigeria. FUTA J Res Sci 12(1):46–54 http://www.fjrs.futa.edu.ng/

Rafnsson V, Gunnarsdottir H, Kiilunen M (1997) Risk of lung cancer among masons in Iceland. Occup Environ Med 54:184–188

Verma SS, Sharma YK, Arora S, Bandopadhyay P, Selvamurthy W (2000) Indirect assessment of peak expiratory flow rate in healthy Indian children. J Trop Paediatr 46:54–55

Vestbo J, Knudsen KM, Raffn E, Korsgaard B, Rasmussen FV (1991) Exposure to cement dust at a Portland cement factory and the risk of cancer. Br J Ind Med 48:803–807

Combined Effect of Effort-Reward Imbalance and Sleep Quality on Depressive Symptoms Risk in Train Drives in China

Shanfa Yu[(✉)], Wenhui Zhou, Guizhen Gu, and Hui Wu

Henan Provincial Institute of Occupational Health, Zhengzhou 450052, China
yu-shanfa@163.com

Abstract. Depressive symptoms is a complex disease caused by the interaction of individual characteristics, environment, lifestyle, environment and genetic factors. In recent two decades, many studies found that workers with high ERI, poor sleep quality were both associated with a higher risk of depressive symptoms. However there is still limited understanding regarding the interrelationships among interrelationships among ERI, poor sleep quality, and depressive symptoms. To this end, we designed combined effect model of ERI and sleep quality to analyze the two variables on depressive symptoms risk. This study showed that the combined effect of ERI and sleep quality was a risk factor for depressive symptoms, and the combined effect was larger than the separate one; Both high ERI and poor sleep quality were the greatest contributors to depression symptoms in combined effect model of ERI and sleep quality, which may supply a clue to explain the link between ERI, sleep quality and increased depression risk.

Keywords: Depression symptoms · Effort-reward imbalance · Sleep quality
Combined effect

1 Introduction

In recent years, psychological problems, suicide and karoshi caused by large work pressure and fatigue were seen frequently in the media [1–4]. People who attempt to suicide were significantly associated with mental illness, especially the depression [5–7]. In 2003, Occupational Safety and Health Administration reported that depression caused by job stress resulted in $ 30.4 billion direct and indirect economic losses in American; In China, the incidence of mental disorders and chronic fatigue syndrome has risen to 30%–40% in some professional population. An investigation showed that depression has become the second disease in the burden of medical expenses, and the proportion accounted for total medical expenses reached 6.2%. Excluding its influence on individual mental health, depression also results in worker absenteeism, leads to the reduced productivity in the workplace [8, 9]. Thus, it is indispensable to pay more attention to the depressive symptoms of workers from a social perspective. In recent two decades, most psychosocial studies in developed countries revealed a strong association between bad work environment and depression [10–12], and some studies

© Springer Nature Switzerland AG 2019
S. Bagnara et al. (Eds.): IEA 2018, AISC 819, pp. 20–30, 2019.
https://doi.org/10.1007/978-3-319-96089-0_3

found that depressive symptoms were more prevalent in workers with high effort-reward imbalance [13–16]. In addition, recently, some studies have been conducted to observe the relationship between sleep quality and depressive symptoms, which suggested that shorter sleeping time and poorer subjective sleep quality were associated with a higher risk of depressive symptoms [17–19]. However, no study has been conducted on the combined effect of job stress and sleep quality on depressive symptoms risk so far.

According to the work characteristics of occupational population, train driver was a kind of occupation population who continued to be in a work state of highly centralized and paramilitary as well as having the characteristics of higher work responsibilities, longer continuous work, faster running speed, irregular working hours, stay up late, mandatory sleep, smaller work space, larger noise intensity, and poorly ventilated environment, etc. These characteristics are more likely to lead poor psychological status during work along with the poor sleep quality. Prolonged poor psychological status at work along with the poor sleep quality may lead to adverse psychological condition, especially to the depression symptoms. Therefore, in this study we selected the train driver as the subjects to explore the effects of Effort-Reward Imbalance and sleep quality on depressive symptoms. Meanwhile, the study attempts to evaluate whether the combination of ERI and sleep quality may induce the greater risk of depression symptoms when compared with ERI or sleep quality alone.

2 Methods

2.1 Study Population

This cross-sectional study was conducted in Henan Province, located in the central region of China, from Mar 2012 to Sep 2012. A total of 1 042 train drivers were recruited from a depot of Railway Bureau, who included 75 China Railway High-speed drivers, 301 passenger train drivers, 678 train freight drivers, and 348 passengers. The informed consents were signed by the subjects and the study protocol was approved by the Medical Ethics Committee of the Henan Provincial Institute of Occupational Health. Each subject was given a questionnaire at his/her workplace and required to complete the questionnaire within 45 min. The questionnaire was designed to collect following information: age, job tenure, education level, smoking and alcohol use histories, effort-reward imbalance, sleep quality and depression symptoms.

2.2 Measurement Methods

Effort-Reward Imbalance. The Chinese version of effort-reward imbalance (ERI) questionnaire was used in this study [20], it consists of the following three scales: extrinsic effort (6 items), occupational reward (11 items), and overcommitment (6 items). The questionnaire was also formulated by means of translation and back-translation. Extrinsic effort was evaluated by measuring the psychosocial workload; occupational reward focuses on the worker's financial status (i.e. salary), self-esteem, and career opportunity (e.g. promotion prospects and job security). Overcommitment as a personal (intrinsic) component was defined as a set of attitudes, behaviors, and emotions reflecting excessive striving along with a strong desire for approval and esteem. Cronbach's alpha of effort, reward, and overcommitment scales were 0.83, 0.71, and 0.65, respectively. Furthermore, according to the theoretical formulation a ratio of effort/(reward × 0.5454) was calculated to assess the degree of imbalance between high effort and low reward at work where a value >1.0 indicates the critical condition. ≥ 1.0 indicates high ERI level and <1.0 indicates low ERI level. In addition, extrinsic effort, occupational reward, and overcommitment were split by the median value.

Sleep Quality. Sleep quality was assessed by using the Chinese language version of the Pittsburgh Sleep Quality Index (PSQI) [21], which is originally developed by the University of Pittsburgh and translated by Yu [20]. The PSQI consists of 14 items including subjective sleep quality, sleep latency, sleep duration, habitual sleep efficiency, sleep disturbances, use of sleeping medication, and daytime dysfunction. Each question uses a 4-point Likert scale (0-1-2-3). Higher score indicates poorer quality of sleep. Cronbach'a of the PSQI in this sample was 0.90. This study used the median value indicating a sleep disorder [21]. People with 19 or more points were defined as poor sleep quality population, whereas people with below 19 points were defined as good sleep quality population.

Depressive Symptoms. Depressive symptoms were measured by the Center for Epidemiological Studies Depression (CES-D) Scale [22]. The CES-D scale consists of 20 items related with characteristic symptoms and behaviors of depression. Each question displayed using a 4-point Likert scale (0-1-2-3), and higher scores indicated higher depression. Cronbach'a of the DES-D in this sample was 0.91. We applied in this study the threshold value 19 recommended for identifying subjects with depressive symptoms by this scale. This scale has been used extensively in China since 1980's [23]. Based on the threshold of 19 points to screen this survey population with depressive symptoms, the subjects with a score 19 or more points were classified as subjects with depressive symptoms, while below 19 points were those subjects no depressive symptoms.

Potential Confounding Variables. In this study, the potential confounding variables included individual factors, (such as driver types, age (≤ 25 years, 25 to 35 years, 35 to 45 years, and ≥ 45 years), job tenure (≤ 5 years, 5 to 10 years, 10 to 15 years, 15 to 20 years, 20 to 25 years, and ≥ 25 years), educational level [high school and below (elementary school, junior high school, high school), college and over], marital status [married, never married, divorced and widowed, remarried, and unmarried cohabitation], smoking habit (current-smoker, non-smoker, and ex-smoker), and drinking habit (current-drinker, non-drinker, and ex-drinker).

2.3 Statistical Analysis

Chi-square test was carried out to analyze the differences of general characteristics in the group with different depressive symptoms. Hierarchical multiple logistic regression analysis [24] was performed to test the relationships between ERI, sleep quality, combined effect of ERI and sleep quality, potential confounding variables (independent variables) and depressive symptoms; Logistic regression analysis was used to estimate the association between ERI, sleep quality, combined effect of ERI and sleep quality and depressive symptoms. Multivariate odd ratios (ORs) and 95% confidence intervals (CIs) were derived from the logistic regression models. In logistic regression analysis, the confounding factors such as driver types, age, job tenure, education level, marital status, smoking habit, and drinking habit were controlled.

In the logistic regression analysis, the variables were indicated as follows: age (1 for less than or equal 25 years, 2 for 25 to 35 years, 3 for 35 to 45 years, and 4 for more than 45 years); Job tenure (1 for less than or equal 5 years, 2 for 5 to 10 years, 3 for 10 to 15 years, 4 for 15 to 20 years, 5 for 20 to 25 years, and 6 for more than 25 years); Educational level (1 for high school and below, 2 for university and over); Marital status [1 for married, 2 for never married, 3 for divorced and widowed, 4 for remarried, and 5 for unmarried cohabitation]; Smoking habit (0 for non-smoker, 1 for current-smoker, and 2 for ex-smoker); Drinking habit (0 for non-drinker, 1 for current-drinker, and 2 for ex-drinker); ERI (0 for low ERI, 1 for high ERI); Extrinsic effort (0 for low extrinsic effort, 1 for high extrinsic effort), Occupational reward (0 for occupational reward, 1 for high occupational reward); Overcommitment (0 for overcommitment, 1 for high overcommitment); Sleep quality (0 for good sleep quality, 1 for poor sleep quality); Depressive symptoms (0 for no, 1 for yes). The effort-reward imbalance indicator was computed by creating four independent categories: (1) low efforts and high rewards, (2) low efforts and low rewards, (3) high efforts and low rewards, (4) high efforts and high rewards; Similar, the combination of ERI and sleep quality was computed by creating four independent categories: (1) low ERI and poor sleep quality, (2) low ERI and good sleep quality, (3) high ERI and poor sleep quality, (4) high ERI and good sleep quality.

All the statistical analyses were performed with SPSS for Windows (version 19.0; SPSS Inc., Chicago, IL), with P values less than 0.05 was considered significant. All significant statements were two-tailed.

3   Results

3.1   Descriptive Statistics

The general characteristics of 1402 participants showed that age ranged from 20 to 59 years (mean 35.0 years) and job tenure ranged from 2 to 37 years (mean 13.1 years). The mean of depression symptoms, ERI, and sleep scores were 24.41 (S.D. = 10.5), 1.27 (S.D. = 0.5) and 19.55 (S.D. = 8.6), respectively. The distribution of depression symptoms was split by 19-point threshold of the total CES-D score, among the subjects, 955 (68.1%, 955/1 402) were defined as with depressive symptoms. Table 1 shows the distribution of depressive symptoms according to the general characteristics. The distribution of depression symptoms score according to driver types, age, job tenure, marital status, smoking habit, and drinking habit did not show statistically significant differences. But a depressive symptoms rate in EMU drivers or high-speed rail drivers (74.7%, 56/75), 35 to 45 year-age group (72.2%, 249/345), 10 to 15 year-job tenure group (73.5%, 150/204), divorced and widowed (82.8%, 24/29), Non-smoker (70.7%, 424/600) and Non-drinker (70.3%, 474/674) were the highest in the same type of group. A depressive symptoms rate was significant differences between the subjects with (college and over) and low educational level group (high school and below), and a depressive symptoms rate in high educational level group (72.0%, 372/517) were higher than low educational level group (65.9%, 583/885).

Table 1. Distribution of depressive symptoms according to general characteristics.

Variables classification	Number (%)	Total CES-D score		χ^2-value	p -value
		<19	≥ 19		
The driver types				1.502	0.682
EMU drivers or high-speed rail drivers	75 (5.3)	19 (25.3)	56 (74.7)		
Passenger train drivers	301 (21.5)	104 (34.6)	197 (65.4)		
Train freight drivers	678 (48.4)	216 (31.9)	462 (68.1)		
Passenger shunting drivers	348 (24.8)	108 (31.0)	240 (69.0)		
Age (years)				5.786	0.123
≤ 25	201 (14.3)	74 (36.8)	127 (63.2)		
25–35	591 (42.2)	185 (31.3)	406 (68.7)		
35–45	345 (24.6)	96 (27.8)	249 (72.2)		
≥ 45	265 (18.9)	92 (34.7)	173 (65.3)		
Job tenure (years)				6.484	0.262
≤ 5	394 (28.1)	126 (32.0)	268 (68.0)		
5–10	267 (19.0)	97 (36.3)	170 (63.7)		
10–15	204 (14.6)	54 (26.5)	150 (73.5)		
15–20	127 (9.1)	34 (26.8)	93 (73.2)		
20–25	176 (12.6)	55 (31.3)	121 (68.8)		
≥ 25	234 (16.7)	81 (34.6)	153 (65.4)		
Education level				5.016	0.025
High school and below	885 (63.1)	302 (34.1)	583 (65.9)		
College and over	517 (36.9)	145 (28.0)	372 (72.0)		
Marital status				8.979	0.062
Married	1057 (75.4)	336 (31.8)	721 (68.2)		
Never married	265 (18.9)	86 (32.5)	179 (67.5)		
Divorced and widowed	29 (2.1)	5 (17.2)	24 (82.8)		
Remarried	17 (1.2)	4 (23.5)	13 (76.5)		
Unmarried cohabitation	34 (2.4)	16 (47.1)	18 (52.9)		
Smoking habit				2.590	0.274
Current-smoker	726 (51.8)	241 (33.2)	485 (66.8)		
Non-smoker	600 (42.8)	176 (29.3)	424 (70.7)		
Ex-smoker	76 (5.4)	30 (39.5)	46 (60.5)		
Drinking habit				2.213	0.331
Current-drinker	682 (48.6)	227 (33.3)	455 (66.7)		
Non-drinker	674 (48.1)	200 (29.7)	474 (70.3)		
Ex-drinker	46 (3.3)	20 (43.5)	26 (56.5)		

3.2   Relationship Between ERI, Sleep Quality and Depression Symptoms

Table 2 displays the associations between ERI, sleep quality, and potential confounding variables (independent variables) and depressive symptoms. The results of the hierarchical multiple logistic regression analysis revealed that after adjusted for potential confounding variables, ERI and sleep quality were shown to be independently associated with depressive symptoms, and combined effect of ERI and sleep quality was also a risk factor for depressive symptoms ($P \leq 0.01$), suggesting that higher ERI and poor sleep were both associated with higher depression symptoms. Importantly, combined action of ERI and sleep quality significantly increased the depression symptoms level. Moreover, education level and drinking habit in the general characteristics were entered into most of Model, and the results showed that high educational level was a risk factor for depression symptoms, but drinking habit was a protective factor for depression symptoms.

Table 2. Hierarchical multiple logistic regression of explanatory for depressive symptoms.

Variables	Model 1[a]		Model 2[b]		Model 3[c]		Model 4[d]		Model 5[e]	
	B	Wald	B	Wald	B	Wald	B	Wald	B	Wald
Job type	0.048	0.438	0.052	0.489	0.053	0.459	0.051	0.426	0.074	0.942
Age	0.156	1.737	0.151	1.565	0.166	1.652	0.165	1.623	0.187	2.240
Job tenure	−0.025	0.164	−0.045	0.527	−0.017	0.063	−0.032	0.225	−0.022	0.113
Education level	0.373	7.740*	0.310	5.015*	0.415	8.163*	0.353	5.794*	0.415	8.460*
Marital status	−0.053	0.487	−0.035	0.199	−0.047	0.324	−0.035	0.169	−0.032	0.159
Smoking habit	−0.118	1.866	−0.174	3.810	−0.058	0.390	−0.095	1.029	−0.077	0.698
Drinking habit	−0.128	1.792	−0.194	3.789	−0.283	7.210*	−0.305	8.205*	−0.284	7.440*
ERI			1.152	76.988*			0.658	21.077*		
Sleep quality					1.816	177.218*	1.630	132.359*		
Combined effect of ERI and sleep quality									1.631	137.571*
Constant	0.166	0.212	−0.472	1.554	−0.620	2.474	−0.870	4.690	−0.563	2.108

Note. [a]Adjusted by job type, age, job tenure, education level, marital status, smoking habit and alcohol drinking; [b]Adjusted for Model 1 plus ERI; [c]Adjusted for Model 1 plus sleep quality; [d]Adjusted for model 1 plus ERI and sleep quality; [e]Adjusted for mode 1 plus combined effect of ERI and sleep quality. *$p < 0.05$ or 0.01.

Table 3 describes the number and percentage of the train drives. The results showed the ERI (i.e. high efforts and low rewards) and three components (efforts, rewards, and overcommitment), sleep quality, combined effect of ERI and sleep quality (i.e. high ERI and poor sleep quality). Furthermore, the association between each of the variables and depression symptoms was analyzed using multivariate logistic regression analysis. After adjusting for the confounding variables, the subjects with high efforts, high rewards and high overcommitment have 2.08, 0.40 and 1.98 times higher depression symptoms than the subjects with low efforts, low rewards and low overcommitment. Similar results were found for sleep quality (Adj OR = 5.82, 95% CI = 4.52−7.49). In addition, the results showed efforts, overcommitment and sleep quality were risk factors for depression symptoms, but rewards were protective factors. Furthermore, the risk of depressive symptoms for train drivers with both low efforts and low rewards, both high efforts and high rewards, both high efforts and low rewards were 2.04, 1.57, and 4.93 times respectively as high as that for train drives with both low efforts and high rewards. With regard to combined effect of ERI and sleep quality on depression symptoms, the results showed the risk of depressive symptoms for train

Table 3. Crude and adjusted ORs for the associations between the ERI model, sleep quality and combined effect of ERI and sleep quality, and depression symptoms.

Variables classification	N(%)	Crude OR(95%CI)	Adj OR(95%CI)*
ERI model			
Efforts			
Low	765 (54.6)	1.00	1.00
High	637 (45.4)	1.98(1.57–2.50)[a]	2.08(1.64–2.64)[a]
		$P < 0.001$	
Rewards			
Low	689 (49.1)	1.00	1.00
High	713 (50.9)	0.41(0.33–0.52)[a]	0.40(0.32–0.52)[a]
		$P < 0.001$	
Overcommitment			
Low	825 (58.8)	1.00	1.00
High	577 (41.2)	1.89(1.49–2.39)[a]	1.98(1.55–2.52)[a]
		$P < 0.001$	
ERI			
Low efforts and high rewards	463 (33.0)	1.00	1.00
Low efforts and low rewards	312 (22.3)	1.86(1.37–2.53)[a]	2.04(1.49–2.80)[a]
High efforts and high rewards	255 (18.2)	1.52(1.11–2.10)[a]	1.57(1.13–2.17)[a]
High efforts and low rewards	372 (26.5)	4.19(3.00–5.86)[a]	4.93(3.46–7.01)[a]
		$P < 0.001$	
Sleep quality			
Poor	712 (50.8)	1.00	1.00
Good	690 (49.2)	5.85(4.50–7.61)[a]	6.18(4.73–8.07)[a]
		$P < 0.001$	
Combined effect of ERI and sleep quality			
Low ERI and good sleep quality	282 (20.1)	1.00	1.00
Low ERI and poor sleep quality	96 (6.8)	9.20(6.62–12.79)[a]	10.09(7.18–14.20)[b]
High ERI and good sleep quality	418 (29.8)	2.33(1.71–3.17)[a]	2.38(1.74–3.26)[b]
High ERI and poor sleep quality	606 (43.2)	14.97(6.63–33.79)[b]	15.40(6.76–35.10)[b]
		$P < 0.001$	

Note. *Adjusted for driver types, age, job tenure, education level, marital status, smoking habit, and drinking habit. **Linear-by-Linear Association. [b]: $P < 0.01$.

drivers with both low ERI and poor sleep quality, both high ERI and good sleep quality, both high ERI and poor sleep quality were 10.09, 2.38 and 15.40 times respectively as high as that for train drives with both low ERI and good sleep quality. Moreover, this study found that sleep quality could estimate the risk of depression symptoms more accurately than that of ERI, the combined effect of ERI and sleep quality was larger than the separate one.

4 Discussion

This study investigated the association between general characteristics, ERI, sleep quality, and depressive symptoms in 1402 male train drivers and the following finding was obtained. The distribution of depression symptoms was split by 19-point threshold of the total CES-D score. Among the subjects, 955 (68.1%, 955/1 402) subjects were classified as people with depressive symptoms, and a depressive symptoms rate in EMU drivers or high-speed rail drivers (74.7%) was the highest in all driver types. Kim et al. [25] have reported a depression symptoms rate of 19.2% in a study involving 17,457 firefighters. Also Lee et al. [26] have reported a depression symptoms rate of 29.4% (50/170) in Korean-Chinese migrant workers in Korea. In China, Yu et al. [27] have reported that a depression symptoms rate of 23.4% (109/466) in teachers and doctors of Beijing. Hu et al. [28] also showed that a depression symptoms rate of 60.5% in a study involves 1048 Prison Guards. Compared to these previous studies, the present study showed a higher depression symptoms rate.

The findings from this study showed that higher educational level (college and over) and no drinking were not the protectors of depression symptoms, which were similar with the reports of Khajehnasiri [29] and Cheng et al. [30]. In addition, the association between other general characteristics and depression symptoms did not show statistically significant differences. But a depression symptoms rate was higher in the 35–45 year-age group (72.2%); 10–15 year-job tenure group (73.5%); divorced and widowed (82.8%) and non-smoker group (70.3%). The results were consistent with the findings of Liu et al. [31] and Gu et al. [32], but opposite with the study involving 4,833 female workers in the manufacturing, finance, and service fields at 16 workplaces in Yeungnam province of Korean [33]. The findings may be related to their family responsibilities, social responsibilities and job stress.

Hierarchical multiple logistic regression analysis explored the associations of ERI, sleep quality and Combined effect of ERI and sleep quality with depression symptoms among train drives in Henan province of China. Multivariate logistic regression analysis further showed that efforts and overcommitment were risk factors for depression symptoms, but rewards were protective factors in ERI model. This result is consistent with previous studies [34, 35]. In addition, in a depression symptoms study conducted among 17 countries with 14236 participants in three continents, odds ratios of depressive symptoms among people experiencing high work stress compared to those with low or no work stress. Adjusted odds ratios vary from 1.64 (95% CI 1.02–2.63) in Japan to 1.97 (95% CI 1.75–2.23) in Europe and 2.28 (95% CI 1.59–3.28) in the USA, which supported our conclusion. This study also showed that after adjusted for confounders, both high efforts and low rewards could accurately assess the risk of depression symptoms (OR = 4.93), which is best in ERI model. Yu [34] have reported that the risk of depressive symptoms for workers with high efforts and low rewards was about two times as high as that for workers with low efforts and high rewards (OR = 1.70), suggesting that both high efforts and low rewards may lead to depressive symptoms more easily, which showed a consistent results with our findings.

With regard to the effect of sleep quality on depressive symptoms, multivariate logistic regression analysis showed that the risk of depressive symptoms for trail drives with poor sleep quality was about two times as high as that for trail drives with good sleep quality (OR = 6.18). The studies of Lee [17], Tan [18] and Choi [19] supported our conclusion, suggesting that poor sleep quality increases the risk of depressive symptoms.

The study on comparison of the risk and the interaction effect of ERI and sleep quality on depression symptoms have not been reported at present. Our results showed that ERI and sleep quality were independently associated with depressive symptoms, and effect of sleep quality (OR = 6.18) was larger than that of ERI; combined effect of ERI and sleep quality was a risk factor for depressive symptoms, and the combined effect was larger than each scale separately; In combined effect model of ERI and sleep quality, both high ERI and poor sleep quality (OR = 15.40) were the greatest contributors to depression symptoms in combined effect model. This result suggested that ERI or sleep quality may increase the risk of depressive symptoms, moreover, both high ERI and poor sleep quality may lead to depressive symptoms more easily.

Several limitations need to be taken into account. Firstly, this study was the cross sectional observations, and this could result in information bias and selection bias, which could lead to either over- or underestimation of the association. Lack of evaluation of the effect of ERI during due to cross-sectional data used could lead to an underestimation of true association. The use of self-reported questionnaire to measure job stress may be subject to response bias. However, recently no reliable objective measurement of job stress was available. Secondly, our sample was a small size in one occupational type and wasn't representative of the total work force in China. To our knowledge, this is the first time to explore the combined effect of ERI and sleep quality on depression symptoms. This findings required replication in other designs, to establish its validity.

5 Conclusion

This study suggested that the combination of ERI and sleep quality is associated with prevalence of depression symptoms, which may supply a clue to explain the link between ERI, sleep quality and increased depression risk.

References

1. Hiyama T, Yoshihara M (2008) New occupational threats to Japanese physicians: karoshi (death due to overwork) and karojisatsu (suicide due to overwork). Occup Environ Med 65 (6):428–429

2. Kondo N, Oh J (2010) Suicide and karoshi (death from overwork) during the recent economic crises in Japan: the impacts, mechanisms and political responses. J Epidemiol Community Health 64(8):649–650

3. Yoon JH, Jung PK, Roh J, Seok H, Won JU (2015) Relationship between long working hours and suicidal thoughts: nationwide data from the 4th and 5th Korean national health and nutrition examination survey. PLoS ONE 10(6):e0129142

4. Jeon HJ, Lee JY, Lee YM, Hong JP, Won SH, Cho SJ, Kim JY, Chang SM, Lee D, Lee HW, Cho MJ (2010) Life time prevalence and correlates of suicidal ideation, plan, and single multiple attempts in a Korean nationwide study. J Nerv Ment Dis 25:643–646

5. World Health Organization (2009) Pharmacological treatment of mental disorders in primary health care. World Health Organization, Geneva

6. Hawton K, Casañas I, Comabella C, Haw C, Saunders K (2013) Risk factors for suicide in individuals with depression: a systematic review. J Affect Disord 147(1–3):17–28

7. Bellini S, Erbuto D, Andriessen K, Milelli M, Innamorati M, Lester D, Sampogna G, Fiorillo A, Pompili M (2018) Depression, hopelessness, and complicated grief in survivors of suicide. Front Psychol 9:198

8. Murray CJ, Lopez AD (1996) Evidence-based health policy–lessons from the global burden of disease study. Science 274(5288):740–743

9. Leykin Y, Roberts CS, Robert J (2011) Decision-making and depressive symptomatology. Cogn Ther Res 35:333–341

10. Saijo Y, Ueno T, Hashimoto Y (2007) Job stress and depressive symptoms among Japanese fire fighters. Am J Ind Med 50:470–480

11. Shields M (2006) Stress and depression in the employed population. Health Rep 17:11–29

12. Bonde JPE (2008) Psychosocial factors at work and risk of depression: a systematic review of the epidemiological evidence. Occup Environ Med 65:438–445

13. Sakata Y, Wada K, Tsutsuki A (2008) Effort-reward imbalance and depression in Japanese medical residents. J Occup Health 50:498–504

14. Niedhammer I, Tek ML, Starke D (2004) Effort-reward imbalance model and self-reported health: cross-sectional and prospective findings from the GAZEL cohort. Soc Sci Mcd 58:1531–1541

15. Kivimaki M, Vahtera J, Elovainio M (2007) Effort-reward imbalance, procedural injustice and relational injustice as psychosocial predictors of health: complementary or redundant models? Occup Environ Med 64:659–665

16. Tsutsumi A, Kawanami S, Horie S (2012) Effort-reward imbalance and depression among private practice physicians. Int Arch Occup Environ Health 85:153–161

17. Lee CM, Chang CF, Pan MY (2017) Depression and its associated factors among rural diabetic residents. J Nurs Res 25(1):31–40

18. Tan Y, Chen Y, Lu Y (2016) Exploring associations between problematic internet use, depressive symptoms and sleep disturbance among southern Chinese adolescents. Int J Environ Res Public Health 13(3):313

19. Choi SH, Terrell JE, Pohl JM et al (2013) Factors associated with sleep quality among operating engineers. J Community Health 38(3):597–602

20. Yu S, Gu G, Zhou W (2008) Psychosocial work environment and wellbeing: a cross-sectional study at a thermal power plant in China. J Occup Health 50(2):155–162

21. Buysse DJ, Reynolds CF, Monk TH, Berman SR, Kupfer DJ (1989) The Pittsburgh sleep quality index: a new instrument for psychiatric practice and research. Psychiatry Res 25 (2):193–213

22. Radloff LS (1977) The CES-D scale: a self-report depression scale for research in the general population. Appl Psycho Meas 1:385–401

23. Zhang MY, Ren FM, Fan B (1987) Investigation of depressive symptoms among normal populations and application of CES-D. Chin J Neurol Psychiatry 20:67–71 (in Chinese)
24. Nakata A, Takahashi M, Irie M, Swanson NG (2010) Job satisfaction is associated with elevated natural killer cell immunity among healthy white-collar employees. Brain Behav Immun 24(8):1268–1275
25. Kim TW, Kim KS, Ahn YS (2010) Relationship between job stress and depressive symptoms among field firefighters. Korean J Occup Environ Med 25(4):378–387
26. Lee H, Ahn H, Miller A (2012) Acculturative stress, work-related psychosocial factors and depression in Korean-Chinese migrant workers in Korea. J Occup Health 54(3):206–214
27. Yu SF, Yao SQ, Ding H (2006) Relationship between depression symptoms and stress in occupational populations. Chin J Ind Hyg Occup Dis 24(3):129–133
28. Hu S, Meng L, Ma L (2012) The interaction of job stress and perceived organizational support on depression symptoms in prison guards. Chin J Health Stat 29(1):155
29. Khajehnasiri F, Mortazavi SB, Allameh A (2013) Total antioxidant capacity and malondialdehyde in depressive rotational shift workers. J Environ Public Health 10 (2013):150693
30. Cheng HG, Chen S, McBride O (2016) Prospective relationship of depressive symptoms, drinking, and tobacco smoking among middle-aged and elderly community-dwelling adults: results from the China Health and Retirement Longitudinal Study (CHARLS). J Affect Disord 10(195):136–143
31. Liu L, Chang Y, Fu J (2012) The mediating role of psychological capital on the association between occupational stress and depressive symptoms among Chinese physicians: a cross-sectional study. BMC Public Health 12:219
32. Gu G, Yu S, Wu H (2015) The effect of occupational stress on depression symptoms among 244 policemen in a city. Chin J Prev Med 49(10):924–929
33. Cho HS, Kim YW, Park HW (2013) The relationship between depressive symptoms among female workers and job stress and sleep quality. Ann Occup Environ Med 25(1):12
34. Yu SF, Gu GZ, Zhou WH et al (2011) Gender difference of relationship between occupational stress and depressive symptoms. Chin J Ind Hyg Occup Dis 29(12):887–892
35. Yu SF, Nakata A, Gu GZ (2013) Combined effect of demand-control support model and effort- reward imbalance model on depression risk estimation in humans: findings from Henan Province of China. Biomed Environ Sci 26(12):962–971
36. Siegrist J, Lunau T, Wahrendorf M (2012) Depressive symptoms and psychosocial stress at work among older employees in three continents. Global Health 8:27

Assessment of Occupational Vibration on Tire × Track Harvesters in Forest Harvesting

Stanley Schettino[1(✉)] , Luciano José Minette[2] ,
Silvio Sérgio Caçador[3] , and Isabela Dias Reboleto[2]

[1] Federal University of Minas Gerais, Av. Universitária, 1000, Montes Claros,
MG, Brazil
schettino@ufmg.br
[2] Federal University of Viçosa, Av. P.H. Rolfs s/n, Viçosa, MG, Brazil
[3] Federal University of Espirito Santo, Av. Gov. Lindemberg 316 J., Monteiro,
ES, Brazil

Abstract. The objectives of this study were to compare the levels of occupational whole-body vibration presented by tire and track harvesters, in order to verify if there are significant differences between the two types of machines, and also if such exposure represents a health risk to the operators. Two models of track harvesters and two of tire harvesters were assessed. The assessment was carried out according to the methodology established in ISO Standard 2631-1: 1997 and the results were compared to the reference values of Directive 2002/44/EC. The tire harvesters presented low vibration levels, not representing disease development risk to the operators. The track harvesters presented about 2.3 higher vibration values than those presented by tire harvesters and above the limit of action foreseen in the reference standard, thus actions to mitigate the deleterious effects of vibration on the health of the workers become necessary. Thus, the results of this study allowed us to conclude that track harvesters have significantly higher occupational vibration levels than those presented by tire harvesters; that such values are above the limit of action established by the reference standard; that immediate interventions in machinery and processes are necessary in order to reduce operator exposure to vibration; and that they are exposed to a high risk of developing occupational diseases resulting from exposure to the physical agent vibration.

Keywords: Ergonomic · Forest mechanization · Workers health

1 Introduction

With an area of more than 500 million hectares of native forests rich in biodiversity, in addition to more than 8 million hectares of reforestation [1], allied to favorable climate and soil, the Brazilian forestry sector has been constantly developing, which increasingly demands forest-based products. To meet this growing demand, in a scenario which labor is scarce, the competitiveness of the sector is increasing and the requirements of globalized consumer markets are each time higher, the mechanization of wood production activities has become imperative for the sustainability of the forestry

© Springer Nature Switzerland AG 2019
S. Bagnara et al. (Eds.): IEA 2018, AISC 819, pp. 31–40, 2019.
https://doi.org/10.1007/978-3-319-96089-0_4

business. Thus, it aims to minimize production costs, decrease labor dependence, increase productivity, reduce environmental damage and ensure a continuous flow of wood supply to the consumer units [2, 3] as well as to reduce the rates of accidents and the risks of developing occupational diseases.

As far as wood harvesting is concerned, the use of harvesters has been the most used by forestry companies, among the available alternatives. According to the literature [4], the harvester or forestry harvester is a type of automotive tractor that has the purpose of cutting and processing trees within the forest, while at the same time carrying out the operations of wood cutting, debarking, delimbing, tracing and stacking. It is composed of a base machine and a head.

The harvesters can be tire or track machines, or adaptations made on track excavators with the placement of a processor head. This option is widely used in the Brazilian forestry sector, mainly due to the lower cost of acquisition and the existence of models manufactured in the country, facilitating technical assistance and the purchase of parts and maintenance. Contrary arguments refer to the high cost of maintaining tracks and a greater restriction to operation on rough terrain compared to tire machines [5]. However, another factor that should not be disregarded is the vibration to which the operators of such machines are subjected, which can be variable depending on the type of machine (tires or tracks), type of terrain and the operation itself.

Tires absorb soil irregularities more effectively than metal tracks, thus reducing traction efforts and providing more comfort to the occupants of the machine or vehicle, as a result of reduced noise and vibration [6]. In turn, tracks are not capable of offering the same result as tires with regard to vibration attenuation, although having other technical and operational advantages, and even low-frequency vibration environments such as those experienced by operators who conduce track forest machines can impose a series of physical constraints to them [7]. Studies have shown that, considering the same machine operating with tires and tracks under the same conditions, track machines can produce up to 2.7 times more vibration than the tire ones [8]. In track machines vibration is produced by their own power transmission system (drive wheel system and tracks) along with the large area of soil contact, whereas in tire machines less vibration is produced due to reduced friction with the soil.

Work-related vibration is an important health risk factor and can lead to serious consequences to the body. It results from one or more sources of mechanical vibration that affect the organism [9], which in the case of forest machines are due to the motor operation, the displacements on irregular floors and the operation itself. Differently from the other agents, on which the worker is exposed to the risks, when the subject is vibration characteristically there must be contact between the worker and the equipment/machine that transmits the vibration. However, even intermittent exposures can bring damage [10].

Harvester operators, whether of tires or tracks, are subject to full body vibrations and mechanical shocks that are uncomfortable and capable of providing the development of occupational diseases, and the part of the body which is most susceptible to damage is the lower portion of the spine, particularly prone, most often caused by mechanical shocks [11]. The effects of vibration on the human body depend on several factors, in particular on the intensity of vibrations, frequency limits, direction, penetration point, time and form of daily application, as well as the time in which the professional has been submitted to exposure [9]. The mechanical vibrations caused by both the machine operation and the roughness of the displacement surface become problematic when the frequency of parts of the human body (e.g., the trunk vibrates at a frequency of 4 to 8 Hz), ends up approaching the frequency of the machine (1–7 Hz), which may increase the chances of health problems in the operator [12].

In the face of the above, the objective of this study is to compare the levels of whole-body occupational vibration presented by tire and track harvesters, in order to verify if there are significant differences between the two types of machines and also if such exposure represents risk to the health of the operators.

2 Materials and Methods

2.1 Characterization of the Study Area

The data were collected in areas of two forest companies located in the interior of the São Paulo State, Brazil, between the meridians of longitude 23°52′26″ and 24°02′44″ west of Greenwich and the parallels of latitude 47°46′38″ to 48°01′02″ south of the Equator. The altitude varies between 600 and 1,000 m. According to Köppen climatic classification, the predominant climate in the region is Cfb - temperate with mild summers, with annual precipitation of around 1,600 mm. The main types of soil that occur in the region are: Dystrophic Red Latosol (LVd), Dystrophic Yellow Latosol (LAd) and Dystrophic Haplic Cambisol (CXBd); with texture varying from medium to very clayey.

In the study area, the forests are, in their totality, cultivated with eucalyptus trees in stands of hybrid clones of different productivities (from 250 to 420 m³/ha), in a first rotation system with 7 years of age, 3 × 2 m spacing and flat to soft wavy terrain. Mechanical harvesting was performed through the cut-to-length system, on which, the tree is processed in the form of logs of less than 6 m of length, in the same place of felling, that is, in the plots [13].

2.2 Machinery Evaluated

In this analysis, the following machines were used:

- M1 = Track harvester, consisting of a Komatsu hydraulic excavator, model PC200-DLC, Valmet SAA6D107E-1 6-cylinder engine, 155 HP of power, operating weight of 22,600 kg and equipped with the Valmet 370-E head;
- M2 = Track harvester, consisting of a Volvo hydraulic excavator, model EC210, Volvo D6D-EAE2 6-cylinder engine, 159 HP of power, operating weight of 22,300 kg and equipped with Ponsse H7 head;
- M3 = Tire harvester, Komatsu, model 941.1, Valmet 74-CW3 6-cylinder engine, 185 HP of power, operating weight of 21,200 kg and equipped with Valmet 370-E head, 6WD configuration, tandem axles;
- M4 = Tire harvester, Ponsse, Ergo model, MB-OM906LA-EU Stage IIIA 6-cylinder engine, 275 HP of power, operating weight of 23,000 kg and equipped with Ponsse H7 head, 8WD configuration, tandem axles.

All machines had enclosed and climatized cabins. The two tracked machines evaluated (M1 and M2) were equipped with cabins resting on hydraulic cushion pads to reduce impact and vibration levels. In all the evaluated machines the seats of the operators had belts, air suspension, adjustment of the calibration according to the weight of the operator and regulation of distance, height and slope of the backrest.

2.3   Vibration Evaluation

In the vibration evaluation a Maestro model 01 dB vibration meter was used, which has a triaxial accelerometer (directions or axes X, Y and Z) and a recording device of the acceleration values in $m.s^{-2}$. The accelerometer was installed on the seat of the operators. The evaluation was performed according to the methodology established in ISO 2631-1: 1997 standard [14]. This standard fixes a coordinate system originating with origin at the point where the vibration is incorporated into the human body, that is, at the interface between the vibrating source and the body. RMS (average vibrating movement energy) values for each axis were measured (Eq. 1); the a_v (vibration weighted total value, resulting from RMS accelerations obtained in the three axes), according to Eq. 2; as well as the aren value (corresponds to the resulting acceleration of exposure converted to a standard 8-h daily journey), according to Eq. 3.

$$a_w = \left[\frac{1}{T} \int_0^T a_w^2(t) dt \right]^2$$
(1)

where:

a_w = acceleration as a function of time (RMS), in $m.s^{-2}$, for each axis; and
t = duration of measurement, in seconds (s).

$$a_v = \sqrt{k_x^2 \cdot a_{wx}^2 + k_y^2 \cdot a_{wy}^2 + k_z^2 \cdot a_{wz}^2}$$
(2)

where:

a_v = weighted acceleration, in $m.s^{-2}$;
a_{wx}, a_{wy} e a_{wz} = weighted accelerations of the x, y and z axes, respectively; and
k_x, k_y and k_z are multiplier factors, where k_x and k_y = 1.4 and k_z = 1.0.

$$aren = are \sqrt{\frac{T}{T_0}}$$
(3)

where:

are = resulting acceleration of exposure, in $m.s^{-2}$;
T = duration of the daily workday, in hours or minutes; and
T_0 = 8 h or 480 min.

Finally, the results of the evaluations were compared with the reference values of the Directive 2002/44/EC [15]. This directive establishes that the daily exposure limit value for triggering preventive actions is set at RMS 0.5 m.s^{-2}, with the daily exposure limit value being RMS 1.15 m.s^{-2}, both normalized to an exposure of 8 h, corresponding to the standard daily working hours.

2.4 Statistical Analysis

The number of sample units to estimate the parameters of an infinite population to a desired precision level, based on the standard error of the mean [16], was given by Eq. 4.

$$n = \frac{t^2 \cdot s^s}{d^2 \cdot m^2} \tag{4}$$

on which:

    n = estimated sample size;
    t = value of Student's t distribution, at 5% probability;
    s^2 = variance;
    d = error in the average estimate, in percentage; and
    m = sample mean.

Considering a 95% confidence level, equivalent to two deviations, and an estimation error of 5%, the minimum sample size was obtained, consisting of 26 measurements for machine M1, 29 for M2, 25 for M3 and 32 for M4. However, 40 measurements for each machine were performed. Each measurement had a duration of four minutes, a time interval capable of absorbing the entire operational cycle of the machine for the slaughter and processing of 10 trees, according to the study of Leite *et al.* [17]. Descriptive statistics with mean and standard deviation was used to characterize the data. The results were submitted to analysis of variance and means test, using Tukey test at 5% of probability. Statistical analyzes were performed using SAS Software version 9.1 [18].

3 Results

The results of the averages of the whole-body vibration assessments in the four assessed machines are presented in Table 1, along with the comparison between the different machines. The values observed for the assessment of occupational vibration are in accordance with ISO 2631-1: 1997 standard, on which the guidelines for measuring the exposure of workers to whole-body vibration are established.

The vibration energy average values (RMS) indicated that there was no prevalence of any axis (x, y and z), and it was verified that harvester operators are subjected to different magnitudes of vibration in different axes, that is, the whole-body vibration is transmitted either longitudinally, transversely or vertically.

The vibration energy average values (RMS) indicated that there was no prevalence of any axis (x, y and z), and it was verified that harvester operators are subjected to

Table 1. Table captions should be placed above the tables.

Parameter	Axis	Machines				P
		M1 (Tracks)	M2 (Tracks)	M3 (Tires)	M4 (Tires)	
RMS (m.s^{-2})	x	0.34 b ± 0.04	0.53 c ± 0.11	0.23 a ± 0.08	0.20 a ± 0.14	0.062*
	y	0.59 d ± 0.06	0.37 c ± 0.30	0.18 b ± 0.12	0.16 a ± 0.07	0.021*
	z	0.46 b ± 0.11	0.35 a ± 0.19	0.25 a ± 0.12	0.28 a ± 0.21	0.001*
a_v (m.s^{-2})		1.06 b ± 0.08	0.97 b ± 0.13	0.48 a ± 0.08	0.45 a ± 0.09	0.005*
aren (m.s^{-2})		0.82 b ± 0.15	0.74 b ± 0.19	0.37 a ± 0.11	0.32 a ± 0.08	0.012*
TLA		02 h 58 m	03 h 40 m	16 h 27 m	19 h 52 m	-
TLM		15 h 45 m	19 h 23 m	87 h 05 m	105 h 10 m	-

* Significant at 5.0% probability, by t-test with n−2 degrees of freedom.
Note: means followed by the same letter in each row do not differ from each other by Tukey test at 5% probability.
TLA = Time to reach the action limit; TLM = Time to reach the maximum tolerance limit.

different magnitudes of vibration in different axes, that is, the whole-body vibration is transmitted either longitudinally, transversely or vertically.

In accordance with ISO 2631-1: 1997 standard, the acceleration value which represents the magnitude of whole-body vibration is the vibration weighted total value, resulting from the RMS accelerations obtained on the three axes. Comparing the exposure between the different types of harvesters, a significant difference was found between them ($p < 0.05$), and the highest values were found in the track machines. Also, by interpreting the Tukey test results ($p < 0.05$), it is possible to separate the assessed machines into two distinct groups in relation to the occupational vibration levels, either those of track in a group and those of tires in the other.

All track machines had vibration weighted total levels above the action limit (0.50 m.s^{-2}). In order for M1 machine operators not to develop disturbances related to vibration exposure, they should only work 02 h and 58 min a day; whereas for M2 operators, this time passes to 03 h and 40 min. Considering that the standard workday is 08 h, actions are necessary to mitigate the deleterious effects of vibration on the health of the workers.

Tire machines, in turn, presented vibration levels below the tolerance limits foreseen in the reference standard for this study, and the times for action limits within the standard workday of 8 h per day were not even reached. Operators of these machines are safeguarded against the physical risk of whole-body vibration, in accordance with Directive 2002/44/EC.

4 Discussion

The vibrations that are transmitted to the human body can be classified into two types according to the part of the body affected: vibration transmitted to the whole-body or localized vibration. The vibrations transmitted to the whole-body, objective of this study, are of low frequency and great amplitude, being situated in the range of 1 to 20 Hz. The vertical oscillations, that is, in the longitudinal direction of the spine, penetrate the body with the person in the sitting or standing position on vibratory bases, often leading to signs of wear on the spine [19].

Agricultural and forestry machine operators are often exposed to high levels of whole-body vibrations during the development of their activities and are therefore affected by spinal disorders and various other types of diseases. The origin of these problems can hardly be eliminated, due to the well-known dynamics of the soil-machine-operator interaction. Machine designers have been adopting many different techniques in order to minimize the vibrations that, from the soil, appear until the human body. However, field activities can take place in very different scenarios: roughness and slope of the soil, machine speed, presence of tires or tracks, maintenance of machines, devices for specific tasks (e.g. harvester head), etc. In addition, the influence of such factors on human health is still not well understood and, therefore, it is not easy to limit the amount of vibrations acting on the design variables [20]. Considering the type of wheel of the machines (tires × tracks), objective of this study, there are important differences between the behavior of each one in relation to the vibration. According to Nguyen and Inaba [21], tires are able to absorb most of the high-frequency vibrations at the interface with the soil; however, low-frequency vibrations are still transmitted through the machine structure to the body of the operator. In track machines, vibration is produced mainly from the driving wheel, the contact of the tracks with the driving wheel and the other elements (rollers, guide wheel and track links) through polygonal action, as well as the impact between the track and the soil during displacement [22].

Mandal et al. [23], when evaluating full-body vibration levels in machines used in coal mines, observed that track machines had 40% higher vibration levels than those of tire machines when operating under the same conditions. In a similar analysis, this time with combat vehicles, Rozali et al. [8] verified that this difference amounts to 2.7 times greater in the vehicles of tracks when compared to the ones of tires. These studies corroborate the results found in this evaluation, on which occupational vibration levels were 2.3 times higher in track machines than in tire machines.

Furthermore, in the case of track machines, the vibration levels are still aggravated in those whose maintenance is deficient, such as the delay in the replacement of the rolling stock elements and the lack of lubrication, besides the low speed during the operation and displacements [8]. Moreover, Mandal et al. [23] state that factors such as rough terrain, speed and seat conditions, among others, are important responsible for the generation and transmission of vibrations; and that parts of the body most likely to be affected by exposure to whole-body vibration depend on the magnitude of vibration, the distribution of movement in the body, body postures and frequency, direction and duration of vibration.

Regarding the occupational risks of exposure to vibration, it was verified that in the track machines the exposure is an important factor to be considered aiming at the preservation of the physical integrity of the operators. In fact, the time to reach the limit of action with respect to vibration was about three hours for the M1 machine and three hours and forty minutes for the M2, worrying values compared to an eight-hour workday. It is important to note that the acceleration of vibration in each axis has a different behavior, linked to the movement of the operator on the seat. On the x axis the operator is receiving the acceleration that enters the chest, passes through the heart and leaves the back, and vice versa, while on the y axis the acceleration enters the shoulder, trespasses the heart and leaves the other shoulder. On the z axis the operator receives the acceleration that enters the buttocks, passes through the spine and heart and leaves the head [24]. In the case of the z axis, the influence of obstacles on the terrain and erosion, together with the metal structure of the machines and in particular the metal track, can result in even higher acceleration values on this axis.

The consequences of repeated and prolonged exposure to whole-body vibrations include spinal degenerative changes that can cause lumbar pain, hernia, and degenerative disorders in the lumbar discs, as well as other musculoskeletal disorders [25]. As track machines had vibration levels above the action limit, measures must be taken to reduce these levels to an acceptable minimum, below the limits established by Directive 2002/44/EC. As such machines are used over long periods of time, such measures should consider at the same time risk reduction, health monitoring and training, and minimization of exposure to vibration.

In order to reduce exposure to whole-body vibration, Bouazara et al. [26] found reduction of vibration exposure with the use of more modern seats with suspension. This type of solution is the most studied aiming at the effective reduction of the vibration magnitude [27]. However, Paddan and Griffin [28] affirm that the seat of a vehicle can both reduce and increase the vibration for the operator, being important to know the dynamics of each one. The suspension of the seats has spring mechanisms and shock absorbers that produce a low resonance frequency and isolate vibrations, which can contribute to the reduction of whole-body vibration.

Little is known about the effects of changes or optimization of machine cab suspensions over the magnitude of vibration. Using a cushion system at the base of a cabin, Lemerle et al. [29] found a reduction of more than 50% in the magnitude of vibration measured at the seat base and at the chassis of the machine. Studies that show the implementation of preventive strategies that aim to reduce the magnitude of vibration are rare, being more common studies that evaluated the effect of a change in a factor which affected the vibration magnitude [27]. Furthermore, according to the authors, the factors that reduce the magnitude of vibration can be categorized as: (1) design considerations or (2) skills and behavior, on which the most convincing evidence for reducing the magnitude of vibration was found.

Based on the studies of Griffin [30], when the issue involves the health and safety of workers, actions should not be based only on a quantitative analysis. It is appropriate that health monitoring actions and other precautions are taken whenever a risk is suspected and not only in the presence of exposure values above the action level.

5 Conclusions

Under the conditions in which this study was conducted, it can be concluded that:

- Harvesters whose base machines are track hydraulic excavators have occupational vibration levels significantly higher than those presented by tire harvesters;
- The whole-body occupational vibration levels presented by track machines are above the limit of action established by the international reference standards;
- Immediate interventions on the machines and the processes are necessary in order to reduce the exposure of the operators to vibration, as well as their deleterious effects on their health;
- The operators of these machines are exposed to a high risk of developing occupational diseases due to exposure to the physical agent vibration;
- The vibration levels presented by tire harvesters do not expose their operators to the risk of developing occupational diseases resulting from that physical agent.

References

1. IBÁ - Industria Brasileira de Árvores (2017) Relatório IBÁ 2017. Pöyry Consultoria em Gestão e Negócios Ltda, São Paulo
2. Oliveira RJ, Machado CC, Souza AP, Leite HG (2006) Avaliação técnica e econômica da extração de madeira de eucalipto com "clambunck skidder". Revista Árvore 30(2):267–275
3. Santos PHA, Souza AP, Marzano FLC, Minette LJ (2013) Produtividade e custos de extração de madeira de eucalipto com clambunck skidder. Revista Árvore 37(3):511–518
4. Lima JSS, Leite AMP (2014) Mecanização. In: Machado CC (ed) Colheita florestal, 3rd edn. Editora UFV, Viçosa
5. Seixas F, Batista JLF (2014) Comparação técnica e econômica entre harvesters de pneus e com máquina base de esteiras. Ciência Florestal 24(1):185–191
6. Kubba AE, Jiang K (2014) A Comprehensive study on technologies of tyre monitoring systems and possible energy solutions. Sensors 14(6):10306–10345
7. Newell GS, Mansfield NJ, Notini L (2006) Inter-cycle variation in whole-body vibration exposures of operators driving track-type loader machines. J Sound Vib 298:563–579
8. Rozali A, Rampal KG, Shamsul Bahri MT, Sherina MS, Shamsul Azhar S, Khairuddin H, Sulaiman A (2009) Low back pain and association with whole-body vibration among military armoured vehicle drivers in Malaysia. Med J Malaysia 64(3):197–204
9. Sebastião BA, Marziale MHP, Robazzi MLCC (2007) Uma revisão sobre efeitos adversos ocasionados na saúde de trabalhadores expostos à vibração. Revista Baiana de Saúde Pública 31(1):178–186
10. Saliba TM (2016) Manual prático de avaliação e controle de vibração, 1st edn. LTr, São Paulo
11. Rehn B, Nilsson T, Lundström R, Hagberg M, Burström L (2009) Neck pain combined with arm pain among professional drivers of forest machines and the association with whole-body vibration exposure. Ergonomics 52(10):1240–1247
12. Zehsaz M, Sadeghi MH, Ettefagh MM, Shams F (2011) Tractor cabin's passive suspension parameters optimization via experimental and numerical methods. J Terrramech 48(6):439–450

13. Malinovski JR, Camargo CMS, Malinovski RA, Malinovski RA, Castro GP (2014) Sistemas. In: Machado CC (ed) Colheita florestal, 3rd edn. Editora UFV, Viçosa
14. International Organization for Standardzation (1997) ISO 2631-1:1997: mechanical vibration and shock: evaluation exposure to whole-body vibration: part 1, general requirements. ISO, Geneve
15. Diretiva 2002/44/CE do Parlamento Europeu e do Conselho (2002) Jornal Oficial das Comunidades Europeias 177:13–19
16. Lemeshow S, Levy PS (2009) Sampling of populations: methods and applications. John Wiley, New York
17. Leite ES, Minette LJ, Fernandes HC, Souza AP, Amaral EJ, Lacerda EG (2014) Desempenho do harvester na colheita de eucalipto em diferentes espaçamentos e declividades. Revista Árvore 38(1):1–7
18. SAS Institute Inc. (2004) SAS/STAT User's Guide, Version 9.1, Volumes 1–7. SAS Institute Inc., Cary
19. Mansfield NJ (2005) Human response to vibration. CRC Press, London
20. Servadio P, Belfiore NP (2013) Influence of tyres characteristics and travelling speed on ride vibrations of a modern medium powered tractor. Part I: Analysis of the driving seat vibration. Agr Eng Int: CIGR J 15(14):119–131
21. Nguyen V, Inaba S (2011) Effects of tire inflation pressure and tractor velocity on dynamic wheel load and rear axle vibrations. J Terrramech 48(1):3–16
22. Liu ZS, Lu C, Wang YY, Lee HP, Koh YK, Lee KS (2006) Prediction of noise inside tracked vehicles. Appl Acoust 67(1):74–91
23. Mandal BB, Sarkar K, Manwar V (2012) A study of vibration exposure and work practices of loader and dozer operators in opencast mines. Int J Occup Saf Health 2(2):3–7
24. Palmer KT, Griffin MJ, Syddall HE, Pannett B, Cooper C, Coggon D (2003) The relative importance of whole-body vibration and occupational lifting as risk factors for low-back pain. Occup Environ Med 60(10):715–721
25. Sherwin LM, Owende PMO, Kanali CL, Lyons J, Ward SM (2004) Influence of tyre inflation pressure on whole-body vibrations transmitted to the operator in a cut-to-length timber harvester. Appl Ergon 35(3):253–261
26. Bouazara M, Richard MJ, Rakheja S (2006) Safety and comfort analysis of a 3-D vehicle model with optimal non-linear active seat suspension. J Terrramech 43(2):97–118
27. Tiemessen IJ, Hulshof CTJ, Frings-Dresen MHW (2007) An overview of strategies to reduce whole-body vibration exposure on drivers: A systematic review. Int J Ind Ergon 37(3):245–256
28. Paddan GS, Griffin MJ (2002) Effect of seating on exposures to whole-body vibration in vehicles. J Sound Vib 253(1):215–241
29. Lemerle P, Boulanger P, Poirot R (2002) A simplified method to design suspended cabs for counterbalance trucks. J Sound Vib 253(1):283–293
30. Griffin MJ (2004) Minimum health and safety requirements for workers exposed to hand-transmitted vibration and whole-body vibration in the European Union: a review. Occup Environ Med 61(5):387–397

Patient Safety Culture in Education and Treatment Centers: Regional Subcultures

Mehdi Mirzaei, Shirazeh Arghami[✉], Ali Mohammadi,
and Koorosh Kamali

Zanjan University of Medical Sciences, 4515786349 Zanjan, Iran
arghami@zums.ac.ir

Abstract. Background: There is increasing interest to prevent patient injury as the main value in healthcare institutes. However, many patients are still in danger while they receive care services. One factor which influences patient safety is national culture. However, there is no evidence of regional subculture contributions. The aim of this study was to investigate the relation of two different regional subcultures on patient safety culture in education and treatment centers (medical universities hospitals).

Methods: This cross-sectional study assessed the relationship of two regional subcultures in Iran (Turk and Kurd) on patient safety culture. Four education and treatment centers affiliated to Zanjan (Turk subculture) and Ilam (Kurd subculture) were investigated. A sample of 421 nurses, laboratory, and radiology staffs participated in this study. Two kinds of data gathering tools were used, including hospital survey on patient safety culture questionnaire (HSOPSC), and Hofstede's cultural values. Besides descriptive statistics, analytical statistics tests were used at .05 confidence level, including Chi-square, Mann-Whitney, and Linear regression.

Results: Positive response of the dimensions of HSOPSC questionnaire has not been more than 75% (patient safety strength). The total scores of HSOPSC for Turks and Kurds were significantly different (p-value = .016). Moreover, comparing the results of HSOPSC and Hofstede's cultural values questionnaires showed significant coloration (p-value = .01) between patient safety culture and three dimensions of Hofstede's cultural.

Conclusions: It seems regional subculture has its own contribution to patient safety culture. Therefore, considering subculture values as predictors will be helpful to improve patient safety culture.

Keywords: Patient safety culture · Regional subculture · Cultural values Nursing · Paramedical staffs

1 Introduction

The quality of care and improvement of hospital and treatment centers performance is a global issue and one of the main concern of health systems [1]. Hence, patient safety is being regarded as the main value in hospitals and healthcare institutes [2]. Patient safety is a rather new arena in the medical domain and referred to all efforts to prevent patient injury caused by care medicine practice [3]. Experts believe hospitals are

© Springer Nature Switzerland AG 2019
S. Bagnara et al. (Eds.): IEA 2018, AISC 819, pp. 41–53, 2019.
https://doi.org/10.1007/978-3-319-96089-0_5

improving the quality of health and care along with changes in structural mechanisms, institutionalization patient safety culture among employees [1]. Establishment of a positive patient safety culture in the organization enhances employee safety behavior can improve organizational performance, as a result [4]. The world health organization (WHO) has also emphasized the importance of patient safety in healthcare organization [5]. Furthermore, National Quality Forum (2010) acknowledged that culture of safety is one of the seven functional categories for improving patient safety [6].

Although the first goal in any healthcare organization is to prevent patient injury, many patients are still in danger while they receive care services. Various studies showed that 2.9 to 16.6% of patients have experienced at least one undesirable consequence during getting care services [7, 8]. According to the report of WHO, annually millions of patients were victims of injuries caused by unsafe and preventable medical errors [9]. Due to lack of effective error recording system in developing countries, determining the occurrence rate of medical error encounter difficulty in these countries. In Iran, however, the pharmaceutical error has been identified as the most common cause of threatening patient safety. So that about 8% of hospital and healthcare services lead to drug complications [10].

While many efforts to improve patient safety are considered as reactive action (responding after injury), identifying and eliminating hazards before incident occurrence are proactive action and have a great potential for significant enhancement in safety [11]. Therefore, safety culture is considered as a leading indicator which, in contrast lagging indicators, is able to provide valuable information about potential hazards before the fact [11]. It is essential for healthcare organizations to evaluate the current status of patient safety culture in their organization and prioritize dimensions required to be improved [12].

A safety approach plays a substantial role to the extent to which staff members comply safety procedures [13]. Whenever employees consider their own responsibilities in mishap prevention, and whenever the organization supports them on their responsibilities, safety culture will be progressed [14].

Hospital Survey on Patient Safety Culture (HSOPSC) is one of a few tools that exists to assess safety culture. This questionnaire was first developed by the US agency for healthcare and quality in 2004. So far, HSOPSC has been used extensively in healthcare centers and hospital around the world [15]. HSOPSC includes 42 items in 12 dimensions. The dimensions consist of: (1) supervisor/manager expectations and actions promoting safety, (2) organizational learning-continuous improvement, (3) teamwork within units, (4) feedback and communication about error, (5) communication openness, (6) non-punitive response to error, (7) management support for patient safety, (8) staffing, (9) teamwork across units, (10) handoffs and transitions, (11) overall perceptions of patient safety, (12) frequency of events reported [15]. The relationships safety culture between hospital patient safety culture and adverse events has been shown. It means the higher HSOPSC scores has been related to the fewer adverse events in hospitals [16]. HSOPSC was translated into Persian and provides for reliability and validity through factor analysis (CFA) in hospitals throughout the country and has become a valid and standard tool for research in Iran [17].

On the other hand, various studies have shown the relationship between national culture and employee behavior [18, 19]. Safety culture can be amplified or weakened based on the characteristics of a national culture. Therefore, improvement of safety culture needs to consider these characteristics. The idea that national culture likely influences on organizational culture and safety culture has supported in safety research [20]. However, some countries include significant differences among regional subcultures. Since organizations are quite open systems, regional subcultures may influence their culture, either [20]. A Subculture is a subdivision of a national culture which may greatly impact the human behaviors. In this secondary societal group, a rather unique shared pattern is demonstrated [21]. Yet, interactions between regional subcultures and safety culture remains unaccounted for. This can be important in countries like Iran that have ethnics with different cultures.

In Iran, Turk and Kurd cultures have different historical origins, so far as each one has literal languages. In such a way that are ethnicity and defense against differences differ from the characteristics of the Kurd subculture and the cultural relativism and adaptation to the differences of the characteristics of Turk subculture [22].

Hofstede's findings around 1967 and 1973 caused cultural values dimensions to be known and consequently, a questionnaire was developed [23]. So far cultural values questionnaire has been adopted in numerous studies. In Iran, psychometric properties of this questionnaire have been studied. The standard Persian version of this tool includes 16 questions which can assess four dimensions of cultural values, consisting of power distance, masculinity/femininity, uncertainty avoidance, collectivism/individualism [24].

The aim of this survey was to study the relevance of the regional subculture to patient safety culture in four education and treatment centers (medical universities hospitals) in centers of two provinces, namely Zanjan (Turk subculture) and Ilam (Kurd subculture), which are about 600 km apart.

2 Methods

2.1 Study Design, Setting and Participants

A cross-sectional study was carried out in four education and treatment centers. Two centers were in Zanjan and two were in Ilam.

A sample size of 421 was selected of nursing and paramedic (laboratory and radiology) staffs in education and treatment centers of Zanjan and Ilam universities of medical sciences. Physicians have not been entered into the study by reason of policy of health and medical education ministry in Iran, which is to distribute physicians in a demand-based pattern across the country. Therefore, the number of non-regional physicians in education and treatment centers was considerable. Yet, this study was designed to determine the potential impact of regional subculture on patient safety culture.

Two-stage stratified sampling scheme was used. The first stratification was based on metropolitan size. By this level, the sample size form Turk subculture (n = 318) and Kurd subculture (n = 103) were determined. In each stratum, we stratified participants by three working groups (nursing, laboratory, and radiology staffs). Then, in each group, simple random sampling was carried out.

Data were collected between April 15 and 27, 2016. In addition to a questionnaire to collect the socio-demographic and professional characteristics data, 42-item HSOPSC and 16-question Hofstede's cultural values questionnaires were used for data gathering. The psychometric properties of both questionnaires have been approved in Iran [17, 24]. In order to ensure the reliability of the Persian version of the questionnaires, however, a pilot test was conducted preceding data collection using coefficient alpha.

2.2 Ethical Consideration

The study was approved by research ethics committee of Zanjan University of Medical Sciences. Prior to questioning, the supervisors of each hospital wards and the participants were acquainted with the study aims. Furthermore, it was assured that the information would remain anonymous and confidential. It helped us provide a platform for cooperation and encouragement. Then, a verbal informed consent was obtained from each participant.

2.3 Scoring Data Gathering Tools

The outcome variables of HSOPSC were calculated based on the agreement rate of each respondent to the items in a five-point scale as "strongly agree to strongly disagree" or "always to never". The two highest response categories (strongly agree/agree and most of the time/always) were combined and regarded as positive responses. Also, the two lowest ones (strongly disagree/disagree and never/rarely) were added together. However, negatively worded items were reversed and considered as positive responses, too. The middle points of the scales (neither or sometimes) were separately calculated. As proposed by the developers of HSOPSC, dimensions with a response rate of 75% or more were considered as "patient safety strength". Dimension with potential for improvement obtained 50–74% positive response. Less than 50% positive response were considered as weakness patient safety culture [15].

In a like manner, the cultural values questionnaire was answered in a five-point scale. The mean score of each dimension was calculated by averaging the summation of scores given on each related item. The score of each dimension was 5 at maximum. Score 3 was considered moderate. So, needless to say, that a score of more than 3 and less than 3 should be considered, respectively, high and low score [24].

2.4 Statistical Analysis

The reliability tests were performed to assess internal consistency for both questionnaires using Cronbach's α.

The data were analyzed using SPSS (version 11.5) and OpenEpi (version 2.2). Descriptive statistics were used to analyze gender and professional characteristics of the respondents. Normality of the data were checked using Kolmogorov-smirnov test. Then, at the .05 confidence level, a few analytical statistics tests were used to compare the results of each questionnaire in education and treatment centers of the cities, as well as, the results of both questionnaires in comparison to each other. These tests were including Chi-square, Mann-Whitney, and Linear regression.

3 Results

3.1 Internal Consistency of the Questionnaires

Internal consistency (Cronbach's α values) obtained 0.94 for HSOPSC questionnaire, and 0.81 for cultural values questionnaire. Since both were above 0.7, the data gathering tools were of enough reliability.

3.2 Respondents' Gender and Professional Characteristics

The findings showed that a majority of the respondents were Female (78.7%). Nurses were the largest working position (81.5%). The highest number of participants were from intensive care unit (25.2%) and emergency ward (15.7%). The lowest number of participants were from maternity and gynecology ward (2.4%). More than 50% of the participants had less than five years of experience in the current education and treatment center. About 90% of the participants worked more than 40 h within a week. More than 90% of the participants stated that they had direct contact with patients (Table 1).

3.3 Comparison of Patient Safety Culture Dimensions

As mentioned before, the score of HSOPSC is calculated based on Positive Response Rate (PRR). Table 2 shows that in both cultures as a whole, the mean PRR was 36.69, which was less than 75% (patient safety strength).

In both subcultures, the highest Mean PRR belonged to organizational learning-continuous improvement (Turk 53%, Kurd 50%) and teamwork within units (Turk 54%, Kurd 60%). Also, the lowest Mean PRR dimensions were related to non-punitive response to error (Turk 15%, Kurd 12%) and staffing (Turk 13%, Kurd 18%).

The mean PRR between two cultures, however, were significantly different. In total, the mean PRR of Turk was higher than Kurd. Four out of 12 dimensions were significantly higher for Turk, including frequency of events reported (p-value = 0.001), communication openness (p-value = 0.001), feedback and communication about error (p-value = 0.001), and management support for patient safety (p-value = 0.001). Yet, teamwork within units (p-value = 0.027) and staffing (p-value = 0.027) obtained a higher mean PRR for Kurd.

Table 1. Respondents' gender and professional characteristics in two culture

Respondents' characteristics		Turk (percent)	Kurd (percent)	Total (percent)
Gender	Female	257 (8.8)	75 (72.8)	332 (78.7)
	Male	61 (19.2)	28 (27.2)	89 (22.2)
Positions at the hospital	Nursing staff	268 (84.3)	75 (72.8)	343 (81.5)
	Laboratory staff	32 (1.1)	12 (11.7)	44 (1.5)
	Radiology staff	18 (5.7)	16 (15.5)	34 (8.1)
Ward	Internal medicine	17 (5.3)	10 (9.7)	27 (6.4)
	Surgery	30 (9.4)	22 (21.4)	52 (14.4)
	Maternity and gynecology	5 (1.6)	5 (4.2)	10 (2.4)
	Neonatal	15 (4.7)	5 (4.9)	20 (4.8)
	Intensive care unit	90 (28.3)	16 (15.5)	106 (25.2)
	Emergency	49 (15.4)	17 (16.5)	66 (15.7)
	Laboratory	32 (1.1)	12 (11.7)	44 (1.5)
	Radiology	18 (5.7)	16 (15.5)	34 (8.1)
	Others	62 (19.6)	-	62 (12.5)
Experience in current hospital (year)	>1	46 (14.5)	12 (11.7)	58 (13.8)
	1–5	129 (4.6)	43 (41.7)	172 (4.9)
	6–10	117 (36.8)	24 (23.3)	141 (33.5)
	11 to 15	15 (4.7)	7 (6.8)	22 (5.2)
	≥ 16	11 (3.4)	17 (16.6)	28 (6.7)
Job experience in current profession (year)	> 1	36 (11.3)	8 (7.8)	44 (1.5)
	1–5	124 (39)	39 (37.9)	163 (38.7)
	6–10	75 (23.6)	24 (23.3)	99 (23.5)
	11–15	49 (15.4)	14 (13.6)	63 (15)
	≥ 16	34 (1.7)	18 (17.5)	52 (12.3)
Working hours within a week (hour)	<20	0	0	(0)
	20–30	36 (11.3)	11 (1.7)	47 (11.2)
	40–59	192 (6.4)	50 (48.5)	242 (57.5)
	60–79	85 (26.7)	40 (38.8)	125 (27.9)
	≥ 80	5 (1.6)	2 (1.9)	7 (1.6)
Direct contact with patients	Yes	293 (92.1)	99 (96.1)	932 (93.1)
	No	25 (7.9)	4 (3.9)	29 (6.9)

3.4 Comparison of Hofstede Cultural Values

The score of all dimensions of Hofstede cultural values, except masculinity/femininity, were higher than cut off point (3.5) in both subcultures. The whole score of Hofstede cultural values for Turk and Kurd subcultures had no significant difference

Table 2. Respondents' gender and professional characteristics in two culture

Dimensions of patient safety culture	Turk (percent)	Kurd (percent)	Total (percent)	p-value
Frequency of events reported	37	22	30	.0001
Overall perceptions of patient safety	47	45	46	NS
Supervisor/manager expectations and actions promoting safety	42	41	41.5	NS
Organizational learning-continuous improvement	53	50	51.5	NS
Teamwork within units	54	60	57	.027
Communication openness	29	20	24.5	.0001
Feedback and communication about error	51	34	42.5	.0001
Non-punitive response to error	15	12	13.5	NS
Staffing	13	18	15.5	.027
Management support for patient safety	39	26	32.5	.0001
Teamwork across units	37	41	39	NS
Handoffs and transitions	42	44	43	NS
Mean PRR	38	34	36	.016

(p-value = 0.137). These dimensions were greater for Kurd, however, only the difference of Power distance was significant (p-value = .017). While, masculinity/femininity dimension was below cut off point, and higher for Turk at a significant level of .0001 (Table 3).

Table 3. Mean scores of dimensions of Hofstede work-related cultural values in education and treatment centers in Turk and Kurd subcultures

Dimensions of Hofstede cultural values	Turk	Kurd	Total	p-value
Power distance	3.8	3.95	3.83	.017
Uncertainty avoidance	3.9	3.92	3.91	NS
Masculinity/Femininity	3.43	3	3.33	.0001
Collectivism/Individualism	3.59	3.74	3.63	NS
Total score	3.68	3.65	3.67	NS

3.5   Patient Safety Culture and Dimensions of Hofstede Cultural Values

Given the results of the dimensions of Hofstede cultural values (Table 3), which showed differences between two subcultures, correlation of total mean PRR of HSOPSC with mean scores of the dimensions of Hofstede's cultural values was analyzed (Table 4). Linear regression analysis was used to determine any potential relationship between patient safety culture and dimensions of Hofstede work-related cultural values for Turk and Kurd subculture.

Table 4. Linear regression of total mean PRR of HSOPSC with mean scores of the dimensions of Hofstede's cultural values

Dimensions of Hofstede cultural values	PRR in Turk subculture				PRR in Kurd subculture			
	B coefficient	B standard deviation	Standard Beta coefficient	P-Value	B coefficient	B standard deviation	B standard coefficient	P-Value
Power distance	−.684	.29	−.131	.019	−.918	.568	−.154	NS
Uncertainty avoidance	−.853	.277	−.173	.002	−.357	.496	−.071	NS
Masculinity/Femininity	−.43	.203	−.119	.035	−.857	.404	−.201	.037
Collectivism/Individualism	.455	.384	.067	NS	−2.22	.728	−.296	.003

For Turk, power distance (p-value = .019), uncertainty avoidance (p-value = .002), and masculinity/Femininity (p-value = .035) exhibited a significant negative relation with total mean PRR of HSOPSC. Whereas for Kurd, masculinity/Femininity (p-value = .037) and collectivism/Individualism (p-value = .003) showed a significant negative relation.

4 Discussion

Although safety culture has become a significant issue for decades, patient safety culture is a rather new emerging concern for healthcare organizations attempting to improve their service. WHO noted that it is essential for organizations to change their culture to achieve "easy to do the right thing, and hard to do the wrong thing" for patient care [25]. In spite of considerable publications on safety culture and national culture, the potential relationship between regional subcultures and safety culture has still remained unexplained. This study is the first of its kind to investigate such a relation in two subcultures (Turk and Kurd) in Iran.

Findings showed that the mean PRR for all education and treatment centers in both subcultures were less than half of "patient safety strength" criterion (75%). Moreover, the percentage of positive responses for all dimensions were less than 50; except two dimensions (teamwork within the unit and organizational learning-continuous improvement) in both cultures, as well as, feedback and communication about the error for Turk subculture. Generally speaking, findings showed that patient safety culture in these centers was far from what it should be.

However, the difference of mean PRR was significant between Turk and Kurd subcultures (P-value = 0.016). In six dimensions of HSOPSC, positive responses of patient safety culture for two subcultures were significantly different (Table 2). Four dimensions obtained higher score in Turk subculture (P < 0.001), including frequency of events reported, communication openness, feedback and communication about error, management support for patient safety. The dimensions with the higher score for Kurd (p = .027) were teamwork within units and staffing.

Teamwork within unit achieved the maximum positive response rate for Turk (54%) and Kurd (60%) subcultures (Table 2), although, there was a significant difference between them (p-value = 0.027). Other studies have shown the various percentage for this dimension. In most cases, however, it has been at the top rate. These studies include the study of El-Jardali et al. [26] in Lebanon, Ballangrud et al. in Norway, and the annual report on the study of patient safety culture (2016) in the United States [26–28]. In the studies conducted in Iran like the study of Baghaee et al. [29], the mentioned dimension was the highest rate. It is the same for the investigation of Abdi et al. (2013). Therefore, one can conclude that apart from geographical location and cultural status, the dimension of teamwork within unit has usually obtained the best score. It could be due to remarkable job demand for team cooperation, or humanitarian aspects.

On the other hand, among the 12 dimensions of patient safety culture, non-punitive response to error and staffing in both subcultures represented the lowest positive points (less than 20%). Authors like Saleem and Hamdan [30] in Palestine and El-Jardali et al. in Lebanon concluded that the dimension of non-punitive response to error is the most fragile point in patient safety culture [26, 30]. In contrast, Ballangrud et al. in Norway showed that the mentioned dimension was in second place with 78% positive answer [27]. The diversity of findings can be attributed to difference of the reaction of the organization to errors in developed and developing countries. In fact, a blame culture causes employees to feel that reporting the errors may be applied against to them, consequently, they will hide their mistakes. Logically, it seems the most relevant cultural value to this factor of HSOPSC is power distance.

The staffing was the other lowest score of HSOPSC dimension in this study. The researchers as Wang et al. in China, and in AL Ahmadi et al. Saudi Arabia reported the somehow same experience [31, 32]. Then, they mentioned that work pressure, shortage of workers and long working hours may influence on positive responses for this dimension.

The results of Hofstede cultural values survey showed that the total mean score of the responses was 3.67 in all four education and treatment centers, which was more than the cut point (3.5). Also, in both subcultures, the scores of all dimensions were higher than the cut point (Table 3). These results are in agreement with the findings of latest Hofstede's study (1984) in Iran. Except that in the mentioned study, Iran was accounted for a country with a femininity culture, which is in contrast with the present study [33]. Over more than three decades, it is not unexpected some kind of changes in cultural values. Since, they could change by varying political, social and economic environment [34].

In both subcultures, all dimensions of cultural values negatively associated with the whole score of HSOPSC questionnaire (Table 4). A similar conclusion has been found in relation to national and international culture in various studies. For example, Mohamad et al. found that power distance and masculinity had a negative relationship with safety culture in the construction industry in Pakistan [18]. Noort et al. showed in a study that there is a negative relationship between uncertainty avoidance dimensions with safety culture in European air traffic management [19]. Therefore, due to the high score of all cultural values dimensions in this study, it is expectable a low score for patient safety culture. In other words, Hofstede's cultural values can negatively use as a predictor of patient safety culture. Therefore, efforts to reduce all Hofstede's cultural values in healthcare centers can possibly result in promoting patient safety culture.

As Table 4 shows, the negative correlation of collectivism/individualism dimension with patient safety culture is clear in Kurd subculture ($\beta = -.296$, p-value = 0.003). Yet, this is not the same for Turk (p-value = 0.237). So one can say that in Kurd subculture, at least one of the predictors of patient safety culture is collectivism/individualism dimension. However, authors like Mohamed et al. found that collectivism as a factor in improving safety culture [18]. The difference may lie rather in the research environment. Mohamed's study was carried out in the construction workshops, where more careers are done individually by employees who may not be well familiar with each other.

As the same, the negative correlation of uncertainty avoidance with patient safety culture is well-represented in Turk subculture ($\beta = -.173$, p-value = 0.002). This means that in this subculture, uncertainty avoidance could be one of the determinants of patient safety culture. Yet in Kurd, no significant correlation was seen (p-value = 0.474). Some studies showed that high rank in uncertainty avoidance indicates the reduction in effectiveness of safety training, which in turn can lead to increase of keeping staff on the same structure that already existed [35].

In this study, power distance was more than cut point for both subcultures. In Turks, however, it was significantly lower than Kurds (P-value = 0.017) (Table 3). On the other hand, four dimensions of HSOPSC, including frequency of events reported, communication openness, feedback and communication about error, and management support obtained the higher score in Turk subculture (Table 2). Considering the concepts of these dimensions, as well as, the higher score of power distance in Kurds; it seems logical to assume power distance as a predictor for the mentioned dimensions of HSOPSC. Other studies showed that increasing power distance can damage and reduce communications between senior executives and employees. The reason has been attributed to developing a closed environment with the potential of decreasing employees' participation, which in turn attenuates safety culture in the organization [36]. For total PRR, however, power distance accompanied by uncertainty avoidance and masculinity/femininity showed a negative correlation with patient safety culture in Turks (Table 4). On the other hand, power distance is not only included in the regression analysis of Turk subculture (Table 4), but also the score for Turk is significantly lower than Kurd culture (P-value = 0.017) (Table 3). So the difference between four mentioned dimensions is quite understandable between Turk and Kurd subcultures.

5   Study Strengths and Limitations

Plentiful of studies were accomplished on patient safety culture in healthcare domain in different countries. However, few of them have examined the relevance of regional subcultures as a predictor of patient safety culture. This study could be regarded as the first one.

Of course, there has been a few limitations. In this study, the participants were from nurses, laboratory and radiology staffs. As mentioned above, physicians have not been entered to the study. Because, the number of non-regional physicians in education and treatment centers was considerable.

The other point that should not be ignored is that masculinity/femininity in the present study may pertain to power distance. Because the physicians and managers in both subcultures were mostly male whereas more than 70% of participants were female.

6 Conclusion

The findings of this study showed that positive response to HSOPSC questionnaire has not extended to patient safety strength (more than 75%) in all four education and treatment centers in both subcultures (Turk and Kurd). Therefore, it seems necessary to improve patient safety culture in support of patients' rights. Considering the negative relationship between positive response to HSOPSC and Hofstede cultural values score questionnaires, the improvement needs to lessen the predictors (cultural values) in all these centers. Since masculinity/femininity dimension was the only common significant cultural value in both subcultures, it can be concluded that weakening the masculinity/femininity dimension would be necessary in order to promote the patient safety culture in both subcultures.

Further research may carry out to comprise patient safety culture and regional subculture in regard to other factors like population of cities, as well as establishment of HSE management system, and accreditation of healthcare centers.

Acknowledgments. The authors would like to express their highest appreciation to Zanjan University of Medical Science which sponsored the study [Grant Number: A-11-44-4], and friendly cooperation of the education and treatment centers personnel participants in the present study.

References

1. Smits M, Christiaans-Dingelhoff I, Wagner C, van der Wal G, Groenewegen PP (2008) The psychometric properties of the 'Hospital Survey on Patient Safety Culture' in Dutch hospitals. BMC Health Serv. Res. 8(1):230
2. Bodur S, Filiz E (2009) A survey on patient safety culture in primary healthcare services in Turkey. Int J Qual Health Care 21(5):348–355
3. Ilan R, Fowler R (2005) Brief history of patient safety culture and science. J Crit Care 20 (1):2–5
4. Clarke S (2006) The relationship between safety climate and safety performance: a meta-analytic review. Educ Publ Found 11(4):315–327
5. World Health Organization (2008) Summary of the Evidence on Patient Safety: Implications for research. World Health Alliance for Patient Safety, Geneva. http://whqlibdoc.who.int/publications/2008/9789241596541_eng.pdf
6. Myer GS, Denham CR, Angood PB et al (2010) Safe Practices for Better Healthcare—2010 Update: A Consensus Report. National Quality Forum, Washington, DC. http://www.qualityforum.org/WorkArea/linkit.aspx?LinkIdentifier=id&ItemID=25690
7. Wilson RM, Runciman WB, Gibberd RW, Harrison BT, Newby L, Hamilton JD (1995) The quality in Australian health care study. Med J Aust 163(9):458–471

8. Thomas EJ, Studdert DM, Burstin HR, Orav EJ, Zeena T, Williams EJ et al (2000) Incidence and types of adverse events and negligent care in Utah and Colorado. Med Care 38(3):261–271

9. World Health Organization. Call for More Research on Patient Safety. http://www.who.int/mediacentre/news/releases/2007/pr52/en/index.html

10. Moghaddasi H, Sheikhtaheri A, Hashemi N (2007) Reducing medication errors: role of computerized physician order entry system. J Health Adm 10(27):57–67 (in Persian)

11. Pfeiffer Y, Manser T (2010) Development of the German version of the hospital survey on patient safety culture: dimensionality and psychometric properties. Saf Sci 48(10):1452–1462

12. Sorra JS, Dyer N (2010) Multilevel psychometric properties of the AHRQ hospital survey on patient safety culture. BMC Health Serv Res 10(1):199

13. US Department of Health, Services H, Human Services US Department of Health, Human Services (2010) Agency for healthcare research and quality. National Healthcare quality report, pp 2635–2645

14. Wiegmann DA, Zhang H, von Thaden T, Sharma G, Mitchell A (2002) A synthesis of safety culture and safety climate research. University of Illinois at Urbana-Champaign, Savoy

15. Sorra JS, Nieva VF (2004) Hospital Survey on Patient Safety Culture. (Prepared by Westat, under Contract No 290-96-0004). AHRQ Publication No 04-0041. Agency for Healthcare Research and Quality, Rockville

16. Mardon RE, Khanna K, Sorra J, Dyer N, Famolaro T (2010) Exploring relationships between hospital patient safety culture and adverse events. J Patient Saf 6(4):226–232

17. Moghri J, Arab M, Saari AA, Nateqi E, Forooshani AR, Ghiasvand H et al (2012) The psychometric properties of the Farsi version of "Hospital survey on patient safety culture" in Iran's hospitals. Iran J Public Health 41(4):80–86

18. Mohamed S, Ali TH, Tam W (2009) National culture and safe work behaviour of construction workers in Pakistan. Saf Sci 47(1):29–35

19. Noort MC, Reader TW, Shorrock S, Kirwan B (2016) The relationship between national culture and safety culture: Implications for international safety culture assessments. J Occup Organ Psychol 89(3):515–538

20. Carnino A (2002) Management of safety, safety culture and self-assessment. Apostilha do Curso Básico de Treinamento profissional em segurança Nuclear da Agência Internacional de Energia Atômica, Instituto Militar de Engenharia, Rio de Janeiro

21. Lenartowicz T, Roth K (2001) Does subculture within a country matter? A cross-cultural study of motivational domains and business performance in Brazil. J Int Bus Stud 32 (2):305–325

22. Ghaderzadeh A (2012) Intercultural sensitivity in ethnic communities of Iran: study of Turks and Kurds in the city of Ghorveh. J Soc Sci 121–156

23. McSweeney B (2002) Hofstede's model of national cultural differences and their consequences: a triumph of faith-a failure of analysis. Hum Relat 55(1):89–118

24. Mansouri Sepehr RMH, Shekarriz J, Nejati V (2013) Evaluation of factor analysis and reliability of questionnaire of cultural values in organization. Q J Career Organ 5(15):57–71

25. Flin R, Winter J, Sarac C, Raduma M (2009) Report for Methods and Measures Working Group of WHO Patient Safety. Human Factors in Patient Safety: Review of Topics and Tools. World Health Organization, Geneva

26. El-Jardali F, Jaafar M, Dimassi H, Jamal D, Hamdan R (2010) The current state of patient safety culture in Lebanese hospitals: a study at baseline. Int J Qual Health Care 22(5):386–395

27. Ballangrud R, Hedelin B, Hall-Lord ML (2012) Nurses' perceptions of patient safety climate in intensive care units: a cross-sectional study. Intensive Crit Care Nurs 28(6):344–354

28. Sorra J, Gray L, Streagle S, Famolaro T, Yount N, Behm J (2016) AHRQ Hospital Survey on Patient Safety Culture: User's Guide. (Prepared by Westat, under Contract No HHSA290201300003C). AHRQ Publication No 2016
29. Baghaee R, Nourani D, Khalkhali H, Pirnejad H (2012) Evaluating patient safety culture in personnel of academic hospitals in Urmia University of medical sciences in 2011. J Urmia Nursing Midwifery Fac 10(2):155–164
30. Hamdan M, Saleem AA (2013) Assessment of patient safety culture in Palestinian public hospitals. Int J Qual Health Care 25(2):167–175
31. Wang X, Liu K, You L-M, Xiang J-G, Hu H-G, Zhang L-F et al (2014) The relationship between patient safety culture and adverse events: a questionnaire survey. Int J Nursing Stud 51(8):1114–1122
32. Alahmadi H (2010) Assessment of patient safety culture in Saudi Arabian hospitals. Qual Saf Health Care 19(5):e17-e
33. Latifi F (2005) Culture of Iranian Organizations. In: 3rd International Conference Management, Tehran (in Persian)
34. Wu M (2006) Hofstede's cultural dimensions 30 years later: a study of Taiwan and the United States. Intercult Commun Stud 15(1):33
35. Burke MJ, Chan-Serafin S, Salvador R, Smith A, Sarpy SA (2008) The role of national culture and organizational climate in safety training effectiveness. Eur J Work Organ Psychol 17(1):133–152
36. Itoh K, Abe T, Andersen H (2002) A survey of safety culture in hospitals including staff attitudes about incident reporting. In: Proceedings of the workshop on the investigation and reporting of incidents and accidents, Glasgow

The Influence of the Metabolism in the PMV Model from ISO 7730 (2005)

Alinny Dantas Avelino[(⊠)], Luiz Bueno da Silva,
and Erivaldo Lopes Souza

Universidade Federal da Paraíba, Cidade Universitária, s/n - Castelo Branco,
João Pessoa, Brazil
avelino.alinny@gmail.com

Abstract. Thermal comfort is one of the most influential variable to the measurable environmental quality. The most used mechanism to measure thermal comfort in a moderate environment is the Predicted Mean Vote (PMV) of the ISO 7730 standard [1]. Studies show discrepancies between the PMV and the occupants' thermal sensation vote (TSV). The purpose of this study is to analyze the influence of the metabolism on the PMV model that represents the real sensation of students while working on a smart environment. Personal and lifestyle data was collected through questionnaires. The metabolism was calculated through six different methods. A comparison found that the study of Gilani et al. [2] was more accurate considering the TSV. Statistical tests were used to analyze the difference and compare the groups of quantitative, binary and nominal variables with significance level of 0,05. Using Generalized Linear Models (GLM) and the metabolism model of Gilani et al. [2] two models were adjusted. It was shown that gender and drug (medicaments) usage have an influence of approximate 10% over the metabolism. The new models were used to calculate PMVx and PMVy. Both PMVx and PMVy were found to be closer to the TSV than the PMV calculated with the metabolism obtained through the activity (1,2 met), though limited by the results from Gilani et al. [2]. These results show that the evolution of the environments such as the new smart teaching environments drive the need for improvement in thermal comfort studies using personal variables.

Keywords: Thermal comfort · Second keyword · Third keyword

1 Introduction

External heating can cause an increase in internal temperature and thus generate discomfort in the work environments. Also, this same increase causes the subjects to acclimate to increasingly higher temperatures and adjust their respective arterial pressures to this new condition.

Thermal comfort is one of the variables that most influences the indoor environment [3–6]. The Predicted Mean Vote (PMV) is the most used measurement mechanism to evaluate thermal comfort in a moderate environment ever since it was proposed by Fanger in 1970. The PMV predicts how the occupants of an environment will vote

© Springer Nature Switzerland AG 2019
S. Bagnara et al. (Eds.): IEA 2018, AISC 819, pp. 54–64, 2019.
https://doi.org/10.1007/978-3-319-96089-0_6

in a scale from −3 to 3 (−3 being too cold and 3 being too hot) based on six physical and environmental variables.

On the other hand, studies [7–14] show that the use of PMV for different environments is often inadequate. ISO 8996 [15] proposes methods to calculate the metabolism, in addition to these, several researches seek a value for the metabolism that is more representative of reality. In this sense, studies have been carried out to find mathematical models that represent more accurately the metabolism of the subjects [2, 16].

Within this context, the purpose of this study is to establish a relationship between metabolism and personal and lifestyle variables of the subjects analyzed to find a metabolism model that approximates the PMV to the real thermal sensation vote of the research's subjects.

2 Materials and Methods

The following steps were performed to carry out this research.

1st step:

Variables were selected from the literature [16–20] as displayed in Table 1.

The environmental variables were collected through a BABUC/A/M and TGD 300 microclimatic station installed in the center of the room, recording the data every 5 min. Personal and subjective parameters were obtained through questionnaires and the heart rate and blood pressure of the students were obtained by pulse frequency meters and tensiometers.

The study was carried out with 23 students in a smart teaching environment of a university in the city of Teresina, State of Piauí, Northeast Brazil, considering that the external temperature was 40 °C and a thermal sensation around 45 °C. The experiment occurred in September, the hottest period of the year and the internal temperature was 20 °C, 24 °C and 30 °C, on three consecutive days, respectively, following the standards related [1, 20, 21] completing a set of 70 samples.

Table 1. Variables and indicators of the study

Type	Indicators
Thermal sensation	Regarding thermal sensation, how are you feeling right now?
Personal	Gender
	Physical activity
	Height
	Drinks alcohol?
	Smokes?
	Ethnical group
	Age
	Body mass index
	Weight

(*continued*)

Table 1. (*continued*)

Type	Indicators
Health	Dry eyes
	Back, wrists or arms pain
	Nasal congestion
	Headache
	Sore throat
	Light sensibility
	Physical discomfort
	Eye sore
	Excessive fatigue
	Illness in the present
	Family illness
	Chest noise
	Dry skin
	Use of medication
	Tearing
	Arterial pressure
	Heart rate

2nd step:

Metabolism was calculated using six methods available in the literature [2, 15, 16].

The first method is called analytical and consists of calculating the maximum working capacity (MWC), the increase in heart rate per unit of metabolic rate (RM) and basal metabolism culminating in a model that relates metabolism to heart rate (HR) as according to Eqs. 1 to 4.

$$HR = HR_0 + RM(M - M_0) \tag{1}$$

$$RM = (HR_{max} - HR_0)/(MWC - M_0) \tag{2}$$

$$MWC = (41,7 - 0,22A)P^{0,666} \quad \text{for male} \tag{3}$$

$$MWC = (35,0 - 0,22A)P^{0,666} \quad \text{for females} \tag{4}$$

$$HR_{max} = 205 - 0,62A \tag{5}$$

The standard also indicates that, with some loss of precision, Table 2 can be used determine the metabolism. However, the standard does not indicate how to use it: whether in bands or by interpolating the values with the collected data.

Table 2. Direct estimations for the metabolism using the relationship with heart rate

Age (years)	Weight (Kg)				
	50 kg	60 kg	70 kg	80 kg	90 kg
Women					
20	2,9xHR-150	3,4xHR-181	3,8xHR-210	4,2xHR-237	4,5xHR-263
30	2,8xHR-143	3,3xHR-173	3,7xHR-201	4,0xHR-228	4,4xHR-254
40	2,7xHR-136	3,1xHR-165	3,5xHR-192	3,9xHR-218	4,3xHR-244
50	2,6xHR-127	3,0xHR-155	3,4xHR-182	3,7xHR-207	4,1xHR-232
60	2,5xHR-117	2,9xHR-145	3,2xHR-170	3,6xHR-195	3,9xHR-219
Men					
20	3,7xHR-201	4,2xHR-238	4,7xHR-273	5,2xHR-307	5,6xHR-339
30	3,6xHR-197	4,1xHR-233	4,6xHR-268	5,1xHR-301	5,5xHR-333
40	3,5xHR-192	4,0xHR-228	4,5xHR-262	5,0xHR-295	5,4xHR-326
50	3,4xHR-186	4,0xHR-222	4,4xHR-256	4,9xHR-288	5,3xHR-319
60	3,4xHR-180	3,9xHR-215	4,5xHR-249	4,8xHR-280	5,2xHR-311

The next model to obtain the metabolism was adjusted by Gilani et al. [2] and calculates the metabolism using mean arterial pressure data, found according to Nácul [22] through Eq. 6.

$$MAP = \frac{2PAD + PAS}{3} \tag{6}$$

In which:

MAP = mean arterial pressure
PAD = Diastolic blood pressure
PAS = Systolic blood pressure

This model relates the metabolism to blood pressure exponentially and it's codified by Eq. 7.

$$\text{Activity Level} = 0,1092e^{0,0296MAP} \tag{7}$$

The last model used calculates the values of metabolism, called "Luo", through Eq. 8 [16]. The PMV first calculated in this model is obtained through ISO 8996's [15] activity tables so the model works as an adjustment to the PMV value.

$$M = 0,4051PMV^2 - 0,7651PMV + 61,49 \tag{8}$$

3rd step:
PMV1, PMVi, PMVf, PMVg, PMVg and PMVu were calculated using the metabolism models named ISO 1, ISO 2, interpolated, lanes, Gilani and Luo,

respectively. This step is necessary to select the metabolism model that best represents the actual metabolic rate of the occupants of the environment.

The formulation of the proposed PMV model, as well as the comparisons between the models, were developed with the resources of the R Project 3.1.1 software [23].

4th step:

The variables collected from the study subjects were classified as quantitative and qualitative. Qualitative were subdivided into ordinal, nominal and binary. This division was needed to select the appropriate test to measure the relationship of the tested variable with the metabolism. For the quantitative and qualitative variables, the Spearman test was used to identify the existence of a correlation between the metabolism and each one of the variables. For the qualitative binary variables, the Mann-Whitney test was used and for the nominal qualitative variable the Kruskal-Wallis test was used both to analyze differences between the classes in relation to the metabolism. All tests used a significance level of 0.05.

This step is necessary to select which variables may be tested to be in the model in the next step.

5th step:

Using generalized linear models, the metabolism model selected in step 3 was adjusted as a function of the variables selected in step 4. This model was used to calculate the PMV value. Finally, the new value of the PMV was compared with the values found throughout the literature and with the thermal sensation to evaluate if the model is better suited to the reality than the others available in the literature, which are used by ISO 7730 [1].

3   Results

The results obtained in the second step showed inconsistency according to the examples in Table 3. The ISO 1 column provides the data obtained through the analytical model established in ISO 8996 [15]. Some of these results are negative which implies a gain of energy by the subject to perform the task. The difference from ISO 1 to ISO 2 is that ISO 1 was calculated using the maximum heart rate data obtained during data collection and ISO 2 was obtained using the maximum heart rate model available in the standard.

The values of the column "interpolated" in Table 3 were obtained through the interpolation of the data in Table 2. The values of the column labeled "bands" of Table 3 were obtained using each column and each row as a weight and age range respectively.

Table 3. Examples of calculated metabolisms

ISO 1	Interpolated	bands	Gilani	Luo	ISO 2
3.150914	0.683172	0.72069	1.025478	1.099586	1.358949
-3.60606	1.106207	1.106897	1.427169	1.099814	0.223306
1.54308	0.828147	0.817241	1.844532	1.096295	1.040636
4.412304	1.511793	1.589655	1.536785	1.102398	1.554086
2.779382	0.781931	0.741379	1.365188	1.106519	1.198105
-1.12367	0.343728	0.237931	1.575164	1.096331	-0.665341
0.948276	0.825759	0.72069	0.888778	1.096446	0.948275
0.948276	0.557909	0.431034	1.056286	1.09995	0.948275
-1.68731	1.385172	1.517241	1.206784	1.098784	0.490787
1.507216	0.544453	0.237931	1.286715	1.100015	1.072161
-1.10697	2.078948	2.265517	0.90203	1.100006	0.389891
-3.27038	1.108966	1.106897	1.427169	1.100012	0.276741
2.194449	1.719871	1.975862	1.230835	1.096334	1.334806
-2.37268	0.807931	0.72069	1.671229	1.096522	0.175340
2.852087	1.751207	1.905172	1.338513	1.101348	1.245255

The comparison between the values of the PMV described in step 3 is shown in Fig. 1. The value of PMV obtained using the methodology of Gilani et al. [2] has the mean and the closest variation of the values obtained through the opinion of the research subjects (TSV) as observed on Fig. 1. Thus, in this paper, that the model of Gilani et al. [2] will be considered the one that best describes the metabolism of the subjects of the research.

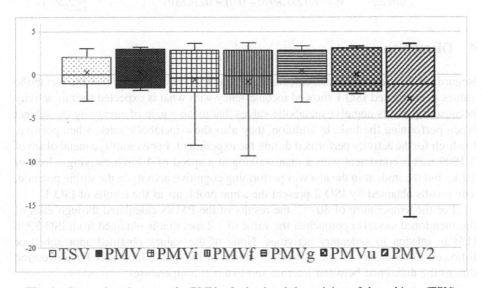

Fig. 1. Comparison between the PMVs obtained and the opinion of the subjects (TSV)

It was observed in the 4th step that there is a relationship between the variables age, pressure, headache, BMI, weight, current illness, family illness, gender, medication use and the metabolism that was calculated by the method of Gilani et al. [2]. The correlations ranged from 0.22 to 0.86, showing that some relationships were weak while others were strong, the relationship between metabolism and headache being the weakest, and the strongest relationship being between metabolism and blood pressure.

The variables that showed a significant relationship with the metabolic variable were: age, diastolic blood pressure measured at the end of the experiment, headache, weight, current illness, family illness, gender and medication use. Then the models were adjusted by testing these variables as shown in Table 4.

The results of the tests show that all the models found represent the metabolism satisfactorily with p-value < 0.05.

Table 4. Development of the generalized linear models from the distributions and bonding functions

Distribution	Bonding function	Model	Likelihood ratio test
Gaussian	Identity	$M = -0.933338 + 0.036013\,\text{FDBP} - 0.085422\text{MED}$	0.0315
	Log	$M = e^{-1.402326 + 0,025408\,\text{FDBP}}$	<2.2e−16
	Inverse	$M = 1/(1,866041 - 0,016657\text{FDBP})$	<2.2e−16
Inverse Gaussian	Identity	$M = -0.863949 + 0.035009\text{FDBP} - 0.090130\,\text{SX}$	0.02055
	Log	$M = e^{-1.546454 + 0.028730\,\text{FDBP} - 0.063320\,\text{SX}}$	0.04097
	Inverse	$M = 1/(2.133796 - 0.020557\,\text{FDBP})$	<2.2e−16
Gamma	Identity	$M = -0.906059 + 0.035682\,\text{FDBP} - 0.090266\,\text{SX}$	0.02069
	Log	$M = e^{-1.489202 + 0.027821\text{FDBP} - 0.061649\,\text{SX}}$	0.04604
	Inverse	$M = 1/(2.028970 + 0.019002\,\text{FDBP})$	<2.2e−16

4 Discussion

Regarding the metabolic rates calculated in step 3, it was observed that the metabolic values denominated ISO 1 showed inconsistency with what is expected for this activity because it presents negative metabolic values that mean a gain of energy by the subject when performing the task. In addition, they also show metabolic rates, when positive, too high for the activity performed during the experiment. For example, a metabolism of 3.1509 met is consistent with a man walking at a speed of 4 km/h carrying a load of 10 kg but the student in the test was performing cognitive activity in the sitting position. The results obtained by ISO 2 present the same problems as the results of ISO 1.

For the temperature of 30 °C, the results of the PMVs calculated through each of the mentioned models approached the value of 1.2 met that is obtained from ISO 8996 [15] in relation to sedentary activities. None of the values obtained approached or followed the tendency of the subjects' opinion of the test, which may have occurred due to the difference between internal and external temperatures.

It is worth of notice that the external temperature was 42 °C at the date of the measurement, so the students are acclimated to a hotter situation which may have influenced their opinion about thermal comfort, despite the acclimation of 30 min before the experiment.

For the temperature of 24 °C, there was no correspondence, but the PMVg, calculated through the method given by Gilani et al. [2], approached more than the result expected by the subjects' opinion. At a temperature of 20 °C, in a given range, the PMV calculated with the metabolism obtained by Table 2, both interpolated (PMVi) and by bands (PMVf), were the same as the responses given by the subjects. However, there were aberrations in the calculation caused by method limitations, which generated a PMVf around −15, for example.

The limitation of this method is that in dynamic work using large muscle groups (as upper limbs) and without mental load and thermal constrictions, there is a linear relationship between heart rate and metabolism [15]. However, the experiment was carried out on subjects performing sedentary activities and who have a mental load component, so this method is not suitable for calculating the metabolism of the subjects in this situation. At the same temperature, although it did not match the opinion of the subjects, the PMVg followed the same tendency of the subjects' opinion.

Preliminarily, it can be concluded that for higher temperatures none of the analyzed methods describes more precisely the opinion of the subjects than the PMV and for lower temperatures, the method developed by Gilani et al. [2] approaches better the opinions of the subjects than the other methods including the use of activity classified tables from ISO 8996 [15]. It happens because the blood pressure of subjects acclimatized at a higher temperature is usually higher than the pressure of subjects acclimated at a lower temperature [26]. Thus, the subjects' pressure is a factor more precise to analyze the metabolism because it accompanies the body regulations caused by acclimatization and thus reduces the influence of acclimatization as a determinant of metabolism.

Considering the above, the model from Gilani et al. [2] was selected as the most representative to be used in this research. The models of Table 4 which contained as only independent variable final diastolic blood pressure (FDBP) had a significant effect, whose p-value < 2,2,10-16. However, the metabolism itself was calculated using the mean pressure that depends on the final diastolic pressure and thus, this model does not bring a relevant relation to this research. Therefore, only the models in function of gender and use of medications will be studied further in this research. Among the three models that relate to gender, the one that presented the best fit according to the likelihood test is model 1 characterized by Eq. 9. Thus, it will be studied as well as model 2 characterized by Eq. 10.

$$M = -0.863949 + 0.035009 \text{FDBP} - 0.090130 \, \text{SX} \tag{9}$$

$$M = -0.933338 + 0.036013 \, \text{FDBP} - 0.085422 \text{MED} \tag{10}$$

The Wald test results for each model showed p-value < 0.05. Thus, model 1 indicates that women have metabolism up to 9.44% lower than that of the men in the study. Model 2 indicates the same way shows that the use of medications can decrease metabolism by up to 10%. Both, for the same blood pressure, have a greater influence on metabolism than gender and the use of drugs.

With the values of the metabolism calculated by models 1 and 2, the PMVs called PMVx and PMVy were calculated. These were compared with the PMVg, PMV and TSV values as shown in Fig. 2. It can be observed that the metabolism elaborated in this study generates a PMV that is better suited to the subjects' opinion than the PMV obtained through the activity in the standard [1]. However, it is also evident that this model is limited by that of Gilani et al. [2] that gave rise to it.

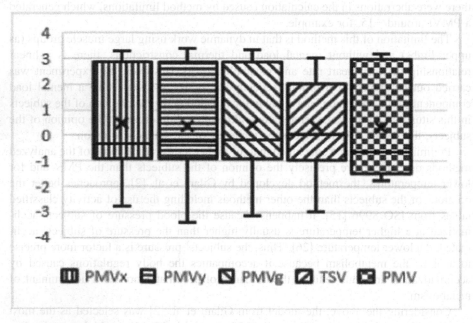

Fig. 2. Statistical comparison of the resulting PMVs with subjects' opinions (SVT)

5   Conclusions

Based on the study carried out it is possible to conclude that metabolism is influenced by gender and the use of drugs that influence the way the subjects perceive thermal comfort. It is also observed that lifestyle habits such as drinking and smoking do not exert a great influence on the metabolism. It is also observed that the improvements in the metabolism calculation make the PMV better fit the reality of the environments, and there is still room for improvement in this work to eliminate the limitation caused using the metabolism of Gilani et al. [2] in the elaboration of the model.

References

1. ISO 7730 (2005) Ergonomics of the thermal environment – analytical determination and interpretation of thermal comfort using calculation of the PMV and PPD indices and local thermal comfort criteria. International Standard Organization, Genève
2. Gilani SI, Khan MH, Ali M (2016) Revisiting Fanger's thermal comfort model using mean blood pressure as a bio-marker: an experimental investigation. Appl. Thermal Eng. 109:35–43
3. Daghigh R (2015) Assessing the thermal comfort and ventilation in Malaysia and the surrounding regions. Renew Sustain Energy Rev 48:681–691
4. Zaki SA, Damiati SA, Rijal HB, Hagishima A, Razak AA (2017) Adaptive thermal comfort in university classrooms in Malaysia and Japan. Build Environ 122:294–306
5. De Dear R, Akimoto T, Arens E, Candido C, Cheong K, Li B, Nishihara N, Sekhar SC, Tanabe S, Toftum J (2013) Progress in thermal comfort research over the last twenty years. Indoor Air 23:442–461
6. Reigner C (2012) Guide to setting thermal comfort criteria and minimizing energy use in delivering thermal comfort. Energy Effi Renew Energy 1–17
7. Attia S, Hensen JLM (2014) Investigating the impact of different thermal comfort models for zero energy buildings in hot climates. In: Proceedings 1st international conference on energy and indoor environment for hot climates, ASHRAE, Doha, Qatar, 24–26 February 2014
8. Auliciems A, Szokolay SV (2007) Thermal Comfort, 2nd and 4th edns. PLEA Notes, Queensland
9. Andreasi WA, Lamberts R, Cândido C (2010) Thermal acceptability assessment in buildings located in hot and humid regions in Brazil. Build Environ 45:1225–1232
10. Dhaka S, Mathur J, Brager G, Honnekeri A (2015) Assessment of thermal environmental conditions and quantification of thermal adaptation in naturally ventilated buildings in composite climate of India. Build Environ 86:17–28
11. Ricciardi P, Buratti C (2012) Thermal comfort in open plan offices in northern Italy: an adaptive approach. Build Environ 56:314–320
12. Ricciardi P, Buratti C (2015) Thermal comfort in the Fraschini theatre (Pavia, Italy): correlation between data from questionnaires, measurements, and mathematical model. Energy Build 99:243–252
13. Humphreys MA, Nicol JF (2002) The validity of ISO-PMV for predicting comfort votes in every-day thermal environments. Energy Build 34:667–684
14. Kim J, Lim J, Cho S, Young G (2015) Development of the adaptive PMV model for improving prediction performances. Energy Build 98:100–105
15. ISO 8996 (2004) Ergonomics of the thermal environment – determination of metabolic rate. International Standard Organization, Genève
16. Luo M, Zhouc X, Zhua Y, Sundell J (2016) Revisiting an overlooked parameter in thermal comfort studies, the metabolic rate. Energy Build 118:152–159
17. Moreira C, Monteiro PRR, Góis J, Miguel AR (2012) Comparative analysis of methods for determining the metabolic rate in order to provide a balance between man and the environment. In: Segurança e Higiene Ocupacionais – SHO. – Livro de Resumos, Sociedade Portuguesa de Segurança e Higiene Ocupacionais (SPOSHO), pp 297–299
18. Choi JH, Loftness V, Aziz A (2012) Post-occupancy evaluation of 20 office buildings as basis for future IEQ standards and guidelines. Energy Build 46:167–175
19. Ruas AC (1999) Avaliação de conforto térmico: contribuição a aplicação pratica das normas interacionais. Campinas. 90f. Dissertação (Mestrado em Engenharia Civil) – Programa de Pós Graduação em Engenharia Civil, Universidade Estadual de Campinas

20. ASHRAE (2003) ASHRAE Standard 55 – Thermal Environmental Conditions for Human Occupancy. Methods. American Society of Heating, Refrigerating and Air-conditioning Engineers, Inc., Atlanta
21. ISO 7726 (2002) Ergonomics of the thermal environment – instruments for measuring physical quantities. International Standard Organization, Genève
22. Nácul FE, O'Donnell JM (2010) Hemodynamic monitoring. In: O'Donnell JM, Nácul FE (eds) Surgical Intensive Care Medicine. Springer, Boston, pp 67–74
23. Garcia R, Ortiz K, Mármol G, Del Cioppo J, Lucio N (2016) Technologies and Innovation. CITI, Guayaquil
24. de Souza AGL (2017) Avaliação da pressão arterial e da frequência cardíaca de alunos na realização de atividades cognitivas em ambientes climatizados. João Pessoa, Paraíba

Perception of Pesticide Contamination Risk in Rural Workers with Low Schooling Level

Luciano José Minette[1] ⓘ, Stanley Schettino[2](✉) ⓘ,
Davi Schettino Mineti[3] ⓘ, and Aracelle Gueler[3] ⓘ

[1] Federal University of Viçosa, Av. P.H. Rolfs s/n, Viçosa, MG, Brazil
[2] Federal University of Minas Gerais, Av. Universitária,
1000, Montes Claros, MG, Brazil
schettino@ufmg.br
[3] Venda Nova do Imigrante College, Av. Ângelo Altoé 888,
Venda Nova do Imigrante, ES, Brazil

Abstract. The rural workers' health can be directly affected by pesticides, through their use in inappropriate amounts and ways, and may have as an aggravating factor the low cognitive level of workers. Therefore, this study aimed to characterize the profile of rural workers with low educational level and to evaluate their perception regarding the pesticide contamination risks. It was carried out in strawberry and tomato producing properties in the State of Espírito Santo, Brazil, on which structured questionnaires were applied, with a sampling of 247 rural workers, containing questions about the perception of pesticide contamination risk, as well as other questions aiming at the characterization of the working process and the morbidity perceived by workers. Through the psychological approach, based on cognitive psychology and using standardized psychometric tests, workers were asked to assign grades on a scale to questions related to reliability, fear, safety, satisfaction and acceptance regarding the use of pesticides. The results evidenced low educational levels of workers, with 23% of them being semiliterate and 12% illiterate, which has led to the following grievances: use of family child labor, lack of use of personal protective equipment, difficulty in interpreting pesticide labels, inadequate disposal of empty packages, deposit of product in inadequate places and lack of training and capacitation. It is concluded that there is a close relationship between exposure to pesticides and the low educational level of rural workers, hampering their perception of occupational problems in the work environment, bringing serious consequences for their health and the environment.

Keywords: Cognitive ergonomics · Social vulnerability · Worker's health

1 Introduction

In the rural environment, the situation of the worker in relation to work safety has become quite worrying. The increasing pressure for production, yield and efficiency has increased the requirement at work and confirmed the insalubrious nature of the activities, demonstrating increased number of accidents, injuries and diseases of all kinds in the rural environment [1, 2]. Agricultural activity has been identified as one of

© Springer Nature Switzerland AG 2019
S. Bagnara et al. (Eds.): IEA 2018, AISC 819, pp. 65–74, 2019.
https://doi.org/10.1007/978-3-319-96089-0_7

the most dangerous and risky in relation to the worker's health and safety [3]. The use of agrochemicals without compliance with aspects related to protection of the worker and safety promotion has also contributed significantly to the development of the risk of diseases resulting from contamination by pesticides, when used in incorrect ways and quantities.

According to Brazilian legislation, pesticides are products and agents of physical, chemical or biological processes, intended for use in the production, storage and processing sectors of agricultural products, in pastures, in the protection of native forests, forest cultures and other ecosystems and urban, water and industrial environments, whose purpose is to change the composition of the flora or fauna, in order to protect them from the damaging action of living beings considered to be harmful, as well as the substances and products used as defoliants, desiccants, stimulators and growth inhibitors [4].

Workers' health can be directly affected by pesticides through contact with contaminated products and indirectly through biota contamination of areas close to agricultural plantations. Contamination can be aggravated by the low cognitive and cultural level of workers. Cognition is the process of knowledge allied to attention, reasoning, thought and memory. Cognitive deficit causes changes in the way the individual processes information, and this deficit is often the result of low educational level, which can hamper performing preventive measures in pesticide contamination [5]. According to Coelho and Coelho [6] there is no doubt that in rural areas the cases of intoxication are intensified due to the lower level of training and education of workers. Education is a socioeconomic variable of extreme importance for cognitive development, since the greater the number of years studied the better the performance in different neuropsychological tasks tends to be [7].

The properly legalized commercial products available in Brazilian market have intrinsic characteristics and meet legal requirements, even though cases of intoxication are registered. It is understood that "a series of precautions should be taken in its use to minimize contact and possible contaminations" [8]. These precautions are related to the procedures of acquiring, transporting, storing, applying and targeting empty packages, and can be defined as well as practices that determine its correct and safe use [9].

Workers do not have the habit of reading the information on the labels of the packages, and when they read it they often interpret it incorrectly, using pesticides in inappropriate quantities and ways. In addition, there is evidence of non-use of Personal Protective Equipment (PPE), which many workers believe to be unnecessary. Lack of knowledge also leads to improper disposal of packages, or their inappropriate washing, which also contributes to contamination of both the environment and man himself. Finally, the lack of information and notion of severity by workers can lead to occupational and environmental contaminations. The fact that most farmers are unaware of the risks they are exposed to results in the negligence of health and safety regulations.

In their studies, Bedor et al. [10] evidenced that low educational level is an important vulnerability for understanding pesticide labeling and its toxicological and environmental implication. In addition to the difficulties in reading, the technical and little understandable language to inform the characteristics of the product and the care with the handling constitute barriers to communication for health and environment.

Farmers have a hard time deciphering the technical terms and even the illustrations inserted in the labels make communication ineffective and dangerous [11].

In face of this, the general objective of this study was to evaluate the perception of pesticide contamination risk in rural workers with low educational level.

2 Materials and Methods

2.1 Characterization of the Study Location

This study was carried out in twelve strawberry and tomato producing properties, located in the municipality of Venda Nova do Imigrante (20°20′59″S and 41°7′49″W), in the south-central region of the State of Espírito Santo, Brazil, with an altitude varying from 630 to 1,550 m, weather with hot and humid summer and dry winter, with average annual temperatures of 19.4 °C and minimum of 5.0 °C. The population is estimated at 20,000 inhabitants, of which about 55% are in the rural area of the municipality. The economic base is in agriculture, with emphasis on the production of vegetable, fruit and farming products, in addition to a smaller scale livestock farming.

2.2 Population and Sampling

The number of sample units to estimate the parameters of an infinite population to a desired level of accuracy based on the standard error of the mean [12], was given by Eq. 1:

$$n = \frac{t^2 . s^2}{d^2 . m^2} \tag{1}$$

where:

n = estimated sample size;
t = value of Student's t distribution, at 5% probability;
s^2 = variance;
d = error in the mean estimate, in percentage;
m = sample mean.

Considering a 95% confidence level, equivalent to two deviations and an estimation error of 5%, the minimum sample size of 225 rural workers was obtained. However, 247 workers were sampled.

All workers were informed about the objectives and methodology of the work, having signed the Free and Informed Consent Term, therefore being in accordance with the Resolution no 466/2012 of the Human Research Ethics Committee of the Ministry of Health [13].

The statistical analyzes were performed using SAS Software version 9.1 [18]. The results found in the work under study do not allow extrapolation to other situations, although they serve as guides for future studies.

2.3   Characterization of the Workers' Profile

A questionnaire in the form of an individual interview was used. The questionnaire was structured in such a way as to avoid problems such as misunderstanding of the questions, allowing clarification at the time of the interview, without mistaken answers. By this method, it was possible to include illiterate people or people with low educational level in the sample. The interviews also helped to know the activities developed in the working process and to familiarize them with the terms used in the work environment.

2.4   Workers' Perception in Relation to Pesticide Contamination Risks

For the study of risk perception, the psychological approach was used, which is based on the opinions expressed by individuals when asked about specific questions related to dangerous activities and/or technologies [14]. It has its foundations in cognitive psychology and standardized tests are often used as an evaluation tool (denominated psychometric), in which the informant is asked to assign grades on a scale to questions related to reliability, fear, safety, satisfaction and acceptance regarding the adoption of a new technology or dangerous activity [14].

As a data collection instrument, a structured questionnaire was used, based on the application of psychometric scales used since the 1970s to evaluate the perception of risks [15, 16], and these scales vary from 1 to 10. The use of a questionnaire in the form of a structured interview, which develops from a fixed list of questions, whose order and writing remain unchanged for all interviewed workers, being in accordance with Britto Junior and Feres Junior [17], is justified as being the most adequate form for the development of social surveys. The questionnaire structured in this way avoids problems such as misunderstood questions, allowing clarification at the time of the interview, and mistaken answers. By this method, it is possible to include illiterate people or people with low educational level in the sample. Interviews also help the researcher to know the activities developed in the working process under study and to become familiar with the terms used in the work environment.

3   Results and Discussion

3.1   Characterization of the Workers' Profile

The study involved 247 workers from the activity of strawberry and tomato cultivation, 65% of which were from the masculine gender and 35% from the feminine gender. The mean age, mass and height of the workers were 35 years old, 67 kg and 167 cm, respectively.

Data from the past decades show that the profile of the rural worker of Espírito Santo did not change, as the surveys of the National Household Sample Survey (NHSS) of 2017 developed by the Brazilian Institute of Geography and Statistics [19] evidenced that the percentage of women involved in rural activities was 35.7% in this period, which does not differ much from the research in question. Both surveys show the incorporation of a significant portion of women working in this productive sphere.

Most workers (57%) were married, 27% were single and the remaining (16%) were cohabiting. The largest contingent of workers had a low average of study years, with 43% not completing elementary school, 23% being semiliterate and 12% unable to read and write. Low educational level may justify the percentage of people involved in strawberry and tomato cultivation, since education is essential for better living conditions. According to NHSS, since the beginning of the research, in the second half of the 60 s of the last century, the educational level of rural workers of Espírito Santo was already low, since most of these workers had no education or studied from 1 to 3 years.

As for working time, 78% perform the function for more than 5 years. However, there were reports of experience in the activity ranging from less than 1 year to 60 years in the investigated properties. This outlook shows that agriculture is one of the oldest economic activities carried out by the human being and involves workers who have dedicated themselves exclusively to work in the field throughout their life [20].

It was observed that most workers (95%) did not receive any type of training, indicating that in the majority of the investigated properties there is no concern with training in any of the labor spheres, like execution of the activity and work safety. The capacitation of workers facilitates the identification of the risks and grievances that they are subjected to. The training of rural workers is essential because, according to Ulbricht and Gontijo [21], agricultural activities were considered as one of the three activities with the greatest occupational risk by the International Labour Organization.

The occupational risk is due to the need for the worker to remain in the standing position, adopting inappropriate postures, to carry out heavy tasks that require handling and transportation of load, exposure to storms, poor use of PPE and especially the use of pesticides. These are easily identified in strawberry and tomato cultivation.

The working day can be considered painful, since 47% of the interviewees affirmed that they work 12 h a day and 33% 10 h. The long working day intensifies occupational risks. According to Labor Law No. 5,889/73, regulated by Decree 73,626/74 of article 7 of the Federal Constitution/1988, the working day of the rural worker is 44 h per week and 220 h per month, and between the two working days a minimum period of 11 (eleven) consecutive hours for rest should be established.

In the properties studied the work activity happens by partnership. In this way the farmer works on lands belonging to other people, with all the production being under his responsibility and distributed with the owners in percentages ranging from 30 to 38%. This fact justifies the long working day of the farmers, who work from 70 to 84 h a week and 280 to 336 h a month, since they work from Monday to Monday. Therefore, the workers have a higher working day than that established by law, being of the order of 59.1% (10 h a day) and 91.0% (12 h a day).

In addition, 60% of workers do not take rest breaks during the activity. According to Salis et al. [20], the concession of breaks throughout the working day is recommended and indispensable.

When questioned about the risks of the work environment, 98% of workers recognized relevantly the heat, cold, rain, noise, dust, falls, machine accidents, repetitive movements, forced postures, physical effort, venomous animals, bacteria, viruses and fungi as factors that could cause harm to health. It is observed that workers recognize well the main agents of risk present in their work environment, despite the low educational level presented by the farmers.

Workers were asked if they felt tired at the start of the working day, 65% answered no, and 35% answered yes. Regarding the physical fatigue after the working day, 87% answered yes, with spine, legs and arms being the most painful parts of the body. The physical fatigue may be related to the extensive working day of the farmers, and the pains may be related to the requirement of using the body parts cited.

It is noted that 54% of workers consider friday the most tiring day of the week and 31% Saturday, due to the accumulation of fatigue of the week, and there is a percentage (18%) who feel tired every day of the week.

It was observed that 79% of workers who perform tomato and strawberry culti-vation considered the work to be harmful to health, although 69% of those interviewed had no health problems at present. According to Veiga et al. [22], a rural worker works up to more than 12 h a day, six times a week, being subjected to a very unhealthy working condition, which can bring serious grievances to his health. The pesticide contamination risk affects not only farmers, but the community in general, since several researches have shown that food contamination by pesticide is alarming [23].

Of the interviewees, 93% usually do not miss work, justifying that tomato and strawberry cultivations require a lot from the worker. Many workers (98%) did not take pre-employment medical examination before starting to work. Only 18% performed periodic examinations, while 82% did not know what they were and what they were for.

The questionnaire used addressed the occurrence of pain and/or musculoskeletal discomfort. It was observed that the main body regions affected were: spine, legs, arms, head, feet/ankles, shoulders and knee.

The greatest complaints were related to legs, spine and arms, members of extreme use during the execution of the activity. The activities carried out manually, especially the agricultural ones, require intense physical effort of the workers. These efforts can have negative effects on the health of the worker.

3.2 Workers' Perception in Relation to Pesticide Contamination Risks

Through the participatory approach of workers, it was observed that the low educa-tional level of the group in question has led to the following grievances: child labor, lack of use of personal protection equipment, difficulty in interpreting agrochemical labels, inappropriate disposal of empty packages, deposit of product in inadequate places and lack of training and capacitation.

It was verified that the majority of the workers involved in strawberry and tomato cultivation developed this function since childhood and adolescence, for this reason they did not have the opportunity to study. This has also been the reality of the children of the workers surveyed, who divide their time between school and plantations. In the holiday months they dedicate themselves fully to work. These children are future candidates for school drop-outs and likely disqualified professionals from the labor market, being forced to accept sub-jobs or continue the poverty cycle of their parents in the future.

The presence of children in plantations is justified by the need to complement the family income. The low educational level of workers does not allow them to understand that school assures the formation of the individual and guarantees his physical, intellectual, emotional and social growth [24].

Workers have an incorrect and distorted view about the future of their children. They believe that if the whole family works every day on plantations they can become a farmer and leave the condition of sharecropper or employee. Therefore, school has been relegated to the second plan. According to Dessen and Polonia [24], school is responsible for the transmission and construction of culturally organized knowledge, modifying the forms of psychological functioning, according to the expectations of each individual. School is the true instrument for changing the financial standards of a family.

The permanence of children in plantations goes through legal context, since child labor is considered illegal in Brazil. According to the Brazilian Ministry of Health [25], under special circumstances the national legislation in force allows work for people over 14 years old as apprentices and for people over 16 years old as workers, protected and with insured labor and social security rights. Unfortunately, this does not happen in tomato and strawberry cultivation. It is known that child labor reduces the child's available time for leisure, family life, education and to establish relations of coexistence with parents, relatives and other people from the community. In addition, children are forced to act as adults, which can generate a source of wear, affecting emotional, cognitive and physical development [25].

Another aggravating factor of the presence of children in plantations is the early exposure to the risks of the work environment. Therefore, the sooner they start working, the worse their health status in adulthood is, since children are more vulnerable to illness and accidents.

When investigated about exposure to pesticides, 91% of workers answered that tomato and strawberry cultivation requires the application of pesticides. According to Barreto [23], the action of pesticides on workers' health can be deleterious and sometimes fatal, provoking from nausea, dizziness, headaches, allergies, renal and hepatic lesions, cancers and other symptoms and pathologies. This action can be noticed immediately upon contact with the product (acute effect) or after some time (chronic effect).

In order to combat the deleterious effects of the action of pesticides, the use of PPE that significantly reduce worker exposure is recommended [11]. However, in the study in question, a considerable percentage of workers (30%) admitted not to use PPE. This is a common reality in Brazilian agriculture, especially in small rural properties, where it is often possible to encounter rural workers without the mandatory PPE during the handling and application of pesticides [22]. Most workers had a low knowledge of the risks of the pesticide use and it is believed that this was one of the factors that contributed to the lack of use of the equipment for their protection.

The studies by Silva et al. [26] showed that low literacy levels contribute to the occurrence of non-use of PPE by many farmers and workers. It was observed that many workers (45%) did not believe in the effectiveness of protection, others (55%) said they did not use it due to the discomfort and hassle caused by the equipment. According to Veiga et al. [22], many PPE may be difficult to use due to the higher resistance of a fabric to permeability, lower thermal comfort, lower portability, anthropometric inadequacies and even environmental characteristics of each locality.

The complaints related to the quality of PPE are constant and can be evidenced in the speech of the workers:

The poison leaks...the whole clothing is soaked.
When we finish spraying, the color of the poison is in the body.
As the visor blurs a lot, we do not wear it and the face is exposed.
It wets the feet, the head...it does not protect anything.
The visor blurs...it's very hot.
It protects little, it's just like ordinary clothing.

Still with regard to the use of PPE, 44% of workers answered that they use the equipment only in situations on which applications are lengthy. Of those who use the security equipment 63% answered that they buy it with their own resource. According to Regulatory Norm No. 31 (RN-31) of the Ministry of Labour and Employment [27], the rural employer must provide personal protection equipment and clothing appropriate to the risks and that do not cause thermal discomfort detrimental to the worker. The equipment and clothing must be in perfect conditions of use and properly sanitized, being replaced whenever necessary.

The RN-31 recommends that the use of personal clothing in the application of pesticides is prohibited, which was commonly observed in the investigated properties. For 60% of workers there is no surveillance body regarding the use of PPE.

During the study, 63% of workers stated that they cannot read and interpret what is written on the packaging of pesticides, thus neglecting the information contained in the labels. According to Bedor et al. [10], the language used to inform the characteristics of the product and handling care constitute barriers to communication of risk to health and the environment. Therefore, pesticide poisoning is associated with low educational level, since workers cannot read the labels.

When asked about the destination of empty packages, 89% answered that most of them are properly discarded and returned to supplying stores. But some packages end up being reused to transport water and residues in the plantation.

Although the majority of workers affirmed that the packages were properly disposed, antagonistic situations were found in the workplace. Thus, the lack of information, education and knowledge of the workers is perceived. For Barreira and Philippi Junior [28], the low level of literacy of workers can be dangerous, and this has led to misuse of packages and incorrect disposal. Low educational level can negatively affect the notion of individuals on environmental conservation. Therefore, workers need to understand that empty packages of pesticides when disposed without control and in inappropriate places can lead to soil and underground water contamination [28].

A large contingent of workers (83%) affirmed that there is a suitable place for pesticide storage in the properties. The lack of perception of these workers was then verified. In practice, inappropriate places without physical infrastructure as required by legislation were observed.

Inadequate storage can cause increased internal pressure in the bottles, thus contributing to the rupture of the package, or even providing the people contamination risk during its opening. In addition, the release of toxic gases can occur, especially from those packages that were not completely emptied or that were externally contaminated by run-off during use. These vapors or gases can endanger the lives of people and even animals from nearby [28]. The greatest aggravating factor is that all of these risks are not perceived by the workers due to the lack of information and knowledge.

Of the workers interviewed, 74% answered that they did not receive guidance on the application of pesticides. In view of this, it is evident the need to carry out a health education work with the workers. It is necessary to implement education strategies involving workers and the question of the use of pesticides [26].

4 Conclusions

In face of the results found in this study, it was concluded that:

- The high rate of pesticide exposure risk (78%) is directly related to the low educational level of rural workers (35% illiterate or semiliterate);
- The low educational level of workers makes it difficult to perceive occupational problems in the work environment, aggravating all pre-existing risks of pesticide exposure;
- The use of pesticides in rural areas has caused a number of serious and irreversible consequences for workers' health and the environment.

References

1. Frank TL, Mcknight R, Kirkhorn SR, Gunderson P (2004) Issues of agricultural safety and health. Annu Rev Public Health 25:225–245
2. Wünsch Filho V (2004) Perfil epidemiológico dos trabalhadores. Rev Bras Med Trab 2 (2):103–117
3. World Health Organization (1995) Global strategy on occupational health for all: the way to health at work. http://apps.who.int/iris/bitstream Accessed 15 Aug 2017
4. Brasil (2002) Decreto nº 4.074, de 4 de janeiro de 2002. Regulamenta a Lei nº 7.802, de 11 de julho de 1989. Presidência da República, Brasília
5. Iida I, Buarque L (2016) Ergonomia: projeto e produção, 3rd edn. Edgard Blucher, São Paulo
6. Coelho EM, Coelho FC (2008) Contaminação por agrotóxicos em São João da Barra, RJ. Rev Perspect Online 2(8):110–116
7. Parente MAMP, Scherer LC, Zimmermann N, Fonseca RP (2009) Evidências do papel da escolaridade na organização cerebral. Rev Neuropsicol Latinoam 1(1):72–80
8. Menten JOM, Canale C, Calaça HE, Flôres D, Menten M (2011) Legislação ambiental e uso de defensivos agrícolas. Citrus Res Technol 32(2):109–120
9. Reccena MCP, Caldas ED (2008) Percepção de risco, atitudes e práticas no uso de agrotóxicos entre agricultores de Culturama, MS. Rev Saúde Pública 42(2):294–301
10. Bedor CNG, Ramos LO, Pereira PJ, Rego MAV, Pavão AC, Augusto LGS (2009) Vulnerabilidade e situações de riscos relacionados ao uso de agrotóxicos na fruticultura irrigada. Rev Bras Epidemiol 12(1):39–49
11. Cerqueira GS, Arruda VR, Freitas APF, Oliveira TL, Vasconcelos TC, Mariz SR (2010) Dados da exposição ocupacional aos agrotóxicos em um grupo assistido por uma unidade básica de saúde na cidade de Cajazeiras, PB. Rev Intertox Toxicol Risco Ambiental Soc 3 (1):16–28

12. Lemeshow S, Levy PS (2009) Sampling of populations: methods and applications. Wiley, New York
13. Brasil (2013) Resolução n° 466, de 12 de dezembro de 2012. Aprova diretrizes e normas regulamentadoras de pesquisas envolvendo seres humanos. Diário Oficial da República Federativa do Brasil, Ministério da Saúde, Brasília
14. Peres F, Rozemberg B, Lucca SR (2005) Percepção de riscos relacionada ao trabalho rural em uma região agrícola do estado do Rio de Janeiro, Brasil: agrotóxicos, saúde e ambiente. Cad Saúde Pública 21(6):1836–1844
15. Fischoff B, Slovic P, Lichtenstein S, Read S, Combs S (1978) How safe is safe enough? A psychometric study of attitudes towards technological risks and benefits. Policy Sci 9 (2):127–152
16. Slovic P (2000) The perception of risk. Earthscan Publishing, London
17. Britto Junior AF, Feres Junior N (2011) A utilização da técnica da entrevista em trabalhos científicos. Evidência 7(7):237–250
18. SAS Institute Inc. (2004) SAS/STAT User's Guide, Version 9.1, Volumes 1–7. SAS Institute Inc., Cary (2004)
19. IBGE - Instituto Brasileiro de Geografia e Estatística (2017) Programa Nacional por Amostra de Domicílios. https://ww2.ibge.gov.br. Accessed 05 Feb 2018
20. Salis HB, Santos JAS, Figueiredo AK, Palhano AN, Diniz RL, Portich P (2002) Apreciação e diagnose ergonômicas no trabalho dos operadores de colheitadeiras de arroz. In: 12th Congresso Brasileiro de Ergonomia, Proceedings, ABERGO, Recife
21. Ulbricht L, Gontijo LA (2003) Os distúrbios osteomusculares relacionados ao trabalho e seus fatores de risco: patologia exclusivamente urbana. Rev Uniandrade 4(1):29–37
22. Veiga MM, Duarte FJCM, Meirelles LA, Garrigou A, Baldi I (2007) A contaminação por agrotóxicos e os equipamentos de proteção individual (EPI). Rev Bras Saúde Ocup 32 (116):57–68
23. Barreto HLP (2008) Manual do trabalhador rural: segurança, saúde e legalidade no uso de agrotóxicos e acidentes com animais peçonhentos. Procuradoria Reg. do Trabalho, Fortaleza
24. Dessen MA, Polonia AC (2007) A família e a escola como contextos de desenvolvimento humano. Rev Paidéia 17(36):21–32
25. Ministério da Saúde (2005) Secretaria de atenção a saúde. Departamento de Ações Programáticas Estratégicas. Trabalho infantil: diretrizes para atenção integral à saúde de crianças e adolescentes economicamente ativos. Ministério da Saúde, Brasília
26. Silva JM, Silva EN, Faria HP, Pinheiro TMM (2005) Agrotóxico e trabalho: uma combinação perigosa para a saúde do trabalhador rural. Rev Ciênc Saúde Colet 10(4):891–903
27. Brasil (2005). Ministério do Trabalho e Emprego. Portaria n° 86, de 3 de março de 2005. Aprova a Norma Regulamentadora de Segurança e Saúde no Trabalho na Agricultura, Pecuária, Silvicultura, Exploração Florestal e Aqüicultura. MTE, Brasília
28. Barreira LP, Philippe Junior A (2002) A problemática dos resíduos de embalagens de agrotóxicos no Brasil. In: Federación Méxicana de Ingenieria Sanitaria y Ciencias Ambientales; AIDIS. Gestión inteligente de los recursos naturales: desarrollo y salud. FEMISCA, Mexico, pp 1–9

Creative Focus Group as an Instrument to Evaluate Work Related Stress

Silvia Gilotta[1]([⊠]), Francesco Deiana[2], Cristina Mosso[2],
Mariangela Ditaranto[3], and Massimo Guzzo[3]

[1] SGS Silvia Gilotta Studio, Turin, Italy
silvia.gilotta@gmail.com
[2] Department of Psychology, University of Turin, Via Verdi, 10, Turin, Italy
[3] Azimut Benetti Group, Via Martin Luther King, Avigliana, TO, Italy

Abstract. From the literature and the outlines guides, it is possible to find different methodologies to collect data from the practice of occupational stress evaluation. Among these, it emerges the Focus Group that Zammuner [24] describes as a "method of qualitative data collection, based on a group talk from which emerge data that the researcher is interested on deeply investigating it". There are several variants of the methodology, that can be used depending on the research purposes. In the case history presented here, there is the need to deal in depth with the aspects concerning the organizational climate and culture. To highlight all the contents, facilitate the sharing of the different points of view, and to ensure the involvement of all participants, it has been chosen a creative alternative inspired by Greenbaum's "expressive drawing" [11]: it is a "projective technique that can be very helpful in eliciting information that might otherwise not be generated in traditional focus group discussions, and that can also energize the group when it's necessary". In practice, it envisages the realization of an artistic artifact in which the workers represent their own perspective, emotional opinion and emotional reaction on topics in question. Compared to a normal "focus group", this variant allows further structuring of the discussion, encouraging participation and comparisons, facilitating the creation of common meanings. Specifically, the operational layout applied is the following: first of all it has been set a short tutorial session related to the constructs of organizational climate and culture; then, it started a warm up phase followed by a creative moment in which the workers have produced some artifacts through which they have described their perception of organizational climate and culture; finally, it has been set a debate in plenary where the participants have talked about their job and its meaning, with the support of a moderator. The data has been collected as notes and processed according to the following categories, obtained from the literature on the organizational climate and culture (James and Jones 1974–1979; Rousseau 1990; Schein [20]): identity-values-ideologies, communication, leadership, rules and incentive, responsibility and freedom, individualism and sense of team, criteria of success. The results obtained have shown that the focus group, in the proposed variant, represents a valid instrument for this activity: the use of the artistic artifact as a way of transmission and sharing the meaning allows a rich and articulated data collection, ensuring a broad and deep vision of organizational reality.

Keywords: Work-related stress · Focus group · Expressive drawing

© Springer Nature Switzerland AG 2019
S. Bagnara et al. (Eds.): IEA 2018, AISC 819, pp. 75–84, 2019.
https://doi.org/10.1007/978-3-319-96089-0_8

1 Context

This research work develops in the context of an assessment of work related stress. Azimut, a well-known company in the nautical production sector, has seized this opportunity as requested by Art.15 of the Legislative Decree.81/2004 [13], to create a chance of listening and dialogue with workers.

Azimut is a large company, a leader in the world market, with several production sites in the Italian territory with more than 1000 employees. In recent years, the company has experienced a period of functional internal reorganization that has also included innovation and changes in the production process.

With a far-sighted intent, the management has hoarded the stress risk analysis activity imposed by the legislator, to start a research process with the aim of understanding in depth the organizational context following this critical period.

This question was addressed to the researchers, authors of this work. In order to respond appropriately to the needs that emerged during the cognitive phase [19], a research activity was initiated aimed primarily at identifying a method of data collection that can give voice to the real experience of workers, and consequently guarantee a deep understanding of organizational reality.

2 Method

The analysis of the literature and of the good operating practices, carried on to identify the most appropriate data collection tool for the specific research context, led to the choice of the "focus group". This tool is particularly suitable for the collection of qualitative data such as the opinions, experiences and narratives of the organizational actors [3, 9].

Zammuner [24] defines the focus group as a qualitative data collection method, based on a group discussion from which emerge data with regards to a topic that the researcher is interested in investigating in depth. It is difficult to trace a single definition in literature, but the researchers agree in defining it as a "discussion involving from 4 to 12 people who, assisted by a moderator, discuss a topic in a permissive and informal environment".

It is a useful method to explore the opinions and attitudes of individuals regarding specific situations and to understand the reasons behind the various human behaviors.

According to Zammuner [24], if we talk about focus groups in the field of social research it is important to use small groups of participants, thus allowing everyone to give a contribution and allowing the moderator to assist the discussion by accompanying them in the exploration of delicate subjects that would not emerge in the confusion of a too large group. As for the duration, it can vary from a minimum of one hour to a maximum of three. The duration will in any case be agreed according to the needs of the client, the participants and the researcher [15].

In general there are four moments that can be identified in a focus group [9]:

(1) Planning and definition: in this first phase the researcher carries on a careful analysis of the context, building on it a working hypothesis, as well as planning the structure of the interview that will be given to the participants.

(2) Conduction: which includes the reception and warm up, in which the moderator introduces himself and explains to the group the aims of the research, after which the moderator will present stimuli that can be visual (films, photos, objects, drawings, vignettes). The following phase is the discussion, which the moderator can decline in a "classical" way by addressing direct questions to the group or by using other methods to facilitate the emergence of the discursive contents. It's right at this stage of the discussion that lies the innovative methodological proposal which is the object of this research work described in the following paragraph.

(3) Data recording and analysis: the data emerged from the focus group can be recorded, by audio/video supports, if in possession of the authorization and the consense of the participants. The analysis phase is very important and is based on the theoretical framework of reference.

(4) Drafting of the final report: in this phase it is possible to follow an ethnographic path, arranging the path according to the treated topic and extending it with the verbalizations of the members, or through a systematic approach, methodically coding the transcription [17].

The literature analysis is also intended to trace other experiences of application of the focus group in the field of risk assessment work related stress, to seek inspiration to define in detail the operating procedure especially for the conduction phase [2–5, 8, 14, 22]. However, in most of the cases collected the practical instructions on the conduction procedure have not been described or explained; in the few cases where they are described it was a simple list of open questions asked to the participants. This lack of literature has offered the opportunity for a creative exercise aimed at developing a new, specific and studied ad hoc technique.

2.1 The Technique of Expressive Drawing

Zammuner [24] shows how in addition to the "classic" open questions that can be asked by the moderator to the group, can be employed other techniques that allow important contents to emerge from the participants: among these alternative or "non-classical" techniques is included the "expressive drawing". Its nature is evidently different from the techniques mentioned above, as participants are required to express their reaction to the subject of study, drawing "how they feel" with respect to it. A different output is required, more significant than just talking. If, in fact, participants are usually called to express themselves through words, and the added value is represented by the interaction between the members and the meanings that emerge, the technique of expressive drawing involves the creation of an artifact, in which it is required to represent your own point of view, opinion or emotional reaction, based on the issues that are being discussed. Warren [23] writes "the individual, be it a professional artist or an inexperienced amateur, enjoys a particular well-being when he creates visual images thanks to their ability to draw from the unconscious and, therefore, to express in colors and shapes an imprint that belongs exclusively to him and that no one else can reproduce in the same way".

Greenbaum [11], inserts the expressive drawing among the projective techniques. These are techniques based on the evocation in individuals of a response to well-known stimuli, in order to elicit the response to new stimuli. According to the author, the technique can be very useful to bring out contents that would be lost by conducing the focus group without the use of stimuli. Participants are asked, individually or in small groups, to create a drawing whose nature depends on the subject of the research, providing them with all the material useful for its creation. When all the drawings have been completed, the information can be elicited in two ways [11]:

1. You can ask each member (or group) to explain to the other participants the meaning of their drawings. The rest of the participants have in this case the task of ensuring that the drawing reflects what they believe "the artist" is thinking, putting him, if necessary, to the test. The important aspect of this exercise is that each artist provides an exhaustive explanation of his own drawing and how it relates to the topic of discussion;
2. You can otherwise leave the participants free to interpret each other's drawings, leaving each one the possibility to correct the others in case their interpretation is different from their own.

It is very difficult to find in literature information on the application of the expressive drawing, for which it seems that don't exist solid procedural indications, so that it remains an exposed area that can be filled only by the professionalism and experience of those who intend to use this approach.

Yuen [26], who used the technique of expressive drawing in research with children and adolescents, underlines several benefits obtained from the use of this technique, including the following:

1. It favors a relaxed atmosphere: the use of figurative art contributes to the creation of a relaxed atmosphere, minimizing the differences among groups of different ages. The technique helps the creation of a common ground, toning down the hierarchical differences and contributing to the creation of shared meanings. By drawing, the participants have the opportunity to reflect on their experience without being influenced by the opinion of others, freely reflecting on the question asked;
2. Useful for describing emotions: drawings are certainly a great way to describe your own emotions, freeing creativity and avoiding the embarrassment of having to describe them in the first person.
3. Useful for understanding the perspective of the participants: in conducting a focus group it is important to encourage the quieter participants. The shyest members can use the drawing as a means of communicating what they would not be able to say with words, adding to the discussion a contribution that would otherwise be lost, and allowing the moderator to elicit thoughts and ideas that would not emerge without the drawings;
4. As a means of structuring and focusing the discussion. Each artifact helps as a focal point in conducting the discussion, avoiding those interruptions that may occur by

conducting the focus group in a classical way. The method helps to focus attention on the topics being discussed.

5. As a means of recognizing and reducing groupthink situations: focus groups are not only conditioned by the relationship between the moderator and members, but also by those among members. As suggested by Fontana and Frey [10], the leader of the focus group must pay close attention to group dynamics, how individual opinions can influence those of others and how relationships outside the focus group can have repercussions within it. Those who participate in focus groups often tend to change their opinions after interacting with others [15]. A sudden change of opinion could depend on the pressure to comply with the idea most shared by the group. This particular type of compliance is called groupthink [16]. The drawings can certainly help to avoid or reduce the emergence of this phenomenon, ensuring the opportunity for everyone to contribute to the discussion. It is certain that groupthink phenomena can also occur in this case (the drawings could be modified after having "peeked" the sheets of others), but recognizing copied drawings is certainly easier than verifying the opinion expressed verbally by an individual is original or less.

The diversity of the application areas shows us how the technique of the focus group is strongly adaptable to different context [6, 18, 25, 26] and the scarcity of indications shows instead how is important that these gaps be filled by the professionalism of those who lead it. Applying the focus group in stress assessment means moving away from the traditional data collection, approaching people instead as carriers of experiences, emotions and opinions. Starting from the indications of Greenbaum, and following these very useful (but almost without operational indications) practical cases, one wondered how this method could be translated within the specific research context. The technique of expressive drawing, although it may be considered with distrust due to its almost playful nature, presents several strengths, including those identified by Yuen [26] and is very useful for overcoming the barriers that can arise in individuals called to expose themselves during data collection for a stress assessment.

This challenging and risky technique, since it's projective and playful, was chosen for its effectiveness in approaching the participants, getting their full involvement, aimed at obtaining in depth information on the organizational context of the research.

2.2 Sample

The project was developed over a period of about five months, involving about 100 workers (belonging to different company partitions) divided between production and administrative departments. Data collection in the field lasted five days.

The evaluation process involved a team composed by the following figures: HR Manager, HR Contact Person, Purchasing Manager, Health and Safety Manager, Health and Safety Psychologist, Doctor in charge, RLS Occupational Psychologist and Ergonomist as External Consultant.

In order to guarantee the richness and variability of data, and the participation of workers that best represent the entire organizational population, the following sampling criteria have been chosen: seniority, which provides a high degree of organizational knowledge; geographical origin (Italian, EU, extra-EU) and gender. It was prepared an information phase that preceded the data collection sessions, which aimed at informing the workers about the nature of the evaluation objectives.

2.3   Operating Procedure

Each session lasted approximately three hours. All 5 days the scheme of activities has been the following:

- Warm up and training on climate and culture. In order to analyze in detail the organizational context, we focused on two fundamental constructs such as culture and organizational climate [18–20]. A frontal training session focused on these topics was then set up, with the aim of creating a theoretical framework and a common language that would facilitate the subsequent discussion. We then used some sequences taken from particularly evocative films, to focus the discussion, capture the attention of the participants and stimulate the comparison among them. The content of the sequences gave the workers the opportunity to recognize elements related to their organizational reality, giving life to a moment of warming and exercise on the issues in question and starting the discussion and exchange of opinions and experiences. This first phase also aims to smooth the differences between researchers and workers, creating a climate of mutual trust.
- Conduction with expressive drawing technique. After the discussion the workers were divided into sub-groups of 4–5 people. They were then asked to represent in an artifact the climate and culture of their company, sharing their experience with their colleagues. Participants were provided with material for the creation of posters: posters, markers, scissors and magazines from which to obtain useful material. After an initial moment of mistrust, everyone took an active part in bringing their own contribution and telling their own experience. Workers have been given complete freedom regarding how to organize themselves (who does what?) and create the billboard: someone used only images, others have written, others have drawn, leaving space for creativity. During the activity, the researchers intervened sporadically only if the participants showed difficulties or were stuck in the realization of the artifact, without interfering in the creative process. The creation of the poster lasted approximately 30 min.
- Group discussion. At the end of the time, the artifacts were collected and presented one at a time by a pair of workers identified internally spontaneously, supported in the arguments by the members of their own subgroup. The participants were asked to present their poster in plenary, telling the contents and explaining the deep motivations that led them to represent certain elements rather than others.

The sharing of artifacts gave rise to a discussion rich in information on organizational reality. The sharing of different business experiences has generated empathy and understanding among the participants of the different groups who have alternated in storytelling and listening.

3 Results

3.1 Data Analysis Procedure

What emerged during the group discussions was transcribed sticking as much as possible to what was expressed by the workers. Non-verbal behaviors were also noted, such as signs and gestures of understanding. Following the directions of Cataldi [6] a qualitative approach was chosen, in order to carry on a critical analysis of the content emerging from the focus groups. For each of the groups, the data collected were inserted within a synoptic framework containing the cognitive categories constitutive of the constructs of climate and culture, to which were subsequently added other categories emerged from the analysis of the data.

The main results in extremely synthetic form are described below.

3.2 Corporate Culture

- Identity-values-ideologies. The identity of the company is very strong and is recognized in the prestige of the brand, in multiculturalism and in the sense of loyalty and belonging, in luxury, in excellence and in the spirit of the group. Despite this, the experiences of the workers are not always positive; in fact, they sometimes feel the failing of the value of craftsmanship, a strong historical component of Azimut ("*nobody seems to remember that the company is artisanal*"). The opening to the world market (the "*big numbers*") has stimulated in the workers the impression of having overshadowed the extreme care of the product (in relation to which they have a strong sense of protection, as they feel it "own"), in favor of an expansion that they say is oriented towards "quantity" rather than "quality". The loss of quality of the product and the scant valorization of the work of the artisans is therefore feared. The company's culture is perceived as a "hybrid" between craftsmanship and industry, an element that could represent a danger if not managed correctly.
- Communication (both bottom-up and top down) is perhaps one of the most critical elements for all groups and this impacts on the knowledge of the phases of the manufacturing process. Moreover, the objectives do not seem to be sufficiently explicit, as are the strategies undertaken by management. In one of the artifacts, some employees draw a boat whose captain (who represents the boss) looks forward, towards "the treasure island", not realizing that there is a fire in the stern. Inside the boat is the crew, represented by three monkeys who "do not see, hear or speak". Employees sometimes, not being aware of the goals, struggle to understand management decisions. It emerged therefore a desire of a clear and concise communication, which supports the production process and helps each worker to raise awareness of his role and that of others, as well as a knowledge of the process that removes the apparent fragmentation in favor of a renewed continuity and simplify information exchange, by eliminating possible ambiguities.
- Leadership. As for the aspects related to leadership, the results have been quite substantial, whether it concerns artifacts or group discussions. There is great nostalgia for the previous leadership, represented by the company's founding founder, an individual capable of transmitting strength, passion and know-how. In one of the

artifacts, he is represented as the "skipper", the one who drives the boat and directs it. Like some sailors who lose their captain, the workers seem to suffer particularly from his absence, repeatedly comparing his figure to the current management. The changes that have characterized the company organization in the two previous years have generated a "gray area", in which middle management is positioned, which has grown exponentially in recent years, but without the necessary integration of skills. At all levels there is a strong stratification, due to an unclear definition of roles that increases the ambiguity of the intermediate positions. This confusion also manifests itself in the figures with responsibilities: they have difficulty in orienting themselves, and often avoid making decisions related to the work of others because they do not possess (or believe not possessing) the right knowledge to do it. This results in a slowdown in production. Workers trust leadership, they ask for more clarity and tools to be more effective and productive.

- Success criteria. All participants agree with the company's success criteria: attention to detail, high quality of work, confidence in the future, love for the product, prestige in the sector, innovation, soundness of the company, passion for craftsmanship, company as a family.

3.3   Organizational Climate

- Responsibility and freedom. The buck-passing is a very strong element: the expectations related to the roles of each do not seem to be sufficiently clear, and this causes a continuous rebound of responsibility. Freedom of decision, understood as the level of discretion that the workers possess, seems to be appreciated at all levels. Despite this, the excessive autonomy with respect to the decisions could turn into a double-edged sword: it is positively perceived when it allows to carry out its own work with good margins of freedom, it is negative when it is a symptom of a scant explicit regulation and clarity of roles. There's a need of clarification of everyone's role; so that everyone knows his responsibilities and cannot avoid facing them.
- Motivation. One of the artifacts is emblematic to describe the motivation in Azimut: bubbles are represented, described as the effervescence, the interest, the motivation to work, conceived as that "complex process of forces that activate, direct and support behavior over time" [1]. The bubbles, however, can not reach the surface, because they are hampered by uncertainty, represented by a suitcase and a directional sign, in which the goals are not specified. Confusion and not always a clear communication have a strong impact on motivational aspects. On the other hand, the group increases the motivation: individuals show a general sense of affiliation towards their colleagues.

4   Conclusions

The results obtained were particularly rich and substantial. The workers, with a great demonstration of trust, expressed the problems of which they are the protagonists every day and their hopes for the future, but also an unconditional bond for their work and for the company that causes joy and pain. The focus group aimed at detecting work-related stress situations was experienced as a moment of listening and communication, in a cognitive moment that allowed us and the workers to obtain a system vision that would not have been obtained with other data collection tools.

All groups of participants showed a strong involvement in the whole process of stress assessment, but the focus group represented in all five days the most emphasized moment especially at an emotional level. In some cases, in fact, the level of involvement has been such deep to reach episodes of outburst, animated discussion, and even emotion.

The focus group, developed with the technique of expressive drawing introduced here for the first time, has achieved the goals of collecting subjective data (the experience of the workers): the people who participated represented with imagination, creativity, transport and freedom their work situation, internal relationships within the specific competence groups and with the rest of the company, through the "organizational narratives" that tell the routine of the company and which took shape and color through the artifacts. The introduction of figurative art has contributed on the one hand to create a relaxed and protected atmosphere, minimizing the expressive and communicative differences between groups, making the process of communication and attribution of meaning more democratic, free and accessible to all, contributing to the creation of a common ground, toning down the hierarchical differences and contributing to the creation and distribution of shared meanings, and on the other, therefore, allowed the researchers to reach the target cognitive objectives useful for the analysis under study.

By drawing individuals see their thoughts taking shape, make connections, and correlations, reflect on their response and introduce elements that would otherwise be lost in the flow of the conversation; they develop complex metaphors together and compare themselves to them, insert the pieces coming from their own daily experience reconstructing processes of which they had no awareness.

The wealth of data emerged, the depth of information obtained on the climate and the organizational culture are a sign of how the stimulus tool chosen has satisfied the requirements of a cognitive tool and a support to the understanding of the dynamics and complex interactions within the organization.

References

1. Avallone F (1994) Psicologia del lavoro. La nuova italia scientifica, Roma
2. Bagnara S, Pietroiusti A, Guidi S (2015) La valutazione approfondita dello stress lavoro correlato in una grande azienda in cambiamento. Med Lav 106(4):250–260
3. Barbour R (1999) Are focus group appropriate tool for studying organizational change? In: Barbour R, Kitzinger J (eds) Developing focus group research: politics, theory and practice, pp 113–126
4. Berland A (2012) Patient safety and falls: a qualitative study of home care nurses in Norway. Nurs Health Sci 14:452–457
5. Berland A, Karvin Natvig G, Gundersen D (2008) Patient safety and job related stress: a focus group study. Intensive Crit Care Nurs 24:90–97

 6. Cataldi S (2012) Come si analizzano i focus group. Franco Angeli, Milano
 7. Cockrell KS, Placier P, Cockrell DH, Middleton J (1999) Coming to terms with "diversity" and "multiculturalism" in teacher education: learning about our students, changing our practices. Teach Teach Educ 15:351–366
 8. Ellis BH, Miller K (1993) The role of assertiveness, personal control and participation in the prediction of nurse burnout. J Appl Commun Res 21(4):327–342
 9. Frisina A (2010) Focus Group: una guida pratica. Bologna
10. Frey JH, Fontana A (1990) The group interview in social research. Soc Sci J 28(2):175–187
11. Greenbaum TL (1998) The handbook for focus group research. Sage, New York
12. Hicks C (2015) Utilization of a focus group to evaluate the percieved stress levels and coping mechanisms of student registered nurse anaesthetist. Doctoral Nursing Capsone Projects. Paper 21
13. INAIL Dipartimento di Medicina del Lavoro - ex ISPESL (2015) Valutazione e gestione del rischio da stress lavoro-correlato, Manuale ad uso delle aziende in attuazione del D.Lgs. 81/08 e s.m.i. Roma
14. Kobasa S, Maddi S, Kahn S (1982) Hardiness and health: a perspective study. J Pers Soc Psychol 42:168–177
15. Krueger RA (1994) Focus group, a practical guide for applied research. Sage, London
16. Janis L (1994) Groupthink: psychological studies of policy decisions and fiascoes. Houghton Mifflin, Boston
17. Morgan DL (1996) Focus groups. Ann Rev Sociol 22:129–152
18. Panteli N (2014) Travelling through Facebook, exploring affordance through lens of age. AIS Electroniv Library (AISeL)
19. De vito Piscitelli P (2002) Clima organizzativo modelli teorici e ricerche empiriche. Franco Angeli, Milano
20. Schein E (1990) Cultura d'azienda e leadership. Guerini
21. Schein EH (2001) La consulenza di processo. come costruire relazioni di aiuto e promuovere lo sviluppo organizzativo. Raffaello Cortina, Milano
22. Warr P (1990) Decision latitude, job demands, and employee well being. Work Stress 4 (4):285–294
23. Warren B (1993) Arteterapia in educazione e riabilitazione. Erickson, New York
24. Zammuner VL (2003) I Focus Group. Il Mulino, Bologna
25. Zhang T, Gobster P (1998) Leisure preferences and open space needs in an urban Chinese American community. J Archit Plann Res 15:338–355
26. Yuen FC (2004) J Leisure Res 36(4):461–482

A Comparison of Sensor Placement for Estimating Trunk Postures in Manual Material Handling

Molly Hischke[✉], Gus Arroyo, Raoul F. Reiser II,
and John Rosecrance

Colorado State University, Fort Collins, CO 80523, USA
hischkmj@colostate.edu

Abstract. Wearable measurement systems have become increasingly more popular in estimating exposures to awkward trunk postures. One limitation in using these systems is the lack of research confirming the optimal placement of the sensors for accurate quantification of trunk postures. The present study explored the effect of sensor placement in estimating trunk postures using Xsens™ (Xsens Technologies, NL) during simulated manual material handling tasks in the laboratory. The researchers found a single IMU on the sternum estimated summary measures and percent time in trunk posture categories similarly to the reference method placement.

Keywords: Low back disorders · Manual material handling · Sensors

1 Introduction

For decades, low back disorders (LBDs) have been recognized as a major cause of injury and disability among many occupational populations (NRC 2001; Marras et al. 2009). A number of systematic reviews have associated awkward postures, specifically postures of the trunk, to the development of LBDs (Andersen et al. 2007; Punnett and Wegman 2004; Putz-Anderson and Bernard 1997; Marras et al. 1995; da Costa and Vieira 2010; Jonsson 1988; Punnett et al. 1991). Certain tasks such as manual material handling (MMH) routinely demand workers to engage in movements inducing awkward trunk postures (Coenen et al. 2013; Putz-Anderson and Bernard 1997; Marras 2010). In an attempt to improve work conditions, NIOSH has called for improved exposure assessment methods, emphasizing on the importance of effectively quantifying exposure to awkward postures in MMH tasks (CDIR 2007).

Direct exposure methods have become common tools used in the quantitative analysis of estimating trunk postures magnitude, duration, and frequency. As a result of recent technological advancements, wearable measurement systems have become more portable for field application, conformable to wear, and cheaper to manufacture (Chaffin et al. 2017).

One limitation of current wearable measurement systems is the lack of knowledge on of the effects of different sensor placement on estimating trunk postures. Previous research has analyzed trunk postures using sensors located on the chest or sternum, lumbar and thoracic regions of the back, shoulders, head, and side of the trunk (Fethke

© Springer Nature Switzerland AG 2019
S. Bagnara et al. (Eds.): IEA 2018, AISC 819, pp. 85–99, 2019.
https://doi.org/10.1007/978-3-319-96089-0_9

et al. 2011; Wong et al. 2009; Faber et al. 2009; Lee et al. 2017; Driel et al. 2013; Graham et al. 2009; Schall et al. 2015a; Yan et al. 2017). No established consensus exists about optimal placement of sensors on parts of the trunk to estimate trunk postures. Estimating trunk postures in the field with wearable measurement systems is challenging because of obtrusive protective equipment preventing sensor placement, sensors being disturbed by thermal, electromagnetic, and mechanical forces, and worker anthropometrics preventing the identification of necessary body landmarks. Determining potential similarities among motion sensors placed on different regions of the trunk can improve the usability of wearable measurement systems in the field. The purpose of the present study was to assess the effect of sensor placement in estimating trunk postures using the Xsens™ (Xsens Technologies, NL) three-dimensional kinematic system.

2   Methods

2.1   Participants

A convenience sample of 30 healthy participants was recruited from Colorado State University. Participants were excluded if they were under 18 years of age or reported experiencing musculoskeletal pain or injury during the time of data collection. After hearing the study protocol and requirements, participants completed and signed forms of consent and photograph release. All procedures in the study were reviewed and approved by the University Institutional Review Board.

2.2   Simulated MMH Tasks

In the lab, participants completed a MMH task which included continuously handling a 1.0 lb. (0.45 kg) cardboard box (length × width × depth = 15 in × 11 in × 2 in) on a table. Data collection began with participants standing upright in a neutral position with arms to the side and feet parallel to one another. Described and depicted in Fig. 1, participants were required to reach for, lower, raise, and push a box in one continuous motion. Participants then returned to neutral position, indicating the completion of one

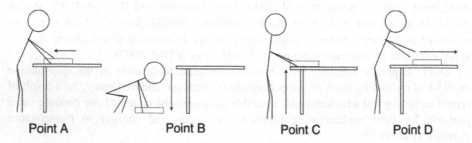

Point A   Point B   Point C   Point D

Fig. 1. The manual material handling tasks completed in the lab by each participant for 10 min (from A to D in one fluid motion). The participants started with (A) reaching for the box and pulling it towards their body (B) lowering the box to the ground without releasing it (C) lifting the box back up to the table and (D) pushing the box across the table.

MMH task cycle, and given five to six seconds of active recovery in the form of walking between MMH task cycles. The frequency of the task was self-paced with participants completing five to eight MMH task cycles per minute for a total of ten minutes.

2.3   Instrumentation

Each participant was fitted with the Xsens™ kinematic system, an inertial measurement system designed for full body and segment motion estimation. The system model used was the Xsens MVN BIOMECH Awinda which consisted of 17 inertial measurement units (IMUs) attached to body segments simultaneously using Velcro straps, a unisex spandex shirt, a headband, and two pairs of gloves (Xsens Technologies B.V. 2015). Each IMU (height × length × width = 55 mm × 40 mm × 10 mm, 16 g) contained a piezoelectric accelerometer (triaxial, ± 16 g), gyroscope (triaxial, ± 2000 deg/sec), magnetometer, and barometer (Xsens Technologies B.V. 2015). The Xsens system estimates velocity, acceleration, and position at a sampling rate of 60 Hz.

Each of the 17 IMUs were secured on the body following anatomical landmarks suggested by the manufacturer. The present study only focused on IMUs on the sternum (2 in Fig. 2), right shoulder (3 in Fig. 2), and sacrum (4 in Fig. 2) but use of additional IMUs was mandatory to execute calibration, data collection, and data processing.

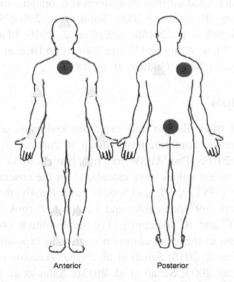

Anterior   Posterior

Fig. 2. Sensor placement for (1) Xsens sensor on sternum, (2) Xsens sensor on right shoulder, and (3) Xsens sensor on sacrum. Grey triangles mark Xsens sensors necessary for system operation but not used to calculate trunk posture estimates.

The Xsens system provided trunk flexion and extension estimates in the sagittal plane based on IMU motion data, Kalman filtering (Xsens Kalman Filter for Human

Movement, Xsens Technologies, NL), body dimensions for each participant, and a built-in biomechanical model. The system uses a right-hand-coordinate system meaning a positive value indicates trunk flexion whereas a negative value indicates trunk extension. Body dimensions were recorded and inputted into supplier-provided software (Xsens MVN Studio 4.0, Xsens Technologies, NL). Prior to the MMH task, the Xsens system was calibrated per manufacturer's instructions.

Estimates of trunk flexion and extension for the sternum, sacrum, and right shoulder segments were recorded in Euler angle form, downloaded in quaternion form using Xsens MVN Studio 4.0 (Xsens Technologies, NL), resampled at 10 Hz, and converted to rotation angles using Matlab (r2016b, The MathWorks Inc., Natick, MA).

Three sensor placement configurations to measure trunk flexion and extension in the sagittal plane were used: (1) the sternum segment values relative to sacrum segment values (X-SST), (2) sternum segment values only (X-ST), and (3) right shoulder segment values only (X-SH). Using trunk flexion and extension estimates derived from IMUs on the right shoulder and sternum was in accordance with manufacturer's requirements of sensor placement and with previous studies in the literature (Plamondon et al. 2007; Foerster et al. 1999; Lee et al. 2017; Fethke et al. 2011; Driel et al. 2012; Graham et al. 2009; Schall et al. 2015a; Yan et al. 2017). Using estimates from the sternum placed IMU relative to the sacrum placed IMU was used as the reference method in the present study because (1) it was a method recommended by the manufacturer and (2) it was similar to comparative studies which have shown this method to be comparable to gold-standard motion systems (i.e. optoelectronic systems) for full body and trunk motions (Roetenberg 2009; Salas et al. 2016; Schepers et al. 2010; Schall et al. 2015a; Schall et al. 2015b; Schall et al. 2016; Plamondon et al. 2007; Robert-Lachaine et al. 2016; Wong and Wong 2008; Van Driel et al. 2009; Bauer et al. 2015; Kim and Nussbaum 2013; Godwin et al. 2009).

2.4 Statistical Analysis

Ensemble averages of trunk flexion and extension estimates were created for each participant (for all sensor configurations) using a custom signal processing tool developed in Matlab (r2016b, The MathWorks Inc., Natick, MA). The arithmetic mean, peak flexion, peak extension values were calculated for the ensemble averages of each measurement method (X-SST, X-ST, and X-SH). Additionally, the 10th percentile, 50th percentile, 90th percentile, 99th percentile, and variation of trunk flexion and extension (difference between 90th and 10th percentiles) of the amplitude probability distribution function were calculated as these are common metrics in exposure assessment studies (Jonsson 1978; Schall et al. 2016; Schall et al. 2015a; Hansson et al. 2010; Lee et al. 2017; Kazmierczak et al. 2005; Schall et al. 2015b; Salas et al. 2016; Howarth et al. 2016). Another metric assessed was the participant's time spent in specific trunk posture categories for the entire ten minutes of simulated MMH task. Based on previous research, the four categories were defined as trunk flexion and extension in the sagittal plane at <0° (Category 1), 0°–30° (Category 2), 31°–60° (Category 3), and >61° (Category 4) (Marklin and Cherney 2005; Hoogendorn et al. 2000; Korshøj et al. 2014; NIOSH 2016; Villumsen et al. 2015; Coenen et al. 2013).

Pearson correlation coefficients and intraclass correlation coefficients were calculated for the mean, 10th, 50th, and 90th percentiles, variation of trunk flexion/extension, and percent time in Category 1 through 4. For the Pearson correlation coefficients, the criteria used to evaluate the strength of linear relationship between metrics was the following: no linear relationship = 0, weak = 0.10 to 0.30, moderate = 0.30 to 0.50, and strong = 0.50 and higher (Taylor 1990). For potential agreement, estimates of ICC and their 95% confident intervals were based on a mean-rating (k = 5), absolute-agreement, two-way mixed-effects model. Agreement level was concluded using criteria by Lee et al. (1989) for the 95% confidence intervals of the ICC: ICC < 0.50 as poor agreement, 0.50 < ICC < 0.75 as moderate agreement, and ICC > 0.75 as strong agreement. Data analysis procedures were conducted using SPSS with a significance level set at 0.05 (Version 21.0, IBM Corp., USA) and graphic procedures were conducted in Excel 2017 (Version 15.36, Microsoft, USA).

3 Results

All participants were recruited from the Colorado State University in Fort Collins, Colorado, USA. The participants (n = 30) were 53% male and 47% female (mean age = 25 years, SD = 4.0; mean height = 452 cm, SD = 27.4 cm).

Estimates of trunk flexion and extension from the reference method, X-SST, and alternative measurement methods X-ST, and X-SH were used to produce ensemble averages of trunk flexion and extension (Fig. 3). The majority of ensemble averages of the participants had three primary peaks characterized by the reaching, pushing, and lowering/lifting motions, respectively, of the simulated MMH task. Ensemble averages typically ranged from approximately three to ten seconds in duration. The largest peak of trunk flexion consistently occurred though the lowering/lifting steps of the MMH task.

3.1 Trunk Flexion and Extension Summary Measures

Summary measures including the mean, peak flexion and extension, 10th percentile, 50th percentile, 90th percentile, and variation (90th–10th percentile) of trunk flexion estimates from X-SH were similar to summary measures from the reference system (Table 1). Similarly, summary measures of X-ST were somewhat comparable, but values for peak flexion and extension, and 90th percentile were higher than estimates from X-SST (Table 1). Pearson correlation coefficients for summary measures for X-ST and X-SH were observed to have strong correlation coefficients, ranging from 0.50 to 0.88 (Table 2). Intraclass correlation coefficients and 95% confidence intervals suggest there was a strong agreement for X-ST and X-SH in estimating trunk flexion and extension observed for the 10th and 50th percentile estimates (Table 3). The 90th, 99th, and variation estimates of trunk flexion for X-ST and X-SH have moderate agreement to the reference method (X-SST) (Table 3).

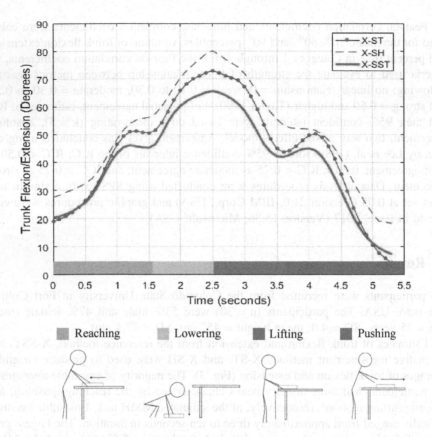

Fig. 3. Example of ensemble average of trunk flexion and extension waveform in sagittal plane by measurement method (X-SST as reference method, X-ST, and X-SH) for one participant.

Table 1. Mean (SD) of summary measures for trunk flexion/extension ensemble averages by measurement method*

Summary measure	X-SST	X-ST	X-SH
Mean (°)	32.0 (9.3)	33.1 (9.9)	30.9 (9.6)
Peak flexion (°)	60.3 (7.8)	69.3 (7.7)	64.2 (9.3)
Peak extension (°)	15.5 (9.0)	11.0 (9.0)	9.8 (9.1)
10th percentile (°)	18.8 (8.6)	14.3 (9.0)	14.3 (9.3)
50th percentile (°)	27.7 (10.3)	27.6 (12.1)	26.0 (11.5)
90th percentile (°)	54.4 (9.8)	62.8 (9.2)	57.7 (9.6)
99th percentile (°)	60.2 (7.8)	69.2 (7.7)	64.1 (9.3)
Variation (90th–10th%)	35.6 (8.3)	48.6 (10.6)	43.4 (11.6)

*X-SST = IMU on sternum relative to IMU on sacrum,
X-ST = estimates from Xsens IMU on sternum,
X-SH = estimates from Xsens IMU on right shoulder.

Table 2. Pearson correlation coefficients (r)* for the mean, 10th percentile, 50th percentile, 90th percentile, and variation of trunk flexion/extension by measurement method**

Summary measure	X-SST	X-ST	X-SH
Mean (r)	REF	0.86	0.68
10th percentile (r)	REF	0.86	0.57
50th percentile (r)	REF	0.88	0.65
90th percentile (r)	REF	0.54	0.55
Variation (90th–10th%) (r)	REF	0.56	0.50

REF = reference method
*Pearson correlation coefficients were statistically significant ($p < 0.05$) unless noted otherwise (two-tailed)
**X-SST = IMU on sternum relative to IMU on sacrum, X-ST = estimates from Xsens IMU on sternum, X-SH = estimates from Xsens IMU on right shoulder.

Table 3. Intraclass correlation coefficients (ICC) and 95% confidence intervals for 10th, 50th, and 90th percentiles and variation of trunk flexion/extension estimates between reference* and alternative methods**

	Intraclass correlation coefficient (ICC)[b]	95% confidence interval	
		Lower bound	Upper bound
10th percentile			
X-ST	0.92	0.84	0.96
X-SH	0.93	0.85	0.96
50th percentile			
X-ST	0.92	0.84	0.96
X-SH	0.78	0.55	0.89
90th percentile			
X-ST	0.70	0.37	0.85
X-SH	0.71	0.39	0.86
Variation (90th–10th%)			
X-ST	0.70	0.37	0.85
X-SH	0.63	0.24	0.82

*Reference method = X-SST, alternative methods = X-ST, X-SH
b = ICC for average measures using a consistency definition, two way mixed models effect
**X-SST = IMU on sternum relative to IMU on sacrum, X-ST = estimates from Xsens IMU on sternum, X-SH = estimates from Xsens IMU on right shoulder.

3.2   Percent Time

On average, the participants spent approximately 60% of the time in Category 2 (0°–30°), about $\sim 20\%$ in Category 3 (30°–60°), and the rest of the time dispersed among Category 1 (<0°) and Category 4 (>60°). Summary measures and Pearson correlation coefficients of percent time in each category by measurement method are presented in

Table 4. Strong correlation coefficients were observed between X-ST and the reference method across all four posture categories. Intraclass correlation coefficients and 95% confidence intervals of the percent time spent in each posture category are provided on Table 5. High intraclass correlation coefficients of X-ST indicated moderate agreement with the reference method in all four posture categories. Percent time estimates from X-SH were a more inconsistent with moderate agreement shown in Category 2 and 4.

Table 4. Summary measures of percent time in Category 1 to 4* per measurement method***

Summary measure	X-SST	X-ST	X-SH
Category 1			
Minimum (%)	0.0	0.0	0.0
Maximum (%)	19.6	21.1	4.6
Mean (%)	1.9	3.0	0.3
Standard Deviation (%)	4.8	5.5	0.9
Pearson correlation coefficient (r)**	REF	0.57	-0.11
Category 2			
Minimum (%)	30.7	44.4	43.8
Maximum (%)	83.8	79.1	78.6
Mean (%)	65.3	66.1	63.4
Standard deviation (%)	13.2	8.1	8.4
Pearson correlation coefficient (r)**	REF	0.60	0.60
Category 3			
Minimum (%)	11.4	7.0	9.2
Maximum (%)	56.0	42.0	43.9
Mean (%)	26.0	21.0	28.2
Standard deviation (%)	11.1	7.3	8.6
Pearson correlation coefficient (r)**	REF	0.58	0.47
Category 4			
Minimum (%)	0.0	0.2	0.0
Maximum (%)	16.8	18.4	21.8
Mean (%)	6.8	9.9	8.2
Standard deviation (%)	4.9	4.0	6.5
Pearson correlation coefficient (r)**	REF	0.50	0.51

* Percent time in Category 1 (>0°), Category 2 (0°–30°), Category 3 (30°–60°), and Category 4 ($\geq 60°$) trunk flexion/extension in sagittal plane
**Pearson correlation coefficients were statistically significant (p < 0.05) unless noted otherwise
*** X-SST = IMU on sternum relative to IMU on sacrum, X-ST = estimates from Xsens IMU on sternum, X-SH = estimates from Xsens IMU on right shoulder.

Table 5. Intraclass correlation coefficients (ICC) and 95% confidence intervals for percent time estimates in Category 1 to 4 between reference* and alternative methods**

	Intraclass correlation coefficient (ICC)[b]	95% confidence interval	
		Lower bound	Upper bound
Category 1			
X-ST	0.72	0.51	0.87
X-SH	0.38	−1.27	0.49
Category 2			
X-ST	0.70	0.52	0.86
X-SH	0.71	0.59	0.86
Category 3			
X-ST	0.69	0.46	0.86
X-SH	0.63	0.52	0.82
Category 4			
X-ST	0.66	0.58	0.74
X-SH	0.65	0.57	0.74

b = ICC for average measures using a consistency definition, two way mixed models effect
*Reference method = X-SST, alternative methods = X-ST, X-SH
**X-SST = IMU on sternum relative to IMU on sacrum, X-ST = estimates from Xsens IMU on sternum, X-SH = estimates from Xsens IMU on right shoulder.

4   Discussion

The present study investigated the effect of sensor placement to estimate trunk postures by comparing an IMU on the sternum (X-ST) and an IMU on the right shoulder (X-SH) to reference method represented by an IMU on the sternum relative to an IMU on the sacrum (X-SST).

4.1   Summary Measures

The findings of the study indicated trunk posture estimates from the sternum IMU were the most comparable to the estimates derived from the IMU on the sternum relative to an IMU on the sacrum. Similar summary measures and strong associations between summary measures were observed between the two measurement methods (Tables 1 and 2). First introduced in Jonsson (1978) for exposure assessments using electromyography, percentiles of exposure from amplitude probability distribution functions have been used extensively as descriptive metrics in occupational studies of biomechanical exposures. Previous literature has shown the use of these descriptive metrics for characterizing jobs and tasks, evaluating effectiveness of interventions, assessing associations between body movements and injury/pain, and comparing exposure assessment tools (Wahlström et al. 2010; Hansson et al. 2010; Kazmierczak et al. 2005; Schall et al. 2015b; Salas et al. 2016; Howarth et al. 2016; Vasseljen and

Westgaard 1997; Bao et al. 1996; Balogh et al. 2006; Unge et al. 2007; Forsman et al. 2002; Jonker et al. 2009; Åkesson et al. 1997). Similar to the present study, other researchers comparing an IMU on the sternum to IMUs on the sternum relative to the sacrum reported comparable measurements of trunk flexion and extension between the two configurations (Schall et al. 2016; Schall et al. 2015a). Because of the strong associations between summary measures, a single IMU placed on the sternum could potentially be used to estimate 10th, 50th, 90th, 99th percentiles, and variation of trunk flexion.

The findings of the study revealed the shoulder IMU was not as comparable to the sternum IMU relative to sacrum IMU. Differences between the methods were the greatest when participants experienced extreme trunk flexion and extension (Table 1). These discrepancies could be due to possible movement artifact from the Xsens shirt, scapular movement, and shoulder posture. Similar to sternum IMU, shoulder IMU estimates for the key percentiles and percent time metrics showed to have acceptable agreement in low flexion variables (10th and 50th percentiles) with estimates from the IMU on the sternum relative to an IMU on the sacrum.

4.2 Percent Time

Assessing time in posture categories has been shown to be practical in a number of industries including manufacturing, nursing, retail, forestry work, military, construction, among others (Wai et al. 2010). The IMU on the sternum showed moderate agreement in measuring time spent in the four different trunk posture categories indicating potential use as a measurement for trunk postures with this system. Similar to sternum IMU, shoulder IMU estimates of percent time metrics showed to have acceptable agreement with estimates from the IMU on the sternum relative to an IMU on the sacrum. Agreement mostly in the time spent in Category 2 (0°–30°), Category 3 (31°–60°), Category 4 (>61°) (Tables 4 and 5). Overall, the sternum IMU had more consistent agreement in estimating the percent of time in each category.

4.3 Limitations

Although the Xsens system has been tested against 'gold-standard' systems for posture analysis, there is not enough consensus in the literature to consider it a 'gold standard' system. Furthermore, the present study was comparing IMUs within the Xsens system in their ability to measure trunk flexion and extension in the sagittal plane for simulated MMH tasks. The system relies on filtering and a biomechanical model, and one cannot generate a full body representation of a participant without using all 17 sensors. The researchers used the system as intended- including all 17 sensors. Although, other research has compared a single IMU on the sternum to an IMU on the sternum relative to an IMU on the sacrum, and reported the two configurations had comparable measurements of trunk postures (Schall et al. 2016; Schall et al. 2015a). Therefore, this suggests if only three sensors from the Xsens system were compared to a validated system (optical motion capture), the sternum IMU and the sternum IMU relative to the sacrum IMU would perform similarly.

Another limitation within the present experiment was the sagittal plane was the only plane analyzed (in terms of trunk flexion and extension). The investigators hypothesize with more complex movements (involving twisting and lateral bending) might require more than a single sensor to produce reliability comparable to the present experiment.

4.4 Implications and Conclusions

Wearable measurement systems, like Xsens, are designed to estimate kinematics based off of sensor placement on anatomical landmarks on the body. The ability to place a single inertial sensor on the sternum or shoulder could serve as an alternative to estimate trunk posture when placing sensors on the sacrum and sternum is not feasible (i.e. in industries such as construction where bulky tool belts, oxygen tanks, and back belts cover parts of the trunk). In situations where worker anthropometrics (e.g. weight, size) make it difficult to locate certain landmarks or are more prone to movement artifact from skin, muscles, or other tissues, having the option to place an inertial sensor on other landmarks can also help assure quality data in exposure assessments (Sazonov et al. 2011; Gemperle et al. 1998; Feito et al. 2011).

The Xsens system requires background knowledge of biomechanics, placement of multiple sensors on different anatomical landmarks, and knowledge of data management and extraction. The complexity of Xsens could discourage occupational safety and health practitioners from adapting them in field applications. Estimating trunk postures with the sternum IMU relative to the sacrum IMU (reference method in the present study) requires a little more data processing than just a single IMU at the sternum. The results of the present study demonstrate the sternum IMU has the capability to measure summary metrics and percent time in posture categories comparable to the reference method which could simplify quantifying trunk posture exposures for occupational professionals.

Future research should include testing wearable devices in a number of simulated and field-based tasks, on workers across different industries, and against properly validated systems. Wearable measurements systems are entering the market quickly, but lacking sufficient research to support their use in daily health and safety practices. The present study adds knowledge on the agreement of estimating trunk postures summary measures and percent time with a single IMU in comparison to an IMU on the sternum relative to an IMU on the sacrum.

This study and publication was supported in part by the Grant T42OH009229 (Mountain and Plains Education Research Center), funded by the Centers for Disease Control and Prevention (CDC). Its contents are solely the responsibility of the authors and do not necessarily represent the official views of the CDC or the Department of Health and Human Services. Additional funding for this project was from the CDC through the National Institute for Occupational Safety and Health (NIOSH) and the High Plains Intermountain Center for Agricultural Health and Safety, grant number U54 OH008085. The content is the responsibility of the authors and does not necessarily represent the official views of the CDC or NIOSH.

References

Åkesson I, Hansson GÅ, Balogh I, Moritz U, Skerfving S (1997) Quantifying work load in neck, shoulders and wrists in female dentists. Int Arch Occup Environ Health 69(6):461–474

Andersen JH, Haahr JP, Frost P (2007) Risk factors for more severe regional musculoskeletal symptoms: a two-year prospective study of a general working population. Arthritis Rheumatol 56(4):1355–1364

Balogh I, Ohlsson K, Hansson GÅ, Engström T, Skerfving S (2006) Increasing the degree of automation in a production system: consequences for the physical workload. Int J Ind Ergon 36(4):353–365

Bao S, Mathiassen SE, Winkel J (1996) Ergonomic effects of a management-based rationalization in assembly work—a case study. Appl Ergon 27(2):89–99

Bauer CM, Rast FM, Ernst MJ, Kool J, Oetiker S, Rissanen SM, Kankaanpää M (2015) Concurrent validity and reliability of a novel wireless inertial measurement system to assess trunk movement. J Electromyogr Kinesiol 25(5):782–790

California Department of Industrial Relations (CDIR) (2007) Ergonomic guidelines for manual materials handling (NIOSH Publication No. 2007-131). U.S. Department of Health and Human Services (DHHS), Center of Disease Control and Prevention, National Institute for Occupational Health and Safety, Washington, DC

Chaffin D, Heidl R, Hollenbeck JR, Howe M, Yu A, Voorhees C, Calantone R (2017) The promise and perils of wearable sensors in organizational research. Organ Res Methods 20 (1):3–31

Coenen P, Kingma I, Boot CR, Twisk JW, Bongers PM, van Dieën JH (2013) Cumulative low back load at work as a risk factor of low back pain: a prospective cohort study. J Occup Rehabil 23(1):11–18

da Costa BR, Vieira ER (2010) Risk factors for work-related musculoskeletal disorders: a systematic review of recent longitudinal studies. Am J Ind Med 53(3):285–323

Driel RV, Trask C, Johnson PW, Callaghan JP, Koehoorn M, Teschke K (2013) Anthropometry-corrected exposure modeling as a method to improve trunk posture assessment with a single inclinometer. J Occup Environ Hyg 10(3):143–154

Faber GS, Kingma I, Bruijn SM, van Dieën JH (2009) Optimal inertial sensor location for ambulatory measurement of trunk inclination. J Biomech 42(14), 2406–2409. 41(Suppl 1), 5527–5528

Feito Y, Bassett DR, Tyo B, Thompson DL (2011) Effects of body mass index and tilt angle on output of two wearable activity monitors. Med Sci Sports Exerc 43(5):861–866

Fethke NB, Gant LC, Gerr F (2011) Comparison of biomechanical loading during use of conventional stud welding equipment and an alternate system. Appl Ergon 42(5):725–734

Foerster F, Smeja M, Fahrenberg J (1999) Detection of posture and motion by accelerometry: a validation study in ambulatory monitoring. Comput Hum Behav 15(5):571–583

Forsman M, Hansson GÅ, Medbo L, Asterland P, Engström T (2002) A method for evaluation of manual work using synchronised video recordings and physiological measurements. Appl Ergon 33(6):533–540

Gemperle F, Kasabach C, Stivoric J, Bauer M, Martin R (1998) Design for wearability. In: Second International Symposium on Wearable Computers, 1998. Digest of Papers. IEEE, pp 116–122

Godwin A, Agnew M, Stevenson J (2009) Accuracy of inertial motion sensors in static, quasi-static, and complex dynamic motion. J Biomech Eng 131(11):114501

Graham RB, Agnew MJ, Stevenson JM (2009) Effectiveness of an on-body lifting aid at reducing low back physical demands during an automotive assembly task: assessment of EMG response and user acceptability. Appl Ergon 40(5):936–942

Hansson GÅ, Balogh I, Ohlsson K, Granqvist L, Nordander C, Arvidsson I, Skerfving S (2010) Physical workload in various types of work: Part II. Neck, shoulder and upper arm. Int J Ind Ergon 40(3):267–281

Hoogendoorn WE, Bongers PM, de Vet HC, Douwes M, Koes BW, Miedema MC, Bouter LM (2000) Flexion and rotation of the trunk and lifting at work are risk factors for low back pain: results of a prospective cohort study. Spine 25(23):3087–3092

Howarth SJ, Grondin DE, La Delfa NJ, Cox J, Potvin JR (2016) Working position influences the biomechanical demands on the lower back during dental hygiene. Ergonomics 59(4):545–555

Jonker D, Rolander B, Balogh I (2009) Relation between perceived and measured workload obtained by long-term inclinometry among dentists. Appl Ergon 40(3):309–315

Jonsson B (1978) Kinesiology: with special reference to electromyographic kinesiology. Electroencephalogr Clin Neurophysiol Suppl 34:417–428

Jonsson B (1988) The static load component in muscle work. Eur J Appl Physiol 57(3):305–310

Kazmierczak K, Mathiassen SE, Forsman M, Winkel J (2005) An integrated analysis of ergonomics and time consumption in Swedish 'craft-type'car disassembly. Appl Ergon 36 (3):263–273

Kim S, Nussbaum MA (2013) Performance evaluation of a wearable inertial motion capture system for capturing physical exposures during manual material handling tasks. Ergonomics 56(2):314–326

Korshøj M, Skotte JH, Christiansen CS, Mortensen P, Kristiansen J, Hanisch C, Holtermann A (2014) Validity of the Acti4 software using ActiGraph GT3X+ accelerometer for recording of arm and upper body inclination in simulated work tasks. Ergonomics 57(2):247–253

Lee W, Seto E, Lin KY, Migliaccio GC (2017) An evaluation of wearable measurement systems and their placements for analyzing construction worker's trunk posture in laboratory conditions. Appl Ergon 65:424–436

Marklin RW, Cherney K (2005) Working postures of dentists and dental hygicnists. J Calif Dent Assoc 33:133–136

Marras WS, Lavender SA, Leurgans SE, Fathallah FA, Ferguson SA, Gary-Allread W, Rajulu SL (1995) Biomechanical risk factors for occupationally related low back disorders. Ergonomics 38(2):377–410

Marras WS, Cutlip RG, Burt SE, Waters TR (2009) National occupational research agenda (NORA) future directions in occupational musculoskeletal disorder health research. Appl Ergon 40(1):15–22

Marras WS, Lavender SA, Ferguson SA, Splittstoesser RE, Yang G (2010) Quantitative dynamic measures of physical exposure predict low back functional impairment. Spine 35(8):914–923

National Institute for Occupational Health and Safety (NIOSH) (2016) Exposure Assessment. https://www.cdc.gov/niosh/programs/expa/. Accessed Aug 2017

National Research Council (NRC) (2001) Musculoskeletal disorders and the workplace: low back and upper extremities. National Academies Press

Plamondon A, Delisle A, Larue C, Brouillette D, McFadden D, Desjardins P, Larivière C (2007) Evaluation of a hybrid system for three-dimensional measurement of trunk posture in motion. Appl Ergon 38(6):697–712

Punnett L, Fine LJ, Keyserling WM, Herrin GD, Chaffin DB (1991) Back disorders and non-neutral trunk postures of automobile assembly workers. Scand J Work Environ Health 17:337–346

Punnett L, Wegman DH (2004) Work-related musculoskeletal disorders: the epidemiologic evidence and the debate. J Electromyogr Kinesiol 14(1):13–23

Putz-Anderson V, Bernard B (1997) Musculoskeletal disorders and workplace factors: a critical review of epidemiologic evidence for work-related musculoskeletal disorders of the neck, upper extremity, and low back. U.S. Department of Health and Human Services, Public Health Service, Centers for Disease Control and Prevention, National Institute for Occupational Safety and Health, (Second, vol. 97–141), Cincinnati

Robert-Lachaine X, Mecheri H, Larue C, Plamondon A (2016) Validation of inertial measurement units with an optoelectronic system for whole-body motion analysis. Med Biol Eng Comput 55:1–11

Roetenberg D, Luinge H, Slycke P (2009) Xsens MVN: full 6DOF human motion tracking using miniature inertial sensors. Xsens Motion Technologies BV, Technical report

Salas E, Vi P, Reider V, Moore A (2016) Factors affecting the risk of developing lower back musculoskeletal disorders (MSDs) in experienced and inexperienced rodworkers. Appl Ergon 52:62–68

Sazonov ES, Fulk G, Hill J, Schutz Y, Browning R (2011) Monitoring of posture allocations and activities by a shoe-based wearable sensor. IEEE Trans Biomed Eng 58(4):983–990

Schall MC, Fethke NB, Chen H, Gerr F (2015a) A comparison of instrumentation methods to estimate thoracolumbar motion in field-based occupational studies. Appl Ergon 48:224–231

Schall Jr, MC, Chen H, Fethke N (2015b) Comparing fatigue, physical activity, and posture among nurses in two staffing models. In: Proceedings of the human factors and ergonomics society annual meeting, vol 59, no 1. SAGE Publications, Los Angeles, pp 1269–1273

Schall MC, Fethke NB, Chen H, Oyama S, Douphrate DI (2016) Accuracy and repeatability of an inertial measurement unit system for field-based occupational studies. Ergonomics 59 (4):591–602

Schepers HM, Roetenberg D, Veltink PH (2010) Ambulatory human motion tracking by fusion of inertial and magnetic sensing with adaptive actuation. Med Biol Eng Compu 48(1):27

Taylor R (1990) Interpretation of the correlation coefficient: a basic review. J Diagn Med Sonogr 6(1):35–39

Unge J, Ohlsson K, Nordander C, Hansson GÅ, Skerfving S, Balogh I (2007) Differences in physical workload, psychosocial factors and musculoskeletal disorders between two groups of female hospital cleaners with two diverse organizational models. Int Arch Occup Environ Health 81(2):209–220

Van Driel R, Teschke K, Callaghan JP, Trask C, Koehoorn M, Johnson PW (2009) A comparison of trunk posture movements: a motion capture system and a new data-logging inclinometer. In: IEA 2009, 17th World Conference on Ergonomics, Beijing, China

Vasseljen O, Westgaard RH (1997) Arm and trunk posture during work in relation to shoulder and neck pain and trapezius activity. Clin Biomech 12(1):22–31

Villumsen M, Samani A, Jørgensen MB, Gupta N, Madeleine P, Holtermann A (2015) Are forward bending of the trunk and low back pain associated among Danish blue-collar workers? A cross-sectional field study based on objective measures. Ergonomics 58(2):246–258

Wahlström J, Mathiassen SE, Liv P, Hedlund P, Ahlgren C, Forsman M (2010) Upper arm postures and movements in female hairdressers across four full working days. Ann Occup Hyg 54(5):584–594

Wai EK, Roffey DM, Bishop P, Kwon BK, Dagenais S (2010) Causal assessment of occupational bending or twisting and low back pain: results of a systematic review. Spine J 10 (1):76–88

Wong WY, Wong MS (2008) Trunk posture monitoring with inertial sensors. Eur Spine J 17 (5):743–753

Wong WY, Wong MS (2009) Measurement of postural change in trunk movements using three sensor modules. IEEE Trans Instrum Meas 58(8):2737–2742

Xsens Technologies B.V. (2015) MVN User Manual. Enschede, Netherlands

Yan X, Li H, Li AR, Zhang H (2017) Wearable IMU-based real-time motion warning system for construction workers' musculoskeletal disorders prevention. Autom Constr 74:2–11

Stochastic Economic Viability Analysis of an Occupational Health and Safety Project

Rogério Miorando[1]([✉]) [iD], Angela Weber Righi[2] [iD], and Priscila Wachs[3] [iD]

[1] Universidade Federal de Santa Catarina, Florianópolis, Brazil
rogerio.miorando@ufsc.br
[2] Universidade Federal de Santa Maria, Santa Maria, Brazil
[3] Universidade Federal do Rio Grande do Sul, Porto Alegre, Brazil

Abstract. This study presents a stochastic analysis for economic viability of an occupational health and safety management systems (OHSMS) project, evaluating the associated risks for a Brazilian electricity distribution company. The method followed four steps: (i) elaboration of discounted cash flow (FC) for the project; (ii) identification of risks associated with the project; (iii) measurement of the risks' financial impact on the project's expected result; and (iv) simulation of the net present value (NPV) considering the impact of the risks using Monte Carlo method. The analysis also performed risk hierarchies to identify the biggest threats and the best opportunities for the project. The risks that presented the greatest impact threaten the benefits: Inability of the affected areas to assimilate the implemented changes, Interruption of the project in the intermediate stage, and Application overhead.

Keywords: OHSMS · Economic viability · Stochastic analysis

1 Introduction

The importance of structured occupational health and safety management systems (OHSMS) is evidenced in the literature by several authors [1, 2]. However, literature is scarce with respect to studies of the uncertainties and risks that threaten the economic and social benefits of these projects, making their economic viability uncertain.

Risk analysis has become an increasingly important area for decision making, since most economic project decisions are made in scenarios involving uncertainty. The sources of uncertainty are multiple and extensive, covering risks associated with markets, suppliers, meteorology, technology, etc. [3].

Authors such as Wu [4] and Pender [5] report the dissatisfaction of project managers regarding the flaws that traditional methods of analysis present in managing uncertainties of projects. Among them is the inability to explicitly address the risks. Regarding OHSMS projects, the reality is no different. Although there is a consensus about the importance of estimating the economic viability of intervention projects in ergonomics, few tools are available for this task [6].

This study presents a stochastic analysis for the economic viability of an OHSMS project, evaluating the risks associated with a Brazilian electricity distribution

© Springer Nature Switzerland AG 2019
S. Bagnara et al. (Eds.): IEA 2018, AISC 819, pp. 100–104, 2019.
https://doi.org/10.1007/978-3-319-96089-0_10

company. The OHSMS project analyzed involves the development of a method to train electrical network electricians in resilience skills [7]. The period of analysis of the project was five years, with one year for implementation and four years of operation before the need for new investments.

The project, encouraged by resources from ANEEL (National Electric Energy Agency - Brazil), left many doubts regarding its possible returns and its economic viability to be reproduced in other electric power companies. In this way, the main contribution of this study is focused on the financial measurement of the risks' impact on the costs and expected returns of the project. This type of analysis in the theme of Ergonomics and Safety allows a greater understanding and convincing of all about the importance of the project and its effective results. Mainly because they are, for the most part, intangible benefits that cannot be quantified by traditional methods of analysis.

2   Method

For this proposed analysis, an adaptation of the Miorando's method [8] was carried out for application in OHSMS projects. The model provides an economic-probabilistic analysis for projects that integrates risk analysis to economic analysis. For this, the model uses probability distributions to quantify the value of possible deviations from the project cash flow, providing an economic-probabilistic analysis of the expected returns.

The method of analysis followed four steps: (i) elaboration of the discounted cash flow for the project, (ii) identification of the risks associated with the project, (iii) quantification of the risks' financial impact on the expected financial result for the project, and (iv) simulation of the project risk-adjusted result. Data for project analysis were collected through meetings with company representatives related to the Occupational Safety and Health Division. Four meetings were held, totaling 9 h, for data collection and project analysis.

2.1   Elaboration of the Discounted Cash Flow

The cash flow of the project has been set up to clearly show related costs and benefits. The costs were divided into five categories: (i) materials and equipment, (ii) consumer material, (iii) human resources, (iv) travel and daily expenses, and (v) others. The benefits were divided, like the costs, in the following categories: (i) reduction of the Accident Prevention Factor, (ii) accidents reduction, (iii) improvement of the indicators with the regulator, (iv) overtime reduction, e (v) reduction of Call Center demand. Costs and benefits were estimated by consensus between the project manager and representatives of the Occupational Safety and Health Division. The discount rate used for the project was 0.8% per month.

2.2   Identification of the Risks

As well as the estimated costs and benefits, the risk identification and assessment were performed jointly by the project manager and the representatives of the Occupational

Safety and Health Division. The option of joint analysis, instead of separate analyzes, aimed to stimulate a greater discussion of the risks involved in the project. The risks were organized within the categories of costs and benefits presented in the cash flow. The risk characterization aimed to minimize possible correlations in the analysis. At the end, 15 risks that impacted the benefits and 22 risks that impacted the project costs were identified.

2.3   Quantification of the Risks' Financial Impact

The financial quantification of the risks impact was carried out by completing a form in which analysts answered the probability of occurrence of each risk for five ranges of values corresponding to their financial impact. The intervals ranged from the worst-case scenario to the best scenario for the presence of risk. For all estimates, the values were determined by consensus among analysts.

2.4   Simulation of the Project Risk-Adjusted Result

The result of the risk-adjusted project was the intersection of the project cash flow information with the quantification of the risks financial impact through a stochastic simulation. As a result, the simulation generated the probability distribution for the Net Present Value (NPV) of the risk-adjusted project. The simulation was performed following the Monte Carlo technique using the @Risk software. At the end of this stage, a hierarchy of the risks was also performed, based on the coefficients of a linear regression analysis of the risks. The hierarchy helped identify the major threats and the best opportunities for the project.

3   Main Results

The result of the simulation showed a considerable reduction of the expected benefits for the project when it is adjusted to the risks. The expected risk-free NPV for benefits was $ 3,738,874 (all amounts are presented in Brazilian Reais - BRL). Considering the impact of the risks present in the project, the average risk-adjusted NPV was $ 1,440,763. This shows that an analysis without considering the risks overestimates the benefits by 160% on average. Even for the implementation costs, which are more easily quantifiable, the risk analysis showed an average variation of 9% when compared to a risk-free analysis. The risk-free NPV of implementation costs was $ 214,997, while the average risk-adjusted NPV was $ 234,096.

The overall project result analysis showed that the risk-free NPV of the project was $ 3,523,878. This value overestimates the project results by 192% on average when compared to the risk-adjusted result. Figure 1 shows the probability distribution for the risk-adjusted NPV of the project. The average NPV value was $ 1,206,667 with a standard deviation of $ 1,431,103. Considering a 95% confidence interval for the project's NPV, it can assume values between - $ 245,567 up to $ 4,470,023. Figure 1 also shows that the probability of the project having a positive NPV is approximately 70%, indicating that it is a project with moderate economic risk.

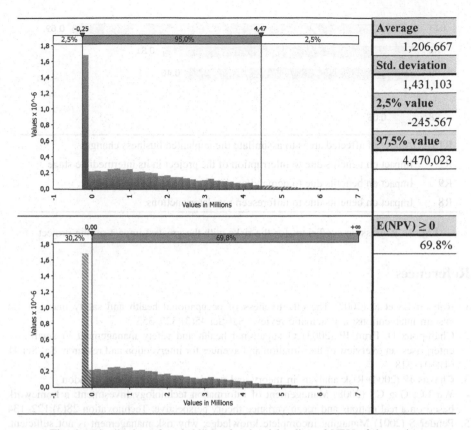

Fig. 1. Distribution of probabilities for the risk-adjusted NPV of the project

The hierarchy of risk made it possible to identify the major threats and the best opportunities available to the project. Figure 2 shows the risk with the greatest impact on results, identified by their respective regression coefficients. The risks that showed the greatest negative impact on the project results were: Ability of affected areas to assimilate the implanted business changes, Impact on benefits due to interruption of the project in its intermediate stage and Impact on benefits due to demand/application overload. On the other hand, the Impact on benefits due to unforeseen regulatory actions presented itself as an opportunity of gain for the project since these possible actions would bring more returns than costs.

That way, despite the moderate probability (30%) of the project does not cover its costs, the analysis points out exactly where managers should focus on developing mitigation plans for the most relevant risks. With this, it is possible to further increase the probability of the project success and its NPV, ensuring an ergonomic intervention with a great chance of financial success. Even knowing that the financial result should not be the only criterion for investments in OHSMS projects, the analysis offers more precise information and arguments that is part of the decision makers language.

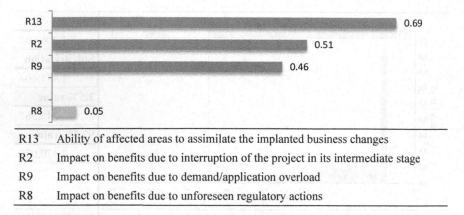

R13	Ability of affected areas to assimilate the implanted business changes
R2	Impact on benefits due to interruption of the project in its intermediate stage
R9	Impact on benefits due to demand/application overload
R8	Impact on benefits due to unforeseen regulatory actions

Fig. 2. Regression coefficient for the risks with the greatest impact on the project

References

1. Robson LS et al (2007) The effectiveness of occupational health and safety management system interventions: a systematic review. Saf Sci 45(3):329–353
2. Champoux D, Brun JP (2003) Occupational health and safety management in small size enterprises: an overview of the situation and avenues for intervention and research. Saf Sci 41 (4):301–318
3. Chavas JP (2004) Risk analysis in theory and practice. Academic Press, London
4. Wu LC, Ong CS (2008) Management of information technology investment: a framework based on a real options and mean-variance theory perspective. Technovation 28(3):122–134
5. Pender S (2001) Managing incomplete knowledge: why risk management is not sufficient. Int J Project Manag 19(2):79–87
6. Hughes RE, Nelson NA (2009) Estimating investment worthiness of an ergonomic intervention for preventing low back pain from a firm's perspective. Appl Ergon 40 (3):457–463
7. Saurin TA, Wachs P, Righi AW, Henriqson E (2014) The design of scenario-based training from the resilience engineering perspective: a study with grid electricians. Accid Anal Prev 68:30–41
8. Miorando RF, Ribeiro JLD, Cortimiglia MN (2014) An economic–probabilistic model for risk analysis in technological innovation projects. Technovation 34:485–498

Case Study in Ergonomics Problem Solving Process at a Beer Distribution Company

Kelsie Daigle[1]([✉]), Colleen Brents[2], Molly Hischke[2],
Rebecca Brossoit[1], Kelly Cave[1], Shalyn Stevens[1],
and John C. Rosecrance[2]

[1] Department of Psychology, Colorado State University,
Fort Collins, CO 80521, USA
kdaigle@rams.colostate.edu
[2] Department of Environmental and Radiological Health Sciences,
Colorado State University, Fort Collins, CO 80521, USA

Abstract. The ergonomic problem-solving process is a uniformly consistent approach to address challenges presented in various environments. Steps include: (1) identification, (2) analysis, (3) brainstorm of possible solutions, (4) implementation (prototypes), and (5) evaluation. This process was conducted at a distribution company that delivered beer to commercial establishments. Less than 200 employees worked at the company. The first three steps of the ergonomic problem-solving process were carried out with warehouse and delivery workers. A safety climate survey was administered. Following data collection (delivery and warehouse tasks observations and a safety climate survey), employees participated in a workshop. The first session introduced work design principles (including product quality, customer service, efficiency, decrease risk of injury, and improve quality of work life) and how they aligned with the company's goals. Concepts were communicated in an interactive setting. During the second session, two separate groups discussed challenges specific to warehouse or delivery workers. Videos gathered during data collection were shared with workers. Researchers facilitated brainstorm sessions to address challenges presented in those videos. Identified challenges included errors in building orders at the warehouse to delivering keg up stairs and around tight spaces. One suggestion included shifting emphasis from individual pick-rates to team accuracy to stress accuracy over speed. Another solution from brainstorming involved standardizing communication policies with clients. This project covered the first three stages of the ergonomics solution development process (identification, analysis, and brainstorming possible solutions). Further collaboration with the company will incorporate the prototype, implementation and evaluation steps.

Keywords: Participatory ergonomics · Beverage distribution
Musculoskeletal disorders · Manual materials handling

© Springer Nature Switzerland AG 2019
S. Bagnara et al. (Eds.): IEA 2018, AISC 819, pp. 105–118, 2019.
https://doi.org/10.1007/978-3-319-96089-0_11

1 Background

1.1 Introduction

The wholesale beer distribution industry, classified as North American Industry Classification System (NAICS) 424810 beer and ale merchant wholesalers, employs approximately 40,000 workers in 1,300 organizations across the United States (North American Industry Classification System 2018). The number of workers might actually be higher, with an estimated 93,000 people in the U.S., averaging about 37 people per organization (McGrath 2003). The NAICS states that "this industry comprises establishments primarily engaged in the merchant wholesale distribution of beer, ale, porter, and other fermented malt beverages" (North American Industry Classification System 2018). The U.S. brewing industry has more than doubled in the last ten years, with over 6,000 breweries in 2016 (Brewers Association 2017). Increases in beer distribution jobs coincide with the brewing industry growth. Job tasks associated with beer distribution include lifting kegs and cases of beer, loading and unloading delivery trucks, carrying kegs and cases up and down staircases, among others.

The job design of beer distribution requires manual materials handling of kegs and cases of beer. Risk factors for musculoskeletal disorders have been associated with manual materials handling tasks (Hagberg et al. 1995; Marras 2000). Work design in warehouses and delivery routes require frequent lifts and repetitive motions that may put the worker at an increased risk of developing low back or upper extremity work-related musculoskeletal disorders (MSDs). There may also be psychophysical risks for MSDs at play in this industry, such as time pressure due to deadlines. The risk for MSDs and costs of such disorders should be of concern to the industry, practitioners, and researchers alike. Established in previous research, MSDs can increase compensation claims, health care costs, and prevent employees from rejoining the workforce due to injury or disability (Hagberg et al. 1995; Institute of Medicine 2001). Low back disorders are considered to be the most common MSD (Marras 2000). Musculoskeletal disorders are the most common work-related illness in the U.S., Japan, and Nordic countries (Putz-Anderson et al. 1997; Institute of Medicine 2001; Pope et al. 1991; Sjøgaard et al. 1993; Punnett and Wegman 2004). In 2011, MSDs accounted for one third of all lost work time (Bureau of Labor Statistics 2011). Importantly, rates of MSDs resulting is lost work time were higher in wholesale and retail trade compared to other industries (Bureau of Labor Statistics 2011). Previous studies have investigated low back motion during pallet loading and unloading (Jorgensen et al. 2005; Marras et al. 1997; Plamondon et al. 2017) While these studies are relevant to distribution industries overall, only a handful of studies are specific to MSD and the beer industry (Abaraogu et al. 2015; Jones et al. 2005; Ramsey et al. 2014).

An interdisciplinary team, consisting of occupational ergonomics and safety and occupational health psychology, applied the participatory ergonomic process at a beverage distribution company. The purpose was to assess the physical and psychosocial risk associated with possible MSDs of the upper extremities and low back with warehouse and delivery workers. This participatory approach utilizes the traditional ergonomic five-step process of (1) identification, (2) analysis, (3) brainstorm possible solutions, (4) implementation (prototypes), and (5) evaluation in conjunction with input directly from the employees. Specifically, this action research process "combines intervention and research components into a systematic and integrated process that contribute to improving the systems performance and expanding general scientific knowledge" (Rosecrance and Cook 2000; Schurman 1996).

1.2 Project Overview

The beverage distribution company serves over 2,000 accounts in Colorado. Kegs and cases may be delivered to a variety of clients, such as restaurants, bars, liquor stores, and grocery stores. The employee base consists of 43.8% sales, 24.8% delivery, 19% warehouse, and 12.4% administration. Overall, the company is 81.9% male, with the majority of females working outside of the warehouse and delivery sectors. About 38.1% of the workforce are supervisors. This company's principles and mission demonstrate their commitment to prioritizing employee safety and well-being. However, the nature of the beverage distribution work presents challenges to this mission. Workers are at an increased risk for developing musculoskeletal disorders throughout the beverage distribution process (Pramitasari et al. 2015). Most manual materials handling tasks occur in either the warehouse filling orders or on delivery routes (shown in Figs. 1 and 2).

Fig. 1. A delivery worker loads a keg onto a dolly to transport it from the delivery truck to the customer's facility. Additional cases and kegs are shown in the truck bays.

The interdisciplinary research team accomplished the first three steps of the ergonomics process during August 2017 through January 2018. Initial meeting with the client help clarify the project goals and introduce the research team's expertise. The primary objectives were to assess the risks associated with beverage handling, conduct

Fig. 2. Warehouse workers build pallets of custom orders by transfer specific beers from large stacks onto a new pallet (shown in the bottom left corner).

an ergonomic assessment of the physical and psychosocial risks, conduct participatory workshops, and collaborate with current employees to develop recommendations to help improve work conditions. The purpose was presenting findings, risks, and possible solutions to the company's top leadership. The primary objectives of the beverage distribution company were to uphold their value of safety in the organization, recognize weaknesses, and identify possible interventions.

The research team observed work tasks, took photographs and videos, applied lift measurement tools, conducted a safety climate survey and unstructured interviews. Based on this data, the team identified risks for MSDs, as well as the safety-related strengths of the company. Employees were invited to participate in workshops where researchers discussed the goals of ergonomics and identifying risk factors in job tasks. Employees were invited to provide input in the solution development process and were encouraged to play an active role in overall company safety. Findings from the observations and safety climate survey were shared with top leadership.

2 Study Design

A participatory action research approach was applied to this study. "Action research is frequently used in applied behavioral sciences when investigating and implementing organizational change" (Rosecrance and Cook 2000). Using this process, investigators hoped to foster collaboration and participation between the research team and employees (Israel et al. 1992; Schurman 1996; Rosecrance and Cook 2000). The research team took this approach for multiple reasons: to provide richer data (about the company, job design, tasks, and safety) and to empower employees to take ownership of the participatory ergonomic process. Chances of a successful ergonomic process are greater with increased employee ownership and participation (Imada 1991; Rosecrance and Cook 2000). Further, using the interdependence between the research team and employees, employee can provide their own insight and ideas, increasing the likelihood of behavior change and buy-in with the process (Elden 1986; Rosecrance and Cook 2000).

3 Ergonomic Process

3.1 Problem Identification

Risks associated with manual materials handling and possible MSDs were identified from four information sources: (1) observation and lifting measurements, (2) unstructured employee interviews, (3) workshops with employees, and (4) safety climate survey. An initial meeting between the safety director of the company and the research team initiated the problem identification process. During this meeting the safety director discussed what he considers to be their safety weaknesses and general safety principles of the organization.

Four site visits to the company's warehouse were conducted to observe job design and conduct lifting measurements. Researchers also photographed and filmed employees performing their assigned tasks. Risk associated with lifting tasks was assessed using the National Institute for Occupational Safety and Health Lifting Equation (NIOSH) lifting equation (Waters et al. 1993).

Researchers rode along with delivery employees on five occasions. During these 'ride-alongs', job tasks were observed to assess manual materials handling risks and to collect lift measurements. The NIOSH lifting equation was used to assess and rank lifting risks during delivery tasks. Details of delivery job tasks and the complex delivery environments were captured using photos and videos.

During these observations both in the warehouse and out on delivery, informal and unstructured interviews were held with employees to gauge their perceptions of specific tasks and possible risk factors. Working tasks were observed to determine where the most obvious risk factors (high forces, high repetition rates, and awkward postures) were present. Once the primary risks were identified, researchers utilized videos, photos, and measurement data to prioritize identified risks.

Safety climate measurements can serve as a leading indicator for safety behaviors and outcomes (e.g., predicting accidents and injuries) in a workforce (Beu et al. 2010; Neal and Griffin 2006). Safety climate, a psychosocial metric, is a measure of the overall perceptions workers may have about the value of safety within their workplace at a given point in time. Safety climate can be considered a 'snapshot' of the broader safety culture. Measuring shared perceptions of safety climate among all employees can help shed light on organizational safety related weaknesses and strengths that may not be visible during observation periods. If one area of safety climate was low, it could indicate an area of safety that may contribute to a higher risk of MSDs. The safety climate survey included items from a widely-used, validated measure of safety climate (Neal and Griffin 2006). All safety climate items were rated on a scale of 1 (strongly disagree) to 7 (strongly agree). Higher scores reflected stronger agreement with the item. The safety climate measure comprised of eight dimensions, which assessed (1) management values, (2) safety communication, (3) training, (4) physical work environment, (5) safety systems, (6) safety knowledge, (7) safety motivation, and (8) safety behavior. The research team created and distributed this online safety climate survey, which was distributed via email to all employees and locations in November of 2017. Two versions of the survey were developed, one for supervisors and one for subordinates (those without a supervisory role). Employees were emailed using the Qualtrics online survey platform. All employees were informed that their participation was voluntary and responses would remain confidential. They were encouraged to complete the survey by management and by members of the research team. Three $50 raffle prizes were randomly awarded as further incentive for employees to complete the survey. Of 155 employees, 105 completed the survey (68% response rate).

3.2   Problem Analysis

The primary goal of the problem analysis step was to better understand job and task requirements and to determine specific circumstances responsible for the identified problem. Researchers focused on the specifics of the work tasks, equipment used, perceptions of safety, and methods of doing work. Videos, photos, observations, safety climate survey, and lifting measurements were used to analyze the problems identified. The team utilized quantitative measures of safety climate, the height of stacking or lifting, size of materials handled, distance traveled with load, reaching distances, and weight of cases or kegs. The team also asked employees and the safety director about the purpose or method of specific tasks and if they have tried other solutions or ways of doing things in the past.

Employees were invited to participate in workshops to discuss challenges in their industry. Due to the scope of risks associated with warehouse and delivery work, two separate workshops were facilitated. The first workshop introduced workers (in both warehouse and delivery areas) to the principles of good work design (efficiency, customer service, quality, and safety). Employees were also briefed on what factors can increase risk in a lifting task or job demand. Videos and photographs from the observational phase were used to demonstrate examples of risk in the beverage distribution industry tasks (such as lifts or job movements). Employees were asked to participate in this discussion. Rather than being directly told what risks were present in the examples, workers were asked to identify what they thought might constitute a risk to their safety. Near the end of these workshops, researchers discussed how employees can play an active role to influence production and safety culture within their own workplace.

After these workshops, the research team combined the qualitative data collected from employee-discussions with quantitative data to reanalyze the challenges to ensure accuracy. One-way quantitative data was analyzed using the NIOSH lifting equation (a tool developed based on biomechanical, physiological, and psychophysical factors to compute a weight limit for manual lifting) (Waters et al. 1993). The computed recommended weight limit represents the maximum weight of an object that should be lifted under the given lifting conditions (e.g., height, distance traveled, twisting, reach, and the type of object being lifted). The lifting index, or measure of how much risk is associated with the lift, is calculated from dividing recommended weight limit by the actual weight of the object. A lifting index less than one indicates low risk, and greater than one indicates higher risk of developing MSDs. Ideally, a task has a lifting index less than one. Therefore, tasks with a lifting index greater than one were considered especially risky. This formula was applied to all the lift measurements the team conducted.

3.3 Solution Development Through Brainstorming

After the initial workshop, employees participated in a subsequent workshop broken into two groups (warehouse and delivery). The purpose of this workshop was to brainstorm possible solutions to challenges identified in the previous workshop. Employees were encouraged to speak freely about their solution ideas, acting as if they had unlimited resources to remedy challenges. The research team also proposed some solutions to further facilitate discussion and employee feedback. During this workshop, the team talked about possible prototypes, potential implementation, and evaluation methods. However, the full exploration of prototypes, implementation, and evaluation was not conducted during the extent of this project due to time and resource constraints.

4   Results

4.1   Observations, Lifting Equations and Discussions with Employees

Warehouse. General areas of improvement focused on MSDs risks and the three goals of ergonomics (quality, efficiency, and safety). The first common warehouse task discussed handling cases of beer. The average case of beer weighs 20 lbs. (9 kg). Warehouse workers move a high volume of cases daily. Due to the high volume, there opportunity for human error when compiling orders. Quality is a concern here because once those orders are compiled, they are loaded on trucks for delivery. Incorrect orders and extra case handling can occur if pallets are incorrectly built. Considering efficiency, after orders are completed on a pallet, they are loaded on a delivery truck. This pallet is weighed to identify filling errors. However, sometimes the weight differs even when orders are correctly picked. This results in pickers and supervisors recounting orders multiple times and handling cases of beer unnecessarily. There were multiple safety concerns regarding case handling in the warehouse. The cases differ in packaging, and therefore degrees of coupling vary between brand packaging. Some cases are challenging to lift and carry because they lack handles. Cases with handles seem to make lifting easier, but they encourage employees to stack them and carry multiples at a time, increasing their risk of developing MSDs. Cases can be stored anywhere between 4.0 in. (10.2 cm) off the ground when stored on a pallet to above the average employees head (approximately 233.8 cm or 8.0 ft. tall). The cases on the ground require the employee to bend down to pick them up. The cases stacked overhead can require employees to climb on other cases of beer to reach them.

An updated order picking software (i.e., Pick Management Software) was one intervention ideas generated by the warehouse employees. Newer software could better tolerate error thresholds of weight per order. Currently, there can be a difference of 16 lbs. (7.25 kg) above or below the expected load weight. Additionally, if the pallet weight does indicate an incorrect amount, that employee must manually recount the order, locate a supervisor to approve the recount, and make appropriate changes to the order.

Another solution was to encourage accuracy over speed by removing picking speed from the order management system. While the company claims to prefer accuracy over speed, employees still feel pressure to work quickly. Employees also suggested hiring additional workers. They suggested this because errors usually occur with larger orders, especially during the busy season around holidays.

Investigators discussed the solution of purchasing a Rotating Load Leveler to ensure products are always at waist-level to endorse safer lifting. Additionally, the research team proposed having adjustable shelves with rollers so that cases were more easily accessible.

The second main task of warehouse workers involves keg handling. Full kegs weigh approximately 160 lbs. (72.6 kg). Warehouse workers pick keg orders based on their label. These labels are 'checked' with a mark to indicate what is inside. If they are labeled incorrectly they can cause a delivery error. The warehouse workers are responsible for loading and organizing pallets of orders for the delivery trucks. Often, kegs were loaded and stacked in ways that delivery drivers would have to reorganize their trucks to get to the kegs they needed first. This is a major efficiency concern. Additionally, kegs exceed the NIOSH recommended weight of 51 lbs. (23 kg.). Further, workers are moving quickly and using forklifts, which can cause potential for accidents and injuries.

Intervention ideas generated by the employees included using pallets to handle kegs. Consistently placing on a pallet guarantees that employees can use forklifts (instead of manually reaching) to transport kegs. However, the employee would also have to stack and organize kegs on the pallet in such a way it was balanced for the forklift. Regarding keg picking errors, employees suggested hiring additional workers to decrease the sense of being rushed. They also suggested rolling kegs off pallets onto the forklift when the pallet and forklift are at the same height. While this solution may decrease risk, it may also decrease efficiency. Further, they suggested kegs are purposefully organized at a safer lifting height for workers building upcoming orders. The research team suggested implementing a lifting partners system so employees would conduct team lifts to handle full kegs.

Delivery. General areas of improvement focused on MSDs risks and the three goals of ergonomics (quality, efficiency, safety). There were more areas for improvement for delivery than warehouse workers. Challenges surrounding maneuvering kegs on client staircases were one major area of potential improvement. Oftentimes, workers deliver to venues that lack elevators. Efficiency is a compromised because transporting kegs up and down stairs in variable conditions is challenging and time consuming. Additionally, keg quality or cases of beer quality can be damaged if the product falls off the dolly. There is a safety concern because there is a high risk for back injury when handling kegs and cases of beer up or down. The company invested in a mechanized stair-climbing-dolly. Employees did not use this tool based on anecdotal evidence that the mechanized stair climber hurts their back more than the traditional dolly. Sometimes employees will carry two kegs stacked up stairs, or a keg with cases of beer on top, increasing the overall weight of the lift. Inclement weather makes this task more dangerous as well.

Intervention ideas generated by employees centered on revisiting the stair climbing dolly (including retraining and re-testing to make sure it is properly calibrated). Workers also suggested retraining on how to squat lift properly up the stairs. Frustration was expressed at the fact that the stairs or delivery site could not be structurally modified. They mentioned that even clients with elevators might not have the elevator functioning or maintained. The research team suggested designing a pulley system to help move kegs up and down stairs, or purchasing a new mechanized stair truck. During discussions between the researchers and employees, investigators learned that some deliveries sites can be mandated as two-man stops. This approach could help reduce the lifting load of the keg by distributing it across two people.

The environment in which workers were delivering the product presented another challenge. Coolers vary in size, organization, and cleanliness. Some employees were forced to reorganize and restack kegs from other companies simply to create space to deliver their own products, sacrificing efficiency. Multiple delivery coolers and storage spaces indicated a NIOSH lifting index greater than three. Lifting full kegs, 160 lbs. (72.6 kg), above the waist greatly increases the risk of developing a low back injury. Lifting cases in tight spaces often requires awkward postures and twisting. Tight delivery spaces can also increase the likelihood of damage to the product, decreasing quality. These tight spaces also have major implications for efficiency. Delivery workers may have to reorganize stacks of beer, which can reach 10 ft. (3.05 m) tall, simply to be able to maneuver their order. Narrow and tight spaces also decrease speed because the worker must cautiously and slowly maneuver the dolly around other beer products. Oftentimes, the walkway is just wide enough to fit the dolly though. Other debris and obstacles may litter the aisle ways, including uneven floors, spilled liquid on the floor, and broken glass, to name a few. Furthermore, these tight spaces limit how and where delivery workers can put their product. Frequently they have to stack cases of beer 10 ft. (3.05 m) or higher. To stack this high, workers craft makeshift stairs with other cases of beer. This strategy requires time and risks compromising product quality if they fall or step on cases of beer at awkward angles.

Solutions from both employees and the research team centered on communicating with clients on characteristics for a safe and efficient delivery space. This included defining a stacking height limit with the client, having a formal policy and procedure for debris (glass, spills, and uneven floors). Specifically, clients notorious for having have multiple obstacles and poor housekeeping practices (that increase the risk of injury to delivery drivers as well as damaged products) could be identified. Then the distribution company could work with these clients to develop a strategy of addressing these housekeeping concerns. New clients could be briefed on what constitutes good, clean housekeeping practices as well. Additionally, the research team suggested that managers remind employees to avoid awkward postures and twisting while lifting.

4.2   Safety Climate

The average age of participants that completed the survey was 35 years old. Regarding gender, 81.9% of participants were male, 14.3% were female, 2.9% identified as "other", and 1% of employees chose to not identify their gender. For supervisor status, 38.1% of participants were in a supervisory role and 61.9% were not in a supervisory role. Next, 77.1% of participants worked at one worksite (where observations and workshops too place) and 22.9% worked at the other location. Finally, 19.0% of participants worked warehouse, 24.8% delivery, 43.8% sales, and 12.4% in administration.

Scale scores were created by calculating the average score of the items within a dimension (e.g., all items pertaining to safety behavior were averaged to create an overall measure of safety behavior). For individual items, percent agree is the percentage of participants who responded with somewhat agree, or strongly agree (i.e., responded with at least a 5 out of 7 for the item). For overall measures, percent agree is the percentage of participants who responded on average with somewhat agree, agree, or strongly agree (i.e., responded with at least an average of 5 out of 7 for all the items in a dimension). Wherever 'employees' are described regarding safety climate, this indicates any worker at the company who does not have a supervisory role. Wherever 'supervisors' are described, this indicates any worker who does have a supervisory role.

Overall, supervisors and employees both perceive very high levels of safety climate (m = 5.92, 96.2% agree). Similar levels of overall safety climate were found regardless of a worker's role (i.e., supervisor status), worksite, and department. Of note is that for all participants, safety motivation was the highest rated dimension; all participants agreed with each of the items related to safety motivation (m = 6.51, 100% agree for employees; m = 6.52, 100% agree for employees). However, the physical work environment dimension was rated the lowest of all dimensions for both employees and supervisors (m = 4.38, 21.5% agree for employees; m = 4.37, 38.5% agree). For example, over half of all participants believed that there were significant dangers inherent in the workplace.

Given that the physical work environment dimension was rated the lowest, further analyses were conducted to determine if this differed based on job title. Participants that work in the delivery department rated the physical work environment lower than participants that work in warehouse, sales, or administration departments. The mean rating of the overall physical work environment (i.e., the average of the items related to physical work environment) for delivery workers was 3.50 out of 7, with 5.6% reporting that they work in a safe environment. The mean rating of the overall physical work environment for delivery supervisors was 3.75 out of 7, with 12.5% reporting that they work in a safe environment. This is not surprising given the dynamic environment that delivery drivers work in. They are required to enter liquor stores, grocery stores, etc. that are not always maintained safely, or have cleared pathways.

4.3 Future Steps

Given the findings, the next steps in the process of ergonomic participatory process would be to specify prototypes and implement the recommended solutions. To do this, buy-in is required from top leadership, supervisors, and employees. Following the implementation of solutions, they should be evaluated. Evaluation should consider if the risks and problems were reduced, and how the solutions influenced the goals of ergonomics (efficiency, customer service, quality, and safety). Additionally, the solutions should be evaluated overtime to explore any longitudinal changes.

The distribution company in the present study wishes to achieve a specific health and safety certification status with the Occupational Safety and Health Administration (OSHA). Starting with findings from the present study and continued collaboration with the interdisciplinary team, the company is working towards becoming part of the Safety and Health Achievement Recognition Program (SHARP) Certified through OSHA.

While the present study covered the first three steps of the participatory ergonomics process, future collaboration to address the final steps (prototype implementation and evaluation) are hopeful for the future. Both the university and distribution company plan to continue interdisciplinary collaboration in the following academic year.

5 Discussion

Both employees and the workplace benefited from worker participation in problem identification, problem analysis, and solution development through participatory action research. Gathering employee perceptions of job tasks and risks helps ensure the researchers target their investigation in relevant and meaningful directions. Second, it helps gain employee buy-in for the implementation of solutions and future organizational changes.

We found multiple inherent risks in the beverage distribution company. One particular example was how employees regularly lift full kegs (160 lbs. or 72.6 kg.), which increases their risk for developing low back MSDs. In addition to the load exceeding NIOSH recommended values, the origin and destination heights are less than ideal.

Overall, the safety climate survey generated desirable results for the company, and highlighted their areas for improvement. These findings suggest that employees place a high value on workplace safety but still perceive aspects of the physical work environment to be dangerous. Workplace safety efforts should emphasize improving the safety of the physical work environment that employees are exposed to. These efforts could be particularly helpful for delivery workers' physical work environment(s), who had the lowest perceptions of safety in their physical work environment. Other target areas to improve workplace safety at this company could include emphasizing safety trainings, clarifying safety systems to prevent workplace safety incidents, and increasing communication about safety-related issues, as these dimensions of safety climate were also rated slightly lower compared to other dimensions (though were generally still highly rated).

6 Limitations

During this project, the first three steps of the ergonomic process were completed. Future collaboration between the interdisciplinary group and beer distribution company hopes to complete the final steps of the participatory ergonomics approach.

During the observational period, investigators were unable to communicate with all employees. Further, not every employee attended the workshops. Scheduling and time conflicts restricted interactions and participation. Therefore, findings may not be generalizable to all employees at the company and may not represent the most accurate picture of the MSDs risks in the organization.

The NIOSH lifting equation was applied to estimate risk associated with different lifting tasks. However the model has limitations and does not apply to one handed lifts, lifting tasks exceeding eight hours, seated or kneeling positions, restricted workspace, unstable objects, unreasonable floor coupling, high speed motion, extreme environments, and unexpected loading (Waters et al. 1993). Furthermore, the NIOSH lifting equation does not apply to combined motions such as lifting and carrying or wheelbarrow or shoveling movements (Waters et al. 1993).

Safety climate perceptions were very high, however, there were still evident risks of MSDs in the company. Additionally, the survey was a one-time cross-sectional measure of safety climate. Safety climate could change day to day, or fluctuate during busier seasons of production. This variation means that current survey findings might not reflect accurate safety culture.

7 Closing Comments

An interdisciplinary group of occupational ergonomics students and occupational health psychology students conducted a participatory ergonomics problem solving process with a beer distribution company in Northern Colorado. The first three steps of the process (1. Identification, 2. Analysis, 3. Brainstorm possible solutions) were completed over the course of an academic semester. Challenges for warehouse and delivery workers were identified and discussed. The interdisciplinary group facilitated workshops to educate employees on basic concepts of identifying risk factors in job environments. Employees were encouraged to contribute and share their perspectives on their own work experience during these workshops. Manual materials handling challenges around transporting cases of beer and full kegs were some situations addressed in the workshops. A safety climate survey was administered to supervisors and employees to gauge perceptions of safety support across the company. While overall high levels of satisfaction and safety support were measured, both supervisors and workers identified environmental challenges to safe distribution practices. The next two phases of the participatory ergonomics approach (4. Prototype and implementation, 5. Evaluation) will be continued during future collaboration between the interdisciplinary group and beverage distribution company.

Acknowledgments. The authors declare no conflict of interest. This study and publication was supported in part by the Grant T42OH009229 (Mountain and Plains Education Research Center), funded by the Centers for Disease Control and Prevention (CDC). Its contents are solely the responsibility of the authors and do not necessarily represent the official views of the CDC or the Department of Health and Human Services. Additional funding for this project was from the CDC through the National Institute for Occupational Safety and Health (NIOSH) and the High Plains Intermountain Center for Agricultural Health and Safety, grant number U54 OH008085. The content is the responsibility of the authors and does not necessarily represent the official views of the CDC or NIOSH.

References

Abaraogu U, Okafor C, Ezeukwu AO, Igwe SE (2015) Prevalence of work-related musculoskeletal discomfort and its impact on activity: a survey of beverage factory workers in Eastern Nigeria. Work 52(3):627–634

Beus JM, Payne SC, Bergman ME, Arthur W (2010) Safety climate and injuries: an examination of theoretical and empirical relationships. J Appl Psychol 95(4):713–727

Brewers Association (2017) Craft beer in review U.S. hits record-breaking 6,000 breweries in operation. Brewers Association [Press release]. https://www.brewersassociation.org/press-releases/2017-craft-beer-review/

Bureau of Labor Statistics (2011) Injuries, illnesses, and fatalities. Nonfatal cases involving days away from work: selected characteristics (2003–2010). US Department of Labor, Bureau of Labor Statistics, Safety and Health Statistics Program, Washington, DC. http://www.bls.gov/iif/data.htm

Elden M (1986) Sociotechnical systems and ideas and public policy in Norway: empowering participation through worker-managed change. J Appl Behav Sci 22:239–255

Hagberg M, Silverstein B, Wells R, Smith MJ, Hendrick HW, Carayonn P, Perussen M (1995) Work related musculoskeletal disorders (WMSDs). Taylor & Francis, London

Imada AS (1991) The rationale and tools of participatory ergonomics. In: Imada AS, Noro K (eds) participatory ergonomics. Taylor & Francis, London, pp 30–49

Institute of Medicine (2001) Musculoskeletal disorders and the workplace: low back and upper extremities. National Academy of Science, Washington, D.C.

Israel BA, Schurman SJ, Hugentoblr MK (1992) Conducting action research: relationships between organizational members and researchers. J Appl Behav Sci 28(1):74–101

Jones T, Strickfaden M, Kumar S (2005) Physical demands analysis of occupational tasks in neighborhood pubs. Appl Ergon 36(5):535–545. https://doi.org/10.1016/j.apergo.2005.03.002

Jorgensen MJ, Handa A, Veluswamy P, Bhatt M (2005) The effect of pallet distance on torso kinematics and low back disorder risk. Ergonomics 48(8):949–963. https://doi.org/10.1080/00140130500182007

Marras WS (2000) Occupational low back disorder causation and control. Ergonomics 43(7):880–902

Marras WS, Granata KP, Davis KG, Allread WG, Jorgensen MJ (1997) Spine loading and probability of low back disorder risk as a function of box location on a pallet. Hum Factors Ergon Manuf 7(4):323–336

McGrath C (2003) 'Family businesses distributing America's beverage': managing government relations in the National Beer Wholesalers Association. J Public Aff 3(3):212–224

Neal A, Griffin MA (2008) A study of lagged relationships among safety climate, safety motivation, safety behavior, and accidents at the individual and group levels. J Appl Psychol 91(4):946–953

North American Industry Classification System (n.d.) NAICS Code 424810 Beer and Ale Merchant Wholesalers. https://siccode.com/en/naicscodes/424810/beer-and-ale-merchant-wholesalers-2

Plamondon A, Larivière C, Denis D, Mercheri H, Natasia I (2017) Difference between male and female workers lifting the same relative load when palletizing boxes. Appl Ergon 60:93–102

Pope MH, Anderson GBJ, Frymoyer JW, Chaffin DB (1991) Occupational low back pain: assessment, treatment, and prevention. Mosby-Year Book Inc., St. Louis

Pramitasari R, Pitaksanurat S, Phajan T, Laohasiriwong W (2015) Association between ergonomic risk factors and work-related musculoskeletal disorders in beverage factory. In: Proceedings international seminar and workshop on public health action. Faculty of Health Sciences Dian Nuswantoro University, Semarang, Indonesia, pp 196–200

Punnett L, Wegman DH (2004) Work-related musculoskeletal disorders: the epidemiologic evidence and the debate. J Electromyogr Kinesiol 14:13–23

Putz-Anderson V, Bernard BP, Burt SE, Cole LL, Fairfield-Estill C, Fine Tanaka S (1997) Musculoskeletal disorders and workplace factors (Rep.). In: Bernard BP (ed.). National Institute for Occupational Safety and Health, Cincinnati, OH

Ramsey T, Davis KG, Kotowski SE, Anderson VP, Waters T (2014) Reduction of spinal loads through adjustable interventions at the origin and destination of palletizing tasks. Hum Factors 56(7):1222–1234. https://doi.org/10.1177/0018720814528356

Rosecrance JC, Cook TM (2000) The use of participatory action research and ergonomics in the prevention of work-related musculoskeletal disorders in the newspaper industry. Appl Occup Environ Hyg 15(3):255–262

Schurman SJ (1996) Making the "new American workplace" safe and healthy: a joint labor-management-researcher approach. Am J Ind Med 29(4):373–377

Sjøgaard G, Sejersted OM, Winkel J, Smolander K, Westgaard J, Westgaard RH (1993) Exposure assessment and mechanisms of pathogenesis in work-related musculoskeletal disorders: significant aspects in the documentation of risk factors. In: Johansen O, Johansen C (eds) Work and health: scientific basis of progress in the working environment. European Commission, Directorate-General V, Employment, Industrial Relations and Social Affairs, Copenhagen, Denmark, pp 75–87

Waters TR, Putz-Anderson V, Garg A, Fine LJ (1993) Revised NIOSH equation for the design and evaluation of manual lifting tasks. Ergonomics 36(7):749–776

The Importance of Identifying Patient and Hospital Characteristics that Influence Incidence of Adverse Events in Acute Hospitals

P. Sousa[✉], A. Sousa-Uva, F. Serranheira, M. Sousa-Uva,
and C. Nunes

Public Health Research Center, NOVA National School of Public Health,
Lisbon, Portugal
Paulo.Sousa@ensp.unl.pt

Abstract. To analyse the variation in the rate of AEs between acute hospitals and explore the extent to which some patients and hospital characteristics influence the differences in the rates of AEs. Methods - Retrospective cohort study. Binary logistic regression models were used to identify the potential association of some patients and hospital characteristics with AEs. A random sample of 4,250 charts, representative of around 180,000 hospital admissions, from 9 acute Portuguese public hospital centres in 2013, was analysed. Results – Main results: (i) AE incidence was 12.5%; (ii) 66.4% of all AEs were related to HAI and surgical procedures; (iii) patient characteristics such as sex, age, admission coded as elective vs urgent and medical vs surgical DRG code, all with p < 0.001, were associated with a greater occurrence of AEs – CCI seems to influence the difference in the rates of AEs; (iv) hospital characteristics such as use of reporting system, being accredited, university status and hospital size, all with p < 0.001, seem to be associated with a higher rate of AEs.

We identified some patient and hospital characteristics that might influence the rate of AEs. Based on these results, more adequate solutions to improve patient safety can be defined.

Keywords: Patient safety · Adverse events · Epidemiologic studies

1 Introduction

In the last 15 years, several studies have been developed with the aim of estimating the rate of adverse events and characterizing their nature, impact and preventability in several countries, including Portugal [1–4]. The identification and measurement of adverse events (AEs) is crucial for enhancing patient safety, hierarchizing priorities for intervention, defining important research areas, and evaluating the impact of the developed solutions in improving safety and quality of care.

The aims of this study were to analyse the variation in the rates of adverse events between acute public hospitals in Portugal and to explore the extent to which some patient and hospital characteristics are associated with rates of adverse events. In this study, we also estimated costs related to the additional length of stay due to the occurrence of AEs.

© Springer Nature Switzerland AG 2019
S. Bagnara et al. (Eds.): IEA 2018, AISC 819, pp. 119–123, 2019.
https://doi.org/10.1007/978-3-319-96089-0_12

2 Methods

This work was based on a retrospective cohort study and was carried out at nine acute public hospital centres in Portugal.

The global sample size was 4,250 medical charts, assuming a margin of error of 1%, a 95% confidence level and an expected AE rate of 11.1% (4,19), representative of around 180,000 hospital admissions (which corresponds to around 24% of all adult hospital admissions in Portuguese public hospitals for the period under analysis – 01 January to 31 December 2013). After that, taking into consideration the proportions of admissions in each type of hospital (small, medium and large), the final sample sizes were: two small hospital centres sampling 290 medical charts each; three medium-size hospital centres sampling 470 medical charts each; and four large hospital centres sampling 565 medical charts each.

A two-stage structured retrospective medical records review was done based on the use of 18 screening criteria. In the first stage, a group of 38 nurses (composed of two to four from each hospital and with a minimum of five years' experience in clinical audits) assessed the medical records, in order to find the presence of at least one of the 18 screening criteria for the presence of a potential adverse event. In stage two, a group of 26 physicians (with a minimum of five years' experience in clinical codes and in clinical audits) reviewed each positive record in order to confirm the presence of an adverse event, estimate its impact and determine its preventability according to a previously established definition. The degree of agreement between the reviewers in each stage was calculated by using the kappa coefficient.

3 Results

A total of 4,225 admissions from nine hospital centres were analysed. One or more of the criteria for an adverse event were identified in 1,029 reviewed medical records (24.3%). Of these 1,029 records that passed to the second screening, a total of 529 (51.4%) were confirmed as an AE, representing an overall incidence rate of 12.5% (529/4,225) with a 95% CI (11.5%; 13.6%). Of the AEs identified, 39.7% (210/529) were hospital-acquired infection, 26.7% (141/529) were related to surgical procedures, 9.8% (52/529) were due to drug errors, and 7% (41/529) were falls. Most of the AEs (66.1%–350/529) occurred in patients aged 65 or over. Comparing groups with and without AEs, age presents a mean difference of 8.2 years (mean ± standard deviation: 68.7 ± 17.2 vs 60.5 ± 20.4 years old, respectively).

Most of the AEs (67.4%–356/529; CI 63.4–71.4) resulted in no or minimal physical impairment or disability, and were satisfactorily resolved during the admission, or within one month of discharge. We estimated that 3.0% (16/529; CI 1.5%–4.5%) of the AEs resulted in permanent disability according to the pre-established definition (1, 4) and 12.5% (66/529; CI 9.7%–15.3%) resulted in death. With regard to preventability, 39.9% (211/529; CI 35.7%–44.1%) were classified as preventable.

We also found that 60.8% (322/529; CI 56.6%–65.0%) of the patients who experienced AEs incurred extra bed days in hospital (a total of 3,091 extra days, at an

average of 9.6 days per patient, ranging from 1 to 73 days), with an additional estimated cost of €1.9 million for the NHS. Of these 1.9 million, around 1.1 million were
associated with AEs considered avoidable (Tables 1 and 2).

Table 1. Association between AE and hospital characteristics

Hospital characteristics		Total (n)	AE n (%)	Without AE n (%)	p-value
Reporting System	Yes	3731	494 (13.2%)	3237 (86.8%)	<0.001
	No	494	35 (7.1%)	459 (92.9%)	
Accreditation status	Yes	2297	314 (13.7%)	1983 (86.3%)	<0.001
	No	1928	215 (11.2%)	1713 (88.9%)	
Type of hospital	University	1356	216 (15.9%)	1143 (84.1%)	<0.001
	Non University	2866	313 (10.9%)	2553 (89.1%)	
Hospital dimension	Small (<10.000 admission/year)	721	93 (12.9%)	628 (87.1%)	<0.001
	Medium (from 10.001 to 18.000 admissions/year)	1321	123 (9.3%)	1198 (90.7%)	
	Large (≥ 18.001 admissions/year)	2183	313 (14.3%)	1870 (85.7%)	
Electronic prescribing drug	Yes	2239	295 (13.2%)	1943 (86.8%)	0.183
	No	1987	234 (11.8%)	1753 (88.2%)	

Table 2. Strength of associations (OR) of patient and hospital characteristics with AE

		OR	95% CI	p
Patients characteristics				
Sex (Female*)	Male	1.36	1.13–1.63	0.001
Age (< 65y*)	≥ 65y	2.10	1.73–2.54	<0.001
Elective*	Urgent	1.83	1.48–2.26	<0.001
Medical*	Surgical	1.17	0.96–1.41	0.105
CCI		1.21	1.15–1.26	<0.001
Hospital characteristics				
Reporting system (No*)	Yes	2.01	1.40–2.86	<0.001
Accreditation status (No*)	Yes	1.26	1.05–1.32	0.014
University hospitals (No*)	Yes	1.54	1.28–1.86	<0.001
Hospital dimension (small*)	Medium	0.69	0.52–0.92	0.012
	Large	1.13	0.88–1.45	0.334
Electronic prescribing drug (No*)	Yes	0.88	0.73–1-06	0–17

*Reference class
OR- Odds ratio; CI- Confidence interval

4 Discussion

In this study, reviews of medical records were performed in order to assess the variation in the incidence, nature, and clinical and economic impact of adverse events between hospitals. The study also aimed to provide some insights into the relationship between the rate of those events and some of the patient and hospital characteristics.

This study gives some insights into patient and hospital characteristics that can influence the rate of AEs.

In the bivariate analysis, all the five patient covariates studied present differences that are statistically significant.

With regard to hospital characteristics, those that have a reporting system, are accredited and have the status of university hospitals present a higher rate of AEs with differences that are statistically significant. One plausible explanation is the fact that those hospitals have a stronger culture of risk awareness and also a systematic practice of data collection and analysis [5, 6]. These results are in line with other studies that state that the estimated rates of AEs are only the "tip of the iceberg", as the magnitude of the problem is much larger [7, 8].

This study presents some limitations, starting with those underlying retrospective studies, such as information bias and hindsight bias.

We assume that the effects of the patient and hospital variables studied partly reflect the most relevant collected characteristics that could influence the rate of AEs. Future research is, therefore, needed in order to provide a stronger body of knowledge that will allow optimal adjustment for the patient case mix and hospital specificities to be obtained.

5 Conclusion

These results could be seen as a small but important step toward enhancing safety in the care provided in Portuguese public acute hospitals. Knowledge of a set of characteristics related to the population treated and to the structure of such hospitals, where care is delivered, is crucial in helping to develop and implement strategies and solutions aimed at reducing AE rates and, therefore, in moving towards a culture and practice of quality and safety of care.

Acknowledgements. We would like to thank everyone who contributed to the study, particularly the nurses and doctors who reviewed the medical charts. We also thank the hospital staff who facilitated the development of the study – Quality and Patient Safety departments and Risk Management cabinets of all participating hospitals. This work was supported by the Fundação Calouste Gulbenkian.

References

1. Baker GR, Norton PG, Flintoft V, Blais R, Brown A, Cox J et al (2004) The Canadian Adverse Events Study: the incidence of adverse events among hospital patients in Canada. Can Med Assoc J 170(11):1678–1686
2. Mendes W, Martins M, Rozenfeld S, Travassos C (2009) The assessment of adverse events in hospitals in Brazil. Int J Qual Health Care 21(4):279–284
3. Rafter N, Hickey A, Conroy RM, Condell S, O'connor P, Vaughan, D et al (2016) The Irish National Adverse Events Study (INAES): the frequency and nature of adverse events in Irish hospitals—a retrospective record review study. BMJ Qual Saf. https://doi.org/10.1136/bmjqs-2015-004828
4. Sousa P, Uva AS, Serranheira F, Nunes C, Leite ES (2014) Estimating the incidence of adverse events in Portuguese hospitals: a contribution to improving quality and patient safety. BMC Health Serv Res 14(1):311
5. Rutberg H, Risberg MB, Sjödahl R, Nordqvist P, Valter L, Nilsson L (2014) Characterizations of adverse events detected in a university hospital: a 4-year study using the Global Trigger Tool method. BMJ Open 4(5):e004879
6. Thornlow DK, Stukenborg GJ (2006) The association between hospital characteristics and rates of preventable complications and adverse events. Med Care 44(3):265–269
7. Classen DC, Resar R, Griffin F, Federico F, Frankel T, Kimmel N et al (2011) Global trigger tool' shows that adverse events in hospitals may be ten times greater than previously measured. Health Aff 30(4):581–589
8. Makary MA, Daniel M (2016) Medical error-the third leading cause of death in the US. Br Med J 353

Ergonomic Issues and Innovations in Personal Protective Equipment for Contaminated Sites Considering European Regulation 425/2016

E. Bemporad, S. Berardi, S. Campanari, and A. Ledda[✉]

Department of Technological Innovations and Safety of Plants,
Products and Anthropic Settlements, INAIL, Rome, Italy
a.ledda@inail.it

Abstract. The Regulation (EU) 425/2016 on Personal Protective Equipment (PPE) repeals the Council Directive 89/686/EEC with effect from the 21st April 2018. Ergonomic aspects are stressed in this Regulation, becoming the basic principle of PPE design. Ergonomics has implications both for the comfort and the safety of workers.

In Italy the Leg. Decree 81/2008, the Code transposing the European Directives on the protection of workers' safety and health, requires for PPE to consider ergonomic and health needs of the worker. The Technical Standard EN 13921 provides guidance on the generic ergonomic characteristics related to PPE while EN 547-3 provides some human body measurements which, although required for openings for whole body access into machinery, can be useful for PPE design.

Regulations, laws or Technical Standards do not exhaustively fulfill the ergonomic requirements, because they cannot take into account the individual characteristics of each worker in PPE design. This is one of the most critical aspects for the adoption of the principles of ergonomics to PPE.

The principle of ergonomics must be carefully evaluated for PPE to be used in contaminated sites, considering the specific risk for the workers (e.g. presence of hazardous pollutants) and working conditions (e.g. in confined space or intercepting groundwater). The aim of this work is to highlight the main ergonomic issues related to the use of specific PPE bearing in mind the new European Regulation. The present work will also illustrate how the use of smart-PPE in contaminated sites could implement workers' safety fulfilling ergonomic principles.

Keywords: Regulation EU 425/2016 · Personal protective equipment
Workers' safety · Smart-PPE

1 Ergonomics in PPE

PPE is to be worn or held by an individual for protection against one or more health and safety hazards, but wearing PPE can also result in unintended negative effects, which can interfere with the user's normal performance of during the risk-related activities.

© Springer Nature Switzerland AG 2019
S. Bagnara et al. (Eds.): IEA 2018, AISC 819, pp. 124–131, 2019.
https://doi.org/10.1007/978-3-319-96089-0_13

Many surveys in the use of PPEs have revealed that they have not been adequately worn by the workers mainly due to mainly the discomfort caused by the lack of-missing wearability needs requirements in their design. During the past 10 years, there has been a continuously increasing interest to improve the human factors or user needs requirements of PPEs.

Ergonomic design assessment in PPE means that when all the protective clothing and equipment worn during a certain task are combined, the clothing still should be functioning properly and provides the basic functions (protection, storage, etc.) to the wearer. Other features should also be provided such as freedom of movement (determining movement speed, flexibility and energy expenditure; any hampering of the wearer's movements increases risk), visibility/conspicuity (exposure during tasks where the wearer's visibility is important; any activity in traffic, or in obscure areas will have a high risk if visibility of the wearer is insufficient).

The principles of ergonomics for PPE are particularly important in the case of workers in a contaminated site. Contaminated sites represent particularly high-risk scenarios (chemical, physical and biological risks). Furthermore, it must be considered that PPE of III category are Complex Design PPE intended to provide protection against mortal danger or protection against serious injury, where the user cannot not identify in sufficient useful time (such as, fall arrest harnesses, respiratory products, head protection).

2 Regulations, Laws and Technical Standards

2.1 Regulation and Italian National Law

The Regulation (EU) 425/2016 on Personal Protective Equipment (PPE) repeals the Council Directive 89/686/EEC with effect from the 21st April 2018. Ergonomic aspects are treated in Annex II "Essential health and safety requirements". Paragraph 1.1.1 "Ergonomics" provides "PPE must be designed and manufactured so that, in the foreseeable conditions of use for which it is intended, the user can perform the risk-related activity normally whilst enjoying appropriate protection of the highest level possible". Therefore, the principles of ergonomics are strictly related to PPE design and manufacturing. Ergonomics has also implications both for the comfort and the safety of workers.

In Italy, the Leg. Decree 81/2008 regulates health and safety at work. This Decree, according to EU Regulation, provides at Article 76 "PPE requirements" that "PPE must take into account the worker's ergonomic or health requirements" and "must fit to the user according to his needs". This Decree at Annex VIII analyzes the risks deriving from protective equipment relating the discomfort to a wrong ergonomic design.

On the other hand many national regulations (e.g. Polish laws) did not still cover the obligation for employers to take into account ergonomic principles when selecting PPE for workers [9].

2.2   Technical Standards

The principles of ergonomics established in the European regulation and acknowledged by the Leg. Decree 81/2008 are supported by the two Technical Standards EN 13921 and EN 547-1.

Technical Standard EN 13921. "Personal protective equipment - Ergonomic principles" [4] provides guidance on the generic ergonomic characteristics related to personal protective equipment (PPE). It specifies for the writers of PPE product standards, principles relating to: anthropometric characteristics related to PPE, the biomechanical and thermal interaction between PPE and the human body, the interaction between PPE and the human senses: vision, hearing, smell and taste, and skin contact.

Essential Factors. This Technical Standard identifies the essential factors to be considered in the determination of the best ergonomic solution for the workers [4]. These essential factors are listed below.

The "optimum level of protection" factors are: the duration of use of the PPE; whether there are different situations in which the PPE is used, requiring different amounts of the body to be covered and for other PPE to be used in addition; the reasonable balance between the severity of the hazard, protection, burden and duration.

The "optimal practicability" factors are identified by tests performed by users to represent a range of the normal movements made while wearing PPE. These tests should always include: understanding instructions and warnings given by the manufacturers; putting on, adjusting and taking off the PPE; general activities, such as moving and communicating; activities specific to situations where the PPE is to be used.

The "physiological impact" factors on workers are: heart rate; oxygen consumption; alveolar gas composition; breathing rate; body temperature change; blood flow; sweat rate; fatigue or strain of muscles.

Factors to be considered for the "adjustability of PPE and its appropriate fixation" to the body are: the information and instructions for fitting and adjusting; the adjustability and stability of adjustments; determining that the PPE has been correctly fitted.

Factors to be taken into account to ensure that PPE does not "irritate or cause discomfort" to the worker are: whether the PPE will be in contact with the skin and how sensitive this particular area of skin is to the effects of rubbing and pressure; for how long the PPE will normally be in contact with the skin; whether the type of PPE may have sharp or hard edges or points; whether the bulk, hardness and position of adjustment and closure mechanisms may have a negative impact on the worker; whether chemical composition of the material used and its by-products may affect the body; whether any materials likely to be used in contact with the skin are known to produce allergic reactions in proportion of the population; whether any closure mechanism, or other feature, might catch up and pull scalp, facial or body hair; whether the outer surface of the PPE could harm other people; whether simple visual and manual examination of the PPE is adequate for an assessment, or whether specific test methods need to be developed for assessing the hardness, roughness or other features of the PPE.

The factors concerning "anthropometric" characteristics of PPE are: the hazard against which the PPE is intended to provide protection; the body part(s) that it will be in contact with or cover; the physical activities expected to be performed during its use; the intended PPE worker group.

Factors to be considered in specifying requirements to take into account the "biomechanical characteristics" of PPE are: the static distribution of mass and consequent load on the worker's body when using different PPE and/or their combinations; the dynamic or inertial forces on the worker's body; optimisation of the influence of biomechanical aspects of PPE on work load and/or task performance.

The factors concerning "thermal characteristics" of PPE are: thermal characteristics of materials and complete PPE (e.g. thermal insulation, water vapor resistance/permeability, air permeability, water absorption/desorption).

Factors to be taken into account for the "sensory effects" of PPE are: vision, hearing, taste or smell, touch or other skin contact.

The features listed above can be used to design and analyze the ergonomics of PPE.

Technical Standard EN 547-3. "Safety of machinery - Human body measurements - Part 3: Anthropometric data", although written for other purposes, provides the standard anthropometric data from European surveys. This standard will have to be considered for the choice and use of right PPE and their ergonomic aspects.

3 Critical Aspects in PPE Ergonomics

Regulations, laws or Technical Standards do not exhaustively fulfill the ergonomic requirements, because they cannot take into account the individual characteristics of each worker in PPE design. This is one of the most critical aspects for the adoption of the principles of ergonomic to PPE. For example, if the values provided in the EN 547-3 are considered to design a protective clothing, the maximum height expected of the worker (95 percentile) is 188.1 cm, inadequate for highest workers in some northern European countries and excessive for other workers of southern Europe [5].

Variability in physical environment (climate) as well as technological and anthropological factors (body sizes) of people between different countries prompted us to believe that PPE technology and laws would result inadequate [2]. This is because technology that does not consider the local human operator within the system of the man-machine-environment can result in unanticipated problems. A Sri Lanka study [1] highlighted that an adjustable helmet designed in Europe did not properly fit 40% of heads. For example a safety clothing designed with insulation properties for cold climates is unsuitable, and extremely uncomfortable for people living in the hot and humid environment of the tropics [3].

Another important consideration in the choice of PPE is that their design is mainly guided by local standards pertaining to these goods (e.g. not EU manufacturer), not considering other countries' standards.

The literature reveals that if PPE is presented to the worker without any regard to the acceptance or wearability factors, it will result in its non-use or wrong use [2, 8]. For example, PPE against falls from a height are dynamic spine loads during fall arresting. Spectacles, goggles, and full-face masks could cause a reduction of the visual field and perception of visual signals, while protective clothing, shoes, and gloves could cause difficulties connected with sweat evaporation (reduction by 60% at 22 °C and by 25% at 40 °C), heat abstraction, movement restriction, and dexterity. Self-breathing apparatus can give an uncomfortable extra weight of about 20 kg. [6].

4   Method for Analyzing the Ergonomics of PPE

In order to meet the ergonomic needs referred to in the European Regulation 425/2016, a method is proposed to be used in occupational contexts to choose the most ergonomic PPE or to determine if the PPE could imply a risk for the worker.

Ergonomics requirements can be taken into account by determination of "performance levels" based on a risk assessment model. A risk analysis related to ergonomic aspects can be performed by assigning a score from 1 to 5 to each of the ergonomic features identified by the EN 13921 standard (Table 1), where "1" means "parameter not respected" and "5" means "parameter fully respected".

For a "quick evaluation" of the PPE ergonomics it is possible to assign a score to each of the 8 macro-areas related the "Effect or characteristic" (considering overall the items in the column "Factor" for each macro-area). For a "detailed evaluation" of the PPE ergonomics it is possible to score each of the 39 items related to the "Factors".

Table 1. Factors to consider in the determination of ergonomics of PPE.

#	Effect or characteristic	Factor		Value [1–5]
1	Optimum level of protection	• Long duration of use of the PPE • Whether there are different situations in which the PPE is used, greater amounts of the body to be covered, for various other PPE to be used in addition • Good balance between the severity of the hazard, protection, burden and duration		…
2	Optimal practicability	• Good understanding instructions and warnings given • Easy to putting on, adjusting and taking off the PPE • Ease of movement for general activities, moving and communicating • Ease of movement activities specific to situations where the PPE is to be used		…
3	Physiological impact	• Low heart rate • Low oxygen consumption • Good alveolar gas composition • Low breathing rate	• Low body temperature change • Low blood flow • Low sweat rate • Low fatigue or strain of muscles	…

(continued)

Table 1. (*continued*)

#	Effect or characteristic	Factor	Value [1–5]
4	Adjustability of PPE and its appropriate fixation	• Completeness of information and instructions for fitting and adjusting • Good adjustability and stability of adjustments • Ease of determination that the PPE has been correctly fitted	...
5	Not to irritate or cause discomfort	• Whether the PPE will be in contact with the skin, low sensibility of this particular area of skin to the effects of rubbing and pressure • Short time for the PPE to be normally in contact with the skin • Absence of sharp or hard edges or points of PPE • Whether the bulk, absence of hardness and good position of adjustment and easy closure mechanisms • Whether chemical composition of the material used does not affect the body • Whether any materials likely to be used in contact with the skin do not produce known allergic reactions • Whether any closure mechanism, or other feature, do not catch up and pull scalp, facial or body hair • Whether the outer surface of the PPE do not harm other people • Whether simple visual and manual examination of the PPE is adequate for an assessment to be developed for assessing the hardness, roughness or other features of the PPE	...
6	Anthropometric	• Appropriateness to the hazard against which the PPE is intended to provide protection • Extended body part(s) it will be in contact with or cover • Easy to perform physical activities during its use • Extended PPE worker group to be protected	...
7	Thermal characteristics of materials and complete PPE	• High thermal insulation • High water vapour resistance/permeability • High air permeability • Low water absorption/desorption	...
8	Sensory effects	• Extended range of vision • Good hearing	• Good taste or smell • Good touch or other skin contact	...

The adopted criterion considers the respect of 70% of the parameters listed as the minimum level to guarantee the ergonomics of the PPE, while the respect of less than 30% defines the PPE as "not-ergonomic". If the PPE obtains in a "quick evaluation" ("detailed evaluation") a score less than or equal to 12 (228) must be considered "not-ergonomic", between 13 (229) and 27 (531) must be considered "ergonomic with changes", greater than or equal to 28 (532) can be considered "ergonomic".

5 Smart-PPE

In addition to the assessment procedure mentioned above, which can lead to a scientific assessment of the ergonomics of PPE, the use of "smart" PPE can be considered.

A smart-PPE is a protective equipment with technologically "advanced" capabilities, also known as wearable technology. The built-in "intelligence" in the PPE can help keeping the worker safe while minimising accidents and hazards. The development of smart-PPE is linked to the advent of the "Internet of Things" and the so-called "Industry 4.0" [8].

The application of ergonomics in a smart-PPE is illustrated by the example of electro-acoustic ear protectors. Ear protectors with an electronic system of active noise reduction put into the ear-cap are a new solution in this field. Active noise reduction ensures greater protection, particularly against low-frequency noise, auditory contact and lower pressure.

Another smart technology used to overcome the difficulties of customizing PPE made on an industrial basis is the 3D printing. For example, as a part of a pilot project for BMW, gloves have been realized by an additive process of 3D printing using a flexible finger-cot to match the form and size of worker's hand. This system measures the worker's thumb with a mobile 3D hand scanner.

A group of other smart-PPE are based on the adoption of smart textiles, with technologically advanced properties. A protective clothing, using a textile that changes colour basing on room temperature, has been developed. This PPE alerts workers to unsafe temperature conditions with a visual indication. This "smart-textile", when applied to the overalls, can "feel" hot surfaces before they actually come into contact with the protective clothing.

Other PPEs use "printed" circuits in the textile itself, decreasing the stiffness of the cloth and the encumbrance of traditional electronic circuits and reducing the likelihood that cumbersome electronic circuits can get caught somewhere.

Another example of "smart" technology applied to PPE is the use of Bluetooth, Bluetooth Low Energy (BLE) or Near-Field Communication (NFC) technologies to provide information about the position, the physical conditions and the type of work performed by the worker. These systems improve ergonomics by reducing the weight of traditional PPE and exchanging real-time information to prevent potential hazardous situations.

6 Conclusions

Applying principles of ergonomics to PPE design helps to ensure that it fits properly and does not interfere with work activities. PPE can only do its job if it is worn, fits correctly, and is fitted properly. Designing, procuring, providing and wearing PPE that has been ergonomically designed to meet the specific needs of workers is vital and helps to ensure the life of the workers (ergonomic saves lives). Therefore, ergonomics, applied to PPE, is about more than comfort and fit (although these are important factors). Ergonomics applied to PPE means increasing workers' safety.

A useful tool to help in the right PPE choice may be the adoption of a scientific procedure. Applying the method proposed for analyzing the ergonomics of PPE can answer to the ergonomics principles provided in the Regulation 425/2016.

The main advantages in using ergonomics criteria, applied to the use and the choice of PPE, are: reduction of chances of injury, making work easier, faster, more comfortable and helping to build relationship.

The use of smart technologies applied to PPE can increase workers' safety and reduce hazards. In this case, the only limit for developing new PPE seems to be the creativity of the designers.

References

1. Abeysekera J, Shahnavaz H (1988) Ergonomic evaluation of modified industrial helmets for use in tropical environments. Ergonomics 31(1):1317–1329
2. Abeysekera J (1987) A comparative study of body size variability between people in industrialized countries and industrially developing countries: its impact on the use of imported goods. In: Proceedings of international symposium on ergonomics in developing countries, OSH series 58, ILO, Geneva CH
3. Abeysekera J, Shahnavaz H (1988) Ergonomics aspects of personal protective equipment: its use in industrially developing countries. J Hum Ergol 17:67–79
4. European Committee for Standardization (CEN) (2007) European standard EN 13921 personal protective equipment - ergonomic principles, Brussels
5. European Committee for Standardization (CEN) (2007) European standard EN 547-3, safety of machinery - human body measurements - part 3: anthropometric data, Brussels
6. Graveling R, Sánchez A, Lewis C, Groat S, Van Tongeren M, Galea K, Cherrie J (2009) Review of occupational hygiene reports on suitability of respiratory protective equipment (RPE). Health and Safety Executive, London, UK
7. National Institute for Working Life (2000) Ergonomics of protective clothing. In: Proceedings of 1st European conference on protective clothing, CM Gruppen
8. National Institute for Working Life (2014) Safe, smart, sustainable new pathways for protective clothing. In: Proceedings of 6th European conference on protective clothing, CM Gruppen
9. Szczecińska K, Łężak K (2000) Review of research studies of ergonomic aspects of selected personal protective equipment. Int J Occup Saf Ergon 6:143–151 Special Issue

The Functional Resonance Analysis Method as a Debriefing Tool in Scenario-Based-Training

Priscila Wachs[1]([⊠]) [iD], Angela Weber Righi[2] [iD],
and Tarcísio Abreu Saurin[1] [iD]

[1] Federal University of Rio Grande do Sul, Rua Osvaldo Aranha, 99,
5°andar, Porto Alegre 90035-190, Brazil
priscilawachs@ig.com.br
[2] Federal University of Santa Maria,
Avenida Roraima, 1000, Santa Maria 97105-900, Brazil

Abstract. The aim of this study is to discuss the use of the Functional Resonance Analysis Method (FRAM) as a debriefing tool in Scenario-Based-Training (SBT). This discussion is based on data collected during a training simulation session carried out as part of a Research and Development Project involving the development of resilience skills of grid electricians. The scenario of this simulation had a client complaining that the power had went off in his residence. The participants of the debriefing identified seven functions performed by the trainees. The use of the FRAM pointed out that there was variability in the outputs of two functions: <to make the repair> and <to check the extent of the power shortage>. Concerning the function <to make the repair> the work constraints, such as time pressure, encouraged workers to make a temporary repair, rather than replacing the cable for a new one. In the debriefing, two actions to re-design the work system were raised: to increase investments in preventive maintenance; and to improve the design of lifting equipment and tools. The instantiation presented showed that using FRAM models and concepts (e.g. output variability, couplings, and functions) can be useful for analyzing workers' and system's performance in the debriefing, since it presents the resonance arising from the variability of everyday performance and lead to recommendations for coping with the variability.

Keywords: Functional Resonance Analysis Method · Training
Debriefing

1 Introduction

Scenario-based-training (SBT) is widely regarded as an effective practice for the development of the necessary skills for working in complex systems. The SBT aims to present realistic scenarios that enables learning, by offering structured and systematic learning experiences, including proper measuring system and feedback [1–3]. To do such, the scenarios present interactive stories in specific contexts. And so, through

S. Bagnara et al. (Eds.): IEA 2018, AISC 819, pp. 132–138, 2019.
https://doi.org/10.1007/978-3-319-96089-0_14

these stories story and the whole scenario presents, participants are introduced to the problem or set of problems and must solve them [4, 5].

The practical experience in the scenario itself is already quite valid for learning, however, when associated with further discussion and feedback with reflection on practice, it becomes even more consistent [6–10]. The debriefing is the SBT stage in which trainees have the opportunity to understand their activity in the context of the broader system performance, and by realizing the gap between work-as-done and work-as-imagined. It also allows for the discussion of the necessary resilience to accomplish the goals and opportunities to improve the socio-technical system (STS) design [11].

Considering the importance of the debriefing, the SBT used in this study, and part of a broader study [11], included the debriefing as one of its four 'SBT stages protocol': (a) theory – basic knowledge on Resilience Skills and the aim of the SBT; (b) simulation briefing (mean time of 10 min); (c) simulation itself (mean time of 77 min); and (d) debriefing (mean time of 69 min). The debriefing started with the electricians describing what they did (a timeline was drawn on blackboard) and identifying incidents and decision-making on working-strategies. It allowed to discuss about the electrician local rationality, the gap between work as done and work as imagined, for example [11].

The aim of this study is to discuss the use of the Functional Resonance Analysis Method - FRAM as a debriefing tool in SBT. This discussion is based on data collected during a training simulation session carried out as part of a Research and Development Project involving the development of resilience skills of grid electricians (details on [11]). It is worth nothing that the development of the SBT used the Resilience Engineering perspective, which is the same perspective used to develop the FRAM. Resilience Engineering understand resilience as "the intrinsic ability of a system to adjust its functioning prior to, during, or following changes and disturbances, so that it can sustain required operations under both expected and unexpected conditions" [12].

2   Method

2.1   The Simulation

The training simulation analyzed was held at the company's training center in a distribution line built specially for the SBT program and the trainees received a vehicle equipped with materials, tools and safety gear (see Fig. 1). The training section happened in a sunny day, during the morning and afternoon, summer time in Brazil with no wind or rain. The scenario of this simulation had the following characteristics: (a) story: the client claims that the power goes 'back and forth'; (b) Service Notification given the to pair of electricians: lack of power in the consumer's unit 10, at Gramado Street; (c) defects on the distribution line: broken connection phase inside the insulation cable; (d) expected solution: replace the cable. The scenario also presented: (i) Difficulty to access to the region; (ii) Lifting weights and need to use a lot of physical strength; (iii) lack of equipment or materials to undertake the activity; (iv) pressure from supervisors, central operations control room or clients; (v) activity carried out previously, at the same location, in an inadequate way.

Fig. 1. Simulation section.

The previously presented SBT protocol was followed: theory (35 min), briefing (5 min), simulation itself (85 min) e debriefing (50 min). The participants in this simulation, and their respective roles, were:

(a) Two members of the research team, who ensured that the simulation protocol was followed. One played the client-role and the other took notes during the simulation, recording information that could support the debriefing;
(b) Two instructors designated by the company, one of the playing the role of the control center, and one Health and Safety Department's member;
(c) The trainees: in accordance with the real work arrangement, a pair of electricians was involved. One of them had 9 years of experience as a front-line electrician, and the other had 11 years of experience.

Details on the SBT development can be read at [11].

2.2   Functional Resonance Analysis Method (FRAM)

The FRAM is a systemic approach to describe activities of a Socio-Technical Complex System, capable of representing the activities of real work (*work as done*) and helping to make sense of human performance variability [13, 14]. The instantiation used to analyse the Functional Resonance was the previously described one and the use of FRAM aimed to better understand the variability present in the real activities, contributing to a deepening discussion in debriefings, which is an important aspect for the knowledge appropriation after simulation sections.

An important FRAM stage is the identification and description of the systems functions. A function refers to the activities, or set of activities, that are required to produce a certain outcome. A function describes what people, individually or collectively, have to do in order to achieve a specific aim [13]. The FRAM defines that each function can be composed by six aspects, which are: (a) Input - starts the function; (b) Output – function's result; (c) Precondition - state or conditions necessary to start the function; (d) Resource - necessary or consumed during the performance of the function; (e) Time - temporal characteristic; (f) Control - supervises or regulates the function [13, 15]. The software FRAM Model Visualizer (version 0.3.2, free and available at www.funciotnalresonance.com) was used to graphically represent the functions and the relationship between them.

3   Main Findings

As mentioned above, the first stage of the debriefing was to identify the 'what they did'. The participants identified seven functions performed by the trainees (Fig. 2): (a) to receive the service order from the control center; (b) to go to the place where the service was requested; (c) to check the extent of the power shortage; (d) to identify the cause of the defect; (e) to make the repair; (f) to turn on the switch trip related to the repaired circuit, re-establishing power to the client; and (g) to inform the central operations control that the repair was made. Those were the main functions identified by the participants and performed by the electricians, but four other functions were also mentioned and played a key role on the activity performed: to provide work instructions and to have the necessary equipment and material (organization's functions); to exchange information with the electrician (client's function); to inform about consumer request and network condition (control room operator's function).

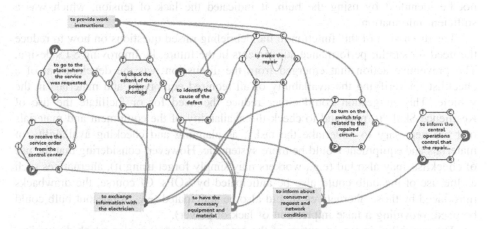

Fig. 2. Identified functions during simulation training. Note: the 'square' functions are not developed by the electricians (green = organization's function; purple = client's function; blue = control room operator) (Color figure online).

The variability of the outputs of two of these functions (c and e) were identified through the FRAM (see in Fig. 1). Concerning the execution of function (c), workers realized that they had not brought with them a voltmeter, the standard equipment to measure tension (work constraint "lack of equipment or materials to undertake the activity"). In addition to this, the client was very upset with the power shortage (work constraint "pressure from supervisors, central operations control room or clients"). As the standardized operational procedure (SOP) did not mention what to do in case of lack of voltmeter, let alone how to deal with upset clients, performance adjustment and use of Resilience Skills[1] was necessary. By discussing the situation with the central operations control room team and within the team in the field, the trainees decided to borrow a bulb from the client and connect it with the switch trip that linked the private and public networks. Since workers had to explain to the client why they needed the bulb for, they had to use Resilience Skill "To discuss, with the consumers and population, the status and hazards of the power line maintenance procedures, as well as the possible causes of defects". If the bulb did not light up, this would indicate that there was a failure in the public distribution network. If the bulb lit up, this would indicate that the problem was in the client private electrical facilities, and thus the electricians should not do any repair; the client should make provisions for the repair himself. It is worth noting that returning to headquarters and bringing a voltmeter was not a reasonable option, since the simulation assumed that the client lived in a rural zone, far away from headquarters (work constraint "difficulty of access to the region").

Although the use of the bulb proved to be good enough to confirm that the problem was in the public distribution network, the variability of the output of function (c) was affected. In fact, the output (i.e., existence of power checked) was too late, since the time taken to make the decision to use the bulb, involving a negotiation with the client and control room team, was longer than the time taken by using a voltmeter. However, the output was acceptable in terms of precision. Although the intensity of tension could not be identified by using the bulb, it indicated the lack of tension, which was a sufficient information.

The discussion of this function in the debriefing raised questions on how to reduce the need for similar performance adjustments in the future, by improving STS design. The preventive action that emerged from the discussion was the development of a checklist for verifying the availability of all required equipment and materials in the vehicle. This suggestion could either reduce the need for or facilitate the use of Resilience Skill "to plan and to check the availability of the equipment and materials that are necessary to undertake the task", as planning and checking availability of materials and equipment would be more systematic. However, considering that the use of checklists may also fail (e.g., workers may simply forget using it), alternatives such as the use of the bulb could also be anticipated by SOPs. Of course, the drawbacks introduced by those alternatives should also be anticipated (e.g., a burnout bulb could be used, providing a false impression of lack of power).

The variability in the execution of the repair (function e) is also worth discussing. The defect was found to be a broken cable, which was inside the insulation layer and therefore was not visible for workers. The SOP recommended the replacement of the damaged cable by a new one. Three work constraints, which were not anticipated by SOPs, were impacting at the moment of the repair: the need for lifting weights and using physical strength; pressure from clients; and activity carried out previously, at the

---

[1] "Individual and team skills necessary to adjust performance, in order to maintain safe and efficient operations during both expected and unexpected situations" [11].

same location, in an inadequate way. This latest constraint was due to the fact that the damaged cable had been amended by another team of electricians before the defect, rather than replaced. Thus, there was evidence that a temporary repair had already been made in the cables.

The work constraints, as well as the variability of the output of function (c) in terms of time, which created time pressure for all downstream functions, encouraged workers to make a temporary repair, by making a new amendment in the cables rather than replacing them. Workers reported in the debriefing that their choice was driven by: cost reduction, since no new materials were used; saving time and effort, since a full repair would be more physically demanding and time-consuming. Moreover, workers reported that they took for granted that making temporary repairs was "normal", in the sense that they and their colleagues used to do that rather than full repairs. As the electricians' decision of making a temporary repair was not in line with SOPs, they adjusted their performance and used the Resilience Skill "to draw up work strategies after the defects have been identified".

While the output of function (e) (i.e., repair was made) happened too early, compensating for the late output of function (c), it was imprecise. Of course, the temporary repair was likely to provoke the same defect in the near future. Moreover, the variability of the output of function (e) is likely to feed back to function (a) (to receive service order), triggering the need for a new repair at the same residence. In the debriefing, two actions to re-design the STS, which could stop that vicious circle, were raised: to increase investments on preventive maintenance, such as by replacing equipment and materials before they fail; in the long-term this would reduce the demand for emergency services and, consequently, the need for using all RSs; to improve the design of lifting equipment and tools, in order to reduce the need for using physical strength. Moreover, SOPs could provide guidance to support workers' decision on whether temporary or definitive repairs are suitable. For instance, there might be strong time-pressure and the materials required for the full repair may not be available. Therefore, it was the trainees and instructors consensus, in the debriefing, that the execution of temporary repairs should be taught in formal training courses. Of course, drawbacks of temporary repairs should also be stated in SOPs.

4   Conclusions

The use of FRAM to analyze simulation allowed to explain the necessary functions to perform the simulation, as well as to analyze the variability of each one of them and the emergent functional resonance. This analysis allows the discussion about opportunities for improvement in the work organization design, improving the safety and efficiency of the activities. In this way, it is worth nothing that FRAM as a tool to conduct the debriefing is opportune both for reflection on the action in the simulation itself, and to identify opportunities for improvement in work organization.

Also, a suggestion for future studies is to deep understand how FRAM contributes to the learning process among the simulation participants. FRAM as a recurrent tool in debriefing and efficient tool for the achievement of the different simulation exercises objectives in the various areas of knowledge, especially human factors.

References

1. Zendejas B, Cook D, Farley D (2010) Teaching first or teaching last: does the timing matter in simulation-based surgical scenarios? J Surg Educ 67(6):432–438
2. Salas E, Guthrie J, Burke S (2007) Why training team decision making is not as easy as you think: guiding principles and needs. In: Cook M, Noyes J, Masakowski Y (eds) Decision making in complex environments. Ashgate, Burlington, pp 225–232
3. Chamberlain D, Hazinski M (2003) Education in resuscitation. Resuscitation 59:11–43
4. Moats JB, Chermack TJ, Dooley LM (2008) Using scenarios to develop crisis managers: applications of scenario planning and scenario-based training. Adv Dev Hum Res 10 (3):397–424
5. Cannon-Bowers J, Bowers C, Procci K (2010) Optimizing learning in surgical simulations: guidelines from the science of learning and human performance. Surg Clin North Am 90:583–603
6. Tannenbaum S, Cerasoli C (2013) Do team and individual debriefs enhance performance? A meta-analysis. Hum Factors: J Hum Factors Ergon Soc 55(1):231–245
7. Paige J (2010) Surgical team training: promoting high reliability with nontechnical skills. Surg Clin North Am 90:569–581
8. Gaba D, Howard S, Fish K, Smith B, Sowb Y (2001) Simulation-based training in Anesthesia Crisis Resource Management (ACRM): a decade of experience. Simul Gaming 32:175–193
9. Alinier G, Hunt W, Gordon R (2004) Determining the value of simulation in nurse education: study design and initial results. Nurse Educ Pract 4:200–207
10. Andersen P, Jensen M, Lippert A, Østergaard D, Klausen T (2010) Development of a formative assessment tool measurement of performance in multi-professional resuscitation teams. Simul Educ 81:703–711
11. Saurin TA et al (2014) The design of scenario-based training from the resilience engineering perspective: a study with grid electricians. Accid Anal Prev 68:30–41
12. Hollnagel E, Paries J, Woods D, Wreathall J (2011) Resilience engineering in practice: a guidebook. Ashgate, Burlington
13. Hollnagel E (2012) FRAM: the functional resonance analysis method e modeling complex socio-technical systems. Ashgate, Burlington
14. Clay-Williams R, Houngsgaard J, Hollnagel E (2015) Where the rubber meets the road: using FRAM to align work-as-imagined with work-as-done when implementing clinical guidelines. Implement Sci 10:125–135
15. Hollnagel E (2014) Safety-I and safety-II: the past and the future of safety management. Ashgate, Burlington

Planning Simulation Exercises as Learning Lab: The Case of Digital Chart Changing Maritime Navigation Activity

Monique Mota Martins[1]([✉])(iD),
Floriano Carlos Martins Pires Junior[1] (iD), José Orlando Gomes[1],
and Angela Weber Righi[2] (iD)

[1] Federal University of Rio de Janeiro, Rio de Janeiro, Brazil
monique250982@hotmail.com
[2] Federal University of Santa Maria, Santa Maria, Brazil

Abstract. The Electronic Chart Display and Information System (ECDIS) has become a key element in ship's bridge, assisting navigators to make the right decision. Despite the technological advance, the ability to make these decisions in the right time is still the determining factor for the safety of navigation, being directly related to the interaction between human and system. The objective of this study was to identify fundamental factors to be incorporated in simulations to analyze the shared cognitive system between the operator and the ECDIS system during navigation, from the perspective of Resilience Engineering (RE). For this, the following methodological strategies were used: (a) CDM method: interviews with navigation specialists, to obtain critical situations experienced by the navigators using the system and the possible actions to be adopted to solve the problems; (b) analysis of accident reports, where inadequate ECDIS configurations were the determining factor. Thus, the research resulted in a scenario elaborated with three critical points for the simulations, which were later validated together with the experts and applied in a course for the qualification of ECDIS operators.

Keywords: ECDIS · Simulation · Resilience Engineering

1 Introduction

Maritime transport is the backbone of trade and the global economy, as more than 90% of freight transport is carried out by international shipping vessels [1, 2].

However, in order to perform a trip, it is essential to determine the position and therefore some equipment has become necessary, such as the compass, the sextant, the radar, the GPS and the maps. However, unlike land navigation, maritime navigation faces many challenges due to the difficulty of obtaining landmarks, and this is the main motive that led man to develop various navigation techniques that would adapt to the reality of his ships and areas of navigation.

Indeed, as the United Nations (UN) [1] points out, ships have at no time been as technologically sophisticated as they are today. However, this technology deserves attention, especially since misuse or lack of knowledge of it can have unintended

© Springer Nature Switzerland AG 2019
S. Bagnara et al. (Eds.): IEA 2018, AISC 819, pp. 139–144, 2019.
https://doi.org/10.1007/978-3-319-96089-0_15

consequences. It is important to note that safety is essential in this area and, for this reason, the maritime industry has been struggling to increase it. However, further progress is needed to prevent accidents or incidents from occurring, such as with Royal Majesty in 1995, CFL Performer in 2008 and CSL Thames in 2011. Such accidents raise the question of how to improve safety.

Nowadays ships use modern equipment such as ECDIS (Electronic Chart Display and Information System), which allows the integration of several sensors to a single screen and obtaining the position of the ship in real time. For the benefits that the system suggests, changes have been made in the current regulations, to make it mandatory for international travel vessels according to the convention [3].

Once the structure of a system changes, so do the possible paths of failure, that is, the ways in which unwanted combinations may occur. Failures are, according to the contemporary view, not the direct consequence of causes, as assumed by the sequential accident model, but rather the result of coincidences that result from the natural variability of human performance and the system [4].

The use of technological development in the maritime industry has brought about major changes in the way of navigation, and consequently the expected actions of those who must work together with modern equipment designed to increase efficiency, accuracy and safety. The navigation traditionally seen using the position of the sun, moon and stars to obtain the ship's position and the use of nautical paper charts is being replaced by the figure of modern ships equipped with ECDIS. With the use of this equipment and the integration of several sensors to it, it is possible to obtain a position in real time and monitoring of hazards to the navigation with generation of alarms. The main function of the system is to increase navigation safety [5].

Such system corroborates the perception that safety has been pursued by the area, but it does not guarantee that it will be fully achieved. The accidents still occurring in maritime transport show that there is a need for some intercurrence. Therefore, it is necessary to know how the system works in practice, so that the human factor is considered, so that it can point to the paths that can be traveled with the purpose of the system being used in favor of maritime transport.

In order to understand the performance of the human element and its importance in the operation of complex systems, concepts derived from the Cognitive Systems Engineering and Resilience Engineering (RE) can be used. And, one way to analyze these factors, based on the paradigms of CSE and RE, is through the use of simulation and training. To do so, aiming at a closer analysis of reality, it is important to adopt a methodological strategy for the elaboration of the simulations. Thus, the objective of this study is to present the methodological tools used to identify fundamental factors to be incorporated in simulations that aim to analyze the relationship between the operator and the ECDIS during navigation, from the perspective of RE.

2   ER in the Context of Maritime Navigation

In the maritime environment, as in many other fields, there is a great deal of human factors contributing to the cause of accidents. According to study [4], "human error" remains the main cause of maritime accidents, although the report highlights other issues such as fatigue, economic problems and lack of training of crew as the main cause for concern. That is, in most accidents that occur, the attributed cause is in one way or another, failure in human performance. The attributed cause may, however, be different from the actual cause [6].

Technological innovations have caused the nature of work to undergo radical changes, modifying the way the task is done or introducing new functions. Usually the use of technology is intended to increase performance and efficiency, reduce costs and increase quality. But the best results are not always achieved and can still generate new and unexpected problems in other areas. Increasing the complexity of the system also increases the chances of failure, either by providing more chances for men to make mistakes, or by situations where actions are unexpected and cause adverse consequences, as well as cases where the system malfunctions. These are some possibilities of problems that may arise [6].

Considering the increase in workload on board, the reduction of crews, simultaneous tasks on bridges and the installation of ever more advanced equipment, such as ECDIS, the scenario proposed for this study is an excellent example in which the application of RE can contribute to changes in the treatment of safety studies. In sociotechnical systems, resilience is the ability to keep the system running and achieve system objectives under a variety of operating conditions, including emergency situations [13]. In RE studies, positive examples of how systems successfully adapt to prevailing requirements are emphasized from four bases: monitoring, response, anticipation, and learning [6].

The recognition of the importance of human cognitive skills in the performance of tasks in the maritime environment can be observed in the insertion of terms such as decision making and situational awareness in the convention [7]. Thus, to meet the amendments to the convention, many seafarers are being trained to meet these requirements. The question is how to train to ensure these skills and how to measure and evaluate the desired competence. It is believed, from the perspective of this study, that qualification involving simulations, using the premises of the CSE and RE, can contribute significantly to the supply of this gap.

3   Tools for Preparing the Simulated Exercise

In order to understand the context in which the simulation would be proposed, as well as to identify possible situations to create scenarios to be used, interviews were conducted with experts from the maritime sector. In these interviews, based on the Critical Decision Method (CDM) [8], the objective was to find critical situations in the use of ECDIS.

In addition to the interviews, data were also collected with the analysis of accident reports, which were detected as the cause of these, mainly, the misuse of the system. The literature on research and applications of knowledge extraction describes document analysis, such as technical reports, as an important and necessary method [9, 10].

3.1   Interview with Experts

This phase, which lasted about four months, twelve seafarers were interviewed. The selection of commanders to be interviewed was initially based on two criteria: (i) commanders who took the ECDIS qualification course; (ii) masters with many years of experience at sea.

Initially, interviews were conducted with specialists based on the CDM technique, following its four phases: event identification, timeline verification, deepening and "what-if" questions [8]. Each step was focused on obtaining specific types of information and gradually deepened the critical cognitive points. In the end, it was possible to have a complete understanding of the incident and to understand the cognitive demand of the task and its configuration. Interviews were transcribed and confirmed and supplemented as necessary.

Despite all the bibliographic research that preceded this phase, the interviews were not easy. Seven attempts were made, although they did not provide interesting data according to the objective of this phase, they motivated changes in the questions, the focus, the specialist profile, as well as they were able to identify the appropriate moment for certain questions.

Thus, after the first interviews, it was observed that the navigators who had been in command for many years did not have much practice in the equipment under study, or it was a more management experience. Then, it was necessary to change the profile of the specialists to be selected for the interview: commanders who had many years of practice in the equipment, especially at the operational level of the system and not only of management. And then, the stories for the elaboration of the exercise were obtained.

Then, from the total of seven initial interviews, in addition to the five subsequent ones, twelve interviews were carried out. Of these, data from the interviews with three of these experts were richer in detail, and these are the main bases for choosing critical events in the use of ECDIS and the procedures followed by operators to maintain the safety of navigation.

3.2   Analysis of Incident Reports

In addition to the interview data, accident reports, in which ECDIS played a major role as the cause of the accident, were analyzed. Three main accidents were analyzed: (i) accident of the CFL Performer vessel [11]; (ii) accident with the CSL Thames vessel [12]; and (iii) an accident with the Ovit vessel [13].

Initially, the entire configuration of the accident scenario was analyzed. The characteristics of the location, the ECDIS, the sensors connected to it, the operators' procedures served as the basis for a selection of data that could be reproduced in the simulated exercise. Then, the critical points of the accidents and their complications were studied for later configuration of the insertion of the critical situations of the simulation.

After this study with the accidents, a selection of the information about the critical events and procedures was made for the elaboration of the simulated exercise. This analysis of the reports occurred in the same period in which the interviews were being conducted.

3.3 Setting the Scenario

In addition to the data from previous steps, other documents about ECDIS and navigation with this equipment - such as standards, resolutions, recommendations, articles, books, journals, model courses and other documents of the International Maritime Organization were also used.

Based on the data collected in the tools, three situations were selected to integrate the simulated scenario: (a) route purposely drawn in an inappropriate manner; (b) loss of position by GPS; and, (c) failure of heading information from gyroscopic compass. The simulation preparation stage lasted about two months, between making and validation by instructors and commanders, and tested by seafarers.

4 Conclusion

This work uses concepts from Cognitive Systems Engineering and Resilience Engineering to guide the search for methodological tools used to identify fundamental factors to be incorporated in simulation to analyze the relationship between the operator and the ECDIS system during navigation, from the perspective of Resilience Engineering.

In addition to obtaining the critical points, the interviews and accident reports also served to better understand the actions that are foreseen, the general context of the accomplishment of the task and the way these navigators perform the task, guiding the elaboration of the simulated exercise and also the subsequent analysis of their results. This "learning lab" shows the essential methodological phase to try to understand the complexity of the systems, as well as the navigation with ECDIS.

References

1. ONU (2016) Maritime Transport Is 'Backbone of Global Trade and the Global Economy', says Secretary-General in Message for International Day. https://www.un.org/press/en/2016/sgsm18129.doc.htm. 28 Apr 2018
2. IMO. IMO's contribution to sustainable maritime development. http://www.imo.org/en/OurWork/TechnicalCooperation/Documents/Brochure/English.pdf. Accessed 26 Feb 2017
3. IMO (2014) International Convention for the Safety of Life at Sea (SOLAS) (1974/1988), 2014 edition
4. Hollnagel E, Woods DD, Leverson N (2006) Resilience engineering: concepts and precepts. Ashgate, Farnham
5. IMO (2006) Resolution MSC 232(82) Adoption of the revised Performance Standards for Electronic Chart Display and Information System (ECDIS). Accessed 5 Dec
6. Hollnagel E, Woods DD (1983) Cognitive systems engineering: new wine in new bottles. Int J Man-Mach Stud 18:583–600
7. IMO (2011) STCW including 2010 Manila amendments, 2011 edition. IMO Publishing, London
8. Crandall B, Klein G, Hoffman R (2006) Working minds: a practitioners guide to cognitive task analysis. The MIT Press, Cambridge

9. Hoffman RR (1987) The problem of extracting the knowledge of experts from the perspective of experimental psychology. AI Mag 8:53–67
10. Hoffman RR et al (1995) Eliciting knowledge from experts: a methodological analysis. Organ Behav Hum Decis Process 62(2):129–158
11. Marine Accident Investigation Branch (MAIB) (2008) Report of investigation of the grounding of CFL Performer in Haisborough Sand, North Sea on 12 May
12. Marine Accident Investigation Branch (MAIB) (2011) Report of investigation of the grounding of CSL Thames in the Sound of the Mull on 9 Aug
13. Marine Accident Investigation Branch (MAIB) (2014) Report of investigation of the grounding of Ovit, in Dover Strait on 18 Sept 2013

Ergonomics and Regulation: The Case of Job Rotation in a Brazilian Slaughterhouse

Iracimara de Anchieta Messias[1](✉) and Adelaide Nascimento[2]

[1] University of the State of São Paulo,
Roberto Simonsen Street, 305, Presidente Prudente – São Paulo, Brazil
iracimara.messias@unesp.br
[2] Cnam-CRTD EquipeErgonomie, 41, rue Gay-Lussac, 75005 Paris, France
adelaide.nascimento@cnam.fr

Abstract. In April 2013, Regulatory Standard 36 was published in Brazil, concerning health and safety at work in slaughtering and meat processing companies, aiming to establish, among other things, standards for the implementation of job rotation. This industry has expanded in Brazil, which is currently the world's largest exporter of beef. This work presents partial data from a survey carried out in a cattle slaughterhouse located in the state of São Paulo, Brazil. The purpose of this communication is to present the acceptance of the slaughterworkers regarding the rotation of posts imposed by the legislation, as well as to identify the determinants of the work. According to the data obtained, the slaughtering sector is composed of 88 workers, of which 91% are male and 47% aged 40 to 54 years. The most common causes of absenteeism in last two years were inguinal and abdominal hernias. The most significant determinants of labor are: product variation (cattle differences), cadence and irregular work rate, and physical and psychological demands. Knives, the main work tools, require time to be cleaned and sharpened. Concerning acceptance of the rotation, 30% of the workers are in favor, as they will learn new activities, and 70% are contrary as they believe they will be more exposed to work accidents in activities for which they do not have specific skills. It is concluded that detailed analysis of the posts, with the effective participation of the workers, are the main steps to comply with current legislation and preserve health at work and reduce the number of accidents and occupational diseases.

Keywords: Ergonomics · Slaughterhouse · Job rotation

1 Introduction

In Brazil, attention to safety and health in the work of the slaughterhouse sector has been a cause for concern. Exportation from the Brazilian meat and processing industry has increased in the last decade, and Brazil is currently the largest exporter of beef and chicken in the world, being in 4th place in pork exportation [1].

Due to this, Regulatory Standard 36 was published in 2013, which legislates on safety and health at work in slaughter and processing companies. This norm, among other issues, establishes standards for the implementation of job rotation of posts in the sector [2].

© Springer Nature Switzerland AG 2019
S. Bagnara et al. (Eds.): IEA 2018, AISC 819, pp. 145–149, 2019.
https://doi.org/10.1007/978-3-319-96089-0_16

According to the inspection body in the area, the Public Ministry of Labor, expansion in the slaughterhouse industry sector has been accompanied by an explosion of illnesses and accidents at work. Payment of labor fines and severances are routine actions decreed by the courts to companies that do not comply with legislation, however, sickness and absenteeism rates increase each year [1].

Among the Brazilian exporter States, the most important are the State of São Paulo, in the southeastern region of Brazil, followed by the States of Mato Grosso and Mato Grosso do Sul, both in central-western Brazil.

Concern regarding risk factors in the performance of work activities in the meat processing industry has been great [4]. The biomechanical and organizational risks in the sector were addressed in the studies of Tappin et al. [5], Farladeau and Vézina [6] and Richard [7]. Psychosocial risks are presented as the main factors responsible for illnesses [8].

According to Neto [9], a working condition in which the ergonomics of the process are not observed leads to low industrial and worker productivity.

Thus, studies are important in order to assess whether the actions implemented and foreseen in the new legislation are effective in their objectives.

This paper presents partial data from a survey carried out in a cattle slaughterhouse located in the State of São Paulo, Brazil, whose objective is to present the acceptance or not of the workers to the implementation of job rotation of posts established by the legislation, as well as to identify the determinants of the work which must betaken into account in this process.

2   Methodology

This is an action – research, in which workers of a medium sized slaughterhouse in the western region of the State of São Paulo, Brazil, were ergonomically analyzed.

The work was initiated due to demand for the company to implement job rotation of worker positions in the slaughtering sector. This sector employs workers who perform complex work activities and, for this reason, it is difficult to carry out rotation.

Firstly, a survey of socio-demographic data and the causes of absenteeism in the previous 24 months was carried out, according to medical data provided by the company.

Subsequently, ergonomic analyzes of the work activities were performed, in areas where the main functions of the sector are carried out: Magarefes and Fakerys.

Systematic observations of the activities were performed using Actogram Kronos® software.

Using a questionnaire adapted from Ouellet et al. [10], semi-structured interviews were conducted with all workers in the slaughtering sector, addressing topics related to the acceptance of job rotation.

3   Results

The company employs approximately 600 workers, 30% in the administrative area and 70% in the operational sectors. The operational sectors are divided into 13 sectors, and those that employ the majority of workers are the Desossa, Meat processing, and Slaughtering sectors with 280, 229, and 88 workers respectively.

The slaughtering sector is composed of 91% men, 47% of whom are between 40 and 54 years of age.

Due to the complexity of the work activities of this sector, workers need prior knowledge and specific training to carry out their work activities, and more than 90% of them are considered experienced in their functions; 48% have been in their current function for from 1 to 5 years and 44% from 5 to 15 years.

The major causes of absenteeism in the previous two years were inguinal hernias and ventral hernias with obstruction (30% of absenteeism), accompanied by carpal tunnel syndrome, responsible for 30% of absenteeism. Small accidents with work tools also appear as a cause of absenteeism, 15% of which are reported as injuries to the fingers and 17% without injuries. Upper limb injuries were responsible for 21% of absenteeism, as observed in Fig. 1.

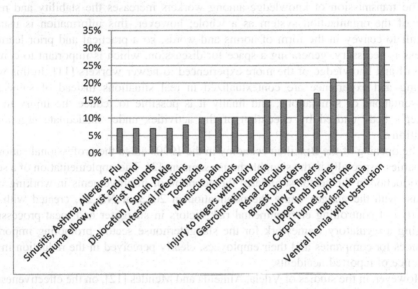

Fig. 1. Causes of absenteeism in slaughterhouse workers in the state of São Paulo, Brazil, in the last two years.

Using Kronos® software and performing ergonomic analysis of the work, it was possible to establish the most important determinants of the work. These are differences in product variation (differences in cattle according to their origin and race), and cadence, which generates an irregular rhythm of work. The frequency of the work is determined by the speed (current velocity ranging from 90 and 120 m/h), which increases the physical, in other words, biomechanical requirements. In addition, the psychosocial demands are caused by pressure stress from the bosses and their peers to perform quality work.

Knives, the main work tools, require time to be cleaned and sharpened, in compliance with the standards of hygiene and quality of the product. However, these activities are not always performed correctly, due to the time demands (cadence and rhythm).

Acceptance of job rotation is low, only 28% of respondents accept carrying out job rotation of posts, while 72% are contrary. Among those who accept job rotation, 30% believe that it will be a good solution to reduce the risks of work, 30% report that they will learn new activities, and 20% report that they will gain relief from physical and repetitive efforts. Among the workers who oppose job rotation, 30% believe that they will be more exposed to accidents at work, and 35% feel insecure in performing activities for which they do not have specific skills.

4 Discussion

In fact, one can observe that the skills and knowledge appropriate to each slaughter post demand learning over time. Aspects linked to competences cannot be overlooked [3].

The transmission of knowledge among workers increases the stability and reliability of the organization system as a whole; however, this information is usually difficult to convey in the form of norms and words, so a practical and prior learning process is necessary, generating a space for discussion, which is important to convey the skill and knowledge of the more experienced to newer workers [11]. In this way, learning and experience are contextualized in real situations instead of subjective representations or simulations, and finally it is possible to reduce the injury to the worker's body caused by execution of the activities under inadequate ergonomic conditions.

The opening of the Brazilian economy in the 1990s aided the professionalization of companies, especially in the slaughterhouse segment [2]. The implementation of a scale of production in the sector stimulated the adoption of improvements in working conditions, with the apogee being the publication in 2013 of NR 36, created with the objective of controlling environmental risk factors in slaughter and meat processing, bringing a regulatory framework for the slaughterhouse sector, promoting important advances for companies and their employees, clearly perceived in the reduction in the incidence of reported accidents.

However, in the studies of Vilela, Almeida and Mendes [12], on the effectiveness of the ergonomic approach in the sector, compared to a standard approach carried out in the same company 10 years earlier, some differences were observed regarding: (1) the understanding of determinants besides physical and biomechanical activities; (2) negotiation for better working conditions; and (3) effective participation of employees in the process.

The results presented here corroborate those already made, which highlight the presence of determinants related to production, leading to effects on the activity and health of employees [13]. Studies [6–8, 10] show that job rotation alone does not favor the prevention of work risks, as one must take into account the different determinants for a job rotation model according to the current activity and health of the workers.

The continuity of our intervention aims to contribute to this type of study, carried out in the Brazilian context. The perspective is to follow the implementation of job rotation in this company, according to the requirement of the NR-36 standard and to evaluate it, carrying out this intervention with a participatory approach between managers and workers within the Brazilian cultural context.

5   Conclusion

As a conclusion, the preliminary data indicate that detailed analysis of the activities of each station where rotation will take place, with the effective participation of the workers, are steps to be taken both to comply with current legislation and preserve health at work and reduce the number of accidents and occupational diseases.

Acknowledgments. The authors thank the Fundação de Amparo à Pesquisa do Estado de São Paulo – FAPESP, for the financial support. Process number: 2017/05299-5.

References

1. ABIEC – Brazilian Association of Meat Exporters (2017). http://www.abiec.com.br
2. BRASIL (2013) Portaria MTE n.º 555, de 18 de abril de 2013. Institui o Ministério de Segurança e Saúde no Trabalho. Diário oficial [da União], Brasília, DF, 19 April
3. Vasconcellos MC, Pignatti MG, Pignati WA (2009) Employment and work accidents in the refrigeration industry in agribusiness expansion areas, Mato Grosso, Brasil. Saúde&Sociedade 18(4):662–672
4. Henchion MM, McCarthy M, Resconi VC (2017) Beef quality attributes: a systematic review of consumer perspectives. Meat Sci 128:1–7
5. Tappin DC, Bentley TA, Vitalis A (2008) The role of contextual factors for musculoskeletal disorders in the New Zealand meat processing industry. Ergonomics 51:1576–1593
6. Falardeau A, Vézina N (2002) Rotationdes postes, assignationtemporaire et impact des absences dans une usined'abattage et de transformation Du porc, Perspectives interdisciplinaires sur Le travail et La santé, 4-2
7. Richard P (2002) Analyse ergonomique et measures biomécaniques dans un abattoir de porcs. Perspectives interdisciplinaires sur Le travail et La santé [Enligne], 4-1
8. Coutarel F, Dugué B, ET Daniellou F (2003) Interroger l'organisation Du travail au regard desmarges de manoeuvre en conception et em fonctionnement. La rotation est-elle une solutionauxTMS?, Perspectives interdisciplinaires sur Le travail et La santé [Enligne], 5-2
9. Neto JMC et al (2011) Avaliação da rotatividade pessoal em indústria de processamento de carnes com intervenção ergonômica no processo. Encontro Internacional de Produção Científica. Acesso em 22 Nov 2013
10. Ouellet S, Vézina N, Malo JL, Chartrand J (2003) L'implantation de larotation de postes: un exemple de démarche préalable, Perspectives interdisciplinairessurletravail et lasanté, 5-2. http://pistes.revues.org/3322
11. Rocha R, Figrueiredo V, Baptista AK (2016) Knowledge Management for Counterbalancing the Process of Loss of Skills at Work: A Practical Study
12. Vilela RAG, Almeida IM (2012) Mendes RWB Da vigilância para prevenção de acidentes de trabalho: contribuição da ergonomia da atividade. Ciência&SaúdeColetiva 17(10):2817–2830
13. Vogel K et al (2013) Improving meat cutters' work: changes and effects following an Intervention. Appl Ergon 44(6):996–1003

Chemical and Biological Risks in Professionals Working in Operating Rooms

Paula Carneiro[1(✉)] [iD], Ana C. Braga[1] [iD], and Roberto Cabuço[2]

[1] ALGORITMI Centre, School of Engineering,
University of Minho, Braga, Portugal
pcarneiro@DPS.uminho.pt
[2] Main Military Hospital/Higher Institute, Luanda, Angola

Abstract. Certain health professionals are at risk of having work-related accidents, mainly due to the contact with biological and chemical materials. The operating room is a place where this type of accident occurs frequently.

This work addresses the problematic of occupational risks associated with operating rooms, aiming to identify and analyze the biological and chemical risks, as well as the professionals' behaviors towards them. The possible operating rooms risks will be presented through the analysis of the results of a questionnaire distributed to operating room health professionals from a hospital in northern Portugal.

Eighty six professionals participated in this study through the completion of the questionnaire. The results revealed the occurrence of 39.5% (95% CI [29.3%, 50.7%]) of accidents involving biological material, and 11.8% (95% CI [6.1%, 21.0%]) of accidents involving chemical material. Not all professionals have had training in occupational risks (63.5% and 47.0% reported having had training in biological risks and chemical risks, respectively). Fisher's exact test did not reveal a statistically significant association ($p > 0.05$) between having had training in biological/chemical risks and having had an accident involving biological/chemical material. The chi-square independence test revealed a statistically significant association ($X^2 = 25.978$, $df = 3$, $p < 0.05$) between the occurrence of accidents involving biological material and the professional category. The occurrence of this type of accident is directly related to the professionals "surgeons".

It is suggested that health professionals and the Institution itself be more involved in risk prevention and, mainly, in promoting a safe working environment.

Keywords: Operating room · Chemical risk · Biological risk

1 Introduction

1.1 Motivation

The health sector employs different types of professionals facing a wide range of occupational hazards. Some of the main problems faced by healthcare professionals are needle stick injuries, musculoskeletal injuries, latex allergies, violence, and stress.

© Springer Nature Switzerland AG 2019
S. Bagnara et al. (Eds.): IEA 2018, AISC 819, pp. 150–159, 2019.
https://doi.org/10.1007/978-3-319-96089-0_17

Although it is possible to avoid or reduce workers' exposure to these risks, the number of non-fatal work-related accidents and illness among healthcare workers is the highest in all industry sectors and have increased over the past decade (NIOSH 2009). In the specific case of the operating room (OR), can exist simultaneously biological, chemical, physical, ergonomic, and psychosocial risks (Morgan et al. 2006). The risks expose OR employees to the occurrence of serious occupational accidents with serious personal and institutional repercussions. Such risks can have consequences with different levels of severity that, in the limit, can lead to the death of the professional (Perdigoto 2012). The operating room is therefore a complex operational structure with a combination of several risk factors, thus defining itself as a high risk area.

1.2 Objectives

The main objective of this work was to characterize the OR of a hospital regarding the occupational risks that the different professionals face, particularly biological and chemical risks. It was also sought to verify if the training professionals have about the risks they face, has some influence on the behaviors they adopt in terms of safety.

2 Operating Rooms

2.1 Characterization of Operating Rooms

The different hospital professionals operate using different instruments and things that can be dangerous. Examples are the different types of puncture-sharp materials, gases for medical use, radiation equipment, pharmaceuticals, and hygiene products, among others (Faria 2008; Kasatpibal et al. 2016).

The OR are services of great differentiation and transversal use by the various surgical specialties. Here can be performed scheduled or urgent surgical interventions. Examinations and other procedures that require a high level of asepsis and/or anesthesia care may also be performed (Lopes 2012). The OR are part of the Operating Theater (OT) which is a service with great technical requirements, at the level of both the facilities and procedures and the technicians who execute those (ACSS 2011). Three access zones must be considered in OT: free; semi-restricted; restricted. In the free area one can move around with outdoor clothing. It includes, e.g., the reception, waiting room, coordination office, and sanitary facilities. In the semi-restricted area, the use of appropriate clothing is compulsory, including a cap. In the restricted area it is compulsory to use clothing appropriate to the OT, including cap and mask. This area includes the preparation room, the anesthesia room, the disinfection area of the personnel, the operating room, the dirty room, the recovery room, and the technical floor.

The surgical team working in OR is generally composed by one main surgeon, two assistant surgeons, one anesthesiologist, one instrumentalist nurse, one circulating nurse, and one anesthetist nurse. Physicians and/or nurses in training can also be present (Lopes 2012).

2.2 Biological and Chemical Risks

Accidents involving exposure to contaminated organic fluids have received special attention from the occupational safety and health services. The main sources of biological risk are personal contact with patients and the handling of biological products: blood and its components, feces, urine, exudates, secretions and vomiting, as well as materials contaminated by them (Lima 2008). The main biological agents are viruses, bacteria, parasites, fungi and genetically modified organisms. The possible routes of entry of biological agents are respiratory, digestive, skin and mucous membranes, and the percutaneous route, which manifests itself through lesions produced by sharp and/or perforating objects, in addition to needle sticks and/or bites (Goméz 2012b). For most studies addressing biological accidents, the diseases of greater impact on workers' health are caused by blood-borne pathogens such as Hepatitis B virus, Hepatitis C virus and human immunodeficiency virus (Canalli 2012).

Some of the referenced chemical risk factors are due to manipulation of cytostatic drugs, exposure to anesthetic gases, exposure to chemical agents in general, allergic reactions to gloves, exposure to bone cement, among others (Xelegati 2003, cited in Faria 2008). The consequences include skin burns, allergic reactions, asthma, eye irritation, headache and neurological signs.

2.3 Personal Protective Equipment (PPE)

PPE used to protect workers from biological agents are mainly gloves, clothing, respiratory protective equipment and eye protection (Goméz 2012a). Gloves are a barrier to direct contact with biological and chemical agents. Facial or surgical masks do not protect against small airborne particles (aerosols). For this there are particulate respirators that provide different levels of protection. The plastic apron or waterproof gown should be worn to protect the gown/uniform during procedures that produce splashes or aerosols of body fluids, secretions or excretions, and should be removed as soon as the contaminating contact is terminated (Lima 2008). The gowns, uniforms and operative fields used in the OR have the function of protecting both the user and the patient against infectious agents, isolating distinct parts of the body from the contact with biological agents. From a chemical protection point of view, the purpose of protective clothing is to prevent chemicals from coming into direct contact with the skin. Regarding the protective footwear, the requirement is that being impermeable or has an impermeable coverage for chemical substances, according to the European standard EN-13832-3: 2006, while for biological agents there is no specific standard of footwear (Márquez et al. 2014). For eyes protection, splashes and aggression by gases/dust in suspension do not require special care in the choice of glasses. Interventions at high risk of blood splashes to the face obey the use of face masks with a visor, or the use of masks with concomitant use of goggles. The surgical cap should be placed so that it completely covers the hair and facial hair, avoiding the dispersion of particles into the atmosphere and the hair loss on the clothing or surgical fields. Uncovered hair houses microorganisms in proportion to their length, curl and oiliness (Duarte and Martins 2014).

3 Methodology and Methods

The present work constitutes a descriptive, analytical and cross-sectional study.

The study was conducted during the year of 2016 in a hospital in northern Portugal, more precisely in its operating theater, which includes six operating rooms. These rooms have direct connection with the room where the material to be used is prepared, with the anesthesia room, with the disinfection room of the surgical staff, and with the cleaning room, all being part of the restricted area of the OT.

The target population of this study consisted of 214 professionals, belonging to the direct and indirect collaborators of the surgical suite of the referred hospital, namely, doctors of various surgical specialties, anesthesiologists, nurses, radiology technicians and operational assistants.

According to the previously defined objectives, a questionnaire was used to collect the necessary information. The questionnaire was developed based on another one used in a similar work, by Perdigoto (2012). However, some issues have been redefined and others have been deleted in order to meet the stated objectives. This adaptation resulted in a questionnaire consisting of four sections. Section 1 consists of questions aiming at characterizing the sample: professional category, demographic data, whether the professional had specific training in biological, chemical, ergonomic, and physical risks, as well as prophylaxis of hepatitis B. This section ends with a question about whether the usual activity causes anxiety or stress to the professional. Section 2 contains a single question, related to the occurrence of occupational accidents involving biological materials. However, this question is subdivided into several sub-sections which address, for example, the level at which the exposure occurred (e.g. mucosa, percutaneous, other), the object and the biological material involved, as well as the activity that the professional was carrying out at the moment, the use of personal protective equipment (PPE), and the behaviors taken after the accident. Section 3 has a question related to the occurrence of occupational accidents with chemical materials. Like the previous question, this also is divided into several sub-sections similar to the described for the previous question. Lastly, the fourth section of the questionnaire consists of a single question relating to the number of working days lost due to the last accident, involving biological/chemical material.

The questionnaire was pre-tested with ten professionals selected by convenience. After this, the questionnaire was able to be applied to all elements of the study population, allowing the collection of various types of information, namely on the occurrence of accidents with biological and chemical materials, information on the main risk factors for such accidents, and also on its consequences. The questionnaires were given to the representatives of the different groups of professionals to make them reach their collaborators, appealing to them to collaborate in the fulfillment of the questionnaires, with the certainty that moral and ethical issues were safeguarded, namely the anonymity of the subjects, and confidentiality of the data.

The data collected was firstly organized in an Excel database. The statistical analysis of these data was performed using the statistical software IBM® SPSS® Statistics, version 25.0. The decision rule used was to detect significant statistical evidence for probability values (p-value) less than 0.05 ($p < 0.05$).

4   Results

The results are distributed in two subsections: descriptive statistics concerning the questionnaire answers; statistical analysis to evaluate comparisons and/or associations between variables.

4.1   Descriptive Statistics

The total number of respondents was 86 professionals. This number represents 40.18% of the target population. Half of the participants were nurses (n = 43, 50%), followed by surgeons and anesthesiologists (n = 19, 22.1%, for both), and by operational assistants (n = 5, 5.8%). Most of the participants were females (n = 58, 71.6%) compared to males (n = 23, 28.4%). The majority of the professionals had no training on chemical risks (53%). Relating to biological risks, most of the professionals had training in this area (63%). In respecting to the moment when they had had the training on biological risks, it was verified that on average it had been 61 months ago with a standard deviation of 58 months (minimum - 12 months, maximum - 336 months). Training on chemical hazards had been done on average 59 months ago, with a standard deviation of 59 months (minimum - 6 months; maximum - 336 months).

It was found that most participants were vaccinated against hepatitis B (89.4%).

The answers to the questions "have you had an accident with biological material in the OR?" and "have you had an accident with chemical material in the OR?" were as follows: 39.5% reported having had at least one biological accident and 11.8% reported having had at least one chemical accident.

Biological Accidents. Regarding the context in which the accident happened, it was verified that the majority occurred in programmed surgeries (n = 22, 66.7%), followed by urgency surgeries (n = 11; 33.3%).

Concerning the level at which exposure to biological material occurred, it was observed that the majority of the professionals reported the percutaneous level (n = 27; 84.4%), followed by the mucosa (n = 4, 12.5%). Only 3.1% (n = 1) referred "not intact skin". In relation to the object involved in the accident, most of the participants reported the needle (n = 24; 75.0%), and a smaller percentage reported the scalpel blade (n = 4, 12.5%). The remains 12.5% referred "other". The answers to the question "what were you doing when the accident occurred?" can be seen in Table 1. The majority reported that they were operating (n = 11, 35.5%).

Figure 1 shows the professionals' responses regarding the PPE that they were using when the accident occurred. The most frequent were: cap, mask, and surgical gloves.

Regarding the behaviors adopted by professionals after accidents with biological material, the most frequently adopted procedures were: make local antisepsis (82.4%), to identify the patient, source of the problem (39.4%); perform medical examinations (32.4%); wash with soap and water (32.4%); perform exams to the patient (29.4%). Only 2.9% reported having done nothing.

Table 1. Information about what was the professional's activity when the accident occurred.

Professional's activity	%
Using the surgical instruments	19.4
Cleaning	6.4
Operating	35.5
Catheterizing vein	22.6
Administer medication	3.2
Suturing	12.9

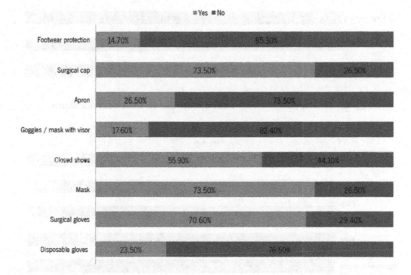

Fig. 1. Answers concerning the PPE they were using when the accident occurred.

Chemical Accidents. The main effects of exposure to chemical agents can be seen in Fig. 2. The most frequent were irritation, dizziness and headache. Vomiting and burns were not reported by any professional.

Figure 3 shows the professionals' responses regarding the PPE that they were using when the chemical accident occurred. Closed shoes, the mask and the cap were the most frequent PPE.

Regarding the behaviors adopted by professionals after the accidents with chemical material, the most frequently adopted were: to identify the product that caused the problem (50.0%); wash with running water (50.0%); do nothing (30%); leave the place (20.0%). No one reported having performed medical examinations or washed with soap and water.

Fig. 2. Distribution of the main effects due to chemical accidents.

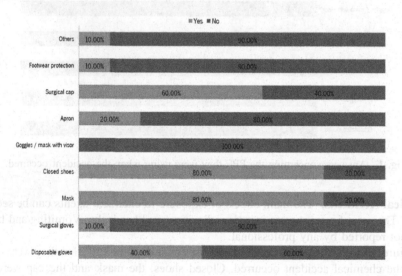

Fig. 3. Distribution of respondents in what concerns the PPE they were using when the chemical accident occurred.

4.2 Statistical Analysis

According to the type of variables to analyze, several statistical tests (Fisher, t-Student, Chi-square) were applied. In Table 2 can be seen all test applied, as well as the results obtained. Only two of them revealed a statistically significant association between the variables: training in biological hazards/wash hands with soap and water after an accident with biological material; professional category of surgeon/occurrence of accident involving biological material.

Table 2. Summary of the statistical tests applied.

Test	Result
Training in chemical risks/accident with chemical material	Fisher Exact test, p = 0.520
Training in biological risks/accident with biological material	Fisher Exact test, p = 0.567
Training in chemical risks/use the adequate PPE	Fisher Exact test, p > 0.05
Training in chemical risks/behaviors after chemical accident	Fisher Exact test, p > 0.05
Training in biological risks/use the adequate PPE	Fisher Exact test, p > 0.05
Training in biological risks/behaviors after biological accident†	Fisher Exact test, p < 0.05*
Professional category/chemical accident	$\chi^2 = 6.939$, df = 3, p = 0.074
Professional category/biological accident	$\chi^2 = 25.978$, df = 3, p = 0.00001*
Seniority in OR/chemical accident	t = −0.482, df = 70, p = 0.632
Seniority in OR/biological accident	t = −0.242, df = 71, p = 0.809
Professional's age/chemical accident	t = −1.104, df = 78, p = 0.273
Professional's age/biological accident	t = −0.982, df = 79, p = 0.329

†The behavior that revealed association was wash hands with soap and water; *Significant at 5%; df: degrees of freedom

5 Discussion of Results

The fact that about 10% of respondents are not vaccinated against hepatitis B is worrisome given that according to some authors (Barash et al. 2014) vaccination is the main strategy to prevent the labor transmission of Hepatitis B virus. The average time elapsed since professionals had training, in both biological and chemical risks, was about 5 years. There was no statistically significant association between has had training and the occurrence of biological or chemical accidents, nor with the use of PPE. The statistically significant association detected between professional category of surgeon and occurrence of accident involving biological material is according to Perdigoto (2012) that points out surgeons as the group of professionals suffering more accidents in the OR. It was found that the occurrence of accidents (biological and chemical) is not dependent on factors such as seniority or age. This is in line with the study by Kasatpibal et al. (2016) which covered more than 2000 operating room nurses, in which age was not identified as a risk factor for accidents. There was a statistically significant association between has had training in biological risks and certain behaviors adopted after an accident with biological material, namely washing with soap and water.

Most of the accidents occurred in the surgical context and at the percutaneous level (84.4%). The needle was the most involved (75.0%) and the blood was the biological material most often implicated (93.5%). Such information is in agreement with other studies that, generally, see a greater risk in the piercing-sharps objects, which are manifested mainly by the skin and mucous membranes (Goméz 2012b). Although professionals understand the importance of using the appropriate PPE to their work, there have been gaps in the use of PPE, namely protection of footwear, apron, glasses or mask and even gloves. Not wearing PPE is a potential risk factor for biological accidents (Kasatpibal et al. 2016).

6 Conclusions

This study aimed at understand the extent to which the training of professionals in biological and chemical risks has an effective influence on the behaviors they adopt in everyday work, in terms of safety. The analysis of the results allowed to verify that in the hospital practice the accidents of biological nature are more expressive and can have serious consequences. Only two statistically significant association between variables were found: between "training in biological risks" and "wash hands with soap and water after an accident with biological material", and between "professional category of surgeon" and "occurrence of accident involving biological material". Although vaccination is not legally enforced, it is recommended that strategies be defined so that all OT collaborators can benefit from the vaccination schedule and consequent immunity to the Hepatitis B virus. It is important to encourage the correct and appropriate use of PPE in the workplace, which could prevent and protect individuals from accidental occurrences. It is suggested that health professionals and the Institution itself be more involved in risk prevention and, mainly, in promoting a safe working environment.

It is relevant to mention that the information collected from the respondents may have been impaired by their memory, since some of them verified that they no longer remembered all the required data.

A more exhaustive, in situ, characterization centered on the type of risk, is suggested so that more specific training programs can be developed, constituting a more effective preventive measure for the reduction of occupational accidents in the OR.

Acknowledgements. This work has been supported by COMPETE: POCI-01-0145-FEDER-007043 and FCT – Fundação para a Ciência e para a Tecnologia - within the Project Scope: UID/CEC/00319/2013.

References

ACSS (2011) Technical Recommendations for Operating Room (in Portuguese). http://www.acss.min-saude.pt/Portals/0/RT_05-2011%20DOC%20COMP%20PDF. Accessed 18 Apr 2016

Barash P, Cullen B, Stoelting R, Cahalan M, Stock M, Ortega R (2014) Manual of Clinical Anesthesia, 7th edn. Wolters Kluwer/Lippincott Williams & Wilkins, Philadelphia

Canalli R (2012) Occupational risks and accidents with biological material in collective health nursing professionals (in Portuguese) PhD thesis. University of São Paulo, Brazil

Duarte A, Martins O (2014) Nursing in Operating Room (in Portuguese), 1st edn. Lidel – Technical editions, Lda, Lisboa

Faria A (2008) Caracterização e Análise dos Acidentes de Trabalho com Profissionais de Enfermagem numa Instituição Hospitalar (in Portuguese). Master's dissertation. University of Minho, Portugal

Goméz EC (2012a) Protective gloves against microorganism (in Spanish). Technical Standards of Prevention. National Institute of Health and Safety at Work

Goméz EC (2012b) Protective clothing against chemical products (in Spanish). Technical Standard of Prevention (NTP) 929. National Center of means of protection. National Institute of Safety and Hygiene at Work

Kasatpibal N et al (2016) Prevalence and risk factors of needlestick injuries, sharps injuries, and blood and body fluid exposures among operating room nurses in Thailand. Am J Infect Control 44:85–90

Lima J (2008) The use of personal protective equipment by nursing professionals - Practices related to the use of gloves (in Portuguese). Master's dissertation. University of Minho, Portugal

Lopes A (2012) Management of Operating Room (in Portuguese). Master's dissertation. University of Minho, Portugal

Márquez J, Carreño M, González J, Díaz R, Otaolarruchi J (2014) Recommendations in the selection of protection and cleaning equipment in a possible Ebola case (in Spanish). http://gruposdetrabajo.sefh.es/gps/images/stories/publicaciones/Documento_WEB_GPS_EPI_14_11_14.pdf. Accessed 12 Apr 2016

Morgan GE, Mikhail MS, Murray MJ (2006) Clinical Anesthesia (in Portuguese), 3rd edn. Revinter, Lda, Rio de Janeiro

NIOSH (2009) State of the Sector - Healthcare and Social Assistance. http://www..cdc.gov/niosh/docs/2009-139.pdf. Accessed 5 Jan 2016

Perdigoto P (2012) Risks in the Operating Room – realities that can influence the management (in Portuguese). Master's dissertation. Polytechnic Institute of Bragança, Portugal

Thermoregulatory Burden from Using Respirators and Performing Composite Movement-Based Exercises of Varying Metabolic Demand

Chen-Peng Chen[1]([✉])(iD), Yi-Chun Lin[2](iD), and Hui-Chen Wei[1]

[1] Department of Occupational Safety and Health, China Medical University,
No. 91 Hsueh-Shih Road, Taichung 40402, Taiwan
chencp@mail.cmu.edu.tw
[2] Department of Public Health, China Medical University,
No. 91 Hsueh-Shih Road, Taichung 40402, Taiwan

Abstract. Using a fitting respirator in exercises of heavy metabolic demand may increase the user's thermal burden, as the metabolic heat typically relieved via respiration is baffled by the mask and less exchanged into the environment. This study investigated the increase in thermoregulatory burden as manifested in skin physiological changes from using N95 filtering-facepiece respirators and simultaneously performing composite movement-based exercises of varying metabolic demand. As the results show, when the participants exercised the transepidermal water loss (TEWL) and skin moisture in the forearm and in the cheek immediately outside the N95 mask rose in accordance with the metabolic intensity of the activity, particularly in the first 10 min (TEWL: 26.9–40.4% increase from the level determined at 0 min; skin moisture: 28.1–250.2% increase). Under high metabolic rate, throughout the entire 30 min of exercise the skin capillary blood flow (SCBF) of the participants increased steadily in both the forearm and cheek (6.4–12.3% increase), signaling a continuous requirement for release of excess metabolic heat. These findings indicated that the respirators and the conditions under which they were used should both be considered when the thermoregulatory burden the users of respirators sustained was evaluated. To prevent potential heat stress in the workplace, it is essential to include in the respirator safety program an ergonomic assessment of workload.

Keywords: N95 masks · Heat stress · Metabolic load · Thermoregulation
Transepidermal water loss

1 Introduction

The air-purifying respirators are often regarded as the last line of defense in protecting the workers from threats of airborne particulate and gaseous contaminants [1]. However, a respirator of sufficient facial fit and faceseal when in use may increase the user's thermal burden, as the metabolic heat typically relieved via respiration is baffled by the mask and the heat flow less exchanged into the environment. As a consequence, the temperature and humidity inside the mask are significantly elevated, increasing thermal

© Springer Nature Switzerland AG 2019
S. Bagnara et al. (Eds.): IEA 2018, AISC 819, pp. 160–165, 2019.
https://doi.org/10.1007/978-3-319-96089-0_18

discomfort among the users. The thermal burden from using a fitting respirator is subject to influences rooted in the user's anthropometric characteristics, the respirator's facepiece design, the metabolic demand of the activity the user is engaged in when using the respirator, the thermal microclimate at the site where the respirator is used, and the duration of respirator use [2]. In the workplace, the respirators are routinely used when the workers handle chemical substances of significant volatility. The buildup of respiratory heat inside the respirator in these labor-intensive tasks may be further amplified, risking a potential thermal imbalance or even thermal stress for the workers. However, when a respirator is selected in the workplace, considerations are often confined to the nature and concentration of the contaminants present in the air.

In evaluating thermo-physiological response of human body to thermal stress, the core temperature, heartbeat rate, systolic/diastolic blood pressure, and sweating rate were often employed as indicators of thermal strain [3]. These indicators observed the exchange of metabolic heat flows between human body and the ambient environment as well as the dynamics underlying thermo-physiological response when the thermal strain was exceeded [4]. In recent years, the advancement in metabolic rate measurement has made available an additional technique by which the metabolic consumption affecting heat strain capacity may be readily gauged [5]. However, in the thermal strain evaluation these indicators were typically limited to describe the net gain or loss of metabolic heat rather than to reveal the thermoregulation taking place in response to demand of balancing metabolic heat when human body was engaged in exercises of significant metabolic demand. To understand the kinetics involved in the balance of metabolic heat flows, skin-based thermo-physiological quantities have been applied to observe and study the thermoregulatory events occurring in the skin microvasculature in response to requirement of heat exchange. Chen et al. [6] investigated the change in skin capillary blood flow (SCBF), transepidermal water loss (TEWL), skin moisture, and skin temperature in the epidermal skin for young workers exercising under different metabolic rates and characterized the sequence of their change. Among these quantities, the SCBF described the capillary flux in the epidermis and represented an early stage of thermoregulation. In comparison, the increase in the TEWL and skin moisture signaled the development of latent heat in the skin and the subsequent loss of invisible water vapor, typically suggesting an advanced requirement of metabolic heat dissipation. The time course and heat-response relationship identified in the change of these quantities also quantitatively described how the acclimatization took place following heat stress.

This current study investigated the thermoregulation occurring in response to the required dissipation of the metabolic heat produced from using N95 respirators and exercising under different metabolic rates. The skin-based thermo-physiological quantities consisting of SCBF, TEWL, skin moisture, skin temperature were adopted to identify and describe the thermoregulation in progress. The findings from this study provided a preliminary assessment on the effect of thermoregulatory adjustment relieving potential thermal stress and on the acclimatization that developed in the course of metabolic heat production and release for the users of filtering-facepiece respirators who worked under strenuous conditions.

2 Materials and Methods

2.1 Study Design

This study was conducted in a microclimatic chamber pre-set at an air temperature of 27 °C and relative humidity of 65%. In the study, the participants performed six specified activities in individual sessions of 30 min wearing a cup-shaped N95 mask equipped with an exhalation valve (Fig. 1). Developing from the maneuvers prescribed in the respirator fit testing procedures established by the US Occupational Safety and Health Administration [7], the specified activities included, in the order of increasing metabolic intensity, talking, bending over, talking while walking on the stairs, walking on the stairs, bending over while jogging, and jogging. The first two, middle two, and final two maneuvers corresponded to exercises of low (70–130 W/m^2), moderate (130–200 W/m^2), and high metabolic rate (200–260 W/m^2), respectively. The ranges of metabolic rate in this study were selected to represent the metabolic consumption of different workloads as prescribed in the International Organization for Standardization Standard 8996 "*Ergonomics of the thermal environment—determination of metabolic rate*" [8]. This study was approved by the China Medical University and Hospital Research Ethics Committee (CMUH 104-REC1-138).

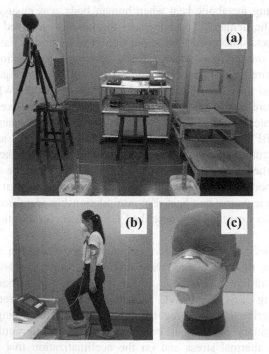

Fig. 1. Evaluation of thermoregulation in response to use of respirators and exercise under varying levels of metabolic demand: (a) microclimatic chamber and equipment used in environmental monitoring; (b) thermo-physiological measurement taken for the participants in standard clothing donning a N95 mask and walking on the stairs; and (c) model of N95 mask tested in this study.

2.2   Participants and Anthropometric Data

A total of 16 participants (8 males and 8 females) were recruited to participate in this study. The participants were of an age between 21 and 25 years old. Table 1 shows the anthropometric data of the recruited participants.

Table 1. Anthropometric data (mean ± standard deviation) of participants in this study.

	Male ($n = 8$)	Female ($n = 8$)
Age (year)	21.4 ± 0.7	21.3 ± 0.4
Height (cm)	174.9 ± 8.0	162.1 ± 3.7
Weight (kg)	71.6 ± 11.6	54.5 ± 8.3
BMI (kg/m^2)	23.4 ± 3.5	20.8 ± 3.8

2.3   Experimental Procedures

The participants entered the chamber in standard clothing (T-shirt and straight trousers; with a clothing insulation of 0.41) and sat for 10 min behind a desk for adaptation to the pre-set thermal environment. The average metabolic rate in this stage ranged from 55 to 70 W/m^2. The readings of SCBF, TEWL, skin moisture, and skin temperature were taken from the upper left forearm and the left cheek outside the N95 mask, starting when the acclimatization finished and ending after 30 min. The participants performed the six specified activities wearing the N95 mask continuously during the course of exercise. The measurement of physiological quantities was conducted at 10, 20, and 30 min into the exercise (Fig. 1).

3   Results and Discussion

In this study the skin-based thermo-physiological response was measured to evaluate the thermoregulation required to balance the metabolic heat produced following the use of respirators and exercise under different metabolic demands. Figure 2 shows the change in skin physiology for the participants wearing the N95 mask and performing specific exercises for 30 min. As the results show, the skin moisture and TEWL in the forearm and in the cheek immediately outside the N95 mask rose in accordance with the metabolic intensity of the exercise. The TEWL increased from 18.3–20.4 g/m^2/hr, the level measured when the participants rested in a sedentary position without wearing a respirator, to 23.9–34.4 g/m^2/hr, the level identified after the participants completed the specified exercise wearing a respirator. Similarly, the skin moisture on the cheek rose from 241.4–307.1 μS to 295.9–1,810.1 μS after the participants wore a respirator and exercised for 30 min. The increases in TEWL and skin moisture were both statistically significant ($p < 0.05$). The most prominent change in these two physiological quantities was observed 10 min into the exercise (TEWL: 26.9–40.4% increase from the level determined at 0 min; skin moisture: 28.1–250.2% increase).

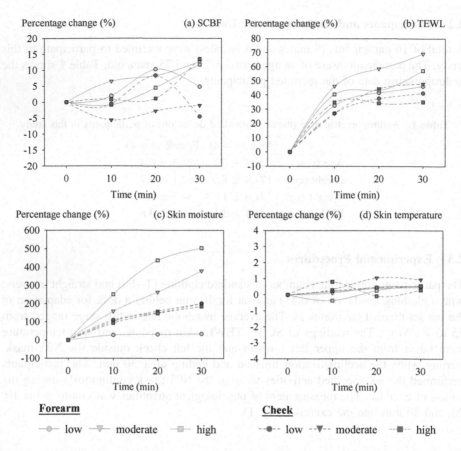

Fig. 2. Change in skin-based thermo-physiological quantities in the forearm and the cheek when the participants wore a respirator and exercised under different metabolic rates (low, moderate, or high) for 30 min: (a) skin capillary blood flow (SCBF), (b) transepidermal water loss (TEWL), (c) skin moisture, and (d) skin temperature. The symbols at each time point represent the mean value of observations.

During the course of exercise, only a minor change was observed in the SCBF when the participants exercised in low and moderate metabolic rates (−0.9–5.4% change from the level determined at 0 min). However, under high metabolic rate, throughout the entire 30 min of exercise the SCBF of the participants increased steadily in both the forearm and the cheek (6.4–12.3% increase), suggesting a continuous requirement of heat dissipation to release excess metabolic heat. Chen et al. [6] described that in the skin-based thermoregulatory events the change in TEWL and skin moisture with a minor adjustment in SCBF often signaled the progress of thermoregulation into the intermediate-to-late stage of heat dissipation, as the dilation of the skin microvasculature as manifested in SCBF increase was an early stage of acclimatization and became less prominent once the metabolic heat flow carried in SCBF was exchanged into latent heat for dissipation via water evaporation. Our findings indicated that the use of respirator and exercise under significant metabolic rate jointly contributed to a thermal burden that demanded a vigorous thermoregulation for the body to regain thermal balance.

4 Conclusions

The findings from this study demonstrated that the respirators and the conditions under which these respirators were used could present a significant thermal burden to the users of respirators and thus should be considered when evaluating the thermoregulation or considering the safety of respirator use. To prevent potential heat stress in the workplace for the workers who must use a respirator and sustain strenuous workload, it is essential to include in the respirator safety program an ergonomic assessment of workload that considers both the musculoskeletal and thermoregulatory stress of the task.

Acknowledgments. This study was financially supported by the Ministry of Science and Technology of Taiwan under Project Number MOST 105-2221-E-039-003-MY3.

References

1. National Research Council (NRC) (2005) Measuring respirator use in the workplace. In: Kalsbeek WD, Plewes TJ, McGowan E (eds) NRC Panel on Review of the National Institute for Occupational Safety and Health/Bureau of Labor Statistics Respirator Use Survey. NRC, Washington, DC
2. Health and Safety Executive (HSE) (2005) Respiratory protective equipment at work: a practical guide, 3rd edn. HSE53. HSE, Sudbury (2005)
3. Ramsey JD, Bishop PA (2003) Hot and cold environments. In: DiNardi SR (ed) The occupational environment: its evaluation, control, and management, 2nd edn. American Industrial Hygiene Association, Fairfax, pp 612–645
4. Yao Y, Lian Z, Liu W, Jiang C, Liu Y, Lu H (2009) Heart rate variation and electroencephalograph - the potential physiological factors for thermal comfort study. Indoor Air 19(2):93–101. https://doi.org/10.1111/j.1600-0668.2008.00565.x
5. Vorwerg Y, Petroff D, Kiess W, Blüher S (2013) Physical activity in 3–6 year old children measured by SenseWear Pro®: direct accelerometry in the course of the week and relation to weight status, media consumption, and socioeconomic factors. PLoS ONE 8(4):e60619. https://doi.org/10.1371/journal.pone.0060619
6. Chen C-P, Hwang R-L, Chang S-Y, Lu Y-T (2011) Effects of temperature steps on human skin physiology and thermal sensation response. Build Environ 46(11):2387–2397. https://doi.org/10.1016/j.buildenv.2011.05.021
7. US National Institute for Occupational Safety and Health (NIOSH) (1987) NIOSH guide to industrial respiratory protection. DHHS (NIOSH) Publication No. 87-116. US Centers for Disease Control and Prevention, NIOSH, Pittsburgh
8. International Organization for Standardization (ISO) (2004) Ergonomics of the thermal environment—determination of metabolic rate. ISO 8996:2004. ISO, Geneva

Equipment Interventions to Improve Construction Industry Safety and Health: A Review of Case Studies

Brian Lowe[1($\boxtimes$)], James Albers[1], Marie Hayden[1], Michael Lampl[2], Steven Naber[2], and Steven Wurzelbacher[1]

[1] National Institute for Occupational Safety and Health, Morgantown, USA
blowe@cdc.gov
[2] Ohio Bureau of Workers' Compensation, Columbus, USA

Abstract. A review was conducted of 153 case studies of construction equipment interventions, representing $6.55 million (2016 USD) of equipment purchases incentivized through the U.S. state of Ohio Bureau of Workers' Compensation (OBWC) Safety Intervention Grant (SIG) program. The source data were drawn from the applications and final reports of employers who received grants between 2003 and 2016. Outcomes were reductions in safety hazards, cumulative trauma disorder risk factors, and a score assessing quality of the intervention evaluative experience as determined through a framework developed by the authors. Items relating to the quality of the evaluative experience were manually extracted from the case study documentation. When aggregated by type of construction equipment, the risk factor reduction and evaluative quality scores were variable within and between equipment types. Equipment for cable pulling, used in the electrical trades, and skid steer attachments for concrete breaking (hydraulic breakers) both emerged as interventions ranked highly for reducing risk factors and for the evaluative quality of their case studies. Other intervention equipment types that ranked highly in both risk factor reduction and evaluative quality were concrete sawing equipment, trailers with hydraulic tilting/ramps, powered hand tools, and man lifts (boom lifts).

Keywords: Construction · Interventions · Case studies

1 Introduction

Construction work can be hazardous, and it subjects workers to a multitude of exposures and injury/illness risk factors and hazards. This is reflected in injury and illness statistics for the sector. In 2010, the rate of days-away-from-work cases in the U.S. Construction sector was 149.6 per 10,000 FTEs, which was 39% higher than the average rate of 107.7 for all private U.S. industries [1]. Though fatal injuries in the U.S. construction industry had decreased in 2010, Construction still accounted for 17.1% of the total fatal work injuries, while the sector comprised approximately 4.6% of the U.S. workforce. In that same year, workers' compensation (WC) insurance covered 124

© Springer Nature Switzerland AG 2019
S. Bagnara et al. (Eds.): IEA 2018, AISC 819, pp. 166–175, 2019.
https://doi.org/10.1007/978-3-319-96089-0_19

million U.S. workers at a total cost of $71 billion to employers, with benefits of $28.1 billion in medical payments and $29.5 billion in benefits to workers [2].

Occupational safety and health (OSH) and risk control programs have recognized the need to address sources of workplace hazards to improve safety and health outcomes [3]. Overexertion due to lifting, being struck by an object, and falls to lower level are the leading causes of nonfatal workplace injury costs [4]. Accordingly, OSH agencies and insurers have interest in assessing the effectiveness of prevention approaches to address the leading causes of workplace injury/illness and to promote the adoption of interventions that are effective in reducing injury/illness burden. The U.S. State of Ohio Bureau of Workers' Compensation (OBWC) Safety Intervention Grant (SIG) program is a grant program in which employers who apply for and are awarded a grant receive matching funds at a multiple of 3:1 (this varied from 2:1 to 4:1 over the program years) for the purchase of equipment to address workplace health/safety [5]. SIG program participant employers must submit a one-year final report describing their experience with the intervention equipment. Thus, the program is one of few that are administratively structured to incentivize employer acquisition of equipment interventions to address safety and health hazards and that systematically collect information about experiences with the equipment. The OBWC SIG program provides an opportunity to assess aggregated employer experiences through individual documented case studies.

In a previous report Wurzelbacher et al. [3] analyzed the accepted claims experience among 468 Ohio employers receiving OBWC SIG program grants from 2003–2009. SIG program participation was demonstrated to reduce workers' compensation (WC) claim rates and costs in most industries. Construction was one industry that was not associated with reduced claims rate in response to SIG participation. However, that analysis [3] examined only the injury claims experience and was a programmatic level assessment that was less able to discern effectiveness of specific types of intervention equipment.

The purpose of the present analysis was to develop and apply a systematic framework by which the SIG experiences could be reviewed and evaluated to assess equipment intervention effectiveness in the Construction industry. The assessment considers additional information beyond the WC claims experience that grantees submitted to OBWC in the reporting of their experience with the equipment. The investigators have interest in the assessment and communication of Construction solutions to industry stakeholders who would benefit by the translation of intervention research findings.

2 Methods

2.1 Source Documentation

Employer-submitted grant applications and final reporting documentation (case studies) for Safety Intervention Grants (SIGs) awarded between 2003–2016 were compiled by the Ohio BWC in March 2017. Final reports are submitted at least one year after the intervention award, so there were few 2016 awards with a final report at that point in time. Case study documentation was organized and keyed by an anonymous application ID number, which served as the linkage between the original application, final reporting documentation, and background information about the grant award: employer category size, amount of grant funding match, and National Council on Compensation Insurance (NCCI) classification code number. The NCCI code is a four-digit classification system, with variation by U.S. state, describing the nature of the business operations for the purpose of assigning collective actuarial risk and for rate administration [6]. The NCCI classification code number of the affected work group, as defined by the employer on the application, was used to crosswalk to an established Construction industry trade/specialty.

At the time that the source documentation was compiled, there had been 368 SIG awards for the purchase of intervention equipment by employers with NCCI classification codes corresponding to Construction sub-industries. However, 134 of those were awarded in 2015 and 2016 alone, and many of these did not yet have final reports submitted by employers at the time of this analysis. (A small percentage of employers fail to submit final reports.) In the present review, there were 224 Construction SIG awards for which reporting materials had been submitted by the grant recipient employer. Of these, 52 grant awards were excluded due to: (1) shop-based equipment/machinery not used on a Construction job site; (2) equipment for lifting construction vehicles and heavy equipment for the purpose of maintenance; (3) equipment for the purpose of hot water heater, furnace, and other home appliance delivery; (4) equipment for landscaping work; and (5) sewer jetting system equipment. Further, 19 of the 224 SIGs were excluded because of significantly incomplete reporting documentation. In total, there were 153 SIG experiences included in the final review.

2.2 Data Extraction

A data extraction form was developed and subjected to peer review by three subject matter experts with background in OSH intervention evaluation, specifically in Construction. The form was based on assessing the quality of evaluative aspects of employers' experience with the intervention equipment in terms of exposure/risk factor reduction, employee acceptance (e.g. usability), and employer return on investment (e.g. productivity, cost/benefit). This information included how the employer's experience demonstrated reductions in risk factors, potential risk transference, employee and management acceptance/adoption of the intervention, and impact of the intervention on productivity/quality. Extracted items were operationalized with the intent to minimize subjective interpretation on the part of the single analyst who read the case studies and performed the data extraction.

A risk factor reduction score for SIG experiences that reported complete baseline and follow-up assessments was based on semi-quantitative instruments for systematically assessing cumulative trauma disorder (CTD) risk factors and safety hazards. Direct subtraction of the follow-up score from the baseline score using these instruments yielded the net change (positive reduction in risk factors). Information on these instruments is available from OBWC.

Equipment purchase costs were extracted from the grant budget and financial documentation in the application, and equipment purchase costs are inflation adjusted to 2016 U.S. dollars (USD) using the Producer Price Index for *Other Heavy Machinery Rental and Leasing: Construction Equipment Rental and Leasing* (Federal Reserve Bank of St. Louis). Intervention equipment cost per affected employee was calculated for single-equipment case studies based on the number of affected employees - those who perform the work for which the new equipment will be used. The affected employee count was a determination made by the employer in the grant application.

2.3   Data Analysis

Primary outcomes were reduction in CTD and safety risk factors and the evaluative evidence quality score, with the individual SIG experiences (case studies) as the unit of analysis. SIG experiences were grouped according to single-equipment SIGs (n = 105) and multi-equipment SIGs (n = 48). Single-equipment SIGs were classified as those in which grant funds were used to purchase a single piece of equipment or an integrated system consisting of a primary piece of equipment and related attachments. Multi-equipment SIGs were those in which multiple pieces of equipment were purchased that were not used as an integrated system in a single construction task. In rare instances did the case studies differentiate risk factor reduction based on individual equipment use. In the single-equipment SIGs, a more straightforward association can be interpreted between the single piece of equipment as an intervention and changes in risk factors.

Change scores for CTD risk reduction and Safety risk reduction between the pre- and post-intervention assessments were z-transformed, and converted percentiles are reported to allow comparisons between the two different scales (CTD risk assessment instrument, Safety assessment instrument). To report on equipment type in aggregate, the mean percentile for the equipment classification was calculated.

Based on the data extraction items, we report Evaluative Evidence Quality Scores for the SIG experiences - intended as a basis for ranking case studies according to attributes believed to be important to establishing evidence of intervention effectiveness (see Table 1). These scores were not considered appropriate for parametric statistics and the median rank was calculated to report on equipment type classifications in aggregate.

3   Results

Equipment purchases were classified according to 13 broad types of construction equipment based on function and mode of operation. Further refined classification of the broader skid steer attachments and walk/ride-behind powered equipment resulted in 25 sub-classifications of equipment (see Table 2). Single-equipment SIGs were then aggregated according to this equipment classification. Multi-equipment case studies were not aggregated according to equipment classification to report quality score or CTD/Safety risk factor reduction. In some multi-equipment intervention grants, there were up to 18 distinct types of equipment purchased spanning multiple classifications. Thus, the row counts by equipment type for multi-equipment SIGs in Table 2 do not

Table 1. Determination of SIG Evaluative Evidence Quality Score. A total of 100 points was possible.

Injury Claims Experience (20%)
- (10%) Was there one, or more, claims indicated for the baseline period?
  No = 0; Yes = 10
- (10%) Was there one, or more, baseline claims that would be *plausibly prevented had the intervention equipment been in place?*
  No = 0; Yes = 10

Risk Factor Abatement Experience (40%)
- Are *CTD, Safety, or IH Risk Factor*s compared in a consistent manner (baseline <u>and</u> follow-up comparable) addressing exposure(s) affected by the intervention?
  none = 0; qualitative description in narrative = 15; BWC instruments (or other quantitative instruments) = 40

Acceptance/Adoption Experience (15%)
- Does the report contain any description of employees' acceptance/non-acceptance of the intervention?
  No = 0; Yes = 15

Work Productivity Experience (15%)
- (10%) Are effects on productivity described?
  none = 0; qualitatively = 5; quantitatively = 10
- (5%) Is a return on investment calculation reported (generally on the CBA form, or in report narrative)?
  No = 0; Yes = 5

Work Quality Experience (5%)
- Are effects on work quality described?
  none = 0; qualitatively = 3; quantitatively = 5

Training (5%)
- Was there evidence of training conducted?
  no description of actual training conducted and no training costs incurred = 0 ;
  actual training described in narrative = 3; CBA lists training costs incurred = 5

---

sum to the total of 48. Multi-equipment SIGs were evaluated differently because of their inherent co-intervention effect, where the effect of individual pieces of equipment cannot be determined. Multi-equipment grants were often a combination of equipment types related to different tasks for a given trade. For example, walk-behind roof cutting systems were frequently purchased in combination with a walk-behind hauling system.

Figure 1 shows raw change scores (pre-intervention minus post-intervention) in Safety and CTD risk assessments by equipment type for single-equipment case studies. Higher change scores reflect greater reduction in risk factors. Table 2 ranks equipment classifications (aggregated across single-equipment SIGs) based on the median evidence quality scores. Columns also list reduction in CTD and safety risk factors by percentile. Table 2 also lists single-equipment SIG costs per affected employee. These summary measures indicate that cable-pulling equipment used in the electrical trades were effective in reducing CTD risk factors and were also associated with grant experiences scoring higher in evaluative quality. This equipment was also associated

Table 2. Summary of Safety Intervention Grant case studies (Construction industry) aggregated by equipment classification.

Equipment classification	Single-equipment						Multi-equipment	
	Sum equip cost (2016 US Dollars)	# Single equip SIGs	Mean equip cost per employee† (2016 USD)	Median evidence quality score (Rank)		Mean percentile CTD risk reduction	Mean percentile safety risk reduction	Sum equip cost (2016 USD)	# Multi equip SIGs associated
Skid steer attachment - other	100,067	1	5,559	85	(1)	28.3	–	47,851	3
Skid steer attachment - augering	90,408	2	3,749	83	(2.5)	34.2	–	31,132	1
Cable pulling systems	174,780	5	800	83	(2.5)	71.0	–	55,698	3
Concrete sawing (not hand tools)	258,521	5	3,068	78	(4)	54.8	34.1	24,686	3
Skid steer attachment - rotary grinding	12,647	1	1,405	75	(5)	40.4	–	0	0
Powered hand tools	136,170	4	1,138	70.5	(6)	27.1	83.1	760,179	26
Vacuum and or hydro excavation	135,403	2	5,043	65.5	(7)	46.9	75.3	0	0
Skid steer attachment - concrete breaking	82,057	4	998	65	(8.5)	62.3	83.1	22,548	2
Trailers (hydraulic tilting/ramps)	67,457	2	5,092	65	(8.5)	84.0	–	65,544	4
Manlift (boom lift)	995,977	14	5,766	63	(10)	59.5	60.9	32,460	1
Conduit bending	29,860	1	459	60	(11)	14.2	–	70,685	5
Lift gates	38,763	4	222	59	(12)	24.1	–	3,151	1
Bulk material transfer/dispensing (tar & adhesive applications)	139,337	4	2,651	58	(13.5)	31.3	21.9	23,415	2
Other	511,337	13	2,468	58	(13.5)	45.1	11.4	218,362	20
Scissor lift/mast lift	700,945	16	2,648	57	(15)	58.7	26.4	169,971	4
Trailer restraints	8,225	1	191	55	(16)	–	–	0	0
Scaffolding/platform	632,305	12	2,945	49	(17.5)	37.5	58.9	69,005	4
Fall protection systems (carts, etc.)	58,308	4	489	49	(17.5)	–	64.0	104,280	7
Walk/ride behind (powered) - trenching	6,736	1	192	48	(19.5)	89.3	32.8	0	0
Walk/ride behind (powered) - screeding	186,316	3	3,047	48	(19.5)	79.7	–	0	0
Cranes/hoists (other than manlifts)	162,802	3	6,103	45	(21)	18.3	–	51,620	4
Skid steer attachment - asphalt cold planer, tiller, brooms, etc.	32,225	2	1,710	44	(22)	63.1	–	0	0
Walk/ride behind (powered) - cutting/removal of roof material	7,657	1	219	30	(23)	–	–	136,355	10
Walk/ride behind (powered) - hauling	–	0	0	–		–	–	76,853	7
Walk/ride behind (powered)- other	–	0	0	–		–	–	17,161	3
	4,568,303	105						1,980,957	48*

*The column total shown does not reflect the sum of rows because these SIGs had multiple equipment classifications associated.
†Mean equipment cost/employee is the total equipment cost for the grant divided by the affected employee count. This is averaged for the equipment classification category.

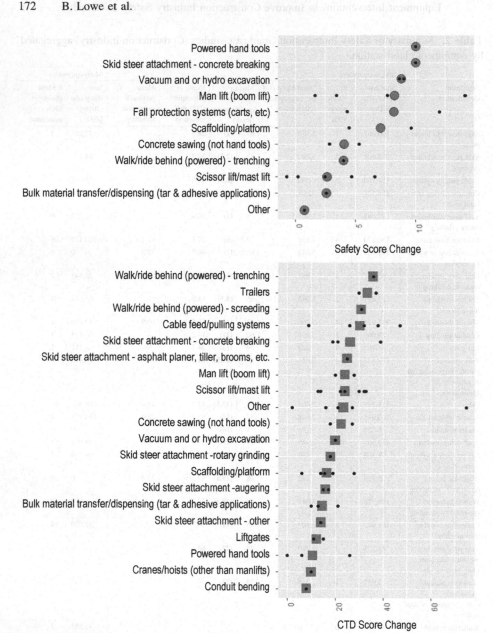

Fig. 1. Changes in safety hazard scores and CTD risk assessment scores (pre-intervention score minus post-intervention) by equipment type. Larger blue points are the group mean by equipment type (smaller black points are individual case studies).

with a low cost per affected employee. Skid steer attachments for concrete breaking (hydraulic breaker) also ranked relatively highly in reducing risk factors (CTD and safety risk factor reduction) and scored highly for evaluative quality. Concrete sawing equipment, trailers with hydraulic tilting/ramps, powered hand tools, and man lifts (boom lifts) were associated with evidence quality scores above the median for all equipment classifications AND above-average scores for either CTD or safety risk factors.

Detailed budget information showed that $6.55 million (in 2016 dollars) in Construction equipment purchases were made from the employer contribution and SIG funds across the 153 grants reviewed (see Table 2). Seventy percent of that total ($4.57 million USD) were single-equipment purchases, and the remaining 30% ($1.98 million) were associated with multi-equipment grants. Forty-three percent of the $6.55 million USD was spent on equipment for work at heights (i.e., scissor lifts, mast lifts, man lifts, boom lifts, scaffolding, non-man lift hoists). This equipment represented 45 of the 105 single-equipment grants, and another 13 multi-equipment grants (of 48) included one of these types of equipment. Equipment for working at heights or transferring of materials to heights tended to be more expensive (in terms of cost per affected employee) in addition to comprising the most commonly purchased construction equipment in the SIG program.

4 Discussion

There are several limitations that affect interpretation of this analysis. The SIG program award eligibility criteria and reporting requirements experienced some changes and were not consistent over 2003–2016. Prior to July 2009, a grant could only be awarded to an employer who had experienced at least one compensable injury claim in the defined affected employee group. As the program was expanded to have a more preventive focus, OBWC revised eligibility requirements so that grants could be awarded to proactively address risk factors, even in the absence of injury claims. In early program years, the application required employers to document all injury claims occurring in the affected employee group, regardless of injury causation. Correspondingly, we observed some baseline injury claim descriptions for which the subsequent equipment intervention was not believed to have any plausible mechanism of prevention. This influenced the decision to place greater emphasis (criteria weighting) on the risk factor reduction experience and less on the injury claim experience.

Employer grant recipients were tasked with determining whether the newly acquired equipment could have potentially introduced new hazards (or exacerbated an existing hazard) in the job/task. Even though employers were asked to anticipate potential new risks in the grant application, newly introduced risks were infrequently described in final reporting, and employer grantees may not have had the necessary health/safety knowledge regarding potential risks to make this determination. It is conceivable that new risks *may* have been introduced for some equipment; for example, new battery-powered hand tools and larger motorized equipment that might be more difficult to lift, carry, or maneuver.

Intervention cost per affected employee was highly, and inversely, related to employer size. Equipment purchased by larger employers was associated with lower cost per affected employee because larger employers had higher numbers of employees in the affected employee group. Equipment interventions that fundamentally alter a construction process, such as the adoption of a walk-behind machine for trenching with a single operator versus multiple employees performing hand digging, may reduce risk factors for employees beyond just the operator of the equipment. However, with other equipment, such as powered hand tools, the beneficial effect may only be realized by the user(s) of the piece of equipment. Defining an affected employee group size may not be straightforward, and there may have been incentive to overestimate the affected employee group size in the application phase. Relatedly, employers documented affected employee *hours* simply as the affected employee group's collective work hours and did not account for actual employee time exposed to the specific tasks and the hazards that were actually mitigated by the intervention.

Cost-benefit analysis (CBA) considerations, as documented by employers, were often incomplete. For example, past leasing costs of equipment were not included in the cost of doing identical work before the new equipment was procured. Equipment depreciation was not factored into the contractors' analyses. A number of reports clearly described a productivity increase in the narrative while not assigning any monetary value to this in the quantitative CBA worksheet. There were case studies in which discrepancies existed between productivity gains described in the narrative and monetary valuation of productivity gains in the CBA worksheet.

5 Conclusion

This review of case studies allowed us to aggregate experiences of multiple employers evaluating similar or identical Construction equipment. Case studies in aggregate provide more compelling evidence than an individual case study in regard to the effectiveness of specific equipment interventions. However, this review illustrates that there are still challenges in demonstrating efficacy of equipment interventions – even from aggregated case study experiences within a program specifically established to improve health/safety outcomes. From the evaluative data extracted in this review of employer case studies, we conclude that electrical cable pulling equipment, skid steer attachments for concrete breaking (hydraulic breakers), concrete sawing equipment, trailers with hydraulic tilting/ramps, powered hand tools, and man lifts (boom lifts) were associated with higher reductions in risk factors and case studies with stronger evaluative quality. In a forthcoming publication, we plan to explore how case study reporting can be improved to strengthen evidence demonstrating equipment intervention effectiveness. A similar methodology is being adopted for case study reviews with other types of workplace equipment.

Disclaimer: The findings and conclusions in this report are those of the authors and do not necessarily represent the official position of the National Institute for Occupational Safety and Health, Centers for Disease Control and Prevention, or the Ohio Bureau of Workers' Compensation.

References

1. CPWR (2013) The construction chart book 5th edn. The U.S. Industry and its Workers. https://www.cpwr.com/publications/construction-chart-book
2. Sengupta I, Reno VP, Burton JF, et al (2012) Workers' compensation: benefits, coverage, and costs, 2010 (2 August 2012). National Academy of Social Insurance
3. Wurzelbacher SJ, Bertke SJ, Lampl MP et al (2014) The effectiveness of insurer-supported safety and health engineering controls in reducing workers' compensation claims and costs. Am J Ind Med 57(12):1398–1412
4. Liberty Mutual Research Institute (2013) 2013 Workplace Safety Index
5. https://www.bwc.ohio.gov/employer/programs/safety/EmpGrants.asp. Accessed 20 Apr 2018
6. National Council on Compensation Insurance (NCCI) (2017) Classification of Industries. https://www.bwc.ohio.gov/downloads/blankpdf/OAC4123-17-04Appendix.pdf

Validation of Computational Fluid Dynamics Models for Evaluating Loose-Fitting Powered Air-Purifying Respirators

Michael Bergman[✉], Zhipeng Lei, Susan Xu, Kevin Strickland, and Ziqing Zhuang

National Personal Protective Technology Laboratory,
National Institute for Occupational Safety and Health,
Pittsburgh, PA 15236, USA
mbergman@cdc.gov

Abstract. Loose-fitting powered air-purifying respirators (PAPRs) are used in healthcare settings to reduce exposure to high-risk respiratory pathogens. Innovative computational fluid dynamics (CFD) models were developed for evaluating loose-fitting PAPR performance. However, the computational results of the CFD models have not been validated using actual experimental data.

Experimental testing to evaluate particle facepiece leakage was performed in a test laboratory using two models of loose-fitting PAPRs. Each model was mounted on a static (non-moving) advanced headform placed in a sodium chloride (NaCl) aerosol test chamber. The headform performed cyclic breathing via connection to a breathing machine. High-efficiency particulate air (HEPA)-filtered air was supplied directly to the PAPR facepiece using laboratory compressed supplied-air regulated with a mass-flow controller. One model was evaluated with six supplied-air flowrates from 50–215 L/min (Lpm) and the other model with six flowrates from 50–205 Lpm. Three different workrates (minute volumes) were evaluated: low (25 Lpm), moderate 46 (Lpm), and high 88 (Lpm). Manikin penetration factor (mPF) was calculated as the ratio of chamber particle concentration to the in-facepiece concentration.

Overall, data analyses indicated that the mPF results from the simulations were well correlated with the experimental laboratory data for all data combined (r = 0.88). For data at the three different workrates (high, moderate, low) for both models combined, the r-values were 0.96, 0.97, and 0.77, respectively. The CFD models of the two PAPR models were validated and may be utilized for further research.

Keywords: Powered air-purifying respirators · PAPR · CFD Manikin penetration factor

1 Introduction

Loose-fitting powered air-purifying respirators (PAPRs) use a battery powered motorized fan to draw air through an air-purifying element (particulate filter and/or sorbent cartridge) and supply it to a loose-fitting facepiece (forming a partial seal to the face) or hood (completely covering the head and neck and may also incorporate a shroud to cover portions of the shoulders and torso). Because loose-fitting PAPRs do not incorporate a tight seal to the face, they do not require fit testing. Interest in their use is becoming more prevalent in healthcare settings [1].

For use in the U.S., PAPRs are designed to meet National Institute for Occupational Safety and Health (NIOSH) certification requirements which were developed based on industrial use [2]. The supplied-air flowrates of loose-fitting PAPRs available on the market are in the range of 170–206 Lpm [3], while peak inhalation flows of adults exercising at heavy workloads can be 255 Lpm [4]. Gao et al. [5] mounted loose-fitting PAPRs on a manikin and measured the protection levels when the manikin breathed at varying workloads (mean inhalation flowrate): low (30 Lpm), medium (55 Lpm), high (85 Lpm), and strenuous (135 Lpm). The protection levels at the high and strenuous workloads were significantly lower than the ones at the other workloads.

Computational Fluid Dynamics (CFD) methods are computerized models which can evaluate the performance of respiratory protective equipment, e.g., N95 filtering facepiece respirators [6, 7]. Lei et al. [8] simulated the performance of loose-fitting PAPRs on a headform using CFD simulations and determined the effects of breathing workloads and supplied-air flowrates on the PAPR's performance. However, the computational results of the CFD models have not been validated using actual experimental data. The objective of this study was to collect and use actual experimental data to validate the CFD PAPR models. The CFD models and experimental methods evaluated PAPR performance at varied workrates and supplied-airflows to the PAPR facepiece.

2 Methods

2.1 PAPR Models

Two loose-fitting PAPR models currently used by healthcare workers were evaluated: MaxAir® 78SP-36 with disposable cuff (size S/M) (Bio-Medical Devices, Inc., Irvine, CA) and the 3M Air-Mate™ with BE-12 facepiece (size regular) (3M Company, St. Paul, MN). Both models have an assigned protection factor (APF) of 25 as designated by the U.S. Occupational Safety and Health Administration (OSHA) [9] (Fig. 1). The MaxAir® is a helmet design, with a blower motor and filter contained within the helmet. The motor draws outside air through a HEPA filter and blows the purified air into the facepiece. It has three airflow settings, tested in our laboratory as low (191 Lpm), medium (214 Lpm), and high (237 Lpm) using a test system described by Bergman et al. [10]. For our experiments, the blower motor was removed from the PAPR and replaced with a tube connected to a source of HEPA-filtered supplied air from the laboratory compressor system (further described in Sect. 2.1). A hole was drilled into the helmet and the tube was inserted through the hole; silicone sealant was then used to seal the interface. The tube bypassed the PAPR's filter and delivered the supplied-air directly into a cavity in the helmet. From this cavity, the air then flowed into the facepiece area.

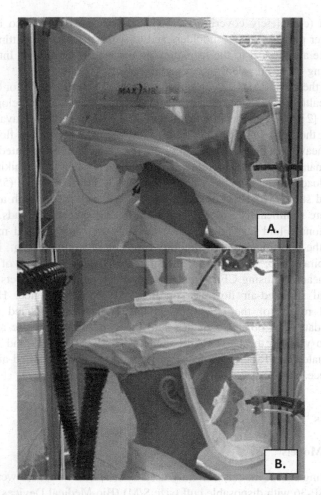

Fig. 1. PAPR models. (A) MaxAir® 78SP-36 Cuff System with disposable cuff (size S/M) (Bio-Medical Devices, Inc., Irvine, CA); (B) 3M Air-Mate™ with BE-12 facepiece (size regular) (3M Company, St. Paul, MN)

The Air-Mate™ is designed with an external blower containing a HEPA filter; a hose connects the blower unit to the facepiece. The facepiece has an elastic neck cuff design. The Air-Mate™ has only one airflow setting (measured at 211 Lpm). For modification of our testing, the exterior blower and filter were not used; the PAPR's hose was connected directly to the filtered laboratory supplied-air system.

CFD Simulations

We modeled the two loose-fitting PAPR systems by 3D scanning and surface processing. In the CFD models, the digital scan of the PAPR was donned on the digital image of the static advanced headform. We then modeled the interior volume of the PAPR. Figure 2 shows the CFD model of the headform and MaxAir® PAPR; the model comprises the breathing zone and the boundaries that include the PAPR interior surface, headform surface, walls of the breathing airway, supplied-air venting holes which direct air towards the face, and the volume of the loose-fitting area around the neck. We further meshed the CFD model into hexahedral cells using ANSYS ICEM software (ANSYS 2018, Ansys, Inc., Pittsburgh, PA). A CFD model using the Air-Mate™ PAPR was also similarly constructed.

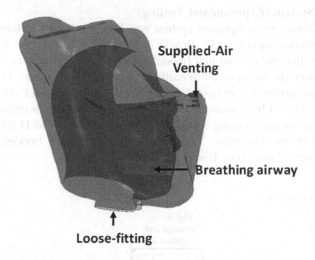

Fig. 2. The CFD model of the headform/MaxAir® PAPR volume

A time-dependent flowrate with a sine wave shape was applied to the inlet of the breathing airway to simulate a cyclic breathing pattern (inhalation and exhalation). A constant flowrate with the direction towards the PAPR breathing zone was applied to the supplied-air venting holes. For both PAPR models, three different workrate minute volumes were tested: low (25 Lpm), moderate (46 Lpm), and high (88 Lpm). One model was tested with supplied-air flowrates from 50–215 Lpm; the other model was tested with supplied-air flowrates from 50–205 Lpm. The loose-fitting area is the pressure outlet boundary where air flowed out of the PAPR. The flow velocity at the surfaces of the headform, breathing airway and PAPR were held at zero (i.e., non-slip boundary conditions were used).

The pisoFoam solver (in OpenFOAM) with the PISO (Pressure Implicit with Splitting of Operators) algorithm was used to perform the CFD simulation because we assumed that the flow is transient and incompressible and has a turbulent effect. Each simulation calculated a 20 s time duration with a 0.001-second time-step; at each time step, the pressure field and the velocity field inside the PAPR breathing zone were determined. Only one simulation run was performed for each supplied-air flowrate/workrate combination; multiple runs would have resulted in negligible variance of results.

The challenge particles, which were the particles outside of the PAPR model, were evenly placed at the gap of the loose-fitting area. At every 0.001-second time-step, particles with size 0.1 μm and concentration 100,000 particles/cm^3 were virtually generated. The tool for the Lagrangian particle tracking in OpenFOAM software calculated the particles that leaked into the facepiece. The interaction between particles and wall boundaries was tracked. We assumed that a particle would stick to a wall boundary after the impact happens between them. Each CFD simulation also determined movements of all particles, which were used to calculate the particle concentration of the inhalation airflow at the opening of the mouth during the entire 20 s simulation time.

Supplied-Air System (Experimental Testing)

Air from the laboratory compressor system was first regulated and then flowed to a mass flow controller (model: 5853, Brooks Instrument, Hatfield, PA) using a microprocessor controller and readout unit (model: 0154, Brooks Instrument, Hatfield, PA). This setup allowed the test operator to easily set a continuous flowrate. Downstream of the mass flow controller, the air passed through a HEPA filter (model: Air-Mate^{TM} filer p/n 451-02-01R01, 3M Inc.) mounted in a test fixture. Before the air entered the PAPR, the flow rate was measured using a mass flow meter (model: HFM-D-301A, Teledyne Hastings Instruments, Hampton, VA) with a digital readout (model: THCD-100, Teledyne Hastings Instruments, Hampton, VA) (Fig. 3).

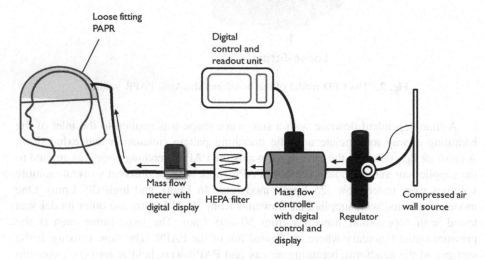

Fig. 3. Supplied-air test setup

NaCl Test Chamber (Experimental Testing)

The facepiece particle leakage for both PAPR models was tested in an acrylic test chamber (l, w, h: 30" × 36" × 72") equipped with mixing fans and an exhaust port. PAPR facepieces were mounted onto a medium-sized static advanced headform described by Bergman et al. [11]. The headform was connected to an external breathing lung with an inflatable bladder by a 22 mm inner diameter tube that travelled from the inside of the headform's mouth to the inflatable bladder. A port on the breathing lung was connected via a hose to breathing simulator (model: BRSS, Koken Ltd., Japan).

The breathing simulator used a cyclic sinusoidal breathing pattern to inflate and deflate the bladder inside the breathing lung by changing the lung's internal pressure.

Sodium chloride (NaCl) aerosol was generated as the challenge aerosol using air from the laboratory compressor system. The airflow was directed through a six-jet Collison (BGI Incorporated, Waltham, MA) with 2 w/v % NaCl solution in deionized water. The aerosol then flowed through a diffusion drier (model: 3062, TSI, Inc., Shoreview, MN) and a Kr-85 neutralizer (model: 3054, TSI, Inc., Shoreview, MN) before entering the chamber. The chamber NaCl aerosol concentration was maintained between 2.0–2.5 × 10^5 particles/cm^3 by adjusting a separate supply line of HEPA-filtered dilution air (Fig. 3).

During five different test periods through the data collection, a size distribution analysis was run on the chamber aerosol using a scanning mobility particle sizer spectrometer (SMPS) (model: 3936, TSI Inc.) system consisting of a classifier controller (model: 3080, TSI, Inc.), a differential mobility analyzer (DMA) (model: 3081, TSI, Inc.), a condensation particle counter (CPC) (model: 377500, TSI, Inc.), and an aerosol neutralizer (model: 3077A, TSI, Inc.). Data from the five SMPS scans were averaged resulting in a count median diameter (CMD) of 60 nm with a geometric standard deviation (GSD) of 1.87 nm (Fig. 4).

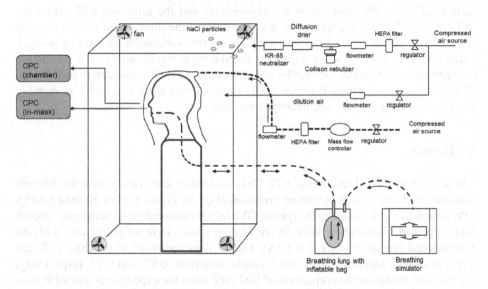

Fig. 4. Sodium chloride aerosol test chamber

Facepiece Leakage Measurements

Leakage measurements were taken using two condensation particle counters (CPC) (-model: 8022A; TSI, Inc.) and two laptop computers with Aerosol Instrument Manager® software (V.9.0.0.0, TSI, Inc.). One CPC sampled the chamber NaCl aerosol concentration (C_{out}) at a location ~2 cm in front of the PAPR hood (i.e., outside of the hood) at the level of the headform's mouth. The other CPC sampled the NaCl aerosol concentration inside the facepiece (C_{in}) via the sample tube penetrating through the lens of the facepeice. The model 8022A CPC has an accuracy of $\pm 10\%$ up to 5×10^5 particles/cm^3 as specified by the manufacturer. Both CPCs were set to sample in low flow mode (300 cm^3/min). Equal lengths of silicon conductive tubing were used to transport the aerosol to each CPC.

For each test, each CPC collected a 2-min sample at the rate of one data point/second. Five 2-min samples were taken in succession for each workrate/supplied-air flowrate combination. Each CPC reported the mean concentration of the 2-min sample which was later used for data analysis.

The average 2-min concentration of the chamber sample [C_{out}] was divided by the average 2-min concentration of the in-mask sample [C_{in}]; (Eq. 1).

$$mPF = [C_{out}]/[C_{in}] \tag{1}$$

Because the two models tested have an OSHA APF of 25, we limited mPF to a maximum of 10,000 for all data analyses. These calculated mPFs were then used to determine geometric mean (GM) mPFs and their geometric standard deviations (GSD) for the various workrate/supplied-air flowrate combinations.

JMP software (V.13, SAS Institute Inc., Cary, N.C.) was used to calculate the correlation coefficient (r-value) and the probability of significance (P-value) for comparison of GM mPF values from the experiments and the estimated mPFs from the CFD models. The r-value quantifies the strength and the direction of the relationship between two variables; if r is approaching +1, the two variables have a strong positive linear correlation. The P-value expresses the statistical significance of the correlation between the experimental GM mPFs and the estimated mPF value determined by the CFD simulation; we chose P-values <0.05 to define that there is a >95% probability that the two variables are significantly correlated.

3   Results

GM mPF experimental results and mPF CFD simulation results are plotted for different supplied-air flowrates and breathing workrates (Fig. 5). Figure 6 plots estimated mPFs determined in CFD simulations against GM mPFs measured in experiments. For all data combined from both models, the simulation results were well correlated with the experimental laboratory data (r = 0.88). For all data combined at the three different workrates (high, moderate, low), the r-values were 0.96, 0.97, and 0.77, respectively. All P-values assessing the correlation of GM mPF from the experiments and mPF from the CFD simulations were significant (P-value < 0.001) except for Model A at the low workrate (P-value > 0.05).

For Model A, the experimental GM mPFs and estimated mPFs were all 10,000 at the low workrate and supplied-air flow rates ≥ 75 Lpm; at supplied-air flow rates <75 Lpm, the experimental GM mPF results decreased to 100 while the simulation results remained at 10,000 (Fig. 5). At the moderate workrate and supplied-air flow rates of 50 and 205 Lpm, the experimental GM mPFs were 9 and 10,000 as compared to the

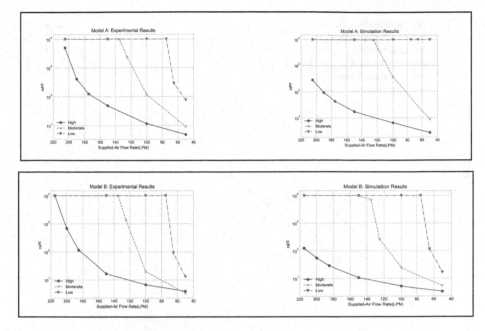

Fig. 5. GM mPF experimental and mPF estimated CFD simulation results by PAPR model at various supplied-air flowrates

estimated mPFs of 9 and 10,000. At the high workrate with supplied-air flow rates of 50 and 205 Lpm, the experimental GM mPFs were 5 and 4990 as compared to estimated mPFs of 3 and 205.

Model B results were similar to the trends for Model A (Fig. 5), with the distinction of Model B having better correlation of results for the low workrate (Fig. 6). The r-values for Model A plots in Fig. 6 are: 0.98, 0.98, and 0.0 for the high, moderate, and low workrates, respectively. The r values for Model B plots in Fig. 6 are: 0.92, 0.98, and 1.0 for the high, moderate, and low workrates, respectively.

4 Discussion

The experimental results validated the CFD simulation results. The turbulent dispersion in airflows with high velocity may cause bias of mPF using the combination of high breathing workrate and high supplied-air flowrates. The particle diffusion in airflows with low velocity may be the origin of the differences in the experimental and estimated mPFs in Model A using the low breathing workrate and low supplied-air flowrates. Future studies will model the turbulent dispersion and the particle diffusion in CFD simulations of loose-fitting PAPRs.

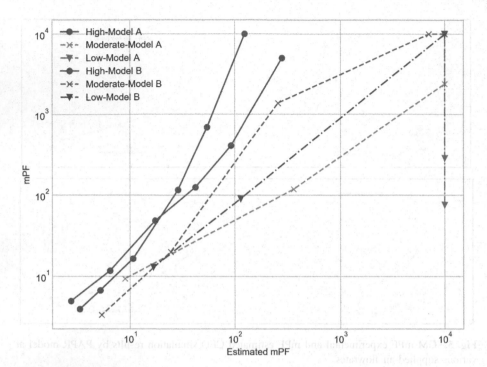

Fig. 6. Estimated mPF determined by CFD simulations versus GM mPF measured experimental results

5 Conclusion

Overall, data analyses indicated that the mPF results from the simulations were well correlated with the GM mPF experimental laboratory data for all data combined (R = 0.88). The CFD models of the two PAPR models were validated and may be utilized for further research. For data at the three different workrates (high, moderate, low) for both models combined, the r-values were 0.96, 0.97, and 0.77, respectively.

Further research can include more PAPR models and consider the impact of dynamic head and mouth movements on PAPR facepiece particle leakage. The CFD models for the two PAPR models were validated and may be applied in future PAPR performance research. Further research is needed to include more PAPR models and to consider the impact of human movements on PAPR particle leakage.

Disclaimer. The findings and conclusions in this report are those of the author(s) and do not necessarily represent the official position of the National Institute for Occupational Safety and Health, Centers for Disease Control and Prevention. Mention of a company or product name does not constitute endorsement by NIOSH. Mention of a product or use of a photo does not constitute NIOSH endorsement.

References

1. Wizner K, Stradtman L, Novak D, Shaffer R (2016) Prevalence of respiratory protective devices in US health care facilities: implications for emergency preparedness. Workplace Health Saf 64:359–368
2. Federal Register (1995) National Institute for Occupational Safety and Health (NIOSH). Respiratory protective devices. US Government Printing Office, Office of the Federal Register, Washington, DC. 42 CFR Part 84
3. Martin S, Moyer E, Jensen P (2006) Integrated unit performance testing of powered, air-purifying particulate respirators using a DOP challenge aerosol. J Occup Environ Hyg 3:631–641
4. Anderson NJ, Casidy PE, Janssen LL, Dengel DR (2006) Peak inspiratory flows of adults exercising at low, moderate and heavy work loads. J Int Soc Respir Prot 23:53–63
5. Gao S, McKay RT, Yermakov M, Kim J, Reponen T, He X, Kimura K, Grinshpun SA (2016) Performance of an improperly sized and stretched-out loose-fitting powered air-purifying respirator: manikin-based study. J Occup Environ Hyg 13:169–176
6. Lei Z, Yang J, Zhuang Z, Roberge R (2012) Simulation and evaluation of respirator faceseal leaks using computational fluid dynamics and infrared imaging. Ann Occup Hyg 57:493–506
7. Lei Z, Yang J (2014) Computing carbon dioxide and humidity in filtering facepiece respirator cavity during breathing cycles. In: ASME 2014 international design engineering technical conferences and computers and information in engineering conference. American Society of Mechanical Engineers, p V01AT02A077
8. Lei Z, Zhuang Z, Bergman M (2017) Application of digital human modeling for evaluating loose-fitting powered air-purifying respirators. In: Proceedings of the 5th international digital human modeling symposium
9. Federal Register (2006) Occupational Safety and Health Administration (OSHA). Assigned protection factors: Final rule. 29 CFR Parts 1910, 1915, 1926 71: 50121–50192
10. Bergman MS, Basu R, Lei Z, Niczgoda G, Zhuang Z (2017) Development of a manikin-based performance evaluation method for loose-fitting powered air-purifying respirators. J Int Soc Respir Prot 34(1):40–57
11. Bergman MS, Zhuang Z, Hanson D, Heimbuch BK, McDonald MJ, Palmiero AJ, Shaffer RE, Harnish D, Husband M, Wander JD (2014) Development of an advanced respirator fit-test headform. J Occup Environ Hyg 11(2):117–125

Sleep and Fatigue in Nurses in Relation to Shift Work

Irina Cekova[(⊠)], Ralitsa Stoyanova, Irina Dimitrova, and Katya Vangelova

National Center of Public Health and Analyses, 1413 Sofia, Bulgaria
i.tzekova@ncpha.government.bg

Abstract. Introduction: Sufficient and good quality sleep is crucial for shift workers because of its implications for alertness, recovery and health. The aim of the study was to follow the effects of night shifts and overtime work on sleep and fatigue of hospital nurses in Bulgaria.

Methods: The study is cross-sectional and comprised 1340 nurses of age 50.1 ± 10.1 years from Sofia hospitals. Anonymous questionnaire survey included Karolinska Sleep Diary, demographic information, work place variables, working hours and shift system. The statistical analyses were carried out with SPSS.

Results: 27.4% of the nurses worked only day shifts, but part of them were with history of night shifts. 18.9% worked from 1 to 4 night shifts, and 46.7% more than 5 night shifts. Great deal of the nurses worked more than 40 h weekly as follows: 34.2% worked 41–50 h per week, 16.8% - 51–60 h and 10.6% >61 h. The reported sleep duration did not differ between the groups of nurses working day shifts and different number of night shifts or with the number of work hours weekly. The quality of sleep, estimated by SQI was worse with the increase of number of night shifts and in comparison to nurses working only day shifts (F = 6.877, p = 0.000) and with increase of the hours worked weekly (F = 5.085, p = 0.002). With the increase of the number of night shifts and the working hours weekly the insufficiency of sleep, fatigue in the morning after awakening and sleep throughout increased highly significantly.

Discussion: Shift working with more night shifts monthly and overtime hours weekly contributed to impaired sleep in the studied group of nurses. The sleep impairment was more evident within the increase in the number of night shifts.

Keywords: Night shifts · Overtime · Sleep quality index

1 Introduction

Sleep is considered as one of the pillars of good health, along with a healthy diet and frequent exercise, and sufficient and good quality sleep is crucial for the workers because of its implications for alertness, recovery and health. Some of the effects of shift work, especially night shift work are related to circadian disruption, sleep impairment and fatigue [1, 2], as well as higher risk for occupational accidents [3], cardiovascular disease [4–6] and peptic ulcer [7]. Shift work may also be related to type 2 diabetes [8], rheumatoid arthritis [9], multiple sclerosis [10], cancer [11, 12], etc.

© Springer Nature Switzerland AG 2019
S. Bagnara et al. (Eds.): IEA 2018, AISC 819, pp. 186–193, 2019.
https://doi.org/10.1007/978-3-319-96089-0_21

Night shift work is more prevalent in healthcare sector, especially hospitals. Individuals working night and rotating shifts rarely obtain optimal amounts of sleep. Night shift workers sleep 1 to 4 h less than normal when they have night shifts, sleep loss is cumulative and by the end of the workweek may contribute to impairment of decision-making, initiative and vigilance [1]. Insufficient sleep and poor sleep quality can contribute to the experienced by the nurses fatigue, a critical issue also in relation to possible medical errors.

Working extended shifts (more than 8 h per day), working overtime (more than 40 h per week), and working both extended shifts and overtime can have adverse effects on sleep, alertness and worker health. Extended shifts have been associated with higher risk of accidents, more psychosocial complaints, more cardiovascular symptoms, higher risk for development of hypertension [1, 13, 14]. Working overtime has also been associated with increased risk of making errors, poorer perceived health, adverse health effects, increased morbidity and mortality [1, 15, 16]. The nurses engaged in rotating night shifts reported the lower quality and quantity of sleep and more frequent chronic fatigue, in comparison with the day shift workers [17].

The nurses in Bulgaria often work overtime due to lack of nurses, as well as second job because of low salaries. Both contribute to long working hours, while performing intense, responsible and decision-rich work in changing clinical environments with frequent interruptions during tasks and irregular working hours, including night work.

The aim of the study was to follow the effects of night shifts and overtime work on sleep and fatigue of hospital nurses in Bulgaria.

2 Methods

2.1 Study Design

A cross-sectional survey was conducted which included the 19 largest hospitals (>150 beds) in Sofia, Bulgaria, as a part of a study of risk factors for health in health care. The survey was anonymous, self-administered questionnaire was filled in by the nurses at their workplaces. 1340 nurses participated in the study, 1330 female and 10 male of average age 50.1 ± 10.1 years and average length of service 27.7 ± 10.8 years.

2.2 Data Collection

The questionnaires gathered information about the demographic, physical and behavioral characteristics as well as the previous and current work schedules of the nurses. Information was collected on the current schedule with the options: only daily shifts, morning/afternoon shifts, rotating shifts or only night shifts. In the study, night shift is defined as hours worked between 24:00 and 08:00. Information is also collected on how many night shifts are worked per month: 0 nights per month, 1–2 nights per month, 3–4 nights per month, more than 5 nights per month. To assess the history of night work, respondents reported the number of years they worked night shifts (including rotating and non-rotating night shifts). The participants also answered question on additional work, worked in the hospital and how often it happened, whether they had a second job, and taking in account the additional and second job how many hours on average they work per week. The average hours worked per week were specified as follows: 21–40 h, 41–50 h, 51–60 h and >60 h per week. In the analysis, we counted over 40 working hours per week for a long working week.

2.3 Sleep Quality and Fatigue

The participants completed a Karolinska Sleep Diary containing questions about "sleep duration", "sleep quality", "insufficiency of sleep", "fatigue in the morning after awakening", "ease of falling asleep", "waking up during sleep", "early awakening" and "sleep throughout". The questions were rated with a 5-point scale (1 - bad sleep, 5 - no sleep problems). A sleep quality index SQI was formed and calculated from questions related to "sleep quality", "fatigue in the morning after awakening", "ease of falling asleep" and "sleep throughout". Fatigue was measured by a questionnaire, consisting of 9 questions concerning the "difficulty to rest after work", "the degree of fatigue at the end of the working day", "the feeling of rest after dinner", "the feeling of rest on the second day off", "the difficulty to concentrate after work"," the communication problems after the end of the working day", "the need for more time to feel rested after work", "the desire for seclusion for about an hour after work", "the lack of willingness to do anything after work due to fatigue", "the inability for optimal work performance at the end of the day due to fatigue" and the fatigue was calculated summing the positive answers.

2.4 Data Analysis

The data was introduced and processed with the IBM SPSS Statistics 15.0 statistical package. ANOVA and Pearson correlation coefficient were applied for statistical analyses, and the significance level was set at $p < 0.05$.

3 Results

During the study data showed that 27.4% of the nurses worked only day shifts, 13.3% worked morning/afternoon shifts and 59.2% worked rotating shifts and only 0.1% worked night shifts. More than 60% of the nurses working only day shifts have history of working rotating shifts with night work. 46.7% from nurses worked more than 5 night shifts, 18.9% worked 1 to 4 night shifts and 34.4% didn't have night shifts.

Great deal of the nurses reported that they had additional work after the end of working time as follows: 55.2% had rarely additional work, 22.1% had additional work 2–3 times per week and 11.3% had additional work every day. Small part of the nurses (11.4%) didn't have any additional work. Also some of the nurses – 27.8% shared that they are working in a second work place. The additional work led to longer working hours per week. Only 38.3% of the nurses shared that they worked 21–40 h per week, while 34.2% worked 41–50 h per week, 16.8% worked 51–60 h per week and 10.6% worked over 61 h per week.

Sleep duration did not differ with the type of working time arrangements and number of night shifts per month, but most of the sleep quality characteristics were significantly influenced by the number of night shifts per month, the most significantly insufficiency of sleep, fatigue in the morning after awakening, sleep throughout and sleep quality index (Table 1), followed by self-rated quality of sleep, ease of falling asleep and waking up during sleep. The differences in fatigue did not reach significance, but the levels of fatigue were higher with the nurses working night shifts.

Table 1. Quantity and quality of sleep (1 = poor sleep, 5 = no problems with sleep) and fatigue (1 = low level of fatigue, 9 = high level of fatigue) according to the number of night shifts per month

Sleep characteristics	$\bar{x} \pm SD$				F
	0 shifts (n = 454)	1–2 shifts (n = 87)	3–4 shifts (n = 162)	>5 shifts (n = 617)	
Sleep duration	7.8 ± 1.2	7.6 ± 1.1	7.7 ± 1.2	7.6 ± 1.4	NS
Quality of sleep	3.3 ± 0.8	3.3 ± 0.7	3.4 ± 0.9	3.2 ± 0.9	4.317**
Insufficiency of sleep	2.6 ± 1.3	2.8 ± 1.1	2.7 ± 1.2	2.3 ± 1.2	6.332***
Fatigue in the morning after awakening	2.8 ± 0.9	3.0 ± 0.7	3.0 ± 0.9	2.7 ± 0.9	6.273***
Ease of falling asleep	3.2 ± 1.0	3.2 ± 0.8	3.2 ± 1.0	3.0 ± 1.0	4.498**
Waking up during sleep	2.9 ± 0.9	3.2 ± 1.0	3.2 ± 0.9	2.8 ± 1.0	5.845**
Early awakening	2.9 ± 1.1	3.1 ± 0.9	3.0 ± 1.1	2.8 ± 1.1	NS
Sleep throughout	3.4 ± 1.1	3.6 ± 1.0	3.4 ± 1.1	3.1 ± 1.1	8.741***
Sleep quality index	3.1 ± 0.7	3.3 ± 0.6	3.2 ± 0.8	3.0 ± 0.7	6.877***
Fatigue	5.1 ± 3.1	5.6 ± 4.0	5.5 ± 3.6	5.4 ± 3.1	NS

*$p < 0.05$ **$p < 0.01$ ***$p < 0.001$

Sleep duration did not differ with the working hours per week, while the characteristics of the sleep showed impairment with increase of hours worked (Table 2), more significant for insufficiency of sleep, fatigue in the morning after awakening and sleep throughout, followed by quality of sleep, waking up during sleep and sleep quality index, especially working 51–60 h and over 61 h per week. The fatigue increased significantly with the number of working hours per week; and the level in the nurses working over 61 h per week is raising concern for health and safety.

The type of work time arrangements and working hours per week correlated negatively with quality of sleep, insufficiency of sleep, fatigue in the morning after awakening, sleep throughout and sleep quality index. Quality of sleep, insufficiency of sleep, fatigue in the morning after awakening and sleep quality index correlated negatively with the additional work and quality of sleep, insufficiency of sleep, ease of falling asleep, sleep throughout and sleep quality index correlated negatively with number of night shifts per month. Sleep duration and early awakening did not show relation to work time arrangements, number of night shifts per month, additional work

Table 2. Quantity and quality of sleep (1 = poor sleep, 5 = no problems with sleep) and fatigue (1 = low level of fatigue, 9 = high level of fatigue) according to working hours per week

Sleep characteristics	x̄ ± SD				F
	21–40 h (n = 478)	41–50 h (n = 427)	51–60 h (n = 210)	>61 h (n = 132)	
Sleep duration	7.6 ± 1.4	7.8 ± 1.2	7.6 ± 1.3	7.6 ± 1.5	NS
Quality of sleep	3.3 ± 0.8	3.3 ± 0.9	3.1 ± 0.8	3.0 ± 0.9	5.782**
Insufficiency of sleep	2.7 ± 1.2	2.4 ± 1.2	2.3 ± 1.2	2.2 ± 1.2	7.801***
Fatigue in the morning after awakening	2.9 ± 0.9	2.7 ± 0.9	2.7 ± 0.8	2.6 ± 0.9	7.762***
Ease of falling asleep	3.1 ± 0.9	3.2 ± 1.0	3.1 ± 1.0	3.0 ± 1.0	NS
Waking up during sleep	3.0 ± 1.0	3.0 ± 1.0	2.8 ± 1.0	2.9 ± 1.0	3.011*
Early awakening	2.9 ± 1.1	3.0 ± 1.1	2.9 ± 1.1	2.8 ± 1.1	NS
Sleep throughout	3.4 ± 1.1	3.4 ± 1.1	3.1 ± 1.1	3.0 ± 1.0	6.423***
Sleep quality index	3.2 ± 0.7	3.1 ± 0.7	3.0 ± 0.7	2.9 ± 0.7	5.085**
Fatigue	4.8 ± 3.2	5.4 ± 3.1	5.4 ± 3.2	6.5 ± 3.7	9.508***

*p < 0.05 **p < 0.01 ***p < 0.001

or working hours per week. The characteristic ease of falling asleep correlated negatively only with the number of night shifts per month and the characteristic waking up during sleep correlated negatively only with the working hours per week (Table 3).

Table 3. Correlation between sleep characteristics and working time arrangements, number of night shifts per month, additional work and working hours per week

Sleep characteristics	Type of work time arrangements (day, morning/afternoon, rotating, night shifts)	Number of night shifts per month	Additional work after the end of working time	Working hours per week
Sleep duration	−.035	−.053	.046	−.003
Quality of sleep	−.059*	−.079**	−.084**	−.116***
Insufficiency of sleep	−.075**	−.097**	−.111***	−.138***
Fatigue in the morning after awakening	−.056*	−.051	−.180***	−.119***
Ease of falling asleep	−.041	−.084**	−.039	−.026
Waking up during sleep	.011	−.042	−.030	−.064*
Early awakening	.014	−.049	−.012	−.024
Sleep throughout	−.063*	−.119***	−.012	−.113***
Sleep quality index	−.069*	−.096**	−.103***	−.114***

*p < 0.05 **p < 0.01 ***p < 0.001

4 Discussion

Shift work, especially the night shift work, and long working hours are known to impair sleep and contribute to fatigue [1, 18, 19]. Quantification of shift work and working hours is quite difficult issue in health care because of the great diversity of shift schedules and changing work time patterns during work experience of the staff. The nurses in the studied hospitals in Sofia work mainly rotating shifts with night work (59.2%), part of the nurses work day shifts, but have history of night shift work. Some of the latter and the ones working morning/afternoon shifts have 1–4 night shifts on duty monthly. Finally 34.4% of the studied nurses didn't have night shifts in the period of the study, but nearly the half of the studied nurses (46.7%) worked more than 5 night shifts.

The present study has demonstrated that the higher percentage of the nurses working in Sofia hospitals had over 40 h work per week, 34.2% worked 41–50 h per week, 16.8% worked 51–60 h per week and 10.6% worked over 61 h per week, perquisite for both sleep impairment and fatigue, especially taking in consideration the job characteristics of caring intense, responsible and decision-making work often in a lack of time.

The sleep quality and fatigue were highly significantly influenced by both night work and long working hours. The quality of sleep was the worst in the nurses, working more than 5 night shifts monthly. These data are in accordance with earlier findings [1, 17, 20]. The disturbed sleep is considered as a common pathway connecting night shift work with adverse health effects and the medium in between is the insufficient recovery [18]. We were not able to differentiate the effect of duration night shift work, as the length of service under night shift work was parallel to age of the nurses.

We consider that one limitation of the study is that it failed to quantify the number of night shifts worked by the former shift working nurses, covering day shifts only during the study. A study of Bjorvatn et al. [21] showed that seventy percent of the nurses (136 women and 14 men) in an intensive care unit in Norway had poor-quality sleep according to the Pittsburgh Sleep Quality Index (PSQI) with shift work including night work. But these sleep problems tended to decrease with lack of exposure and recur with re-exposure. Another study showed that the effect of shift work experience was the greatest among people in their forties and reduced as participants left shift work (i.e. when they transferred to day working or retired), suggesting that the effects of shift work did not persist once people quit shift work [19]. Another limitation of the study is the lack of data on the available periods for recovery and their effect on sleep and fatigue because of the complexity and differences in schedule arrangements between the different hospitals and departments, as well as individual schedules of the nurses.

In conclusion, our results strongly support the suggestion that shift work with 5 and more night shifts monthly and overtime work contribute to impaired quality of sleep and fatigue among hospital nursing staff. Further research is needed to explore the effect of particular schedules and periods for recovery, as well as the possible relationship between quality of sleep and fatigue of the studied nurses and risk of accidents and health impairment. Considering the precaution principle shift work schedules of the nurses should be arranged on the bases of good practice in the ergonomics and overtime work should be avoided when possible.

References

1. Hughes R (2008) Patient safety and quality: an evidence-based handbook for nurses. AHRQ Publication No. 08-0043. Rockville, MD
2. Zhang L, Sun DM, Li CB, Tao MF (2016) Influencing factors for sleep quality among shift-working nurses: a cross-sectional study in China using 3-factor Pittsburgh sleep quality index. Asian Nurs Res 10(4):277–282. https://doi.org/10.1016/j.anr.2016.09.002
3. Wagstaff AS, Sigstad Lie JA (2011) Shift and night work and long working hours-a systematic review of safety implications. Scand J Work Environ Health 37(3):173–185. https://doi.org/10.5271/sjweh.3146
4. Frost P, Kolstad HA, Bonde JF (2009) Shift work and the risk of ischemic heart disease - a systematic review of the epidemiologic evidence. Scand J Work Environ Health 35(3):163–179. https://doi.org/10.5271/sjweh.1319
5. Puttonen S, Härmä M, Hublin C (2010) Shift work and cardiovascular disease - pathways from circadian stress to morbidity. Scand J Work Environ Health 36(2):96–108. https://doi.org/10.5271/sjweh.2894
6. Vyas MV, Garg AX, Iansavichus AV, Costella J, Donner A, Laugsand LE et al (2012) Shift work and vascular events: systematic review and meta-analysis. BMJ 345:e4800. https://doi.org/10.1136/bmj.e4800
7. Knutsson A, Bøggild H (2010) Gastrointestinal disorders among shift workers. Scand J Work Environ Health 36(2):85–95. https://doi.org/10.5271/sjweh.2897
8. Pan A, Schernhammer ES, Sun Q, Hu FB (2011) Rotating night shift work and risk of type 2 diabetes: two prospective cohort studies in women. PLoS Med 8(12):e1001141. https://doi.org/10.1371/journal.pmed.1001141
9. Puttonen S, Oksanen T, Vahtera J, Pentti J, Virtanen M, Salo P et al (2010) Is shift work a risk factor for rheumatoid arthritis? The Finnish public sector study. Ann Rheum Dis 69 (4):779–780. https://doi.org/10.1136/ard.2008.099184
10. Hedstrom AK, Akerstedt T, Hillert J, Olsson T, Alfredsson L (2011) Shift work at young age is associated with increased risk for multiple sclerosis. Ann Neurol 70(5):733–741. https://doi.org/10.1002/ana.22597
11. Ijaz S, Verbeek J, Seidler A, Lindbohm ML, Ojajärvi A, Orsini N et al (2013) Night-shift work and breast cancer – a systematic review and meta-analysis. Scand J Work Environ Health 39(5):431–447. https://doi.org/10.5271/sjweh.3371
12. Åkerstedt T, Knutsson A, Narusyte J, Svedberg P, Kecklund G, Alexanderson K (2015) Night work and breast cancer in women: a Swedish cohort study. BMJ Open 5(4):e008127. https://doi.org/10.1136/bmjopen-2015-008127
13. Dembe AE, Erickson JB, Delbos RG, Banks SM (2005) The impact of overtime and long work hours on occupational injuries and illness: new evidence from the United States. Occup Environ Med 62(9):588–597. https://doi.org/10.1136/oem.2004.016667
14. Folkard S, Tucker P (2003) Shift work, safety, and productivity. Occup Med 53(2):95–101. https://doi.org/10.1093/occmed/kqg047
15. van der Hulst M (2003) Long workhours and health. Scand J Work Environ Health 29 (3):171–188. https://doi.org/10.5271/sjweh.720
16. Yang H, Schnall PL, Jauregui M, Su TC, Baker D (2006) Work hours and self-reported hypertension among working people in California. Hypertension 48(4):744–750. https://doi.org/10.1161/01.HYP.0000238327.41911.52

17. Ferri P, Guadi M, Marcheselli L, Balduzzi S, Magnani D, Di Lorenzo R (2016) The impact of shift work on the psychological and physical health of nurses in a general hospital: a comparison between rotating night shifts and day shifts. Risk Manag Healthc Policy 9:203–211. https://doi.org/10.2147/RMHP.S115326

18. Härmä M (2006) Workhours in relation to work stress, recovery and health. Scand J Work Environ Health 32(6):502–514. https://doi.org/10.5271/sjweh.1055

19. Tucker P, Folkard S, Ansiau D, Marquié J-C (2011) The effects of age and shiftwork on perceived sleep problems: results from the VISAT-combined longitudinal and cross-sectional study. J Occup Environ Med 53(7):794–798. https://doi.org/10.1097/JOM.0b013e318221c64c

20. Sveinsdóttir H (2006) Self-assessed quality of sleep, occupational health, working environment, illness experience and job satisfaction of female nurses working different combination of shifts. Scand J Caring Sci 20(2):229–237. https://doi.org/10.1111/j.1471-6712.2006.00402.x

21. Bjorvatn B, Dale S, Hogstad-Erikstein R, Fiske E, Pallesen S, Waage S (2012) Self-reported sleep and health among Norwegian hospital nurses in intensive care units. Nurs Crit. Care 17 (4):180–188. https://doi.org/10.1111/j.1478-5153.2012.00504.x

Ergonomics in Agriculture: Critical Postures, Gestures, and Perceived Effort in Handling Foldable Roll-Over Protective Structures (ROPS) Fitted on Tractors

Federica Caffaro[1] , Margherita Micheletti Cremasco[2] ,
Ambra Giustetto[2] , Lucia Vigoroso[1] ,
Giuseppe Paletto[1], and Eugenio Cavallo[1(✉)]

[1] Institute for Agricultural and Earthmoving Machines (IMAMOTER),
National Research Council of Italy (CNR), Turin, Italy
eugenio.cavallo@cnr.it
[2] Department of Life Sciences and Systems Biology, University of Torino,
Turin, Italy

Abstract. Tractor overturn is the main cause of injuries and fatal accidents in agriculture. Rollover Protective Structures (ROPS) showed to be effective in reducing fatalities during tractor overturn. Foldable ROPS (FROPS) have been developed to offer greater mobility when working in low overhead clearance zones and more storage options. However, many fatalities and serious injuries in tractor overturn accidents occur for a misuse of the FROPS. The study proposed a multidimensional ergonomic investigation of FROPS handling to identify criticalities in the human-machine interaction which prevent farmers from raising the roll-bar. An observation of users' behaviors while operating the FROPS and a collection of subjective ratings about perceived effort were performed, considering also tractor objective features. Eleven operators and nine tractors were involved in the study. The results showed that the participants exposed themselves to different safety (e.g. falls) and health (e.g. biomechanical overload of the spine) risks when raising the FROPS, even though they did not perceive any intense effort in handling the roll-bar. The vertical distance between the FROPS pivot pin and the ground affected participants' gestures and postures. To facilitate the actual use of the FROPS and to prevent injuries or fatal consequences in case of tractor overturn, some technical solutions to enhance the accessibility of the FROPS may be developed. Furthermore, the correct postures and gestures to handle the roll-bar should be addressed during safety and health training courses.

Keywords: Agriculture · Foldable ROPS · Occupational risk

1 Introduction

Agriculture is one of the most dangerous occupational sectors in both developing and industrialized countries [1]. Machinery is the main source of accidents, and tractors are involved in the highest rates (46%) of injuries [2]. In particular, since the 1950s, tractor

© Springer Nature Switzerland AG 2019
S. Bagnara et al. (Eds.): IEA 2018, AISC 819, pp. 194–202, 2019.
https://doi.org/10.1007/978-3-319-96089-0_22

overturning/rollover has been reported as the main cause of fatal and non-fatal accidents [3, 4]. The most effective way to prevent deaths during overturning accidents is the combined use of a Rollover Protective Structure (ROPS) and a seatbelt [5], which has been proven to reduce 99% of deaths when tractor overturn occurs [6]. ROPS are structures that absorb a portion of the impact energy generated by the tractor weight in an overturning accident. They decrease the risk of severe injury by providing the operator with an adequate clearance zone [7]. The first mandatory requirement for ROPS to be fitted on newly manufactured agricultural tractors was introduced in Sweden in the 1950s [4]. Following this experience, the 1980s were characterized by the adoption of safety structures to protect operators from injuries caused by vehicle overturns/rollovers in most of the developed countries [8]. In Europe, ROPS have been mandatory on new standard tractors since 1976 in France, 1977 in Italy and 1980 in Spain [6]. In the United States, the U.S. Occupational Safety and Health Act (OSHA) requires approved ROPS to be fitted on agricultural tractors over 20 horsepower (15 kW) manufactured after October 25, 1976, when operated by hired employees [9].

In the Scandinavian countries a significant reduction in the frequency of fatal rollovers for ROPS-equipped tractors was observed [4]. However, other European countries do not report the same positive trends and registered a high number of tractors without an adequate structure to protect the driver during overturning accidents [10]. For instance, in Italy, in the period June 2008–June 2011 there were more than 400 fatal accidents related to tractor overturning and more than 50% of tractors involved were not equipped with ROPS [11]. As concerns the United States, in 2012, only 59% of agricultural tractors were equipped with ROPS [12].

Overhead obstacles were reported as the most important reason for the operators not to install a ROPS [13]. To facilitate tractor operation in low overhead clearance zones, foldable ROPS (FROPS) have been developed. They are made of two parts: the upper and folding frame and the lower part, the support, fixed to the tractor body or chassis (Fig. 1). The foldable frame is connected to the lower part by a pivot point and a pin, or a bolt, to keep it upright. By this construction, the height of the FROPS – which is similar to that of a conventional ROPS when it is in upright position – can be significantly decreased by folding the upper foldable part downward. Tractors equipped with foldable ROPS offer greater mobility and more storage and transport options compared to fixed ROPS.

However, recent surveys have revealed a new issue. The number of fatal accidents and severe injuries in tractor rollover accidents with folded-down FROPS has increased in the last years [4, 12, 14]. Research showed that after lowering the FROPS to pass an obstacle, many operators prefer to leave it in the folded-down position. One possible explanation for this behavior is that raising and lowering the FROPS is a time-consuming and strenuous process [15]. These issues have been typically investigated by means of subject-based techniques [16], as questionnaires or interviews, to collect data about farmers' perceptions of the barriers hindering FROPS utilization; however, in an ergonomic analysis of the work activities, a conjoint adoption of both subject-based and observation-based techniques is recommended, to integrate and compare different information, and to provide a more comprehensive analysis of the topic under investigation [17].

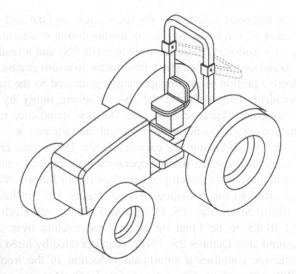

Fig. 1. Example of a tractor fitted with rear-mounted foldable ROPS.

Based on these considerations, the present study proposed a multidimensional ergonomic analysis of FROPS operation, to identify critical gestures and postures, and related operators' subjective assessment of the effort required to raise the FROPS. Pointing out the critical issues in the handling of FROPS will allow to suggest targeted preventive interventions, in terms of both technical improvements of the machine and training actions, to facilitate the actual use of the foldable ROPS, thus preventing injuries and fatalities in case of overturning accidents.

2 Method

2.1 Participants

Eleven farmers participated in the study, and nine models of tractors of different brands equipped with rear-mounted foldable ROPS were analyzed. The participants were all males, because of the predominance of male workers among Italian farming population [18]. The machinery considered for the present study were standard-track tractors (i.e. track width larger than 1150 mm, as stated by OECD Tractor Codes, [7]) fitted with rear-mounted two-pillar FROPS.

2.2 Instruments

For the present investigation we adopted a mixed-method exploratory approach [19]. Both machine-related and subject-related measures were taken, to identify potential mismatches between the two components of the human-machine interaction, which may hinder FROPS operation.

The data collection is ongoing, as this study is part of a wider project related to safe FROPS handling. Table 1 reports the variables and the instruments we considered for the purposes of this paper. The Borg CR10 scale [20] is a subjective scale ranged from 1 to 10 indicating the perceived level of effort when performing a task. In this study, it was used to collect subjective ratings of perceived exertion during the manual handling of FROPS.

Table 1. Type of measurements, variables considered in the study and respective instruments of data collection.

Type of measurement	Variable	Instrument
Machine objective features	Overall height from ground-to-top of folded ROPS	Measuring tape
	Vertical distance ROPS pivot pin-to-ground	
Participants' subjective evaluation	Perceived exertion	Borg CR10 scale [20]
Participants' actual behavior	Gestures, postures and place to stand	Video-recorded observation

2.3 Procedure

To be included in the study, the farmers had to own a tractor satisfying the inclusion criteria listed in Sect. 2.1 (i.e. track width larger than 1150 mm, fitted with rear-mounted two-pillar FROPS). A list of possible participants was provided by the dealers of various brands of agricultural machinery in the province of Cuneo and Asti, Piedmont Region, Northwestern Italy. Respondents were contacted by telephone and, if willing to participate, were met at their own farm. Each visit lasted about 20 min. The participation was voluntary and all the farmers gave their written informed consent prior to their inclusion in the study.

2.4 Data Analysis

Data collected through the abovementioned instruments and techniques were analyzed by means of a concurrent procedure -i.e. in which both qualitative and quantitative data are collected at the same time and then the information is integrated in the interpretation of the overall results [19] –, to point out the most critical issues in FROPS operation. Two independent judges analyzed the videos using an observational grid. The grid provided different postural and behavioral categories which are known to be critical variables when assessing postural comfort/discomfort and unsafe behaviors [21, 22], adjusted for the type of task considered, i.e. handling the FROPS. The placement and movements of operators' head, trunk, left and right upper and lower limbs, hands and feet were annotated and considered for subsequent analysis.

3   Results

The farmers had a mean age of 48.64 years (SD = 12.79) and have been working in agriculture for 25.17 years on average (SD = 18.45). As concerns the frequency of FROPS operation, many of the interviewees reported a seasonal handling of this device: they typically had to move it several times in different periods of the year, to work under hazelnut trees or in the wood. Two farmers reported a frequent folding down of the FROPS, to work in greenhouses, or to store the tractor in the warehouse.

As regards the characteristics of the investigated tractors, the distance between the crossbar in lowered position and the ground was between 1335 and 1385 mm (M = 1346.11, SD = 31.50), whereas the distance between the ground and the FROPS pivot pin, ranged from 1840 to 1955 mm (M = 1900.00, SD = 37.50).

Regarding the placement of the participants and the gestures performed during the FROPS-raising task, all the interviewed operators performed the task by standing in front of the FROPS trajectory, on the back of the tractor.

The observations showed that operation of the FROPS folding frame may be split in two different phases, with regard to critical gestures and postures required: (1) rotation from the lowered position to the horizontal position; (2) rotation from the horizontal position to upright position.

Two main types of operators' behaviors were detected to grasp the ROPS when it was fully lowered: 6 farmers used either some parts of the machine or a ladder as a platform to reach and operate the ROPS, whereas other 5 participants raised the roll-bar by standing on the floor (Fig. 2a and b). These two different placements were maintained also in the second phase of FROPS operation, with one farmer moving from the ground to the tractor lift arm to accompany the roll-bar in its upright position.

Two main types of hand gestures and placement were also observed when the participants moved the roll-bar to the upright position: 4 participants finished the raising task by using only one hand (the other one was used just as a support) while 7 farmers by pushing the roll-bar with both hands (Fig. 2c and d).

The support of the feet induced postural criticalities which mainly concerned the shoulders and the spinal column. The shoulders were asymmetrical if the lifting operation wasn't finished with both hands but by accompanying the roll-bar toward its upright position with just one hand. This type of movement determines a unilateral lengthening of the muscular bundles of the back and it is often associated with a redistribution of weight on the lower limbs, moving the feet or raising the heels, thus decreasing their area of support, which may therefore create a risk for the operator's safety and health. Among the operators who lifted the FROPS with two hands, a posterior hyperextension of the back and the neck can occur, determined by the lack of a standing surface for operators who use parts of the tractor as a support for the feet. Even this movement can cause health risks for the operator, such as contractions at lumbar and neck level, but also safety risks, such as the risk of falling. The observed behaviors were affected also by the objective features of the FROPS: higher distances between the crossbar in the lowered position and the ground and between the FROPS pivot pin and the ground affected the grasping point of the crossbar in lowered position, requiring more awkward postures and uncomfortable gestures.

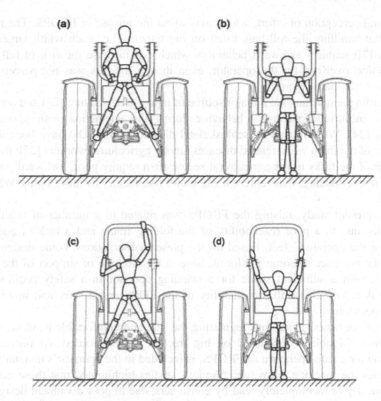

Fig. 2. Two typical placements and postures adopted by the operators to grasp the roll-bar in its lowered position and to move it to the upright position. Initial position: (a) standing on some parts of the machine, or (b) standing on the ground. Final position: (c) standing on some parts of the machine, or (d) standing on the ground.

It was also observed another incongruous movement, i.e. the torsion of the column during the pin locking and unlocking phases, sometimes accompanied by a slight forward bending of the operators, who carried out this activity by standing on the two lift arms of the tractors and not from the ground.

Despite the adoption of critical postures and gestures required by FROPS operation, the average rating of perceived effort reported by the participants was 3.00 (SD = 1.81), corresponding to a moderate effort.

4 Discussion

Raising a foldable ROPS is reported in the literature as a time-consuming and strenuous process. For these reasons, several farmers keep the roll-bar in a folded-down position, thus making it ineffective in preventing injuries during tractor overturn accidents [15]. In this study, an ergonomic analysis of the interaction between the operators and the foldable ROPS fitted on their own tractors was performed, to identify critical postures, gestures and perception of effort, which may affect the misuse of FROPS. The results showed that handling the roll-bars fitted on big tractors (i.e. track width larger than 1150 mm [7]) requires awkward behaviors which can increase the risk of fall or of biomechanical overload for the operator, even though the task was not perceived as strenuous by the participants.

Falls from the machine are a major source of injury in agriculture [23] and are often caused by incautious operator's behavior during the interaction with agricultural machinery [24]. Work-related musculoskeletal disorders (WMSDs) have been described as one of the main work-related diseases among agricultural workers [25]: the type and nature of the tasks in the agricultural sector often require poor, awkward postures and muscle overloading, that represent the major risk factors for developing WMSDs [26].

In the present study, raising the FROPS was related to a number of health and safety risks, due to a poor reachability of the folding frame and a lack of adequate support for the operators' feet. Based on the present observations, some design solutions, as the presence of some platforms, able to lift the base of support of the operators' feet, with a sufficient space for a standing operator in a safety position, are recommended, to increase the reachability of the ROPS and a safer and more comfortable operation.

A set of operators' guidelines regulating the safe and comfortable handling of the rear-mounted foldable ROPS fitted on big tractors is also needed. At present, the information for a safe operation of FROPS, is included in the operator's manual which accompanies the machine. However, previous studies highlighted that these manuals are infrequently or incompletely read by consumers, due to poor document design and excessive information content [27, 28]. The recommended grasping areas and the places to stand should be clearly identified and reported onto the manual, and simple drawings may be used to illustrate the correct gestures for a safe and comfortable handling of the rear-mounted foldable ROPS. Moreover, the safe and comfortable manual handling of FROPS may be specifically addressed during safety training courses and activities.

5 Conclusions

The ergonomic assessment showed that the majority of farmers experienced some criticalities related to the height of the grasping point of the folding frame of the FROPS and had some difficulties to complete the task without some kind of support, adopting uncomfortable and unsafe postures and gestures (leading for instance to the risk of pinching or falling from improvised places to stand). Specific design interventions, operators' guidelines and training activities may be developed to overcome these issues and to encourage the correct use of FROPS for safe tractor operations.

References

1. ILO. Agriculture; plantations; other rural sectors. http://www.ilo.org/global/industries-and-sectors/agriculture-plantations-other-rural-sectors/lang–en/index.htm. Accessed 15 Apr 2018
2. Kogler R, Quendler E, Boxberger J (2015) Occupational accidents with agricultural machinery in Austria. J Agromedicine 21(1):61–70
3. Abubakar MS, Ahmad D, Akande FB (2010) A review of farm tractor overturning accidents and safety. Pertanika J Sci Technol 18(2):377–385
4. Pessina D, Facchinetti D, Giordano DM (2016) Narrow-track agricultural tractors: a survey on the load of the hand-operated foldable rollbar. J Agric Saf Health 22(4):275–284
5. NIOSH. Preventing death and injury in tractor overturns with roll-over protective structures. https://blogs.cdc.gov/niosh-science-blog/2009/01/05/rops/. Accessed 02 May 2018
6. Cavallo E, Langle T, Bueno D, Tsukamoto S, Görücü S, Murphy D (2014) Rollover protective structure (ROPS) retrofitting on agricultural tractors: goals and approaches in different countries. J Agromedicine 19(2):208–209
7. OECD (2017) OECD standard codes for the official testing of agriculture and forestry tractors. Organisation for Economic Co-operation and Development, Paris
8. Cavallo E, Ferrari E, Coccia M (2015) Likely technological trajectories in agricultural tractors by analysing innovative attitudes of farmers. Int J Technol Policy Manag 15(2):158–177
9. Murphy DJ (2018) Rollover protection for farm tractor operators. http://extension.psu.edu/business/ag-safety/vehicles-and-machinery/tractor-safety/e42. Accessed 15 May 2018
10. Mangado J, Arana JI, Jarén C, Arazuri S, Arnal P (2007) Design calculations on roll-over protective structures for agricultural tractors. Biosys Eng 96(2):181–191
11. Pessina D, Facchinetti D (2011) Il ruolo del web nel monitoraggio degli incidenti mortali dovuti al ribaltamento dei trattori agricoli [Web monitoring of fatal accidents due to tractor overturns]. In: Convegno di Medio Termine dell'Associazione Italiana di Ingegneria Agraria, Belgirate, Italy, pp 1–7
12. NIOSH. Fatality assessment and control evaluation (face) program. http://www.cdc.gov/niosh/face/inhouse.html. Accessed 04 May 2018
13. Spielholz P, Sjostrom T, Clark RE, Adams DA (2006) A survey of tractors and rollover protective structures in Washington State. J Agric Saf Health 12(4):325–333
14. Hoy RM (2009) Farm tractor rollover protection: why simply getting rollover protective structures installed on all tractors is not sufficient. J Agric Saf Health 15(1):3–4
15. Khorsandi F, Ayers PD, Jackson D, Wilkerson J (2016) The effect of speed on foldable ROPS actuation forces. J Agric Saf Health 22(4):285–298
16. Kirwan B, Ainsworth LK (eds) (1992) A guide to task analysis. Taylor and Francis, London
17. MacLeod IS, Wells L, Lane K (2000) The practice of triangulation. In: McCabe PT, Hanson MA, Robertson SA (eds) Contemporary Ergonomics, 2000. Taylor & Francis, London, pp 244–248
18. ISTAT (2013) Atlante dell'agricoltura italiana [Atlas of Italian agriculture]. https://www.istat.it/it/files/2014/03/Atlante-dellagricoltura-italiana.-6%C2%B0-Censimento-generale-dellagricoltura.pdf
19. Creswell JW (2013) Research design: qualitative, quantitative, and mixed methods approaches, 4th edn. Sage Publications, Thousand Oaks
20. Borg G (1990) Psychophysical scaling with applications in physical work and the perception of exertion. Scand J Work Environ Health 16(1):55–58
21. ISO (2000) ISO 11226: Ergonomics - evaluation of static working postures. International Organisation for Standardisation, Geneva, Switzerland

22. Kroemer KH, Grandjean E (1997) Fitting the task to the human: a textbook of occupational ergonomics. Taylor & Francis, London
23. Bancej C, Arbuckle T (2000) Injuries in Ontario farm children: a population based study. Inj Prev 6(2):135–140
24. Caffaro F, Roccato M, Micheletti Cremasco M, Cavallo E (2017) Falls from agricultural machinery: risk factors related to work experience, worked hours, and operators' behavior. Hum Factors: J Hum Factors Ergon Soc 60(1):20–30
25. Walker-Bone K, Palmer KT (2002) Musculoskeletal disorders in farmers and farm workers. Occup Med 52(8):441–450
26. Naeini HS, Karuppiah K, Tamrin SB, Dalal K (2014) Ergonomics in agriculture: an approach in prevention of work-related musculoskeletal disorders (WMSDs). J Agric Environ Sci 3(2):33–51
27. Caffaro F, Cavallo E (2015) Comprehension of safety pictograms affixed to agricultural machinery: a survey of users. J Saf Res 55:151–158
28. Tebeaux E (2010) Improving tractor safety warnings: readability is missing. J Agric Saf Health 16(3):181–205

Theoretical Requirements and Real Benefit of Cold Protective Clothing for Order-Pickers in Deep Cold

Sandra Groos[✉] [iD], Mario Penzkofer[iD], Helmut Strasser,
and Karsten Kluth[iD]

Ergonomics Division, University of Siegen,
Paul-Bonatz-Str. 9-11, 57068 Siegen, Germany
groos@ergonomie.uni-siegen.de

Abstract. In a deep-cold working environment with variable physiological stress and considering individual thermoregulatory capacities, cold protective clothing must maintain the users' body core and skin surface temperature on an acceptable level. In order to evaluate the cold protective clothing used, physiological effects of order-picking in a cold store (-24 °C) were examined. 60 subjects (Ss) participated in the study, classified in 4 groups with 15 Ss each (younger/older males, younger/older females). The Ss had to work under predetermined, realistic working conditions with modified working phases with a duration of 80, 100, and 120 min, separated by identical warming-up breaks of 20 min. Additionally, 128 professional order-pickers employed in deep cold-storage depots had been systematically interviewed with regard to different topics of the cold protective clothing. The results show that the cold protective suit appears to protect sufficiently, whereby it should be noted that the insulation value is 2 times higher than required by an international standard. However, there is a need for improvements of the cold insulating boots and gloves to enable preventive occupational health and safety. In addition, a complete revision of the standards which are based on theoretical deliberations and assumptions is strongly recommended.

Keywords: Cold protective clothing · Working in deep cold · Field study

1 Introduction

The sales figures of frozen food have been rising steadily over the years [12, 22]. Due to this trend, this has also led to an increase in the number of jobs, such as order-pickers in cold-storage depots which are working at approximately -24 °C. As a cold environment may be a significant health risk, among other things cold protective clothing should be designed on the basis of the results of ergonomic measurements. However, the theoretical requirements of cold protective clothing which are grounded on national and international laws and standards are usually not based on scientific and practical studies, even though the large number of legal and normative requirements may create the impression of high accuracy.

© Springer Nature Switzerland AG 2019
S. Bagnara et al. (Eds.): IEA 2018, AISC 819, pp. 203–212, 2019.
https://doi.org/10.1007/978-3-319-96089-0_23

Nevertheless, there are still problems with the test methods and the selection of cold protective clothing, since a natural working environment cannot be completely and adequately simulated in a laboratory. However, field studies can be expected to deliver objectively determined cold-induced strain, e.g. the decrease of the skin or body core temperature from which statements can be derived regarding the real benefit of cold protective clothing. The results should help to recommend improvements of the garment with the aim of enabling preventive occupational health and safety and, ultimately, also to increase the efficiency of the work process.

1.1 Legal and Normative Requirements

Cold protective clothing must meet certain general requirements which are specified in the standard EN ISO 13688 [8]. Additionally, the cold-specific standard EN 342 [3] deals with the special demands of protective clothing for cold environmental conditions. It also serves as the basis for requirements and test methods for ensembles and garments against cold. The classification of the measured characteristics of cold protective clothing intends also to ensure an adequate level of protection in consideration of different conditions of use. The thermal insulation and the air permeability are the characteristics to be examined which must be specified on the label of the garment. The thermal insulation as the most important characteristic of cold protective clothing, is measured via using a thermal manikin whose size and form is equal to an adult person. However, the air permeability is of similar significance, as wind can substantially increase the convective heat losses [3]. The measurement of the thermal insulation by means of a thermal manikin, which has movable arms and legs to simulate air movements, is described in detail in EN ISO 15831 [9]. When selecting the materials for protective clothing, it should also be regarded that the water-vapor resistance must be low, particularly in conjunction with physically demanding work. The standard EN ISO 11092 [7] describes the measurement of thermal- and water-vapor resistance of textiles.

As noted by Wang et al. [25], cold protective clothing has to fulfill two essential requirements which can influence each other negatively: On the one hand, the clothing should provide an adequate isolation against cold. On the other hand, the garment must transfer the water vapor from the skin to the environment, which means, that a high water-vapor permeability of the clothing must be ensured. As EN 342 [3] points out, instead of the maximum isolation always the optimum isolation of the garment has to be selected. Since a very high isolation can lead to increased perspiration, the isolation of the garment will be increasingly affected by the moisture absorption. Therefore, the set of standards recommends determining the required isolation of the garment, which should ensure a thermal equilibrium of the human body, based on a specific level of physiological strain [6]. This value is known as $IREQ_{neutral}$, determined through the endogenous heat production (energy expenditure in Watts per m^2 body surface) and the ambient temperature (in °C). The resulting thermal insulation value is expressed through the unit "clo" (clothing) or in the common SI unit "m^2 K W^{-1}", whereby "1 clo" is equivalent to "0.155 m^2 K W^{-1}". According to the standard EN ISO 9920 [5], 1 clo represents a clothing combination consisting of underwear (short-sleeved and -legged), trousers, shirt, coat, socks and shoes.

1.2 Standardized Test Methods and Improper Selection of Cold Protective Clothing

To determine the minimum required isolation (IREQ$_{min}$) or the isolation which ensures a thermal equilibrium (IREQ$_{neutral}$) according to EN ISO 11079 [6], the level of physiological strain must be identified. EN ISO 8996 [4] delivers the generally valid values for the physical activity of "typical working tasks" which have to be determined in a laboratory under thermo-neutral conditions. The maximum value under cold conditions and during cold shivering is indicated here with 200 W/m². The German Standard DIN 33403-5 [2] specifies the absolute metabolic rate for manual palletizing and order-picking of "moderately heavy goods" and "heavy goods" to 360 W and 420 W, respectively. The energy expenditure measured under realistic working conditions in a cold store at −24 °C was 473/470 W for younger/older male Ss [22] and 377/387 W for younger/older female Ss [12]. These values related to the body surface result in 223/230 W/m² and 218/220 W/m² for the younger/older males and females, respectively. The values determined exceed clearly the 200 W/m² in the EN ISO 8996 [4]. As illustrated in Fig. 1, according to the EN ISO 11079 [6] the minimum required isolation (IREQ$_{min}$) is between 1.2 and 1.5 clo when considering the energy expenditure measured under realistic working conditions in the cold. The required isolation which ensures a thermal equilibrium (IREQ$_{neutral}$) with 1.4 to 1.8 clo is slightly higher. These values are significantly lower than the values of the used cold protective suit with 2.85 and 2.98 clo (see Fig. 4).

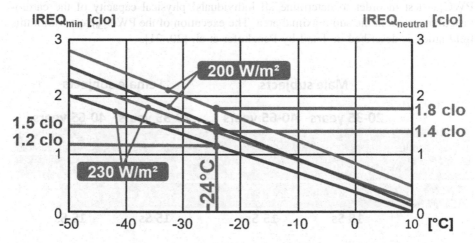

Fig. 1. IREQ$_{min}$ [clo] and IREQ$_{neutral}$ [clo] for a given metabolic rate of 200 and 230 W/m² at a surrounding temperature of −24 °C according to EN ISO 11079 [6].

Furthermore, the test procedure for the cold protective clothing may have faults since the specifications from EN ISO 15831 [9] do not consider the moisture from sweating, which to a great extent can affect the isolation of the garment. As noted by Fukazawa et al. [11] and Meinander and Hellsten [16] a sweating thermal manikin allows to adapt the test procedure to reality. This has already been taken into account by the American Society for Testing Materials (ASTM) which inter alia published the ASTM F 2370 [1]. Also the test procedure for cold protective shoes and boots is not feasible as Kuklane et al. [14] found out, that even a thin textile shoe and a sandal would pass the test described by EN ISO 20344 [10]. This, in turn, means that an open sandal can be certified as cold protective shoe. The explanations above indicate, that in particular the Standards from the present Comité Európen de Normalisation (CEN) and the International Organization for Standardization (ISO) are based more on theoretical calculations and assumptions rather than on scientific and practical studies.

2   Methods

2.1   Test Subjects

To comply with absolutely real working conditions, 60 subjects (Ss) were asked to carry out whole working day tasks in a cold store (at approximately −24 °C). The layout of a typical cold store is described in detail in [12, 22]. As visualized in Fig. 2, depending on age and gender, the Ss were separated in four different groups (20–35 or 40–65-year olds males and females), with 15 Ss each, representing one group. In addition to the described characteristics, the body mass index (BMI) was calculated and further behavioral attributes such as physical activity or the consumption of nicotine and alcohol were recorded. Furthermore, each subject had to carry out a standard PWC_{130} test in order to determine all individuals' physical capacity of the cardio-vascular system in the sub-maximal area. The execution of the PWC_{130} test within this field study is described in detail by Penzkofer et al. [20, 21].

	Male subjects		Female subjects	
	20-35 years	40-65 years	20-35 years	40-65 years
Age [Years]:	25.9 ± 3.5	55.4 ± 6.7	23.7 ± 3.4	53.3 ± 6.8
Size [m]:	1.83 ± 0.1	1.80 ± 0.1	1.69 ± 0.1	1.67 ± 0.1
Weight [kg]:	79.0 ± 12.4	89.8 ± 17.2	64.3 ± 8.1	68.9 ± 16.4
	15 Ss	15 Ss	15 Ss	15 Ss

Fig. 2. Characteristics (age, size and weight) of the 60 Ss, separated in four groups depending on age and gender with 15 Ss in each group.

2.2   Test Procedure

Test Performance and Work Physiological Measurements. The left part of Fig. 3 shows the test layout which was developed to enable the measurement of the physical strain while working in a cold store at −24 °C. First of all, the physical and mental conditions of the participants were registered. Afterwards, the Ss were "instrumented" with the measurement systems shown in the upper right part of Fig. 3. Utilizing these devices, the relevant physiological parameters "body core temperature" and "blood pressure" were recorded discretely every 15 min, and the "heart rate" and "skin surface temperature" were measured continuously. As illustrated in the lower right part of Fig. 3, the skin surface temperature was taken at various positions of the body. Through a mobile spirometry system, the energy expenditure was quantified for the identification of the intensity of the physical work and heat production. Afterwards the Ss had to work in the cold at −24 °C for three periods with a duration of 80, 100 and 120 min. The working phases were interrupted by warming-up phases of 20 min, each, at about +21 °C. A work load adapted to the real job with an average of 1.6 t/h of goods which had to be order-picked, should guarantee comparable results.

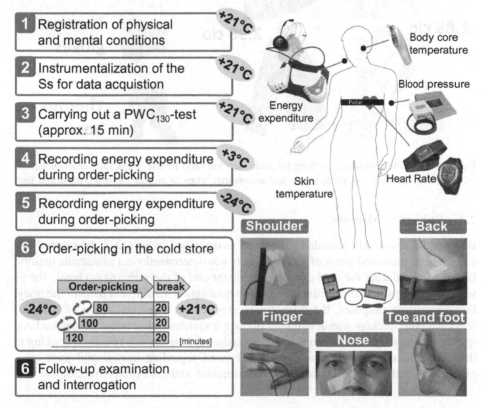

Fig. 3. Diagram of the test procedure (left part) and the measurement equipment for the determination of stress and strain of the Ss in the cold store at −24 °C (upper right part) including the measuring points for the recording of the skin temperature at various measuring points (lower right part).

Cold Protective Clothing. The cold protective clothing consisted of different layers of clothing and special accessories. Besides normal everyday wear (warm underwear, a pullover and a pair of trousers), the Ss were also protected with cold protective clothing. As illustrated in Fig. 4, the cold protective suit consisted of a thick jacket and long trousers with an isolation of 2.85 or 2.98 clo. Furthermore, a thick wool hat, thermo-gloves – normally made of fleece – and thermo-boots (with thermo-socks) were worn.

Fig. 4. Used cold protective clothing for male (left suit) and female (right suit) order-pickers at −24 °C with the isolation value in clo and accessories worn to protect the head, hand and feet.

2.3 Subjective Methods

Besides the objectively recorded parameters as described in Sect. 2.2, the subjectively experienced stress and strain of all 60 subjects was determined via a systematic inquiry. Every 15 min during the cold exposure and at the end of the warming-up break, the Ss were asked in particular regarding their cold sensations at 12 different parts of the body. For comparison purposes, the subjectively experienced stress and strain of 128 professional order-pickers was determined through a systematic interview. A standardized questionnaire comprised 57 items referring to the subjectively felt effects of working in the cold. The questionnaire, amongst other items regarding muscle and joint complaints, covered cold sensations, frostbite-symptoms and cold protective clothing.

3 Results

3.1 Work Physiological Parameters

While working in the cold store at a surrounding temperature of −24 °C, the core temperature decrease, depending on age and gender, differed between 1.3K (younger females) and 2.1K (older males) compared to the temperature at the outset taken in the morning. In individual cases decreases of up to 3K (older males) and 2.3K (older females) could be determined which means, that the body core temperature dropped to an intolerable value of approximately 34 °C.

As to be expected, the temperature sensors in the area of the shoulder and kidney recorded inconspicuous values. Due to the thick cold protective clothing, the skin surface temperature decreased only slightly, and the value of the outset was reached after a very short time during the warming-up break. The skin surface temperature of the unprotected nose dropped to 12–15 °C on average with slightly lower values for the female Ss. The low temperature of the nose could be compensated very quickly during the warming-up breaks, supported by a more intensive blood circulation of the face after leaving the cold store. The skin temperature of the fingertip, with average temperatures of 15° to 20 °C, exhibits a similar trend as the nose. Individual declines of values below 5 °C could be measured which resulted in severe pain in one case of a younger female, persisting even after the warming-up break. The skin temperature profile of the toe showed a somewhat smaller but continuous decrease with values between 20 °C and 30 °C. The temperature at the outset was mostly not reached after a 20 min warming-up break. The sensor under the sole of the foot measured nearly constant temperatures throughout the entire working day.

3.2 Subjective Assessments

In order to identify the subjective experiences of the cold exposures, the 60 Ss were also interviewed regarding the cold sensations. The majority of the Ss experienced cold sensations at the nose, the mouth, the fingers and the toes. The ratings of cold sensations on the 4-step scale from "0" to "4" (unbearably cold) are between "chilly" (1.0) and "very cold" (3.0). Especially the toes were still rated as "chilly" after the 20 min warming-up breaks.

The 128 professional order-pickers employed in 24 deep cold-storage depots, among other things, evaluated their subjective experiences regarding the cold protective clothing in the topics "degree of restraint", "comfort", "cold protection" and "perspiration". The "degree of restraint" of the cold protective suit and the accessories described in Fig. 4 was rated between −1 and −1.6 on the 4-step scale from "0" (no restraint) to "−4" (can hardly fulfill the task). Nearly the same values were registered for the "comfort", which means the order-pickers experience their cold-protective clothing as almost comfortable. With exception of the thermo-boots (−0.7) and thermo-gloves (−0.4), the cold protection was rated as nearly perfect. Clearly worse was the rating about the "perspiration" which could be rated from "0" (no perspiration) to "−4" (extremely elevated perspiration). All garments were assessed with values between −1.5 to −1.9. Overall, the cold protective clothing used in the cold store was rated with +2.7 on a bipolar scale from "−4" (very bad) to "+4" (very good).

The work physiological results of the field study and the results of the interviews of the 128 professional order-pickers are described in detail by Groos [12] and Penzkofer [22].

4 Discussion

The results from the field study show that order-picking in a deep cold environment results in elevated stress and strain for the human organism due to the superposition of the extremely low temperatures and the high physical workload. But as noted by Strasser and Kluth [24], order-picking itself is a benefit rather than an additional stress since the numerous changes in the body posture and movements of the arms and legs improve the blood circulation, and associated heat production and transfer to the fingers and the toes is increased. The cold protective suit (comprising a thermal jacket and thermal pants) appears to protect sufficiently, even though it was not possible to keep the body core temperature on a constant level. Nevertheless, it must be noted that the isolation value is nearly two times higher as the IREQ$_{neutral}$ required by the international standard EN ISO 11079 [6]. In contrast to the protective suit there is a need for improvements of the cold-insulating boots and gloves as the skin surface temperature of the fingers and the toes while working in the cold showed a substantial decrease. Also the subjective assessments of cold sensations during the field study and the evaluated subjective experiences of 128 professional order-pickers showed not satisfying results of the boots and gloves.

In order to promote occupational health, it is mandatory to take care of the workers' needs and high quality of the garments when selecting cold protective clothing. The responsible persons should know about the weaknesses of the laws and standards for cold protective clothing and the risks of inadequate cold protective clothing. Yet, a deviation from the comfortable thermic zone may result in decreased physical and mental performance [17, 19]. In addition to the negative impacts on the physical and mental performance, working at −24 °C with inadequate cold protective clothing can also have an impact on an individual's health. Besides an increased muscular strain and other musculoskeletal complaints and symptoms [18, 23], everyday cold exposure may aggravate the symptoms of prevailing chronic diseases [15]. Also Hassi et al. [13] mentioned that cardiovascular disease, respiratory symptoms, musculoskeletal diseases, peripheral circulation problems, and skin diseases are associated with cold exposures.

In summary, it has to be emphasized that order-picking in deep cold leads to high physical stress and strain which requires adaptations and improvements on both individual as well as operational levels. Besides a mandatory optimization of the length of working and break times, there is a need for improvements of the cold-insulating boots and gloves. These adjustments may not only create preventive occupational health and safety but, ultimately, should increase the efficiency of the work process for every individual and support the working ability of employees in the long term.

5 Conclusion

- In order to evaluate the cold protective clothing used in a cold store at −24 °C, ergonomic measurements and subjective assessments were carried out.
- While working in the cold store, the body core temperature, in comparison to the value at the outset, decreased up to 3K.

- The skin surface temperature of the finger, toe and nose decreased significantly during the cold exposure, and reached values which may result in pain sensations in the short term and possible health risks in the long run.
- The results of the subjective assessments confirmed the objectively measured values.
- A complete warming-up during the breaks of 20 min at about +21 °C, was not possible for all individuals.
- Based on the results there is a need for improvements of the cold insulating boots and gloves to ensure preventive occupational health and safety.
- A complete revision of the theoretical requirements based on national and international laws and standards is strongly recommended.

Acknowledgement. This research project was financially supported by the "German Research Foundation" (Grant no. STR 392/5-1 and KL 2067/1-2).

References

1. ASTM F 2370 (2016) Standard test method for measuring the evaporative resistance of clothing using a sweating manikin. Beuth Verlag, Berlin
2. DIN 33403-5 (1997) Climate at workplaces and their environments – part 5: ergonomic design of cold workplaces. Beuth Verlag, Berlin (in German)
3. EN 342 (2017) Protective clothing – ensembles and garments for protection against cold. Beuth Verlag, Berlin
4. EN ISO 8996 (2004) Ergonomics of the thermal environment – determination of metabolic rate. Beuth Verlag, Berlin
5. EN ISO 9920 (2009) Ergonomics of the thermal environment – estimation of thermal insulation and water vapor resistance of a clothing ensemble. Beuth Verlag, Berlin
6. EN ISO 11079 (2008) Ergonomics of the thermal environment – determination and interpretation of cold stress when using required clothing insulation (IREQ) and local cooling effects. Beuth Verlag, Berlin
7. EN ISO 11092 (2014) Textiles – physiological effects – measurement of thermal and water-vapor resistance under steady-state conditions (sweating guarded-hotplate test). Beuth Verlag, Berlin
8. EN ISO 13688 (2013) Protective clothing – general requirements. Beuth Verlag, Berlin
9. EN ISO 15831 (2004) Clothing – physiological effects – measurement of thermal insulation by means of a thermal manikin. Beuth Verlag, Berlin
10. EN ISO 20344 (2013): Personal protective equipment – test methods for footwear. Beuth Verlag, Berlin
11. Fukazawa T, Lee G, Matsuoka T, Kano K, Tochihara Y (2004) Heat and water vapor transfer of protective clothing systems in a cold environment, measured with a newly developed sweating thermal manikin. Eur J Appl Physiol 92(6):645–648
12. Groos S (2018) Alters- und geschlechtsdifferenzierte Objektivierung von Belastung und Beanspruchung bei berufsbedingten Kälteexpositionen unter Berücksichtigung eines variablen Arbeitszeit-Pausenzeit-Regimes. Doctoral thesis, University of Siegen (in print)
13. Hassi J, Rytkönen M, Kotaniemi J, Rintamäki H (2005) Impacts of cold climate on human heat balance, performance and health in circumpolar areas. Int J Circumpolar Health 64 (5):459–467

14. Kuklane K, Ueno S, Sawada S-I, Holmér I (2009) Testing cold protection according to EN ISO 20344: is there any professional footwear that does not pass? Ann Occup Hyg 53 (1):63–68

15. Mäkinen TM (2007) Human cold exposure, adaptation, and performance in high latitude environments. Am J Hum Biol 19(2):155–164

16. Meinander H, Hellsten M (2004) The influence of sweating on the heat transmission properties of cold protective clothing studied with a sweating thermal manikin. Int J Occup Saf Ergon 10(3):263–269

17. Oksa J (1998) Cooling and neuromuscular performance in man. PhD thesis, University of Jyväksylä

18. Oksa J, Ducharme MB, Rintamäki H (2002) Combined effect of repetitive work and cold on muscle function and fatigue. J Appl Physiol 92(1):354–361

19. Palinkas LA (2001) Mental and cognitive performance in the cold. Int J Circumpolar Health 60(3):430–439

20. Penzkofer M, Kluth K, Strasser, H (2008) Physiological responses of male subjects to cold exposures at +3 °C and −24 °C during warehouse commissioning work. In: Proceedings of the 2nd international conference on applied human factors and ergonomics, Las Vegas, Nevada, USA, 8 pp

21. Penzkofer M, Kluth K, Strasser H (2008/2009) Heart rate and work pulses of two age groups associated with working in the cold at +3 °C and −24 °C. Occup Ergon 8(4):135–145

22. Penzkofer M (2013) Feldstudien zur Objektivierung von Belastung und Beanspruchung jüngerer und älterer Arbeitspersonen bei berufsbedingten Kälteexpositionen. Ergonomia Verlag, Stuttgart

23. Sormunen E, Oksa J, Pienimäki T, Rissanen S, Rintamäki H (2006) Muscular and cold strain of female workers in meat-packing work. Int J Ind Ergon 36(8):713–720

24. Strasser H, Kluth K (2006) Sensations of cold and physiological responses to groceries handling in cold-storage depots. In: Pikaar RN, Koningsveld EAP, Settels PJM (eds) Proceedings of the xvith triennial congress of the international ergonomics association. Proceedings-CD, Maastricht, The Netherlands, 6 p

25. Wang SX, Li Y, Tokura H, Hu JY, Han YX, Kwok YL, Au RW (2007) Effect of moisture management on functional performance of cold protective clothing. Text Res J 77(12):968–980

Readiness to Change: Perceptions of Safety Culture up and down the Supply Chain

Shelley Stiles[1,2(✉)], Brendan Ryan[1], and David Golightly[1]

[1] Human Factors Research Group, University of Nottingham, University Park,
Nottingham NG7 2RD, England
Shelley.stiles@2020she.com
[2] 2020 SHE Solutions Ltd., 58-60 Wetmore Road,
Burton on Trent DE14 1SN, England

Abstract. Safety culture research tends to treat organisations as a single body, with less focus on understanding how perceptions vary in a multi stakeholder environment. One such example of a multi-stakeholder environment is a construction project. The success of safety interventions must be sensitive to the interfaces and relationships, and different perceptions, between Principal Contractors and their Supply Chain, particularly for Small to Medium Enterprises (SMEs) that may have fundamentally different safety management systems and culture. This paper explores whether there is a difference in perception between project members. It tests whether perceptions are driven by a perceived hierarchy of greater maturity for Principal Contractors or whether different organisational layers of the project rate themselves highly in comparison to others, as a form of self enhancement. 17 workshops were undertaken across four different Principal Contractors, and their respective Supply Chains, comprising a total of 367 participants (Principal contractor n = 114; supply chain n = 253). Participants were asked to rate the safety culture maturity of their organisation and the safety culture maturity of the other group using Hudson's safety culture maturity model. The results identified a significant difference in the perceived safety culture maturity of the Principal Contractor and Supply Chain, with Principal Contractors perceived by all parties as more mature than the supply chain. This suggests that the power structure across a project has more of an effect on perceptions than self-enhancement by any organizational type. The divergent power relationship between Principal Contractor and their Supply Chain may influence the reported levels of safety culture maturity for the project as a whole, and has a bearing on how safety culture interventions should be delivered to effect change.

Keywords: Safety · Culture · Supply chain · Construction

1 Safety in the Construction Industry

The UK Construction Industry has a workforce of over two million people working across 170-200,000 firms [1] with one of the highest accident rates of all industries in the UK [2]. In the 2014/2015 period there were 35 fatal injuries. 25% of all UK fatal injuries were in the Construction Industry.

© Springer Nature Switzerland AG 2019
S. Bagnara et al. (Eds.): IEA 2018, AISC 819, pp. 213–223, 2019.
https://doi.org/10.1007/978-3-319-96089-0_24

This paper focuses on the civil engineering sector which covers general construction for civil engineering works, including road and railway construction, and utility projects. Whilst there has been an overall downward trend in the rate of all workplace injury in the Construction sector since 2001, there is recognition across industry that safety performance has reached a plateau and therefore there is a need to change and improve safety [3].

The common measure of safety is derived from lagging indicators such as total days lost (absence), accident incidence and frequency rates, although more recently techniques focused on leading indicators are being developed within the sector [4]. Leading indicators are more focused around safety management arrangements and their effectiveness at improving safety, and can include the numbers of training days undertaken, senior management safety tours and workforce engagement sessions. Since Ladbroke Grove and the Cullen Report [5], high risk industries, including construction, have been encouraged to reduce their reliance on lagging indicators and consider leading indicators to measure their safety performance [6]. One aspect of this is how the importance of understanding safety culture has become more prevalent in the civil engineering sector of the Construction Industry.

1.1 Project Delivery Organisations

The Construction Industry's safety management arrangements in the UKare governed by the Construction Design and Management (CDM) Regulations 2015. These regulations outline interfaces between organisations for project delivery and the management of safety. By definition a project is *'a temporary endeavour with a defined beginning and end undertaken to meet unique goals and objectives'* [7].

The Client is *'the person or company, with the controlling interest in the project'* [8]. The Client is usually the funder of the project and specifies the project remit. Even though the CDM Regulations 2015 do not expect that Clients have detailed safety knowledge, their impact is through the selection and appointment of competent organisations to deliver the project (the Project Delivery Organisation), and the allowance of adequate time and resources to deliver the project safely [9]. There have been limited studies undertaken regarding the role of Client and their impact on safety performance on construction projects. The Client has been excluded from this study as there is usually little or no Client involvement in project-based safety improvement programmes.

A Project Delivery Organisation (referred to as 'PDO' for the remainder of this paper) is established with a number of companies co-ordinated via contractual obligations, for a determined period of time. A PDO is made up of numerous organisations, and typically comprises a main Principal Contractor to manage and deliver the work. The rest of the PDO is then comprised of a supply chain to perform general or specialist work, the majority of whom are small to medium sized enterprises (SMEs), and account for 80% of the cost of a construction project [10]. The industry's Supply Chain is made up of 98% SMEs with less than 60 employees [11], which is common within the civil engineering sector [12]. When these aspects are combined with the lack of standardisation renown on construction projects, the challenge of developing a mature safety culture becomes significant [13].

This is an important characteristic of the industry as research has shown that SMEs typically have less adequate control and arrangements for safety, which is exacerbated in the Construction Industry [14, 15]. The management of safety issues in the Construction Industry is influenced by the size of organisations involved, with the safety performance of SME's working on larger projects aligned to the prevailing culture under that Principal Contractor [16]. Vickers et al. [17] state two areas of concern; price competition amongst contractors gives advantage to those less diligent in safety, and lengthy subcontractor chains dilutes responsibilities of the Principal Contractor. Therefore, the contribution and significance of Supply Chain companies to overall project safety culture is worthy of further exploration.

1.2 Safety Culture and the Impact of Contractorisation

A safety culture is described as the values and beliefs of the organisation which govern how people should work and behave [18] or *'how safety is placed as a priority within an organisation'* [19].

Westrum [20] identified three levels of safety culture maturity, which was enhanced by Reason [21] to five levels. A study by Parker et al. [22] identifies these levels as:

- Level 1 Pathological: Who cares about safety as long as we are not caught?
- Level 2 Reactive: Safety is important: we do a lot after an accident.
- Level 3 Calculative: We have systems in place to manage all hazards.
- Level 4 Proactive: We try to anticipate safety problems before they arise.
- Level 5 Generative: HSE is how we do business around here.

Each of the five levels include 18 different organisational safety management arrangements; including management interest and communication with the workforce, safety management system, incident reporting, balance between safety and productivity, contractor management, competence and training, trends and performance statistics. It can be concluded that a combination of safety arrangements, and how effectively these are deployed, provides information on the effectiveness of an organisation's safety arrangements and culture; collectively assessed as its safety performance [22, 23].

As it has already been identified that a PDO has numerous companies brought together under contract to deliver a defined product/service for a client, the project safety culture would be derived from a combination of each respective company's set of values and beliefs. According to literature, safety culture is established at company level. Therefore, the safety culture of each member of the PDO makes a contribution (either positive or negative) to the project's safety culture [24]. Each member company may have varying levels of impact dependent on the size of their organisation, prevailing safety culture, the level of significance given to a project by their company in proportion to their overall turnover and workload, and their respective position in the project social structure. Therefore, the project safety culture would be derived from a combination of each respective company's culture.

The power relationship between Principal Contractors and their Supply Chain can also be a factor [25–28]. The Supply Chain are beholden to the Principal Contractor for awarding them contracts, hence the power of the relationship lies with the Principal

216 S. Stiles et al.

Contractor. Such studies have shown that this is often accompanied by an adversarial attitude from the Principal Contractor. The significance of these aspects for improving project safety culture are not well understood.

1.3 Perception of Safety Culture

Organisations within the Construction Industry recognise the need to improve safety performance [3]. Organisational change is dependent on a number of factors, which includes the commitment to and perceived need for change [30, 31]. Bandura [32] identified the importance of self-efficacy for achieving specific goals, which was developed further by Weiner [31] in relation to change efficacy. According to Herscovitch and Meyer [33] organisational commitment to change can be in three forms; because they want to, they have to or they ought to. The first of these is most powerful level of commitment. Kotter [34] identifies that many change programme fail due to the lack of obtaining such commitment, amongst several other considerations. If an organisation perceives that their safety culture is poor they may be more likely to be motivated to make improvement. An organisation that perceives that they have an acceptable or good culture may be less likely to focus on making improvement.

Readiness to change involves all parties being prepared to take action and be committed to the change [32]. For a PDO, each member organisation will have variable motivation and degrees of commitment to improve safety (want to, have to or ought to). If the whole project safety culture is to improve, each member organisation will need to be motivated and committed to change. However, within a PDO there are potentially many different organisations, each with their own perception of prevailing safety culture maturity. It could be argued that a lack of consistency amongst PDO members could inhibit the change efficacy for the PDO as a whole [31].

Therefore, how an organization perceives its safety culture is important to achieving change. However, the complexity of a PDO is that it is unclear whether safety culture is perceived uniformly across the PDO, or diverges because of organizational type – Principal contractor or Supply Chain.

There are two potential and potentially competing mechanisms that would influence perceptions of safety culture in a multi-stakeholder environment configuration like a PDO; self-enhancement and organizational size/type.

Literature has shown that when individuals rate themselves they have a tendency to rate their own safety culture and performance higher than other parties; this is referred to as self-enhancement [35]. Self-enhancement is particularly relevant in the context of safety, as it makes people feel good about themselves, giving a further level of complexity for consideration when initiating a change programme to improve safety culture. If self enhancement is the prevailing mechanism, then all parties are likely to give themselves positive and preferential ratings of their own safety culture in comparison to others.

The alternative mechanism is that organizational type influences perception, in that larger organisations (typically the Principal Contractor) and the power relation of the Principal Contractor driving safety culture, shapes perception. Walters and James [14] showed that there was a difference in safety culture and performance for larger organisations in comparison with smaller organisations, with larger organisations being more mature. The Supply Chain, mostly made up of SMEs, are typically found to have less financial security which impacts on money allocated to safety (e.g. resources, training, investment in safety equipment, with a lack of full time internal health and safety personnel, resulting in a reduced awareness of safety requirements, plus informal safety arrangements rather than formal safety management systems; which should direct the management of safety throughout day to day operations [14, 26, 35–37]). In relation to a construction project it could be expected that the Principal Contractor's safety culture would be more mature than their supply chain as they tend to be the larger organisations. Thus, the Principal Contractor is perceived by all parties as having a superior safety culture.

2 Study Aim

This study seeks to determine the perceived level of safety culture maturity across a sample PDO's within the Construction Industry (civil engineering sector), considering:

- How do Principal Contractors and the Supply Chain rate themselves and others?
- Are there differences in these ratings?
- If so, is there evidence to support self-enhancement, where project members score themselves higher than they score other parties, or a power relationship where all parties perceive the Principal Contractor to have higher perceived safety maturity?

3 Method

3.1 Participants

To evaluate the relevant organisation's perceptions of safety culture maturity, a series of 17 workshops were undertaken across four different Principal Contractors, and their respective Supply Chains. Supply Chain members were engaged in this study through their existing communication arrangements with the Principal Contractors. Individuals were invited to participate as part of a wider behavioural safety training course that was being undertaken on the same day. All participants were managers who worked within PDOs as part of the day to day activities. The participants comprised a total of 367 participants including 114 Principal Contractor representatives and 253 Supply Chain representatives.

Workshops were delivered to the Principal Contractor representatives only as a group, then repeated for the Supply Chain of the Principal Contractor. Separate workshops were held for the Principal Contractor and Supply Chain to minimise peer pressure. However, each Principal Contractor had requested and arranged for the supply chain to attend the workshops. There were a maximum number of 25 participants in each workshop with a mean of 21 participants per workshop.

3.2   Procedure

The procedure outlined below was repeated for all of the workshops for both the Principal Contractor and Supply Chain participants.

For each Principal Contractor, workshops on assessing safety culture maturity were conducted based on the Hudson Model [22]. At each workshop participants received a brief tutorial on safety culture maturity and were provided written guidance on the characteristics associated with the different maturity levels; level 1 pathological, level 2 reactive, level 3 calculative, level 4 proactive, level 5 generative.

Every participant was asked to rate safety culture maturity. Ratings were against the five levels of maturity on the model: 1 to 5. Each participant provided two ratings;

- One for the maturity level of Principal Contractor
- One for the Supply Chain as a whole.

The ratings were recorded by individuals on paper and then collated. Confidentiality and anonymity of ratings was maintained during this process.

For each workshop the rankings for both Principal Contractor and Supply Chain were plotted anonymously on a flip chart and shared with the group. The summary results were shown on the flip chart. These results were discussed as a group to confirm the data listed on the flip chart was in-line with how participants perceived safety culture maturity of the Principal Contractor and Supply Chain.

3.3   Analysis

Following each workshop, the paper records were transferred into an excel spreadsheet which identified rater group (independent variable; Principal Contractor (PC) or Supply Chain (SC), who the score was for (independent variable) and what the score was (dependent variable). All of the rankings collected were interval data, based on the Hudson Model 1–5 maturity levels. Regarding the data collected from each participant, it is important to note the following:

- Each data point for a Principal Contractor represents a different individual from one of four Principal Contractor organisations, whereas
- Each data point for a Supply Chain represents a different organisation

Once all of the workshops had been completed, the data were collated and analysed via independent t-tests using SPSS.

4   Results

4.1   Overall Ratings of Maturity

Across all ratings given (N367) the mean perceived safety culture maturity of the Principal Contractor was 3.29 (st Dev 0.732) and the mean perceived safety culture of the Supply Chain was 2.80 (st Dev 0.817). The difference between these means was found to be significant ($t(734) = p < 0.05$).

4.2   How Do Principal Contractors and Supply Chain Rate the Maturity of Themselves and Others?

Figure 1 illustrates the differences in perception between the Principal Contractor (PC) and Supply Chain (SC), whether they were giving a score for themselves or the other group. There was no significant difference between the ratings of the Principal contractor provided by the Principal Contractor and those provided by the Supply Chain (PC = 3.28; SC = 3.30, t(365) p = −0.268). There was a significant difference between the ratings of the Supply Chain provided by the Principal Contractor and those provided by the Supply Chain (PC = 2.27; SC = 3.03), t(365) = −9.133 p = < 0.00).

The Principal Contractor perceived the Supply Chain cultural maturity to be significantly lower than their own maturity level (mean score given for PC = 3.28; SC = 2.27; t(226) = 12.48 p = < 0.00). The Supply Chain perceived the Principal Contractor cultural maturity to be significantly higher than their own maturity level (means PC = 3.30; SC = 3.03); t(504) = 3.803 p = < 0.00).

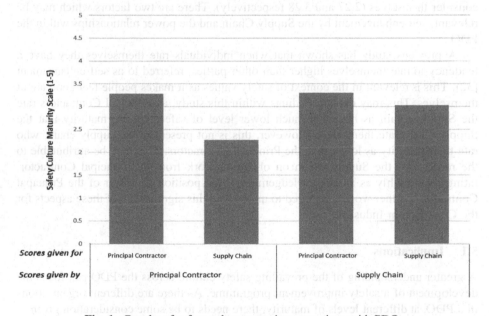

Fig. 1. Results of safety culture maturity perceptions with PDO.

5   Discussion

This study has evaluated perceptions of the safety culture maturity of both Principal Contractors and the Supply Chain to establish whether or not there is a difference in perceptions of cultural maturity level for a given PDO.

This study has found significant differences in perceived safety culture maturity between Principal Contractors and their Supply Chain. Across all participants the perceived safety culture maturity of Principal Contractors was found to be higher than that of the supply chain. This is supportive of a study by Walters and James [14] where larger organisations were found to have a more mature safety culture. Organisational, size affects the awareness of safety requirements and there is more access to specialist competent safety resource and extensive safety management systems [14, 26, 35–37], which has an impact on safety culture maturity. As a large number of the Supply Chain are SMEs, it is often found that SMEs have fewer safety arrangements due to lack of commitment, awareness and resources [26, 35, 38], and therefore may have a less mature safety culture. Therefore, as a Principal Contractor looking to implement safety improvements, due consideration should be given to the organisational size of the members of the PDO, with particular attention given to the Supply Chain. This in turn may influence the selection of interventions which the Principal Contractors applies on the project.

This study also found that there were differences in how the Supply Chain rate themselves and how they were rated by the Principal Contractor, with the Principal Contractor perceiving the Supply Chain maturity to be lower than the Supply Chain consider themselves (2.27 and 3.28 respectively). There are two factors which may be relevant; self-enhancement by the Supply Chain and the power relationships within the PDO.

A previous study has shown that when individuals rate themselves they have a tendency to rate themselves higher than other parties; referred to as self-enhancement [32]. This is relevant in the context of safety values as it makes people feel good about themselves. This may explain findings within this study, as Principal Contractors rate the Supply Chain as having a much lower level of safety culture maturity that the Supply Chain rate themselves. However, this is not present in the supply chain, who rate their maturity as lower than the Principal Contractors. This may be attributable to the reliance of the Supply Chain on obtaining work from the Principal Contractor, rating them highly as an acknowledgement of the position of power of the Principal Contractor. Further work is required to understand this significance of these aspects for the Construction Industry.

5.1 Implications

A greater understanding of the prevailing safety culture across the PDO can focus the development of a safety improvement programme. As there are different organisations in a PDO, at different levels of maturity, there needs to be some consideration given to how safety improvements are implemented. This is particularly relevant as there is a substantial range of ratings given for the Supply Chain (PC 3.29 (st Dev 0.732), SC 2.80 (st Dev 0.817). Any programme to improve safety may have to be developed, bearing in mind that the Principal Contractor and the Supply Chain may be starting from different levels of maturity.

Organisations also need to have sufficient change efficacy for such a change programme to be effective [32]. For each PDO change efficacy will be dependent on the organisation's motivation to do something differently. How organisations rate themselves may impact on their commitment to any change programmes.

An accurate understanding of current safety culture maturity within the PDO and recognition of the need to improve will facilitate a readiness for change which would otherwise be lacking and hinder the commitment to and future embedment of improvements in safety culture [30, 34]. Therefore, PDO's change efficacy could be limited by the Supply Chains perception of their own safety culture i.e. they think they are better than the Principal Contractors consider them to be, and resistance to change for improvement programmes may be higher.

It is advisable that wider assessments of safety culture maturity across project members are undertaken to establish a clearer baseline of the prevailing safety culture. This could include ratings provided by the Client. Further studies could explore whether there were specific aspects of safety culture that were perceived to be more or less mature; for example, leadership, communications, training, openness and approach to learning etc. as detailed within the 18 steps of the maturity model developed by Parker et al. [22]. A more in-depth understanding would provide further information on which to focus a safety improvement programme.

5.2   Potential Study Limitations

Regarding potential study limitations, the main consideration is the collection of data points. For each Principal Contractor, the number of data points is provided by a number of different people representing the Principal Contractor organisation. In contrast the data points provided by the Supply Chain are from one individual per supplier/subcontractor organisation. Therefore, the results for the Principal Contractor are based on a number of ratings for four Principal Contractor organisations and may not be truly representative of the whole industry. However, this method was chosen as it is indicative of the organisational structure of a PDO and the Construction Industry (fewer larger organisations, many smaller organisations). It is recognised that there will always be intercompany differences which may not have been addressed through the purposive sampling technique undertaken within this study, although this is addressed to some extent with the large sample size (N367).

The safety culture maturity model that has been used within the study has ratings against 5 levels. There may be a central tendency evident in the results, as the mean maturity level for all PDOs was 3.04. However, the standard deviation was 0.814, indicating that there was a reasonable variance of responses amongst a high number of participants (N367).

A further potential limitation is the study's reliance on perceptions of safety culture without further evaluation of safety culture using evidence of safety management arrangements with safety performance data. There is recognition that evaluating interventions aimed at improving safety culture are difficult to carry out [38]. It is recommended that future studies could use methods of data triangulation to address this; where evaluation of safety culture will combine multiple methods to gather data.

6 Conclusion

This study explores perceived safety culture maturity of members of PDO's within the civil engineering sectors of the Construction Industry. There is a perceived difference in safety culture maturity of organisations within a construction project. Principal Contractors are perceived to have a more mature safety culture than their Supply Chain, which is also supported by the Supply Chain. Self-enhancement may be evident within the results, but only for the Principal Contractors.

The findings are important for those who are working to improve safety performance within this sector of industry, as this study identifies typical safety culture maturity in relation to a PDO and recognises that Principal Contractors are perceived to be more mature that the Supply Chain. Therefore, the design of specific interventions within a safety improvement programme should be cognisant of the maturity level of all organisations working within the PDO to be most effective.

This study has raised a number of points worthy of further exploration in future studies. The difference in perception amongst key organisations (excluding the client) within the PDO poses a challenge to the current safety culture maturity models; in particular whether they provide sufficient scope to assess the safety culture maturity of member organisations within a project with the additional complex relationships this involves. A greater understanding of change efficacy would also be beneficial where one member of the PDO scores themselves high, and another low, to determine the impact this may have on implementing a successful safety improvement programme.

References

1. DTI (2006) Construction statistics annual report 2006. The Stationary Office, London
2. HSE. HSE construction intelligence report - analysis of construction injury and ill health intelligence. Accessed 2009
3. HSE. http://www.hse.gov.uk/statistics/industry/construction/construction.pdf. Accessed 2015
4. DTI (2010) Key performance indicators and benchmarking: construction industry KPI's. The Stationary Office, London, Chap 16
5. Cullen Report (2000) The ladbroke grove rail inquiry. HSE Books, London
6. Flin R, Mearns K, O'Connor P, Bryden R (2000) Measuring safety climate: identifying the common features. Saf Sci 34:177–192
7. Oxford Brookes. http://www.brookes.ac.uk/services/hr/project/pm_at_brookes/definition.html. Accessed 2011
8. Health and Safety Authority: Clients in construction: best practice guidance. Health and Safety Authority, Metropolitan Building, James Joyce Street, Dublin http://www.hsa.ie/eng/Publications_and_Forms/Publications/Construction/Clients_in_Construction_Best_Practice_Guidance.pdf. Accessed 2009
9. Bednarz J. One of our health & safety manager's Speller Metcalfe. http://www.spellermetcalfe.com/blog/2015/health-safety-and-clients-responsibility. Accessed 2016
10. Constructing Excellence (2004) Supply chain management, constructing excellence
11. DTI (2011 July) Structure of the construction industry. The Stationary Office, London, Chap 3

12. Office for National Statistics. http://www.ons.gov.uk/businessindustryandtrade/cons tructionindustry/datasets/constructionstatisticsannualtables. Accessed 2014
13. Duff AR, Robertson IT, Cooper MD, Phillips RA (1993) Improving safety on construction sites by changing personnel behaviours, Report number 51
14. Walters D, James P (2009) Understanding the role of supply chains in influencing health and safety at work IOSH, Leicester
15. Winkler C, Irwin JN (2003) Contractorisation – aspects of health and safety in the supply chain. HMSO, London
16. HSE: Construction sector strategy 2012–15 http://www.hse.gov.uk/aboutus/strateg iesandplans/sector-strategies/construction.htm. Accessed 2012
17. Vickers I, Baldock R, Smallbone D, James P, Ekanem I (2003) Cultural influences on health and safety attitudes and behaviour in small businesses, HSE. HMSO, London
18. Schein EH (1992) Organisational culture and leadership, 2nd edn. Jossey-Bass, San Francisco
19. Farrington-Darby T, Pickup L, Wilson J (1997) Safety culture in railway maintenance
20. Westrum R (1993) Culture with requisite imagination cited. In: Wise J, Stager P, Hopkins J (eds) Verification and validation in complex man-machine systems. Springer, New York
21. Reason J (1997) Managing the risks of organisational accidents. Ashgate, Aldershot
22. Parker D, Lawrie M, Hudson P (2006) A framework for understanding the development of organisational safety culture. Saf Sci 44:551–562
23. Fleming M (2001) Safety culture maturity model. HMSO, London, pp 5–6
24. Ekvall, G.: Climate, structure and innovativeness of organisations. In: Working paper for Swedish council for management and organisational behaviour (1983)
25. Cox A (2002) Ireland, P.: Managing construction supply chains: the common sense approach. Eng. Constr. Archit Manag 9(5/6):409–418
26. Briscoe G, Dainty A, Millett S (2001) Construction supply chain partnerships: skills, knowledge and attitudinal requirements. Eur J Purch. Supply Manag. 7:243–255
27. Hinze J, Gambatese J (2003) Factors that influence safety performance of speciality contractors. J Constr Eng Manag 2003:159–164
28. Briscoe G, Dainty A (2005) Construction supply chain integration: an elusive goal? Supply Chain Manag 10(4):319–326
29. Burnes B (1996) Managing change – a strategic approach to organisational dynamics, 2nd edn. Pitman Publishing, London
30. Weiner BJ (2009) A theory of organizational readiness for change. Implement Sci 4:67
31. Bandura A (1997) Self-efficacy: the exercise of control. W.H. Freeman, New York
32. Hoorens V (1993) Self-enhancement and superiority biases in social comparison. In: Stroebe W, Hewstone M (eds) European review of social psychology, 4th edn. Wiley, Hove
33. Herscovitch L, Meyer JP (2002) Commitment to organisational change: extension of a three-component model. J Appl Psychol 87:474–487
34. Kotter JP (1996) Leading change. Harvard Business School Press, Boston
35. Cooper D (2001) Improving safety culture – a practical guide. Applied Behavioural Sciences, Hull
36. Arditi D, Chotibhongs R (2005) Issues in subcontracting practice. J Constr Eng Manag 131:866–876
37. Akintoye A, McIntosh G, Fitzgerald E (2000) A survey of supply chain collaboration and management in the UK construction industry. Eur J Purch. Supply Manag. 6:159–168
38. Hale AR, Guldenmund FW, van Loenhout PLCH, Oh JIH (2010) Evaluating safety management and culture interventions to improve safety: effective intervention strategies. Saf Sci 48(8):1026–1035

Acoustic Assessment of Rifle Shots While Hunting by Audiometric Tests, Interviews and Online Survey

Karsten Kluth[⊠] and Dennis Wurm

Ergonomics Division, University of Siegen, Paul-Bonatz-Strasse 9-11,
57068 Siegen, Germany
kluth@ergonomie.uni-siegen.de

Abstract. Hunters expose their hearing – without suitable precautions – with each shot an extreme sound level. Shots with up to 160 dB close to the muzzle are particularly harmful to the ears and can sooner or later cause permanent hearing loss. However, the ability to wear a hearing protection when firing a shot is perceived by many hunters as disturbing and therefore often encounters rejection. Initially, a subjective survey was conducted. Thus, information about noise exposure of the hunters in everyday life, about their shooting behavior, the dangers of hunting, the level of knowledge of the hunters regarding possible hearing hazards and existing safety precautions could be obtained. The survey was attended by 74 active but non-professional hunters of different age and occupational groups. The evaluation of the survey revealed that the danger to hearing caused by the pop of a shot is largely known to the hunters. Nevertheless, the use of hearing protection on the hunt is largely neglected. Additionally, audiometric investigations were carried out. The results of 20 surveyed hunters show in some cases dramatic changes in the audiogram at 6000 Hz. This can be interpreted as an indication of an existing bang trauma. Furthermore, the evaluation of the measurement revealed that there is a clear differentiation concerning hearing loss between the ear which is facing the weapon and which is on the far side.

Keywords: Noise exposure · Hearing loss · Audiometric investigation

1 Introduction and Problem Definition

The beauty and diversity of nature often becomes a very special experience through the perception of sounds. These sounds bring many people the feeling of relaxation, leisure and rest and thus often an increased level of quality of life. Due to the increasing life expectancy, it is important to be able to enjoy these acoustic perceptions even in old age.

The importance of hearing for humans becomes apparent by the fact that our ears are always ready to receive. Earlier, the ever-awake ear had a significant function of protecting people in a quiet environment. Attention and vigilance were often necessary to preserve one's life if necessary. On the one hand, the things heared warned in time of an approaching danger and on the other hand allowed hunting of prey for one's own

© Springer Nature Switzerland AG 2019
S. Bagnara et al. (Eds.): IEA 2018, AISC 819, pp. 224–236, 2019.
https://doi.org/10.1007/978-3-319-96089-0_25

nutrition. Today, in an environment strongly influenced by noise, this function of the ever-awake ear also poses a danger to humans due to overstraining or excessive demand, since today's conditions do not even allow the ear to take a break. Therefore, it is essential to be careful with the hearing, not only in professional life, but also in leisure time.

In the past, the game was hunted silently with bow and arrow. This has changed in the last few centuries. Now firearms are used, which produce an extremely loud shot. Both the professional hunters and the leisure hunters hearing is thereby exposed extreme stress. Even a single shot on the hunt can cause a huge, long-term or irreparable impairment of hearing. For this reason, it is very important to be aware of the danger of an acoustic shock. No one should unconsciously expose themselves to an additional harmful stress on the body.

Impulse sound, i.e. sound with high sound pressure peaks, has a much higher hazard potential than continuous noise but the subjective perception is different. The shortness of the sound effect often leads to false perceptions and this is doubly problematical. In the case of impulse noise, the hearing risk is underestimated due to the short duration of the event. As a result, ear protectors are often not used, although the exposure time of the sound of 1–2 ms may result in significant hearing loss at 6000 Hz [1]. Added to this is the age-related hearing loss, which can become a problem with increasing life expectancy [2]. In this case, hearing first decreases in the high frequencies (see Fig. 1) and the result is a dramatic hearing loss. Only an early education can raise the awareness of the high potential danger of impulse sound.

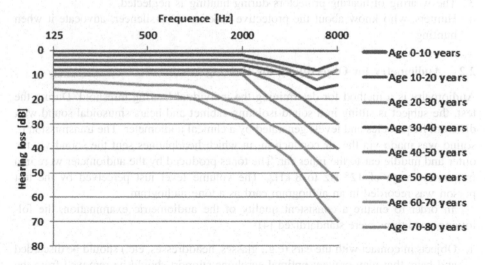

Fig. 1. Comparison of age-related hearing threshold levels [3]

2 Methods

2.1 Questionnaire for the Determination of Hunting and Shooting Behavior

For the collection of information on the hunting behavior a questionnaire with the topic "Danger of the hearing impairment during hunting" was created. This questionnaire was intended to clarify the subjective perception of a possible hearing impairment caused by the use of firearms in leisure or work. A total of 73 German hunters took part in the survey. The questionnaire was classified into five sections:

1. Details of the person as well as the currently practiced or formerly practiced occupation. Here, important findings could be obtained on any other causes of hearing loss, for example, whether the subject was regularly exposed to noise in the profession and which age-related hearing threshold level had to be used.
2. Information on the shooting behavior and the time spent on the hunt. Furthermore, it was recorded whether it was a right-handed or left-handed shooter.
3. The personal hearing as well as experiences and habits that have been made in the context of hearing protection.
4. Opinion of the German hunters to the silencer as a primary sound protection.

The following hypotheses have been made in connection with the topic of hearing impairments during hunting:

1. The risk of hearing loss from the acoustic shock caused by the shot is not known to every hunter.
2. The hunter feels after the hunt harmful influences on hearing.
3. The wearing of hearing protectors during hunting is neglected.
4. Hunters, who know about the protective function of a silencer, advocate it when hunting.

2.2 Audiometry for Objectifying the Hearing

Audiometry is a method for determining the individual hearing threshold. During the test, the subject is sitting in a sound isolating cabinet and hears sinusoidal sound with different frequencies and levels generated by a clinical audiometer. The transmission of sound was made via the air conduction, in which headphones sent the sound via the outer and middle ear to the inner ear. The tones produced by the audiometer were in a frequency range of 125 Hz to 8 kHz. The volume level just perceived by the test person was recorded in an audiogram card as a tone audiogram.

In order to ensure a consistent quality of the audiometric examination, the following conditions were standardized [4]:

1. Objects in contact with the ears (e.g., glasses, headdresses, etc.) should be discarded and hairs that may prevent optimal earphone support should be removed from the support area.
2. The headphones should be positioned exactly and then not be changed.

3. The positioning of the air conduction headset should be done just above the opening of the ear canal.
4. The subject should keep the body position as unchanged as possible, so that disturbing noise is prevented.
5. The subject's response should be given for the duration of the test tone for better results.
6. The subject hears the sound before the actual measurement in order to sensitize him for the subsequent test.
7. The better hearing ear should be measured first.
8. To avoid a routine, the tones should vary between 0.5 s and 3 s and have no specific regularity.

3 Results

3.1 Subjective Assessment

73 people from a hunting district in Germany took part in the survey. 85% were male and 15% of the respondents were female. There was a strong difference in age between participants (see Fig. 2).

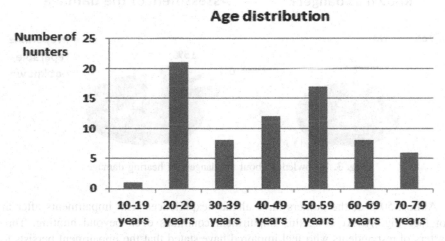

Fig. 2. Age distribution of hunters (n = 73)

The age structure shows that the largest number of participants is in the age group 20–29 years old and the second strongest group is the age group 50–59 years. At the age of 19, the youngest hunter is almost 60 years younger than the oldest hunter at the age of 78. The average age is 44.6 years.

The survey indicates that no typical occupational field is recognizable for those who hold a hunting license, but that hunting license holders from all professions are recruited. 17% of the participants have a technical job (6% engineers, 11% employees on the shop floor), 16% from trade and 13% from commercial professions. Medical professions and teachers are represented with 6% each.

For 87% of the participants the hunt is only a hobby, for 7% both hobby and profession and for 6% the hunt is the profession. Furthermore, very different is the weekly time for hunting. The time required for almost half of the hunters (49%) is up to 5 h/week. For 4% of the hunters the hunt even takes more than 20 h. A hobby hunter spends as much as 40 h per week on the hunt, although this high time expenditure is probably due to the age and his retirement. 92% of the hunters spend up to 5 h a month on the shooting range and 1 hunter even stays there for more than 20 h.

The number of shots also shows an extreme range. The maximum shooting performance showed two hunters with approximately 100 shots in the weekly average. The hunter with the lowest shooting performance, however, shot on average less than one shot a week. On average, all hunters participating in the survey shoot 3.4 shots per week. A clear assignment was found on the body side, where the rifle is applied. At 88%, there were significantly more right-hand than left-hand shooters.

Figure 3 shows the feedback on hearing and hearing protection. The group of people questioned (98%) is largely aware of the danger caused by shots. After all, almost 80% know that the damage in the ear is an irreparable damage. In addition, there is knowledge about the serious hearing damage.

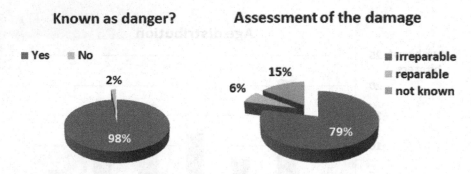

Fig. 3. Knowledge about the dangers of hearing damage

Almost 50% of the hunters have already experiences with impairments after the shot (see Fig. 4). At 14%, this impairment continues even beyond hunting. Three quarters of respondents who feel impaired have stated that the impairment persists for up to one hour after the shot. For 17% it was felt after 6 h and for 6% even the next day.

75% of the hunters are not personally affected by serious noise. However, 17% are exposed to noise at work, 5% at work and leisure and 5% only during leisure time. Here, however, the sensation of the individual with respect to noise has to be taken into account. Also sound which is not perceived as noise, e.g. music, can be harmful. About one third of respondents regularly have a hearing examination.

When wearing the hearing protection, a distinction must be made between wearing the protection on the shooting range or during hunting (see Fig. 5). Since most hunters

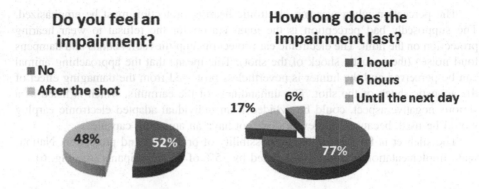

Fig. 4. Individual impairment due to the acoustic shock by the shot (left) with indication of the duration of the impairment (right)

are aware of the risk of a possible hearing damage, only a few of them wear not an earmuff when shooting on the shooting range but only a quarter of hunters wear hearing protection when hunting. The main reason given here is the impairment of perception. In addition, the handling is described as impractical. Another aspect is the temporal factor. So is criticized that there is not enough time to raise the ear protection when a wild animal is targeted. In addition, the habit of not wearing ear protection plays a decisive role. Younger hunters (<45 years) protect themselves rather from the acoustic shock of the shot. All of them wear earmuffs on the shooting range and 35% on the hunt. For the older hunters (>45 years) 85% do not protect their hearing at all.

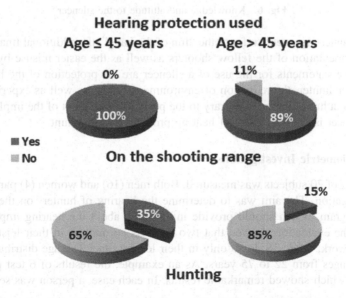

Fig. 5. Age-related use of hearing protection on the shooting range and during hunting

The perception when wearing electronic hearing protection must be emphasized. The supposedly bad perception is the main reason for the refusal to wear hearing protection on the hunt. The electronic ear protection amplifies soft sounds and dampens loud noises (the acoustic shock of the shot). This means that the approaching animal can be perceived and the hunter is nevertheless protected from the damaging effect of the acoustic shock of the shot. The unhandiness of the earmuffs, which is listed as a serious negative aspect, could be avoided if an individual adapted electronic earplug would be used, because this device does not have an annoying capsule.

The silencer is for 92% a known possibility of primary sound protection. Nation-wide implementation would be welcomed by 75% of the participants (see Fig. 6).

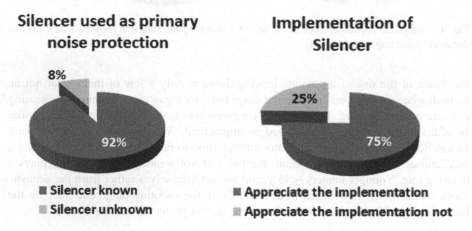

Fig. 6. Knowledge and attitude to the silencer

The counter-arguments refer to the "fun of the bang", the additional financial costs, the better orientation of the fellow shooters as well as the easier misuse by poaching. The positive statements for the use of a silencer are the protection of the hearing and health of the hunter, the reduction of environmental noise as well as experiences with the silencer at hunts abroad. Contrary to the positive assessment of the implementation of the silencer is the actual use of hearing protection on the hunt.

3.2   Audiometric Investigations

The hearing of 20 subjects was measured. Both men (16) and women (4) participated in the investigation. The aim was to determine the hearing of hunters on the basis of a tone audiogram, which should provide information about the hearing impairment by hunting. The evaluation showed that two subjects go hunting in their leisure time as well as at work, while 18 do so only in their leisure time. The age distribution of the subjects ranges from 22 to 75 years. As an example, the results of 6 test persons are presented, which showed remarkable results. In each case, a person was selected who

1. had a pronounced hearing loss in the region around 6 kHz,
2. in conversation explicitly described damage to the hearing by a shot,

3. explicitly showed effects of noise in the form of communication problems,
4. appeared interesting due to its old age,
5. pursues the hunt both as a profession and as a hobby and thus has a very high priority, and
6. had the supervision of a shooting range and was thus exposed to high numbers of shots.

All individual results of the measurement can be taken from the audiogram cards in Figs. 7, 8, 9 and 10.

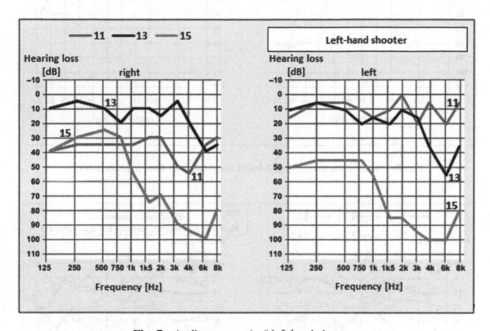

Fig. 7. Audiogram card of left-hand shooters

Evaluation of the Audiometry of Subject No. 1

The 61-year-old hunter has his hunting license for 19 years, is a right-handed shooter and has with 20 shots per week a relatively high shooting performance. He wears hearing protection on the shooting range. However, if he pursues his hobby - hunting - he only wears hearing protection occasionally. His very good hearing at the beginning of hunting has diminished greatly, which seems relatively normal considering the practiced profession (foreman). Noticeable is the hearing loss at 6 kHz, which may be related to a blast trauma (see Fig. 10). At the right ear, this threshold shift is absolutely 75 dB, at the left ear 40 dB. This impairment at 6 kHz can be detected in several subjects in different degrees of severity. As an example, test person no. 13 should be mentioned, who has such a detectable hearing loss on the left ear (see Fig. 7).

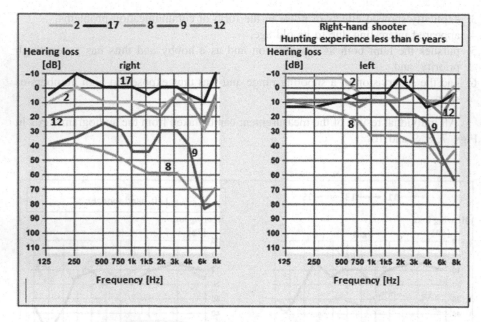

Fig. 8. Audiogram card of right-hand shooters, less than 6 years hunter

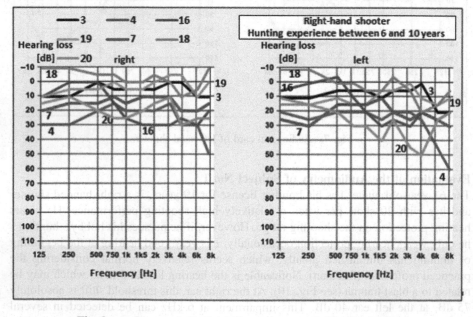

Fig. 9. Audiogram card of right-hand shooters, 6–10 years hunter

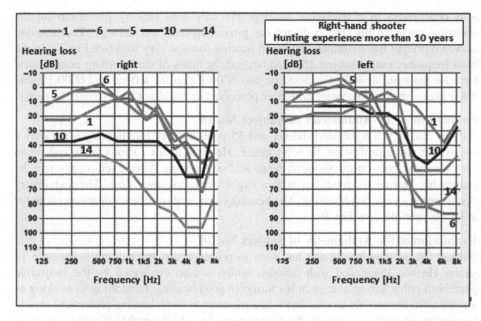

Fig. 10. Audiogram card of right-hand shooters, more than 10 years hunter

Evaluation of the Audiometry of Subject No. 6

In the hearing examination shown in Fig. 10, an auditory threshold shift in the treble range was determined. The 58-year-old person, who has been a civil engineer for 36 years, has his hunting license for 27 years and a shooting performance of 10 shots per week. The right-hand shooter assesses his hearing as mediocre. In addition, it has changed quite a bit over time. Furthermore, tinnitus was diagnosed. Nevertheless, subject 6 uses the hearing protection only on the shooting range and not on the hunt. When looking at the figure, it can be seen that his hearing performance on his left ear is significantly lower than that of the right ear. From a frequency of 1000 Hz, the hearing decreases clearly (2000 Hz: 35 dB; 3000 Hz: 85 dB) and remains approximately constant at the other frequencies by 85 dB–90 dB.

Evaluation of the Audiometry of Subject No. 11

The 31-year-old hunter, a left-hand shooter, works as an electrical engineer and is therefore regularly exposed to machine noise. His audiogram card shows a strong threshold shift on the right ear. During the questioning, the subject rated his hearing as good and not changed. Contrary to these answers subject 11 notes that he has some problems with conversations. According to his own statement, the subject hears better with the right ear and not with the left ear as shown by the values determined by audiometry (see Fig. 7). It also becomes clear that at 6000 Hz a second maximum of the hearing threshold shift is reached for the left ear, which may be related to a blast trauma.

Evaluation of the Audiometry of Subject No. 15

Subject 15 is at the age of 75 and the oldest of the test persons. He has his hunting license for 51 years and is a left-hand shooter. The profession as a welder has certainly also contributed to a damaging hearing. He only uses hearing protection on the shooting range. In his case, a doctor has proven an inner ear damage. The recorded curves represent this extremely advanced hearing damage very well (see Fig. 7). In the high frequency range, subject 15 came beyond the limits of the recording possibilities, because the tones at 4000 Hz, 6000 Hz and 8000 Hz on the left ear and 6000 Hz and 8000 Hz on the right ear were no longer perceived or could not be determined exactly.

Evaluation of the Audiometry of Subject No. 18
Subject No. 18 is a right-hand shooter and 35 years old. For him hunting is not only a hobby but also a profession, he is a forester. He gives about 1 shot a week and wears ear protection both on the shooting range and on the hunt. After a self-assessment, he judges his hearing as good, but with a slight worsening over time. The audiometry documented a very good hearing. The hearing curve is above the comparison curve for a 30–40 year-old (see Fig. 9).

Evaluation of the Audiometry of Subject No. 20
Subject 20 is 59 years old and has been in possession of his hunting license for 10 years. He was diagnosed with tinnitus, which is also confirmed by the subjective perceived rather strong change in his formerly good hearing. In addition to working as a machinist, the activity as shooting range supervision (with hearing protection) should be mentioned. In the activity of shooting supervision a high number of shots have to be considered. This subject is very careful with his hearing, because he always uses hearing protection, both on the shooting range as well as on the hunt. The audiogram shows that the threshold shift in the left ear of the subject is more pronounced than in the right ear (see Fig. 9).

Evaluation of the Audiometric Analyses in an Overall Comparison
To make the data clearer, the hearing curves were arranged in four diagrams. The consideration of the shooting orientation (left-hand or right-hand shooter) provides information whether there is a different stress on the two ears. It turned out that some subjects had more hearing loss on the ear facing the gun. Thus, the right ear of the subjects 1 and 14 is more damaged (see Fig. 10). Subject 14, a 65-year-old hunter, was not under professional harmful influence of noise. Compared to the two previous right-hand shooters, the left-hand shooters 13 and 15 show worse hearing on the left ear (see Fig. 7).

Furthermore, it can be seen in the diagrams that several subjects have a hearing loss at 6000 Hz, which may point to the already described phenomenon of a blast trauma.

The results of the audiometric studies showed that the use of firearms without hearing protection further increases the declining hearing in old age and increases the threshold shift. This result is also confirmed by the study by Counter and Klareskov [5]. The two researchers at Harvard Medical School examined the hearing of Greenlandic residents who are not exposed to industrial noise. They found that about 75% of men who go hunting with firearms have hearing loss. In comparison, women who do not go hunting and are thus not exposed to a shooting blast had an age-related normal hearing [5].

4 Discussion

Each single shot exceeds the maximum sound pressure level in Europe of 137 dB (C). Since hunting can be both a hobby and a profession, the limit does not apply to both areas. The limits are only for the professional hunter obligatory and serve the hobby hunter merely as a recommendation. Since hunting is a hobby for 87% of respondents, the majority of hunters are not subject to the general laws of noise.

Basically, 72% would like a better informative education about the consequences of a shot and the possibilities of using hearing protection. The question concerning the selection of a hearing protection shows that 43% of the used hearing protectors are electronically controlled, which is certainly related to the positive perception characteristics. However, it should be considered that in the relevant frequency range of 1000 to 2000 Hz only a noise dampening of 25 to 30 dB(A) is present, depending on the quality of the hearing protector. The awareness of the danger from the shot is also confirmed by the knowledge of the silencer. As a possibility of primary sound protection, the silencer is known by 94% of the hunters. 75% would welcome an implementation. But even a silencer has a limited performance, as Fig. 11 illustrates. Even with it only sound reductions of 30 dB are possible. Overall, this would result in a reduction of 60 dB. At a sound level of about 160 dB during shooting, about 100 dB remain on the eardrum. All in all, the risk is clearly alleviated, but the situation is still not good.

Weapon	Caliber	Noise at the muzzle undamped	Noise at the ear undamped	Noise reduction at the muzzle	Noise reduction at the ear
Pistol SIG-Sauer P226	9 mm	160,5 dB	157,7 dB	33,1 dB	28,1 dB
Pistol Glock 21	.45 ACP	162,5 dB	162,5 dB	30,7 dB	33,9 dB
Assault rifle Colt M4	.223 Rem	164,0 dB	155,0 dB	26,6 dB	29,8 dB
Rifle Remington 700	.308 Win	165,7 dB	157,2 dB	26,8 dB	26,0 dB

Fig. 11. Exemplary silencing of a weapon by using a silencer [6]

In the evaluation of the audiometry results a connection between shooting orientation and threshold shift could be determined. The damaged hearing was usually on the ear facing the weapon. Significant drops in the curves at 6000 Hz were shown by the audiometry results, which can be the result of a blast trauma. Overall, the hearing loss increased above average in the age comparison.

However, these results are just a documentation of the moment. In order to gain more detailed insights, long-term studies with hunters would have to be carried out in the future. Interesting would be a comparison measurement between today's results of the hunters with audiometry results in a few years, possibly in a 5-year cycle. Correlations between the wearing of hearing protection and hearing loss as well as between the number of shots and hearing loss would be of interest to science and health care policy.

Finally, the great interest of the hunters in the issue should be mentioned, as many consider the reduction of the volume of the acoustic blast and the associated health protection as important. This is evident by the individual's knowledge of the danger and the use of hearing protection at least on the shooting range. It remains to be seen how the case law concerning the permission of the silencer for hunting develops. On the basis of the findings gained from the surveys and the investigations carried out, a clear recommendation for shooting with a silencer is ultimately to be made.

5 Conclusion

- In order to evaluate the danger for the hearing of the acoustic shock of the shot while hunting, audiometric measurements and interviews were carried out.
- The exposure time of the acoustic shock of the shot of 1–2 ms may result in significant hearing loss at 6000 Hz.
- The danger of the acoustic shock of the shot on the hearing is widely known and impairments are felt.
- Wearing a hearing protector on the shooting range is a matter of course for many hunters, but the protectors are often rejected in the hunt. This is justified by an impairment of perception and impractical handling.
- The implementation of silencers is for many hunters a welcome measure of primary sound protection.
- The use of firearms without hearing protection further increases the declining hearing in old age and increases the threshold shift.
- Significant drops in the audiometric curves at 6000 Hz can be the result of a blast trauma.

References

1. Twardella D (2018) Lärm: Grundlagen. https://www.lgl.bayern.de/gesundheit/arbeitsplatz_umwelt/physikalische_umweltfaktoren/laerm_grundlagen.htm. Accessed 20 Apr 2018
2. Bachmann K-D, Vilmar K (2018) Gehörschäden durch Lärmbelastungen in der Freizeit. http://www.bundesaerztekammer.de/fileadmin/user_upload/downloads/Gehoers.pdf. Accessed 20 Apr 2018
3. Boenninghaus H-G, Lenarz T (2007) HNO, 13th edn. Springer, Heidelberg
4. Böhme G, Welzl-Müller K (1984) Audiometrie: Hörprüfungen im Erwachsenen- und Kindesalter. Hans Huber Verlag, Bern
5. Counter SA, Klareskow B (2018) Hypoacusis among the Polar Eskimos of northwest Greenland. https://www.researchgate.net/publication/20919326_Hypoacusis_Among_the_Polar_Eskimos_of_Northwest_Greenland. Accessed 20 Apr 2018
6. Neitzel C (2014) Jagd mit Schalldämpfer. Selbstverlag, Westerstede

CEN/TR 16710 Feedback Method: A Tool for Gathering the Creative Contribution of End Users to Improve Ergonomics of Standards, Design, Construction and Management of the Machines

Massimo Bartalini[1], Alessandro Fattorini[1], Claudio Stanzani[2], and Fabio Strambi[3(✉)]

[1] PISLL AUSL SUD EST Toscana, Siena, Italy
[2] SindNova, Rome, Italy
[3] EUR ERG, Genoa, Italy
fabiostrambi@gmail.com

Abstract. The CEN–TR 16710-1 "Feedback method", approved by CEN on November 2015, "…is addressed to standards writers, designers and manufacturers, employers-buyers, end users, craftsmen and workers, market surveillance and authorities.". It represents a valid instrument for the systematic and reproducible collection of knowledge, experience and creativity of end users of machines.

Keywords: Feedback · End users · Machinery · Standardization

1 Introduction

The use of machines is still one of the most risky activities for the health and safety of workers.

Accidents at work with machines, such as tractors in agriculture and fork lifts in many different sectors, are still frequent and sometimes fatal. Occupational diseases such as MSD, musculo-skeletal diseases, can be caused or worsened by the incorrect use of machines designed and manufactured without taking due account of ergonomic principles.

The social Directive 89/391/CEE and daughters state that:

- workers use machines regularly maintained, complying with the essential safety requirements of the European directive 2006/42;
- workers are skilled and well trained in the use of each specific machine whose consented use and the eventual residual risks are very well known.

The intrinsic safety level of each machine is in any case the central element for an effective prevention. [1, 2] The essential health and safety requirements, EHRS, relating to the design and construction of machinery are exhaustively reported in the Annex 1 of the Machinery Directive 2006/42/EC [3].

© Springer Nature Switzerland AG 2019
S. Bagnara et al. (Eds.): IEA 2018, AISC 819, pp. 237–246, 2019.
https://doi.org/10.1007/978-3-319-96089-0_26

Other relevant requirements are contained in technical, private and voluntary documents, the technical standards, prepared by standardization bodies on the basis of a request by the European Commission. The standards, when adopted and published on the European Official Journal are harmonized and guarantee compliance with safety requirements (EHSRs).

The harmonized standards, representing the state of the art, are the results of a standardization process that involve many stakeholders, public authorities, businesses and consumers and social associations; a standard periodic review is normally scheduled every 5 years. Respecting a harmonized standard guarantees a presumption of conformity with the EHSRs but standards are voluntary documents so that a manufacturer is not obliged to follow them and can use other solutions to demonstrate that his product meets the EHSRs.

The European Standardization organisations, ESOs (the European Committee for Standardization (CEN), the European Committee for Electro-technical Standardization (Cenelec) and the European Telecommunications Standards Institute (ETSI)) prepare the European technical standards and cooperate increasingly with other international standardization bodies such as ISO/IEC to share the standards.

"Finally, standards can help to bridge the gap between research and marketable products or services" [4]. As mentioned above the process of preparing a standard must consider the state of the art intended as the balance between a technical knowledge, feasibility and be cost effective: so the state of the art is a dynamic process. The design of a new standard must be based on a risk assessment, that provides the collection of the experience of users of similar machines and a risk reduction process following an iterative method, EN ISO 12100 [5].

The revision of an existing standard also requires the gathering of the experience of the end users of the studied machine. Many standards [6–8] provide for the involvement of end users of a machine both in the design and in the revision phase through the use of simulation with prototypes and models and collecting the operators feedback in various ways, as described in the EN 614-2 [9]: group discussions, interviews, questionnaires, checklists, and observational studies.

Despite the description of these various investigative ways, no specific validated method was, for long time, published and available to collect the contribution of the end users of the standard (of the machine manufactured in conformity with the standard) to understand the limits and the critical aspect arising from its daily use in the real work cycles in the different contexts. There was an evident need to have a tool for gathering the creative contribution of end users to improve ergonomics of standards, design, construction and management of the machines.

Finally, on 17 November 2015 it was approved by CEN The Technical Report CEN/TR 16710-1: Ergonomics methods - Part 1: Feedback method - A method to understand how end users perform their work with machines [10].

1.1 CEN/TR 16710-1 Feedback Method

The strategy of the European Trade Union Institute, ETUI, for improving machinery standards through users' feedback [11] is described in its web site: "based on their experience and practical knowledge of the equipment, users constitute a precious source of information on the adaptability of the technical solutions provide by the manufacturer". Developing standards is a long process that often takes several years before their publication. As a result, some standards may mention solutions which have since been superseded by recent technological advances, even if these standards may be revised every five years" [11].

In 1997 the European Trade Union Confederation, ETUC, through ETUI, commissioned SindNova, to research and develop a method capable of collecting the knowledge of workers who are expert users of machinery through job reconstruction with the various machines in the different micro and small European enterprises – a method that would yield concrete results within a reasonable period, with limited resources and with validated, verifiable and updatable instruments. The project was carried out in 1999 in Tuscany, Italy, by the Siena Local Occupational Health & Safety Unit (PISLL/USL). The results of this research were published [12] and described the "Feedback method". In 2003 the Bilbao Agency sponsored the "Feedback method" this methodology in the context of the SME Funding Scheme 2003–2004, when Italy (Regione Toscana) and Germany (GrolaBG) decided to apply the "Feedback method" to study the forklift trucks (FLT).

During the ETUI seminar in Brussels on 27th march 2006 based on the results of the two "Feedback method" projects on wood working machines and forklifts, the representative of the Enterprise and Industry Directorate General of the European Commission affirmed that: "The ETUI-REHS experiment has so far produced an impressive amount of useful data: we believe that if the methodology is applied to other classes of machinery, similar good results will be achieved.". In December 2006 the same Directorate General of the Commission wrote in the M/396 EN mandate for the standardization in the field of machinery: "When executing the standardization tasks covered by this mandate, CEN and CENELEC are requested to take due account of feedback from the end-user of the machinery concerned" [13, 14].

In September 2008 in Solna, Sweeden, the resolution n. 318 was written: "CEN/TC 122 asks ETUI to submit a proposal for a Technical Report based on the "Feedback method" designed by Fabio Strambi. This proposal should be developed in close cooperation with CEN/TC 114 [15].

The interesting and challenging work within the Work Group 2 of the CEN/TC 122 was supported by the collection and presentation of further important results achieved by the application of the "Feedback method" to the study of many different machines in many member states [16–20] with the collaboration of many associations, institutions and the POLO for the promotion of health, safety and ergonomics in the Micro, Small and Medium Sized Enterprises, SMEs, in province of Siena. For the first time, in the standardization system a tool is available, the CEN/TR 16710-1, to collect the "feedback from the end-user of the machinery concerned" as requested by the Commission mandate M/396 EN.

1.2 How to Gather the Creative Contribution of End Users of Machine: The Feedback Method

The "Feedback method" has been "designed specifically to collect the contribution of end-users of machinery by reconstructing and understanding how work is actually performed (i.e. the real work)". The acquired knowledge "can help to improve technical standards, as well as the design", [10] manufacturing and use of machinery.

The seven main steps of the "Feedback method" may be summarized as follow:

- Choice of the machine to study
- Collection and evaluation of data on the machine
- Identification of companies to be involved in the project
- Inspection of workplace
- The Feedback method's work groups, FMWGs, and work analysis with skilled end users of the machine
- The validated Report of the FMWG
- The Technical Report

Choice of the Machine to Study

Some good motivations for choosing the machine to study by using the "Feedback method" are:

- the definition of a new standard or a revision of an existing one;
- the presence of a lack of the safety systems or of the ergonomic requirements of a machine together with the knowledge of the causes of major or fatal accidents during its use;
- The large diffusion of the machine and the specific interest of some stakeholders to acquire information on how the real work with a specific type of machine is performed.

Collection and Evaluation of Data on the Machine

A technical dossier is collected to give evidence of the existing knowledge and experience concerning ergonomics and safety of the machine, e.g. technical standards and studies, reports, guidelines and any literature available, data on accidents at work and relevant cases.

Identification of Companies to Involve in the Project

Trade Unions, Employer Associations, safety experts and ergonomists collaborate to identify the companies that use the machine and are interested in participating in the study.

Inspection of Workplaces

Specifically designed forms are compiled during inspection, containing data of the company and all the interesting information on the machine and its use. The data collected is used during the work analysis that is carried out in working groups.

The Feedback Method Work Groups, FMWGs, and Work Analysis with Skilled End Users of the Machine

Work groups are then formed, each group being made up of 5 to 9 skilled end-users.

Each FMWG reconstructs work activities in detail and then carries out a systematic analysis of each work activity/phase under the "coordination" of a facilitator.

Using an ergonomic analysis the group describes both the daily work activities and each single identified work activity/phase in detail with the help of the reported "Feedback method" sheet, Fig. 1 below.

n._____		Work Phase: _____		
	Operating Procedure	Competence	Critical aspects: hazards/risks; disorders/diseases/ injuries.	Solutions, suggestions for prevention; need of further research
Sequence of tasks/ activities	[Detailed description of each action, procedure and method of executing each task/activity, with information on the equipment used, safety devices and personal protective equipment (PPE)...]	[Information about the competence required for: (1) optimal execution of the task/activity and each action (use of equipment; choice, use, and handling of materials); (2) the organisation and disposition of work/ workplace and layout and environment; (3) understanding and applying the instruction handbook]	[Identification of: (1) the critical aspects affecting the health and safety of workers or limiting the efficient performance and reliability of tasks and actions; (2) every hazard and risk; (3) intrinsically safe machinery and equipment; (4) awkward postures, incorrect work practices, environmental conditions (microclimate, dust lighting, layout, etc.); (5) fatigue, complaints, occupational diseases, accidents or injuries; (6) work related stress or problems linked to organizational aspects (rhythm, shifts, etc.).]	[Identification of solutions/suggestions on how to eliminate or minimize the identified problems, hazards and risks and apply the relevant ergonomic principles to: machines, equipment, safety devices, PPE, work procedures, work organisation, environment, etc.; Guidance on: Training, Inspection, Instruction handbooks. Proposals for further research to find new solutions]

Fig. 1. "Feedback method" sheet

For each activity/work phase, the job tasks are identified, and for each of them the consensus on the following elements are put in writing by the facilitators: operating procedure, competence, hazards/risks and suggestions.

The Final Validated Report and the Technical Report

At the end of the meeting the facilitators write a draft of the final report that is submitted to all participants. They are requested to give their feedback with an approval or a proposal for amendments.

Once all participants have validated the draft, the researchers/ergonomists prepare a final report on the study and a technical proposal for the amendment of the standards.

Suggestions for designers, manufacturers, employers- buyers, end-users and market surveillance authorities are also drafted.

2   Results

During the years 2000–2018, 410 end users participated in 39 work groups of the "Feedback method" (FMWGs) in 7 European member states to describe how they perform their work with 9 different types of machines, identifying critical aspects and ergonomic shortcomings and provide their own contribution to possible preventive solutions (Table 1).

Table 1. Applications of the "Feedback method". 2000–2018

Machinery	Country	Companies	Users	Feedback WGs	N. machinery	Years
Woodworking machinery	1	14	28	4	58	2000–2001
Forklift	5	45	60	11	1658	2003–2005
Angle grinder	1	19	19	3	85	2005
Telehandler	5	35	35	5	39	2006–2008
Combine harvester	4	46	110	6	117	2009–2013
Agricultural tractors	1	74	110	7	87	2012–2015
Towed grape harvester	1	34	48	3	36	2016–2018
Total	7	267	410	39	2080	2000–2018

The first FMWGs analyzed the work activity with the woodworking machines and produced interesting and useful information [21, 22] for manufacturers, employers and employees, market surveillance authorities and standardizers.

In particular, taking into account the risk of accidents linked to an incorrect use of the existing protection devices the need was underlined to improve the usability of protection devices, the visibility of the line of cut and to forbid performing some tasks with the riving knife mounted saw guards. The same FMWGs suggested providing the machine with an extraction system of the wood dust, a classified carcinogenic agent [23, 24], during production, maintenance and cleaning operations. The use of compressed air to clean the machine and the workplace must be forbidden. In 2004 the harmonized standards EN 861; EN 1870-4 and EN 1870-15 finally provided specific indications for the cleaning of some woodworking machines.

The "Feedback method" has demonstrated its validity to study the machines fixed in a work place, such as the woodworking machines, the small hand held machines, such as the angle grinder, and the big self propelled machines, like the fork lift truck, the telehandler and the combine harvester [24–27].

The main proposals of the FMWGs for the improvement of the standard [28] of the angle grinder (Fig. 2) may be summarized as follows:

• characteristics of the main switch to avoid accidental operation, inadvertent starting (switch with minimum voltage), exclusion of locking device in the "on" position (in 2012, these improvements were included in the IEC 60745-2-3:2006/AMD2:2012, the standard concerning angle grinder and other types of hand held machines);

- stability of the side handles
- declaration of noise and vibration emission at work
- improvement of the instruction sheets readability and entirety. Definition of the limits of use of the machine. Instructions on polluting substances produced.

The final reports of the international FMWGs organized to analyze the work with the fork lift truck, the telehandler and the combine harvester contributed in a significant way to the writing of the CEN/TR 16710-1, "Feedback method".

In 2012 a "Feedback method" study was carried out to improve understanding the work performed with agricultural tractors gathering experience, creative contribution of the end users and to write a technical report to add to the table of the extensors of the delegated acts of Regulation 167/2013.

Some recommendations by the FMWGs for agricultural tractors:

- working on sloping ground, the slope of the ground on which the tractor is working must be known, to verify if it is suitable or not to the tractor and the carried and/or towed equipment. The presence of an inclinometer on the control panel is desirable the maneuver for coupling the equipment must be facilitated with automatic coupling systems (quick coupling devices), thus avoiding the driver getting in and out of the cab or the assistance of another operator at ground level;
- the visibility of the coupling area from the driving position is inadequate for many models. The visibility of the coupling area should be improved with systems equipped with camera and monitor or with properly positioned rear-view mirrors;
- in some conditions (eg. during chemical treatments or special dusty conditions or bad weather) the windows of the cab need to be carefully cleaned in order to maintain an adequate visibility. Minimum requirements for the equipment supplied to the tractors for glass cleaning should be set.

By using the "Feedback method", ETUI has contributed to the elaboration of the delegated acts concerning the Regulations: (1) on vehicle functional safety requirements; (2) on vehicle construction requirements; (3) general recommendations for improving the safety of tractors [29].

After the agricultural tractor and the combine harvester another very complex agricultural machine was studied by using the "Feedback method": the towed grape harvester. No existing standard specifically helps to design and manufacture this machine in conformity with the EHSRs. The diffusion of the towed grape harvester is constantly growing especially in regions with a large production of grapes and high quality wines such as the south of France, Tuscany in Italy and so on.

Manufacturers are also interested in gathering the contribution of the workers/end users to understand any critical aspects related to both the aspects of health, safety and well being of the operators and the quality and efficiency of the production of their machine. Some of the main critical aspects were highlighted by the FMWGs during the following activities:

- disassembly, cleaning, maintenance and reassembly of machine parts
- hooking the machine to the tractor;
- driving the grape harvester in the field.

For example. During the daily cleaning and maintaining of the machine workers are frequently obliged to assume poor postures especially when reaching the various parts of the machine in the elevated and badly accessible area.

The installation of centralized greasing points and of a platform protected with railings when in elevated position are recommended by the FMWGs.

The operators are requested to keep their head turned backwards during the many operations of driving and harvesting the grapes.

The installation of video cameras on the rear of tractor can attenuate the upper mentioned critical aspect allowing the drivers to diminish the maintenance of incorrect posture and the risk of musculoskeletal disorders.

Workers have referred to the difficulty of driving these machines in vineyards that do not have adequate space characteristics: the spaces between the rows and the maneuvering spaces should be suitable to the reduced maneuverability of tractor with the towed grape harvester.

Further studies by using the TR 16710-1 are in progress regarding both machines used in the mechanical and agricultural sector and different work processes as scaffolding in the building industry and assembling in caravan production.

Indeed the "Feedback method" has proved, gathering the contribution of workers, to be an helpful tool to analyze and better understand a whole work process and not only the use of a machine.

3   Conclusion

One key factor for any improvement of the ergonomic and health and safety requirements of both activities and work instruments/machinery, is the monitoring of the work process, of the regular development of the productivity and the registration and understanding of all the critical aspects emerging during daily work. Knowledge of the failure and/or adverse effect of: an organizational choice and/or a design, of manufacturing and use of an instrument/ machine allows searching for the cause and to try to eliminate the risk and the incongruity. There are instruments to collect this information: quantitative and qualitative data on production, accidents at work, sick absence, information received from the operator (EN 614-2) and so on. The "Feedback method", is an ergonomic method which, considering work as an articulated, unitary process carried out in consecutive phases and activities, assumes the centrality of the worker, especially the skilled worker, to understand how the work is really performed in each different work place. In a few words, the "Feedback method", as reported in the scope of TR 16710-1:

"… is designed to minimize the influence of the subjectivity of the facilitators and researchers in reconstructing and describing the reality of work, and to maximize the "objective" contribution of the skilled users of the machine." This creative contribution may influence the state of art and consequently "is addressed to the standard writers, designers and manufacturers, employers-buyers, end users, craftsmen and workers, market surveillance and authorities."

References

1. GROWTH. http://ec.europa.eu/growth/sectors/mechanical-engineering/machinery/. Accessed 22 May 2018
2. Tool#18. https://ec.europa.eu/info/sites/info/files/better-regulation-toolbox.pdf. Accessed 22 May 2018
3. Directive 2006/42/EC –new machinery directive of 17 May 2006
4. Eur Lex. http://eur-lex.europa.eu/legal-content/EN/TXT/?uri=CELEX:52011DC0311. Accessed 22 May 2018
5. EN ISO 12100:2010: Safety of machinery - General principles for design - Risk assessment and risk reduction
6. CEN Guide 414 (2004) Safety of machinery - Rules for the drafting and presentation of safety standards
7. EN 614-1:2006+A1:2009 Safety of machinery - Ergonomic design principles - Part 1: Terminology and general principles
8. EN ISO 6385:2004 Ergonomic principles in the design of work systems
9. EN 614-2:2000+A1:2008 Safety of machinery - Ergonomic design principles - Part 2: Interactions between the design of machinery and work tasks
10. CEN TR 16710-1:2015 Ergonomics methods - Part 1: Feedback method - A method to understand how end users perform their work with machines
11. ETUI. https://www.etui.org/Topics/Health-Safety-working-conditions/Safety-of-machinery-directives-and-standards/Standardisation/Knowledge-base-for-an-active-ETUI/The-strategy-for-improving-machinery-standards-through-users-feedback. Accessed 22 May 2018
12. Strambi F, Stanzani C, Bartalini M, Cucini M (2001) Ergonomia e norme tecniche di sicurezza: il contributo degli utilizzatori. Editor Franco Angeli, Milano
13. Boy S (2006) A European system to improve machinery safety by drawing on users' experience. ETUI, Brussels
14. Mandate to CEN and CENELEC for standardisation in the field of machinery. M/396 EN. Brussels, 19 December 2006
15. ISO/TR 22100-3:2014 Safety of machinery – Relationship with ISO 12100 Implementation of ergonomic principles in safety standards
16. Strambi F, Bartalini M, Cianotti R, Tini MN, Stanzani C (2006) Feedback: a method to collect the contribution of machinery users in order to improve the quality of design standards. In: Proceedings ISSA, Nice
17. Strambi F, Bartalini M, Boy S, Gauthy R, Landozzi R, Novelli D, Stanzani C (2012) End users "Feedback" to improve ergonomic design of machinery. J Prev Assess Rehabil 41 (Supplement 1):1212–1220. https://doi.org/10.3233/wor-2012-0305-1212
18. Strambi F, Bartalini M, Boy S, Stanzani C (2015) The "Feedback method" a tool to better understand the real work activities with the contribution of end users of machinery. In: 8th international conference of the safety of industrial automated system SIAS, 18–20 November 2015, Konigswinter, Germany, pp 169–219
19. Bartalini M, Fattorini A (2016) Applicazione del "Metodo Feedback" alle trattrici agricole nell'Area Vasta Sud della regione Toscana. Toscana RLS anno, vol 1
20. Strambi F, Bartalini M, Boy S, Stanzani C (2017) The "Feedback method" - A standardized method to give workers a voice. CLR NEWS – European Institute for Construction labour research, EFBWW, Brussels, no 02, pp 30–39
21. EN 848-1:2007+A1:2009 Safety of woodworking machines - One side moulding machines with rotating tool - Part 1: Single spindle vertical moulding machines

22. EN 1870-1:2007+A1:2009 Safety of woodworking machines - Circular sawing machines - Part 1: Circular saw benches (with and without sliding table), dimension saws and building site saws
23. Directive 2004/37/EC carcinogens or mutagens at work of 2004/04/29 amended by Directive (EU) 2017/2398 of 12 December 2017
24. EN 1459:1998+A2:2010 Safety of industrial trucks - Self-propelled variable reach trucks
25. EN 1726-1:1998+A1:2003 Self-propelled trucks up to and including 10000 kg capacity - Part 1: General requirements
26. EN ISO 4254-7:2009 Agricultural machinery - Safety - Part 7: Combine harvesters, forage harvesters and cotton harvesters
27. ERGOMACH. https://ergomach.wordpress.com/information-and-solutions/good-solutions-from-end-users-feedback/. Accessed 22 May 2018
28. EN 50144-2-3:2002 Safety of hand-held electric motor operated tools - Part 2-3: Particular requirements for grinders, disk type sanders and polishers
29. ETUI. https://www.etui.org/Topics/Health-Safety-working-conditions/Safety-of-machinery-directives-and-standards/Machinery-Legislation/ETUI-Seminar-Machinery-safety-in-agriculture-better-ergonomics-for-better-working-conditions/The-technical-recommendations-from-the-Feedback-study-on-agricultural-tractors. Accessed 22 May 2018

Shaping Future Work Systems by OSH Risk Assessments Early On

Peter Nickel[1](✉), Markus Janning[2], Thilo Wachholz[3],
and Eugen Pröger[3]

[1] Institute for Occupational Safety and Health of the German
Social Accident Insurance (IFA), Sankt Augustin, Germany
peter.nickel@dguv.de
[2] German Social Accident Insurance Institution of the Federal Government
and for the Railway Services (UVB), Münster, Germany
[3] German Federal Waterways and Shipping Administration (GDWS),
Bonn, Germany

Abstract. With future work systems becoming more interactive, dynamic, and
flexible, occupational safety and health (OSH) calls for prospective assessments
of hazards and risks to facilitate prevention through design. German river locks
for inland waterways will be composed of standardised objects representing a
high level of OSH. A research project aimed at conducting risk assessments of
river locks of the future; early in their planning stage, with standardised objects
and referring to different EU Directives addressing safety and health at work.
About 150 work scenarios across operational states and variations in river lock
standardisation have been compiled and supplemented for instructing risk
assessments. They also conveyed design requirements for setting up dynamic
virtual reality (VR) planning models in future contexts of use. Initial feedback
about VR simulation use for scenario-based risk assessments is positive and a
design review suggested minor adjustments before risk assessments can be
conducted that should facilitate OSH improvements and templates for future
assessments. The project motivates applications in similar contexts.

Keywords: Occupational safety and health · Risk assessment
Modelling and simulation · Design requirements · Virtual reality

1 Introduction

1.1 Purposes of Risk Assessments

Future work systems often are characterised as becoming more interactive, dynamic,
and flexible. Occupational safety and health (OSH) therefore has a growing interest in
prospective assessments of hazards and risks to enable improvements early in design
[1, 2]; i.e. human-system interactions regarding work tasks, organisation, equipment,
and environment. With its design strategies (e.g. task orientation) and principles (e.g.
compatibility) human factors and ergonomics is well prepared for prospective design as
it provides a broad range of recommendations to meet requirements in human centred
design and OSH; already part of standards [e.g. 1, 3] and legislation (e.g. [4–6]).

© Springer Nature Switzerland AG 2019
S. Bagnara et al. (Eds.): IEA 2018, AISC 819, pp. 247–256, 2019.
https://doi.org/10.1007/978-3-319-96089-0_27

In the German transport sector high priority is given to standardisation of river lock design, because it has the potential to expand transport resources, to reduce costs across the life cycle, to simplify the planning process, to speed up the construction process, to ease maintenance and operation, and to improve OSH [7]. An expert group on standardisation of river locks has been appointed by the German Federal Ministry of Transportation and Digital Infrastructure (BMVI) to identify objects yielding good practice in river lock design, operations and maintenance and match experience and expertise in river lock construction and use for inland waterway transportation. Standardisation in the given context comprises a kit of most components of river locks such as gates, inspection safety closings, and floating bollards. Standardised objects have already been agreed upon and several are already going to be used for future river lock construction [8].

Improvement of OSH in the context of standardisation requires hazards and risks to be assessed. Among them are those, already taken into account by the expert group and others, having not yet taken into account across the river lock life cycle. The endeavour is challenging, because hazard and risk assessment of machinery components itself does not result in machinery safety; i.e. the whole is greater than the sum of the parts. In addition, the future river lock in its future context of use across the life cycle should be included in OSH assessments; even though it is not yet available.

Improvement of OSH is seen most effective early in design, because re-design due to safety and health issues would be resource-demanding, if not impossible, when river lock construction has already been completed. Therefore, OSH risk assessments will be conducted early on for future river locks; with a specific focus on the kit of standardised objects resulting in several variations for standardised river locks.

1.2   Simulation for Risk Assessments

Over the past decades simulation techniques have cleared their way for applied use in industry and services [9, 10], however, not necessarily to substitute traditional techniques for human factors and ergonomics analysis, design, and evaluation in laboratories or in the field, but to bridging the gap between them [11]. Simulation opens up new and effective methods to face some problems in work systems analysis, design and evaluation, in that it allows for systems investigation across the life cycle from construction and development, over application in the context of use, up to modification and recycling [9, 12].

There is a long tradition also in OSH to use simulation techniques. Applications may refer to (a) procedures agreed on and established for testing product safety (e.g. laboratory heat stress testing for hydraulic pipes simulates product use), (b) role plays in OSH trainings intending to simulate safety behaviour at the workplace, (c) retro-perspective accident analysis (e.g. cause effect analyses and if what/when reasoning), and (d) investigations of unavailable, undesirable or future work scenarios (e.g. safety concepts development in virtual reality (VR) simulation for human work in smart manufacturing). Simulation should support work systems design in early stages of development and illustrate dynamic and close to reality scenarios across the life cycle of future river locks.

1.3 Risk Assessments for Improving OSH at Future Work Systems

The German Social Accident Insurance Institution of the Federal Government and for the railway services (UVB) launched a project to strengthen the impact of OSH on the standardisation of river lock components in their future context of use. IFA and UVB in cooperation with partners from BMVI, German Federal Waterways and Shipping Administration (GDWS) and German Social Accident Insurance Institution for Transport and Traffic (BG Verkehr) are working together to prepare for risk assessments for future contexts of use according to different EU Directives addressing safety and health at work and being early in design (see www.dguv.de/ifa/sutave; [13]). The aim of the research is to perform risk assessments of river locks of the future; including standardised components and referring to different European Directives addressing safety and health at work. In parallel with the planning process of first river locks adjusted to standardisation requirements, risk assessments should be conducted based on scenarios presented in VR planning models for current and future contexts of use.

2 Methods

2.1 Different Types of Risk Assessments

At work and across the life cycle of work systems, in the present context i.e. human-system interactions at river locks, several types of risk assessments are mandatory according to different European Directives transferred into member countries legislation [4–6]. While hazards are intrinsic characteristics of objects or procedures such as work materials, equipment, procedures and practices with the potential to cause harm, risk is the likelihood that the potential harm will be attained under the conditions of use and/or exposure, and the possible extent of the harm [14]. Risk assessments should cover all risks arising out of working conditions which are reasonably foreseeable and should also include reasonably foreseeable misuse. Assessments referring to early stages of development or planning are called preventive and prospective risk assessments [15], while for iterations in risk assessments or concluding assessments terms like amendatory or corrective are used.

Among others, the EU Directive on Safety of Machinery [5] requires to perform risk assessments before machinery is placed on the European market and put into service. This type of risk assessment follows a three step procedure for risk reduction [16]; starting with risk analysis, followed by risk evaluation, and resulting in risk reduction according to OSH design requirements. Risk analysis includes determination of the limits of the machinery, hazard identification and risk estimation. This taken together with risk evaluation forms the risk assessment. Risk assessments as iterative processes should be performed at different stages of the machine life cycle (i.e. construction, modification), levels of maturity (i.e. mock-up, prototype) and degrees of detail (e.g. components, whole machinery). Following [8], this type of risk assessment is mandatory at the planning stage as well as at final construction stages before putting the river lock into operation. Performing this risk assessment for variations of standardised river locks may also further improve existing templates and samples to guide future risk assessments according to [5].

Another type of risk assessment is required according to the EU OSH Framework Directive [4]. This type often is focusing on safety and health with regard to human activities during operations; i.e. normal, maintenance and repairs after machinery have been put into service. Assessments of hazards and risks and implementation of measures for risk reduction should be done before operator task performance in order to ensure safe operations. The purpose of carrying out this type of risk assessment is to enable the employer to effectively take the measures necessary for the safety and health of workers. Measures to prevent from occupational risks according mainly refer to the hierarchy of controls [4]. There is no given procedure to follow for conducting this type of risk assessment, however, a structured procedure including documentation should be established that ensures that all relevant hazards and risks will be identified and addressed. Hazards identified should be matched against criteria to ensure health and safety based on legal requirements, published standards and guidance and principles of the hierarchy of controls seeking to ensure an improvement in the level of protection [14, 17]. The focus in the present context will be risk assessments during maintenance operations with several operators involved (e.g. draining of river lock and gate maintenance) and on safety issues with regard to the lock superstructure (e.g. guard railing, signalling, lightning and camera systems repair). In reality, this type of risk assessment should be performed before human-system interaction starts; i.e. to prepare for lock gate maintenance. In the present context, however, it will already be done during planning stages of work systems to affect construction with the consequence to prevent from safety and health issues in future operations. Performing this risk assessment for variations of standardised river locks will establish safe and healthy work procedure for maintenance activities as well as facilitate templates and samples to guide future risk assessments according to [4].

With regard to the EU Construction Site Directive [6] risk assessments are mainly related to safety and health plans and safety and health files to identify and prevent OSH risks during on-site construction as well as during subsequent construction work across the facility life cycle. The preparation of the open and dynamic documents should start with construction preparation, always go along with the facility and be updated throughout the life cycle [18]. The *safety and health plan* aims to arrange for OSH prevention during construction stages, to place risk assessment and risk management at the core of improved safety and health performance and to manage safety and health issues on construction sites; referring to all activities including those relevant next to the site and to all contractors involved. *Safety and health files* are the principal documents for consideration of risks for a timespan from completion of initial construction to design and construction work throughout the life cycle until eventual demolition. Files should contain relevant OSH information including a description of the project, most important drawings and contact details of stakeholders for potential access to more detailed design information. Also of importance are hazards not obvious to others (e.g. surveys on rock samples of the facility), those that might foreseeably arise during further construction work (e.g. loadbearing capacities of soil of the facility, limitations of floor loadings) and those that refer to future construction activities and operations (e.g. façade cleaning, roof repairs, annexe to existing building, heavy or bulky goods delivery, evacuations, restauration, conversions of use). In the present context this type of risk assessment is required in early planning stages to anticipate future work activities at river locks, to collect relevant information for future use of river locks and to adapt construction in a way that lower constrains for OSH at later stages of use. Performing this type of risk assessment for variations of standardised river locks will also serve to develop templates and sample safety and health plans and files to guide future risk assessments according to [6].

Common among all risk assessments is that they all refer to human-system interaction, aim at risk reduction at source, and apply similar procedures to conduct the assessments. Unfortunately, with some of the standardised objects being new, there is not much work experience available for some of the working procedures required. As a consequence, it will not always be possible to perform assessments based on sound levels of experience. Since all assessments will be done at early planning stages, the full hierarchy of controls will be available for developing measures for risk reduction.

2.2 Scenario Compilation and Inspection Activities

Hazards and risks at work situations should be assessed in static and dynamic scenarios to allow for analyses of human-system interactions during task performance in present and future contexts of use. A scenario approach is used in the present context, since scenarios cumulate situations of relevance for risk assessments and for improving OSH of working conditions at river locks. Although scenarios neither have the same impact nor likelihood to achieve its goals, they serve as focal points representing a range of variations and possibilities. The different types of risk assessments at river lock sites determine scenario selection and development. Potentially relevant scenarios at river locks including further explanations about their integration in risk assessments were compiled based on different criteria [19–21]:

- scenarios usually taken into account at former risk assessments in reality,
- scenarios showing standardised objects of river locks in contexts of use,
- typical work tasks conducted at river locks during normal operations; i.e. upstream and downstream locking of different barges,
- typical work tasks conducted at river locks during operations deviating from normal operations; i.e. maintenance and repairs,
- user requirements for operating the river lock and for shipping,
- accident prone situations or core hazardous situations with regard to river lock use or during operational states across the life cycle, and
- work activities at different operational concepts and operating modes (e.g. automatic, manual and special), to name but a few.

Risk assessments usually are on-site investigations to determine the presence, type, severity, and location of hazards and provide suggested ways to control them. Assessments are often conducted by groups of inspectors having expertise in different domains for scenarios under investigation, i.e. OSH experts in cooperation with experts e.g. for the tasks at hand, machinery construction, production requirements. Conducting risk assessments in scenarios is investigation and inspection work that relies on static and dynamic scenario design, especially if scenarios are available as VR models only. VR model construction was informed about design recommendations and requirements by the above mentioned criteria [19–21], the relevant literature [e.g. 21–

23], and expertise of members of the project group having already been involved in scenario based risk assessments.

3   Results on Design of Risk Assessment

3.1   Scenarios Relevant for Different Types of Risk Assessment

A list of scenarios has been circulated among members of the project group and among relevant working groups in the GDWS for completion and verification. In total, 150 scenarios have been listed for to be taken into account in one or more types of risk assessments. The scenarios were supplemented with descriptions and explanations for instructing risk assessments. In addition, building the VR planning model for standardised river locks in current and future contexts of use was informed (see below). Scenarios were structured according to operational states (i.e. normal, maintenance, other operations) and variations in standardisation of river locks (e.g. elastic ship impact protection devices independent of the gates or use of saving basins). Substructures were organised along standardised objects and work activities related to components of the river lock (Fig. 1).

Normal operations	automatic operating mode	...
	manual operating mode	...
	special operating mode	...
	...	...
Maintenance and Repairs	chamber draining/flooding	...
	head/aft gate operations	...
	elastic ship impact protection	...
	...	...
Other operations	frost operations	...
	ship evacuation	...
	reasonably foreseeable misuse	...
	...	...
Variations in standardisation	savings basins	...
	floating bollards	...
	elastic ship impact protection	...
	...	...

Fig. 1. List of scenarios for different types of risk assessments hierarchically structured according to operational states, related standardised objects and work activities required.

3.2   Scenario Requirements for Risk Assessments

Scenario-based risk assessments to be conducted at a VR planning model refer to four dimensions of requirements. First of all, it is required to present the *scenario environment*. The scenario consists of the river lock and its immediate environment. All standardised objects should already work in its future context of use. Therefore, components of the river lock should allow for opening and closing of gates, signalling or floating bollard movements adjusted to water level. In addition, barges of different sizes, truck-mounted cranes, mobile elevating work platforms, other transportation equipment, illumination and weather conditions should be available for dynamic operations in specific scenarios.

Second, *basic human-system interaction* is required for scenarios in terms of simple walks at a river lock site to observe objects, places and work processes. This also refers to relocating position, selecting among relevant positions or moving to places usually not easily accessible or available, such as lock chamber ground level, cable channels, video masts, or even bird's eye positions above the river lock. In addition, this is relevant in dynamic scenarios triggered by personnel working at river locks. This allows observing the locking of ships organised by the control room operator in cooperation with the barge captain, although personnel is not necessarily seen when following dynamics in scenarios.

Third and most importantly, the scenarios should guide conducting risk assessments. This requires visualising static objects and dynamic working conditions, i.e. moving parts (e.g. gates, barges), hiding or showing river lock components (e.g. cover plates for caverns, lock superstructure) and switching between scenarios representing alternative objects. Among others, the standard refers to variants of elastic ship impact protection devices independent of the gates, one with a restraining heavy steel cable and another with a beam. Control of natural speed of movements would be helpful to support realisms while at the same time benefit from simulation systems. Options like speeding up or replaying sequences have been shown advantageous for effectively performing risk assessment [24]. Effective risk assessments also require measures to be taken in scenarios, e.g. by flexible tape measure or laser distance measure, and specifications about installations, e.g. size and weight of floating bollard components or hoist cylinders.

Finally and for *documentation purposes* of risk assessments, it is necessary to take pictures and videos of scenarios under investigation or specific hazards identified. Documentation should also include adding notes to specific components in scenarios and sum up activities across scenarios considered during risk assessments.

3.3 Scenario-Based Requirements for the Design of the Virtual River Lock

A dynamic VR planning model with different layout options for standardised components has been developed [13]. This was based on exact planning information and technical drawings available for a standardised river lock currently in early planning stage (i.e. Wanne-Eickel Nordschleuse) for barges up to 190 m long and 12.5 m wide. Since the risk assessments refer to standardised river locks in general, the VR planning model represents variations in standardisation. It therefore will serve assessments of standardised objects for future river locks [8] and address variations with regard to e.g. saving basis, floating bollards, chamber walls, length of river lock, lift height, road bridge across aft of river lock, elastic ship impact protection devices, camera masts, and illumination plans. A final design review [25] has been conducted and requires minor adjustments in design, such as cover plates for caverns, sealing, lock superstructure. A two-day session for risk assessments according to the OSH Framework Directive [4] has already been scheduled and preparations are under way. This type of risk assessment will be conducted by external inspectors supported by members of the project group; i.e. river lock experts regarding OSH, maintenance and repairs, river lock planning and experts from accident insurance institutions as well as scenario-based modelling and simulation.

Initial feedback from members of the project group and others visiting the VR planning model in general favoured the use of VR in risk assessments. As exceptions they mentioned that it is not possible to address safety issues of the control software and that it will only refer to common knowledge with no simulation possible for unknown situations or effects. Advantages of simulations using 3D CAD or VR were seen in that they may:

- provide a fully functioning work system in a future context of use in 1:1 scale,
- allow for dynamic simulations, i.e. dynamic movements of different parts of machinery (e.g. locking if ships) or human-system interaction (e.g. tape measure use),
- integrate fragmented planning information (e.g. drawings, maps, photos, listings, machinery components),
- are visible, accessible, walkable, assessable and trigger group discussions,
- provide views from different perspectives and locations (e.g. quay, barge, mast),
- enable access of the inside model structure (e.g. caverns, saving basins),
- support identification of testing modes and logic reasoning for dimensioning functional safety,
- support identification of undetected hazards (e.g. by dynamics of moving parts of whole machinery),
- allow for multiple repetitions of the same scenario (e.g. scenario play, pause, stop),
- serve assessments of various modes of operation (e.g. normal operation, maintenance, repairs), and
- support the selection of provisions and measures.

4 Discussion and Conclusions

With the aim to improve OSH for future standardised river locks about 150 scenarios have been developed for different types of risk assessments. Scenarios are transferred into a 3D CAD and VR planning model that simulates river locks and work processes in a wide context of use across operational stages and the life cycle. Outcomes of risk assessments will result in documentations on:

- design improvements for river locks currently under planning,
- design improvements of standardised objects to better fit OSH requirements in standardisation of future inland river locks,
- improved templates and samples for conducting risk assessments according to [5],
- new templates and samples for risk assessments according to [4] and [6],
- basic information for operation instructions for technical components,
- basic information for working instructions for maintenance and repairs, and
- references for risk assessments to be conducted after successful construction.

Results of risk assessments early on have the potential to improve OSH in the process of standardisation while at the same time immersing into future work scenarios in the context of use over next decades [13].

Scenarios for conducting risk assessments have been collected and elaborated and, among others, refined to be set up in 3D CAD and VR simulations. There are, however, two methodological issues why care must be taken, when results related to simulations should be applied to practice. One refers to scenario selection and the other to 3D CAD and VR as simulation techniques. Although most important scenarios have been selected, neither all work situations potentially taking place in the future nor all elaborated in particular detail will be considered in risk assessments for river lock standardisation.

However, scenarios provided a rationale and supported systematic development of 3D CAD or VR simulations. In addition, the simulation as built in the present project also extends the effective range of prevention through design by inclusion of scenarios not yet available and those not desirable or too dangerous to face in reality [26]. Simulation techniques such as 3D CAD and VR may be seen as representations of reality or imitations of real processes in the past, present or future [12]. Complexity in reality is always reduced for modelling and simulation purposes and therefore will never fully replace reality, but may to a certain extent represent and augment risk assessments as suggested above [9, 27]. The use of 3D CAD and VR in the near future will neither replace final risk assessments in reality in late stages of machinery design nor should it replace alternative means for support. But as could be demonstrated, 3D CAD and VR planning models serve risk assessments as a beneficial complements and make risk assessments have the potential to shape OSH in future work systems early on.

References

1. EN ISO 6385 (2016) Ergonomic principles in the design of work systems. CEN, Brussels
2. Kantowitz BH, Sorkin RD (1983) Human factors: understanding people-system relationships. Wiley, New York
3. EN 614-2 (2008) Safety of machinery – Ergonomic design principles – Part 2: Interactions between the design of machinery and work tasks. CEN, Brussels
4. EU OSH Framework Directive 89/391/EEC of 12 June 1989 on the introduction of measures to encourage improvements in the safety and health of workers at work (with amendments 2008). Off J Eur Union L 183:1–8 (1989)
5. EU Machinery Directive 2006/42/EC of the European Parliament and the Council of 17 May 2006 on machinery, and amending Directive 95/16/EC (recast). Off J Eur Union L 157:24–86 (2006)
6. EU Construction Directive 92/57/EEC on the implementation of minimum safety and health requirements at temporary or mobile construction sites. Off J Eur Union L 245:6–22 (1992)
7. BMVI (2014) Verkehrsinvestitionsbericht für das Berichtsjahr 2012. Deutscher Bundestag, Drucksache 18/580, 18 February 2014
8. BMVI (2016) Standardisierung von Binnenschiffsschleusenanalgen bis zu 10 m Fallhöhe (Erlass vom 20.12.2016 – WS 10/ 5212.4/1-2 (2576192)). BMVI, Bonn
9. Meister D (1999) Simulation and modelling. In: Wilson JR, Corlett EN (eds) Evaluation of human work. A practical ergonomics methodology. Taylor & Francis, London, pp 202–228

10. Miller C, Nickel P, Di Nocera F, Mulder B, Neerincx M, Parasuraman R, Whiteley I (2012) Human-machine interface. In: Hockey GRJ (ed) THESEUS cluster 2: psychology and human-machine systems – report. Indigo, Strasbourg, pp 22–38

11. Chapanis A, van Cott HP (1972) Human engineering tests and evaluations. In: van Cott HP, Kinkade RG (eds) Human engineering guide to equipment design. AIR, Washington, pp 701–728

12. Stanton N (1996) Simulators: a review of research and practice. In: Stanton N (ed) Human factors in nuclear safety. Taylor & Francis, London, pp 117–141

13. Nickel P, Lungfiel A (2018) Improving occupational safety and health (OSH) in human-system interaction (HSI) through applications in virtual environments. In: Duffy VG (ed) Lecture Notes in Computer Science, vol 10917. Springer, Cham, pp 1–12

14. Commission European (1996) Guidance on risk assessment at work. ECSC-EC-EAEC, Luxemburg

15. Merdian J (1995) Risikobeurteilung in Arbeitssystemen. Die BG 10:518–524

16. EN ISO 12100 (2010) Safety of machinery – general principles for design – risk assessment and risk reduction. CEN, Brussels

17. Główczyńska-Woelke K, Łyjak G, Gruber H, Vlková S, Nagy K, Schenk C, Šmerhovský Z (2010) Guide for risk assessment in small and medium enterprises. Verlag Technik & Information, Bochum

18. European Commission (2010) Non-binding guide to good practice for understanding and implementing Directive 92/57/EEC 'Construction Sites'. Common, Frankfurt

19. International Commission for the Study of Locks (1996) Final report of the international commission for the study of locks. PIANC, Brussels

20. Dutch Ministry of Transport, Public Works and Water Management (2000) Design of locks (Report by the Civil Engineering Division). Bouwdienst Rijkswaterstaat, Utrecht

21. Nickel P, Kergel R, Wachholz T, Pröger E, Lungfiel A (2015) Setting-up a virtual reality simulation for improving OSH in standardisation of river locks. In: Proceedings of the 8th international conference on the safety of industrial automated systems (SIAS 2015). DGUV, Berlin, pp 223–228

22. Chun CK, Li H, Skitmore RM (2012) The use of virtual prototyping for hazard identification in the early design stage. Constr Innov 12(1):29–42

23. Zhou W, Whyte J, Sacks R (2012) Construction safety and digital design: a review. Autom Constr 22:102–111

24. Pröger E, Nickel P, Lungfiel A (2015) Risikobeurteilung nach Maschinenrichtlinie an einer virtuellen Neckarschleuse. Der Ing 54(2):9–13

25. EN IEC 61160 (2005) Design review. CEN, Brussels

26. Wickens CD, Hollands JG, Banbury S, Parasuraman R (2013) Engineering psychology and human performance. Pearson, Upper Saddle River

27. Nickel P, Nachreiner F (2010) Evaluation arbeitspsychologischer Interventionsmaßnahmen. In: Kleinbeck U, Schmidt K (eds) Arbeitspsychologie (Enzyklopädie der Psychologie, D, III, 1). Hogrefe, Göttingen, pp 1003–1038

Improvements of Machinery and Systems Safety by Human Factors, Ergonomics and Safety in Human-System Interaction

Michael Wichtl[1,2], Peter Nickel[1,3(✉)], Urs Kaufmann[1,4],
Peter Bärenz[1,5], Luigi Monica[1,6], Siegfried Radandt[1,8],
Hans-Jürgen Bischoff[1,8], and Manobhiram Nellutla[1,7]

[1] ISSA Section Machinery and Systems Safety, WG Human Factors,
Ergonomics, Safe Machine, Mannheim, Germany
peter.nickel@dguv.de
[2] Austrian Workers' Compensation Board (AUVA), Vienna, Austria
[3] Institute for Occupational Safety and Health of the German Social Accident
Insurance (IFA), Sankt Augustin, Germany
[4] Swiss Insurance Institution for Occupational Safety and Health (SUVA),
Lucerne, Switzerland
[5] Research Centre for Applied System Safety and Industrial Medicine (FSA),
Mannheim, Germany
[6] Italian Insurance Institution for Occupation Safety and Health (INAIL), Rome,
Italy
[7] Manufacturing Safety Alliance of BC, Chilliwack, Canada
[8] International Social Security Association, Section Machinery and Systems
Safety (ISSA), Mannheim, Germany

Abstract. The International Social Security Association (ISSA) Section for Prevention of Occupational Risks in Machine and System Safety established a working group on Human Factors, Ergonomics and Safe Machines for reviewing, selecting and presenting design requirements and recommendations according to occupational safety and health (OSH) and human factors and ergonomics (HFE). An internet platform will inform, motivate and support machinery manufacturers, users and OSH experts to apply OSH and HFE when constructing, risk assessing, setting-up and operating machinery. Concepts like hierarchy of controls and work system design guide along relevant issues. Besides traditional approaches a special emphasis is given to smart manufacturing calling for design solutions for human-system interaction according to OSH and HFE.

Keywords: Occupational safety and health · Work systems design
Design principles · Design requirements · Hierarchy of controls

List of Technical Abbreviations

- OSH: Occupational Safety and Health
- MSS: Machinery and System Safety
- HFE: Human factors and Ergonomics
- HFESM: Human factor, Ergonomics and Safe machines
- CPS: Cyber Physical Systems

© Springer Nature Switzerland AG 2019
S. Bagnara et al. (Eds.): IEA 2018, AISC 819, pp. 257–267, 2019.
https://doi.org/10.1007/978-3-319-96089-0_28

1   Introduction

1.1   Tasks and Objects of the ISSA Section of Machinery and Systems Safety

The International Social Security Association (ISSA) with its headquarters in Geneva is the principal international institution bringing together social security agencies and organisations with representatives from the employers' and employees' sides. The ISSA's aim is to promote dynamic social security as the social dimension in a globalising world by supporting excellence in social security administration.

The ISSA Section for the Prevention of Occupational Risks in Machine and System Safety as one among 13 sections on prevention is committed to improving safety and health at work in the field of machine and system safety worldwide. Based on the fact that each individual component of the system Human-System-Environment may contribute to total system weakness and thus lead to risks, the MSS section has different tasks and objectives (see www.issa.int/en/web/prevention-machines/about).

1.2   Human Factor, Ergonomics and Safe Machines

The MSS section currently has four international working groups, with the working group "Human factor, ergonomics and safe machines" (WG HFESM) targeting on:

- Describe the requirements of the EU Machinery Directive [1] on ergonomics (i.e. minimum goal setting for safety) while using relevant CEN and ISO standards (e.g. on anthropometric measures) and add information of practical relevance and provide useful links for users.
- Develop practical working aids for designers and construction engineers on how to integrate safety as well as human factors and ergonomics (HFE) into the design and construction of machinery.
- Design the necessary steps according to complexity, provide applicable practical aids (i.e. indicate relevant standards that may serve as a theoretical basis).
- Deal specifically with ergonomics of software and user interfaces.
- Reflect requirements from emerging technologies; i.e. cyber-physical systems (CPS) as discussed with regard to Industry 4.0 and smart manufacturing.

1.3   Current Project of the WG HFESM

Machinery and systems safety aims at prevention of occupational hazards and risks with the consequence of improving system availability and reliability as well as operational safety. Taking a comprehensive approach requires addressing different stake-holders (e.g. manufacturers, users, OSH experts) and considering risk assessments for manufacturers and users less a legal obligation but a tool to make human interaction with machines safer, ergonomic and in consequence more reliable and productive.

With some future work systems remaining unchanged, others in the context of digital manufacturing may develop into cyber-physical systems (CPS). New challenges will therefore arise for HFE and safety disciplines as dynamics and interactions will be more predominating in function allocation, human-centred design requirements, safety measures, and intelligent environments. Therefore, the working group should inform about how to integrate HFE design requirements into machinery construction, integration in workplace design and use at the shop floor level.

2 The Strategy: Work Systems Design

2.1 Work System Design and Occupational Safety and Health

Work system design aims at improving overall system performance while at the same time optimising operator physical and mental workload as well as facilitating high level OSH [2]. According to the World Health Organisation (WHO) OSH refers to all aspects of health and safety in the workplace and should have a strong focus on primary prevention of hazards. Concepts on OSH may vary since OSH legislation differs across countries; however, concepts usually refer to measures for safety and health of employees at work and other forms of activity and include prevention of occupational accidents, occupational diseases and work-related health risks as well as human centred design of working conditions [3, 4].

OSH usually adapts to conventions of the ILO and unlike legal requirements for OSH, the hierarchy of controls remains fairly similar across countries and application contexts. The hierarchy provides some guidance for selecting effective measures for risk reduction and prevention in systems design. Traditionally, a hierarchy of controls may follow levels according a "STOP!" principle [5]; with OSH interventions along subsequent levels of the hierarchy assumed to be less effective, albeit required:

- (S) substituting component (e.g. eliminating hazard),
- (T) technical measures (e.g. safeguard),
- (O) organisational measures (e.g. job rotation),
- (P) personal measures (e.g. personal protective equipment, PPE), and at low level
- (!) information by instructional measures (e.g. warning sign) [4, 5].

Especially effective and sustainable efforts in OSH facilitate requirements and guidance on primary prevention. Manufacturers are required to design safe machinery that meet a set of minimal health and safety requirements and employers are held responsible for providing safe work equipment to employees. At the same time, employees are responsible to work according to the requirements. The concept of primary prevention is a guiding principle for priority consideration of high level measures to combat hazards and risks at work across the life cycle from early on with the consequence of lowering rates of occupational accidents and diseases [4, 6]. In this context the human being at his work has to be seen with the general claim of complete physical, mental and social well-being (see Sect. 2.5).

2.2   Work Systems Design Criteria

HFE in work system design contributes to OSH in that it traditionally applies design principles referring to four criteria; with an increasing impact on OSH and similar to those relevant in human centred design of working conditions [7, 8].

The most basic criterion in work system design is *feasibility of work* for the human being. It describes generic human abilities and refers to human physical (e.g. body postures, movements and strength) and mental dimensions (e.g. cognitive and emotional aspects in human information processing and task performance). It addresses human capabilities across the life cycle (e.g. hand grip, pattern recognition), limitations (e.g. reach height, no visibility for infrared light) and needs (e.g. training physical fitness and situational awareness). This criterion already determines fundamentals in the design of human working conditions (i.e. maximal physical load to be carried depends on power and endurance of performance; acoustic information for humans shall be in their hearing range of 2 Hz to 20 kHz, preferable in range of 2–5 kHz).

Freedom from harm goes beyond feasibility of work in that it specifically refers to prevention of severe occupational accidents, occupational diseases and work-related health risks by design of safe and healthy working conditions. Despite human abilities and capabilities e.g. (a) to avoid obstacles, provide ground floor without obstacles to reduce trip, slip, and fall accidents, (b) to carry asbestos materials, avoid use of asbestos to prevent from cancer diseases, and (c) to pay attention, time restrict monitoring tasks to avoid vigilance decrements.

Freedom from physical and mental impairment refers to working conditions which neither initially nor all of a sudden result in harm, but do impair human activities and conscious experience that may result in accidents. However, impairments of human performance capabilities over longer periods of time or occurring frequently may manifest in harm. HFE requirements therefore aim at designing human tasks and conditions for human task performance as well as variations in tasks or temporal restrictions in task performance. Effective OSH interventions may be illustrated by setting up weight or frequency limits for handling of physical load, providing parallel information presentation when the human task requires comparing different sources of information, or as a short term measure recommend application of task enrichment, enlargement or rotation strategies as well as time restrictions for vigilance tasks such as quality control tasks or control room operations.

Work and working conditions are among the most important factors having an impact on *development of learning, health and personality*. Therefore, fostering the latter refers to the highest level criterion in the design of work systems.

Workload assessment is considered crucial for assessments referring to the four criteria and it is guiding for HFE as well as OSH [7, 9]. Since ergonomic design strategies (e.g. task orientation) and principles (e.g. compatibility) aim at optimising the level of workload, it should be avoided to generally reduce physical or mental workload for unimpaired performance [10]. Deviations of workload from an optimal level will cause either facilitating or impairing effects (e.g. fatigue, reduced vigilance), with the latter closely related to safety and health issues.

The process of work system design comprises workers and work equipment acting together to perform the system function in the workspace, in the work environment, under the conditions imposed by the work tasks [7]. The project group aims at instructing and supporting prospective manufacturers and users of machinery to improve OSH in machinery design by addressing the following topics:

- work place design (e.g. biomechanical requirements, display and control design),
- software design (e.g. task, interaction and information design),
- psychological issues in systems design (e.g. health, safety, behaviour), and
- CPS design (e.g. smart manufacturing and challenges in risk assessments).

2.3 Work Place Design Requirements and Recommendations

With regard to HFE requirements on work place design, biomechanical aspects such as anthropometry and physiology are chosen for presentation. Requirements for the layout of manipulation areas at industrial work places are chosen to illustrate given information. Figure 1 shows areas for main, frequent, and occasional grasp activities in sitting and standing postures.

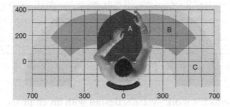

Fig. 1. Areas for main, frequent, and occasional grasp.

HFE requirements for the selection, design and location of displays and hand control actuators shall be taken into account to avoid potential hazards associated with their use (e.g. operating machines) [11]. Figure 2 provides information on recommended fields of vision for locating visual displays when performing detection and monitoring tasks in order to avoid ergonomic hazards and potential health and safety impairments (e.g. neck pain and detection/monitoring errors).

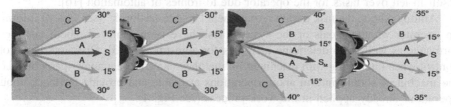

Fig. 2. Vertical and horizontal fields of vision for detection and monitoring tasks with (location in zones A is recommended, B is acceptable, and C is not suitable for the task).

2.4   Work Equipment and Software Design Requirements and Recommendations

Design of human-system interaction for work system design with an emphasis on design of equipment and software refers to three different but interrelated interfaces; i.e. the task, the interaction and the information interface (Fig. 3). HFE and OSH in work system design set the operator task in the centre of the design strategy (primacy of task design [12]) that should be supported by organisational and technical components to serve task completion and overall system performance.

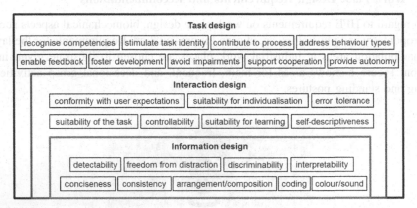

Fig. 3. Design requirements for work systems design with an emphasis on design of equipment and software at task, interaction, and information level [11, 13, 17].

Designing the task *interface* means analyses of system goals, functions as well as activities and information required before allocating functions to the operator and/or to technical components of a work system such as machinery [13, 14]. Taking into account task design requirements (see Fig. 3) should result in tasks for the operators which they can perform at least to avoid physical and mental impairment (see above) [e.g. 10, 13, 15]. As regards human-automation interaction this also refers to flexible adaptation to adequate degrees of automation in machinery design, which must not result in left over tasks for the operator due to ironies of automation [16].

Among others, the task design requirement "enable feedback" (see Fig. 3) informs an operator whether goals of the systems task have been achieved, about task performance as well as to determine activities for adjustments [13]. Operators have job control when interacting with machinery in that they can choose among operations; i.e. the machinery should inform about operations available and provide feedback about operations already applied in the work process (see Fig. 4).

Designing the interaction interface is to be done appropriate for the tasks; i.e. to interface human operators for task interactions with technical system components while taking into account human performance capabilities and limitations. Principles of human information processing (perception, decision making, action implementation) are transformed into design requirements (e.g. feedback, compatibility, function) [11,

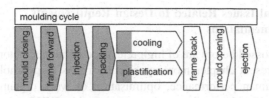

Fig. 4. Batch procedure of injection moulding machine provides feedback to operator about production process

18, 19]. The design of human-system interaction interfaces according to requirements facilitates operator interactions to perform assigned tasks safely and efficiently.

Interaction design requirements (see Fig. 3) should foster co-working with machinery. Interactions conform to user expectations (Fig. 3) if they correspond to predictable contextual needs of the user and to commonly accepted conventions; i.e. interfaces used in Europe must follow European conventions for reading from left to right [11]. Interaction interfaces are controllable when modes are consistently defined. Valves may either be set in manual or automatic mode; as in Fig. 5. If any is indicated, there shall be no undefined mode to avoid potential hazards.

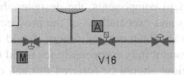

Fig. 5. Inconsistent valve indications, with control valve (left) in manual mode, block valve (centre) in automatic mode, and undefined control valve (right). [Picture: GAWO e.V.]

Designing the information interface refers to modalities (e.g. visual, auditory, haptic), objects (e.g. text, graphic, list, display) and may be passive or active (e.g. label, control, feedback). It is closely linked to human information processing as it impacts reasoning and decision making while directly linking to perception and action implementation (feedback).

Conciseness as an information design requirement (see Fig. 3) is of special importance for information presentation when an information flow should be guided, e.g. by moving his/her gaze in a direction. According to the example given above in Fig. 4, steps in moulding are supported by gaze flow in horizontal sequence representing the production process within the machine.

All task, interaction and information design requirements according to HFE and OSH (see Fig. 3) will be illustrated and presented on the information platform.

2.5   Psychological Issues Related to Design Requirements and Recommendations

Though issues discussed before are of high relevance in human factors and engineering psychology, the present chapter more specifically refers to psychological issues; i.e. workers health, human performance, optimisation of work organisation, and safety behaviour at work. *Health* is understood as mentioned by WHO principles as a state of complete physical, mental and social well-being and not merely the absence of disease or infirmity. Promoting workers health is therefore more than prevention of disease as it goes beyond, addresses risks and includes measures that allow for optimising physical activity at work through systems design, promoting more healthy diets, and protecting cognitive and emotional wellbeing at work.

Human performance is relevant in physical and mental terms and is closely linked to the four criteria for work systems design and workload assessment (see above).

Optimisation of *work organisation* will be addressed with topics such as improvement of social relations at work, work-life balance, and organisational safety culture, as well as reduction of risks of impairment by multiple factors of work stress and working time risks.

Safety behaviour at work is often discussed in OSH, however, sometimes erroneously used for ensuring safety when no technical measure has been found in time. When designing work system components, the designer should take due account of abilities, skills, experiences and expectations of the prospective user population. System design omissions leading to sub-optimal design should not be replaced by calling for operator training, as it cannot fully compensate for it and may impair overall system performance. Safety behaviour should be fostered through corrective and prospective design of human-system interaction, but may require support through qualification, training and instruction. System redesign according to HFE design requirements shall always have priority.

2.6   Cyber Physical Systems Design Requirements and Recommendations

CPS are characterised by technology driven interactive processes in industry and services that will mainly be monitored and controlled by software algorithms referring to sensor technologies and machine-to-machine communication in networks. Although not mentioned in definitions, it can reasonably be assumed that human-system interaction will always be involved, e.g. when setting-up or maintaining CPS. Therefore, development and use of CPS calls for appropriate implementation of OSH and HFE requirements. Smart manufacturing has a strong impact on human-system interaction and seems to foster shifts in hazards (e.g. mechanical superimposed by informational) and in human task requirements (e.g. action implementation superimposed by perception and reasoning). Perspectives for prevention should extend and adapt accordingly, in that design of human task and interaction interfaces referring to HFE principles become key performance indicators in mediating OSH [6]. It is required to interface to human information processing, because the basis for interactive processes in smart manufacturing and Industry 4.0 is exchange of information variable and dynamic in time, quantity and quality. The design of interfaces is challenging as has already been demonstrated in solutions on human-automation interaction according to HFE [24]. With new technologies available, the variety of interfaces for human-system interaction increases (e.g. natural, tangible, organic user interfaces) with the potential to improve HFE and OSH in work system design.

New technologies are meant to tap into, analyse and predict aspects of human behaviour in complex systems by e.g. input monitoring, language or gesture recognition for human inclusion into technological systems or with a HFE perspective, in order to assist (tool), partially supplement (prosthesis) or temporarily represent (agent) human interaction with technical system [16, 26]. This requires OSH and systems design to address challenges in human-system interaction [16].

- Feedback challenges: High degrees of automation provide humans with insufficient information for reasonable participation. Operators then find changes and errors in automation difficult to identify and impossible to compensate for.
- Human task structure challenges: Functions are taken away from human operators by automation, and their work is consequently impaired or even hindered.
- Relationship structure challenges: Technology-centred design of automation leads to unintended uses by operators.

With the aim to improve safety of function and result in smart manufacturing, risk assessments are required to disclose potential hazards and risks in human-system interaction with regard to human tasks, hardware and software as well as all system processes. In the given context this refers to safety in terms of functional, process and operational safety and therefore extends to security as well as design according to HFE requirements across all operational states.

A potential solution to tackle the complexity of CPS due to variability and dynamics in interactive processes is seen in modular approaches as is common e.g. in the process industries. The overall system is separated in modules with clearly defined interfaces. Risk assessments should then be organised for individual modules and their combinations since the whole system is greater than the sum of its parts. In addition, risk assessments should also become flexible and dynamic and may even take advantage of emerging technologies in that draft assessments become an integrated component in smart manufacturing.

3 Conclusions

Information will be made available through a specific internet platform located at the Machinery and System Safety Section of the ISSA. Most common design requirements and recommendations in HFE for improving safety in machinery and system design have already been identified based on reviews of relevant standards, OSH research and OSH expertise in machinery design. So far individual sheets on biomechanical issues and on software design issues in HFE are available (see Sects. 2.3 to 2.6). An internet presentation has been chosen since OSH and HFE recommendations and requirements can be updated and adapted to new challenges and solutions in smart manufacturing. All information will be provided to guide designers and safety and health experts at early stages in machinery and systems design and use as well as to provide input for training courses by the Section for the Prevention of Occupational Risks in Machine and System Safety of the ISSA.

References

1. EU Machinery Directive 2006/42/EC of the European Parliament and the Council of 17 May 2006 on machinery, and amending Directive 95/16/EC (recast). Off J Eur Union L 157:24–86 (2006)
2. Nachreiner F (1998) Ergonomics and standardization. In: Stellmann JM (ed) ILO encyclopaedia of occupational health and safety. ILO, Geneva, pp 29.11–29.14
3. Lehto MR, Cook BT (2012) Occupational health and safety management. In: Salvendy G (ed) Handbook of human factors and ergonomics. Wiley, Hoboken, pp 701–733
4. Nickel P (2016) Extending the effective range of prevention through design by OSH applications in virtual reality. In: Nah FF-H, Tan C-H (eds) Human computer interaction in business, government and organization: information systems (HCIBGO 2016), Part II, LNCS 9752. Springer International Publishing, Cham, pp 325–336
5. EU OSH Framework Directive 89/391/EEC of 12 June 1989 on the introduction of measures to encourage improvements in the safety and health of workers at work (with amendments 2008). Off J Eur Union L 183:1–8 (1989)
6. Nickel P, Lungfiel A (2018) Improving occupational safety and health (OSH) in human-system interaction (HSI) through applications in virtual environments. In: Duffy VG (ed) Lecture Notes in Computer Science, vol 10917. Springer, Cham, pp 1–12
7. EN ISO 6385 (2016) Ergonomic principles in the design of work systems. CEN, Brussels
8. Hacker W (1998) Mental workload. In: Stellmann JM (ed) ILO encyclopaedia of occupational health and safety. ILO, Geneva, pp 29.41–29.43
9. EN ISO 10075-2. Ergonomic principles related to mental workload – Part 2: Design Principles. CEN, Brussels
10. Sanders MS, McCormick EJ (1993) Human factors in engineering and design. McGraw Hill, New York
11. EN 894 Series (ISO 9355 Series) (2010) Safety of machinery – Ergonomics requirements for the design of displays and control actuators – Part 1: General principles for human interactions with displays and control actuators, Part 2: Displays, Part 3: Control actuators, Part 4: Location and arrangement of displays and control actuators. CEN, Brussels
12. Ulich E, Schüpbach H, Schilling A, Kuark J (1990) Concepts and procedures of work psychology for the analysis, evaluation and design of advanced manufacturing systems: a case study. Int J Ind Ergon 5(1):47–57
13. EN 614-2 (2008) Safety of machinery – Ergonomic design principles – Part 2: Interactions between the design of machinery and work tasks. CEN, Brussels
14. EN 16710-2 (2016) Ergonomics methods – Part 2: A methodology for work analysis to support design. CEN, Brussels
15. Nachreiner F, Nickel P, Meyer I (2006) Human factors in process control systems: the design of human–machine interfaces. Saf Sci 44:5–26
16. Lee JD, Seppelt BD (2012) Human factors and ergonomics in automation design. In: Salvendy G (ed) Handbook of human factors and ergonomics. Wiley, Hoboken, pp 1615–1642
17. EN ISO 9241-112 (2017) Ergonomics of human-system interaction – Part 112: Principles for the presentation of information. CEN, Brussels

18. Wickens CD, Hollands JG, Banbury S, Parasuraman R (2013) Engineering psychology and human performance. Pearson, Upper Saddle River
19. Cosmar M, Nickel P, Schulz R, Zieschang H (2017) Human-machine-interaction (oshwiki. eu/wiki/Human_machine_interface). EASHW, Bilbao
20. Sheridan TB, Parasuraman R (2005) Human-automation interaction. Rev Hum Factors Ergon 1(1):89–129
21. Miller C, Nickel P, Di Nocera F, Mulder B, Neerincx M, Parasuraman R, Whiteley I (2012) Human-machine interface. In: Hockey GRJ (ed) THESEUS cluster 2: psychology and human-machine systems – report. Indigo, Strasbourg, pp 22–38

Can Interventions Based on User Interface Design Help Reduce the Risks Associated with Smartphone Use While Walking?

Jun-Ming Lu[✉] and Yi-Chin Lo

Department of Industrial Engineering and Engineering Management,
National Tsing Hua University, No. 101, Section 2, Kuangfu Road,
Hsinchu 30013, Taiwan
jmlu@ie.nthu.edu.tw

Abstract. In response to the accidents resulting from smartphone use while walking, this study aims to propose interventions based on user interface design and evaluate whether they can help reduce the behavioral changes and associated risks. First, four types of walking-sensitive changes in user interface including a pop-up reminder (reminding not to use the smartphone while walking; disabled only when tapping on it), a blank screen (disabled only when double-tapping the screen), simplified push notifications (from other applications), and hidden push notifications (from other applications), were proposed to solve problems resulted from over-concentrating on the screen and switching between multiple mobile applications. Subsequently, the proposed interventions were evaluated through the simulated tasks of smartphone use while walking along a 5.1 m × 3.6 m rectangular track in the laboratory environment. 10 males ranging from 20 to 24 years old who use smartphone for one hour per day or longer were recruited. Along with the case of no intervention, a total of five conditions of smartphone use were considered. Two conditions without smartphone use, including free walking and walking while holding a smartphone, were assessed as well. Over the seven conditions, walking speed, detection rate of unexpected stimulus, and perceived level of situation awareness were compared. Results suggested that the risks related to impaired visual attention became even higher when given any of the four interventions based on user interface design, whereas significant changes were not observed in other measures no matter whether the interventions were taken or not.

Keywords: Distracted walking · Multitasking · Pedestrian safety

1 Introduction

As more than three quarters of the population owns a smartphone [1], one of the challenges following the increasing user dependence is distracted walking while using a smartphone. Among all the fatal traffic accidents of pedestrians, the percentage of the cases caused by using a cellphone/smartphone rose from 1.0% in 2004 to 3.6% in 2010 [2]. In the United States, the number of pedestrians transported to emergency rooms due to falls or collisions associated with smartphone use increased by 124% from 2010

© Springer Nature Switzerland AG 2019
S. Bagnara et al. (Eds.): IEA 2018, AISC 819, pp. 268–273, 2019.
https://doi.org/10.1007/978-3-319-96089-0_29

to 2014 [3]. Considering the causal factors, distractors from the smartphone can reduce one's risk awareness of potential hazards, leading to the danger for not only the user him-/her-self but also other road users.

The major challenge caused by distracted walking is the need for multitasking, which results in the behavioral changes in body movements [4], gaze patterns [5], allocation of attention resource and reaction time to unexpected stimulus [6], and so on. Besides, Hyman et al. reported that the impaired situation awareness and the occurrence of inattentional blindness could be also some of the reasons of accidents [7]. Further, the analysis of gaze behavior showed that the ability of change detection is reduced as one uses the cellphone while walking [8].

Even though the statistics have highlighted the problems with distracted walking due to smartphone use, there have been limited effective solutions proposed. Warning signs on the road could be the most common way being adopted [9, 10]. Unfortunately, the prevalence of such behaviors did not decline much. Hence, some ideas involving user interface design were proposed. For example, Kane et al. developed the so-called "walking user interface" for cellphones, in which the font size and menu design will be changed automatically as the walking behavior is detected [11]. In addition, DoCoMo provides an alternative in which a pop-up message will appear for every 10 s while one tries to use the smartphone while walking [12]. However, the effectiveness of such countermeasures hasn't been well verified.

Therefore, in response to the accidents resulting from smartphone use while walking, this study aims to propose interventions based on user interface design and to evaluate whether they can help reduce the behavioral changes and associated risks.

2 Development of Interventions Based on User Interface Design

In order to identify what kinds of user interface design may be helpful for reducing the risks associated with distracted walking due to smartphone use, one male (24 years old) and one female (25 years old) were recruited for analyzing their mental activities involved in smartphone operation. Each participant was asked to sit in front of a desk and play a typing game on the specified smartphone, during which messages in Facebook were randomly sent by the test giver. Gaze data was collected by an eye-tracking system, whereas smartphone operations were obtained using a screen recording application. This enables the observation of how a user allocates his/her visual attention on the screen while interacting with multiple mobile applications.

Greater mental efforts were found while over-concentrating on the screen to read or send messages, as well as switching between multiple mobile applications. Thus, walking-sensitive changes in user interface were proposed to solve these problems. First, as shown in Fig. 1(a), a pop-up reminder will be presented every 10 s as long as the walking behavior is detected (when walking speed is greater than 3.49 km/hr). A second option is to activate a blank screen every 10 s as long as the walking behavior is detected, as shown in Fig. 1(b). Both of them were designed to prevent the user from starring at the screen long and to adequately draw his/her visual attention to the surrounding road conditions. The only difference is how the intervention can be disabled. The pop-up reminder is disabled temporarily by tapping the "understood" button in the dialog box, whereas the blank screen is removed temporarily by double-tapping the screen.

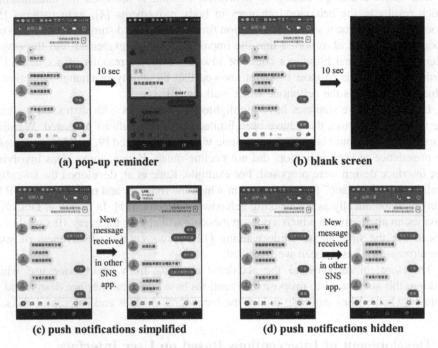

(a) pop-up reminder

(b) blank screen

(c) push notifications simplified

(d) push notifications hidden

Fig. 1. The four types of walking-sensitive interventions based on user interface design.

Further, push notifications from other mobile applications will be simplified or hidden, in other not to encourage the user to switch between applications actively. The two designs are shown in Fig. 1(c) and (d), respectively. In the case of simplified notifications, it is assumed that users may hence decide to ignore the push notifications (from LINE, in this experiment) since the information about the sender is not displayed. As for the case of hidden notifications, the user will not be able to capture the push notifications directly, unless he/she checks the application voluntarily.

3 The Evaluation of Walking-Sensitive Interventions

3.1 Experimental Design

Following the development of four walking-sensitive interventions, it is necessary to assess the effectiveness in terms of the reduction in behavioral changes. In addition to the distracted walking without interventions, comparisons were also made against free walking and walking while holding a smartphone (so as to identify whether the changes are purely due to smartphone operation). Hence, a total of seven conditions were considered.

Ten males ranging from 20 to 24 years old were recruited. All of them use the smartphone for one hour per day or longer. Each participant was instructed to walk counter-clockwisely along a 5.1 m × 3.6 m rectangular track for four laps, as shown in Fig. 2. Under the five conditions with smartphone use (four interventions and the one without any intervention), the participant was required to use Facebook to interact with the test giver, during which push notifications from other applications might be presented. In the meantime, he/she needed to walk along the 0.3 m-wide pathway as stably as possible. These are the primary tasks. At the end of the two longer sides, a monitor was placed to randomly show a red, green, or yellow color lasting for 0.6 s. Besides, as passing through the shorter sides, the recorded sound of a 65-decibel car horn was played through the wireless speaker located at the center of the track. The participant was required to speak out loud the color he/she saw or whether he/she heard the car horn. These are treated as the performance measures of the secondary task. During the experiment, an optical motion capture system and an eye-tracking system were used to collect the gait and gaze data. Responses to the unexpected visual and auditory stimulus were collected by the test giver, whereas perceived level of situation awareness was obtained using the Situation Awareness Rating Scale. It took about 90 min for each participant. Prior to the experiment, all participants read and signed the informed consent approved by the Research Ethics Committee, National Tsing Hua University.

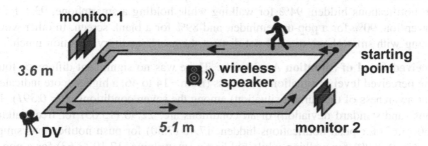

Fig. 2. The rectangular track for conducting simulated tasks. (Color figure online)

3.2 Results

Repeated Measured ANOVA was adopted to compare the walking speed, detection rate of unexpected stimulus, and perceived level of situation awareness among the seven conditions, with the significance level at 0.05. The significance of main effects and associated post-hoc analyses are presented in the following paragraphs.

Walking Speed. The walking speed was first normalized by the walking speed under the condition of free walking. If the number is smaller than 1.00, it means the walking speed is reduced under a specific condition. There was no significant difference found in the normalized walking speed among the seven conditions ($p = 0.322$). The means (and standard deviations) of all conditions other than free walking are: 0.93 (0.23) for walking while holding a smartphone, 0.90 (0.28) for push notifications simplified, 0.89 (0.25) for push notifications hidden, 0.88 (0.32) for a blank screen, 0.87 (0.29) for a pop-up reminder, and 0.79 (0.21) for no intervention. In other words, walking with smartphone use did not significantly slow down the user, no matter whether the interventions were taken or not. However, it seems that the two interventions with push notifications might help prevent the user from slowing down, making the behavior slightly closer to the safer conditions.

Detection Rate of Unexpected Stimulus. There was a barely significant difference found in the percentage of unexpected visual stimulus detected among the seven conditions ($p = 0.056$). The post-hoc analysis determined three groups in terms of detection rate. One is free walking (80%), which is significantly higher than those including a pop-up reminder (55%), push notifications simplified (51%), push notifications hidden (50%), and a blank screen (48%). The third group contains walking while holding a smartphone (70%) and no intervention (61%), in which no significant difference was found against the other two groups. In other words, the interventions did not help enhance the ability to notice unexpected visual stimulus. Instead, the performance is even worse than the condition without any intervention. It seems that the intervention designed to draw the user's attention from the smartphone to the environment eventually leads to more attention allocated on the smartphone. On the other hand, there was no significant difference found in the percentage of unexpected auditory stimulus detected among the seven conditions ($p = 0.268$). The means of all conditions are: 100% for free walking, 98% for push notifications simplified, 94% for push notifications hidden, 94% for walking while holding a smartphone, 93% for no intervention, 90% for a pop-up reminder, and 89% for a blank screen. In other words, walking with smartphone use did not distract the user's auditory attention much.

Perceived Level of Situation Awareness. There was no significant difference found in the perceived level of situation awareness (from −14 to 46, a higher score indicates a better awareness of the current situation) among the seven conditions ($p = 0.391$). The means (and standard deviation) of all conditions are: 22.90 (13.05) for free walking, 19.60 (3.63) for push notifications hidden, 17.40 (7.50) for push notifications simplified, 17.30 (8.80) for walking while holding a smartphone, 17.10 (3.63) for a pop-up reminder, 16.10 (6.92) for no intervention, and 15.80(7.34) for a blank screen. In other words, the interventions did not reduce the perceived situation awareness. The user may still be able to understand the situation as clearly as when not distracted by smartphones, no matter whether the interventions are adopted or not. However, there is a slight trend that interventions with push notifications might better help maintain situation awareness, making the behavior slightly safer.

4   Conclusion

In this study, it was found that the use of smartphone while walking did impair the visual attention to the surrounding environment, whereas walking speed, residual auditory attention, and overall situation awareness were not affected significantly. Considering the reduced visual attention, it became even more risky as walking-sensitive interventions based on user interface design were adopted. Follow-up studies will be hence needed to confirm whether it is possible to allow smartphone use while walking in a safe manner, by adopting countermeasures other than interface redesign.

References

1. Institute for Information Industry (2015) Innovative service experience ecosystem research and development project. http://www.find.org.tw/market_info.aspx?k=2&n_ID=8482/. Accessed 24 May 2018 (in Chinese)
2. Nasar JL, Troyer D (2013) Pedestrian injuries due to mobile phone use in public places. Accid Anal Prev 57:91–95
3. Smith DC, Schreiber KM, Saltos A, Lichenstein SB, Lichenstein R (2013) Ambulatory cell phone injuries in the United States: an emerging national concern. J Saf Res 47:19–23
4. Lamberg EM, Muratori LM (2012) Cell phones change the way we walk. Gait Posture 35 (4):688–690
5. Kim R, Lester BD, Schwark J, Cades D, Hashish R, Moorman H, Young D (2016) Gaze behavior during curb approach: the effect of mobile device use while walking. In: proceedings of the human factors and ergonomics society annual meeting, vol 60, no 1. SAGE Publications, Los Angeles, pp 1580–1584
6. Jeon S, Kim C, Song S, Lee G (2016) Changes in gait pattern during multitask using smartphones. Work 53(2):241–247
7. Hyman IE, Boss SM, Wise BM, McKenzie KE, Caggiano JM (2010) Did you see the unicycling clown? inattentional blindness while walking and talking on a cell phone. Appl Cognit Psychol 24(5):597–607
8. McCarley JS, Vais MJ, Pringle H, Kramer AF, Irwin DE, Strayer DL (2004) Conversation disrupts change detection in complex traffic scenes. Hum Factors 46(3):424–436
9. Iyengar, R (2014) Chinese city sets up 'no cell phone' pedestrian lanes. TIMES. http://time.com/3376782/chongqing-smartphone-sidewalk-meixin-group/. Accessed 24 May 2018
10. Seoul Metropolitan Government (2016) New traffic signs for smartphone users. http://english.seoul.go.kr/new-traffic-signs-smartphone-users/. Accessed 24 May 2018
11. Kane SK, Wobbrock JO, Smith IE (2008) Getting off the treadmill: evaluating walking user interfaces for mobile devices in public spaces. In: Proceedings of the 10th international conference on human computer interaction with mobile devices and services. ACM, pp 109–118
12. DoCoMo. About the approach to prevent from smartphone use while walking. https://www.nttdocomo.co.jp/info/news_release/2013/12/03_00.html/. Accessed 24 May 2018 (in Japanese)

Proposal and Verification of a Method for Maintaining Arousal by Inducing Intrinsic Motivation: Aiming at Application to Driving of Automobiles

Yuki Mekata[1]($\boxtimes$), Shuhei Takeuchi[2], Tsuneyuki Yamamoto[2],
Naoki Kamiya[2], Takashi Suzuki[2], and Miwa Nakanishi[1]

[1] Keio University, 3-14-1 Hiyoshi, Kohoku-ku Yokohama,
Kanagawa 223-8522, Japan
wyume7921@gmail.com
[2] Tokai Rika Co., Ltd., 3-260 Toyota, Oguchi-cho,
Niwa-gun, Aichi 480-0195, Japan

Abstract. With the advance of automation in recent years, the interaction between systems and users has become less intense and more monotonous. Thus, there is some concern that the arousal level of users is liable to decrease while interacting with a system. The decrease in the arousal levels of users causes drowsiness in their bodies and deterioration of their performance, potentially leading to serious accidents. Thus, it is necessary to establish a method for maintaining arousal level. At present, external-stimulation methods such as sounds or vibrations are popular; on the other hand, some psychophysiological studies have suggested a relationship between motivation and arousal, but there has been little research on how to apply these studies, as well as only a few examples of practical application. Therefore, in this research, we propose and verify a method for maintaining arousal by inducing intrinsic motivations. This arousal maintenance was found to work well during manual operations, but to be less effective by influence of stress during automatic operations. In addition, the varieties of presentation of feedbacks and the arousal level during operations were found to be influenced the effectiveness.

Keywords: Maintaining arousal · Intrinsic motivation · Driving

1 Background and Purpose

Since automation has advanced in recent years, the interaction between systems and users has become less intense and more monotonous. Thus, there is concern that the arousal level of users may decrease while interacting with systems. If this level cannot be maintained, users may get into serious accidents due to drowsiness in their bodies and deterioration of their performance. Thus, it is required to establish a method for maintaining arousal level.

Currently external stimulations such as sounds and vibrations are popular for maintaining arousal level [1]. However, these methods have problems such as causing

© Springer Nature Switzerland AG 2019
S. Bagnara et al. (Eds.): IEA 2018, AISC 819, pp. 274–284, 2019.
https://doi.org/10.1007/978-3-319-96089-0_30

discomfort and sleep rebound [1, 2]. On the other hand, some psychophysiological research has suggested a relationship between motivation and arousal [3], but only a few preceding studies have examined applying these results to maintaining arousal. There have also been only a few examples of practical applications. Therefore, in this research, we propose and verify a method of maintaining arousal level by inducing intrinsic motivations. Here, we focus specifically on the driving of automobiles.

2   Method

2.1   Experimental System and Environment

The experimental system was created using Unity (ver.2017.1.0f3, Unity Technologies). Participants carried out experiments while sitting facing the desk on which the display (50 inch, NEC), steering wheel (G29 Driving Force, logicool), and gaze-measuring devices (EMR-AT VOXER, NAC) were installed (see Figs. 1 and 2).

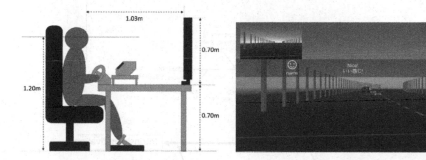

Fig. 1. Experimental system   **Fig. 2.** Experimental screen

2.2   Experimental Task

Participants controlled a simulated car on the screen using the steering wheel and performed two types of tasks: driving on a course of a certain width for ten minutes (manual operation) and monitoring automatic driving for ten minutes (automatic operation).

In manual operation, participants operated the steering wheel to advance the course, and when other cars approached, took appropriate countermeasures, such as overtaking or giving way. Also, when branching the course, participants determined the direction according to instructions presented on the screen with arrow symbols.

In automatic operation, participants only selected the direction of branching; all other operations were performed automatically.

2.3   Experimental Conditions

For each case of manual and automatic operation, three conditions were prepared: imposition of only simple driving without giving meaningful information

(Condition 1); provision of only messages as navigational aids (Condition 2); and provision of presentation for inducing motivation (Condition 3). Participants performed the experiment in one set of six trials, which comprised two types of operations, each under three conditions. The trial order was random with ten minutes of resting time between tasks, during which participants closed their eyes. To consider the influence of the circadian rhythm, the same participants conducted one set of experiments during the morning, daytime, and at night.

Condition 3 was established based on a cause-and-effect diagram of intrinsic motivation (Fig. 3). Presentation of feedback, presentation of information on other cars, and communication with other cars were set as mechanisms to satisfy desires for competence, autonomy, and relationship. Feedback concerning the task performance by participants was presented on the screen, and verbal rewards or countermeasures were presented on the screen by button operation by the participant. The driver's hometown and history were presented when participants saw other cars on the screen using the results of the gaze-measuring device. Greetings and thanks were presented from the other cars, and the participant was able to send messages from their own car.

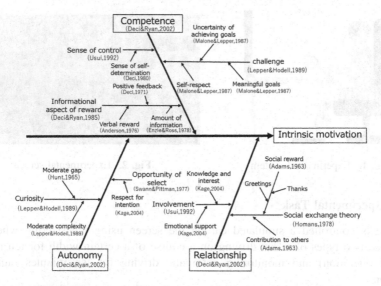

Fig. 3. Cause-and-effect diagram of intrinsic motivation [4–10]

2.4 Measurements

Subjective assessments and physiological-index measurements (including electroencephalograms (EEG), skin-conductance level (SCL), respiration (RSP), electrocardiograms (ECG), skin temperature (SKT), and changes in the volume of blood in the brain by fNIRS(functional near-infrared spectroscopy)) were recorded.

For subjective assessment, the mental work strain (MWS) checklist [11] was used. The checklist obtains six category scores from answers of 12 questions on a scale of one to seven: between agree and disagree. Table 1 shows the categories and questions of the MWS checklist.

Table 1. The MWS checklist

Category	Question
Sleepiness	Eyelids feel heavy
	Sleepy
General engagement	Feeling energetic
	Feeling active
Relaxation	Feeling relaxed
	Feeling calm
Mental strain	Feeling nervous
	Feeling anxious
Difficulty in concentration and attention	Dullness in thoughts
	Difficulty paying attention
Reduced motivation	Do not feel motivated
	Do not feel determined to do something?

In this research, the axis was reversed for the categories of Sleepiness, Reduced motivation, and Difficulty in Concentration and Attention, so that they could be analyzed as Arousal, Motivation, and Concentration and Attention, respectively. In addition, the axis was reversed for the category of Mental strain and coupled with the existing category of Relaxation.

EEG, SCL, RSP, ECG, and SKT were measured using MP150 (BIOPAC). The sampling rate was 2 kHz. Table 2 shows the position of the electrode, the measurement filter, and each index's relationship with the arousal level.

Here SCL, RSP, ECG, SKT, and RSA reflect sympathetic activity, which is used as an index of stress; Thus, it is necessary to consider the influence of stress.

Changes in blood volume in the brain was measured using OEG-16 (Spectratech). A band was attached over the forehead, and the oxygenated-hemoglobin-concentration change of 16 CH was measured. The mean of values of CH corresponding to the dorsolateral prefrontal cortex (DLPFC), medial prefrontal cortex (MPFC), and orbitofrontal cortex (OFC), which are thought to be related to motivation, were compared between conditions.

Table 2. Physiological indices

Index	Position of electrode	Filter	Relationship with the arousal level
EEG	A1, A2, and C3 of the 10–20 system	0.5 Hz ~ 35 Hz	Calculate amplitude of α, β, θ, and δ waves the value of $(\alpha+\beta)/(\theta+\delta)$ is proportional to the arousal level
SCL	Index finger and middle finger of left hand	1.0 Hz ~	Standardize the mean values of each task the value is proportional to the arousal level
RSP	Wear a band on the abdomen	0.05 Hz ~ 10 Hz	Calculate breathing interval, and standardized the mean values of each task the value is inverse proportional to the arousal level
ECG	Left and right clavie and abdomen	0.5 Hz ~ 35 Hz	Calculate heart rate, and standardized the mean values of each task the value is proportional to the arousal level
			Calculate RSA component of heart rate variability, and standardized the mean values of each task the value is inverse proportional to the arousal level
SKT	Nose tip	1.0 Hz ~	Standardize the mean values of each task the value is inverse proportional to the arousal level

2.5   Participants and Ethics

Participants were nine university students aged twenty-one to twenty-five. Participants did not consume alcohol or caffeine, and had an adequate amount of sleep the night before the experiment. This research was conducted after receiving approval from the Research Ethics Review Committee of the Keio University Faculty of Science and Technology.

3   Results and Discussion

3.1   Maintaining Arousal During Manual Operation

Figure 4 shows the mean scores in each category of the MWS checklist for manual operation. Statistical testing of the difference in score between the experimental conditions shows that the score of Condition 3 is highest in the categories of Arousal, Concentration and Attention, General Engagement, and Motivation. On the other hand, no significant difference between the conditions is found in the category of Relaxation.

Figure 5 shows the mean values of SKT during manual operation. There is found to be no significant difference in score between experimental conditions, but the value of Condition 3 is lower than that of Condition 1, meaning that the sympathetic system is activated in Condition 3. No significant differences between the conditions is found in terms of EEG, SCL, RSP, ECG, or RSA.

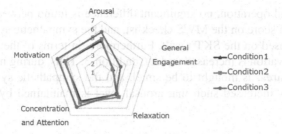

Fig. 4. The scores of MWS checklist for manual operation

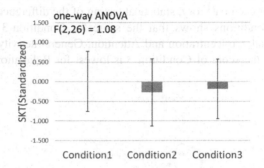

Fig. 5. The values of SKT for manual operation

Figure 6 shows the mean values of change in the volume of blood in the OFC during manual operation. The OFC is suggested to be related to motivation, and a large change in the value of oxygenated-hemoglobin concentration in the OFC indicates that motivation is high [12]. The value of Condition 3 is shown by statistical testing to be highest, meaning motivation is high in this case. No significant difference between the conditions is found in the DLPFC or MPFC.

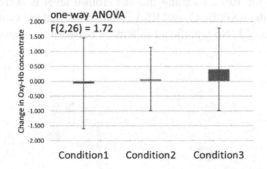

Fig. 6. The values of change in the volume of blood in the OFC during manual operation

During manual operation, no significant difference is found between the conditions for the Relaxation score on the MWS checklist, and the sympathetic system is activated in Condition 3 based on the SKT result. Furthermore, in terms of the volume of blood in the brain, motivation is increased under Condition 3. Thus, during manual operation, the influence of stress is thought to be small, and the sympathetic system is activated and motivation is increased such that arousal level is maintained by the motivation-inducing effect.

3.2 Maintaining Arousal During Automatic Operation

Figure 7 shows the mean scores in each category of the MWS checklist during automatic operation. For each category, statistical testing of the difference in score between the experimental conditions shows that the score for Condition 3 is highest in the categories of Arousal, Concentration and Attention, General activity, and Motivation. On the other hand, the score of Condition 3 is lowest for Relaxation.

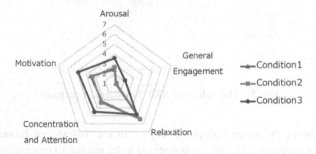

Fig. 7. The scores on the MWS checklist during automatic operation

Figures 8, 9, 10, and 11 show the mean values of EEG, RSP, ECG, and RSA during automatic operation. Statistical testing of the difference in score between the experimental conditions shows that the value for Condition 3 is lower than that for the other conditions in the EEG, meaning that the arousal level is decreased in that case. On the other hand, the RSP, ECG, and RSA results show that the sympathetic system is activated in Condition 3. No significant difference between the conditions is found for SCL and SKT.

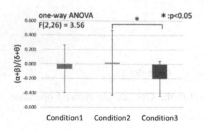

Fig. 8. EEG during automatic operation

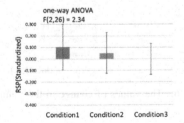

Fig. 9. RSP during automatic operation

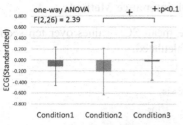

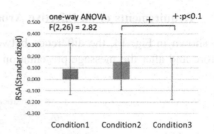

Fig. 10. ECG during automatic operation

Fig. 11. RSA during automatic operation

Figures 12 and 13 show the mean values of change in the volume of blood in the DLPFC and MPFC during automatic operation. DLPFC and MPFC are thought to be related to motivation, and a large change in value of oxygenated-hemoglobin concentration there indicates that motivation is high [13, 14]. The value of Condition 3 is found to be lowest, meaning motivation is low under Condition 3. No significant

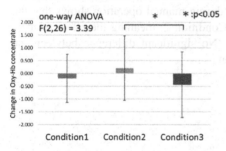

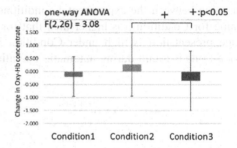

Fig. 12. The values of change in the volume of blood in the DLPFC during automatic operation

Fig. 13. The values of change in the volume of blood in the MPFC during automatic operation

differences between the conditions are found in the OFC.

During automatic operation, the Relaxation score is low in Condition 3 and the sympathetic system is activated under this condition according to RSP, ECG, and RSA; however, EEG suggests that the arousal level is low under Condition 3. In addition, based on the change in volume of blood in the brain, motivation is low under Condition 3. Thus, the activation of the sympathetic system and the decrease in motivation are thought to be caused by stress. In Condition 3, a response by a button was requested to confirm that participants accepted the presentation. It is thought that requiring button operation increases the stress of participants in automatic operations that require monitoring only, leading to a decrease in motivation.

3.3   Requirements of an Effective Arousal-Maintenance Method

As shown in Fig. 14, the difference between the mean SCL values over ten seconds before and after the message presentation was calculated.

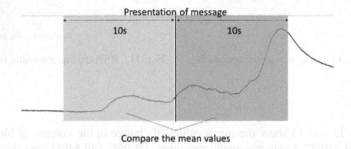

Fig. 14. Calculation of the change amount in SCL

Figures 15 and 16 show the change in the amount of SCL before and after message presentation during manual and automatic operation. Statistical testing of the difference in value between the experimental conditions during manual operation shows that the value for Condition 3 is higher than that for Condition 2, meaning the effect of message presentation is high under Condition 3. No significant difference between the conditions is found during automatic operation.

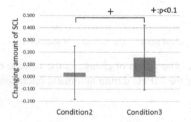

Fig. 15. The values of change in SCL during manual operation

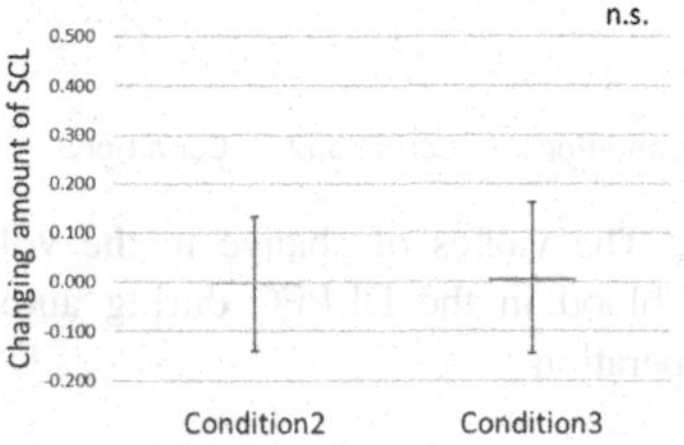

Fig. 16. The value of change in SCL during automatic operation

The change in SCL before and after message presentation is higher for manual operation than automatic. Thus, the effect of message presentation is greater during manual operation than automatic. One difference is the target of feedback. For manual operation, there are four targets: operation of the steering wheel, driving position, distance to other cars, and gaze transition. In contrast, for automatic operation, the target of feedback is only gaze transition. Thus, it is thought that a variety of forms of feedback enhances the motivation-inducing effect.

Beyond the content of the messages, there is a difference in the level of decrease of arousal between manual and automatic operators. During automatic operation, the arousal level has a higher tendency to decrease compared with manual operation. This can also be seen from the fact that the Arousal score on the MWS checklist during automatic operation is lower than that during manual operation. In order to examine how the arousal level at the time of the tasks affects the motivation-inducing effect, the results of manual operation were analyzed by classifying them into groups with high arousal level (group 1) and low arousal level (group 2). This classification was made according to the arousal level immediately after the start of the task. Figures 17 and 18 show the mean values of SCL and SKT in group 1. As a result of a statistical test of the difference in value between the conditions, the result indicates that the arousal level is high in Condition 3. No significant difference is found in the EEG, RSP, ECG, and RSA values of group 1 and all physiological indices of group 2.

From the result, the effect of maintaining arousal level is seen in the high-arousal-level group only; the effect tends to be seen in the case where arousal level is high. It is thought that the arousal level (when the effect occurs) affects the motivation-inducing effect.

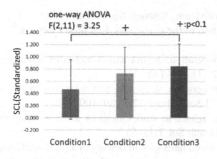

Fig. 17. The values of SCL for group 1   **Fig. 18.** The values of SKT for group 1

4 Conclusions

In this research, the effect of maintaining arousal level by inducing intrinsic motivation was considered.

Our experiment showed that the motivation-inducing effect maintained arousal level during manual operation, but not during automatic operation. Since the production of this experiment is seeking responses from participants, it is possible that this caused stress during automatic operation, lowering the intrinsic motivation. In addition, in the case of automatic operation, since the variety of feedback is poor, and the arousal level during the task is low, and the possibility is considered wherein the motivation-inducing effect in such a situation is low.

In the future, we plan to propose a method that will be effective even during automatic driving.

References

1. Yoshiyuki M et al (2016) Arousal retention effect of magnetic stimulation to car drivers preventing drowsy driving without sleep rebound. IEEJ Trans Electron Inf Syst 136(3):383–389
2. Yukie H (2000) A study on the effective method of aroma release program for keeping alertness. In: Proceedings of social automotive engineers of Japan, 61-00, pp 5–8
3. Psychology Notes HQ. The Arousal Theory of Motivation. https://www.psychologynoteshq.com/arousal-theory-of-motivation. Accessed 30 Jan 2018
4. Deci EL, Ryan RM (2002) Handbook of self-determination research. University of Rochester Press
5. Deci EL, Ryan RM (1985) Intrinsic motivation and self-determination in human behavior. Springer, Heidelberg
6. Mafumi U (1992) The effects of perceived competence and self-determination on intrinsic motivation. Jpn J Soc Psychol 7(2):85–91
7. Shigeo S (1984) The comparison of the effects of language and token rewards on intrinsic motivation. Jpn J Educ Psychol 32(4):286–295
8. Masaharu K (2004) Forfront of research of motivation. Kitaooji Shobo Publishing
9. Lepper MR, Hodell M (1989) Intrinsic motivation in the classroom. Res Motiv Educ 3:73–105
10. Kenichiro F (2002) Intrinsically motivated cooperation: merits and demerits of monetary reward in social survey. Infrastruct Plan Rev 19(1):137–143
11. Makoto T (1996) Analysis of the relationship between sleepiness and relaxation using a newly developed mental work strain checklist. J Sci Labour 72(3):89–100
12. Yoshihiko A, Kenichi K (2012) Brain activation related to work efficiency while listening to background music. Bull Gunma Prefect Coll Health Sci 7:45–53
13. Hiroshi S, Toshihiro K (2006) A change of the blood hemoglobin density in a prefrontal area at the time of attention control task enforcement: examination with "Kanahiroi" multi-cancellation test. Annual reports of School of Health Sciences Faculty of Medicine. Kyoto University, vol 3, pp 7–15
14. Taketoshi O, Hisao N (2005) Mechanism of emotion and intelligent information processing. High Brain Funct Res 25(2):116–128

Psychological Hardiness and Coping Strategies to Deal with Traumatic Event at Algerian Firefighters

Lahcene Bouabdellah[1]([⊠]) [iD], Idir Bensalem[1] [iD],
Mohamed Mokdad[2] [iD], and Houda Kherbache[1] [iD]

[1] Sétif University 2, Sétif, Algeria
doylettres@yahoo.fr, idir.hope@live.fr,
houdakhe@yahoo.fr
[2] University of Bahrain, Sakhir, Bahrain
mokdad@hotmail.com

Abstract. The purpose of this study is to investigate the psychological hardiness and coping strategies among Algerian firefighters.

The study used the quantitative method (cross-sectional survey).

Data were collected using the Scale of Psychological hardiness [1], and Ways of Coping Check List (WCC) [2].

The study sample consisted of (99) firefighters.

It has been found that both psychological hardiness, and coping strategies increased over time among Algerian firefighters.

Keywords: Algerian firefighters · Psychological hardiness · Coping strategies

1 Introduction

Civil protection is the state body responsible for helping members of society during disasters and accidents. It is important to carry out in-depth studies on the agents of this organization in order to understand the different factors that contribute to the formation of their personalities, and to have practical results that help to increase their professional skills and their psychological preparation. The help of this organization means the help of society as a whole.

Civil protection response officers are subjected to violent scenes, and traumatic events that can have an impact on all aspects of the firefighter's life, as shown in the Bos et al. [3]. Who have shown that the prevalence rate of complaints or disabilities among firefighters was high. Bailly found that if an individual faces a traumatic event or sees a traumatic scene, he will be subjected to psychological stress [4], but the reaction is different from one individual to another. This hypothesis is supported by Pierron [5] who concluded that an individual's exposure to a traumatic event produces a stressful energy that may exceed the individual's ability to withstand it, causing psychological trauma. Further, it is supported by the study of Kobasa [6] on psychological and social variables which to the individual to maintain his psychological and physical health. The results of his research have shown that the variable of

© Springer Nature Switzerland AG 2019
S. Bagnara et al. (Eds.): IEA 2018, AISC 819, pp. 285–292, 2019.
https://doi.org/10.1007/978-3-319-96089-0_31

psychological hardiness is one that acts as a psychological variable that reduces the impact of stressful events on the physical and psychological health of the individual. If people who have high psychological hardiness are subjected to psychological pressure, they do not get sick because they are more resistant. This indicates a dynamism, between the force of traumatic events and the strength of the individual's resistance called psychological hardiness. The confrontation of these two strengths requires the use of coping strategies to overcome the psychological stress. This is supported by Ray et al. [7] saying that when faced with stressful life events, the individual uses a wide range of coping strategies to modify the impact of stress. Research on coping over the past 30 years has been dominated by contextual models that emphasize a particular stressful encounter [8].

From the foregoing, we ask the following question:

• What is the level of psychological hardiness of firefighters?
• What are the coping strategies used by firefighters to curb work stress?

2 Methodology

Method: Researcher adopted the survey approach (the cross-sectional survey).

Sample: The study sample consisted of (99) firefighters. The age is determined between 24 and 45 years, which corresponds to the age of adults. Firefighters were divided into (4) groups; (30) firefighters with 05 months of experience, (28) firefighters with 10 months of experience, (25) firefighters with 15 months of experience and (16) firefighters with 20 months of experience.

Data collection tools: The following scales have been used:

• **Scale of Psychological hardiness** [1]. It is a tool that quantifies a psychological hardiness, consisting of 47 items distributed on the three dimensions of psycho-logical hardiness as follows:

1. Commitment: consists of 16 items and refers to the degree of obligations of the individual towards himself and his goals and others.
2. Control: consists of 15 words and refers to the degree of belief of the individual ability to control events.
3. Challenge: consists of 16 items and refers to the degree of individual challenges.

  The researchers confirmed the validity of the scale and its stability before using it.

• **Ways of Coping Check List (WCC)** [2]. The (WCC-R) has 27 items. It is a revised version of the older version of Lazarus and Folkman [8], the (WCC-R) was developed in 1996 by Cousson et al. [2].

  It measures 3 dimension:

1. Problem-focused coping strategies.
2. Emotion-focused coping strategies.
3. Seeking social support.

The researchers confirmed the validity of the scale and its stability before using it.

3   Results

3.1   Psychological Hardiness

The results show that the arithmetic mean of psychological hardiness of the firefighters with experience between 5 months and 20 months ranges from 98.84 to 106.09, meaning that it has increased in the medium level of psychological hardiness. It was found that psychological hardiness has increased with a high frequency in (5–10 month), thereafter it continued to increase with a less frequency till 20 months. This means there is a probability to increase after 20 months of experience to reach a higher level of psychological hardiness.

Firefighters had to increase their effectiveness and ability to use their available psychological and environmental resources to effectively cope with stressful events, and to adapt with their colleagues as it was found in the study of [9].

It is obvious that at point 5 months to 10 months, the firefighters acquired a great psychological hardiness in a short time, what corresponds to the early stages of joining the profession (Fig. 1).

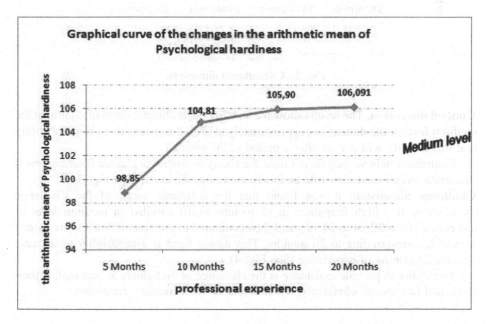

Fig. 1. Development of hardiness overtime.

Commitment dimension. It was found that the arithmetic mean of the commitment increased with a high frequency in (5 months to 10 months) from the medium to the high level (from 34.42 to 38.03), and then continued to increase with a less frequency to 38.39, corresponding to 20 months, This means there is a probability to increase beyond 20 months of experience. (See Fig. 2).

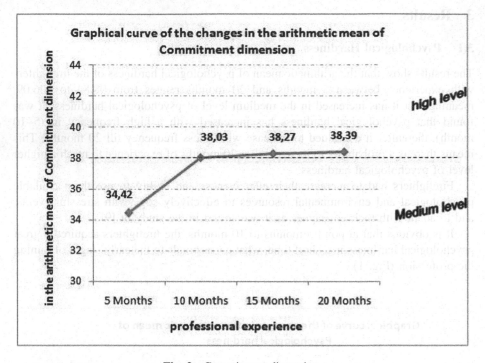

Fig. 2. Commitment dimension.

Control dimension. The results show the stability the arithmetic mean of control at the medium level of the dimension, approximately at (32), this means there is a probability that this stability will persist after a period of 20 months.

Firefighters believe they do not have the ability to control or change what happened (traumatic event), and their role is intervention after the event (See Fig. 3).

Challenge dimension. It was found that the arithmetic mean of the Challenge increased with a high frequency in (5 months to 10 months) in medium level of dimension (from 32.24 to 34.72), and then continued to increase with a less frequency to 35.52, corresponding to 20 months, This means there is a probability to increase beyond 20 months of experience (See Fig. 4).

The ability to provide assistance is the challenge of firefighters to accomplish their work and face events effectively, which explains the increasing dimension.

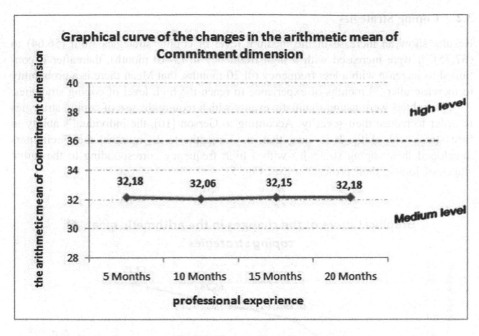

Fig. 3. Control dimension.

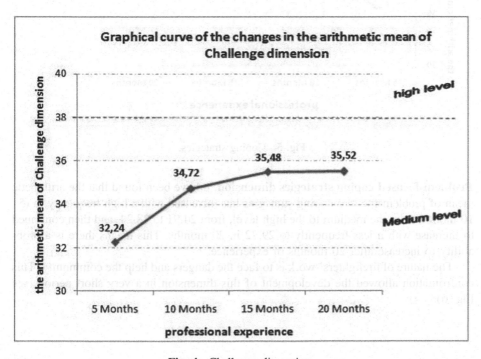

Fig. 4. Challenge dimension.

3.2   Coping Strategies

Results show an increase in the medium level of coping strategies from (56.64) to (67.75), it were increased with a high frequency in (5–10 month), thereafter it continued to increase with a less frequency till 20 months, that Mean there is a probability to increase after 20 months of experience to reach the high level of coping strategies.

Firefighters were facing traumatic events which require the use of coping strategies in order to reduce their severity, According to Gerson [10], the individual's ability to face stress with skills. It is clear that in (5 months to 10 months) the Firefighters developed their coping strategies with a high frequency corresponding to the initial stages of joining their profession (See Fig. 5).

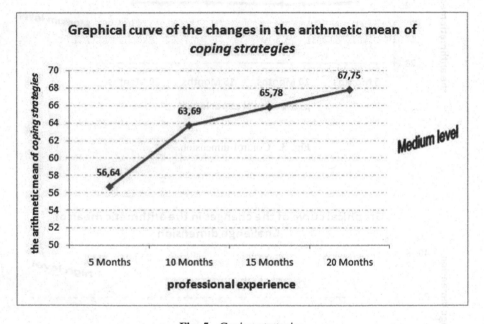

Fig. 5. Coping strategies.

Problem-focused coping strategies dimension. It have been found that the arithmetic mean of problem-focused coping strategies has increased with a high frequency in (5–10 month) from the medium to the high level, from 24.91 to 28.24, and then continued to increase with a less frequently to 29.72 in 20 months. This means there is a probability to increase after 20 months of experience.

The nature of firefighters' work is to face the dangers and help the community. This confrontation allowed the development of this dimension in a very short period (see Fig. 6).

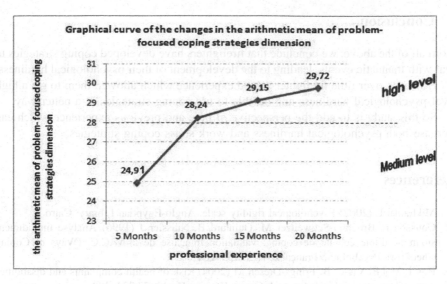

Fig. 6. Problem-focused coping strategies.

Emotion-focused coping strategies dimension. It was found that the arithmetic mean of Emotion-focused coping strategies dimension increased in the medium level from 15.21 to 19.85 in the 20 months, which means there is a probability to increase after 20 months of experience.

To face traumatic events, firefighters develop their strategies to control their emotions (See Fig. 7).

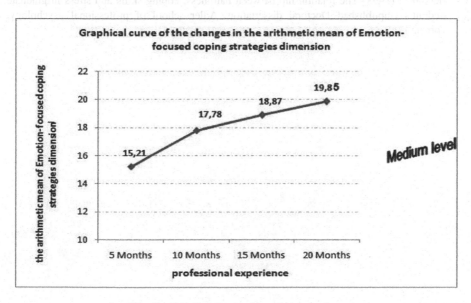

Fig. 7. Emotion-focused coping strategies.

4 Conclusion

From all of the above, we conclude that firefighters have developed coping strategies to deal with traumatic events, leading to the development of their psychological hardiness.

However, over time firefighter gained experience which allowed them to get a high level psychological hardiness and be able to use coping strategies in a better way,

So this study is to add the perspective of time and previous experience which can increase both psychological hardiness and work stress coping strategies.

References

1. Mikhamar E (2002) Psychological rigidity scale. Anglo-Egyptian Library, Cairo
2. Cousson F, Bruchon-Schweitzer M, Quintard B, Nuissier J (1996) Analyse multidimensionnelle d'une échelle de coping: validation française de la W.C.C. (Ways of Coping Checklist). Psychologie Française 41(2):155–164
3. Bos J, Mol E, Visser B, Frings-Dresen M (2004) Risk of health complaints and disabilities among Dutch firefighters. Int Arch Occup Environ Health 77:373–382
4. Bailly D (1996) The role of β-adrenoceptor blockers in the treatment of psychiatric disorders. CNS Drugs 5(2):115–136
5. Pierron R (1983) l'agression c'est les autres. édition Bordas, Paris
6. Kobasa SC (1979) Stressful life events. J Personal Soc Psychol 37(1):01–11
7. Ray C, Lindop J, Gibson S (1982) The concept of coping. Psychol Med 12(2):385–395
8. Lazarus RS, Folkman S (1984) Stress, appraisal and coping. Springer, New York
9. Sawhney G, Jennings KS, Britt TW, Sliter MT (2017) Occupational stress and mental health symptoms: examining the moderating effect of work recovery strategies in firefighters. J Occup Health Psychol, 1–14. Advance online publication, 12 June
10. Gerson M (1998) The relationship between hardiness, coping skills and stress in graduate student's Unpublished Doctoral dissertation. Adler school of professional psychology, Torronto

Effective Factors on the Occurrence of Falling from Height Accidents in Construction Projects by Using DEMATEL Method

Reza Gholamnia[1], Mobin Ebrahimian[1(✉)],
Saeid Bahramzadeh Gendeshmin[1], Reza Saeedi[2], and Sina Firooznia[3]

[1] School of Health, Safety and Environment,
Shahid Beheshti University of Medical Sciences, Tehran, Iran
m95.ebrahimian@sbmu.ac.ir
[2] Department of Health Sciences, School of Health, Safety and Environment,
Shahid Beheshti University of Medical Sciences, Tehran, Iran
[3] Islamic Azad University North Tehran Branch, Tehran, Iran

Abstract. In construction industries, falling from height recognized is one of the main factors in the occurrence of construction accidents. In this study, 10 construction blocks were selected in Tehran that were under construction. Three groups of organizational, individual and environmental factors and sub-factors were considered as effective factors on the falling from height. Using the experts' opinion, the main factors were divided into three groups, whereas their sub-factors were divided into 15 cases. Then, DEMATEL method was used to determine the most important factor and it's sub-factors influencing the occurrence of accidents falling from a height and their relationship. The matrix of the Decision Making Trial and Evaluation (DEMATEL) method shows both the causal relationship between the factors and the effectiveness of their sub-factors variables. The results showed that individual factors were considered as the most important factors and their sub-factors had the most effect on the occurrence of falling from height accidents. It can be said that in determining the relationships of sub-factors, three out of six individual sub-factors including motivation, training hours, and age/experience and two out of five organizational sub-factors including management commitment and safety culture have the most impact on the occurrence of falling from height accidents. Paying attention to safety training and also increasing the motivation of workers can be effective in reducing occupational/psychological stress and accidents of falling from height.

Keywords: Falling from height · DEMATEL method · Construction project

1 Introduction

Construction industries had been known as one of the most dangerous industries in the world and falling from height has been reported as one of the most cause to death in building industry [1]. The complexity of construction, short-term organizational structure, working site altering, complexity of working environment [2, 3], and special behaviour of workers in different working condition is the most important characteristic

S. Bagnara et al. (Eds.): IEA 2018, AISC 819, pp. 293–305, 2019.
https://doi.org/10.1007/978-3-319-96089-0_32

of construction industries which distinguish it from other industries [4, 5]. According to studies, the highest number of falling from height accidents is related to the Construction industry, about 50–60% of mentioned accidents in this industry contribute to death [6].

In recent years industrial and developed countries' statistics like USA, China and Hong Kong did not show downward trend in the number of construction mortality rate [7]. Chi and woo showed that about 30% of work-related death is due to falling [8]. Falling from height leads to deaths and injuries among the workers of construction industries in the USA which include 40% of deaths and 20% of working days lost in 2010 [9].

Some studies have shown that different factors such as; organizational, individual, and environmental factors; plays a role in occurring the constructional accidents, especially in the industry of construction [10–12]. Other studies have shown that organizational factors play a key role in safety attitude [13–17], the severity of accidents outcome and the number of the accidents [18–20]. For instance, after Bubal accident, organizational factors singled out as a latent cause in accidents which can be used in surveying the safety function of organizations [21, 22]. Organizational factors do not have the same impact on safety activity in different countries because it is related to some outer factors like social, technical, economic and cultural situation [23]. Some researchers have reported organizational factor such as communications, supervising, inspecting and safety culture as identifying an effective factor in preventing of constructional accidents and falling from a height [24–26]. On the one hand, to tackle work-related accidents and hazards, individual factors like hours of training [25]; personal protective equipment [27]; age, experience [28]; and lack of balance is very important and effective [30, 31]. Nil and griffin have introduced motivation as an effective factor in occurring accidents especially construction industry which motivation of workers to the safety factors results in increasing safe behaviour and decreasing work-related accidents [32]. On the other hand, in case of the positive attitude of workers on up and running safety plans and principles in the organization will enhance their partnership in implementing the safety plans and safe behaviour [33]. Environmental factors which are effective to falling from height accidents include the height of work platform [34]. Management commitment to safety is an effective factor in successfully implementing of safety plans [25].

With regard to the high number of organizational accidents and in particular falling from height in Iran, in this study, the factors affecting the falling from height accident were identified by experts and questionnaires and the causal relationships between factors affecting high-altitude incidents in construction projects of west Tehran were investigated.

2   Method and Material

After studying the accidents happened from 1992 to 1395, 92 out of 421 work-related accidents allocated to the accidents of falling from height. According to integrated information reported from health and safety departments of mentioned projects and case study of similar studies, 82 factors and sub-factors of construction accidents especially falling from height accidents were identified [14, 25, 27, 30]. In this research, identified factors were survived by a group of safety managers and experienced safety experts in related fields. Finally, with regard to the aims of the study and the project condition, 3 main factors categorized in 15 sub-factors (Table 1).

Table 1. Factors and sub-factors

Factors	Sub-factors
Organizational factors	*Management commitment:* management commitment is one of the most important organizational factors in implementing safety plans and safety culture and plays a key role in accidents [36–40]
	Safety culture: safety culture is derived from a series of individual and group values including believes, understanding, attitude, qualification as well as behavioural patterns which determine commitment, style, and skills of health and safety management and safety of organization, affecting the rate of accidents [41]
	Supervision and Inspection: both supervision and inspection decrease the number of work-related and falling from height accidents leading to safety [42, 43]
	Organization and project size: size of organization of projects is one of the main contributing factors of accidents
	Rules and regulations: rules and regulations of safety is one of significant factors and should be taken into account as the main factor [43, 45]
Individual factors	*Education:* illiteracy is effective in accident rate [33].
	Training hours: attending safety training classes related to workers job as well as being familiar with hazards, decrease injuries and accident rate [46]
	Personal Protective Equipment: lack of using fall arrest equipment has been titled as the main contributing factor in many studies. Supplementary to this, PPE is a preventive approach [34, 42]
	Age/experience: age and experience of workers is one effective factor in accidents [28]
	Lack of balance: lack of maintain of worker's balance resulting from the physical ability of workers, affect the rate of accidents occurrence [29, 43]
	Motivation: existence of motivations in workers create a commitment to work in a safe way [33]
Environmental factors	*Climate condition:* workers are in touch with a range of weather condition such as cold weather, hot weather, wind, rainy and etc., and these conditions cause falling in some cases [35]
	Interference: Interference between different work situations sometimes results in the presence of two executives from two different companies along with interference within the management of work
	Level Smooth: The purpose of the level smooth in this study is the lack of housekeeping on the workplace platform at the height. In many cases, the falling is due to the lack of individual's attention for discipline of the workplace safety which results to the collapse of personal balance caused by this neglect
	Work platform height: Height is one of the most important parameters in falling from height, which has been studied by many researchers. Huang and Hinze recommended that more attention is needed to be paid to workers who work at an altitude of more than 30 ft (9 m) due to the high accident rate in this industry [47]

DEMATEL technique was used to identify the causal and relationships between factors and sub-factors. This technique is operated by a step-by-step approach as follows [47, 48]:

Step1- Calculation the direct relation matrix (M): At first, a group of experts investigate the relationship between sets of factors based on paired comparison scale. Each respondent produces a direct relation matrix. Then, an average matrix is derived using the mean of the same factors/sub-factors in the various direct matrices of the respondents.

Step2-Calculation of the normal direct-relation matrix (N): The sum of all rows and columns was computed and then the inverse of the largest number is formed (k). Then, all values of M are multiplied by the inverse of this number (K), so that normal matrix is obtained.

Step3- Calculate the total-relation (T) matrix: Once the normalized M matrix, is obtained to calculating the T matrix, first the identity matrix was formed (I). Then, the normal matrix (N) is subtracted from the identity matrix and the results are reversed. Finally, the normal matrix is multiplied by the inverse matrix.

Step4- Calculation threshold intensity value and display of the network relationships map: The threshold intensity was calculated to determine the network relationship map (NRM). In order to explain the structural relationship among the factors while keeping the complexity of the system to a manageable level, it is necessary to set a threshold value to filter out the negligible effects in matrix T. The threshold value was computed by the average of the elements in the total-relation matrix. Once threshold value was calculated only the effects greater than the threshold value, and significant would be chosen and shown in a digraph. At this step D and R vectors were calculated that demonstrating the sum of rows and columns of the total-relation matrix, respectively. D vectors demonstrating the factor attribute influencing other factors of the model. R vectors criteria attribute over which other criteria exert an influence. Once D and R vectors were calculated, the diagram can be acquired by mapping the data set of (D + R, D − R). (D + R) is defined as the prominence, show the influence of factor and total extent of being influenced. D + R is called prominence, which indicates the influence of element's degree. If it is positive, the factor tends to fall under the effect category. If it is negative, the criterion tends to fall under the causal category.

In this study, factors and sub-factors are named by the numerical index according to Table 2.

Table 2. Factors and sub-factors of the study

Sub-factors	Markers	Factors	Markers
Education	SS1	Individual factors	C1
Training hours	SS2		
PPE	SS3		
Age/experience	SS4		
Lack of balance	SS5		
Motivation	SS6		
Management commitment	SS7	Organizational factors	C2
Safety culture	SS8		
Inspection/investigation	SS9		
Project and organization size	SS10		
Rules and regulations	SS11		
Climate condition	SS12	Environmental factors	C3
Interference	SS13		
Level smooth	SS14		
Work platform heights	SS15		

3 Results

The matrix derived from the Dimethyl technique (internal communication matrix), which presents both the causal relationship between the factors and the effect and variability of the variables, has been presented. The following steps were taken in order to present and obtain the results:

- Calculation of direct relationship matrix (M):

Since comments of a few experts were used, the simple arithmetic mean of the comments was used and the direct contact matrix (M) was obtained as follows (Table 3)

Table 3. Direct contact matrix (M)

Total of rows	Environmental factors	Organizational factors	Individual factors	
4.8	2.3	2.5	0	Individual factors
4.60	2.2	0	2.5	Organizational factors
4.50	0	2.00	2.5	Environmental factors
0	4.5	4.6	5.0	Total of columns

- Calculation of normal direct-relation matrix (N = K*M);

The sum of all rows and columns was calculated. Inverted the largest k number has been made. Based on the table, the largest number is 5.0. and all table values are multiplied by the inverse of this number so that the following matrix is obtained (Table 4)

$$k = \frac{1}{\max \sum\limits_{j=1}^{n} a_{ij}} = 0.20$$

Table 4. Normalized matrix (N)

Environmental factors	Organizational factors	Individual factors	N
0.460	0.500	0.000	Individual factors
0.440	0.000	0.500	Organizational factors
0.000	0.400	0.500	Environmental factors

- Calculation of complete-communication matrix

To calculate the complete matrix, at first the identity matrix (I) was formed. Afterwards, normalize matrix was subtracted from the identity matrix and derived matrix has been inverted. Finally, we multiplied the normal matrix by the inverse matrix (Table 5).

$$T = N(I - N)^{-1}$$

Table 5. Complete-communication matrix (T)

Environmental factors	Organizational factors	Individual factor	T Matrix
6.113	5.612	6.245	Individual factors
5.459	5.351	6.112	Organizational factors
5.069	5.452	6.119	Environmental factors

- Showing the network relationships map

To determine the network relationships map (NRM) The severity of the threshold was calculated. In this study, the severity of the threshold was determined 6,287. Therefore, a significant relationship pattern was determined as follows (Table 6).

Table 6. The pattern of significant relationship of main factors

Environmental factors	Organizational factors	Individual factor	
6.113	0	6.110	Individual factor
6.440	0	6.422	Organizational factors
0	0	6.259	Environmental factors

In the table above, all zero values indicate that the causal relationship between the factors is not considered.

Table 7. The pattern of causal relationships of main factors

D−R	D + R	R	D	
0.909	36.599	17.845	18.754	Individual factor
−0.907	34.907	17.907	17.000	Organizational factors
−0.680	36.123	17.024	17.411	Environmental factors

According to Table 7, the sum of elements of each row (D) presents effectiveness rate of that factor on other model factors. On this basis, improving individual factors has the greatest impact, then organizational factors fall to the next level. In addition to this, Environmental factors have the least influence.

Sum of column elements for each factor indicates susceptibility rate of that factor from other system factors. Therefore, organizational factor has the greatest susceptibility (Table 7).

Horizontal vector (D + R) is impressionability and effectiveness rate of the factor on the system. In other words, more D + R rate of a factor means more interaction of the factor with other factors of the system.

Based on this, the individual factors' criterion has the highest interaction with other criteria and the organizational factors criterion has the least interaction with other criteria. Vertical vector (D−R), indicates effectiveness power of every factor. This means that, if D−R is positive the variable will causal factor and if it is negative the variable will be impacted factor. So on this basis, individual factors are causal and environmental and organizational factors are impacted. Identify the inner relations of the sub-factors:

In this research, relationships between the sub-factors have also been studied. First, the matrix of the relationship between the elements and the relationship between the elements were measured by using expert's comments. Subsequently, the decision matrix was built by computing the average judgment of the experts. Finally, the matrixes of inner relationships of sub-factors in Table 8 have been presented.

After calculating the matrix of complete relationships, the threshold value was determined to be 0.551. Then significant relationship pattern was obtained and according to its relations.

Table 8. Matrixes of inner relationships of sub-factors (Complete communication matrix)

ss15	ss14	ss13	ss12	ss11	ss10	ss9	ss8	ss7	ss6	ss5	ss4	ss3	ss2	ss1	Matrix
0.401	0.402	0.461	0.449	0.418	0.446	0.514	0.450	0.424	0.441	0.425	0.430	0.465	0.429	0.402	ss1
0.408	0.410	0.463	0.449	0.416	0.456	0.513	0.438	0.421	0.464	0.441	0.428	0.468	0.370	0.468	ss2
0.410	0.418	0.463	0.440	0.426	0.448	0.514	0.442	0.444	0.456	0.428	0.419	0.410	0.424	0.464	ss3
0.325	0.335	0.376	0.347	0.42	0.410	0.420	0.345	0.347	0.348	0.344	0.295	0.377	0.339	0.366	ss4
0.460	0.470	0.514	0.492	0.473	0.510	0.542	0.512	0.499	0.522	0.419	0.411	0.545	0.482	0.537	ss5
0.455	0.465	0.522	0.502	0.472	0.519	0.578	0.491	0.481	0.439	0.470	0.481	0.541	0.479	0.537	ss6
0.412	0.422	0.475	0.401	0.438	0.466	0.515	0.459	0.293	0.454	0.441	0.441	0.501	0.475	0.469	ss7
0.448	0.453	0.520	0.499	0.472	0.510	0.573	0.426	0.478	0.511	0.478	0.480	0.526	0.477	0.534	ss8
0.445	0.451	0.540	0.521	0.494	0.517	0.509	0.478	0.489	0.511	0.493	0.494	0.550	0.497	0.545	ss9
0.421	0.441	0.499	0.486	0.410	0.427	0.561	0.471	0.466	0.496	0.466	0.471	0.516	0.469	0.508	ss10
0.313	0.333	0.423	0.396	0.333	0.410	0.470	0.403	0.402	0.412	0.389	0.385	0.438	0.487	0.424	ss11
0.366	0.388	0.427	0.353	0.384	0.406	0.483	0.410	0.392	0.419	0.398	0.402	0.423	0.400	0.439	ss12
0.305	0.325	0.396	0.428	0.411	0.442	0.520	0.443	0.428	0.452	0.422	0.410	0.471	0.432	0.468	ss13
0.315	0.338	0.426	0.411	0.408	0.423	0.490	0.424	0.414	0.431	0.420	0.406	0.443	0.396	0.433	ss14
0.312	0.328	0.422	0.401	0.406	0.413	0.480	0.420	0.412	0.421	0.418	0.4-6	0.440	0.376	0.413	ss15

Table 9 indicates that to determine the relations of sub-factors, four out of five organizational sub-factors, namely, Inspection/investigation, management commitment, safety culture and organization/project size, and three sub-factors of the individual factors, namely, training hours, motivation and age/Experience have the most impact and fall into the causal variables and others sub-factors are affected. According to Table 9, mutual understanding has the most interaction with other sub-factors.

Table 9. Causal relationships pattern of research sub-factors

D−R	D + R	R	D	Markers	Sub-factors
−0.439	12.749	6.594	6.115	ss1	Education
0.345	12.214	6.008	6.981	ss2	Training hours
−0.480	12.869	6.674	6.195	ss3	PPE
1.115	10.936	6.036	7.476	ss4	Aga/experience
0.228	13.054	6.035	4.785	ss5	Lack of balance
0.845	12.889	6.018	6.881	ss6	Motivation
0.551	13.335	6.357	5.421	ss7	Management commitment
0.359	12.399	6.067	4.333	ss8	Safety culture
0.651	13.140	6.201	6.939	ss9	Inspection/investigation
0.456	13.077	6.339	6.738	ss10	Project and organization size
−0.107	14.345	7.226	4.229	ss11	Rules and regulations
−0.283	11.594	5.939	5.656	ss12	Climate condition
−0.400	11.941	6.217	5.724	ss13	Interference
−0.227	12.641	6.497	6.145	ss14	Level smooth
0.021	11.675	5.808	5.866	ss15	Work platform heights

4 Discussion

The main purpose of this study was to identify the factors affecting the incidence of falling from the height and to determine the inner relations of these factors using the DEMATEL method. The results of this research showed that individual factors are among the other main factors causal factors. This shows the high importance of individual factors and effectiveness of that on other factors of falling from height. Staff training, training hours, age/experience is important factors of effective individual factors on falling from height [49]. Most of the organization and HSE managers have acknowledged that training employees and paying attention to their level of awareness can be important in promoting safety culture to better control workers' behaviors and protecting safety awareness. On the other hand, organizational factors and its sub-factors such as management commitment and safety culture have been confirmed as key elements of organizational success in competitive areas, regardless of the role of certain aspects such as quality, production, job satisfaction and safety [50].

The results show that three sub-factors i.e., training hours, motivation and age/experience related to individual factors and also 2 sub-factor like management commitment and safety culture related to organizational factor has the greatest effect and are causal. And other sub-factors are affected. Various studies have shown that management commitment to safety leads safety to be seen as a value in projects and also it cause staffs to feel that safety attitudes toward their managers are positive and supportive attitudes [51]. On the other hand, it should be taken into account that the adherence of low-level managers to the programs and guidelines of safety in the organization is achieved if the high-level managers of the organization pay attention to the safety and training of its principles [52].

In a study carried out to identify effective factors in improving the safety culture, the sub-factors of management commitment, training, motivation, communication and employee participation were announced as the most important factors. In current study, safety culture as one of the organizational factors has the most impact and is considered as causal variables. The importance of safety culture is not an unfamiliar concept to any of the professionals in the safety and health field. In construction projects, work is often done at different locations, and the mix of contractors and employees changes throughout the project's progress which poses many problems in implementing safety and safety culture promotion programs [53]. Safety culture affects the general habits and beliefs of individual and is highly effective in occupational accidents and injuries [54].

Mutual understanding is another organizational sub-factor that has the most effective in this study and has been identified as causal variables. The existence of a significant relationship between the mutual understanding in the organization and the adaptation of activities and obedience of employees has been emphasized in many studies [55–58]. The results of using DEMATEL technique in this study showed that training hours and psychological/occupational stresses related to individual factors have the most impact on other factors and are considered a causal variable. Recent studies have shown that safety education is one of the most important tools for preventing injuries, risks and occupational diseases [59, 60] since it is a prerequisite for improving safety and health at work [61].

Finally, this study showed that individual factors like motivation among workers and increasing their awareness of safety principles can be effective in promoting the organization's safety culture and reducing incidents of falling from height accidents. On the other hand, organizational factors such as management commitment can be effective in improving the safety of the organization and encourage workers to fallow these principles as much as possible.

5 Conclusions

According to the results of the study, construction projects should always considered as a strategy to reduce the unsafe conditions of the environment. Also, specialized safety training programs and stress control management should be implemented on the site which can play an effective role in reducing the incidence of falls from height accident in the construction industry and increase management efficiency. Individual factors and their sub-factors, based on the results of this research, have the most impact on the other sub-factors. It means that the training and control of the personnel, supervision of the personnel should be taken into consideration by the managers of the organization and health and safety experts. Therefore, more attention to individual factors and its dimensions in the construction industry is necessary to prevent incidents of falling from height.

Acknowledgements. The authors gratefully acknowledge SAMAN MOHIT Construction Company for being partners in the research team.

References

1. Sorock GS, Smith EO, Goldoft M (1993) Fatal occupational injuries in the New Jersey construction industry, 1983–1989. J Occup Med 35:916–921
2. European Commission Building (1987) Construction safety. Building Advisory Service, London
3. Fang D, Wu H (2013) Development of a safety culture interaction (SCI) model for construction project. Saf Sci 57:138–149
4. Geller ES (2001) Working safe. how to help people actively care for health and safety. CRC Press, Boca Raton
5. Geller E (2001) Sustaining participation in a safety improvement process: ten relevant principles from behavioral science. Prof Saf 46(9):24–29
6. Falls from height - Prevention and risk control effectiveness. Prepared by BOMEL Limited for the Health and Safety Executive (2003)
7. Zhang M, Fang D (2013) A continuous behavior-based safety strategy for persistent safety improvement in construction industry. Autom Constr 34:101–107
8. Chi C-F, Wu M-L (1997) Fatal occupational injuries in Taiwan—relationship between fatality rate and age. Saf Sci 27:1–17

9. Bureau of Labor Statistics (2014a) U.S. Department of Labor. Table A-9. Fatal Occupational Injuries by Event or Exposure for all Fatalities and Major Private Industry Sector. All United States, 2014. http://www.bls.gov/iif/oshcfoi1.htm#2014
10. Wagenaar WA, Groeneweg J (1987) Accidents at sea: multiple causes and impossible consequences. Int J Man-Mach Stud 27:587–598
11. Reason, J. Human error. Cambridge University Press, Cambridge
12. Drury CG, Brill M (1983) Human factors in consumer product accident investigation. Hum Factors 25(3):329–342
13. Pidgeon N, O'Leary M (2000) Man-made disasters: why technology and organizations fail. Saf Sci 34:15–30
14. Neal A, Griffin MA, Hart PM (2000) The impact of organizational climate on safety climate and individual behavior. Saf Sci 34:99–109
15. Oliver A, Cheyne A, Tomas JM, Cox S (2002) The effects of organizational and individual factors on occupational accidents. J Occup Organ Psychol 75:473–488
16. Seo DC (2005) An explicative model of unsafe work behavior. Saf Sci 43:187–211
17. Tomas JM, Melia JL, Oliver A (1999) A cross-validation of a structural equation model of accidents: organizational and psychological variables as predictors of work safety. Work Stress 13:49–58
18. Hunag YH, Ho M, Smith GS, Chen PY (2006) Safety climate and self-reported injury: assessing the mediating role of employee safety control. Accid Anal Prev 38:425–433
19. Siu OL, Philips DR, Leung TW (2004) Safety climate and safety performance among construction workers in Hong Kong: the role of psychological strains as mediators. Accid Anal Prev 36:359–366
20. Varonen U, Mattila M (2000) The safety climate and its relationship to safety practices, safety of the work environment and occupational accidents in eight wood-processing companies. Accid Anal Prev 32:761–769
21. Flin R, Mearns K, O'Connor P, Bryden R (2000) Measuring safety climate: identifying the common features. Saf Sci 34:177–193
22. Takano K, Kojima M, Hasegawa N, Hirose A (2001) Interrelationships between organizational factors and major safety indicators: a preliminary field study. In: Wilpert B, Itoigawa N (eds) Safety culture in nuclear power operations. Taylor & Francis, London, pp 189–205
23. Helmreich RL, Merritt AC (1998) Culture at work in aviation and medicine: national, organizational, and professional influences. Ashgate, Aldershot
24. Simard M, Marchand A (1994) The behaviour of first-line supervisor in accident prevention and effectiveness in occupational safety. Saf Sci 17:169–185
25. Cohen A (1977) Factors in successful occupational safety programs. J Saf Res 9:168–178
26. Embrey DE (1992) Incorporating management and organizational factors into probabilistic safety assessment. Reliab Eng Syst Saf 38:199–208
27. Drury CG, Brill M (1983) Human factors in consumer product accident investigation. Hum Factors 25(3):329–342
28. Chau N, Pe'try D, Gavillot C, Guillaume S, Beaucaillou C, Bourgkard E, Gruber M, Monhoven N, Andre' JM (1995) Implications professionnelles des le'sions se've`res du member supe' rieur. Arch Mal Prof 56:12–22
29. Simon-Rigaud ML (1995) Physical activities and health: results of a national survey of workers. Bull Acad Natl Med 179:1429–1436
30. Weick KE (1990) The vulnerable system: analysis of the Tenerife air disaster. J Manag 16:571–593

31. Forcier BH, Walters AE, Brasher EE, Jones JW (2001) Creating a safer work environment through psychological assessment: a review of a measure of safety consciousness. In: Stuhlmacher A, Cellar D (eds) Workplace safety: individual differences in behavior. Hayworth Press, Chicago, pp 53–65

32. Neal A, Griffin MA. A study of the lagged relationships among safety climate, safety motivation, safety behavior, and accidents at the individual and group levels. J Appl Psychol

33. Rahimi Pordanjani T, Mohammadzadeh Ebrahimi A. The mediating role of safety motivation in relationship between of safety management practices and unsafe behaviors. Received 21 Aug 2015 Revised 24 Dec 2015 Accepted 07 May 2016

34. Huang X, Hinze J (2003) Analysis of construction worker fall accidents. J Constr Eng Manag 129(3):262–271

35. Nadhim EA, Hon C, Xia B, Stewart I, Fang D (2016) Falls from height in the construction industry: a critical review of the scientific literature

36. Hadikusumo BHW, Jitwasinkul B, Memon AQ (2017) Role of organizational factors affecting worker safety behavior: a Bayesian belief network approach. Procedia Eng 171:13–19

37. Rahimi Pordanjani T, Mohammadzade EA (2015) The effect of employees' management commitment to safety and consciousness on unsafe performance: the mediating role of safety self-efficacy. J Health Saf Work 4(4):69–80

38. McKeon C (2004) Psychological factors influencing unsafe behaviour during medication administration. University of Southern Queensland

39. Vinodkumar M, Bhasi M (2011) A study on the impact of management system certification on safety management. Saf Sci 49(3):498–507

40. Fernández-Muñiz B, Montes-Peón JM, Vázquez-Ordás CJ (2012) Safety climate in OHSAS 18001-certified organisations: antecedents and consequences of safety behaviour. Accid Anal Prev 45:745–758

41. Gadd S, Collins AM (2002) Safety culture: a review of the literature. Health & Safety Laboratory, pp 1–36. (HSL/2002/25)

42. Occupational Safety and Health Administration (OSHA) (1998) Fall protection in construction. Publication 3146, Washington, D.C

43. Lubega HA, Kiggundu BM, Tindiwensi, D (2000) An investigation into the causes of accidents in the construction industry in Uganda. In: 2nd International conference on construction in developing countries: challenges facing the construction industry in developing countries, Botswana, 15–17 Nov 2002, pp 1–12. http://buildnet.csir.co.za

44. Simon-Rigaud ML (1995) Physical activities and health: results of a national survey of workers. Bull Acad Natl Med 179:1429–1436

45. Tam CM, Zeng SX, Deng ZM (2004) Identifying elements of poor construction safety management in China. Saf Sci 42(2004):569–586

46. Chalak MH, Shabahang H, Laal F, Almasi Z (2016) Evaluating the risk of cement open mining activities and its health hazards by FMEA method in Zabols cement mines before and after training interventions 2015. Eur J Biomed 3(12):623–628

47. Lin Y-T, Yang Y-H, Kang J-S, Yu H-C (2011) Using DEMATEL method to explore the core competences and causal effect of the IC design service company: an empirical case study. Expert Syst Appl 38(5):6262–6268

48. Bacudio LR, Benjamin MFD, Eusebio RCP, Holaysan SAK, Promentilla MAB, Yu KDS et al (2016) Analyzing barriers to implementing industrial symbiosis networks using DEMATEL. Sustain Prod Consum 7:57–65

49. Byrd TC (2014) Factors impacting safety climate in a small, rural hospital setting. Union University

50. Wang R-C, Chuu S-J (2004) Group decision-making using a fuzzy linguistic approach for evaluating the flexibility in a manufacturing system. Eur J Oper Res 154(3):563–572

51. Farh J-L, Cheng B-S (2000) A cultural analysis of paternalistic leadership in Chinese organizations. In: Li JT, Tsui AS, Weldon E (eds) Management and organizations in the Chinese context. Springer, London, pp 84–127. https://doi.org/10.1057/9780230511590_5

52. Hsu SH, Lee C-C, Wu M-C, Takano K (2008) A cross-cultural study of organizational factors on safety: Japanese vs. Taiwanese oil refinery plants. Accid Anal Prev 40(1):24–34

53. Jeschke KC, Kines P, Rasmussen L, Andersen LPS, Dyreborg J, Ajslev J, et al (2017) Process evaluation of a toolbox-training program for construction foremen in Denmark. Saf Sci 94:152–160

54. Chib S, Kanetkar M (2014) Safety culture: the buzzword to ensure occupational safety and health. Procedia Econ Finan 11:130–136

55. Bies RJ, Moag JS (1986) Interactional justice: communication criteria of fairness. Res Negot Organ 1(1):43–55

56. Bies RJ, Shapiro DL (1987) Interactional fairness judgments: the influence of causal accounts. Soc Justice Res 1(2):199–218

57. Shapiro D, Brett J (eds) (1991) Comparing the instrumental and value-expressive models of procedural justice under conditions of high and low decision control. National Academy of Management meeting, Miami, FL

58. Shapiro DL (1993) Reconciling theoretical differences among procedural justice researchers by re-evaluating what it means to have one's views" considered": Implications for third-party managers

59. Kinoshita T, Kanehira E, Matsuda M, Okazumi S, Katoh R (2010) Effectiveness of a team participation training course for laparoscopy-assisted gastrectomy. Surg Endosc 24(3):561–566

60. Faigenbaum AD, Myer GD (2010) Resistance training among young athletes: safety, efficacy and injury prevention effects. Br J Sports Med 44(1):56–63

61. Vidal-Gomel C (2017) Training to safety rules use. some reflections on a case study. Saf Sci 93:134–142

Comprehensive Rehabilitation for Accident and Occupational Diseases in Colombian Workers

Juan Carlos Velásquez V[1] and Diana Marcela Velasquez B[2(✉)]

[1] Universidad del Valle, Cali, Colombia
juan.carlos.velasquez@correounivalle.edu.co
[2] Universidad Libre, Cali, Colombia
diana.counsel88@gmail.com

Abstract. The numbers of accidents and occupational diseases in Colombia are high. In 2014, 13,918 workers received compensation for permanent partial disability or invalidity. The Social Security provides rehabilitation services to workers who require it. Its objective is the social integration and recovery of productive capacity.

However, the results of the rehabilitation is poor. In recent years a medical center in southwestern Colombia implements a model of ergonomic rehabilitation aimed at improving care for workers and optimize resources, achieve better results in functional recovery and employment integration in less time, positively impacting the quality of life and labor productivity in the región.

The aim of the study was to determine characteristics and results of a comprehensive rehabilitation process for workers who suffered an accident or illness.

Were characterize the population and quality of health indicators. 6336 cases were analyzed, 74% were men. 80% of the population was between 30 and 59 years, the group between 40 and 49 years was the most representative. The most frequent pathologies requiring rehabilitation were: STC (90%), lumbar disc disorders, mental disorders, Quervain tenosynovitis and epicondylitis.

Reduction in the number of days, from 96 days of incapacity average in 2013 to 50 days in 2014 were observed. Similarly comportment in the Rehabilitation times showed a downward trend from 230 days in 2013 to 43 days in 2014. The sector with more cases is the agricultural sector with 46%, followed by the industrial sector with 30%. return to the job was 95% of cases.

Keywords: Occupational health · Ergonomics · Rehabilitation Disability

1 Introduction

The rehabilitation defined as the set of social, therapeutic, educational and training actions, of limited time, articulated and defined by an interdisciplinary team, which involve the user as an active subject of their own process, to the family, to the labor community and the social community, in the fulfillment of the objectives set, which aim to achieve changes in the worker, his environment, which allow him to return to

© Springer Nature Switzerland AG 2019
S. Bagnara et al. (Eds.): IEA 2018, AISC 819, pp. 306–315, 2019.
https://doi.org/10.1007/978-3-319-96089-0_33

work and experience a good quality of life [1]. It requires a series of actions that are integrated systematically and systemically.

This document considers that in addition to the individual approach should take into account the behavior of groups of workers and its determinants where the environment plays an important role. For authors and from the perspective of ergonomics, in addition to work processes, consumption processes and productsmust be taken into account [2].

This vision of rehabilitation (ergonomic rehabilitation) does not deny the importance of the clinical, epidemiological and environmental approach, but, on the contrary, it broadens and strengthens it. In the ergonomic approach, the rehabilitation of workers involves two fundamental aspects: the work system and the analysis of human activity. This conception understands that the intervention must be carried out in both components. Some authors even privilege the component of the work system, over the eminently therapeutic component [2].

The rehabilitation process means a failure in primary prevention, to a certain extent, rehabilitation cases are an indicator of results in the processes of health promotion and prevention of occupational morbidity and accidents; and at the same time, the rehabilitation processes are an opportunity for the promotion and prevention in occupational health and safety systems. Without a doubt, a rehabilitation case should be considered a sentinel marker for prevention.

The epidemiological trend in the General System of Labor Risks of Colombia (GSLR) where only the diseases and accidents (WA) suffered by formal workers are registered (informality is not covered by the GSLR), shows that there is a marked increase in the diagnosis of occupational diseases and a stable but high trend in occupational accidents. Table 1.

Table 1. Work accidents statistics in Colombia Source FASECOLDA

Year	Affiliated workers GSLR	Work accidents	N° deaths
2010	6.813.660	450.564	689
2011	7.499.489	555.479	692
2012	8.430.802	659.170	676
2013	8.271.920	622.486	706
2014	8.936.937	688.942	564
2015	9.656.829	723.836	563

The Ministry of Labor of Colombia in 2013, was make a comparison between the results obtained by the first survey of occupational safety and health conditions (year 2007) and by the second (year 2013). An increase of 19% in the permanent partial disability was found.

This impact, in addition to affecting individuals, families and society as a whole, leads to economic losses that have not yet been quantified. There is only information about direct costs, including economic benefits.

The costs for disabilities and temporary or definitive disability subsidy add up to 40% of the total costs paid, however, this apects does not represent the highest costs, it is the benefits that demand the most, in summation (both) they constitute the highest direct cost of the Labor Risk System.

The indirect costs far exceed the direct costs, therefore, it is understandable the figure posed by the Pan American Health Organization, up to 7 points of the Gross Domestic Product of the country, by accident and labor illness [3].

At present, the rehabilitation processes have marked weaknesses such as low opportunity in the identification of cases, inadequate quality in the diagnosis, classification and definition of rehabilitation plans; problems in integration and cohesion of the rehabilitation process between entities of health services and companies. Added to this, the late reincorporation of the worker to job, increases costs, worsens the prognosis and the results of the rehabilitation process.

There are a few practical tools for employers, there is a lack of awareness over of the disabled worker, which reveals a significant lack of knowledge over policies and actions that employers should by have make in to the rehabilitation system for work.

Therefore, a rehabilitation program must be comprehensive, not only for the disciplines that comprise it, but also for the approach and model that supports it.

2 Materials and Methods

A model of integral rehabilitation for work was designed, developed and implemented, which was applied in 6336 workers, during the period 2013–2014.

This paper presents the rehabilitation model proposed by the authors and the results of its implementation in the population subject to comprehensive rehabilitation during the study period.

2.1 Rehabilitation Model Stages

In ergonomic terms, rehabilitation involves two stages, the first between the therapy team and the patient (worker) that influence the effectiveness of the treatment. The second between the patient and the work system which is the determining result.

In the process of rehabilitation in the patient interaction and the therapist participate two fields of dominance, the psychological and the physical.

In the psychological field, cognitive and affective aspects play an important role. The knowledge, and the expectations of the therapist model the, motivations and commitment of the patient. In turn, it is the responsibility of the therapist to promote autonomy and independence of the patient with the rehabilitation process and equipment.

In this area, aspects such as worker perception, cognition, judgment or integration criteria, the patient's decision-making capacity, speed and response time, operational capacity, fatigue and motivation among other aspects must be taken into account.

At the physical level, above the technology, the specificity and precision of the thechnique is of paramount importance for the effectiveness of the treatment.

In this field, physiological, biomechanical, anthropometric aspects such as strength, flexibility, proprioception, speed, joint amplitude, aerobic capacity, resistance, precision, dexterity, tolerance of tissues, tolerance to pain, among others, must be taken into account.

Additionally, it is important to consider social aspects such as commitment to work or activities, goal orientation, confidentiality, self-esteem, self-acceptance, social support, etc.

2.2 Patient and Task, Interface

Components of the work environment participate in the interface of the patient and work system; technical and technological, organizational and socio-cultural aspects of work. Here it is necessary to take into account the prescribed task (task) and the real task (activity) as well as the mechanisms that regulate it and its effects on the worker and the work system.

Among the environmental factors, the conditions of the micro climate and other factors must be considered; between the technological and technical factors, the modes of operation, the operation techniques, the standards, types of controls, the type of instruments, equipment and tools, interfaces, etc.

The system of shifts, schedules, breaks, management modality, leadership, forms of communication, etc., should be considered among the organizational aspects. The culture of the organization is a key factor and determinant for the rehabilitation process of there that must be implemented programs that prepare employers for the labor reintegration and the continuity of the labor rehabilitation in the company [4].

2.3 Operation of the Model, Implementation

Early detection, adequate diagnosis of cases. The appropriate diagnosis requires a timely identification of the health event, therefore, an active search of the cases and the timely report to the rehabilitation center is necessary.

Once the case is identified, the diagnosis is confirmed by the intervention of the medical team, composed of the work physician or the physiatrist, who will classify the case and make the initial definition of the prognosis. It is very convenient to make the clinical diagnosis in an early and adequate way and to define the occupational criterion of it.

Classification, reclassification of the patient. It is given by the functional state of the patient and/or the severity of the injury. This classification is made by the physiatrist, a key strategy to decide on the appropriate rehabilitation plan.

Diagnosis of the rehabilitation team. Patients with a lesion classified as mild should be assessed by the functional rehabilitation team, occupational medicine and physiatry; critics patients, by the integral rehabilitation team, the pertinent surgical clinical specialty (orthopedics, neurosurgery, psychiatry, and others), and chronic patients by the team of occupational rehabilitation.

In order to specify the functional rehabilitation process, an evaluation of the work system where the worker carries out his/her habitual work activity must be done, this aspect is key for the definition of the rehabilitation plan, the therapy must be develop with purpose. At this stage, the analysis of the activity must be done, the diagnosis of the conditions of the patient (worker) system interface, in order to have an inventory of the environmental, technical and technological, organizational and sociocultural conditions of the system. work and the company. This way, the company is included in the rehabilitation and return at work program.

2.4 Definition of the Rehabilitation and Forecast Plan

The ergonomic rehabilitation plan should consider the two steps, the first step, patient and the therapy equipment, where goals, specific rehabilitation actions, times and resources are defined and second step where the activity analysis, the diagnosis of the effects of the activity on the worker and the work process, are basic to define the actions of preparation for the reincorporation to the work and the actions of engineering in the system of work (redesign design).

2.5 Implementation of the Specific Rehabilitation Program

The rehabilitation approach must integrate the physiological components (physical conditioning, improvement of abilities), biomechanical (posture, strength, flexibility, arcs of joint movement, etc.), psychological (self-determination, self-esteem, self-care, confidence, etc.). and articulate the functional restoration with the exploration and training of the potentialities. It should be seeking early reincorporation, as a principle of quality and oportunity in the process.

The employer (boss, middle managers, managers) and workers (coworkers) must be prepared for the process of reinstatement of the worker in the process of rehabilitation. In addition, the support team that will prescribe design and redesign actions will advise the employer on the administrative and engineering changes to be implemented so that the worker who returns to work can perform adequately (design and redesign).

This team, made up of an ergonomist, engineers, designers, physicians, psychologists, social workers and therapists, advises the employer on the changes that must be made in the work system for the purpose of labor reincorporation. In addition, when the worker returns to work during the rehabilitation phase, the company's health team must accompany the actions prescribed by the support team for the reinstatement of work. epidemiological criteria.

The intervention of groups of workers with similar activitiesshould be considered, for, in order to optimize the rehabilitation resources in the aforementioned interfaces and promote through the unions collective actions that articulate the model of integral rehabilitation for work. Additionally, an information system should be built that generates the indicators for the surveillance of the most prevalent pathologies object of rehabilitation and the rehabilitation process itself, that is, the effectiveness and efficiency of the process.

2.6 Monitoring of the Rehabilitation Process

Individual, and collective follow-up (with an ergonomic and clinical epidemiological approach) is alwais necesary. The condition of the rehabilitated worker is determined, the functional (physical and psychological) and productivity aspects are evaluated. Reinforcement is made to the preparation program for labor reintegration and rehabilitation in the company. It is at this moment where the actions of the rehabilitation program and the impact of the programs and the promotion and prevention actions of the company and the ARL can be evaluated.

2.7 Evaluation and Audit of the Process

The model and the rehabilitation process must be audited and evaluated, with process and result indicators. In terms of quality, opportunity, costs, efficiency and effectiveness.

The diagnosis of origin also requires the clinical certainty, the contribution of the tests that confirm the relationship with the occupation, for this purpose an interdisciplinary team is formed, specialized in the evaluation of jobs with ergonomic, epidemiological and clinical criteria, which technically support the causal relationship with the risk factors of the task.

3 Results

6336 workers were evaluated who entered and completed the integral rehabilitation process, during the years 2013 to 2014. 70% of the cases that were admitted were classified as severe. 26% of the rehabilitated workers were women, 74% were men. 77% of workers were under 49 years old, the average age was 41 years. The highest proportion of patients presented levels of secondary and primary education with 44.4% and 26.2% respectively, this is equivalent to approximately 70.6% of the population. El 90% de los pacientes vivían en el área urbana y solo el 10% en el área rural.

The agricultural sector 46%, industrial 30%, temporary services 10% were the sectors with the largest number of workers in the rehabilitation processes and the occupations were trades services 13%, sugarcane cutters 10%, operators 10%, maintenance and mechanical services 8%, cleaning services 6% and administrative staff 5%.

The parts of the body with the most affections were the upper limbs with the 50% of the cases by work accidents (WA), fingers 18%, lumbo sacral column 12%, knee 12%, shoulder and arm 10%, wrist and hand 9% the most frequent injuries were Contusions 28%, fractures 24%, sprains 9%, muscular contractures 9% mono neuropathies of upper limbs (CTS) 6%, were the most prevalent lesions by WA.

70% of the total cases were classified as severe, 29% as mild and 1% chronic. No significant differences were found by gender, economic activity, or type of activity.

11% of the cases for rehabilitation were occupational diseases, 89% were sequelae of work accidents. The most prevalent lesions were carpal tunnel syndrome, lumbar disc disorder, mental sphere disorders, Quervain's tenosynovitis and epicondylitis. these differences were statistically significant.

There was an increase of 28% in the rehabilitation times of patients who presented symptoms or emotional disorders compared with the group that did not present it, this difference were considered statistically significant p = 0.000.

After applying the model, there was a significant decrease in the tendency of the days of permanence in the rehabilitation process.

The average number of days of permanence in rehabilitation decreased from 196 average days of permanence in rehabilitation in the first quarter of 2013 to 22 days average in rehabilitation in the fourth quarter of 2014. This trend can be seen as a slope evidently negative (See Graph 1).

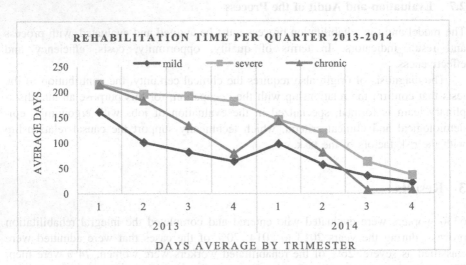

Graph 1. Rehabilitation times, severity in the injury, years 2013–2014. IPS. Cali

A marked decrease was identified in days of incapacity in workers rehabilitated due to work accidents and occupational illnesses during the study period (Graphics 2 and 3).

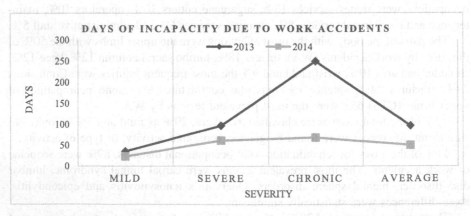

Graph 2. Behavior and trend of patient disabilities rehabilitated by work accidents 2013–2014. Cali

Workers who comeback to job were very important percentage (95% of the cases), 5% continued with inability.

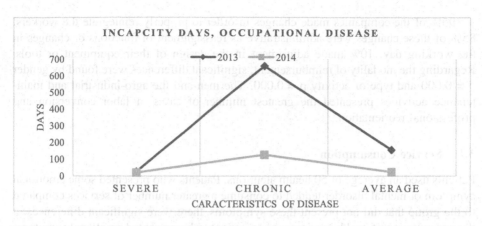

Graph 3. Behavior and trend of patient disabilities rehabilitated by ocupational disease 2013–2014. Cali

The number of workers without modifications in the jobs and in the tasks, (OJM) was higher as expected in the mild injuries patients (88.2%), while the modifications in the Jobs and in the tasks (JTM) it were higher in the severes injuries patients (55.5%), the great majority of the chronic injuries patients were comeback without task modifications. A smaller proportion of the seriously ill patients had comeback with temporary relocation (TR), definitive relocation (RD), and a smaller percentage could not continue performing their tasks and had to receive professional reorientation (PR) reconversion of work forcé (RWF) or were declared disabled-invalid (DI) due to the severity of the injuries (Graphic 4).

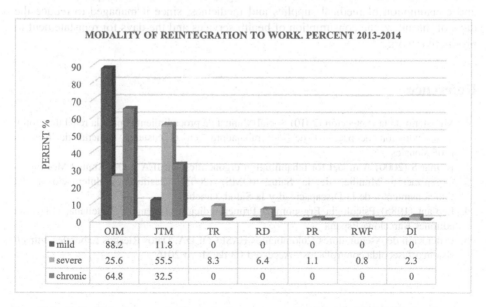

Graph 4. Behavior of the modalities of reinstatement of the integral rehabilitation program during the period 2013–2014. IPS. Cali

95% of the companies made changes in order to properly reintegrate the workers. 85% of these changes were administrative as reassignment of functions or changes in the working day. 10% made adjustment in the design of their equipment or tools. Regarding the modality of reimbursement, significant differences were found by gender p = 0.000 and type of activity p = 0.000. The men and the agro-industrial and maintenance activities presented the greatest number of cases of labor conversion and professional reorientation.

3.1 Service Consumption

Patients used an average of 50 health atentions. Patients who presented some emotional symptom or mental disorder tended to consume a greater number of services compared to the group that did not present these symptoms. there were significant differences in the consumption of health services among patients who presented emotional symptoms and/or mental disorders of those who did not, p = 0.00000 with the consumption of health services of the patients being higher. patients who presented these symptoms.

4 Conclusions

The rehabilitation model for patients with musculoskeletal injuries proposed by the authors, shows that ergonomic rehabilitation with integral management reduces the days of permanence in the rehabilitation process, reduces the days of disability, reduces complications, improves to the work return process, adequately prepare companies and work groups for the acceptance of the worker with some degree of disability. The model also proved to be successful in reducing direct costs for consumption of services and consumption of medical supplies and medicines, since it managed to reduce the days of disability, the consumption of health services and the days for reinstatement to productive work.

References

1. Ministerio de la Protección (2010) Social Manual de procedimientos para la rehabilitación y reincorporación ocupacional de los trabajadores en el Sistema General de Riesgos Profesionales
2. Kumar S (2000) A model for rehabilitation ergonomics. In: IEA 2000, Annual Meeting, 4
3. Organización Mundial de la Salud (OMS) (2001) Clasificación Internacional del Funcionamiento, de la Discapacidad y la Salud, Ginebra
4. Leplant (1985) Psicología Ergonómica Primera edición el Lengua Castellana, Oikos-tau ediiones Barcelona España
5. Federación de Aseguradores Colombianos FASECOLDA. Datos Riesgos Laborales. https://sistemas.fasecolda.com/rpDatos/. Accessed 03 de Mayo 2016

6. Hernández E, Velásquez JC (tutor) (2015) Comportamiento de los tiempos de reincorporación y modalidades de reintegro laboral según la extremidad afectada, el cargo, la escolaridad y el manejo medico recibido, en un grupo de trabajadores que presentaron fractura de las extremidades por accidente laboral y que fueron usuarios de un programa de rehabilitación integral en la ciudad de Cali durante el año 2014. Tesis Maestría Salud Ocupacional, Universidad del Valle
7. Bonafede M, Espindle D, Bower AG (2013) The direct and indirect costs of long bone fractures in a working age US population. J Med Econ 16(1):169–178
8. Corral M, Vargas Prada S, Gil JM, Serra C (2015) Reincorporación al trabajo tras un episodio de incapacidad temporal por trastornos músculo-esqueléticos: revisión sistemática de guías de buenas prácticas. Arch Prev Riesgos 18(2):72–80
9. Hou WH, Chi CC, Lo HL, Kuo KN, Chuang HY (2013) Vocational rehabilitation for enhancing return-to-work in workers with traumatic upper limb injuries. Cochrane Database Syst Rev
10. Lopez Sullaez Lía LC, Estrada Ruiz R (2009) Repercusión Ocupacional de las Amputaciones Traumáticas en Dedos de la Mano por Accidente de Trabajo. Med Segur Trab 55 (217):41–48
11. Ustun TB, Sartorious N (1995) Mental illness in general health care: an international study. Wiley, Chichester
12. Waddell G, Burton AK (2006) Is work good for your health and well-being?. TSO, London
13. Vargas Porras PA, Orjuela Ramirez ME, Vargas Porras C (2013) Lesiones osteomusculares de miembros superiores y región lumbar: caracterización demográfica y ocupacional: Universidad Nacional de Colombia, Bogotá 2001–2009. Enferm Glob 12(32):119–133
14. Martin Alfonso L (2003) Aplicaciones de la psicología en el proceso salud enfermedad. Rev Cuba Salud Pública 29(3):275–281
15. Carrillo Castrillo JA (2014) Caracterización de la Accidentalidad Laboral en el Sector Industrial Andaluz en el Período 2003–2008. Aplicaciones en el Diseño y Evaluación de Programas de Intervención. Tesis doctoral, Universidad de Sevilla

A Safety-II Approach on Operational Maneuvers of a Hydropower Plant

Juliano Couto Portela[1(✉)] and Lia Buarque de Macedo Guimarães[2]

[1] Itaipu Binacional, Av Tancredo Neves 6731,
Foz do Iguaçu, PR 85867-000, Brazil
juliano.portela@gmail.com

[2] Graduate Program in Industrial Engineering, Federal University of Rio Grande
do Sul, Av. Osvaldo Aranha 99, 5° andar, Porto Alegre, RS 90035-190, Brazil
liabmg@gmail.com

Abstract. This work investigates, under the Resilience Engineering (RE) approach, the environmental conditions that lead to accidents at the Itaipu Binacional hydroelectric power plant (Itaipu HPP). The socio-technical importance of the installation and the technological update of all control, supervision and monitoring systems in a near future justify the work. The study focused on the impacts of variability in the normal operation of four typical maneuvers representing the four quadrants of a periodicity-complexity matrix, and the variability influencing each step of the maneuvers. The opinions from the operational staff were raised, organized, and the RE principles, techniques and heuristics were applied to them. Although the selected maneuvers might not represent the whole universe of more than five hundred maneuvers executed by the Itaipu HPP, the results indicate that some variability types such as "maneuver environment", "the need to confirm the maneuver steps with another operator", "situations that draw attention from the operator", and "the similarity with another operating environment" act decisively in practically all maneuvers. The operator knowledge was not mapped as a fundamental variability impacting the failures. The results were compared with historical data and presented to the staff, who proposed necessary actions to increase the operational safety under the RE perspective. Overall results imply that there is room for moving from a reactive traditional safety management (Safety-I), based on retrospective analysis of accidents and "what went wrong," to an effective proactive one (Safety-II), based on the variability of normal operation and therefore "what goes right."

Keywords: Safety-II · Resilience Engineering · Hydroelectric power plant

1 Introduction

Accidents in critical infrastructures, although rare, cause serious social, environmental and economic impacts in their area of influence. Therefore, they must be avoided even if a "normal" accident rate is expected due to the operation complexity. Accidents in Three Mile Island [1], Chernobyl [2], Sayano-Shushenskaya [3], Deepwater Horizon (2010) [4] and Fukushima [5] showed that the variability that affected the operational

© Springer Nature Switzerland AG 2019
S. Bagnara et al. (Eds.): IEA 2018, AISC 819, pp. 316–326, 2019.
https://doi.org/10.1007/978-3-319-96089-0_34

process had become out of control, impacted vital functions of the process, and eventually lead to the fatal failures.

The approach for understanding accidents, no matter the type (industrial or services) or complexity of the system, is often "Safety-I" using techniques that search for contributing factors to the event, focusing on "what goes wrong". This traditional approach understands safety as an absence of unwanted results in normal operation; therefore the goal is to maintain a safe condition where the number of unwanted outcomes is as small as possible [6, 7]. "New" accidents occur because their cause was not eliminated, as it was unknown. Safety-I thus promote a bimodal way of seeing how work is done: things work out because the system works as it should and because people work in the prescribed way, as one imagines; things go wrong because something goes wrong or it fails [6, 7].

Hollnagel [6] explains the distinction between "Work-as-Done" and "Work-as-Imagined" (where the latter is generally considered in Safety-I) and how mistaken it is thinking of human beings as machines capable of responding in infinite ways to infinite possibilities. Besides, Safety-I mainly refers to the evaluation of the lack of safety rather than the presence of safety.

Under the reactive approach of Safety-I, even if subsequent analysis had been successful in pointing out the root causes of the accidents and new safety standards were established, it's clear that if the conditions that led to the accidents had been anticipated, they probably had been avoided. However, for perceiving the hazards of the process before they get out of control, it is necessary a different approach to normal operation and a different proactive way of thinking about safety. This new approach, called "Safety-II" [6, 7] as proposed by the Resilience Engineering (RE) [8] considers important focusing on adaptive capacity in order to maintain control over unforeseen disturbances or events [9]. In addition, safety should not focus on the rare cases of failures because they do not explain why performance is almost always satisfactory and how it helps to meet organizational goals. The statistical probability of a failure is 1 out of 10,000 while things go right 9,999 times out of 10,000 [7].

Based on the principle of equivalence between "successes" and "failures", Safety-II assumes that both normal operation and accidents emerge from the same source (variability of normal operation), and safety is a consequence of the way the complex system behaves, not a static property. Under this view, the outcomes may differ from what was expected, and this might be harmful or beneficial. The efficiency of the system depends on the adaptability and flexibility of human work and, although very rarely, failures might result from this behavior [7].

Adopting Safety-II premises, this study investigates the conditions leading to operational failures in the Itaipu Binacional hydroelectric power plant (Itaipu HPP), the largest hydroelectric power generator of the world, having produced nearly 2.5 billion MWh since 1984. It is responsible for 17% of the Brazilian's energy consumption and 75% of the Paraguayan's. Therefore, shutdowns at Itaipu HPP have the potential to cause production loss and instability to both electrical systems. As most companies, Itaipu HPP adopts a reactive approach to safety, and introducing the proactive approach might improve its operational practices. Furthermore, adopting Safety-II practices meets its vision of sustainable production and high operational performance, represented by a sequence of world records of energy production, the last one achieved in 2016 (more than 103 million of MWh).

2   Method

The study followed the Safety-II management approach [6, 7] as proposed by Resilience Engineering (RE) [8]. It is based on the variability of the normal operation and, therefore, "in the many things that goes right", in contrast to the traditional and reactive view of Safety-I, based on the retrospective analysis of accidents and "the few things that went wrong" [6, 7]. Inspired by the Functional Resonance Analysis Method (FRAM) [10] the study mapped the functions of typical maneuvers under normal operation (Work-as-Done) at the Itaipu HPP, from 2006–2015, their variability and interfering conditions on normal operation.

The FRAM is used in RE to describe outcomes using the idea of resonance arising from the variability of everyday performance. This idea of resonance replaces the traditional cause-effect relation in Safety-I. The following steps are used to describe functional variability and resonance, and generate recommendations for damping unwanted variability: (i) identify and describe essential system functions, and characterize each function using the six basic characteristics (aspects); (ii) check the completeness/consistency of the model; (iii) characterize the potential variability of the functions in the FRAM model; (iv) define the functional resonance based on dependencies/couplings among functions and the potential for functional variability; (v) identify ways to monitor the development of resonance either to dampen variability that may lead to unwanted outcomes or to amplify variability that may lead to wanted outcomes.

Due to the significant number of maneuvers executed by the Itaipu HPP operational team, each one with several steps with different levels of complexity and attention demand, a matrix of four quadrants was created with categories "complexity" and "periodicity", where four maneuvers represent the universe for the purpose of this study. The periodicity refers to the number of times the maneuver was performed during the period. The complexity refers to the number of steps to be taken and the number of factors (vulnerabilities/noises) to influence its success. One of the criteria for choosing the maneuver was that it should cause an impact on the energy production or safety of human resources, facilities or the surroundings of the plant if unsuccessful.

Using a participatory ergonomics method [11, 12], 31 technicians (mean age 44.7 years; mean time in company 21.8 years) and eight engineers (mean age 35.9 years; mean time in company 7.1 years) from the operational staff: (i) chose the maneuvers representing the four quadrants of the periodicity-complexity matrix; (ii) evaluated their normal operation; (iii) determined the variability that influences each step of the chosen maneuvers; (iv) evaluated the impact of each step of the maneuver on the overall result – success or failure.

The Itaipu HPP operational staff is composed of 19 engineers, 97 operational technicians and six technical and administrative assistants. The twenty-four hours work is done with a six hours rotating shift of groups composed of 11 operational technicians responsible for supervising and controlling the plant. They inspect the equipment, perform maneuvers, analyze and monitor more than 20,000 digital or analogical statuses of the displays that inform the conditions of installations, equipment and systems of the plant. The engineers give technical and managerial support for pre and post operation.

3 Results

Table 1 shows the four typical maneuvers selected for analysis by the operational staff that participated in the study.

Table 1. Maneuvers considered in the study (period 2006–2015).

Maneuver	Periodicity	Complexity	Total maneuvers
Select generating units to conventional control	Low	Low	553
Separation of generating units for ANDE	Low	High	27
Reversal of speed regulator pumps	High	Low	10,400
Start and stop of generating unit	High	High	5,322
Total			16,302

3.1 Identifying Variability

The variability types that influence the four operational maneuvers were identified in the interviews with the 39 professionals. Regardless of the complexity and periodicity of the maneuver, 16 variability types, grouped into six categories, are common to the four types of maneuvers, as shown in Table 2.

From the start to the end of a maneuver, several steps ("functions" in the FRAM [10]) are performed. For the purpose of the study, there are two types of steps: (1) key step: central function from which the purpose of the maneuver is achieved; an error in a key step most probably causes an error in the maneuver itself. Examples: breaker closure, unit synchronization. It is similar to the functional resonance of the FRAM [10]; (2) relevant step: although it is not a key step, it contributes to the normal flow of the maneuver in a decisive way; a failure in this step can influence the key step, and therefore causes error in the maneuver. Examples: trigger starting preparation, confirm voltage bar without voltage before switch closure.

Table 3 shows the seven steps of one of the maneuvers ("select generating units to conventional control") with some of the variability types – see category "operator" in Table 2 – to illustrate how the general table was assembled. It was considered less important to understand the technical meaning of each step than to understand how each variability influences each step. The "X" implies that the variability influences the step.

Table 2. Variability influencing the four operations under study.

Category	Variability
Operator	Knowledge/experience
	Ability to adapt to unexpected situations
	Necessity to confirm each step with another operator
Communication	Availability of communication equipment
	Operator/Supervisor reports each step
	Quality of operational communication (e.g. A clear and objective communication between O&M)
Human resources	Availability of human resources
	Hierarchy in the execution of step
Operational instruction	Quality of the operational instruction
	Current operational restrictions
Maneuver environment	Situations that take the operator's attention (e.g. telephone/alarm)
	Urgent/ emergency situations or concomitant work
	Similarity with another maneuver environment
Maneuver equipment	Normal operation of the maneuver equipment
	Availability of the computerized system
	Tacit peculiarities of the equipment (not described in any instructions

Table 3. Example of maneuver steps and their variability.

Maneuver step	Key step	Operator knowledge/ experience	Operator ability to adapt to unexpected situations	Operator necessity to confirm each step with another operator
1. Request to the dispatch center to turn off the Automatic Control of Generation	N			
2. Set reference power in the Turbine Joint Control	S	X	X	X
3. Confirm that the voltage value is zero in the Joint Control	S			
4. Set reference voltage in the Voltage Joint Control	S	X	X	
5. Select generating units, transmission lines and voltage bars from Scada 1 to conventional	S	X	X	X
6. Confirm the switch key 43JCS in the Joint Control in the "conventional" position	N			
7. Switch generating units on conventional panel to the same position as in the digital system	N	X	X	

In the example from Table 3, it was considered that the "operator knowledge" variability does not influence the "request to the dispatch center to turn off the automatic control of generation" step, as the knowledge is not relevant to the outcome of this step. On the other hand, it was considered that "ability to adapt to unexpected situations" influences the "set reference voltage in the voltage joint control" step because, if there is an unexpected situation at the time of this maneuver - for example, disturbance in the electrical system - the operator must be able to adapt the maneuver to this situation. Based on the evaluation of the operation team, the question "which variability in Table 2 influences this step?" was asked for each of the steps. The overall result is shown in Table 4, which gives an overview of the specific variability for each of the four quadrants of the complexity-periodicity matrix.

Table 4. Variability frequencies on the maneuver key steps.

Low complexity	%	High complexity	%
Maneuver environment – Similarity with another maneuver environment	85	Maneuver environment – Urgent/emergency situations or concomitant work	76
Maneuver environment – Situations that take the operator's attention (e.g. telephone/alarm)	62	Maneuver equipment – Normal operation of the maneuver equipment	74
Instruction – Current operational restrictions	62	Maneuver environment – Situations that take the operator's attention (e.g. telephone/alarm)	74
Instruction – Quality of the operational instruction	54	Human Resources – Availability of human resources	67
Operator – Knowledge/experience	46	Operator – Necessity to confirm each step with another operator	64
Operator – Ability to adapt to unexpected situations	46	Instruction – Current operational restrictions	64
Low periodicity		High periodicity	
Maneuver environment – Situations that take the operator's attention (e.g. telephone/alarm)	93	Maneuver equipment – Normal operation of the maneuver equipment	61
Human Resources – Availability of human resources	85	Operator – Knowledge/experience	57
Instruction – Quality of the operational instruction	85	Maneuver environment – Similarity with another maneuver environment	57
Operator – Necessity to confirm each step with another operator	81	Maneuver environment – Urgent/emergency situations or concomitant work	50
Instruction – Current operational restrictions	78	Maneuver environment – Situations that take the operator's attention (e.g. telephone/alarm)	50
Maneuver environment – Similarity with another maneuver environment	74	Instruction – Current operational restrictions	50

The following considerations may be drawn from Table 4:

(a) The "maneuver environment" is significant in all situations of complexity and periodicity;
(b) The "knowledge/experience", despite its central importance, does not appear as a main variability in the maneuvers of high complexity and low periodicity;
(c) Most of the steps are affected by "similarity with another maneuver environment"; it is usual in plants with several generating units that each environment where the operator performs the maneuver is exactly equal as the adjacent, except for the operational identifications;
(d) The category "instruction" affects significantly the low complexity and low frequency maneuvers;
(e) In the case of low periodicity maneuvers, many of the variability types affect almost all the steps. The main variability is the "situations that take the operator's attention". This is because as the maneuver is rarely executed, the operator tends to turn all the attention to the execution, and all the attention-taking conditions influence the normal operation;
(f) In maneuvers of high complexity, the two variability types that most affect the functions ("maneuver environment-urgent/emergency situations or concomitant work" and "normal operation of the maneuver equipment") are external to the operation, so there is no much to do as a concrete action to decrease their influence on the maneuver, but emphasize the importance of the experience as well as the knowledge management as a way to increase the operation capacity of adaptation and response to eventual disruption, one of the primal precepts of RE.

3.2 Comparison Between the Raised Variability Types and Actual Failures

Figure 1 (right) shows the number of well-succeeded operational maneuvers performed by the Itaipu HPP in the period 2006–2015 and the failures resulting in loss of energy production, reliability or damage to equipment during that time. Four failures occurred in the period, representing 0.025% of the universe, while the successful maneuvers represent 99.975% confirming [7] as shown in Fig. 1 (left).

Fig. 1. (Left) from [7], the imbalance between things that go right (the gray area in Fig. 1) and things that go wrong (the thin line); and (right) well-succeeded x failed maneuvers performed by the Itaipu HPP operation between 2006 and 2015.

The variability types raised by the group were used as a basis for analysis of the four operational failures that occurred between 2006 and 2015. The variability types that influenced each one are summarized in Table 5. The first operational process failure (in 2010) occurred in a maneuver of low complexity and high periodicity. The analysis indicated process failures in five out of six steps. There was a coincidence between the activation of an alarm and an unrelated work on the equipment, where the latter required special attention of the operator. When analyzing the alarm in front of the equipment, the operator missed the comprehensive understanding of the operational situation, performing an undesired maneuver of pump transfer.

The second failure (in 2008) occurred in a maneuver of high complexity and low periodicity. The analysis indicated process failures in 2 out of 17 steps. In the occasion, the generating unit U9A should be switched off for maintenance. However, at the same time, the dispatcher asked to shut down the U03 along with the U9A in order to meet electrical systems demands. Upon entering the digital system screen to open the U9A circuit breakers, the operator was induced by the information that U03 should be also stopped. Thus, mistakenly the operator accessed the control panel of the U03 instead of the U9A and opened the circuit breakers with U03 still at full power, causing load rejection in the U03.

Table 5. Summary of the variability that influenced the four operational failures in Itaipu HPP (period 2006 – 2015).

Failure 1: 2010 (Low complexity/High periodicity)	Failure 2: 2008 (High complexity/Low periodicity)	Failure 3: 1st 2014 (High complexity/High periodicity)	Failure 4:2nd 2014 (High complexity/High periodicity)
Maneuver environment Situations that take the operator's attention (e.g. telephone/alarm)	**Maneuver environment** Similarity with another maneuver environment	**Maneuver environment** Urgent/emergency situations or concomitant work	**Maneuver environment** Similarity with another maneuver environment
Maneuver environment Urgent/emergency situations or concomitant work			
Operator Knowledge/experience	**Operator** Necessity to confirm each step with another operator	**Operator** Necessity to confirm each step with another operator	**Operator** Necessity to confirm each step with another operator
Maneuver equipment Normal operation of the maneuver equipment	**Maneuver equipment** Normal operation of the maneuver equipment	**Maneuver equipment** Normal operation of the maneuver equipment	
		Human Resources Availability of human resources	

The third failure (in 2014) occurred in a maneuver of high complexity and high periodicity. In this, failure was found in 3 out of the 22 steps. The generating unit U09 should start and synchronize to the electrical system. In order to run the start-up sequence automatically, it would be necessary for an operator be locally on the unit's local control panel to switch the control mode to automatic. Since the local operator wasn't found, the operator in the Control Room decided to perform the maneuver even so. With concomitant work in progress in the control room, the maneuver was performed by a single operator. In this process, the U09 circuit breaker was closed out of ideal synchronization conditions, resulting in mechanical and electrical damage to the unit.

The fourth failure (in 2014) occurred in a maneuver of high complexity and high periodicity. In this, failure was found in 3 out of the 22 steps. During the stopping process of generating unit U05 to meet electrical system requirements, the operator mistakenly transferred the power supply selection of the unit U04 instead of U05. It was necessary to stop the U04, which was operating normally at the time.

Despite the different underlying problems contributing to the four analyzed failures, it is possible to note some similarities that corroborate conclusions drawn from Table 5:

(a) The "maneuver environment" is important in all situations of complexity and periodicity;
(b) The "operator knowledge" is not a main variability influencing maneuvers of high complexity and low periodicity. In fact, the reports pointed out that in the cases of faults 2, 3 and 4 the operators were fully aware of the operational procedure;
(c) Most of the steps are affected by the variability "similarity with another maneuver environment". It was present in two out of the four failures;
(d) The category "instruction" affects more significantly the low complexity/periodicity maneuvers.
(e) In low periodicity maneuvers, many variability types affect practically all steps. The main variability affecting normal operation comes from situations that get the attention of the operator.
(f) In high complexity maneuvers the two main variability types ("maneuver environment urgent/emergency situations or concomitant work" and "maneuver equipment – operation") affecting the operation are external conditions, and the analysis of the reports has proved that malfunctioning of the maneuvering equipment and concomitant emergency/work situations are basic failure conditions. Operators should prioritize their knowledge of systems management and training to decrease the influence of these types of variability on the maneuvers.

4   Conclusion

Based on the precepts of Safety-II, this study focused on the impacts of variability on the normal operation of typical maneuvers at the Itaipu HPP. Inspired by the Functional Resonance Analysis Method (FRAM), and using participatory ergonomics, the opinions of 39 people from the operational staff were raised, organized, and the RE principles, techniques and heuristics were applied to a set of four maneuvers characteristic of the operation of the plant. They were categorized in high and low complexity and periodicity, in order to understand which variability influences normal operation. Maneuvers were detailed in steps; by identifying the variability that acts on each step, it was clear that the variability types influencing normal operation are common to all studied maneuvers.

These same variability types were found in the reports of the four operational failures that occurred between 2006 and 2015 (which represents only 0.025% of the operations performed in the period). Therefore, the conclusion is that although it is fragile to say that the four maneuvers represent the whole universe of more than five hundred maneuvers executed by the Itaipu HPP, they provided valuable inputs for comparing what was verified in the study with the actual cases of failure. An important result is that some variables act decisively in practically all maneuvers: the "maneuver environment", "the need to confirm the steps of the maneuver with another operator", "situations that draw attention from the operator" and the "similarity with another operating environment". Another important finding is that the operator knowledge was not mapped as a fundamental variability leading to failures. Once identified the maneuver functions, it was possible to understand which would be the most important variability influencing normal operation, providing subsidies to decision-making before an event of failure. Comparisons between normal operation and the four failure cases showed that success and failure have a same source (variability of normal performance), which is adherent to FRAM and Resilience Engineering principles.

The main contribution of this study for the development of Resilience Engineering and Safety-II approaches is that the same variability types influence the operational steps, regardless of the complexity or the periodicity of the maneuver.

The operational team was surprised and excited with the "focusing on success" approach and the participatory method used in the study, and proposed actions to increase the operational safety of the plant. Adopting Safety-II practices meets the company's vision of sustainable production and high operational performance.

References

1. Kemeny J, Babbitt B, Haggerty P, Lewis C, Marks P, McBride L, McPherson H, Peterson R, Pigford T, Taylor T, Trunk A (1979) Report of the President's Commission on the accident at Three Mile Island: the need for change, Washington, DC
2. IAEA (1992) Report INSAG-7 Chernobyl accident: updating of INSAG-1, Vienna
3. Belash IG (2010) Hydrotechnical construction: causes of the failure of the no. 2 hydraulic generating set at the Sayano-Shushenskaya HPP: criticality of reliability enhancement for water-power equipment. Power Technol Eng 44(3):165–170
4. BP (2010) Deepwater horizon accident investigation report, London. https://www.bp.com/content/dam/bp/pdf/sustainability/issue-reports/Deepwater_Horizon_Accident_Investigation_Report.pdf. Accessed 21 May 2018
5. IAEA (2011) IAEA International fact finding expert mission of the nuclear accident following the great east Japan earthquake and IAEA expert mission to Japan, Vienna
6. Hollnagel E (2014) Safety-I and Safety-II, the Past and Future of Safety Management. Taylor & Francis, London

7. Hollnagel E, Wears RL, Braithwaite J (2015) From Safety-I to Safety-II: a white paper. The Resilient Health Care Net: Published simultaneously by the University of Southern Denmark, University of Florida, USA, and Macquarie University, Australia

8. Hollnagel E, Woods D, Leveson N (eds) (2006) Resilience Engineering: Concepts and Precepts. Ashgate Publishing Co., Aldershot

9. Lundberg J, Johansson BJE (2015) Systemic resilience model. Reliab Eng Syst Saf 141 (special issue on resilience engineering):22–32

10. Hollnagel E, Hounsgaard J, Colligan L (2014) FRAM – the functional resonance analysis method - a handbook for the practical use of the method, 1st edn, no june. Middelfart, Denmark

11. de Macedo Guimaraes LB, Fogliatto FS (2000) Macroergonomic design: a new methodology for ergonomic product design. In: Fourteenth triennial meeting of the international ergonomics association. International Ergonomics Association, San Diego, p 328

12. de Macedo Guimaraes LB (1999) Ergonomic approach: the macroergonomic method. In: de Macedo Guimaraes LB (ed) Ergonomics in production, 4th ed, vol 1. (Chap 1.1) FEENG, Porto Alegre

Working Postures and 22-Year Incidence of Acute Myocardial Infarction

Niklas Krause[1(✉)], Onyebuchi A. Arah[1], and Jussi Kauhanen[2]

[1] University of California, Los Angeles, CA 90095, USA
niklaskrause@g.ucla.edu, arah@ucla.edu
[2] University of Eastern Finland, Kuopio, Finland
jussi.kauhanen@uef.fi

Abstract. *Introduction*: In contrast to body postures during the entire day or during leisure, associations of work postures with acute myocardial infarction (AMI) have not been examined while adjusting for socioeconomic status, leisure time physical activity, fitness, and other CVD risk factors.

Methods: The associations between work postures (assessed by questionnaire at baseline) and 22-year incidence of first AMI (ascertained via record linkage with national registries) were estimated by Cox regression models adjusting for 19 confounders among 1831 Finnish men, separately for those with (n = 1515) and without (n = 316) pre-existing ischemic heart disease (IHD).

Results: In the full sample, work postures showed no substantial associations with AMI risk except for "standing quite a lot" (HR 1.26; 95%CI 0.94–1.70). The effects of work postures differed by baseline IHD status. Among men *without* IHD, sitting showed 7–16% reduced risks, standing 4–21% increased risks, and walking had no impact. Among men *with* IHD, walking showed 18–23% reduced risks, sitting "quite a lot" or "very much" showed 28% and 67% increased risks, respectively. Standing "a little" (HR 1.38, 95%CI 0.79–2.42) and "quite a lot" (HR 1.33, 95%CI 0.73–2.41) were positively but standing "very much" (HR 0.68, 95%CI 0.36–1.32) inversely associated with AMI.

Conclusions: Considering previously reported strong positive associations between standing and progression of atherosclerosis the latter finding is probably due to healthy worker selection bias. It is therefore prudent not to recommend standing at work for men with IHD and to always consider IHD status when assessing risks and benefits of different work postures.

Keywords: Physical activity · Work posture · Cardiovascular disease

1 Introduction

Prolonged sitting may increase risk for cardiovascular disease (CVD), however, in contrast to total daily sitting time, occupational sitting has not been linked to CVD in prospective epidemiologic studies. [9, 11, 13, 14, 18] Standing at work predicted venous diseases, [15–17] atherosclerosis, [7] incident heart disease, [13] and all-cause mortality, [4, 8] with strongest effects seen for men with pre-existing ischemic heart disease (IHD). [7] Total and leisure but not occupational walking time reduced risk. [10] The associations between work postures and acute myocardial infarction

© Springer Nature Switzerland AG 2019
S. Bagnara et al. (Eds.): IEA 2018, AISC 819, pp. 327–336, 2019.
https://doi.org/10.1007/978-3-319-96089-0_35

(AMI) while adjusting for socioeconomic status, conditioning leisure time physical activity, cardiorespiratory fitness, and other CVD risk factors have not yet been investigated.

2 Methods

Associations of work postures with first AMI were investigated in a representative random population sample of 1831 middle-aged Finnish men enrolled in the Kuopio Ischemic Heart Disease Risk Factor Study, and separately for 1515 of them with and 316 without pre-existing IHD. Details of the study population have been described previously. [5] Typical work postures (sitting, standing, walking) were each assessed for all 1831 men at baseline, and again for 591 men still working at 11-year follow-up, by questionnaire using four rank-ordered exposure categories: not at all, a little, quite a lot, and very much. Participants were considered to have IHD at baseline if they had a history of myocardial infarction or angina pectoris, took currently anti-angina medi- cation, or had positive findings of angina from the London School of Hygiene car- diovascular questionnaire. [12] Cardiorespiratory fitness (also known as aerobic capacity or maximum oxygen uptake or VO2max) was assessed by a maximal but symptom-limited exercise test on an electrically braked bicycle ergometer. The assessment of all potential confounders followed standard methods and has been described in detail elsewhere. [5] AMI during follow-up was ascertained via record linkage with national hospitalization registries and review of respective medical records by a cardiologist. Multivariate Cox regressions models were used to estimate hazard ratios with 95% confidence intervals separately for each working posture, stratified by baseline IHD, and with incremental adjustment for 19 potential confounders listed under "Covariates" in Table 1. This secondary data analyses was exempt from human subjects institutional review. The original study was approved by the University of Eastern Finland and by the University of California.

3 Results

The distribution of all variables across baseline IHD status is shown in Table 1. Results from Cox regression analyses with incremental adjustment for 19 potential confounders are shown in Tables 2, 3 and 4. Compared to those working "not at all" in the respective posture, baseline postures showed no substantial associations with AMI risk in the full sample of all men with the exception of standing "quite a lot" (HR 1.26; 95% CI 0.94–1.70) (Table 2). Results differed by baseline IHD: Among men *without* IHD (Table 3), sitting was associated with a 7–16% reduced risk, standing with a 4–21% increased risk, and walking did not influence risk. Among men *with* IHD (Table 4), sitting "quite a lot" or "very much" was associated with a 28% and 67% increased risk, respectively (HR 1. 28, 95%CI 0.74–2.21; and HR 1.69, 95%CI 0.98–2.85). Standing "a little" (HR 1.38, 95%CI 0.79–2.42) and "quite a lot" (HR 1.33, 95%CI 0.73–2.41) was positively but standing "very much" inversely (HR 0.68, 95%CI 0.36–1.32) associated with AMI. Walking was associated with 18–23% reduced risks. 11-year work postures with additional adjustment for baseline work postures showed similar but less precise estimates due to a much smaller sample size (data not shown). For example, standing "very much" at 11-years was associated with a 41% increased AMI risk (HR 1.41, 95% 0.56–3.58) for all men.

Table 1. Characteristics of the study population: distribution of independent variables by ischemic heart disease status at baseline. Kuopio Ischemic Heart Disease Risk Factor Study 1984–2011 (N = 1831).

Independent variables	Men without IHD (N = 1515)		Men with IHD (N = 316)		Difference
	N or mean	% or (SD)	N or mean	% or (SD)	P-value (t-test or chi-square test)
Working postures					
Sitting					
- Not at all	376	24.8	105	33.2	
- A little	474	31.3	109	34.5	
- Quite a lot	310	20.5	49	15.5	
- Very much	355	23.4	53	16.8	0.001
Standing					
- Not at all	238	15.7	49	15.5	
- A little	665	43.9	105	33.2	
- Quite a lot	339	22.4	85	26.9	
- Very much	273	18.0	77	24.4	0.002
Walking					
- Not at all	137	9.0	26	8.2	
- A little	798	52.7	139	44.0	
- Quite a lot	394	26.0	102	32.3	
- Very much	186	12.3	49	15.5	0.019
Covariates					
Age and technical factors					
Age (years)	51.5	(5.1)	53.6	(3.8)	0.001
Participant in placebo group of lipid lowering drug trial	134	8.8%	28	8.9%	0.993
Participant in treatment group of lipid lowering drug trial	127	8.4%	26	8.2%	0.928
Biological factors					
Blood glucose (mmol/l)	4.70	(0.9)	4.83	(1.22)	0.067
Plasma fibrinogen (g/l)	2.95	(0.52)	3.10	(0.57)	0.001
Body mass index	26.7	(3.4)	27.1	(3.8)	0.054
LDL-cholesterol (mmol/l)	4.00	(0.97)	4.18	(1.07)	0.007
HDL-cholesterol (mmol/l)	1.31	(0.29)	1.28	(0.32)	0.230
Systolic blood pressure	134.0	(16.2)	133.0	(18.3)	0.393
	4	0.3%	4	1.3%	0.014

(*continued*)

Table 1. (*continued*)

Independent variables	Men without IHD (N = 1515)		Men with IHD (N = 316)		Difference
	N or mean	% or (SD)	N or mean	% or (SD)	P-value (t-test or chi-square test)
Lipid-lowering medication (yes)					
Anti-hypertensive medication (yes)	176	11.6%	109	34.5%	0.001
Behavioral factors					
Alcohol consumption (g/week)	71.3	(111)	88.9	(199)	0.128
Cigarette/day*years	140	(289)	214	(351)	0.001
Conditioning LTPA (hours/year)	91	(102)	90	(110)	0.962
Cardiorespiratory fitness (VO_2max in ml O_2/kg/min)	32.8	(7.0)	27.5	(6.9)	0.001
Socioeconomic status					
Personal income (1000 FIM/yr)	93.1	(57.3)	72.9	(37.6)	0.001
Psychosocial job factors					
Social support score at work (range 0–12)	6.5	(2.5)	6.5	(2.5)	0.994
Mental strain at work index (range 0–37)	11.5	(6.3)	13.3	(7.1)	0.001
Stress from work deadlines (yes)	347	22.9%	101	32.0%	0.001

Table 2. Work postures and 22-year incidence of acute myocardial infarction among *all* men. Hazard ratios and 95% confidence intervals from incrementally adjusted Cox regression models. Kuopio Ischemic Heart Disease Risk Factor Study, N = 1831, 1985–2011.

Working posture	Model 1 Age-adjusted		Model 2 Adjusted for traditional CVD risk factors[1]		Model 3 Adjusted for all covariates in Table 1 excl. fitness[2]		Model 4 Adjusted for all covariates in Table 1 incl. fitness[3]	
	HR	95% CI	HR	95% CI	HR	95% CI	HR	95% CI
Sitting								
- Not at all	1.00		1.00		1.00		1.00	
- A little	0.81	0.64–1.03	0.86	0.68–1.09	0.89	0.70–1.13	0.91	0.71–1.15
- Quite a lot	0.93	0.72–1.21	1.01	0.78–1.32	1.04	0.79–1.36	1.05	0.79–1.38
- Very much	0.89	0.69–1.14	0.96	0.74–1.24	1.04	0.79–1.38	1.01	0.77–1.33
Standing								
- Not at all	1.00		1.00		1.00		1.00	
- A little	0.97	0.75–1.27	1.01	0.77–1.32	1.07	0.82–1.41	1.12	0.85–1.48
- Quite a lot	1.17	0.88–1.56	1.26	0.95–1.69	1.22	0.91–1.65	1.26	0.94–1.70
- Very much	0.96	0.71–1.31	0.95	0.70–1.30	0.99	0.72–1.37	1.00	0.73–1.38

(*continued*)

Table 2. (*continued*)

Working posture	Model 1 Age-adjusted		Model 2 Adjusted for traditional CVD risk factors[1]		Model 3 Adjusted for all covariates in Table 1 excl. fitness[2]		Model 4 Adjusted for all covariates in Table 1 incl. fitness[3]	
	HR	95% CI	HR	95% CI	HR	95% CI	HR	95% CI
Walking								
- Not at all	1.00		1.00		1.00		1.00	
- A little	0.94	0.68–1.29	0.92	0.67–1.26	0.96	0.69–1.33	1.00	0.72–1.39
- Quite a lot	0.97	0.69–1.36	0.92	0.65–1.29	0.92	0.65–1.31	0.97	0.68–1.38
- Very much	1.03	0.71–1.50	0.92	0.63–1.35	0.89	0.60–1.31	0.92	0.62–1.35

[1]Adjusted for traditional CVD risk factors (age, bmi, LDL-cholesterol, HDL-cholesterol, systolic blood pressure, alcohol, smoking, and conditioning leisure time physical activity)
[2]Adjusted for all covariates listed in Table 1 excl. cardiorespiratory fitness (VO_{2max})
[3]Adjusted for all covariates listed in Table 1 incl. cardiorespiratory fitness (VO_{2max})

Table 3. Work postures and 22-year incidence of acute myocardial infarction among men *without* ischemic heart disease at baseline. Hazard ratios and 95% confidence intervals from incrementally adjusted Cox regression models. Kuopio Ischemic Heart Disease Risk Factor Study, N = 1515, 1985–2011.

Working posture	Model 1 Age-adjusted		Model 2 Adjusted for traditional CVD risk factors[1]		Model 3 Adjusted for all covariates in Table 1 excl. fitness[2]		Model 4 Adjusted for all covariates in Table 1 incl. fitness[3]	
	HR	95% CI	HR	95% CI	HR	95% CI	HR	95% CI
Sitting								
- Not at all	1.00		1.00		1.00		1.00	
- A little	0.81	0.61–1.07	0.83	0.62–1.09	0.84	0.63–1.11	0.85	0.64–1.13
- Quite a lot	0.88	0.65–1.19	0.93	0.68–1.27	0.93	0.67–1.28	0.95	0.68–1.31
- Very much	0.83	0.61–1.12	0.86	0.63–1.16	0.87	0.63–1.21	0.85	0.61–1.18
Standing								
- Not at all	1.00		1.00		1.00		1.00	
- A little	0.91	0.66–1.24	0.91	0.67–1.25	1.00	0.72–1.38	1.04	0.75–1.44
- Quite a lot	1.11	0.79–1.56	1.16	0.82–1.63	1.19	0.83–1.69	1.21	0.85–1.72
- Very much	1.04	0.73–1.48	1.03	0.72–1.47	1.12	0.77–1.63	1.13	0.78–1.65
Walking								
- Not at all	1.00		1.00		1.00		1.00	
- A little	1.01	0.69–1.47	0.98	0.67–1.43	1.00	0.68–1.46	1.03	0.70–1.51
- Quite a lot	1.02	0.68–1.52	0.99	0.66–1.49	0.99	0.65–1.50	1.02	0.67–1.56
- Very much	1.04	0.66–1.63	0.97	0.62–1.53	0.95	0.59–1.52	0.97	0.61–1.55

[1]Adjusted for traditional CVD risk factors (age, bmi, LDL-cholesterol, HDL-cholesterol, systolic blood pressure, alcohol, smoking, and conditioning leisure time physical activity)
[2]Adjusted for all covariates listed in Table 1 excl. cardiorespiratory fitness (VO_{2max})
[3]Adjusted for all covariates listed in Table 1 incl. cardiorespiratory fitness (VO_{2max})

Table 4. Work postures and 22-year incidence of acute myocardial infarction among men with ischemic heart disease at baseline. Hazard ratios and 95% confidence intervals from incrementally adjusted Cox regression models. Kuopio Ischemic Heart Disease Risk Factor Study, N = 316, 1985–2011.

Working posture	Model 1 Age-adjusted		Model 2 Adjusted for traditional CVD risk factors[1]		Model 3 Adjusted for all covariates in Table 1 excl. fitness[2]		Model 4 Adjusted for all covariates in Table 1 incl. fitness[3]	
	HR	95% CI	HR	95% CI	HR	95% CI	HR	95% CI
Sitting								
- Not at all	1.00		1.00		1.00		1.00	
- A little	0.88	0.57–1.35	0.98	0.63–1.51	0.90	0.57–1.41	0.92	0.59–1.46
- Quite a lot	1.39	0.86–2.26	1.57	0.95–2.59	1.39	0.81–2.40	1.28	0.74–2.21
- Very much	1.40	0.87–2.25	1.67	1.03–2.72	1.72	1.01–2.93	1.67	0.98–2.85
Standing								
- Not at all	1.00		1.00		1.00		1.00	
- A little	1.32	0.79–2.23	1.51	0.88–2.59	1.34	0.77–2.34	1.38	0.79–2.42
- Quite a lot	1.21	0.71–2.08	1.46	0.83–2.56	1.24	0.69–2.23	1.33	0.73–2.41
- Very much	0.69	0.38–1.25	0.69	0.37–1.27	0.67	0.35–1.28	0.68	0.36–1.32
Walking								
- Not at all	1.00		1.00		1.00		1.00	
- A little	0.72	0.39–1.30	0.67	0.36–1.23	0.78	0.41–1.47	0.83	0.44–1.57
- Quite a lot	0.67	0.36–1.25	0.57	0.30–1.07	0.69	0.35–1.36	0.77	0.39–1.52
- Very much	0.77	0.39–1.51	0.66	0.33–1.31	0.78	0.38–1.58	0.82	0.40–1.66

[1]Adjusted for traditional CVD risk factors (age, bmi, LDL-cholesterol, HDL-cholesterol, systolic blood pressure, alcohol, smoking, and conditioning leisure time physical activity)
[2]Adjusted for all covariates listed in Table 1 excl. cardiorespiratory fitness (VO_{2max})
[3]Adjusted for all covariates listed in Table 1 incl. cardiorespiratory fitness (VO_{2max})

4 Discussion

Key strengths of this study include its prospective design, long and virtually complete follow-up includings medical specialist certification of AMI diagnosis, comprehensive control for possible confounders including virtually all traditional CVD risk factors plus important psychosocial risk factors, and stratified analyses by baseline IHD status avoiding the exclusion of the most vulnerable high risk worker population usually missed by most other studies of chronic cardiovascular disease outcomes that limit enrollment or analyses to healthy subjects only. It is noteworthy, that differences existed in observed risk patterns between men with and without baseline IHD and that these would have been missed had we excluded workers with pre-existing IHD from out study. It is generally advisable to study both subgroups because men with IHD are an integral part of middle-aged and an increasing part of older aging working populations. It is further advisable to examine their risks in stratified analyses so that different subgroup effects do not cancel each other out as happened to a certain extent in the combined analyses of the entire sample.

Two main potential sources of bias need to be considered: exposure misclassification and potential confounding via healthy worker selection effects.

One important limitation of this study is the measurement of working postures by self-report in four broad categories. The correlation between self-report and direct observation for physical work activities is often only moderate. However, questions about static postures involving the whole body such as sitting or standing have been shown fair to high reliability and validity in self-reported contexts. Workers tend to underestimate the amount of standing which would introduce a conservative bias. The potential for bias due to measurement error has been shown to be lowest for prospective studies. No direct job observations were available in this population-based study but we had the opportunity to check current occupational titles for all subjects and found that occupational titles were consistent with reported working positions. For example, men doing administrative work, operating stationary machinery, or driving motor vehicles were the three occupational groups most often reporting no standing at work, while physicians, sales people, metal workers, and forest and construction workers typically reported high amounts of work in a standing position. These occupations are heterogeneous in terms of socioeconomic status and other relevant work characteristics. For this reason we believe that it is the standing position at work per se that has a detrimental effect on the cardiovascular system. Measurement bias is also possible because the amount of time spent in the different work postures in our analyses was based on a single measurement at baseline and did not reflect possible changes in exposure during the 22-year follow-up period. For example, of those who were still employed at 11-year follow-up and who were standing very much at baseline, 53% reported standing very much, 28% quite a lot, 14% little, and 5% not at all, respectively at 11-year follow-up. The Pearson correlation coefficient between the four categories of standing at baseline and at 11-year follow-up was 0.61 (p < 0.001). Although these figures indicate a modest stability of exposure over the follow-up period, some exposure changes have definitely occurred and this most likely has resulted in an underestimation of the magnitude of the effects associated with prolonged standing. In general, all sources of exposure misclassification in this study probably led to an attenuation of effect estimates, a conservative bias towards no effect.

In contrast to our earlier study on prolonged standing and non-symptomatic progression of carotid atherosclerosis measured as change in intima media thickness by ultrasound, [7] this study of AMI is vulnerable to health-based selection in and out of physically demanding work because myocardial infarctions usually occur in persons who suffer from chronic coronary heart disease that is typically characterized by activity-related symptoms such as chest pain or dyspnoe prompting affected workers not to seek and perform physically demanding work and work postures and/or to migrate into less demanding jobs and/or to less demanding tasks within a given job. Therefore, the higher risk of a sitting posture among men with baseline IHD observed in our study may in part reflect this selection effect rather than a detrimental effect of sitting per se. Similarly, the increased risks observed for a standing work posture at baseline in this same subgroup of workers has probably been underestimated in this study because workers with symptomatic IHD at baseline may have already changed to a less physically demanding sitting position at work thus inflating the risks associated with sitting and deflating the risks associated with standing or walking for the same reason. The apparent lower risks of AMI among men with IHD in the highest standing category ("very much") or were walking at work should therefore be interpreted with caution because these estimates may not indicate any true protective effect but need instead to be understood as a spurious effect that is the result of an inherent bias, the so-called "healthy worker effect". [1] The very strong positive associations between standing very much and 4-year asymptomatic progression of atherosclerosis especially among these men with IHD as reported previously, [7] and consistent risk patterns observed for occupational physical activity measured as energy expenditure and relative aerobic strain in this same study population with regard to cardiovascular disease incidence [6] and both CVD and all-cause mortality [5], support the presence of such a conservative bias due to healthy worker effects in this study.

Our findings among healthy men (without IHD at baseline) showing a reduced risk for AMI for sitting work postures and increased risks for standing work are consistent with a recent 12-year prospective cohort study of a representative Canadian population-based cohort study that found a doubling of IHD incidence among both men and women without IHD at baseline even after adjustment for a wide range of potential confounders including comorbidity and several psychosocial and traditional CVD risk factors. [13] That study used a job exposure matrix that identified four different exposure groups: those with primarily (nearly exclusive) sitting work postures served as a reference for a group with primarily standing work postures, a group with a combination of sitting, walking, and standing, and a group with other work postures. Interestingly, only the primarily standing group showed a substantially increased IHD risk of about 100% while the other groups' estimates hovered around zero, ranging from minus 3–7% for those with mixed postures to plus 4–9% for those in other work postures, depending on the number of controlled variables in the model.

5 Conclusions

This study alone does not provide conclusive evidence about the health effects of certain working postures, however, point estimates and confidence intervals of the results are most compatible with the hypothesis that prolonged standing increases the risk of AMI in both healthy workers and workers with IHD. Work posture associations with AMI risk differed between men with and without IHD. Among men *without* IHD, sitting reduced risk, standing increased risk, and walking had no impact. Among men *with* IHD, sitting more than a little and standing "a little" or "quite a lot" increased risk, while standing "very much" and walking tended to reduce risk. Considering the previously reported strong positive association between standing and progression of atherosclerosis among these men with IHD, [7] and a doubling of IHD risk among primarily standing workers without IHD in the recent prospective Canadian mentioned above, [13] it seems that the apparently reduced risk among workers with IHD who stand very much may actually be a due to healthy worker selection bias. We therefore conclude that it is currently prudent for clinicians and occupational health professionals not to recommend a predominantly standing work posture, and instead to advocate for the provision of seating options for those who currently work for substantial amounts of time in a standing work posture.

That recommendation is also consistent with the goal to prevent other common health conditions that have been associated with prolonged standing at work such as venous diseases and musculoskeletal fatigue, discomfort, and pain of the extremities and lower back that are more frequently observed among those who predominantly work in a standing position. Evidence from both laboratory and epidemiological studies indicate that relative modest amounts of standing at work of less than one or two hours may already increase the risk of musculoskeletal symptoms. [2, 3] While it is important for future research to reduce exposure misclassification through better objective and repeat measures of work postures and to determine detailed dose-response relationships between working postures and cardiovascular disease and mortality outcomes, a proactive workplace health promotion and disease prevention strategy may in the meantime be well guided by more readily available evidence from such musculoskeletal health research. Another possible strategy for more rapidly assess cardiovascular health risks of working postures may include using changes in levels of traditional cardiovascular risk factors as a surrogate indicator for hard clinical cardiovascular morbidity or mortality outcomes. However, the health effects of prolonged standing at work reported in the literature, for example, have been consistently and to a very large extent been independent from other known cardiovascular disease risk factors and therefore these cardiovascular disease risk factors may not be important mediators or valid indicators of the CVD risks associated with a standing working posture. On the other hand, most past research was not designed to formally investigate mediation effects and future research needs to fill this gap. Any future research on the health effects of working postures or respective worksite interventions should also include workers with pre-existing chronic musculoskeletal or cardiovascular diseases and to always consider baseline health status and possible healthy worker effects when assessing risks and benefits of different work postures.

References

1. Arrighi HM, Hertz-Picciotto I (1994) The evolving concept of the healthy worker survivor effect. Epidemiology 5:189–196
2. Coenen P, Willenberg L, Parry S, Shi JW, Romero L, Blackwood DM, Maher CG, Healy GN, Dunstan DW, Straker LM (2016) Associations of occupational standing with musculoskeletal symptoms: a systematic review with meta-analysis. Br J Sports Med 0:1–9. https://doi.org/10.1136/bjsports-2016-096795
3. Coenen P, Willenberg L, Parry S, Shi JW, Romero L, Blackwood DM, Maher CG, Healy GN, Dunstan DW, Straker LM (2018) Associations of occupational standing with musculoskeletal symptoms: a systematic review with meta-analysis. Br J Sports Med 52:174–181
4. Hayashi R, Iso H, Cui R, Tamakoshi A (2016) Occupational physical activity in relation to risk of cardiovascular mortality: the Japan collaborative cohort study for evaluation for cancer risk (JACC study). Prev Med 89:286–291. https://doi.org/10.1016/j.ypmed.2016.06.008

5. Krause N, Arah OA, Kauhanen J (2017) Physical activity and 22-year all-cause and coronary heart disease mortality. Am J Ind Med 60:976–990. https://doi.org/10.1002/ajim.22756
6. Krause N, Brand RJ, Arah O, Kauhanen J (2015) Ocupational physical activity and 20-year incidence of acute myocardial infarction: results from the Kuopio Ischemic Heart Disease Risk Factor Study. Scand J Work Environ Health 41:124–139. https://doi.org/10.5271/sjweh.3476
7. Krause N, Lynch JW, Kaplan GA, Cohen RD, Salonen R, Salonen JT (2000) Standing at work and progression of carotid atherosclerosis. Scand J Work Environ Health 26:227–236
8. Kristal-Boneh E, Harari G, Melamed S, Froom P (2000) Association of physical activity at work with mortality in Israeli industrial employees: the CORDIS study. J Occup Environ Med 42:127–135
9. Møller SV, Hannerz H, Hansen AM, Burr H, Holtermann A (2016) Multi-wave cohort study of sedentary work and risk of ischemic heart disease. Scand J Work Environ Health 42:43–51. https://doi.org/10.5271/sjweh.3540
10. Murtagh EM, Murphy MH, Boone-Heinonen J (2010) Walking – the first steps in cardiovascular disease prevention. Curr Opin Cardiol 25:490–496. https://doi.org/10.1097/HCO.0b013e32833ce972
11. Rempel D, Krause N (2018) Do sit-stand workstations improve cardiovascular health? JOEM Publ Ahead Print. https://doi.org/10.1097/jom.0000000000001351
12. Rose GA, Blackburn H, Gillum RF (1982) Cardiovascular survey methods. World Health Organization, Geneva
13. Smith P, Ma H, Glazier RH, Gilbert-Ouimet M, Mustard C (2017) The relationship between occupational standing and sitting and incident heart disease over a 12-year period in Ontario, Canada. Am J Epidemiol 187:27–33. https://doi.org/10.1093/aje/kwx298
14. Stamatakis E, Chau JY, Pedisic Z, Bauman A, Macniven R, Coombs N, Hamer M (2013) Are sitting occupations associated with increased all-cause, cancer, and cardiovascular disease mortality risk? A pooled analysis of seven British population cohorts. PLoS One 8: e73753
15. Tabatabaeifar S, Frost P, Andersen JH, Jensen LD, Thomsen JF, Svendsen SW (2015) Varicose veins in the lower extremities in relation to occupational mechanical exposures: a longitudinal study. Occup Environ Med 72:330–337. https://doi.org/10.1136/oemed2014-102495
16. Tüchsen F, Hannnerz H, Burr H, Krause N (2005) Prolonged standing at work and hospitalization due to varicose veins: a 12-year prospective study of the Danish population. Occup Environ Med 62:847–850
17. Tüchsen F, Krause N, Hannerz H, Burr H, Kristensen TS (2000) Standing at work and varicose veins. Scand J Work Environ Health 26:414–420
18. van Uffelen JG, Wong J, Chau JY, van der Ploeg HP, Riphagen I, Gilson ND, Burton NW, Healy GN, Thorp AA, Clark BK, Gardiner PA, Dunstan DW, Bauman A, Owen N, Brown WJ (2010) Occupational sitting and health risks: a systematic review. Am J Prev Med 39:379–388. https://doi.org/10.1016/j.amepre.2010.05.024

Proposal of New Map Application for Distracted Walking When Using Smartphone Map Application

Tomoki Kamiyama[1(✉)], Mitsuhiko Karashima[2],
and Hiromi Nishiguchi[2]

[1] Graduate School of Information and Telecommunication Engineering,
Tokai University, 2-3-23 Takanawa, Minato, Tokyo, Japan
7bjnm011@mail.u-tokai.ac.jp
[2] School of Information and Telecommunication Engineering, Tokai University,
2-3-23 Takanawa, Minato, Tokyo, Japan

Abstract. This study aimed to propose a new map application that would reduce the time of looking at the map while walking to the degree of the time of reading the paper map and an experiment was conducted to confirm the effectiveness of the proposed map application.

The proposed map application had a difference from Google Maps that the walker's current position was shown on the map only when he/she stopped. In the experiment, the participants were required to walk three randomly assigned routes in KOMAZAWA Olympic Park General Sports Ground, respectively, by using the paper map, Google Maps, and the proposed map application. They equipped a wearable camera for recording the gaze direction during the experiment. After walking each route, they evaluated subjectively their workload while walking the route by using Japanese version of NASA-TLX.

The results of the experiment revealed that the number of stops walking to look at the proposed map application was significantly more than Google Maps. The results also revealed that the time of looking at the proposed map application while walking the route was significantly reduced in comparison with Google Maps. These meant that the proposed map application induced the walker to stop walking to look at the map and could reduce the distracted walking in comparison with Google Maps. The effectiveness of the proposed map was confirmed through the experiment.

Keywords: Distracted walking · Map application · Smartphone

1 Introduction

The danger of distracted walking when using smartphone has become a social problem in many countries, including Japan, as smartphones spread. According to the reports by Tokyo Fire Department between 2012 and 2016, 145 distracted walkers were injured and transported to hospitals by ambulance [1]. Telecommunications Carriers Association reported that the top four most frequent reasons for distracted walking in the principal cities of Japan were the use of map, SNS, website, and email applications [2].

© Springer Nature Switzerland AG 2019
S. Bagnara et al. (Eds.): IEA 2018, AISC 819, pp. 337–346, 2019.
https://doi.org/10.1007/978-3-319-96089-0_36

Previous researches have mainly focused on distracted walking when texting and reading messages and talking on the smartphone [e.g. 3–12]. However, there were few studies focusing on distracted walking when using a map application [13]. Our previous study aimed to find the difference of walking behaviors between the paper map and the map application (Google Maps). The results of the previous study revealed the difference of walking behaviors that the ratio of the time of looking at the map to the walking duration and the time of looking at the map while walking the route with using Google Maps were larger than the paper map. The study also revealed that the number of times to look around while walking the route with the paper map was larger than Google Maps. From the results of the study, the new function was conceived that the walker's current position was shown on the map only when he/she stopped in order to reduce the time of looking at the map while walking [14].

This study aimed to propose the new map application that would reduce the time of looking at the map while walking to the degree of the time of reading the paper map and an experiment was conducted to confirm the effectiveness of the proposed map application.

2   Proposed Map

As mentioned above, our previous study revealed the difference of walking behaviors that the ratio of the time of looking at the map to the walking duration and the time of looking at the map while walking with using Google Maps were larger than the paper map. The study also revealed that the number of times to look around while walking with the paper map was larger than Google Maps. From the viewpoint of the safety for the walking with using smartphone map application, these results suggested that it should be needed to decrease the time of looking at the map while walking and to increase the number of times to look around while walking. So, the new map application was proposed that the walker's current position was shown on the map only when he/she stopped. By using the proposed map application, as the walker cannot confirm the current position on the map at the time of walking, he/she would be expected to memorize the route information at the time of stopping, to guess the current position from the memorized route information, and to confirm the current position with looking around while walking. From this expectation, in comparison with Google Maps, the proposed map application would be expected to be able to decrease the time of looking at the map while walking and increase the number of times to look around while walking.

3   Experiment

3.1   Participants

Twelve university students (6 males and 6 females, aged 20 to 22) participated in the experiment. This study has been approved by the research ethics committee of Tokai University (No. 177077). They were required to answer a questionnaire for checking the lack of sense of direction before the experiment in order to confirm that the participants didn't have a poor sense of direction. This questionnaire was structured from Sense of Direction Questionnaire-Short Form (SDQ-S), also added two questions ("Have you ever been told that you have no sense of direction?" and "Do you think you have no sense of direction?") [15]. The participants were required to equip a wearable camera (Panasonic HX-AIH-D) for recording gaze direction while walking the route.

3.2 Experimental Task

The participants were required to walk three randomly assigned routes (each distance was about 650 meters) in KOMAZAWA Olympic Park General Sports Ground, respectively, by using the paper map, Google Maps, and the proposed map application. Moreover, they were required to think aloud while walking the route (Fig. 1).

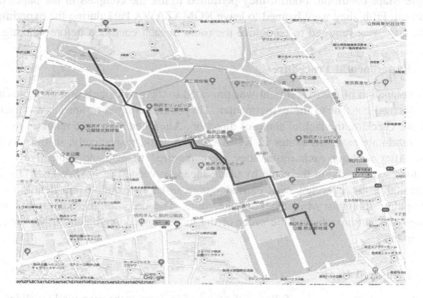

Fig. 1. Map of KOMAZAWA Olympic Park General Sports Ground and routes of the experiment.

3.3 Maps

Paper Map: KOMAZAWA Olympic Park General Sports Ground on Google Maps was printed on the paper map, which consisted of 4 A4 size sheets and on which page numbers and compass points were written. The route was written by hand on the paper map.

Google Maps: Google Maps (version 4.43.15) was used in the experiment.

Proposed Map Application: This application was created by Android Studio (version 2.3.3). Though the proposed map application constructed of the same map as Google

Maps, the application had a difference from Google Maps that the walker's current position was shown on the map only when he/she stopped walking. The on/off of the indicator of the walker's current position was judged from his/her speed calculated by programing function. The location information was updated once per 5 ms. The routes were drawn by polyline, but the route searching function was not used.

3.4 Conditions

The participants were instructed the following 6 rules as follows in order to ensure the same condition in all three maps. First, the participants were banned to shrink and expand the map scale in conditions of Google Maps and the proposed map application. Second, they were banned to push the current position button in conditions of Google Maps and the proposed map. Third, they were banned to use navigation function in Google Maps condition. Fourth, they permitted to use the compass in the paper map condition. Fifth, they were banned to leave KOMAZAWA Park during the experiment. Sixth, they were required to recognize the route in each condition before walking.

3.5 Experimental Procedure

The explanation of the experiment was provided to the participants and they provided the informed consent to the experimenter. They were required to answer a questionnaire for checking the lack of sense of direction. After then they were required to walk three randomly assigned routes in KOMAZAWA Park, respectively, in three conditions. After walking each route, they were required to evaluate subjectively their workload while walking the route by using Japanese version of NASA-TLX [16, 17]. Additionally, the interview was carried out in order to examine the details of the participants' thoughts while walking the route. The order of the routes was fixed through the participants and the order of the maps was counterbalanced between the participants.

3.6 Measured Items

The measured items were as follows: the time of looking at the map while walking the route (second), the ratio of the time of looking at the map to the walking duration while walking the route, the number of times to look around while walking the route, the walking speed (km/h), the number of stops walking, and the subjective workload score (NASA-TLX AWWL(Adaptive Weighted Workload) score).

The data of each item were analyzed statistically by using one-way within-subjects analysis of variance (ANOVA) with post-hoc Bonferroni multiple comparison test.

4 Results

4.1 Time of Looking at the Map While Walking the Route

The result of ANOVA for the time of looking at the map while walking the route revealed that the mean time was significantly difference between three maps

(F(2,22) = 4.5409, p < 0.05). As shown in Fig. 2, the results of the post-hoc test revealed that the mean time of looking at the proposed map application was signifi-cantly less than Google Maps (t = 2.8747, p < 0.05). On the other hand, the results also revealed that there was no significant difference between the proposed map and the paper map (t = 2.2206, p > 0.10). These mean that the proposed map contributes to reducing the time of looking at the map while walking the route in comparison with Google Maps and that the time at the proposed map was similar to the paper map.

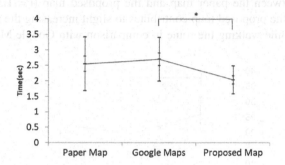

Fig. 2. The time of looking at the map while walking the route

4.2 Ratio of the Time of Looking at the Map to the Walking Duration While Walking the Route

The results of ANOVA for the ratio of the time of looking at the map to the walking duration while walking the route revealed that the mean ratio was significantly dif-ference between the maps (F(2,22) = 7.2604, p < 0.01). The results of the post hoc test revealed that the mean ratio of the time of looking at the proposed map application was significantly less than Google Maps (t = 2.9824, p < 0.05). These results also revealed that the mean ratio of the time of looking at the proposed map was significantly less than the paper map (t = 3.5453, p < 0.01). These mean that the proposed map con-tributes to reducing the ratio of the time of looking at the map while walking the route in comparison with Google Maps (Fig. 3).

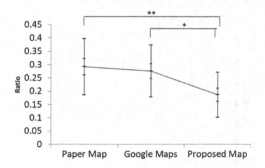

Fig. 3. Ratio of the time of looking at the map to the walking duration while walking the route

4.3   Number of Times to Look Around While Walking the Route

The results of ANOVA for the number of times to look around while walking the route revealed that the mean number of times was significantly difference between the maps (F (2,22) = 3.4891, p < 0.05). The results of the post hoc test revealed that the mean number of times with using Google Maps was significantly less than the paper map (t = 2.6416, p < 0.05). On the other hand, the results revealed that there is no significantly difference between Google Maps and the proposed map application (t = 1.3384, p > 0.10) nor between the paper map and the proposed map (t = 1.3032, p > 0.10). These mean that the proposed map contributes to slight increasing the number of times to look around while walking the route in comparison with Google Maps (Fig. 4).

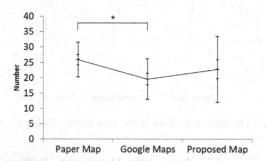

Fig. 4. The number of times to look around while walking the route

4.4   Walking Speed

The results of ANOVA for the walking speed revealed that mean speed was significantly difference between the maps (F(2,22) = 4.9316, p < 0.05). The results of the post hoc test revealed that the mean speed with using Google Maps was significantly more than the paper map (t = 2.9247, p < 0.05). On the other hand, the results revealed that there is no significantly difference between Google Maps and the proposed map application (t = 0.4712, p > 0.10) nor between the paper map and the proposed map (t = 2.4534, p > 0.10). These mean that the proposed map doesn't cause a reduction in walking speed in comparison with Google Maps (Fig. 5).

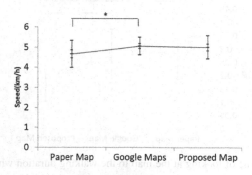

Fig. 5. Walking speed

4.5   Number of Stops Walking

The results of ANOVA for the number of stops walking revealed that the mean number was significantly different between the maps ($F(2,22) = 17.8254$, $p < 0.01$). The results of the post hoc test revealed that the mean number with using the proposed map application was significantly more than Google Maps ($t = 5.5072$, $p < 0.01$). The results also revealed that the mean number with using the proposed map was significantly more than the paper map ($t = 4.7513$, $p < 0.01$). These mean that the proposed map contributes to increasing the number of stops walking (Fig. 6).

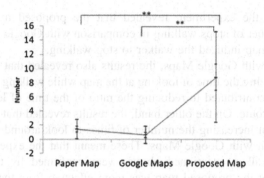

Fig. 6. The number of stops walking

4.6   AWWL Scores

The results of ANOVA for the AWWL scores revealed that mean AWWL scores was significantly difference between the maps ($F(2,22) = 9.6644$, $p < 0.01$). The results of the post hoc test revealed that the mean AWWL scores with using Google Maps was significantly less than the paper map ($t = 4.1857$, $p < 0.01$). The results also revealed that the mean AWWL scores with using the proposed map application was significantly less than the paper map ($t = 3.2576$, $p < 0.05$). On the other hand, the results revealed that there was no significant difference between the proposed map and Google Maps ($t = 0.9281$, $p > 0.10$). These mean that the proposed map does not cause an increase in the subjective workload in comparison with the Google Maps (Fig. 7).

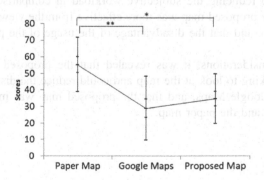

Fig. 7. The AWWL scores

5 Discussions

As aforementioned, by using the proposed map application, as he/she cannot confirm the current position on the map at the time of walking, it would be expected that the walker memorized the route information at the time of stopping, guessed the current position from the memorized route information, and confirmed the position with looking around while walking. From this expectation, in comparison with Google Maps, the proposed map would be expected to be able to decrease the time of looking at the map while walking and increase the number of times to look around while walking.

The results of the experiment revealed that the proposed map contributed to increasing the number of stops walking in comparison with Google Maps. This meant that the proposed map induced the walker to stop walking.

In comparison with Google Maps, the results also revealed that the proposed map contributed to reducing the time of looking at the map while walking the route, and that the proposed map contributed to reducing the ratio of the time of looking at the map while walking the route. On the other hand, the results revealed that the proposed map contributed to slight increasing the number of times to look around while walking the route in comparison with Google Maps. These meant that the expectations about the behavior of the walker, mentioned above, were confirmed by the results of the experiment and that the proposed map was more effective from the viewpoint of the safety than Google Maps.

Moreover, the results revealed that the proposed map did not cause a reduction in walking speed in comparison with Google Maps. The results revealed that the proposed map did not cause an increase in the subjective workload in comparison with Google Maps. Therefore, the disadvantage of the usage of the proposed map was insignificant in comparison with Google Maps.

The results of the experiment revealed that the proposed map contributed to increasing the number of stops walking in comparison with the paper map and that the time of looking at the proposed map while walking the route was similar to the paper map. In comparison with the paper map, the results also revealed that the proposed map contributed to reducing the ratio of the time of looking at the map to the walking duration. Moreover, the results revealed that the proposed map did not cause a reduction in walking speed in comparison with the paper map, and that the proposed map contributed to reducing the subjective workload in comparison with the paper map. Therefore, the proposed map was more effective from the viewpoint of workload than the paper map, and that the disadvantage of the usage of the proposed map was insignificant.

From these considerations, it was revealed that the proposed map induced the walker to stop walking to look at the map and could reduce the distracted walking in comparison with Google Maps, and that the proposed map was more effective than both Google Maps and the paper map.

6 Conclusion

This study aimed to propose a new map application that would reduce the time of looking at the map while walking to the degree of the time of reading the paper map and an experiment was conducted to confirm the effectiveness of the proposed map application.

The proposed map application had a difference from Google Maps that the walker's current position was shown on the map only when he/she stopped.

The results of the experiment revealed that the number of stops walking with using the proposed map application was significantly more than Google Maps. The results revealed that the proposed map contributed to reducing the time of looking at the map while walking the route, the reducing the ratio of the time of looking at the map to the walking duration, and slight increasing the number of times to look around while walking in comparison with Google Maps. The results also revealed that the proposed map contributed to be similar to the time of looking at the map while walking the route and reducing the subjective workload in comparison with the paper map.

From mentioned above, it was revealed that the proposed map induced the walker to stop walking to look at the map and could reduce the distracted walking, and that the proposed map was more effective than both Google Maps and the paper map.

References

1. Tokyo Fire Department. http://www.tfd.metro.tokyo.jp/lfe/topics/201602/mobile.html. Accessed 20 Nov 2017
2. Telecommunications Carriers Association distracted walking survey. http://www.tca.or.jp/press_release/pdf/170308sumahochosa.pdf. Accessed 20 Nov 2017
3. Shigeru H, Ayaka S, Yuri S, Hideka S, Saki Y, Kosuke M (2015) Effects of using a smart phone on pedestrians' attention and walking. Procedia Manuf 3:2574–2580
4. Shigeru H, Kanae F, Eri K, Yoshinori S, Azuri Y (2016) Effects on auditory attention and walking while texting with a smartphone and walking on stairs. In: HCII 2016 posters, part I, CCIS 617, pp 186–191
5. Syuji Y, Hiroshi T, Kayoko T, Toru M, Takuya F (2017) Effects of smartphone use on behavior while walking. Urban Reg Plan Rev 7:138–150
6. Aoi N, Shin M, Kouhei I, Toshiki I, Keita S, Aya N, Daiki N, Kayoko S, Teppei A, Kunihiko A, Katsuyuki M, Jun H (2016) Influences of smartphone use while walking on gait parameters. Jpn J Health Promot Phys Ther 6(1):35–39
7. Kazuhiro K (2016) Gaze measurement of a pedestrian texting while walking and a rider texting while riding a bicycle, and verification of dangerousness of texting. IEICE 10 (2):129–136
8. Nasar J, Hecht P, Wener R (2008) Mobile telephones, distracted attention, and pedestrian safety. Accid Anal Prev 40(1):69–75
9. Schildbach B, Rukzio E (2010) Investigating selection and reading performance on a mobile phone while walking. In: Proceedings of the 12th conference on human-computer interaction with mobile devices and services, mobile HCI 2010, pp 93–102
10. Stavrinos D, Byington K, Schwebel D (2011) Distracted walking: cell phones increase injury risk for college pedestrians. J Saf Res 42(2):101–107

11. Schwebel D, Stavrinos D, Byington K, Davis T, O'Neal E, Jong D (2012) Distraction and pedestrian safety: how talking on the phone, texting, and listening to music impact crossing the street. Accid Anal Prev 45:266–271

12. Lu J-M, Lo Y-C (2017) Investigation of smartphone use while walking and its influence on one's behavior among pedestrians in Taiwan. In: HCI 2017: HCI international 2017 - posters' extended abstracts, pp 469–475

13. Beeharee A, Steed A (2006) A natural wayfinding-exploiting photos in pedestrian navigation. In: MobileHCI 2006. ACM Press, pp 81–88

14. Tomoki K, Mitsuhiko K (2017) Effects of using smartphone map application on walking behavior. Jpn J Ergon 53(Supplement1):S308–S309

15. Yoshiaki T (1992) Sense of direction and its relationship with geographical orientation, personality traits and mental ability. Jpn J Educ Psychol 40(1):47–53

16. Hart SG, Staveland LE (1988) Development of NASA-TLX (task load index); results of empirical and theoretical research. Adv Psychol 52:139–183

17. Shigeru H, Naoki M (1996) Japanese version of NASA task load index: sensitivity of its workload score to difficulty of three different laboratory tasks. Jpn J Ergon 32(2):71–79

Defining an Occupational Risk Prevention Approach for Networked Organisational Configurations: The Case of Road Haulage and Logistics

Liên Wioland(✉) and Virginie Govaere

Working life Department, Preventive Ergonomics and Psychology Laboratory,
INRS, 1 rue du Morvan – CS 60027, 54519 Vandœuvre Cedex, France
lien.wioland@inrs.fr

Abstract. All the functions of the logistics chain, from the producer to the end customer, are supported by a multitude of companies connected to each other. These companies form a complex network in which each one of them is dependent on what happens upstream and downstream of its activity. In this context, the INRS conducted an ergonomic study whose objective was to determine a prevention approach specific to this type of network organization. The purpose of this paper is to present this preventive approach entitled PRO-PAGIR in the sector of road transport of goods and logistics. This approach has been tested in a company, the results are also presented. This study is an ergonomics analysis and not a business case.

Keywords: Prevention · Road haulage and logistics · Network organisation

1 Context and Objectives

1.1 Context

Road Haulage and Logistics companies form a vital sector of the French economy, with more that 87% of all goods transported in France going by road and with the sector employing a total of 1.8 million employees. The sector is characterised by just-in-time organisation, very dense regulation, high competitiveness, growing pressure from clients and suppliers, and tight deadlines. Added to this, the competitiveness and flexibility factors are increasing, due, in particular to the e-commerce boom (online sales have grown by 675% in 10 years).

Road Haulage and Logistics fits more broadly into the concept of "Supply Chain". The supply chain covers all of the functions from supplier to customer, all of the activities (delivery, storage, sorting, handling, etc.), the resources (material, human, financial, technical, and informational) and the methods enabling the company to make available to the customer the required economic quantity of product and/or of service with the best possible productivity and profitability and at the right time. The product and/or service should be of expected quality in an environment of safety, regulations, and contingencies. All the way along the logistics chain, the flow of goods goes

© Springer Nature Switzerland AG 2019
S. Bagnara et al. (Eds.): IEA 2018, AISC 819, pp. 347–356, 2019.
https://doi.org/10.1007/978-3-319-96089-0_37

through various handover points that take charge of some of the handling and processing it undergoes [1]. The literature in the logistics field emphasises that an issue to be addressed is to optimise and impart smoothness to the logistics chain as a whole, and not segment by segment [2, 3]. To this end, information should be managed optimally, both in terms of how it is distributed and processed, and in terms of how it is shared inside and outside the company. This should enable the companies not only to be fast and flexible, but also to trace and rationalise their activity [4, 5]. Ultimately, the idea is to integrate all of the logistics functions by synchronising all of the flows of goods and of information, while also co-ordinating all of the players involved in these processes.

The logistics chain is generally simplified and represented linearly, from supplier to customer. In reality, the flows of goods and of information are manifold, take different paths, and form complex networks made up of multiple companies (or entities) having distinct functions. These interconnected companies are therefore dependent on what happens upstream and/or downstream from their activities.

1.2 Health, Safety, and Prevention

1.2.1 A Few Figures

Health and safety appear as forming a stake and a challenge in their own right. In terms of accident rate, employees in the sector suffer accidents 2 to 3 times more often than employees in the other sectors of activity, and the seriousness of the accidents is higher than average. It is observed that the employees essentially concerned by ergonomic and prevention studies are drivers and goods handlers [6, 7]. As regards health, the risks usually identified for such employees concern both the physical activity (risks of musculoskeletal disorders (MSDs) related to goods handling, same-level falls (trips and slips), exposure to vibration, etc. [8, 9]) and also the mental aspects of the work (high mental workload, segmentation of the work, intensification of the pace of work, management of contingencies, reinforcement of the inspection and monitoring of the work ...) [10]. As regards safety, the accident risks are of the collision and/or crushing types related to using mechanised goods-handling vehicles or plant, or else road accidents related mainly to phenomena of drowsiness and/or loss of vigilance [11, 12].

It must be admitted that those studies are focussed on single activities of the supply chain, and that the analyses are conducted independently and in a manner isolated from the other activities of the supply chain despite it having a networked organisational configuration.

1.2.2 Specificities Related to Networked Organisational Configurations

Earlier work has highlighted the phenomenon of "propagation" [13, 14]. That phenomenon raises questions about the extent of the effects related to a technological transformation among the operatives from various departments that are interconnected within a single company or between different companies. That work shows that not only the users of a new technology (Information and Communication Technology (ICT) in that case) but also non-users should be incorporated into the analysis and the development of risk prevention avenues. Otherwise, even if they contribute to improving the work situation of the direct users of the new ICT, the prevention solutions that are proposed might also degrade the working conditions of the non-users.

For example, an improvement in the work situation of order pickers who use ITC (voice picking) was accompanied by a deterioration in the working conditions of the forklift drivers who stocked up the picking racks, due to an increase in the pace of their work and to tensions appearing with the order pickers, for example. Thus, implementing a solution that is targeted on a single one of these work situations does not prevent the propagation of the degradation factors to the connected situations from continuing.

Its counterpart in terms of prevention actions, namely the "act elsewhere" principle, aims to determine that work situation on which it is most appropriate to act in order to improve prevention, while taking account of two aspects:

- the prevention action that focuses directly on the work situation is not systematically the most effective or the most appropriate; and
- direct prevention action may be limited or impossible to implement.

In the earlier work on transport [15], it has been shown that prevention measures that were focused directly on drivers were difficult to implement, given the diversity of work situations and of the operatives involved. The daily work of the driver takes place to a large extent on the road, but also on the sites of multiple customers, and in interaction with other operatives. One way of responding to that difficulty was to act on the organisational aspects and working conditions of the platform with which the drivers were strongly connected. Ultimately, it was by improving that situation, which was not the initially targeted situation, that the situation of the drivers was improved.

1.2.3 Objectives

In an occupational risk prevention context, the objective of this study was to test a prevention approach that was specific to the networked organisational configuration and that incorporated the specificities of "propagation" and of "act elsewhere". Thus, the idea was to incorporate into this approach:

- analyses of work situations involving the operatives who are directly concerned; and
- analyses of work situations connected to the preceding situations and that therefore involve operatives who are indirectly concerned. Here, the idea was to analyse the relationships existing between these interconnected situations so as to define action logics adapted to the chain of work situations, and so as to bring some answers about the relevant entities on which action was possible.

2 Method

The study was conducted in a transport (haulage) company. That company is presented below in such a manner as to highlight the specific points related to the prevention approach that was tested.

The company has a national agency where the customer orders arrive. The orders are then prepared by the logistics branch of the company and are then transferred to the transport unit of the national agency. The goods are then transported every night by truck to the local platforms close to the urban centres of various regions. The study focused on the national agency and on the largest of the local platforms that was concerned by the main questions raised by the company in terms of working conditions. The goods arrive at about 4 a.m. and are checked, sorted, and re-consolidated by delivery round. Two delivery rounds are scheduled per day for each vehicle and driver. After the first rounds (starting at about 4:30 a.m. and finishing about 8 a.m.), the drivers return to the platform to organise the goods they are going to deliver on the second rounds. The company has an Information System (IS) making it possible to track the goods all the way to delivery to the customers. Depending on the tasks to be performed, the IS is in the form of software on a computer, or in the form of a Personal Digital Assistant (PDA) for the drivers.

Although the request from the company was focused on the working conditions of the operatives at the platform, the spectrum of the analysis was broadened, consistent with the tested approach. Thus, analyses in the real working situation concerned the platform, the national agency (entity upstream from the platform), and also the deliveries to customers (entity downstream from the platform). To conduct these analyses, "paper & pencil" type observation techniques and video recording observation techniques were used. Then interviews were conducted with the management, the first-line supervisors, and the employees of the national agency and of the platform, and the drivers. The resulting data was qualitative. After a pre-observation stage at the national agency and at the platform for 4 days, data collection was organised in the 3 phases that were spaced apart in time:

- at the national agency, a campaign of observations and interviews was conducted for 3 days;
- at the platform, two campaigns of observations and interviews were conducted; the first lasted 5 days in succession, while the platform was operating normally, and the second lasted 4 days in succession during peak activity (complex situation); and
- at customer delivery level, a campaign of observations and interviews was conducted for 3 days in succession (i.e. 6 drivers monitored during their working days).

3   Results

3.1   The Platform and the National Agency

The platform management players were constituted by the platform supervisor and his assistant. The observations and interviews made it possible to identify their main functions. The supervisor constituted the "interface" between the platform and the national agency. He acted as an "orchestra conductor": he distributed the assignments to be done, co-ordinated all of the activities of the operatives, and made sure that the assignments were completed properly by coping with all contingencies and incidents. He made all the preparations necessary so that the next day's deliveries went as well as possible (preparing the parcels, and organising the storage in rack bays). For all these tasks, a certain number of procedures were defined by the national agency but relative autonomy existed, in particularly as regards defining the resources to be implemented.

Difficulties, stemming from the national agency, could be observed as having propagated to the activity of all of the operatives at the platform. For example, delays in deliveries of goods coming from the national agency were observed. Such delays created "uncertainty" for the operatives at the platform. They found themselves waiting "tensely", not knowing whether the goods would be able to be delivered within the allotted times. In order to remove the uncertainties and anticipate the organisational changes to be implemented, the supervisor regularly asked the national agency for information. When the goods arrived, he reorganised the activity of all of the operatives. Thus, regardless of their functions, the operatives helped with unloading and managing the goods. Such reactive organisation reduced the time available for doing their usual activities.

Other difficulties resulted from the interactions between the national agency and the platform. The geographical distance tended to cause their communications and coordination to deteriorate. For example, the national agency required a new administrative task to be performed (inputting indicators into new dashboard-type progress charts) without any support being given to the remote sites. The work to start implementing this new task at the platform was felt to be an additional task that was pointless and that was complex to perform. The interviews showed that the platform operatives, across all of the functions, had the impression they lacked recognition from the operatives at the national agency. Such degraded communication and coordination damaged social relations and undermined inter-site team spirit.

In a peak activity situation (complex situation), all of the difficulties that were propagated or that were related to the interactions were exacerbated. During peak activity, the volume of goods to be delivered doubled. The first consequence was saturation (overloading) of the platform, reinforcing the risks related to moving around the site (collisions, being crushed or run over, etc.). The second consequence was an increase in the risks of errors in handling and processing the goods (parcels lost or mixed up, for example). Although reinforcements were provided, in the form of a team of temporary workers, intensification in the mental and physical activities could be observed in all of the operatives (additional and complex checking and vigilance was necessary because the goods were unknown and very heterogeneous). That intensification also resulted in an increase in the number of goods-handling operations (looking for parcels, sorting, goods to be wrapped in film, etc.), and some of those operations interfered with other work.

The causes of that saturation of the platform were related not only to factors inherent to the platform (platform under-sized) but also to the factors identified above and that were found to a more sustained extent. Thus, the lack of communication with the national agency reduced the possibility of anticipating certain difficulties related to the platform being saturated, or of thinking about how to change the way work was organised in order to cope with peak activity. The operatives worked and organised themselves even more reactively as the day and the events went by. In this tense and strained context, new difficulties were also added, such as those related to material resources. For example, the quantity of goods-handling equipment remained identical, regardless of whether or not the activity at the platform was complex, thereby giving rise to competition between the employees for accessing the equipment and thus weakening the team spirit.

Finally, the difficulties identified were either inherent to the platform, or they stemmed from the national agency or from interactions between the two entities.

3.2   The Platform and the Clients: Deliveries to the Downstream Node

The delivery activity handled by the drivers broke down into three main activities.

One part involved managing the goods during the goods loading and unloading operations. The observations showed that drivers faced various difficulties related, in particular, to the availability of the goods (goods not delivered or impossible to find at the platform) and to the stability of the goods (e.g. goods improperly packaged or prepared). These difficulties were related to how the platform was organised (e.g. storage rules unclear or choice of storage zone not optimum) and to its structure being under-sized compared with the flow of goods. They had a direct impact on one of the main assignments of the driver that was to comply with delivery deadlines. A risk of not meeting deadlines generated tension and strain in the drivers because some customers could refuse the goods or claim financial compensation. Since the drivers could be considered to be liable, they tried to make up for these disruptions by making use of a certain amount of autonomy that they had. For example, some questioned their colleagues to find the goods, others arrived on site earlier so as to have more time to find them, or indeed repacked pallets so that they were more stable or easier to handle. When the organisational configuration so permitted (time available and autonomy enjoyed), some drivers supported their colleagues by helping them in their tasks. When such making up for disruptions took place under time pressure, e.g. in peak activity situations, such situations were then found difficult to bear.

The second activity related to managing the goods deliveries proper. Difficulties were observed that were related to access to the sites (e.g. absence of dedicated delivery areas, or indeed delivery in complex zones such as shopping centres), to delivery hours constraints dictated by shop opening hours not necessarily consistent with the hours at which traffic flows more smoothly. Such constraints constituted an indicator of driver activity smoothness and revealed the relationship of dependence between driver and customer availability (shop opening hours, availability of delivery-taking warehouse staff). Constraints related to the interactions with the customers were observed (unpleasant reception from the customer, refusal to take the goods, and distribution of the work between driver and customer sometimes "vague"). The driver finding himself at the interface with the customer did not necessarily have the means to negotiate and did not always have the information necessary to do so.

These difficulties made the driver's task complex, in particular when he had to cope with more than one of them at the same time. In order to mitigate them, and accomplish their assignments within the allotted time, the drivers developed anticipation strategies. For example, they planned various scenarios on the way in which the goods could be taken charge of, they included their experience of the encountered situation in such scenarios, and they built networks of "good relations" for their regular customers, enabling them to deliver more easily and quickly. They could also re-plan their route or the order in which they made their deliveries if organisational constraints so permitted. These cognitive activities, some of which took place while the drivers were driving had a cost in terms of the mental workload and could give rise to tiredness, loss of vigilance at the wheel, or tension and strain.

The last activity of the drivers involved managing the flow of information that enabled the goods to be traceable and the regulatory aspects to be complied with (managing delivery notes, for example). In practice, they had to fill in documents, check the transported goods corresponded to what was expected (use of lists or of management tools such as the personal digital assistant (PDA)). During that task, drivers were expected to detect discrepancies, report them, and correct them. Such discrepancies could involve inconsistencies between computerized information and the reality of the goods (22 parcels were listed in the documents, but only 20 parcels were present on the pallet because 2 were identified as weighing 0 kg), or indeed they could involve malfunctions related to the traceability systems (incomplete bar code or PDA failure). Drivers made up for such discrepancies and inconsistencies by doing searches, making additional checks, performing extra handling operations that interfered with their other work, and communicating to additional extents with their colleagues. Making up for such discrepancies and inconsistencies could rapidly become complex, especially when the incoming and outgoing flows of information were numerous, heterogeneous, and, in some cases required by the customer. If the organisational configuration required a fast pace of work, if the tools were not appropriate or defective, and the discrepancies were manifold, drivers were then under pressure.

In a peak activity situation, all of these difficulties were exacerbated with less room for manoeuvre to absorb them, in particular due to the increase in the number of customers to deliver to and in the volume of activity.

Finally, the difficulties or disruptions with which the drivers had to cope were inherent to the platform and related to the customers to deliver to (downstream entity). The difficulties or disruptions relating to managing the flow of information stemmed both from the IS of the national agency (entity upstream from the platform) and from the ISs of the customers (downstream entity).

It was observed that psychosocial risk factors were present for all of the operatives (uncertainty, tensions, lack of recognition, deterioration in social relations and weakening of inter-site team spirit, work and attention overload, and time pressure). However, these risks were partly offset by the operatives having a relatively large amount of autonomy. The increase in the number of handling operations generated mainly MSD risks. Traffic, site access, and vehicle manoeuvring difficulties, as well as the multiple information acquisitions or checks, and organisational and re-organisational operations gave rise to accident risks at the platform. In terms of difficulties or disruptive factors, they stemmed from the national agency, from the platform itself and from its resources, and from the customers.

3.3 Avenues for Improvement

The diagnostic findings were shared with all of the operatives in question, whether they belonged to the national agency or to the platform, in order to define avenues for improving working conditions and risk prevention.

As regards the platform, it was decided with the company not only to put in place clear organisation of the flows of goods and of the circulation paths and gangways on the platform, but also to put in place organisation of storage so as to facilitate goods handling. These avenues for risk prevention are crucial to implement during peak activity. It was also decided, with the management of the company, to inform the employees about the risks and the prevention measures related to their activity. The management of the company also chose to extend its platform, and that became effective the following year through purchasing additional floor area.

As regards the difficulties stemming from the national agency and that propagated to the platform, a more in-depth diagnostic analysis was made of the situation to identify their own difficulties and to define appropriate levers for action. For example, the trucks leaving late to go to the platform was pinpointed as being due to organisational difficulties that were discussed and thought about with them so that those difficulties could be improved. Thus, for example, the roles of the operatives interacting with the platform were redefined and clarified. As regards the difficulties relating to the IS, although they have been established and admitted by the company, to date no solution has yet been found because thinking about them and the changes to be made appear complex and, in particular, involve other players and departments from the logistics branch of the company.

As regards the difficulties relating to interactions between the national agency, the platform, and deliveries, the main problems concerned communications and coordination. The main avenue for finding solutions to these problems involved improving organisation so as to make inter-site synergy possible (beyond mere juxtaposition). Such synergy mainly required needs to be better explained and better listened to, everyone's roles and responsibilities to be redefined clearly, and information to be passed around and shared more smoothly. A platform manager was hired to take charge, in particular, of these communication and coordination aspects. This action should make it possible to reduce uncertainties, and to anticipate and to be better organised to cope with urgent situations.

As regards the difficulties relating to deliveries to customers, it was agreed that the drivers should feed back to the management of the platform (platform supervisor or manager) any difficulties and malfunctioning encountered with the customers. The management should then talk to the customers and find compromises for improving the situation. Discussions were also initiated by the company with the suppliers to smooth out the sorting process at the agency (e.g. the labelling of the goods was changed to reduce goods-handling operations).

The advantages perceived by the company about this approach concern awareness of the manifestation of degradation factors in certain work situations that have their sources in other work situations, on other sites of the company (local and remote). Deployment of a "win-win" approach with the customers and the suppliers was also emphasized. This approach contributed to increasing the overall performance of the company and of its partners, while also incorporating the health and safety of the employees.

4 Discussion and Conclusion

The three categories of difficulties identified, namely the ones specific to each of the levels, the ones relating to the interactions between those levels, and the ones related to the interdependence with the upstream and downstream entities, lead to avenues for recommendations that come under different approaches:

The first one comes under an approach usually deployed in ergonomics, where detection of a problematic work situation leads to an analysis of that situation, and then to proposed prevention solutions appropriate to it.

The other two (interactions and interdependence with upstream and downstream) require understanding of how these various levels interact and have impacts on one another, and therefore also need to be analysed at least to a minimum extent. The intra-site interactions between operatives on the same site are often taken into account. However the inter-department interactions are taken into account to a lesser extent, and the inter-site ones are almost never taken into account. The difficulty lies in that fact that implementing a solution that is targeted on a single one of these work situations does not prevent the propagation of the degradation factors to the connected situations from continuing. For example, in this study, it would have been vain to implement only solutions targeted on the platform in order to solve the difficulties related to its inter-dependence with the national agency.

To conclude, regardless of the supply chain segment in question, addressing the health and safety of the operatives as comprehensively as possible cannot be done solely from a single level or from a single work situation. It is crucial to take into account what happens upstream and downstream. Analyses of the situation should be conducted with particular attention being given to the interactions between the operatives of the other departments of the company or of external companies so as to distinguish between the constraints/resources specific to the work situation and those coming from other situations. By distinguishing the "origins" of the constraints/resources, the level on which action should be taken is identified more appropriately (on the situation itself? On a connected situation?). In terms of the approach for preventing occupational accidents and diseases, this approach, called "PROPAGIR", is intended to be deployed in a work situation so as to improve not only that work situation but also the work situations that are connected to it. This prevention approach appears to a promising one for a sector such as the transport and logistics sector, where enterprises are interconnected, and in which health and safety problems are very marked. This approach would be supplementary to the actions conventionally implemented, but it remains to be developed in more depth, in particular regarding the methodological and deployment aspects. In particular it needs to be lightened in order to make it more accessible to occupational safety and health specialists.

References

1. France Stratégie, Dares (2015) Les métiers en 2022. Rapport du groupe Prospective des métiers et qualifications. France stratégie et Dares. http://dares.travail-emploi.gouv.fr/dares-etudes-et-statistiques/etudes-et-syntheses/synthese-stat-synthese-eval/article/les-metiers-en-2022
2. Dornier JP, Fender M (2007) La logistique globale et le supply chain management. Editions Eyrolles
3. Camman C, Livosi L, Roussat C (2007) La logistique simplement, activités. Editions Liaisons
4. Medan P, Gratacap A (2008) Logistique et supply chain management. Editions Dunod
5. Pache G, Spalanzani A (2007) La gestion des chaînes logistiques multi-acteurs: perspectives stratégiques. Edition PUG
6. Apostolopoulos Y, Shattell MM, Soenmez S, Strack R, Haldeman L, Jones V (2012) Active living in the trucking sector: environmental barriers and health promotion strategies. J Phys Act Health 9:259–269
7. Hamelin P (2001) La durée de travail des conducteurs professionnels comme enjeu de la flexibilité et da la compétitivité des transports routiers de marchandises et de voyageurs. BTS newsletter, no 15–16, pp 42–51
8. Bovenzi M, Hulshof CTJ (1986–1987) An updated review of epidemiologic studies on the relationship between exposure to whole body vibration and low back pain. Int Arch Occup Environ Health 72:351–365
9. Thierry S, Chouaniere D, Aubry C (2008) Conduite et santé, une revue de la littérature. Documents pour le médecin du travail. INRS, 1er trimestre, no 113
10. Klein T, Ratier D (2012) L'impact des TIC sur les conditions de travail. Rapport et documents Centre d'Analyse Stratégique
11. Shibuya H, Cleal B, Mikkelsen KL (2008) Work injuries among drivers in the good-transport branch in Denmark. Am J Ind Med 51(5):364–371
12. Schulz JD (1998) A fatal combination. Traffic World, p 16
13. Govaere V (2013) Changements de l'activité de travail liés aux NTIC: intérêts et limites du phénomène de propagation pour la prévention. Références en santé au travail 133:43–58
14. Govaere V (2013) Une approche exploratoire pour la prise en compte de la transversalité inhérente à un système d'Information (SI). Epique 2013, Bruxelles, pp 275–284
15. Wioland L (2012) Ergonomic analyses in the transport and logistics sector. Reflection on developing a new prevention approach: "Act Elsewhere". Accid Anal Prev 59:213–220

sEMG Activity Contribution to Risk Assessment for PRM Assistance Workers

Alessio Silvetti[1(✉)], Lorenzo Fiori[1], Giorgia Chini[1],
Alberto Ranavolo[1], Antonella Tatarelli[1], Massimo Gismondi[2],
and Francesco Draicchio[1]

[1] INAIL Lab of Ergonomics and Physiology - DiMEILA,
Monte Porzio Catone, Italy
al.silvetti@inail.it
[2] Rome, Italy

Abstract. Aim of this study is to analyze the task of pushing Passengers with Restricted Mobility (PRM) on three different wheelchairs currently supplied in an Italian airport. The wheelchairs differed in their width, weight and wheel dimensions. We investigated the task with two different PRMs weight (100 and 55 kg) and three different caster wheels positions (0°, 90° and 180°). We computed the Average Rectified Value, as percentage of maximum voluntary contraction, recorded from Erector Spinae and Anterior Deltoid muscles bilaterally in the starting phase of pushing. We can conclude that by means of sEMG it is possible to obtain useful data about the risks of pushing and pulling tasks in addition to those obtained by measuring the applied forces. In future research, it could be useful to analyze also muscle co-activation to better understand the biomechanical risks of pushing and pulling tasks

Keywords: Airport · Pushing and pulling · Wheelchair · Electromyography

1 Introduction

Instead of a large diffusion of automation and mechanization, pushing and pulling tasks remain a considerable part of many activities. In a previous report [1] Klein et al. showed that workers' compensation claims for low back injuries related to pushing or pulling task were 9% of all claims. Other epidemiological studies indicate pushing/pulling activities as an occupational risk factors for low back disorders (LBDs) [2–9]. However, this evidence is still considered weak or inconsistent [10]. Van den Heuvel et al. [11] demonstrated no significant associations between pushing/pulling activities and low back and shoulder complaints in some different work activities.

These conflicting epidemiological data about the association between LBDs and pushing and pulling tasks, could be dues to the fact that pushing and pulling are often not independent causative factors [5, 6, 12]. Furthermore in several studies causal associations resulted difficult because of the limitations imposed by specific research designs and study populations [13].

The systematic review of Roofey et al. [12] has been criticized [14, 15] for not taking into account biomechanical studies.

© Springer Nature Switzerland AG 2019
S. Bagnara et al. (Eds.): IEA 2018, AISC 819, pp. 357–362, 2019.
https://doi.org/10.1007/978-3-319-96089-0_38

The current pushing and pulling risk prevention policies are based on psychophysical studies. The proposed limits are founded on the assumption that subjective perception of an individual's maximum acceptable external forces corresponds to biomechanical tolerance [16]. The psychophysical approach has been applied in largely used guidelines [17, 18]. The proposed limits, however, does not necessarily correspond to biomechanical load, particularly at lumbar spine level as reported in other studies [19–21].

Aim of this paper is to analyze muscle activity, by means of surface electromyography (sEMG) from the shoulder and the back in the task of assisting passengers with reduced mobility with different combinations of task constraints: pushing three different wheelchairs (A, B, C), in three starting position of the caster wheels (180°, 90°, 0°) and with two weights of passengers (55 kg and 100 kg).

1.1   Task Description

The workers provide assistance to PRMs accompanying them and any baggage: (i) in departure, from their arrival at the air terminal (at the Terminal or at the railway station or at a multi-floor carpark, etc.) to their seat on the airplane; (ii) in arrival, from their seat on the airplane to their ground destination; (iii) in transit between an arriving flight and subsequent departure on another flight. The activities of the operators are performed throughout the whole airport using a wheelchair (made available by the assisting company or, rarely, the personal wheelchair of the PRM), with the exception of some special kind of assistance for which there are minivans equipped to carry disabled passengers.

2   Materials and Method

Preliminary recordings of initial and sustained forces, by means of a digital dynamometer, according to the annex D of ISO 11228-2 [22], were made. We decided to investigate through EMG only the initial phase because sustained forces values we recorded were far lower than the limits proposed by Snook and Ciriello tables [16].

Five male subjects were enrolled. Their average (SD) age, height, and weight were respectively 41.8 years (3.5), 180.2 cm (6.6) and 83.8 kg (11.3). None of them had any history of either musculoskeletal disorders or neurological diseases.

Participants had to perform pushing activities with different combinations of task constraints: pushing two dummy patients of 55 and 100 kg, in three starting position of the caster wheels (180°, 90°, 0°) for three different wheelchairs (A, B, C).

Recordings were made on a flat surface with the same characteristics as the airport floor. All three wheelchairs (WhCh) models were in excellent working order and all the wheels were at the pressure recommended by the manufacturer.

Five measurements were taken under all wheel starting positions, for each condition and for each WhCh.

Table 1 shows technical features of the three tested WhChs. Table 2 resumes the experimental setup and images of the three tested wheelchairs.

Table 1. Technical features in cm (weight in Kg) of the three tested wheelchairs

WhCh	Handle height	Handle width	Anterior wheel Ø	Posterior wheel Ø	Weight
A	89	50	20	60	24
B	91	53	18	60	22.8
C	89	39	19	30	16.9

Table 2. experimental setup

Wheelchair A						Wheelchair B						Wheelchair C					
55 Kg			100 Kg			55 Kg			100 Kg			55 Kg			100 Kg		
180°	90°	0°	180°	90°	0°	180°	90°	0°	180°	90°	0°	180°	90°	0°	180°	90°	0°

Electrical muscle activity, of the initial phase, was recorded using a 16 channel Wi-Fi surface electromyography system (FreeEMG, BTS SpA, Milan, Italy) at a sampling frequency of 1 kHz. After skin preparation, surface electromyographic (sEMG) signals were detected from each muscle by two Ag/AgCl pre-gelled disposable surface electrodes (H124SG, Kendall ARBO, Donau, Germany) which had a detection surface of 10 mm (gelled). Electrodes were placed bilaterally over the muscle belly of Deltoideus Anterior and Erector Spinae in the direction of the muscle fibers, according to the Atlas of muscle innervation [23].

Data were processed using MATLAB R2017b software (vers. 9.3.0, MathWorks, Natick, MA, USA). The sEMG signals were filtered with a digital filter IIR bandpass Butterworth of 5° order (20–450 Hz). Afterward, to obtain linear envelope, signals were rectified and filtered with a digital filter IIR bandpass Butterworth of 3° order with a cut frequency of 10 Hz.

The Average Rectified Value (ARV) of each muscle in each condition was measured as percentage of Maximum Voluntary Contraction (%MVC).

Statistical analysis was performed using SPSS Statistics 17.0 software (SPSS Inc., Chicago, IL, USA). A multivariate analysis of variance (MANOVA) was performed to determine whether there was a significant effect of the wheelchair and of the starting position of the steering wheels for both investigated weight. A post-hoc Tukey's test between each different starting position was used to detect any significant differences between the three wheelchairs for both investigated weights. P-values lower than 0.05 were considered statistically significant.

3 Results

3.1 ARV

Tables 3, 4, 5 and 6 show, for the three wheelchairs tested (A, B, C), mean (±SD) ARV values of the four muscles investigated with two weights (55 and 100 kg) and three positions of the caster wheels (180°, 90°, 0°). Surface electromyography results showed no statistical differences.

Wheelchair A showed lowest ARV values in all the muscles investigated for the most demanding task constraints (pushing 100 kg with caster wheels at 180° and 90°).

Wheelchair C showed highest ARV values for the most demanding task constraints in both Erector Spinae and in right Deltoideus Anterior muscles.

Wheelchair B showed highest ARV values for the most demanding task constraints in left Delotideus Anterior muscle.

The results of the less demanding task (55 kg) didn't show a clear trend.

Table 3. Table show mean (±SD) ARV values for right Deltoideus Anterior (%MVC)

Weight (Kg)	Position	A	B	C
55	180°	11 ± 2	12.7 ± 1.7	11.1 ± 1.2
55	90°	13 ± 4.2	18.3 ± 8.1	13.7 ± 2.4
55	0°	10.4 ± 3.5	10.7 ± 2.1	10.7 ± 3.3
100	180°	14.2 ± 4.1	15.4 ± 2.9	16.4 ± 2.9
100	90°	16.7 ± 3.9	18.2 ± 2.8	20.2 ± 4
100	0°	14.2 ± 3.7	14.6 ± 4.9	15.8 ± 3.4

Table 4. Table show mean (±SD) ARV values for left Deltoideus Anterior (%MVC)

Weight (Kg)	Position	A	B	C
55	180°	12.9 ± 3.9	12.1 ± 3.6	11.2 ± 2.9
55	90°	11.6 ± 4.2	16.6 ± 4.3	13.1 ± 2.7
55	0°	12.6 ± 5.4	10.6 ± 2.5	11 ± 2.7
100	180°	15.4 ± 6.3	15.8 ± 5.5	15.4 ± 4.5
100	90°	18.2 ± 5.2	23 ± 6.2	21.5 ± 6.4
100	0°	17.3 ± 6.5	17.9 ± 4.1	14.9 ± 4.8

Table 5. Table show mean (±SD) ARV values for right Erector Spinae (%MVC)

Weight (Kg)	Position	A	B	C
55	180°	10.8 ± 4.9	11.2 ± 7.1	10.7 ± 4.1
55	90°	10.9 ± 4.2	13.7 ± 6.6	13.2 ± 5.3
55	0°	10.6 ± 5.2	9.6 ± 5.3	11.3 ± 4.2
100	180°	12 ± 4	12.8 ± 4.3	13.4 ± 4.5
100	90°	12.8 ± 5	13.4 ± 6.2	14.4 ± 4.7
100	0°	13 ± 5.4	11.9 ± 4.7	12.8 ± 5.1

Table 6. Table show mean (±SD) ARV values for left Erector Spinae (%MVC)

Weight (Kg)	Position	A	B	C
55	180°	10.4 ± 4	11.7 ± 2.5	10.7 ± 4
55	90°	10.4 ± 2.6	12.1 ± 2	12.4 ± 2.1
55	0°	11.3 ± 2.7	9.9 ± 2.6	11.9 ± 3.1
100	180°	9.5 ± 4.5	10.9 ± 3.6	11.9 ± 4.6
100	90°	11.7 ± 3.3	14 ± 4.6	15.1 ± 4.2
100	0°	9.9 ± 4	9.9 ± 4.7	11.9 ± 3.9

4 Discussion

Surface electromyography didn't show statistical differences in all investigated muscles. This could be due to the small sample size and to operators different motor strategies. Some wheelchairs technical features, particularly weight and handle width (Table 1), could also affect the data. In spite of that, it is possible to observe some characteristic trends. Wheelchair A, despite its greater weight (Table 1), showed the lowest values in the most demanding configurations (pushing 100 kg with the steering wheels at 180° and 90°). Pushing 55 kg didn't show any trend probably because the different weights of wheelchairs (Table 1) influenced largely the whole weight when the passenger weighted only 55 kg.

The muscle activities together with applied forces could give data to select wheelchairs and to prepare recommendations for their use.

In order to have a better understanding of the pushing and pulling tasks, it could be useful in future research to investigate muscle co-activation in trunk and shoulder as highlighted from several authors [24–26].

References

1. Klein BP, Jensen RC, Sanderson LM (1984) Assessment of workers' compensation claims for back strains/sprains. J Occup Med 26:443–448
2. Clemmer DI, Mohr DL, Mercer DJ (1991) Low-back injuries in a heavy industry. I. Worker and workplace factors. Spine 16:824–830 (Phila Pa 1976)
3. Riihimaki H (1991) Low-back pain, its origin and risk indicators. Scand J Work Environ Health 17:81–90
4. Hoozemans MJ, Van Der Beek AJ, Frings-Dresen MH, Van Dijk FJ, Van Der Woude LH (1998) Pushing and pulling in relation to musculoskeletal disorders: a review of risk factors. Ergonomics 41:757–781
5. Hoozemans MJ, Van Der Beek AJ, Fring-Dresen MH, Van Der Woude LH, Van Dijk FJ (2002) Low-back and shoulder complaints among workers with pushing and pulling tasks. Scand J Work Environ Health 28:293–303
6. Hoozemans MJ, Van Der Beek AJ, Frings-Dresen MH, Van Der Woude LH, Van Dijk FJ (2002) Pushing and pulling in association with low back and shoulder complaints. Occup Environ Med 59:696–702

7. Yip YB (2004) New low back pain in nurses: work activities, work stress and sedentary lifestyle. J Adv Nurs 46:430–440

8. Andersen JH, Haahr JP, Frost P (2007) Risk factors for more severe regional musculoskeletal symptoms: a two-year prospective study of a general working population. Arthritis Rheum 56:1355–1364

9. Plouvier S, Renahy E, Chastang JF, Bonenfant S, Leclerc A (2008) Biomechanical strains and low back disorders: quantifying the effects of the number of years of exposure on various types of pain. Occup Environ Med 65:268–274

10. Kelsey JL, Golden AL, Mundt DJ (1990) Low back pain/prolapsed lumbar intervertebral disc. Rheum Dis Clin North Am 16:699–716

11. van den Heuvel SG, Ariens GAM, Boshuizen HC et al (2004) Prognostic factors related to recurrent low-back pain and sickness absence. Scand J Work Environ Health 30:459–467

12. Roffey DM, Wai EK, Bishop P, Kwon BK, Dagenais S (2010) Causal assessment of occupational pushing or pulling and low back pain: results of a systematic review. Spine J 10:544–553

13. Bombardier C, Kerr MS, Shannon HS, Frank JW (1994) A guide to interpreting epidemiologic studies on the etiology of back pain. Spine 19:2047S–2056S

14. Ward A (2009) The role of causal criteria in causal inferences: bradford hill's "aspects of association. Epidemiol Perspect Innov 6(2):1–22

15. Takala E (2010) Lack of "statistically" significant association does not exclude causality. Spine J 10:944–946

16. Snook SH, Ciriello VM (1991) The design of manual handling tasks: revised tables of maximum acceptable weights and forces. Ergonomics 34(9):1197–1213

17. Mital A, Nicholson AS, Ayoub MM (1997) A guide to manual materials handling. Taylor & Francis, London

18. Garg A, Waters T, Kapellusch J, Karwowski W (2014) Psychophysical basis for maximum pushing and pulling forces: a review and recommendations. Int J Ind Ergon 44:281–291

19. Jorgensen MJ, Davis KG, Kirking BC, Lewis KEK, Marras WS (1999) Significance of biomechanical and physiological variables during the determination of maximum acceptable weight of lift. Ergonomics 42(9):1216–1232

20. Davis KG, Jorgensen MJ, Marras WS (2000) An investigation of perceived exertion via whole body exertion and direct muscle force indicators during the determination of the maximum acceptable weight of lift. Ergonomics 43(2):143–159

21. Le P, Dufour H, Monat J, Rose Z, Huber E, Alder RZR, Radin Umar B, Hennessey MD, Marras WS (2012) Association between spinal loads and the psychophysical determination of maximum acceptable force during pushing tasks. Ergonomics 55(9):1104–1114

22. International Organization for Standardization (ISO) (2007) Ergonomics – manual handling – part 2: Pushing and pulling. ISO, Geneva. Standard No. ISO 11228-2:2007

23. Barbero M, Merletti R, Rainoldi A (2012) Atlas of muscle innervation zones: under-standing surface electromyography and its applications. Springer, Heidelberg

24. Lee KS, Chaffin DB, Waikar A, Chung MK (1989) Lower back muscle forces in pushing and pulling. Ergonomics 32:1551–1563

25. Thelen DG, Ashton-Miller JA, Schultz A (1996) Lumbar muscle activities in rapid three-dimensional pulling tasks. Spine 21:605–613

26. Nussbaum MA, Chaffin DB, Baker G (1999) Biomechanical analysis of materials handling manipulators in short distance transfers of moderate mass objects: joint strength, spine forces and muscular antagonism. Ergonomics 42:1597–1618

Juvenile Workers or Pocketmoney Precariat? – The Process and Effect of Establishing Health and Safety Committees for Juvenile Workers in Danish Retail Sector

Hans Jørgen Limborg[2(✉)], Kristina Karstad[1], Karen Albertsen[2],
Dorte Ekner[1], Anders Ørberg[1], Sisse Grøn[2],
Charlotte D. N. Rasmussen[1], Andreas Holtermann[1],
and Marie B. Jørgensen[1]

[1] National Research Centre for the Working Environment (NRCWE),
Lersø Parkallé 105, 2100 Copenhagen Ø, Denmark
[2] TeamArbejdsliv, Høffdingsvej 22, 2500 Valby, Denmark
hjl@teamarbejdsliv.dk

1 Introduction

In the Danish retail industry, young workers employed in temporary positions for few hours a week constitutes more than half of the employees. For the main part of the young workers this employment is their first meeting with the labor market, they lack knowledge and awareness of the work environment, and at some workplaces the introduction is limited, 'ad-hoc' or 'fast-track' [1]. Often the young worker feel only limited involved in decisions regarding their work [2]. It may be beneficial both for short- and long-term performance and security to increase the young workers knowledge of and involvement in the work environment issues, as well as the employer's awareness on the needs of the young workers.

In the literature, young people are often considered difficult to 'reach' through traditional information campaigns [3, 4]. Some have stressed that young workers perceives the risk taking as a natural part of their identity formation [5–7] and this combined with a sense of 'inferiority' [8], making the young a difficult group to reach. Most often, young people are not aware that their work contains dangers or see them as conditions at work [9, 10].

Other researchers have emphasized the position of young people in the workplace as being at the bottom of the hierarchy [11]. Therefore, young people's accidents and unsafe behaviors are also dependent on the workplace they are part of [7, 10]. The 15–17 year-olds raise a special issue. Unlike, for example, apprentices and students, young people in leisure jobs do not aim to become part of a subject and industry, and do not feel the same connection with the workplace as older colleagues. They work a limited weekly hour, are only for a short period in their jobs, and the workplace is not the arena they primarily identify with [7]. There are two positions in security research: A social-psychological perspective, focusing on the individual, i.e. the young people's special risk awareness, knowledge and behavior. And a culture/organizational theory

S. Bagnara et al. (Eds.): IEA 2018, AISC 819, pp. 363–372, 2019.
https://doi.org/10.1007/978-3-319-96089-0_39

approach, focusing on the workplace culture, norms and work routines [12, 13]. Each position produces its prevention strategy. The former will change the knowledge/awareness of the young people and the ability to act through 'information' and 'attitude processing'. The latter will orient themselves towards the collective level: the group, the organization and the work itself.

The means of this project was aimed at both, as it had an upbringing function, but was also integrated into the organization. The project's thesis was that there is no incompatible contradiction but that the two approaches complement and enrich each other [14].

2 Aim

The aim of the present study was to evaluate the implementation process of an intervention where health and safety committees for young workers were established in Danish supermarkets with the purpose to improve the young workers influence on and awareness of their work environment.

3 Method

The study was originally designed with a cluster-randomized one-way crossover design [15]. However, due to dropout, it became necessary to change the design to a multiple case study with matched control worksites. For the present purpose, only qualitative data from the implementation and evaluation at the intervention worksites were used. Quantitative analysis of effects is reported separately.

3.1 Intervention

The intervention was implemented in two rounds between April 2016 and June 2017. Six supermarkets were allocated to the first round (of which two continued with a second committee for young workers) and five new supermarkets participated in the second round. In total eleven worksites participated in the intervention, and of those, two worksites participated in both rounds. Each intervention lasted 8 weeks.

First, a worksite based youth health and safety committee was established including two young workers, and at least one adult (safety representative or manager) and a researcher.

Next, the committee performed a workplace analyze to evaluate hazardous work situations for the youth workforce. The young workers were encouraged to suggest possible solutions and improvements to the work environment, and the solutions were then analyzed and rated according to the possible impact and the practical opportunities for implementation.

Finally, the committee implemented selected changes and disseminated information regarding the actions taken, e.g. on local social media.

The young workers were supported toward the process by the safety representatives and by the researchers.

3.2 Data Were Collected in the Form of

Researchers and participants notes and photo-documentations from the meetings.

Researchers logbooks. After each communication (either meeting, phone or on-line), the researchers took notes about time, content and form.

Posts from the social media (Facebook), messages from SMS, messenger and e-mails.

Group interviews at the end of the intervention process with the participants from the health and safety committees for young workers.

3.3 Quantitative Evaluation

Between March and June 2017, we conducted a controlled before-and-after study with eight of the supermarkets allocated to either intervention (n = 5) or control (n = 3). A total of 251 juvenile workers (age 15–17 yrs.) employed at the supermarkets were allocated for intervention (n = 197) and control (n = 54), respectively. Before and after the intervention, an online questionnaire was sent as a link through text messages. The primary outcome was the degree of involvement in decisions about work environment measured by the question "Do you get involved in decisions that affect your work environment?" (Response option on 5-point Likert scale).

4 Results

4.1 Implementation

Intervention plan for intervention covered in both rounds:

Time	Activity
Week 1	Start-up meeting with planning and adjustment of expectations, with participation of researcher, manager and safety representative
Week 1	Selection of two young workers for the H&S committee for young workers, buy the manager and H&S representative
Week 2	Course for the H&S committee for young workers, including presentation about work environment, selection of focus for a H&S activity for committee and development of an action plan
week 2–6	The H&S committee for young workers meet, check out the prevalence of the problem in the shop and work with the activities in action plan
Week 6	Follow-up meeting with dialogue of the experiences so far between researcher and representatives in the H&S committee for young workers

(*continued*)

(continued)

Time	Activity
Week 6–10	The H&S committee for young workers meet, and work with the activities in the action plan
Week 10	Final meeting between researcher and H&S committee for young workers with evaluation of the course and progress

In the second intervention round the interventions were strengthened by a more close contact and follow-up between the responsible researcher and H&S committee for young workers, and by intensified use of social media in the communication process.

All of the elements in the intervention plan was conducted at all the intervention workplaces. However, at some workplaces the follow-up meeting and the final meeting were held as phone or skype-meetings.

4.2 Activities Started by the Health and Safety Committees for Young Workers

The committees choose to focus on a range of issues concerning their daily working life and performance as workers in the supermarkets. The topics covered among others:

Procedures for Introduction to the New Young Employees: In four of the supermarkets it was decided to focus on the procedures for introduction of new employees. Different models for an introduction procedure were developed; lists with tasks that should be presented for the new, a quiz with questions to practice and follow up on the learned, and in one place a video was produced by the young committee members and shown on Facebook and on a service-meeting for all the young workers.

In one of the supermarkets they had developed a written procedure for introduction of new employees, a plan for the training in different areas, a pop-quiz for practicing and fun and a checklist to be used by the person responsible for the introduction. They summarized the immediate benefits: "The new knows what she has to go through, the responsible knows what he or she shall do, and NN [The manager] knows what he can expect that the new employee knows."

Communication and Guidelines for Acute Situations: In some other supermarkets the young workers became aware that they missed guidelines in acute situations as e.g. robbery, stealing, acute illness among costumers, fire. In one shop, information was made available by posters and posts at Facebook and supported by a quiz with gifts to two young employees every week for right answers. They further printed a small calling card for the pocket to all employees with internal 'Nice to know' numbers on the one side and with numbers for help in acute situations on the other side.

Strategy Toward Angry Costumers: The young health and safety committee members had in one supermarket chosen to focus on the challenges they sometimes experienced in the meeting with aggressive costumers. They started with a small investigation among the colleagues about their experiences with surly costumers and how they tackled it. They found that a common problem was the situations sitting at the cash, where costumers had a different opinion of the prize than the one on the bar code. As a result, the management decided on a general rule, that the costumer's prize should always be followed for differences below 30 D.kr [appr. 4 Euro]. And this prevented a lot of uncomfortable situations for the employees and for the costumers.

Accessibility and Functionality of Tools: In some supermarkets the young workers experienced problems finding the stools used e.g. for trimming on the high shelves. This led to buying of new stools and to agreements about the right fixed places for the stools. In one shop, the trolleys (T-vogne) were not in place where they should be used, and they started an information campaign about availability of the trolleys (T-vogne).

In another shop the young committee members pointed on a perennial problem with the bottle refund machine. The boy in the committee was very familiar with the machine, and he decide to write a trouble-shooting manual and make it available beside the machine preventing future break downs.

Norms for Exchange of Rosters/Shift Among the Young Workers: In one supermarket the committee took initiative to invite all the young employees to discuss the norms and implicit rules for exchanges of shifts; appropriate time for response to a request and that it was okay to refuse to take a shift and to remind each other of answering. The initiative made it much easier to ask for exchanges.

The locker rooms: One of the committees decided to gain keys to all closets in the locker rooms so that their belongings could be safe during working hours and further, they decided to get rid of the mess and increase the cleaning standard, particularly in the boy's locker room.

In Addition to the Activities Conducted Directly as Part of the Project: In one shop, the retail manager arranged a large meeting in the shop for the parents of the young workers. On this meeting, at the same time advertisement were made for the super-market and the parents were informed about the workplace of their kids and the efforts made toward the young workers.

The Danish trade unions for employees and employers within trade and retail sponsored the production of three small films with the young workers and the managers as target groups [16].

4.3 The Experiences of the Young Participants

The experience from this project was that, in general, the young workers were happy for the opportunity to participate in the health and safety committees and expressed satisfaction with the process. A young worker expresses with proud: *"It's so very good to see what it is possible to do with so little work - how little is needed to make changes"*.

Some of them had gained more awareness of safety and quality of work by the process: *"Sometimes it happens, that those who need to trim - they may not trim properly. I have, for example, often experienced, that they are trimming with one hand and have their mobile in the other hand"*.

And some would like to learn and do more to improve the work environment: *"Somehow you want to know more about the working environment as it really concerns us. You may also be interested in being active in the work environment again, as you can really see that you are making a difference in your workplace"*.

For some the committee went out to be an important channel for influence: *"I've felt that I've been listened to in another way because I've had the role you've been given me.... It also gives you another desire when you feel that there are some who listen to you"*. Another says: *"Before, it was just work, but now it's a bit like we have a voice. And I think I'm daring to say something without fear of getting stuck on."*

The new possibilities for influence gave rice to the satisfaction of problem solving: *"It has become much easier to solve problems after we have to make some decisions ourselves, and it is also very nice to solve a problem that I know how to solve"*.

However, the committees had not been equally successful in all places. In some of the workplaces the committees didn't function. A young participant tells: *"It failed like we could not convey anything to NN [the manager] and everything should be done now and here. But we also did not really investigate where to place the tools, and what should happen"*.

4.4   The Experiences of the Work Environment Representatives and Managers

In the shop, where the committee of young workers had made a video to their colleagues, the safety representative was clearly proud: *"...the film they have made – I've also said to them – I'm really proud of you when I see that movie. I really think it was a crazy good idea, and I also said so across the house..... I am so impressed of their movie and what they've got out of seven hours of work"*.

In some other shops the project had been an eye opener toward trouble shooting and increased effectivity: *"I have not really thought so, but it's a waste of time that they go around and spend a lot of time on finding the stools...... what are the tools needed to perform their work as effectively as possible?"*

The young workers had in some supermarkets also challenged the usual habits at the workplace e.g. by using Facebook, video and pop-quizzes actively in the communication process.

One of the managers expressed that the project had added a lot of value for a relatively low cost. Another posed the question, whether the workplace did benefit optimally of the young workers ressources. She had experienced how much value the young workers could add to the workplace through their huge engagement in the H&S committe.

Many of the supermarkets decided to continue the involvement of young workers in the work environment after the end of project. Some workplaces would invite the young workers to participate in the ordinary work environment system. Other decided to arrange specific evening-meetings about work environment issues for the young workers. For some, the project had raised the awareness toward the young workers: *"We should be much better at communicating with the young people and making them respond. But they will never be part of the H&S group - they are never allowed to that"*.

There was a clear, and not surprising pattern, that the workplaces experiencing the most successful H&S committees for young workers were also the most dedicated in order to continue the involvement of the young workers in the future work environment work.

4.5 Difficulties in Raising Awareness Among the Peers

As part of the project, the young committee members had to involve, or as a minimum inform, their young colleagues about the work in the committee. This was in some cases difficult and sometimes quite disappointing for some of the young committee members. After posting news at Facebook with only few 'likes' from the peers, a young worker said: *"I felt like talking to a wall when I tried to spread the message, and after some time you lose a bit of courage"*. In a vulnerable position and in a vulnerable age, it may be very demanding for a young person not to receive 'likes' from activities at Facebook the peers.

Also at workplaces with successful results from the work and with highly supportive managers and safety representative, a committee member tell: *"The response from the managers and the adult collegues were super, but our peers did not say so much"*. The peers may have felt disregarded themselves.

4.6 Decreasing Commitment at Some of the Workplaces

Even though the management in most supermarkets found the project relevant, it was not easy to maintain the commitment toward the intervention process. It was particularly difficult for supermarkets facing challenges e.g. with shifting managers or problematic issues with the psychosocial work environment. At worksites where they had difficulties with the H&S work in general and lacked continuity in the H&S group, it was not possible to support the committee for the young workers either.

Some workplaces from the first round of intervention decided to stop their participation in the second round because they found it to demanding to participate. The manager and safety representative explained that the results obtained from the work did not mach the resources it had required to support the group. Especially one of the members had been difficult to pull through. They believed that the project in general was too ambitious. They had thought that the young workers would have been more self-propelled than they were.

4.7 Motivation, Support and Encouragement

At the workplaces where the intervention proceeded best, the young workers in the H&S committees were clearly motivated and simultaneously they reviewed efficient both practical and mental support from the management and safety representatives and researchers, demonstrating organizational ressources and a supportive culture. If either one or both of these two factors were not present, the intervention was perceived as less successful by the participants.

In some of the committees, the young workers were either not as motivated or not as competent to work by their own with the action plans, even though there was available support from the workplace and the researchers. At a single workplaces, one of the young delegates continued and fulfilled the aim of the action plan, even though he received only very limited support and encouragement form the worksite. At another workplace the young participants in the H&S committee received support from the safety representative, but not at all from their peers, thus lowering the motivation and the perception of the process.

4.8   Questionaire on Involvement in Decisions About Work Environment

At baseline, response rates to the questionnaire were 65% (intervention) and 56% (control) and at follow-up 52% (intervention) and 43% (control) of the eligible youth workers. At baseline, 41% (intervention) and 40% (control) of the youth workers reported to have some or high degree of involvement in decisions regarding their work environment. At the intervention sites, 41% of the youth workers had heard of the youth health and safety committee, 39% had seen posts on Facebook, and 52% had noticed a specific intervention activity. At the time of conference, analysis of the effectiveness of the intervention conducted will be finalized and results presented.

5   Discussion

The results from this qualitative evaluation first of all point to highly different perceptions of the processes at the different workplaces. In some shops the intervention was perceived by the participants as a huge success, while in other shops it was perceived as a failure. This is not at all an unusual finding from organizational intervention studies [17, 18], and it may suggest that the intervention may be efficient under certain circumstances and not under other.

Both organizational/culture and social psychological issues seems to have been crucial for the proceeding of the intervention process, supporting the thesis that there is no incompatible contradiction between the two approaches but that they may complement and enrich each other [14]. The approach of the young workers toward involvement and working with the work environment issues seems to be highly linked to the workplace practices and culture related to the work environment organization. This parallels the notion from previous research that risk perception and unsecure behaviors are inseparable from daily work practices and workplace culture [19–22], which not least the experienced employees are carriers of.

6   Conclusion

The implementation of health and safety committees for juvenile workers within the retail industry have potential for increasing awareness and involvement in the work environment among the young workers and for being beneficial for the supermarkets. The group of juvenile workers in the retail industry is generally perceived as a very difficult group to reach. With a participatory approach and use of social media, we expect the intervention to be effective in improving the youth workers' influence on and knowledge of their work environment. However, the implementation requires continuous support and commitment from the management and safety representatives. In some shops the costs may overrule the benefits, while other shops experienced high value for the costs.

References

1. Nielsen ML, Jørgensen A, Grytnes R, Nielsen KJ, Dyreborg J (2017) Sikkert arbejde for unge gennem læring og instruktion (SAFU Læring)
2. Nielsen ML, Görlich A, Grytnes R, Dyreborg J (2017) Without a safety net: precarization among young danish employees. Nord J Work Life Stud 7(3):3–22
3. Aftaletekst (2012) En strategi for arbejdsmiljøindsatsen frem til 2020
4. Regeringen (2010) Nye veje til et bedre arbejdsmiljø. Regeringens strategi for arbejdsmiljøindsatsen frem til 2020
5. Austen L (2009) The social construction of risk by young people. Health Risk Soc 11 (5):451–470
6. Lupton D, Tulloch J (2002) 'Life would be pretty dull without risk': voluntary risk-taking and its pleasures. Health Risk Soc 4(2):113–124
7. Mitchell WA, Crawshaw P, Bunton R, Green EE (2001) Situating young people's experiences of risk and identity. Health Risk Soc 3(2):217–233
8. Illeris K, et al (2009) Ungdomsliv: Mellem individualisering og standardisering. Forlaget Samfundslitteratur
9. Breslin C et al (2007) Workplace injury or "part of the job"?: towards a gendered understanding of injuries and complaints among young workers. Soc Sci Med 64(4):782–793
10. Nielsen ML (2012) Adapting the normal - examining relations between youth, risk and accidents at work. Nord J Work Life Stud 2(2):71–85
11. Baarts C (2004) Viden og Kunnen – en antropologisk analyse af sikkerhed på en byggeplads. Arbejdsmiljøinstituttet
12. Dejoy DM (2005) Behavior change versus culture change: divergent approaches to managing workplace safety. Saf Sci 43:105–129
13. Tharaldsen JE, Haukelid K (2009) Culture and behavioural perspectives on safety – towards a balanced approach. J Risk Res 12(3–4):375–388
14. Dyreborg J, et al (2013) Review af ulykkesforebyggelsen – reivew af eksisterende videnskabelige litteratur om effekten af forskellige typer tiltag til forebyggelse af arbejdsulykker. Det Nationale Forskningscenter for Arbejdsmiljø
15. Reich NG, Milstone AM (2014) Improving efficiency in cluster-randomized study design and implementation: taking advantage of a crossover. Open Access J Clin Trials 6:11–15
16. http://bfahandel.dk/Fysisk-arbejdsmiljoe/Unge#35461
17. Nielsen KM, Fredslund H, Christensen K, Albertsen K (2006) Success or failure? Interpreting and understanding the impact of interventions in four similar worksites. Work Stress 20(3):272–287
18. Albertsen K, Garde AH, Nabe-Nielsen K, Hansen ÅM, Lund H, Hvid H (2014) Work life balance among shift workers – results from an intervention study about self-rostering. Int Arch Environ Health 87(3):265–274

19. Gherardi S, Nicolini D (2000) The organizational learning of safety in communities of practice. J Manag Inq 9(1):7–18
20. Gherardi S, Nicolini D (2000) To transfer is to transform: the circulation of safety knowledge. Organization 7(2):329–348
21. Gherardi S, Nicolini D, Odella F (1998) What do you mean by safety? Conflicting perspectives on accident causation and safety management in a construction firm. J Conting Crisis Manag 6(4):202–213
22. Grøn S, Richter L (2013) Navigating safety – second report from safety culture and reporting practice on danish ships in the danish international ship register. Center of Maritime Health and Safety, University of Southern Denmark

Extending Participatory Ergonomics to Work Stress Prevention Adapted to Local Situations

Kazutaka Kogi[1(✉)], Yumi Sano[1], Toru Yoshikawa[2],
and Etsuko Yoshikawa[3]

[1] Ohara Memorial Institute for Science of Labour, Tokyo 151-0051, Japan
k.kogi@isl.or.jp
[2] National Institute of Occupational Safety and Health,
Kawasaki 214-8585, Japan
[3] Japanese Red Cross College of Nursing, Tokyo 180-8618, Japan

Abstract. Recent trends in extending participatory ergonomics to work stress prevention were examined. The practical ways to plan and implement broad-ranging workplace risks were compared between customary participatory ergonomics steps and participatory job stress prevention processes. Participatory stress prevention programs reviewed included those for small enterprises, health care facilities and local government offices. Key common steps similarly comprised application of action checklists listing typical local good practices as well as workplace-level group work for proposing multifaceted improvements, followed by reporting of benefits achieved. These steps were relatively simplified in the case of stress prevention programs so as to address broad-ranging stress-related factors such as internal communication, work planning and social support. These steps were usually preceded by short-term training of facilitators selected from among local workers. The trained facilitators played a key role by assisting co-workers in conducting locally feasible improvements within a broad mental health scope. It was confirmed that extending participatory ergonomics to stress prevention could be effectively organized by applying the simplified procedures addressing broad psychosocial factors. The participatory steps taken with the support of trained facilitators were shown to have real impact on stress prevention at work. It is suggested to involve locally trained facilitators and organize simplified participatory steps focusing on feasible improvements reflecting local good practices.

Keywords: Participatory ergonomics · Work stress prevention
Trained facilitators

1 Introduction

One of recent trends in preventing stress at work is to extend participatory ergonomics by addressing broad-ranging workplace risks by means of participatory steps involving workers [10, 12]. The adopted participatory activities have led to multifaceted improvements related to stress at work. This new strategy focuses on locally feasible improvements that are planned and implemented at the initiative of workers as in the case of serial activities undertaken in applying participatory ergonomics [1, 6–9, 13].

© Springer Nature Switzerland AG 2019
S. Bagnara et al. (Eds.): IEA 2018, AISC 819, pp. 373–381, 2019.
https://doi.org/10.1007/978-3-319-96089-0_40

International cooperation has been active in promoting participatory programs for improving workplace environment in small industries, agriculture, construction, fishing, health care and informal workplaces [2, 4, 5, 8, 11, 17]. A new impetus is in progress to extend this cooperation to job stress prevention [3, 12, 15].

It is useful to know how stress-reducing participatory steps can be effectively organized in varied local situations. Attention is drawn to the types of workplace improvements undertaken, practical means of selecting multifaceted improvements and the roles of facilitators acting through the participatory steps.

2 Methods

Participatory steps taken in organizing workplace-level activities for reducing work stress in small enterprises, health care facilities and local government offices [9, 10, 12] were compared with sequential steps undertaken in customary participatory ergonomics programs [2, 4–7, 17]. In particular, the ranges and types of workplace improvements conducted, the group work steps leading to planning of locally appropriate improvements and the action-oriented tools facilitating the workplace-level activities were taken into account [1, 7, 9, 10, 12].

The roles of volunteer facilitators who supported the participatory steps leading to concrete results in the reviewed programs [6, 10, 13] were extracted and compared. How these facilitators could assist participatory activities aimed at locally feasible improvements particularly in stress prevention programs was examined and discussed.

3 Results and Discussion

3.1 Types of Improvements Achieved by Participatory Steps

The main features of participatory activities undertaken to achieve meaningful workplace improvements were similar in both customary participatory ergonomics programs and work stress prevention programs. As shown in Table 1, both these programs relied on local good practices as immediate goals and aimed at locally feasible improvements in multiple technical areas of workplace conditions. Both the programs made use of 'action checklists' listing locally appropriate improvement actions. In facilitating the planning and implementation of available multifaceted improvements, trainers or volunteer facilitators selected from among workers assisted co-workers in group work activities.

In stress prevention programs, the ranges of workplace improvements were obviously broadened to cover in particular workplace-level communication and social support measures. This reflected the emphasis placed on work organization and psychosocial factors in these programs. As a result, the technical areas covered by action checklists used in the programs were wider than in the case of those for customary participatory ergonomics. Table 2 compares the technical areas covered by the action checklists used.

Table 1. Main features of customary participatory ergonomics and recent stress prevention programs.

Features	Participatory ergonomics	Stress prevention programs
Setting goals	Learning from local good examples	Learning from local good practices
Aimed improvements	Immediate improvements in work methods and physical environment	Immediate improvements in work planning, work methods and mental health
Action tools used	Ergonomic action checklist	Mental health action checklist
Facilitators	Trainers, occupational safety and health specialists	Trained facilitators, occupational health specialists
Basic references	[2, 4, 5, 11, 14, 17]	[9, 12, 15, 16]

Table 2 indicates the numbers of check items in different technical areas covered by the action checklists used in customary participatory ergonomics and stress prevention programs. While the action checklists used for stress prevention also covered basic ergonomic principles, they had a wider scope to cover extended actions related to improved communication, work schedules and social support measures.

Examples of the workplace improvements achieved by the reviewed work stress prevention programs are listed in Table 3. While the majority of improvements were usually done in work methods and physical environment, many improvements were also conducted to improve internal communication, work schedules and social support. It is of interest that the participatory steps aimed at stress prevention at work could substantially address these areas related to organizational and psychosocial aspects.

3.2 Sustained Improvements Achieved by Participatory Steps

The sequence of participatory steps taken by workers was similar in either case of customary participatory ergonomics and stress prevention programs. These similar steps are shown in Fig. 1. Following the previous training sessions of trainers or facilitators, the workers learned local good practices and organized group discussion of feasible options. In the case of stress prevention, this was done by a workshop-style meeting involving workers in the same workplace. After identifying locally feasible improvement options, action plans were agreed on and then implemented. Usually, the implemented improvements were reported and disseminated. This feedback of achieved improvements was favourable for continuing the workplace actions, particularly in stress prevention programs.

These participatory steps were commonly facilitated by the use of an 'action checklist' that listed typical low-cost improvements feasible in the local context. The technical areas covered by the action checklist in the case of customary participatory ergonomics usually included work methods (in particular, materials handling and workstation design), physical environment, welfare facilities and teamwork arrangements.

Table 2. Numbers of check items in different action areas covered by action checklists applied in customary participatory ergonomics and in recent stress prevention programs.

Technical areas for improvement actions	Participatory ergonomics				Stress prevention at work		
	Small enterprises	Agriculture	Construction sites	Trade union initiative program	Mental health action checklist	Health care work	Local government offices
Materials handling	8	9	4	6	1	3	1
Workstations	8	7	4	6	3	3	2
Machine safety	7	5	5	5	1	1	1
Physical environment	13	10	6	6	3	8	2
Welfare facilities	3	4	3	3	1	3	1
Sharing information	3	2	2		8	2	5
Work schedules		2		1	4	1	3
Social support					5	1	4
Emergency preparedness	3	3	2	1	4	3	4
Other technical areas			4	2		5	
Total number of items	45	42	30	30	30	30	23

The technical areas covered in the case of stress prevention were broader and included communication and information sharing, work schedules, social support and emergency preparedness measures, thus reflecting the broader scope of improvement actions. Table 4 indicates typical improvement actions in the six areas of the mental health action checklist. The actual action checklists used were its modified versions according to each local situation, and these six areas were equally covered by the action checklists used in all the reviewed stress prevention programs.

The broad-ranging improvements in these six areas were confirmed to be sustained in subsequent years when the participatory procedures were simplified in the form of brief workplace-level workshops, e.g., repeated once every year. The improvements reported in consecutive years by stress prevention activities in local government offices are shown in Fig. 2. In the case of local government K, over 90% of about 170 workplaces made improvements each year, and their distribution in the six areas was similar, those in areas A, B, E and F amounting to 27–35% through 2011–2015. In the case of local government H, the rate of workplaces doing improvements increased from 49% in 2014 to 89% in 2017 among about 630 workplaces, while their distribution in the six areas was also similar each year. It was notable that the improvements in areas A, B, E and F similarly amounted to 26–39% in these years. Thus, the workplaces making improvements in internal communication, work schedules and social support,

Table 3. Improvements achieved in different technical areas after stress prevention programs.

Technical areas	Glass recycling factory	General hospital	Offices in local government K	Offices in local government H
A. Communication	6	4	44	92
B. Work schedules	3	15	8	26
C. Work methods	23	26	91	198
D. Physical environment	22	19	58	255
E. Social support	15	8	11	34
F. Preparedness/training	15	13	16	30
Total	84	85	228	635

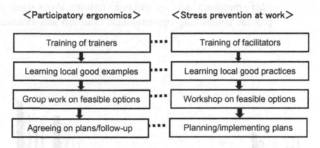

Fig. 1. Sequence of participatory steps taken by participatory ergonomics and stress prevention at work.

i.e., those contributing to improvements in organizational and psychosocial aspects, increased annually.

It was obvious that the simplified participatory procedures in the case of stress prevention programs contributed to sustaining the workplace-level actions for reducing stress at work from year to year. It was striking that about 60% of participating workplaces conducted improvements in two or more areas.

3.3   Roles of Trained Voluntary Facilitators in Supporting Workplace Improvements

The experiences in the stress prevention programs in varied occupations clearly demonstrated the active roles of volunteer facilitators trained in supporting the whole process of planning and implementing valid improvements at each participating workplace. Similar roles used to be played by trainers also in the case of customary participatory ergonomics. The simplified procedures of organizing workplace-level activities in the reviewed stress prevention programs evidently contributed to sustaining the improvement actions in participating workplaces. The specific roles played by the facilitators could be summarized as in Table 5. Easy-to-apply support measures

Table 4. Examples of improvements achieved in different technical areas after stress prevention programs.

Technical areas	Examples of improvement actions
A. Internal communication	Hold brief meetings; avoid excessive tasks; share information (e.g., weekly schedules); utilize notice boards and shared files
B. Work schedules	Limit overtime (e.g., no-overtime days); reduce night tasks; secure breaks; jointly plan paid leaves
C. Work methods	Provide labelled storages; use mobile racks; classify files; work at elbow height; use colour coding; use safety devices
D. Physical environment	Use daylight; provide local lights; Improve ventilation; label containers of chemicals; indicate places requiring PPE; provide resting facilities
E. Social support	Encourage supportive climate; enhance co-worker support; create manuals for absentees' tasks; organize informal gatherings
F. Preparedness measures	Plan emergency actions; organize training about stress prevention; provide counselling opportunities; establish anti-violence procedures

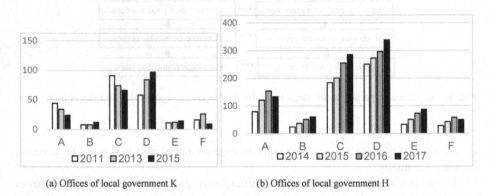

(a) Offices of local government K   (b) Offices of local government H

Fig. 2. Improvements done in successive years in the six areas of stress prevention at work in the offices of local governments K (about 170 workplaces) and H (about 630 workplaces). In local government H, the percentage of workplaces doing improvements increased from 49% (2014), 66% (2015) and 77% (2016) to 89% (2017).

conducted with the help of 'action tools', such as local good examples, an action checklist and a simple report form, apparently helped doing these roles.

The training of these volunteer facilitators selected from among workers of each workplace was conducted as a rule by a brief training session of 2–2.5 h. The facilitators were usually trained by going through the actual participatory steps that they could do with their co-workers. Thus, they learned the validity of simple group-work steps that comprised learning local good examples in the six technical areas, applying the locally adapted action checklist and conducting group discussion to agree on existing good points and feasible improvement actions. They thus learned the merits of simple steps of making proposals for immediate actions as well as encouraging co-workers to plan, implement and report a few feasible improvements within the deadline of about 2–3 months.

As main contributing factors for facilitating the whole participatory steps at the guidance of these facilitators, three points illustrated in Fig. 3 may be mentioned. The main contributing factors were (1) the overall orientation aiming at locally feasible plans, (2) the use of action-oriented tools such as action checklists and (3) the simplified procedures for planning and implementing immediate changes.. These factors obviously made it easy for participating workplaces to achieve improvements available in varied occupations through the workplace-level Plan-Do-Check-Act cycle for risk management.

Table 5. Roles of volunteer facilitators supporting the participatory steps in stress prevention programs.

Aims of facilitation	Support measures	Action tools utilized
(a) Emphasizing locally feasible improvements reducing stress	Present local good practices having impacts on reducing stress	Locally achieved examples in multiple areas of stress prevention
(b) Support joint planning of immediate improvements	Organize a brief workshop at each workplace on feasible options	A locally adapted 'action checklist' listing feasible improvements
(c) Encourage sustained actions involving workers/managers	Assist in agreeing on action plans and reporting improvements done	A simple-format 'report form' describing joint actions taken

In the case of 458 workplaces of local government H that reported the technical areas of improvements conducted in 2016, 59% conducted the improvements in two or more areas. Of these 59% workplaces, 38% took actions in two technical areas, 16% in three areas, 5% in four or five areas. It was thus striking that the majority of workplaces could plan and implement improvements in multiple areas. This clearly showed the active roles of facilitators encouraging workers to conduct feasible improvement actions taking advantage of the simplified procedures. Intervention studies have confirmed these multifaceted actions through participatory stress prevention programs had real impact on stress reduction of workers [12, 14].

Among these 458 workplaces, 56% did some improvements in work methods (area C) and 54% in physical environment (area D). It was noteworthy that 34% did improvements in internal communication (area A), 9% in work schedules (area B), 18% in social support (area E) and 13% in preparedness measures (area F). While 26% of the workplaces took combined actions in areas C and D, 18% did so in areas A and C, 11% in areas A and E, and 10% in areas A and D. Apparently, many workplaces included improvements in internal communication or sharing work tasks as important areas for stress reduction.

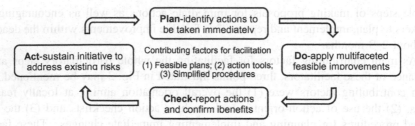

Fig. 3. Three main contributing factors that facilitated the Plan-Do-Check-Act cycle to reduce stress at work.

The improvements in work schedules (area B) were considered important for preventing overwork, and it was noteworthy that almost all the workplaces that took action in this area made improvements also in other areas, such as areas A, C, D and E. It was of interest that the workplaces that tried to limit overtime, secure breaks or jointly plan paid leaves also took actions about communication, work methods or social support.

4 Conclusions

Extending participatory ergonomics to stress prevention is effectively in progress in various occupations commonly applying simplified procedures. Many improvements valid for addressing psychosocial factors, such as better communication and social support, could be planned and implemented. The participatory steps taken with the support of trained facilitators can have real impact on stress prevention in varied occupations. It is suggested to involve locally trained facilitators and organize simplified participatory steps at each workplace focusing on feasible improvements reflecting local good practices.

References

1. Dul J, Bruder R, Buckle P, Carayon P, Falzon P, Marras WS, Wilson JR, van der Doelen B (2012) A strategy for human factors/ergonomics: developing the discipline and profession. Ergonomics 55:377–395
2. International Labour Office (2010) Ergonomic checkpoints, 2nd edn. International Labour Office, Geneva
3. International Labour Office (2012) Stress prevention at work checkpoints. International Labour Office, Geneva
4. International Labour Office (2014) Global manual for WIND: work improvement in neighbourhood development. International Labour Office, Geneva
5. Kawakami T, Kogi K (2005) Ergonomics support for local initiative in improving safety and health at work: international labour organization experiences in industrially developing countries. Ergonomics 48:581–590

6. Khai TT, Kawakami T, Kogi K (2011) Participatory action oriented training, ILO DWT for east and south-east asia and the pacific, Hanoi
7. Kogi K (2002) Work improvement and occupational safety and health management systems: common features and research needs. Ind Health 40:121–133
8. Kogi K (2008) Facilitating participatory steps for planning and implementing low-cost improvements in small workplaces. Appl Ergon 39:475–481
9. Kogi K (2010) Roles of occupational health good practices in globalization. J Occup Saf Health 18:172–181
10. Kogi K (2012) Practical ways to facilitate ergonomics improvements in occupational health practice. Hum Factors 54:890–900
11. Kogi K, Kawakami T, Itani T, Batino JM (2003) Low-cost work improvements that can reduce the risk of musculoskeletal disorders. Int J Ind Ergon 1:179–184
12. Kogi K, Yoshikawa T, Kawakami T, Lee MS, Yoshikawa E (2016) Low-cost improvements for reducing multifaceted work-related risks and preventing stress at work. J Ergon 6(1):147. https://doi.org/10.4172/2165-7556.1000147
13. Noro K, Imada A (1991) Participatory ergonomics. Taylor and Francis, London
14. Thurman JE, Louzine AE, Kogi K (1988) Higher productivity and a better place to work: practical ideas for owners and managers of small and medium-sized industrial enterprises: action manual. International Labour Office, Geneva
15. Tsutsumi A, Nagami M, Yoshikawa T, Kogi K, Kawakami N (2009) Participatory intervention for workplace improvements on mental health and job performance among blue-collar workers: a cluster randomized controlled trial. J Occup Environ Med 51:554–563
16. Yoshikawa T, Kawakami N, Kogi K, Tsutsumi A, Shimazu M, Nagami M, Shimazu A (2007) Development of a mental health action checklist for improving workplace environment as means of job stress prevention. Sangyo Eiseigaku Zasshi 49:127–142
17. Yu I, Yu W, Li Z (2011) The effectiveness of participatory training on reduction of occupational injuries: a randomized controlled trial. Occup Env Med 68(Suppl 1):A24–A25

Touchless Access Control Using iBeacons in Norwegian Hospitals

Sebastian Brage Hansen[1] and Sashidharan Komandur[2(✉)]

[1] Evry Inc., Oslo, Norway
[2] NTNU, Oslo, Norway
sash.kom@gmail.com

Abstract. Contact with surfaces is one of the leading causes of infections and cross contaminations in hospitals. Although some efforts in design in addition to developing process routines to prevent/limit pathogen transmission exist, there may be lapses and this calls for better technological intervention to prevent such transmission. The less number of touchpoints the hospital employees need the greater the possibility of decreasing cross contamination. This paper introduces a touchless access control based on beacon technology for a regular hospital setting. Much of the data used in this paper are obtained from an ethnographic approach which also is the basis for the three use cases introduced in the paper. A prototype consisting of three Estimote Bluetooth low energy beacons/ transmitters; one Android smartphone receiver; a cloud based API and one Raspberry Pi 3 mini computer was developed as a part of this work. It was found that the system's classification model provided accurate positioning in a 2D space.

Keywords: Beacons · Healthcare · Universal design

1 Introduction

A report indicated that nearly 6% of patients admitted to Norwegian hospitals were affected by a disease which was not related to the initial cause for which they were admitted (Aftenposten [1]). This figure was around 10% for Denmark (Burcharth et al. [2]). It is estimated that the general average for this kind of occurrence is around 5 to 10% (D'Agata et al. [3]). Good hand sanitation practices by healthcare professionals, researchers and specialists is believed to be the most important contributor to limiting the spread of pathogens (D'Agata et al. [3]). Some research shows that despite there are lapses in maintaining personal hand hygiene by healthcare professionals, for example forgetting to wash their hands after visiting bathrooms (Burcharth et al. [2]). Much has been done to improve sanitation and prevent spread via surface contacts, but there is still much to desire from the design of things healthcare workers interact with on daily basis. There is a substantial scope of work to be undertaken to improve sanitation and prevent spread of pathogens via surface contacts through efforts such as better design of interaction touchpoints.

© Springer Nature Switzerland AG 2019
S. Bagnara et al. (Eds.): IEA 2018, AISC 819, pp. 382–386, 2019.
https://doi.org/10.1007/978-3-319-96089-0_41

1.1 Problem Statement

The main reason for the spread of pathogens within hospitals is surface contact. There have been substantial efforts in creating sanitation routines and designing touchless interfaces. For example automation for sanitation dispenser and automated doors already quite a lot in use but there is still room for improvement. There are many touchpoints such as keychains, levers to open mechanical doors, keyboards for punching keycodes and all of these maybe contributing to the transmission of pathogens, thus it is absolutely important that such areas of contact are absolutely minimized in a hospital environment. Access control is a major reason why hospital staff require to contact items such as keys, keycards or keychains when locking or unlocking doors. It is possible with newer technologies available these touchpoints can be re-envisioned and in its place a touchless interface is created.

2 Methods

The project aims to triangulate a user's position using a personal marker. Literature review of practical ways of positioning users within 2D space suggest using bluetooth beacons [4]. For exploring relevant methods of implementing the beacons in a work environment, a "fast and dirty" ethnographic study adapted for HCI as per Preece, Sharp and Rogers recommendations was conducted [5]: "[…] it takes place in the natural setting of the people being studied; it involves observation of user activity, listening to what user says, asking questions, and discussing the work with the people who do it. […] In contrast to long-term 'pure' ethnography, […], the 'quick and dirty' version of ethnography has been adapted for HCI. Although involving significant shorter time with subjects and correspondingly less of analysis, this version still requires observation of subjects in their own environment and still requires attending to sociality of subjects in their work context". Based on the ethnographic finding, a smartphone app utilizing three beacons for triangulation was built and used to test the human ergonomic factors of controlling doors using iBeacons. The app was built in Java on Android Studio with the Android SDK, and ran on a Sony Xperia Z1 Compact (Xperia) and a Nexus 6 (Nexus) both using Android 6 Marshmallow. The beacons were Estimote second generation beacons set to transfer at 100 ms intervals, the shortest available for the particular beacons. The smartphone used trigonometry to calculate a user's travel vector in a 2D space, then contact a RESTful API and trigger a routine for opening a door if the user was on a collision course with the door. The RESTful API was built using Python running on a Raspberry Pi 3 mini computer connected to the internet using direct, high capacity cable. Following Steve Krug's guidelines for "recruiting loosely and grade on a curve" [6], subjects were recruited by convenience sampling from users already acquainted with operating security doors, and the prototype tested in two locations: One test with two users at an environment approximating those of the target environment, and one test in a lab like environment using the author as subject. To mimic the behavior of an access control, a lamp was connected to the output of the API and triggered to light up when the API sent the signal for opening the door. Qualitative data from observation and unstructured interviews was recorded for test using subjects from the convenience sampling. When the author acted as subject, notes on first hands experience and evaluation based on best practices was recorded (Fig. 1).

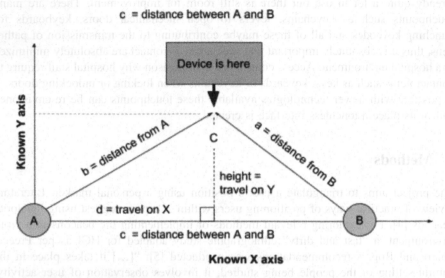

Fig. 1. Diagram depiction of how the smartphones position is triangulated on a 2D axis using distance from two known points.

3 Results

Prototype 1 consisted of the smartphone app, the Raspberry Pi polling every 10 ms, a light bulb connected to a 220 V circuit, the cloud API and a cluster of three beacon transmitting at 950 ms. When the phone was in range of all beacons a signal was sent to the API telling the Raspberry to turn on the light. When first prototyped, the iBeacons were found to have a less than optimum response rate for perceived instantaneous interaction. Where the Raspberry Pi polls a doors state every 10 ms, the iBeacons relative position to the phone are only updated every 950 ms. Using Estimotes developer app on the smartphone, the beacon's transmission interval was decreased to the allowed minimum of 100 ms. Although the lowest setting available it is not low enough to meet the planned 10 ms response time.

Prototype 2 was the same as prototype 1 but with beacon's transmission speed adjusted to the minimum available 100 ms. The second iteration adjusted for transmission broadcasting cycles that granted a faster response. Still, some of the latency persisted. When deploying the initial prototype for a proof of concept run, a Sony Xperia Z1 Compact (Xperia) smartphone was used. The phone was found not to be powerful enough to continuously monitor all beacons at optimal efficiency. For every "loop" of the monitored beacons – where the program iterates over every single beacon in range one by one and checks if the current beacon is one of the three used for triangulation – the Xperia would run significantly slower than a more powerful Nexus 6. When test users were given the Nexus 6 they would describe the system as "more responsive" and "faster" than for the Xperia. Whether or not this subjective feeling of "slowness" was due to hardware or software issues was not investigated. Later when looking over timestamps of when beacons were read while scanning, on both phones separately, the Xperia was shown to read beacons at a slower rate (field notes, timestamps of measurements from the two phones in debug log).

Prototype 3 used the grid design for positioning the user and calculating valid approach vectors. When calculating grid positions, a new issue was discovered: The beacons vary in transmission strength even when stationary throwing the triangulation off and making it impossible to calculate an exact vector or path in high resolution/small areas. Based on 5 sets of 100 distance measurements at increasing distances the app is shown to report an average inaccuracy of $\sim \pm 30\%$ (inaccuracy = standard deviation/mean).

4   Discussion

This paper's research questions are mainly about door automation in a hospital environment using bluetooth based low energy beacons, and the use cases were obtained after an ethnographic study which was designed according to the guidelines of Hartson and Pardhaa [7]. Due to technical and practical limitations the desired level of high fidelity prototype could not be created. Despite this substantial data was collected through this low fidelity prototype.

Access control will to some extent vary from hospital to hospital. In the hospital for this study there was the usage of keycards, automated doors and the use of old fashioned keys and pull doors. This is representative in general of other hospitals as well. Some doors are redesigned upon request by lab technicians, such as a western style saloon door going into the lab allowing for hands free movement both ways through the door.

This study found the most challenging obstacle to be technical limitations affecting the speed of operations. Whenever a door did not open as intended the users would be frustrated and proceed to open the door manually or by keycard. As expressed by McEwan et al. [8], cues is a fundamental part of any interaction. When they are missing or hard to interpret the experience will feel broken or unpleasant. For a door to be well designed, good cues must be present, a thought also entertained by notable designer Norman [9].

To assure the proper delivery of cues to the user, the designed could be altered with a feedback loop following best practice guidelines as Norman Nielsen's 10 heuristics for system design to offer affordance as described by Norman [9]. This feedback loop could be in form of a light signal describing the doors different states, a mechanical flag or something similar clarifying when the door is active and operational, occupied performing an action or in another state. However Norman has a valid point in stating: "The violation of the simple use of constraints on doors can have serious implications." [9] so the chosen cues should be investigated, prototyped and user tested.

The beacons afford the ability to track a single user in 2D space inside the hospital. This tracking is however not accurate enough to give further advantages to fully automated door in its current form. Alternatively the devices could be used in a simpler form, as checkpoints or broad term locators for personnel, but not for the pin point precision needed to fully replace existing door sensory – like infrared sensors, cameras or indeed push buttons.

2D tracking can help locating personnel throughout keypoints and greater areas [10]. The ranging of the beacons afford for a close, medium, far classification of user positions. Gathering of which rooms healthcare workers have been in and thereby which doors they have passed could therefore strengthen the general knowledge of pathogen spread.

References

1. Aftenposten (2011) Smittefare ved norske sykehus. http://www.aftenposten.no/meninger/leder/Smittefare-ved-norske-sykehus-6597907.html. Accessed 6 Nov 2015
2. Burcharth J, Pommergaard H, Alamili M, Danielsen AK, Rosenberg J (2014) Hver femte kirurg vasker ikke hænder efter toiletbesøg–et etnografisk feltstudie. Ugeskrift for læger 176 (25):2364–2367
3. D'Agata E, Horn MA, Ruan S, Webb GF, Wares JR (2012) Efficacy of infection control interventions in reducing the spread of multidrug-resistant organisms in the hospital setting. PLoS ONE 7(2):e30170
4. Gast MS (2014) Building applications with ibeacon: proximity and location services with bluetooth low energy. O'Reilly Media, Inc., Sebastopol
5. Preece J, Sharp H, Rogers Y (2011) Interaction Design-beyond human-computer interaction, 3rd edn. Wiley, Hoboken
6. Krug S (2005) Don't make me think: A common sense approach to web usability. Pearson Education India, Noida
7. Hartson R, Pardha SP (2012) The UX book: process and guidelines for ensuring a quality user experience. Elsevier, New York City
8. McEwan B, Westerman D, Sumner E, Bowman N (2016) Let's interface: connecting social research to UXD. Panel Debate at South by Southwest Interactive 2016, Austin, TX, 12 March 2016
9. Norman DA (2002) The design of everyday things. Basic books, New York City
10. Hatlevoll DB (2011) Using Bluetooth low energy in sensor devices: possible applications in welfare technology. NTNU

Smart Planning - Approaching the Characteristics of a Valid, Balanced Transport Round

Virginie Govaere[1]([⊠]), Liên Wioland[1], Julien Cegarra[2], Didier Gourc[3], and Antoine Clément[3]

[1] INRS, Working Life Department, 1 rue du morvan – CS 60027, 54519 Vandoeuvre Cedex, France
{virginie.govaere,lien.wioland}@inrs.fr
[2] Université de Toulouse, INU1 Champollion, SCoTE EA7420, Place Verdun, 81100 Albi, France
julien.cegarra@univ-jfc.fr
[3] Université de Toulouse - IMT Mines Albi 2, Campus Jarlard, 81013 Albi Cedex 09, France
{didier.gourc,antoine.clement}@mines-albi.fr

Abstract. Road Freight Transport (RFT) companies represent 37,200 companies and roughly 420,000 employees. This sector is confronted by strong competition and growing pressure from customers and suppliers, tight delivery times, exacerbated flexibility, etc. In parallel, they are required to fulfill performance duties in terms of preventing risks of occupational accidents and diseases. In 2016, CNAM statistics reports 70 deaths per year, 3,000,000 work days lost, an average 6 work days lost per employee, an index of frequency (73‰). The planner builds the transport rounds by integrating at best all dimensions (regulation, economic, environmental and prevention of health and safety of their employees). In this context, the Smart Planning project aims to develop a computer system to help create more balanced planning. The purpose of this paper is to present the first results. It proposes, with an ergonomic analysis, to identify the prescribed and tacit constraints manipulated by the planners in two companies. A questionnaire is drawn up to validate and enrich the data on the health and safety dimension.

This study is not a business case; it is ergonomic analysis to validate different determinants identified (health and safety) and investigates the assessment of these determinants and their possible consideration during planning.

Keywords: Prevention of health and safety · Balanced planning Compromise

1 Introduction

1.1 Business Sector Background

Road Freight Transport (RFT) companies represent an annual turnover of 44 billion euros, which is generated by more than 37,200 companies. This sector is confronted by

S. Bagnara et al. (Eds.): IEA 2018, AISC 819, pp. 387–396, 2019.
https://doi.org/10.1007/978-3-319-96089-0_42

strong competition and growing pressure from customers and suppliers, tight delivery times, exacerbated flexibility, and many other challenges [1]. Financial issues impose additional constraints upon the already complex problem of planning transport operations. Procurement, production and distribution are indeed all organized based on stockless production involving "just-in-time" (JIT) deliveries. In transport terms, this is reflected by smaller freight consignments to be moved at higher frequency for recipients requiring delivery as quickly as possible or even by appointment. Ensuring RFT company survival implies taking these constraints into account by optimizing delivery rounds. RFT companies are also subjected to specific national and community standards and regulations (regulatory and environmental issues). Alternative and combined modes are emerging in rail, sea, river transportation on environmental grounds. However, whatever the energy combination, certain delivery modes are rapidly beset by limits in terms of flexibility, delivery times, customer proximity, etc., and the last kilometers of a journey are necessarily traveled by road. Optimization of the number of kilometers traveled and optimum vehicle selection, based on transport round constraints, are levers for responding to part of the financial and environmental issues. At the same time, companies are committed to fulfilling performance duties in relation to preventing risks of occupational accidents and diseases under Article L4121-1 of the French Labor Code governing occupational health-safety issues. CNAM national health insurance fund statistics and other institute show that the claims ratio in this branch in France remains very much higher than the national average for all branches in that employees are twice as often victims of accidents and of more serious accidents [2, 3]. The commonly identified risks for employees in this sector involve both the physical aspects (musculoskeletal disorders related to handling, falls on the level, exposure to vibration, etc.) and mental aspects (high mental load, activity segmentation, higher pace of work, dealing with unforeseen events, etc.) of the work activity. Multiple research studies and actions have mobilized stakeholders from different areas and have contributed to improving driver safety. Nevertheless, this problem remains an issue. Furthermore, RFT planner activity is performed within a specific context involving segmentation of various operations, performance of dual tasks [4]. Interruptions, segmentations and performance of several tasks in parallel constitute risk factors for the planner and factors that impact on work quality and the compromises to be adopted during planning. These compromises affect driver working conditions: managing an urgent situation in a tense, stressful environment leads operators to protect themselves less and is often associated with increases in accident risk. We defend a dual viewpoint in this paper in order to act upon the sector's claims ratio:

- Integration of the health-safety issues right from delivery round planning stage. Planning stage consideration of the frequency of handling operations to be performed by the driver, of available aids (pallet trucks, bay operators, etc.) and of infrastructures (stairs, lifts, etc.) at the customer, alternation of more and less demanding activity phases or balanced allocation of "difficult" delivery points among drivers may represent a solution to alleviating driver levels of stress.
- Provision of assistance suited to planner activity. This could entail planning support in a context, in which the mental aspects of these employees' work are strongly present.

Ultimately, it is up to the planner to integrate all the relevant and often antagonistic financial, environmental and social issues, when designing transport rounds. Under these complex conditions full of uncertainty, the planner must adopt a compromise between these constraints. The quality of the compromise depends on the time the planner can give to this operation, on his/her representation of each constraint and on the margins for maneuver available to him/her.

1.2   Research Context

Aids to managing transport or Transport Management Systems (TMS) are available to the planner, but these solutions focus only on the financial issues. The Smart Planning project aims to overcome this limitation by integrating financial, environmental and health and safety aspects for both drivers and planners into robust planning aids for transport operations. The project is financed by the ANR and features an interdisciplinary consortium. The primary scientific aim of this project is to model the constraints integrated by the planner in order to propose valid delivery rounds from financial, environmental and social standpoints, while integrating operator (planner and driver) health and safety aspects [5]. The objective is also to model the dynamics of the compromise adopted by the planner which respond differently to both the challenges and the transport company strategy. These models should be subjected to break-down into operational criteria, which can be rated and are: (1) financial, (2) environmental and (3) health and safety-related.

To achieve this primary project aim, a number of stages are unavoidable, including listing and categorizing the constraints to be managed by the planner, defining transport round performance indicators from health and safety, financial and environmental standpoints, identifying constraint prioritization strategies to be adopted by planners, etc. Ergonomic analysis is conducted to identify the planners' constraints (formulated, released or satisfied constraints). The purpose of this paper is to present these results, which contribute to defining the initial components of a valid, balanced transport round in financial, ecological and health and safety terms.

The second project aim is to introduce the approach to validate the list of determinants identified in relation to the health and safety dimension and to investigate rating of these determinants by planners during their work. Preliminary results illustrate the significance of this approach.

2   Methodology

The ergonomic analysis involved the two industrial partners of the Smart Planning project. These partners are freight transporters. These SMEs employ 195 and 60 people. Identical methodology was deployed at both companies.

2.1   Stage 1: Description of Planning Activity

Analysis of the planning activity was conducted using instrumented observations (video and audio recordings). The video recordings were made using cameras to observe two planners/company performing a usual operation during their work shifts. Observation-based data were collected from the start to the end of the work shift on full working days during a single week. This operation provided 24 h of recordings (2 planners/company on morning and afternoon shifts, excluding breaks, on 3 full working days). The observations were formalized in the form of an activity code, which subsequently enabled us to determine the time and frequency of the operations performed, the various tools used, the communications, the purpose of these communications and the contact persons involved. This analysis was complemented by interviews with participants both directly (planners) and indirectly (operations managers, drivers and general managers) involved in the planning activity. The latter management personnel establishes the company strategy and hence guidelines in terms of planning criterion selection and prioritization). An interview questionnaire was drawn up and systematized. This included 5 main categories of information: constraints related to the client and its requirement, constraints related to human and equipment resources, constraints related to infrastructures and regulations, company guidelines and strategic decisions and the different types of unforeseen events encountered.

2.2   Stage 2: Constraint Identification and Prioritization

Following the interviews and observations, we identified the concepts used by the planners for building transport rounds. For the stakeholders, this involved prioritizing characteristics for each concept based on their importance, when planning transport rounds. We collected the verbal accounts concomitant with this prioritization task. A card game was developed for this prioritization exercise; it included the concepts identified during Stage 1 and their characteristics. For example, there was the "vehicle" concept along with its typical attributes: "tailgate present", "bed length" or "side loading possible", etc. At each company, seven employees took part in interviews and prioritization. Data acquisition required a week-long visit to each company.

2.3   Stage 3: Collection of Sociotechnical System (STS) Determinant Ratings by Sector Professionals

A questionnaire was edited based on the Stage 1 interview findings and earlier work in the sector [ref lien]. This questionnaire was validated by the project steering committee, tested by the 2 industrial partners (7 planners), then circulated online via professional federations within the RFT sector. The purpose of circulation was to generalize the results obtained with the project industrial partners. The questionnaire included the following sections:

- A section identifying the respondent's job and the company profile
- A section describing the planning activity
- A section focusing on factors causing difficulty or arduousness experienced by drivers and their consideration during planning. For each factor, this involved determining its presence or absence in the activity, rating its degree of difficulty or arduousness on a scale of 1 to 10 (1 very low and 10 very high) and stating whether this factor is considered during planning.

Three categories of factor were envisaged in the last section: those causing difficulty or physical arduousness, those causing mental arduousness and those causing emotional arduousness (see Table 1). A free data entry zone for factors complementing this list closed out this section.

Table 1. Example of questioning for each factor category

	Physical	Mental	Emotional
Factor present or absent	Do you reckon handling is a factor of arduousness in driver activity? • Yes • No	Do you reckon your drivers encounter driving situations that require a high or prolonged level of attention? • Yes • No	Can the client relationship be a factor causing difficulty in driver activity? • Very often • Often • Sometimes • Rarely • Never
Definition	Handling: carrying packages, moving pallets without mechanical equipment, filming, sorting, etc.	This may involve driving at night, driving electrical silent vehicles or maneuvers requiring constant vigilance	Factor related to welcoming, assisting or collaborating, possible relational difficulties, etc.
Rating (for each factor)	1--10 Very minor Very major		
Consideration (for each factor)	If it exists, do you take this factor into account during planning? • Yes • No		

3 Results

3.1 Description of Planning Activity

The planners managed a client relationship (11% of operator activity time), available physical and human resources (21% of activity time) and transport schedules (16% of activity time). They established the traceability (36% of activity time) and provided an interface between the different stakeholders (management, clients, drivers, etc.). Comparing activity and operation frequency therefore highlighted the fragmented nature of the planning activity. This activity fragmentation was evidenced by the brevity of the operations as shown by their average times (approximately 1 min. 30 secs. per operation). The planners issued or received oral communications (face-to-face or by telephones) for 40% of the activity time (i.e. 28 h of communications).

These communications occurred in parallel with the transport round planning activity. Planning represented 21% of the planners' total activity time. Typically, it could be divided into collection of data, allocation of resources and optimization of various criteria. Observation of the planning activity revealed a succession of back-and-forth operations on the planned transport round or order addition or deletion.

3.2 Identification, Categorization and Prioritization of Constraints

The interviews and card game were run with planners, operations managers and company managers: a total of 14 study participants. The data exhibited no variations in participant status. The interviews and observations revealed that the concepts used by the different players in planning transport rounds were the delivery sites, the vehicles, the client, the order, the driver and the round. These concepts were characterized with attributes. For example, the Site (warehouse, store, private individual) was characterized with address, loading location, unloading location, delivery time window, flexibility, type of vehicle, accessibility, presence of bay, driver contribution, handling equipment. The planners handled these concepts (site, vehicle, client, order, driver and round), when building their planned transport rounds. The concepts and their attributes created resources and constraints, when performing a planning operation. Planners also integrated constraints involving regulations applicable to work, driving or the environment. However, these depended on concepts (e.g. downtown delivery is regulated in some cases): delivery times, type of vehicle authorized, parking location…. and through the planner's handling of the concepts that these constraints covered. Categorization of these constraints therefore corresponded to the concepts dealt with. Interview analysis allowed us to define 4 levels of constraint:

- Major constraints, for which it was imperative that the planner ensure their integration into his/her planning operation. These constraints arose from "regulation" or minimum characteristics to be implemented to fulfill the transport operation. In the former case, regulations were applicable, when they involve the highway code, the labor code, the environment or Heavy Goods Vehicle (HGV) driving. With regard to minimum characteristics to be implemented, these also represented major constraints because they could not be adapted. Thus, if the delivery site imposed maximum vehicle dimensions (e.g. underground passage access, hence maximum vehicle height and width), one major planning constraint would be the height and width of the vehicle to be used for this site
- Moderate constraints were to be integrated into transport round planning and their integration was indeed necessary. Adaptations could nevertheless be implemented, if needed.
- Minor constraints were more recommendations than necessities.
- "Absent" constraints are those that were not spontaneously signaled as planning constraints. Thus, health and safety determinants were associated with specific situations or points requiring vigilance and were not related to risks. These situations were to be carefully watched, when planning the transport round, but they did not constitute declared constraints for planners during construction of the schedule.

Except for major constraints, managing levels of constraint remained dependent on the situation. For example, a moderate constraint could be relaxed in order to provide an acceptable planning solution. In a low activity situation, driver inexperience could be ranked at a level of moderate constraint and could lead to calling in a more experienced driver. In a high activity situation, round preparation was more constrained. Planners then had to abandon consideration of this constraint.

The card game allowed us to see that the constraints were prioritized by grouping them together or at least by processing aggregated attributes since constraints and attributes were not independent of each other. Type of vehicle, its capacity, driver's license requirement were aggregated and considered globally in some situations. In other situations, such as the "exceptional" request described in the remainder of this text, each attribute and constraint was considered separately.

The expression most often encountered in the interviews or in the card game with planners was "It depends!"..... it depended not only on the client, the site, the driver but also on the company strategy, the company's financial health, etc. The level of constraints and their prioritization were therefore contingent. Moreover, the level of constraints was variable, depending on the planner's knowledge of the situation; in cases involving a new client, an exceptional order or an unusual type of goods, all the constraints would be prioritized differently by globally considering their levels higher than usual. Under these circumstances, what would be a valid, balanced transport round in financial, ecological and health and safety terms? For the project industrial partners, a valid round was a round that, ideally, combined two components, two types of performance:

- Performance internal to the company: suitable usage, from a financial standpoint, of human and physical resources and of the delivery time. This usage was intended to generate a positive financial margin.
- Performance external to the company: adequate response to client requirements. This response constituted quality of service to the client.

The "valid" transport round therefore depended on proportioning the quality of service to clients located in the round, while controlling costs and profitability. These two components and their "proportioning" had to be weighted according to company strategy. Round validity was assessed independently of the "ecological" and "health and safety" dimensions. The project industrial partners integrated these dimensions into the balanced round. The balanced round depended on mobilizing human and physical resources that were adequate for the company's objectives. Nevertheless, this mobilization had to conserve a margin for maneuver in terms of managing clients and unforeseen events during round performance.

3.3   Rating of STS Determinants by Sector Professionals

It should be remembered that the planners described health and safety determinants with respect to situations and points requiring vigilance and not with respect to risks or constraints. Thus, site accessibility problems, complex loading/unloading locations, handling and stripping and unusual or extended working hours were all situations identified for their physical aspects. A new client, heavy traffic, complex infrastructures, an unforeseen wait at the client causing time-related stress were all situations identified as hazardous for mental health. Emotional aspects did not give rise to specific situation identification.

The Questionnaire

The findings presented are from the questionnaire's experimental stage with the project industrial partners (7 planners); they provide elements of information in terms of STS determinant rating. These elements need to be supported by data from the questionnaire's wide circulation. Two general issues were included prior to presenting the results based on factors causing difficulty or arduousness.

With regard to safety, defined by 4 items (theft and holdup, assault, accidents when stationary or driving), 57% of participants asserted that they took it into account by providing technical solutions such as secure vehicles (holdup and assault) and assignment to rounds based on driver skills and potential hazards. Planners, who did not integrate this aspect into their planning, asserted that this was dealt with by corridoring or watchmen at client sites to prevent assault, by driver training and compliance with regulations preventing accidents, when stationary or driving.

With regard to health, 87% of participants stated that they took this into account during planning by providing technical solutions to alleviate certain constraints (e.g. electrical pallet trucks), driver assignment (attention given to driver age, known health problems...) and order management (high volume orders broken down). One of the planners did not integrate health into planning because he considered health to be managed by applying regulations and by equipment selection by company management. Following these two general issues, more specific issues were raised (Table 1). These are based on physical, mental and emotional aspects.

Physical Aspects

These aspects were based on situations involving site access difficulties and handling situations. 87% of the participants considered that handling and site accessibility were factors causing difficulty and arduousness. However, all participants took them into account in their planning. Site accessibility-related difficulty was rated at between 5 and 7 on a 1-to-10 scale with 1 very low and 10 very high.

Mental Aspects

"Driving" and "Prolonged attention" were envisaged as mental aspects. With regard to these aspects, the planners considered that all their drivers were affected. Hence, these aspects were taken into account during planning and were rated between 4 and 8 with 42% rated 8).

Emotional Aspects

Assaults and relationship with the client were situations identified as causing emotional difficulties. All the planners reckoned that drivers were exposed to such situations. With regard to relationship with the client, emotional difficulty was "often" effective in 28% of cases, "sometimes" in 43% of cases and "rarely" in other situations. 71% of planners integrated this difficulty into their planning. However, this difficulty was rated moderate to low (between 1 and 5).

Other Difficulties

The 7 planners referred to difficulties or situations considered tough:

- Lifestyle (sleep, food, adapting to variable work rhythms, etc.) was quoted 3 times with a rating of 9
- Meteorological conditions were quoted twice with ratings of 8 and 7, Traffic was quoted twice with a rating of 8, Time-related stress (lateness) was rated 8, Bay entry and driving maneuvers were rated 7
- Lack of knowledge of the client was a difficulty rated 5.

Physical, mental and emotional dimensions were all present within these "other difficulties".

4 Discussion and Conclusion

The findings of this study highlight that transport round planning activity is segmented, that situations involving dual tasks are usual and frequent, and that usage of tools is intensive and brief. Work intensity and information load handled by various tools are constants of the planner's activity. We also recall that planning is spread right across the entire daily activity.

Planners manage multiple constraints, which arise from outside the company or depend on decisions made by the company. The former are constraints related to regulations, infrastructures or clients. Constraints inherent to the company are related to the choice and sizing of the vehicle fleet, the strategy of the company and its human resources (drivers) for managing the activity. Constraints external to the company are considered major constraints to be overcome whatever the situation. Constraints inherent to the company have no status a priori; depending on the situation, they can be major, moderate or minor. Planners consider these constraints to be mutually dependent and they therefore deal with them globally. However, when the situation, context or demand is exceptional, constraints are prioritized differently and processed separately. With the exception of major constraints, prioritization cannot be performed a priori. It must be contingent on both the context and the situation.

It should be noted that employee health and safety aspects during planning were not spontaneously referred to during the study interviews. Furthermore, when the employees were questioned, they spoke mainly about situations requiring special care by the project industrial partners and not about exposures or risks. The physical dimension was referred to within this framework. When the mental and emotional dimensions were raised, they were not dismissed but were considered inherent to the driver's job or to the driver's responsibility. However, with the questionnaire, consideration of the health and safety dimension in planning was effective for a large majority of planners, even though a variability in dimension integration was noted: the emotional dimension being less integrated by planners. It is also the dimension that was rated as having a lower intensity. It is not because the health and safety dimension was not stated as being systematically integrated during planning that it was not necessarily taken into account. The planners considered that certain factors, such as the company's choice of handling equipment, vehicle fleet deployment, corridoring or compliance with regulations on driving time, safety training for drivers, etc., effectively meet the requirements of driver health and safety.

We also noted that the RFT context and the corporate strategy led planners to adapt constantly their management of constraints and consideration of the different dimensions. The next stage of this research will therefore be to focus on designing a compromise development system rather than a constraint prioritization system. Compromise development analysis conducted by planners during their activity would appear to represent an interesting route to preparing a delivery round that is both valid and balanced for industrial stakeholders.

References

1. Camman C, Fiore C, Livolsi L, Querro P (2017) Supply chain management and business performance: the VASC model, 1st edn. Wiley-ISTE, London
2. Joling C, Kraan K (2008) Use of technology and working conditions in european union. European foundation for improvement of living and working conditions, Luxembourg
3. Bilan de l'Observatoire social des transports (2013) Commissariat général du développement durable, service de l'observatoire et des statistiques
4. Wioland L (2013) Ergonomic analyses in the transport and logistics sector. Reflection on developing a new prevention approach: "Act Elsewhere". Accid Anal Prev 59:213–220
5. Apostolopoulos Y, Shattell MM, Soenmez S, Strack R, Haldeman L, Jones V (2012) Active living in the trucking sector: environmental barriers and health promotion strategies. J Phys Act Health 9:259–269

EEG Based Assessment of Pedestrian Perception of Automobile in Low Illumination Road

Rahul Bhardwaj[1] and Venkatesh Balasubramanian[2(✉)]

[1] Haria Seating Systems Limited, Hosur, India
rb@haritaseating.com
[2] RBG Lab, Engineering Design Department, IIT Madras, Chennai, India
chanakya@iitm.ac.in

Abstract. Pedestrian involvement is a major subset of road crashes. It is estimated that pedestrian road crash is about 22% of all road traffic related deaths.
Since pedestrians share road space and traffic, they are susceptible to the crashes
especially where a large number of pedestrians are seen on roads. In this study,
we have estimated the pedestrian's response and time taken to estimate the
correct recognition of the vehicle approaching them while crossing the road.
EEG analysis has been performed to estimate cognitive response of pedestrians.
Thirty volunteers participated in this study. Six scenarios were presented as were
shown to the participants. Analysis was performed based on EEG acquired
activity. It was observed that less beta activity (p > 0.0%) was estimated when
participants were shown the video of "low beam car with active light source"
and "high beam car with active light source". This clearly indicates that active
light source in addition to headlights of car make pedestrians less confused
about the oncoming traffic. This helps them to cross especially in low illumination allies.

Keywords: Pedestrian perception · Pedestrian safety · Pedestrian crossing
Low illumination crashes · Active light source

1 Introduction

According to the recent report by Ministry of Road Transportation and Highways [1] of
India, it has been found that the pedestrian death was 15,746 of total road accidents
deaths (150,785) in the year 2016 which was 10.5% of total road accidental death. [2]
Reported that among traffic crashes in the United States of America, 14% (4,735) of
total fatalities (33,719) are pedestrian. Among that 66,000 were injured in 2013. Older
people belonging to the age group 65 and above were more involved in the pedestrian
fatalities (19%) of all (896 of 4,735) and 10% of all pedestrians injured (7,000 of
66,000). According to World Health Organization [3], total no of death of pedestrians
are 275,000 globally and it is 22% of the total road accidental death. Pedestrian-vehicle
collision at nighttime is common resulting in high loss of life.

There might be several reasons for pedestrian-vehicle collision in night time condition. However a common thread in most of them would be that pedestrians having

© Springer Nature Switzerland AG 2019
S. Bagnara et al. (Eds.): IEA 2018, AISC 819, pp. 397–405, 2019.
https://doi.org/10.1007/978-3-319-96089-0_43

not noticed vehicle [4] or vehicles are insufficiently visible to pedestrian. Both the possibilities eventually result in road accident. So far, many researchers have been performed to analyze pedestrian crashes and the severity level. They had recognized the possible risk aspects and suitable countermeasures based on different statistical methods [5–8].

Some researchers have suggested using of reflective clothing to increase the contrast visibility of pedestrians to the drivers and the use of reflective clothing has shown increased pedestrian visibility [9, 10]. [11] Had used retro–reflective material which was applied to the major joints of cyclist and similar material was mounted on a vest to recognize the pedestrian using video based technique.

It is very difficult in the nighttime for drivers to see and recognize the pedestrians not wearing conspicuity enhancing clothing. Especially in developing economies, such as India, the use of nighttime clothing is not affordable or adhered by pedestrians in both professional and non-professional environments. It has been observed that the old age pedestrians take more time to decide while crossing the road [12]. Delay in taking decision is due to slow brain processing speed due to old age (Salthouse 1996).

Recently one study has been performed by [13] in which authors estimated pedestrian's response towards an approaching vehicle in night time conditions based on signal detection theory. Authors estimated time taken to correctly recognize the type of vehicle while placing active light source in between the headlights of the car. It was found that pedestrians were quick and comfortable to recognize the approaching vehicle when active light source was used.

In recent years in the India, a huge penetration of two-wheelers and four-wheelers has been observed. According to Investment Information and Credit Rating Agency of India [14], the trend of two wheeler segment growths has been observed to be increasing i.e. in 2016, it was 3.0% whereas, in 2017 it was increased to 4–6%. Trending on same path, the growth of four-wheeler is also expected to increase in next decade in India. Massive growth of four wheelers and two wheelers in India is a serious threat for pedestrian safety and road safety.

The inability of the pedestrian to judge and discriminate between four-wheeler and two two-wheeler may lead to confusion among the pedestrian. This confusion results in increased reaction time for decision making among the pedestrian while crossing the road especially in the rural area and highways where the facilities of the street light or high way light is minimal. In this study, we are evaluating pedestrian's brain processing and decision taking ability using electroencephalography (EEG) while placing active light source in between the headlight of the car while crossing the road in night time or low illuminative conditions.

2   Methods and Materials

2.1   Subject Details

Thirty male participants volunteered in this study. All the participants were screened for physical disability and corrected vision. Their mean age was 22(±5.6) years, height 1.70(±0.07) m and weight was 62.3(±6.3) kgs. It was ensured before start of the experiment that all the participants took 8 h of sleeping rest and it was also ensured that they are refrained from tea, coffee, tobacco and alcohol. Study protocol and nature of the experiment were explained to all the participants before the start of the experiment. Written consent was also taken from all the participants.

2.2 Design of Experiment

The accident scenarios of interest are simulated using a dynamic scene. Video of multiple sub-modalities of the accident scenarios were simulated and recorded using a high quality commercially available camera (Nikon inc.) on a five hundred meters long straight road. Six scenarios were taken in this study which is following. These are presented in sequence in Figs. 1, 2, 3, 4, 5 and 6.

- Low beam car (LBC)
- Two bikes low beam (BLB)
- High beam car (HBC)
- Two bikes high beam (BHB)
- Low beam car with LED strip (LBCS)
- High beam car with LED strip (HBCS)

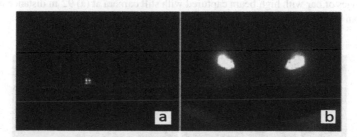

Fig. 1. Images of car on low beam captured with still camera at (a) 92 m distance and (b) 2 m from observation point.

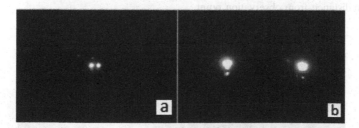

Fig. 2. Images of 2 bikes with low beam captured with still camera at (a) 92 m distance and (b) 2 m from observation point.

Fig. 3. Images of car with high beam with LED strips captured with still camera at (a) 92 m distance and (b) 2 m from observation point.

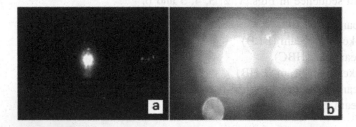

Fig. 4. Images of car with high beam captured with still camera at (a) 92 m distance and (b) 2 m distance from observation point.

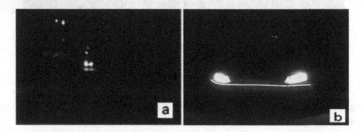

Fig. 5. Images of car with low beam captured with still camera with strips at (a) 92 m distance and (b) 2 m distance from observation point.

Fig. 6. Images of 2 bikes with high beam captured with still camera at (a) 92 m distance and (b) 2 m distance from observation point.

Design of experiment and Subjects were explained the protocol in a standardized format and the dynamic scene was simulated. Pedestrian perceive a set of two lights heading towards them with limited visibility in night. Due to lack of further spatial information, they assume this to be as lights of two motorcycles as shown in Fig. 1. Their reasoning is supported by the common observation of motorcycle driving on narrow roads. Therefore, they drive straight into the perceived juxtaposed 'two wheelers'. Sometimes, pedestrians get wrong perception of the situation and this causes an accident as represented in Fig. 2.

Video recording of the above six modules were projected to the subjects in a temperature and noise controlled environment to give the real feeling of the road environment to the subjects. Other protocols and directions were adopted from [13].

2.3   EEG Data Collection and Processing

EEGs data were recorded from frontal, temporal, parietal and occipital lobes of the brain using scalp electrodes using the 10–20 lead position/location system via a commercially available EEG machine (Mitsar-EEG-201) at a sampling frequency of 500 Hz. The electrode locations were, frontal (F3, F4), parietal (P3, P4), temporal (T3, T4) and occipital (O1, O2) lobe of the brain. EEG signals were filtered using sixth-order band pass filter with a pass band range of 0.5 Hz and 30 Hz. Wavelet packet decomposition of Daubechies 8 (db8) family was adopted as the mother wavelet for EEG data analysis. Based on the frequency ranges, the root mean square (RMS) of beta (13–30 Hz) band were obtained. The ratio between $((\alpha + \theta)/\beta)$ were also estimated.

2.4   Statistical Analysis

To perform the statistical analysis, a normality of distribution was checked using Shapiro-Wilk normality test on all the subjective and objective data acquired. It was found that the subjective data were normally distributed whereas objective data were not normally distributed. Friedman test, a non-parametric was performed to determine the significant differences of RMS values of EEG signals of four lobes (frontal, temporal, parietal and occipital) of the brain. Post-hoc analysis was performed using Bonferroni correction at significance levels of $p < 0.05$.

3   Results

In the objective evaluation part, the RMS values of beta activity from all the four lobes of the brain were estimated based on EEG. It has been observed in the frontal part of the brain that the RMS value of beta activity was least and shown significant difference $(p < 0.05)$ for the video of low beam car with strip (LBCS) and high beam car with strip (HBCS) in frontal, temporal and occipital lobes as compared to rest of the scenario. Parietal lobe did not show any significant difference in this scenario. RMS value of beta activity was highest for high beam car (HBC) and two bikes high beam (BHB). However, more or less same trend was followed by the parietal, temporal and occipital lobes of the brain (Figs. 7 and 8.). RMS value of the ratio index $((\alpha + \theta)/\beta)$ been

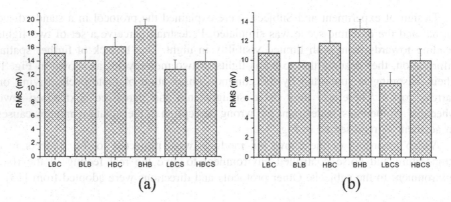

Fig. 7. Mean RMS values and standard error of relative wavelet packet coefficient in beta band on (a) frontal lobe and (b) parietal lobe of brain plotted against the six video (low beam car (LBC), two bikes low beam (BLB), high beam car (HBC), two bikes high beam (BHB), low beam car with strip (LBCS), high beam car with strip (HBCS)) which was shown to the subject on the projector.

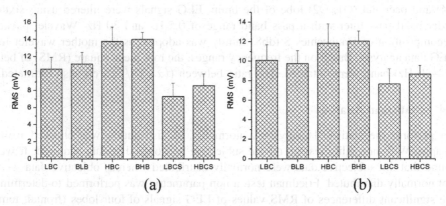

Fig. 8. Mean RMS values and standard error of relative wavelet packet coefficient in beta band on (a) temporal (b) occipital lobe of brain plotted against the six video (low beam car (LBC), two bikes low beam (BLB), high beam car (HBC), two bikes high beam (BHB), low beam car with strip (LBCS), high beam car with strip (HBCS)) which was shown to the subject on the projector.

observed high in the frontal part of the brain and shown significant difference ($p < 0.05$) for the video of low beam car with strip (LBCS) and high beam car with strip (HBCS) in frontal, temporal and occipital lobes as compared to rest of the scenario.

4 Discussion

Minimal lighting condition on the road in night time leads pedestrians difficult in taking decision while crossing the road especially in rural areas. It becomes very tough for the pedestrian to guess whether a car or two two-wheeler coming in parallel approaching them. According to a study, it is found that pedestrians are 3 to 6.75 times more at risk to accidents in the dark than in daylight [4]. Place of occurrence has also the impact on the road accidents. According to [15], it was found that 15.9% of fatal road accidents were reported near residential area in mega city in India among those 10.3% accidents reported due to pedestrian crossing and 5.4% accident reported near schools/college/other educational institutions. These statistics are also the concern for researchers from pedestrian safety point of view.

Several studies have been performed so far to address the difficulty faced by drivers in recognizing pedestrian in night time but surprisingly to our knowledge, no study has been done to address the pedestrian dilemma towards type of vehicle approaching towards pedestrian [16]. Focus of this study is on the pedestrian's dilemma and difficulties while crossing road in night condition scenario.

Fear of misconstruing the bikes as four wheelers may be a factor that would have caused these errors. This fear may be due to the increased occurrence of such accident scenarios. On the other hand, the misidentification of the high beam car with strips may be due to one of two reasons. The reason could be either that there is a reduced number of four wheelers using such strips, or the lack of knowledge on part of the observer of such lighting methodologies on four wheelers. It is also interesting to note that drivers were not confused in the 'low beam car' context and the 'low beam car with strips' context. These two observations may have implications for road safety as the high beam of four wheelers is activated on all highways in many countries and particularly in India.

Placing active light source in between the headlights of the car not only make pedestrian to recognize correctly but also helps in speeding up the decision making process. Glare also plays a vital role in taking decision and makes correct response by the pedestrian. It has been observed in this study that pedestrian's correct response towards the low beam car and two parallel bikes with low beam was high and time taken to make this correct response were less as compared to the high beam car and two parallel bikes with high beam.

All the four lobes of the brain have distinct and special function. Movement, emotion and problem solving is related to the frontal lobe whereas, parietal lobe is linked with recognition and perception of stimuli. The temporal lobe is associated to memory, interpreting sounds and languages whereas occipital lobe is related with vision responses [17]. EEG was used to track the brain activity of the pedestrian during visualization of six different videos.

In this study, we investigated EEG activity in frequency bands, such as beta, in four lobes of the brain. Beta activity was taken in this study since beta waves can be observed under a state of exhaustion and an ineffective state of cognition [18, 19]. Beta wave is related to concentrated mental activity such as nervousness, anxiety and it has been traced mostly during alert and wakeful state. It has been observed that the beta activity of the frontal lobe was highest for "two bike with high beam" whereas beta activity of the "low beam car with LED strip" and "high beam car with LED strip" was seen lowest among all the scenario of the video shown to the participants (Fig. 7.). Continuous guessing about the vehicle approaching to the pedestrian would have increased the cognitive demand of the pedestrian.

Higher beta activity when the scenario "two bike with high beam" was presented clearly indicates that the pedestrian's brain was highly active, nervous and unease which makes the pedestrian confuse. Lower beta activity during exposure to "low beam car with LED strip" and "high beam car with LED strip" clearly indicates that the pedestrian did not take much time and effort to take right decision. However same trend has been observed in all the three lobes (parietal, temporal and occipital) of the brain in which the beta activity was highest for "two bike with high beam" whereas beta activity of the "low beam car with LED strip" and "high beam car with LED strip" was seen lowest.

In this study, videos were recorded for different scenario and were shown to the subjects on projector. Subjects were conditioned and their focusing effort was only on this activity. However in real world there could be multiple distractions which will make it even worst. Therefore it is important to fix illuminators or LED strip in between headlights. Currently the law requires getting illuminators or LED strip for commercial vehicle (CV). This works well if the opposite vehicle is able to illuminate the marker to alert the pedestrian and hence to reduce the pedestrian accidents.

5   Conclusion

Drivers may not be the cause of accidents all the time. Sometimes, accident happens due to confused pedestrian too. There is a possibility of a mixed problem of both. Pedestrian taking more time to cross the road than estimated by the driver.

This study highlights the pedestrian perception towards juxtaposed vehicles. There is a need of fixing LED strip between head light of the car or other commercial vehicle. It was observed that the pedestrian took very less time to recognize the vehicle (car or two-wheeler) approaching him when LED strip was fixed in between head light of car. From the EEG study, it was observed that the beta activity was less when participants were shown the video of "low beam car with LED strip" and "high beam car with LED strip" which confirms that the pedestrian was less confused, nervous and calculative while crossing roads.

There should be LED strip between headlights of vehicle but not to be enforced. Large chunk of public have less awareness. This paper gives specific details of the importance of putting LED strip. These can eventually help in avoiding road accidents thus escalating transportation safety.

References

1. Road Accidents in India – 2015 (2016) Ministry of Road Transport and Highways, Government of India
2. National Highway and Traffic Safety Administration's (NHTSA) (2015) National Center for Statistics and Analysis, Washington, D.C. www.nhtsa.gov
3. World Health Organization (WHO) (2013) Global status report on road safety: supporting a decade of action: summary
4. Sullivan JM, Flannagan MJ (2002) The role of ambient light level in fatal crashes: inferences from daylight saving time transitions. Accid Anal Prev 34(4):487–498
5. Roudsari BS, Mock CN, Kaufman R (2005) An evaluation of the association between vehicle type and the source and severity of pedestrian injuries. Traffic Inj Prev 6(2):185–192

6. Nasar JL, Troyer D (2013) Pedestrian injuries due to mobile phone use in public places. Accid Anal Prev 57:91–95
7. Oh C, Kang YS, Kim B, Kim W (2005) Analysis of pedestrian-vehicle crashes in Korea. In: 84th annual meeting of the transportation research board, Washington, D.C.
8. Mohamed MG, Saunier N, Miranda-Moreno LF, Ukkusuri SV (2013) A clustering regression approach: a comprehensive injury severity analysis of pedestrian–vehicle crashes in New York, US and Montreal, Canada. Saf Sci 54:27–37
9. Shinar D (1984) Actual versus estimated night-time pedestrian visibility. Ergonomics 27 (8):863–871
10. Blomberg RD, Hale A, Preusser DF (1986) Experimental evaluation of alternative conspicuity-enhancement techniques for pedestrians and bicyclists. J Saf Res 17(1):1–12
11. Owens DA, Antonoff RJ, Francis EL (1994) Biological motion and nighttime pedestrian conspicuity. Hum Factors J Hum Factors Ergon Soc 36(4):718–732
12. Oxley JA, Ihsen E, Fildes BN, Charlton JL, Day RH (2005) Crossing roads safely: an experimental study of age differences in gap selection by pedestrians. Accid Anal Prev 37 (5):962–971
13. Balasubramanian V, Bhardwaj R (2018) Pedestrians' perception and response towards vehicles during road-crossing at nighttime. Accid Anal Prev 110:128–135
14. Investment Information and Credit Rating Agency of India (ICRA) (2016)
15. National Crime Research Bureau Ministry of Home Affairs (NCRB) (2016) Accidental deaths and suicide in India. www.ncrb.nic.in
16. Theeuwes J, Alferdinck JW, Perel M (2002) Relation between glare and driving performance. Hum Factors J Hum Factors Ergon Soc 44(1):95–107
17. Kar S, Bhagat M, Routray A (2010) EEG signal analysis for the assessment and quantification of driver's fatigue. Transp Res Part F: Traffic Psychol Behav 13(5):297–306
18. Eoh HJ, Chung MK, Kim SH (2005) Electroencephalographic study of drowsiness in simulated driving with sleep deprivation. Int J Ind Ergon 35(4):307–320
19. Klimesch W (1999) EEG alpha and theta oscillations reflect cognitive and memory performance: a review and analysis. Brain Res Rev 29(2):169–195

Sit-Stand Workstation for Office Workers: Impact on Sedentary Time, Productivity, Comfort and Feasability

Claire Baukens[1,2]([⊠]), Veerle Hermans[3,4], and Liesbeth Daenen[5,6]

[1] University College Odisee, Brussels, Belgium
claire.baukens@idewe.be
[2] Department of Ergonomics, Group IDEWE (External Service for Prevention and Protection at Work), Louvain, Belgium
[3] Department Head of Ergonomics, Group IDEWE (External Service for Prevention and Protection at Work), Louvain, Belgium
[4] Department of Experimental and Applied Psychology, Work and Organisational Psychology (WOPS), Faculty of Psychology and Education Sciences, Vrije Universiteit Brussel, Brussels, Belgium
[5] Department of Rehabilitation Sciences and Physiotherapy, Human Physiology and Anatomy (KIMA), Faculty of Physical Education and Physiotherapy, Vrije Universiteit Brussel, Brussels, Belgium
[6] Knowledge, Information and Research Center, Group IDEWE (External Service for Prevention and Protection at Work), Louvain, Belgium

Abstract. Sedentary behavior, also at work, increases the risk of cardiovascular disease, obesity, diabetes and even earlier mortality (e.g. Biswas et al. 2015). Employees should be encouraged to change and break up their sitting behavior. Therefore, the objective of this study is to determine if a sit-stand workstation can reduce sitting time at work. Secondary, feasibility and usability of the sit-stand workstation is evaluated.

A pre-post study was conducted with 20 office workers. After receiving an information session on the importance of good ergonomics and optimizing the ergonomic conditions, a sit-stand workstation was installed. Sitting time was measured using ActivPAL[TM] accelerometers at T0 and after 4 weeks of using the sit-stand workstation (T1). A questionnaire regarding sitting and movement behavior and performance indications was filled out at T0 and T1.

The results revealed a significant difference in sitting time between T0 and T1: sitting time decreased with almost 10% and standing time increased with 6%. Subjects revealed that they were more active at work and sat less at work due to using the sit-stand workstation.

The majority of the subjects (respectively 74% and 79%) mentioned that the use of the sit-stand workstation had no effect on their work productivity and quality of work. In addition, respectively 79% and 69% of the subjects revealed that they could perform easy computer tasks and read comfortably while using the sit-stand workstation. Otherwise, half of the subjects agreed that the sit-stand workstation is easy to use and they would use it in the future. Only 32% of the subjects found it useful in a typical office environment.

The results indicate that a sit-stand workstation is a helpful tool to decrease sitting time and increase standing time during working hours, although possible

© Springer Nature Switzerland AG 2019
S. Bagnara et al. (Eds.): IEA 2018, AISC 819, pp. 406–414, 2019.
https://doi.org/10.1007/978-3-319-96089-0_44

future use of the sit-stand workstation was questioned by half of the subjects, possibly due to its design.

Keywords: Ergonomics · Sit-stand workstation · Sedentary work

1 Relevance

With the increasing amount of office workers, more and more people increase their sitting behaviour. It is proven that increased sitting time at work, increases the risk of cardiovascular disease, obesity, diabetes and even earlier mortality (e.g. Biswas et al., 2015). Therefore, employees should be encouraged to change and break up their sitting behavior. There are several devices on the market to promote this: sit-stand tables, ergobikes, treadmill desks, etc. The effectiveness of these interventions is a popular theme in recent research. In the review study of Shrestha et al. (2016), it was concluded that a sit-stand desk can reduce sitting between a half to two hours per day. However, after 6 months the reduction in sitting is less than one hour during work. So the proposed guideline towards accumulating 2 h/day of standing and light activity (e.g. light walking) during working is not achieved (Buckley et al. 2015). Therefore, it is important to know why people do not reduce their sitting time sufficiently and what the feasibility is of implementing solutions to decrease sitting in practice.

2 Objectives

With this study, we want to (1) study **if a sit-stand workstation can reduce sitting time at work** and (2) evaluate the feasibility and use of a sit-stand workstation. We hypothesize that (1) participants will decrease their sedentary time at work as a result of using the sit-stand workstation and (2) participants will find the sit-stand feasible and acceptable for use in the work environment.

Research questions:

- Are workers sitting less using the sit-stand workstation?
- Are users changing the number of changes between sitting and standing per day using the sit-stand workstation?

What is the influence of using the sit-stand workstation on performance indicators (productivity and quality), comfort and physical activity.

3 Methodology

3.1 Subject Group

Twenty office workers from the same company (PSA, Antwerp, Belgium) volunteered to take part in the study. The sample size was based on other studies focusing on interventions to reduce sedentary behavior (e.g. Carr et al. 2012).

Inclusion criteria were:

- 20–65 years
- at least 30 h/week employment contract
- general good health situation according to the individual medical records
- no sit-stand table or dynamic office workstation at present

All subjects signed an informed consent, in which anonimity and confidentiality was guaranteed (approved by ethical committee of IDEWE, 14/12/2016).

3.2 Experimental Condition

At baseline (T0), subjects were asked to wear an activity tracker (ActivPal™ Micro) for one week to analyze their usual daily sitting and movement pattern. A questionnaire was filled in regarding sitting and movement behavior, physical discomfort and performance indications. All subjects received an information session on the importance of good ergonomics (including general advice on reducing sitting behavior). It was explained how to use the sit-stand and an individual workplace visit was arranged to optimize the working environment. Subjects used the sit-stand for 4 weeks. Post-measurements were done (T1) including wearing the activity tracker for one week and filling in the self-reported questionnaires (identical to those at baseline) and a feasibility questionnaire.

3.3 Outcome Measures

Mean sitting time (in min and in % of workday) and mean standing time (in min and in % of workday) were measured, during 7 days at week 1 (T0) and week 4 (T1), using the ActivPal™ Micro. Also the number of changes between sitting and standing were measured. The Occupational Sitting and Physical Activity Questionnaire (OSPAQ) was used to evaluate time to sit, stand and move (Chau et al. 2013). The OSAPQ has good test-retest reliability to evaluate sitting time (ICC = .89) and standing time (ICC = .90) at work, and sufficient validity. The OSAPQ is a useful instrument to measure sitting, standing and walking time at work in addition to an objective measurement such as the activity tracker (van Nassau et al. 2015).

Feasibility of the sit-stand was evaluated using a questionnaire based on the study of Carr et al. (2012). In this questionnaire, also productivity and comfort indicators were questioned. The results of his/her questionnaire were discussed with the subject at the end of T1. Additional remarks were noted.

3.4 Statistical Analysis Techniques

Statistical analysis was performed using SPSS for Windows (version 23, SPSS Inc; Chicago, IL). Baseline data were analyzed by using descriptive statistics. Normality of the variables was tested with the Kolmogorov-Smirnov test. A Paired samples t-test was used to compare the mean of sitting time between T0 en T1. A Wilcoxon signed rank test was used to compare the number of changes between sitting and standing between T0 en T1. The significance level was set at 0.05.

4 Results

4.1 Demographic Data

Twenty subjects participated in this study, including thirteen woman and seven men. Fourteen participants had a fulltime work schedule.

4.2 Sitting and Standing Behavior Measured with ActivPAL

A significant difference in sitting time was found between T0 (without the sit-stand workstation) and T1 (with the use of the sit-stand workstation): sitting time decreased with 7.95% and standing time increased with 6.75% at T1 compared to T0 (see Table 1). In other words, workers sit on average 25 min less on a working day of 8 h when using the sit-stand workstation.

Table 1. Mean sitting and standing time (% per workday) at T0 and T1 (activPAL^{TM})

	T0	T1	Difference (T1 − T0)	
Sitting time X (SD)	76.75 (6.82)	68.81 (8.62)	−7.95 (9.48)	p = .007
Standing time X (SD)	24.01 (7.15)	30.76 (8.61)	6.75 (8.96)	p = .003

Regarding the number of changes from sitting to standing, no significant differences were found (p = .464). Subjects changed 3.4 times/hour from sitting to standing at T0 and 3.2 times/hour at T1 (see Table 2). This means respectively 27 and 25 times on an average workday of 8 h. Similar results were found for the changes from standing to sitting (see Table 2).

Table 2. Mean sit to stand and stand to sit changes (per hour per workday) at T0 and T1 (activPAL^{TM})

	T0	T1	Difference (T1 − T0)	
Sit to stand X (SD)	3.46 (0.94)	3.27 (1.00)	−0.19 (1.04)	p = .464
Stand to sit X (SD)	3.46 (0.94)	3.28 (1.00)	−0.17 (1.05)	p = .506

4.3 Sitting and Standing Behavior Measured with OSPAQ

The results of the OSPAQ questionnaire show that the participants also indicate that they reduced their sitting time significantly and increased their standing time significantly at T1 compared to T0, with differences of 10.05% and 8.32% respectively (see Table 3).

Table 3. Mean sitting and standing time (% per workday) at T0 and T1 (OSPAQ)

	T0	T1	Difference (T1 − T0)	
Sitting time X (SD)	84.37 (8.23)	74.32 (13.23)	−10.05 (11.15)	p = .001
Standing time X (SD)	6.42 (3.93)	14.74 (10.03)	8.32 (9.17)	p = .001

4.4   Performance

The majority of the subjects mentioned that the use of the sit-stand workstation had no effect on their work performance such as work productivity (i.e. 74% of the subjects) and work quality (i.e. 79%) (see Fig. 1).

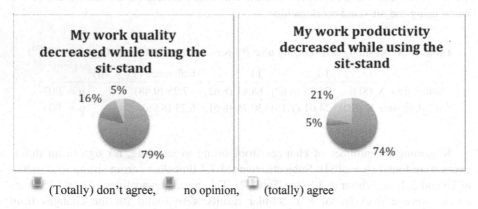

(Totally) don't agree,   no opinion,   (totally) agree

Fig. 1. Work productivity and work quality using the sit-stand workstation

Furthermore, subjects revealed that they could perform easy computer tasks such as typing simple text and data input (i.e. 69% of the subjects) and read comfortably (i.e. 79%) while using the sit-stand (see Fig. 2).

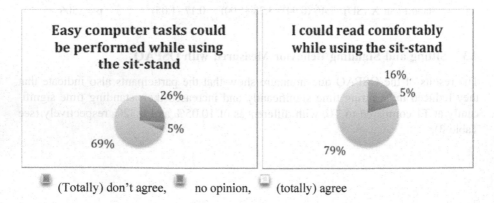

(Totally) don't agree,   no opinion,   (totally) agree

Fig. 2. Performance of computer tasks using the sit-stand workstation

4.5   Physical Comfort

Regarding physical comfort, the use of the sit-stand was not related to physical discomfort: the majority of the subjects didn't have more back, neck/shoulder, hip/legs, muscle/joint pain on days using the sit-stand, and 68% of the subjects didn't feel more tired on days using the sit-stand (see Fig. 3).

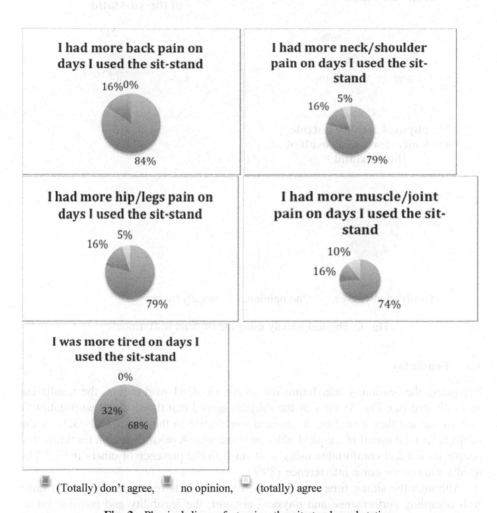

I had more back pain on days I used the sit-stand
16% 0%
84%

I had more neck/shoulder pain on days I used the sit-stand
16% 5%
79%

I had more hip/legs pain on days I used the sit-stand
5%
16%
79%

I had more muscle/joint pain on days I used the sit-stand
10%
16%
74%

I was more tired on days I used the sit-stand
0%
32%
68%

▪ (Totally) don't agree,   ▪ no opinion,   ▫ (totally) agree

Fig. 3. Physical discomfort using the sit-stand workstation.

4.6   Physical Activity

Most subjects mentioned that they were more active at work and sat less due to the use of the sit-stand workstation. However, there was no effect on the amount of physical activity out of work. Only 2 subjects mentioned that they were more physically active outside work as a result of the sit-stand workstation (see Fig. 4).

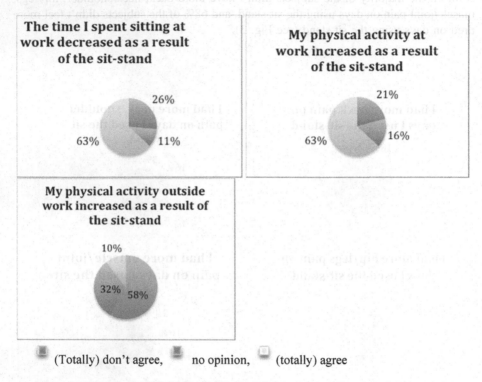

The time I spent sitting at work decreased as a result of the sit-stand

26%
63%   11%

My physical activity at work increased as a result of the sit-stand

21%
63%   16%

My physical activity outside work increased as a result of the sit-stand

10%
32% 58%

▣ (Totally) don't agree, ▣ no opinion, ▢ (totally) agree

Fig. 4. Physical activity using the sit-stand workstation.

4.7   Feasibility

Regarding the feasibility and future use of the sit-stand workstation, the results are more divided (see Fig. 5). Half of the subjects agreed that the sit-stand workstation is easy to use and they would use a sit-stand workstation in the future. Only 32% of the subjects found it useful in a typical office environment. A possible reason for this is that people do not feel comfortable using a sit-stand in the presence of others (63%). The results showed no noise interference (89%).

Although the sitting time was decreased and subjects responded in general quite well regarding performance and physical comfort, the feasibility and possible future use of the workstation was questioned by half of the subjects. Therefore, at the end of T1, the results of the feasibility were discussed with each subject individually and additional remarks regarding feasibility were noted. Figure 6 gives an overview of the mentioned remarks.

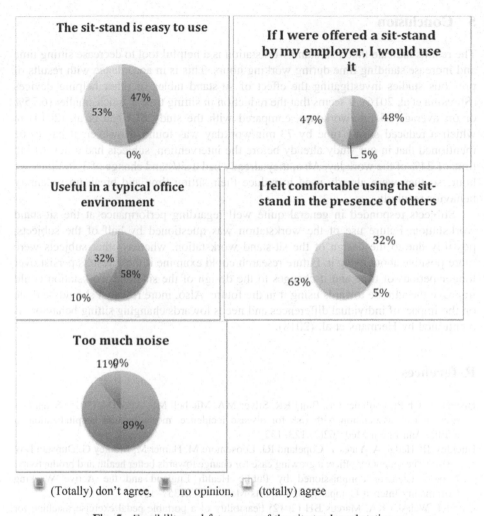

(Totally) don't agree,   no opinion,   (totally) agree

Fig. 5. Feasibility and future use of the sit-stand workstation

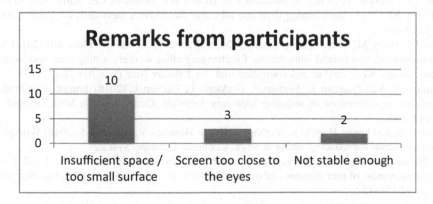

Fig. 6. Most common remarks regarding the sit-stand workstation

5 Conclusion

The results indicate that a sit-stand workstation is a helpful tool to decrease sitting time and increase standing time during working hours. This is in accordance with results of previous studies investigating the effect of sit-stand tables or other helping devices (Shrestha et al. 2016). It seems that the reduction in sitting time is much smaller (6.75% or on average 25 min/workday), compared with the study of Chau et al. (2014) in which a reduced sitting time by 73 min/workday was found. However, it has to be mentioned that in our study already before the intervention, subjects had a non-sitting time of 24% of the workday. Also, they already had a sit/stand change of 3.46 times per hour, so they were already used to reduce their sitting time and break up sedentary behavior regularly.

Subjects responded in general quite well regarding performance at the sit-stand workstation. Future use of the workstation was questioned by half of the subjects, possibly due to the design of the sit-stand workstation, whereas other subjects were more positive about using it. Future research could examine if these results persist over longer periods of time and if changes to the design of the sit-stand workstation could improve the attitude towards using it in the future. Also, more research should be done on the impact of individual differences and needs towards changing sitting behavior, as mentioned by Hermans et al. (2018).

References

Biswas A, Oh PI, Faulkner GE, Bajaj RR, Silver MA, Mitchell MS, Alter DA (2015) Sedentary time and its association with risk for disease incidence, mortality, and hospitalization in adults. Ann Intern Med 162(2):123–132

Buckley JP, Hedge A, Yates T, Copeland RJ, Loosemore M, Hamer M, Bradley G, Dunstan DW (2015) The sedentary office: a growing case for change towards better health and productivity. Expert statement commissioned by Public Health England and the Active Working Community Interest Company. Br J Sports Med 49(21):1357–1362

Carr LJ, Walaska KA, Marcus BH (2012) Feasibility of a portable pedal exercise machine for reducing sedentary time in the workplace. Br J Sports Med 46:430–435

Chau JY, Grunseit AC, Chey T, Stamatakis E, Brown WJ, Matthews CE, Adrian EB, van der Ploeg HP (2013) Daily sitting time and all-cause mortality: a meta-analysis. PLoS ONE 8 (11):1–14

Chau JY, Daley M, Dunn S, Srinivasan A, Do A, Bauman AE, van der Ploeg HP (2014) The effectiveness of sit-stand workstations for changing office workers' sitting time: results from the Stand@Work randomized controlled trial. Int J Behav Nutr Phys Act 11:127–137

Hermans V, Van Naemen L, Seghers J, Godderis L, Daenen L (2018) Impact of individual motivation differences on reducing sedentary behavior. Occup Environ Med 75(suppl 2): A47–A48

Shrestha N, Kukkonen-Harjula K, Verbeek J, Ijaz S, Hermans V, Bhaumik S (2016) Workplace interventions for reducing sitting at work. Cochrane Database Syst Rev (3)

van Nassau F, Chau JY, Lakerveld J, Bauman AE, van der Ploeg HP (2015) Validity and responsiveness of four measures of occupational sitting and standing. Int J Behav Nutr Phys Act 12:144–152

Analyses of Time Use in Informal Economy Workers Reveals Long Work Hours, Inadequate Rest and Time Poverty

Jonathan Davy[1]($\boxtimes$), Didintle Rasetsoke[1], Andrew Todd[1],
Tasmi Quazi[2], Patric Ndlovu[2], Richard Dobson[2], and Laura Alfers[3]

[1] Department of Human Kinetics and Ergonomics, Rhodes University,
Grahamstown, South Africa
j.davy@ru.ac.za
[2] Asiye e Tafuleni, Durban, South Africa
[3] WIEGO, Durban, South Africa

Abstract. Workers in the informal economy operate under difficult and unsafe working conditions. This and other systemic challenges can negatively affect their work ability. One aspect of informal work that is not well understood is working time, an important element of any work system as it dictates how much time is available for rest and how it contributes to time poverty. Therefore, the aim of this study was to determine the working time of informal workers through a time use survey. This cross sectional, descriptive study recruited ten informal worker participants from Warwick Junction, in Durban, South Africa. Traders completed a participatory time-use survey method and an activity clock, reporting on activities over a normal and busy 24-h period. More specifically, participants reported total working time, commuting time, personal time, time spent doing domestic chores and reported sleep duration, while accounting for the time allocated to each activity. On average, workers reported working 6 to 7 days per week and extended work hours. Reported sleep time varied between 6 ($\pm$1.4) and 7.2 ($\pm$1.9) h per night on busy and normal nights respectively, which provided evidence of time poverty. Female workers reported having higher demands on their time than men. While the sample was small and the data self-reported, the study provides evidence of extended work hours, inadequate rest and time poverty faced by informal workers and in particular, women. Further research should empirically explore the effects of the extended work hours and reduced time for recovery amongst informal workers.

Keywords: Informal economy · Working time · Time use
Occupational hazards

1 Background

Time use surveys can provide a comprehensive and specific overview of human activities that are sometimes difficult to obtain through other surveys [1]. While they provide insights into formal work and its associated activities, time use surveys play an important role in identifying and estimating the value and impact of other types of

© Springer Nature Switzerland AG 2019
S. Bagnara et al. (Eds.): IEA 2018, AISC 819, pp. 415–424, 2019.
https://doi.org/10.1007/978-3-319-96089-0_45

work, including that which is informal, unpaid or revolves around care work, voluntary work or the household or reproductive economy [1–3]. Such information is vital to understand the extent and characteristics of the workforce in focus, to measure socio-economic changes, determine the wellbeing and level of poverty of a particular group, assist with the valuation of unpaid work and contribute to macro policy making that supports work activities not formally recognized [1]. Although research in this area is limited, [4] reported that the average working hours in a week was 53 h (±9 h a day, for 6 days) and the working day length was controlled by the workers, with more than half of the workers reporting that they sometimes worked extended hours (over 12 h). Other studies [5] have reported work duration of between 10.56 and 11.78 h, which includes both paid and unpaid work.

In the context of low to middle income countries, time use data has and continues to highlight how the poor and in particular, women, suffer time poverty [3, 6]. According to Kes and Swaminathan [2], time poverty refers to how competing claims on an individual's time limits the choices they make (around how time is allocated), which is usually accompanied by some compromise, resulting in either increased work intensity or the neglect of other important activities, such as those that are economically pro-ductive, leisure or recreation focused or those dedicated to sleep and rest [3]. While difficult to define, Gammage [3] uses the threshold of 12 h (combined paid and unpaid work) as indicator of time poverty. The work hours in this case (paid and unpaid work) are a necessity, not a choice, an important distinction made by Bardasi and Wodon [7]. While time poverty is experienced by both genders, particularly those from an impoverished background, certain social and cultural norms tend to dictate, rather rigidly, the expected roles of men and women [8]. In this context, women are restricted to activities that relate to the domestic sphere, while men work beyond this and gen-erally engage in more economically productive activities [2, 3]. This trend is evident in the context of the informal economy, where women, largely, are restricted to activities that have a low earning potential (over and above their domestic responsibilities) and further increase the risk of remaining in poverty [3, 9].

In addition to providing important insights into different types of activities, time use data can provide important information on how and why work (and other activities) are arranged the way they are [1, 3], the constraints that drive this time allocation and the hazards and risks that workers may be exposed to as a result. This is of particular importance in the informal economy, which, for the purposes of this paper, refers to activities that are regulated, semi-regulated or not regulated at all, characterized by an ease of entry, that have small scale operations which are labour intensive and make use of adapted technology [10, 11]. Understanding time use in this sector and its drivers is important, given that those involved in this sector face many challenges. These include low income generating activities, seasonal, irregular or sporadic work, limited job security and little to no social protection [11–13]. From an infrastructural perspective, street traders, which refers to operating businesses in public spaces (also referred to as informal traders) more often than not have limited access to basic amenities, such as shelter for their goods, toilets, water, reliable, safe sources of energy and fire safety equipment [14]. Not only does this affect their productivity, but it also has implications for their health. The challenges associated with poor infrastructure are accentuated by the plethora of occupational hazards faced by street traders [15, 16]. These include exposure to physical, environmental, psychosocial, organizational and chemical hazards [12–14, 16]. As a result, physical injuries, reduced health status, work-related musculoskeletal disorders, chemical exposure and psychosocial stress are commonly reported challenges amongst this group [12, 13].

An important aspect of any working environment is how work is arranged and scheduled. More specifically, this refers to the working time (and non-working time) arrangements. Working time, if atypical (such as night, early morning or extended work) has been associated with reduced alertness, performance and productivity [17, 18] and can result in reduced health [19] and an increased risk of accidents [18, 20]. Ultimately, extended or abnormal work hours affect the ability for individuals to recover, most importantly through sleep (an often compromised activity which is necessary for health and general functionality). Reduced sleep, which refers to sleep that is less than between 7 and 9 h in length per night, can have a negative effect on alertness, neuro-behavioral functioning, mood and general motivation [21]. This can, in turn, affects work performance and safety while at work [12, 22, 23] and can contribute to a reduction in health status [24]. Furthermore, there is evidence that individuals from a lower socioeconomic status also generally experience poorer quality sleep, the reasons for which are discussed in Grandner et al. [25].

To date, there has been extensive research into how different communities allocate their time, particularly those involved in informal economy activities. However, there is limited research that has investigated how the time allocated to work and other activities may affect sleep and recovery in the context of the informal economy. Furthermore, it is not clear whether this would differ between male and female traders and whether it would contribute to time poverty. Therefore the purpose of this study was firstly, to analyze time usage in a select informal economy setting, to be able to quantify working and recovery time and secondly, to determine if there were differences between the genders and whether there was evidence of time poverty.

2 Method

2.1 Experimental Design

This study adopted a cross-sectional, descriptive and quantitative time use survey that was field based at Warwick Junction, a transport hub in the city of Durban in South Africa.

2.2 Time Survey Method

To determine the time spent on different activities during a normal 24-h day, an activity clock was adopted. The clock was divided up into 24 h and provides a useful, visual method to determine how time is allocated to different activities over a day [26]. This method was minimally intrusive and did not require constant monitoring on the part of participants or researchers but merely required an account of time spent on different activities carried out on typical days [26]. For the purposes of this study, traders were asked to indicate when in a day and for how long they performed the following activities: work and work-related activities, commuting, household and child-care activities, personal activities (grooming, bathing, using the toilet, eating) and sleep. The traders were further asked to distinguish between normal and busy days, while also reporting on when in a month and a year they were busiest. These data were supplemented with field notes obtained through semi-structured interviews that provided important details on the factors that influenced the reported allocation of time to the different activities.

2.3 Participants Characteristics and Recruitment

Upon ethical research clearance by the Human Kinetics and Ergonomics Department at Rhodes University, the data collection commenced. Research staff from a collaborating Non-Governmental Organisation (Asiye e'Tafeleni), a community partner that works alongside the informal traders at Warwick Junction, assisted by approaching each of the traders through a purposive sampling approach. There were no restrictions or limitations on who could participate. There was also no attempt at selecting participants on any basis other than that they worked in the informal economy at Warwick Junction and were willing to participate in learning about the study. Ten traders (5 males and 5 females) who were workers from the different sectors of the informal economy at Warwick Junction were recruited.

The specific jobs of these individuals were one traditional hat maker, one storage operator, one Bovine head cook (this refers to a cows head, which is a Zulu delicacy), one Mealie cook and vendor (Mealies are also referred to as maize or corn, the cooked mealies are conveyed in and sold from supermarket trolleys), two cardboard and plastic recyclers (who use trolleys to collect and convey recyclables to central areas where they are purchased by middle-agents), one water transporter (who transports up to 400 L of water nearly 5 km every day), one bead maker and seller and two barrow operators (who use self-made carts to transport informal traders goods from the storage facilities to their trading posts and back again every day). In addition to these jobs, most of the traders performed additional jobs (such as farming, carpentry and clothing and apparel sales.

2.4 Approach

Prior to consenting to participate in the study, all traders were verbally and in writing, given an overview of the aims and objectives of the study in a language of their choice (IsiZulu, IsiXhosa or English). The semi-structured interview and time usage determination was administered by researchers from Asiye e'Tafuleni who were fluent in the three languages. Following consent, basic demographic information (age, sex, sector of employment in Warwick, number of years in informal employment and the number of dependents of the participants) was obtained. Following that, participants were provided with a daily activity clock. Participants were required to visually quantify (draw out) the amount of time spent on activities in various activity categories during a 24-h period (for normal and busy days), while the interviewer wrote down the details and duration of the activities. Traders were then asked follow-up questions regarding the different activities

Following data collection, all data were reduced. Descriptive statistics (mean and standard deviation) were determined, with male and female data being compared. Common themes that emerged from the field notes were identified to determine common challenges and patterns of behaviour related to the traders reported time use.

3   Results

3.1   Basic Demographic Information

The study consisted of an equal number of male and female participants (n = 10) who worked in various sectors of the informal economy at Warwick Junction. The age range of participants was 27–61 years with a mean of 47.10 ± 11.84 years. The mean number of dependents was 6.20 ± 2.68 and participants had spent 16.70 ± 7.87 years working in the informal economy at Warwick Junction. The activities performed by these traders in this study ranged from regulated (bovine head cook, Hat maker, bead maker, storage operator, informal recycler), to semi-regulated (mealie cook, water transporter) and unregulated (barrow operators).

3.2   Time Use Survey Results

Most traders reported that their busiest work periods occurred in the lead up to and over the festive holidays (September to December and over Christmas and Easter) or other important cultural holidays (Heritage Day, which falls in September). Busy periods were also linked to the availability of goods – the Mealie cooks reported their busiest periods to be between December and April, which coincides with the season when maize is available. Monthly, business varied between the different sectors, but most reported month end to be busiest periods. This was not the case for the hat maker and bead seller, who found that at month end, consumers favoured necessities (groceries) over their luxury items.

All but two of the traders worked 7 days a week (the others worked 6 days). With respect to reported working time, on average, traders reported a total working time that ranged from 53–87 h a week. The Bovine head cook reported the highest weekly working time (87 h per week), while the water transporter had the least with 53 h per week. This translates to a mean reported, daily working duration of 12 (±2.15 h).

Table 1. Summary of reported time use during normal and busy periods (means ± standard deviation). It also includes the most extreme time usage reports out of the whole sample.

Activity	Busy	Normal	Extreme
Work (h)	13.5 (2.13)	10.7 (2.78)	15.5
Day start	04h12	04h33	03h00
Day end	22h27	21h45	00h00
Commute (h)	1.1 (1.32)	1.35 (1.64)	4.5
Household (h)	0.975 (1.53)	1.32 (1.44)	0
Personal (h)	2.35 (1.23)	3.55 (1.81)	0
Sleep (h)	6.1 (1.35)	7.15 (1.93)	4

Table 2. Comparison between genders regarding time use around work-related activities (reported mean duration, start and end times), commuting time, household/childminding activities, activities around personal care and reported sleep during normal and busy periods (means ± standard deviation).

	Male		Female	
	Busy	Normal	Busy	Normal
Work (h)	12.94 (2.3)	10 (2.9)	14 (2.17)	11.4 (1.1)
Day start	04h00	04h00	04h24	05h06
Day end	22h00	21h18	22h54	22h12
Commute (h)	1.24 (1.5)	1.5 (1.8)	1 (1.3)	1.2 (1.7)
Household (h)	0.3 (0.67)	0.43 (0.6)	1.65 (1.9)	2.2 (1.6)
Personal (h)	2.7 (1.7)	4.4 (2.3)	2 (0.61)	2.7 (1)
Sleep (h)	6.7 (1.78)	7.4 (2.1)	5.5 (.5)	6.9 (2.1)

Reported working time differed between busy and normal days (13.5 vs 10.7 h), with day start (waking) times on busy days being 21 min earlier than those on normal days (Table 1). Activity reportedly ended 42 min later on busy days, relative to normal days. With respect to reported sleep, traders reported obtaining 7.15 (±1.93) h during normal working periods, and 6.1 (±1.35) h during busy periods. As a result, this sample of traders was on average awake for between over 16 and 18 h each working day. While not all traders commuted (as some lived at the station), one trader (the Bovine head cook) reported spending 4.5 h a day commuting, over and above working hours which, during busy periods, was over 15 h.

Comparison Between Male and Female Traders
Female traders reportedly worked longer hours than the males during both normal and busy work periods respectively (Table 2). As a consequence, reported sleep time for females was generally shorter than that reported by the males for both normal and busy periods respectively (Table 2). Female traders also generally spent more time doing household related activities than male traders. This meant that female traders had less personal time than males irrespective of the demands on them from work.

4 Discussion

This study aimed to explore time use in a small cohort of informal workers at Warwick Junction in Durban with the intention of understanding the time allocated to work, personal and recovery related activities and secondly, to determine whether there were differences in time use between male and female traders. Despite the small sample and the descriptive nature of this study, the results of this study have highlighted that reports of extended work shifts (over 12 h), inadequate sleep, lack of personal time and long commutes are commonplace among traders in the informal economy at Warwick Junction. Furthermore, the majority of the traders worked seven days a week, with most having more than one income generating activity, thus moonlighting was common.

All informal traders in this study reported working extended shifts (which refer to work shifts longer than 8 h). While in some instances, extended work shifts are pre-ferred over conventional 8-h shifts the preference revolves around the fact that the total working time within a week will be restricted through legislation – in South Africa, this is 45 h, with 10 h over time, as outlined in the Basic Conditions of Employment Act [27]. In the context of the informal economy, no such legal limits exist, resulting in some traders working in excess of 80 h per week regularly, largely by necessity. Extended workdays and weeks can result in an increased risk of accidents at and away from work, sleepiness and reductions in alertness while at work, reduced quality and quantity of sleep, increased exposure to risks within the working environment and ultimately, reductions in health [20]. While the prevalence of accidents related to informal work was not explored in this study, there is ample of evidence that informal workers face extensive physical, chemical, social and occupational hazards while working in public spaces [16]. The necessity to generate income, therefore, necessitates the risks associated with extended work shifts in these public spaces.

This study also found evidence that inadequate sleep was common amongst the traders, particularly during busy periods, which occurred each week and towards month end and due to the unconducive working/living environment. The contributing factors to reduced and poor quality of sleep corresponded with previous research and included; having to work extended work shifts [20], early starts [28] and late finishes [29], the need for extended commuting [30], noise from the immediate living environment, particularly in lower socioeconomic areas [25] and musculoskeletal pain [31]. This was also affected by whether traders slept at their workplace, more specifically Berea Station or not. Not obtaining adequate sleep is associated with reduced alertness and increased sleepiness during the day, which can accumulate over time in the form of sleep debt [32]. This may be associated with an increased risk of accidents at work [19], but, more importantly, chronic inadequate sleep can affect individual health status, which may, in turn, affect workability [33]. While this was not part of this study, it would be important for future research to empirically quantify sleep in informal traders and understand the prevalence and effects of long-term sleep loss on trader health and workability.

There was also evidence of time poverty in this small cohort of traders. On average, irrespective of whether it was a normal or busy period, the traders reported extended work shifts, reduced time for domestic and personal activities and a reduced time allocation for sleep (Table 1), criteria that supports the definition of time poverty outlined in previous research [2, 3, 7]. Irrespective of gender differences, the extended working time appears to be by necessity rather than by choice, particularly amongst female traders, observations supported by previous literature [1, 3, 8]. With reference to Table 2, irrespective of the intensity of work (normal or busy), female traders tended to work longer hours, have less time for personal activities, have a higher domestic workload and consequently, less time for sleep. While not a focus of this study, this disparity between male and female traders likely highlights the predominating social and cultural norms around gender roles, which imposes a heavy domestic load on top of extended work hours, which, as a result, increases their exposure to the various hazards and risks inherent in the working environment [15]. Furthermore, this time poverty limits sleep time, particularly in female traders, which may point to the fact that this cohort, and perhaps women in the informal sector in general, are chronically sleep deprived.

While this study highlights an important occupational and personal challenge faced by informal traders, specifically at Warwick Junction, there were some limitations associated with the study. The sample was small and the data self-reported, with some of the interviews lasting over three hours. However, the incorporation of the activity clock facilitated the collection of time use in a participatory manner. It was the first time that a study of this nature had been performed in this context and despite these limitations; the findings align with previous research.

5   Conclusion

Through this small time use study, we found that informal workers in the informal sector at Warwick Junction in Durban work extended shifts during both normal and busy periods, and that this has reduced the time available for personal and domestic activities, particularly sleep, all of which points to evidence of time poverty in this cohort. While the results are not generalizable to all informal workers at Warwick Junction, the overall implications of the extended working time limits recovery and increases the exposure of the workers to existing occupational and environmental hazards that are inherent to this sector. Over time, this may affect the health, well-being and the income generating ability of these workers, most especially for female traders. In order to better understand and improve working conditions (with regard to working time), more research is required, particularly to understand the South African context. Such findings are important for the formulation of policies and programs to promote, ensure important for the formulation of policies and programs to promote, ensure and to advocate for better working conditions in the informal economy. Furthermore, organizing the informal economy and acknowledging its role may ultimately help workers meet their basic needs by increasing their income generating ability and recognizing them as a vital part of the economy in countries like South Africa.

References

1. Hirway I (2009) Mainstreaming time use surveys in national statistical system in India. Econ Polit Wkly 44(49):56–65
2. Kes A, Swaminathan H (2006) Gender, time use, and poverty in Sub-Saharan Africa. World Bank Working Paper 73
3. Gammage S (2010) Time pressed and time poor: unpaid household work in Guatemala. Feminist Econ 16(3):79–112
4. Devey R, Skinner C, Valodia I (2003) Informal economy employment data in South Africa: a critical analysis. TIPS & DPRU Forum, 8–10 September, Johannesburg
5. Encuesta Nacional sobre Condiciones de Vida (ENCOVI) [National Living Conditions Survey] 2000, 2006. National Institute for Statistics, Guatemala. http://www.ine.gob.gt/. Accessed Jun 2010 – cited from Gammage 2010
6. Blackden CM, Wodon Q (eds) (2006) Gender, time use, and poverty in sub-Saharan Africa, vol 73. World Bank Publications, Washington, D.C.

7. Bardasi E, Wodon Q (2010) Working long hours and having no choice: time poverty in Guinea. Feminist Econ 16(3):45–78
8. Cagatay N (1998) Gender and poverty. UNDP, Social Development and Poverty Elimination Division
9. Chen MA (2012) The Informal economy: definitions, theories and policies. WIEGO Working Paper 26
10. Njenga TM, Ng'ambi FE Women working in the informal economy. http://www.sexrightsafrica.net/wp-content/uploads/2016/11/OSISA-womenworkinginformaleconomy-copy.pdf. Accessed 20 Mar 2017
11. Smit N, Mpedi L (2010) Social protection for developing countries: can social insurance be more relevant for those working in the informal economy? Law Democr Dev 14(1):1–33
12. Ametepeh RS (2011) Occupational health hazards and safety of the informal sector in the Sekondi-Takoradi metropolitan area of Ghana. Res Humanit Soc Sci 3(20):87–99
13. Naidoo RN, Kessy FM, Mlingi L, Petersson N, Mirembo J (2009) Occupational health and safety in the informal sector in Southern Africa – the WAHSA project in Tanzania and Mozambique, pp 46–50
14. Lund F, Alfers L, Santana V (2016) Towards an inclusive occupational health and safety for informal workers. NEW SOLUT: J Environ Occup Health Policy 26(2):190–207
15. Alfers L (2009) Occupational health & safety for informal workers in Ghana: a case study of market and street traders in Accra. WIEGO, Manchester
16. Basu N, Ayelo PA, Djogbénou LS, Kedoté M, Lawin H, Tohon H, Oloruntoba EO, Fobil J, Fayomi B, Robins T (2016) Occupational and environmental health risks associated with informal sector activities—selected case studies from West Africa. NEW SOLUT: J Environ Occup Health Policy 26(2):253–270
17. Åkerstedt T (1998) Shift work and disturbed sleep/wakefulness. Sleep Med Rev 2(2):117–128
18. Folkard S, Tucker P (2003) Shift work, safety and productivity. Occup Med 53(2):95–101
19. Knutsson A (2003) Health disorders of shift workers. Occup Med 53(2):103–108
20. Knauth P (2007) Extended work periods. Ind Health 45(1):125–136
21. Goel N, Basner M, Rao H, Dinges DF (2013) Circadian rhythms, sleep deprivation, and human performance. In: Progress in molecular biology and translational science, vol 119, pp 155–190. Academic Press
22. Sonnentag S, Fritz C (2007) The recovery experience questionnaire: development and validation of a measure for assessing recuperation and unwinding from work. J Occup Health Psychol 12(3):204–221
23. Van Dongen HPA, Baynard MD, Maislin G, Dinges DF (2004) Systematic inter individual differences in neurobehavioral impairment from sleep loss: evidence of trait-like differential vulnerability. Sleep 27(3):423–433
24. Kinnunen U, Mauno S, Siltaloppi M (2010) Job insecurity, recovery and well-being at work: recovery experiences as moderators. Econ Ind Democr 31(2):179–194
25. Grandner MA, Patel NP, Gehrman PR, Xie D, Sha D, Weaver T, Gooneratne N (2010) Who gets the best sleep? Ethnic and socioeconomic factors related to sleep complaints. Sleep Med 11(5):470–478
26. Seymour G, Malapit H, Quisumbing A (2016) Measuring time use in development settings, (1), 1–29. Annual World Bank Conference on Development Economics, Washington, D.C., 20–21 June 2016
27. Department of Labour. Basic Guide to Working Hours. http://www.labour.gov.za/DOL/legislation/acts/basic-guides/basic-guide-to-working-hours. Accessed 19 Mar 2017

28. Rosa RR, Härmä M, Pulli K, Mulder M, Näsman O (1996) Rescheduling a three shift system at a steel rolling mill: effects of a one hour delay of shift starting times on sleep and alertness in younger and older workers. Occup Environ Med 53(10):677–685

29. Kecklund G, Åkerstedt T (1995) Effects of timing of shifts on sleepiness and sleep duration. J Sleep Res 4(s2):47–50

30. Tucker P, Folkard S (2012) Working time, health and safety: a research synthesis paper. ILO

31. Harrison L, Wilson S, Munafò MR (2014) Exploring the associations between sleep problems and chronic musculoskeletal pain in adolescents: a prospective cohort study. Pain Res Manag 19(5):e139–e145

32. Van Dongen H, Maislin G, Mullington JM, Dinges DF (2003) The cumulative cost of additional wakefulness: dose-response effects on neurobehavioral functions and sleep physiology from chronic sleep restriction and total sleep deprivation. Sleep 26(2):117–126

33. Hillman DR, Lack LC (2013) Public health implications of sleep loss: the community burden. Med J Aust 199(8):7–10

Usability Evaluation of an Online Workplace Health and Safety Return on Investment Calculator

Olivia Yu[1,2(✉)] ⓘ, Kelly Johnstone[1] ⓘ, and Margaret Cook[1] ⓘ

[1] School of Earth and Environmental Sciences,
The University of Queensland, Brisbane, QLD, Australia
o.yu@uq.edu.au
[2] The Office of Industrial Relations, Queensland Government,
Brisbane, QLD, Australia

Abstract. The online Return on Investment (ROI) calculator developed by the Queensland Government (Australia) is a financial evaluation tool that is publicly available and captures the direct and indirect costs and benefits of implementing a workplace health and safety (WHS) intervention to estimate a comprehensive ROI figure. Whilst cost is an important factor in decision-making, how useful is such a tool to WHS professionals in deciding whether to implement a WHS control? **Aim:** This study aimed to evaluate user perceptions of the online ROI calculator's usefulness, ease of use, and value. **Methods:** Google Analytics and the ROI calculator's results were obtained via the Queensland Government database and a usability questionnaire was administered to capture how users perceive the online calculator's ease of use and usefulness. **Results:** Google Analytics recorded 18,633 sessions, 12,803 new users, average of 1.12 page views per session, average session duration of 3 min 30 s and a bounce rate of 86.24%. The ROI calculator yielded 548 observations with a 13.50% conversion rate (74 completions). Overall, users (n = 17) perceived the ROI calculator to be average to slightly positive in terms of usefulness and ease-of-use. **Conclusions:** The results indicate effective site promotion or interest around ROI in WHS, however, the calculator has poor ability to retain users to complete the tool. Future research is currently being conducted to gain a better understanding of the potential roadblocks to greater tool use and user design and experience of the tool.

Keywords: Return on investment · Usability evaluation · Cost benefit analysis Occupational health and safety

1 Introduction

Evaluations are performed to help investigators make informed decisions about a system or program. Evaluations of evaluations (EOE) are generally deemed not as important and thus less research on this topic is available. However, this may seem somewhat ironic, as determining an evaluation's validity and reliability ensures that decision makers receive and act upon relevant high-quality information [1]. This enhances the degree of 'how informed' the decision maker is and can help determine

S. Bagnara et al. (Eds.): IEA 2018, AISC 819, pp. 425–437, 2019.
https://doi.org/10.1007/978-3-319-96089-0_46

whether the cost of the evaluation was 'worth it' [2]. The evaluation is considered worthwhile if the marginal information value exceeds the cost threshold, where the threshold is often arbitrarily decided [1].

One of the most widely used methods for evaluating user acceptance of tools and technology is usability testing. It is an approach focussed on evaluating the users' attitudes and perceptions gathered by investigators to determine behavior about the user while interacting with technology [3]. As Nielson [4] observes, "users experience usability of a site before they have committed to using it". That is, first impressions of a site's ease of use and usefulness plays a large role on a user's final perceived attitude; therefore, design is a major indicator of intention to use and a predictor of future use [5].

Usability testing requires real people who actually use the technology as participants in the test. It is important to differentiate between user feedback and how technology performs technically and functionally [6]. There are no purely objective measures in usability testing that researchers can use to conclude whether a tool is 'good' or 'bad' in the eyes of the user, therefore, subjective measures must be used to determine users' perceptions towards a system. These indicators are supported by underlying theories with validated and reliable measures that explain the psychology surrounding people's attitudes and behaviors. Usability testing, just like any evaluation, is a process that requires relevant inputs or information to yield outcomes in order to allocate or stop allocating resources [1].

Carroll [7] explains that "(u)sability is the extent to which this interactivity is accomplished effectively, efficiently and satisfactorily", where interactivity is between the user and the technology or tool. Usability describes the methods that users will be satisfied when using technology, as measured by ease of use, ease to learn and ability to perform required tasks. Alternatively, Benbunan-Fich [8] defines usability as a concept of "how well and how easily a user, without formal training, can interact with an information system of a website". In other words, a usable system or tool is one that not only understands human behavior and attitudes, but also takes into account how humans remember, understand, problem solve and communicate with a system [9]. Usability is not just about functionality; allowing a user to perform their task efficiently, but also about the perceived effort and enjoyment of doing the task [5].

The Technology Acceptance Model (TAM) is one of the most widely used and well-established models evaluating usability of technology. It purports that the two major factors of a user's behavior are ease of use and usefulness, where perceived usefulness is defined as *"the degree to which a person believes that using a particular technology would enhance his or her job performance"* and perceived ease of use as *"the degree to which a person believes that using a particular system would be free of effort"* [10]. Ease of use and usefulness are theorised to determine predicted future use, or the intention to use the technology in the future. Davis [10] justified the model with a number of experiments concluding that perceived ease of use and usefulness have the greatest impact on user acceptance. Numerous studies show that TAM explains about 40% of the variance of user behavior and intention, which is a lot higher than the models it was based on (Theory of Reasoned Action and Theory of Planned Behavior) [11].

As discussed by Turner [12], the original TAM measured four variables: perceived ease of use (PEU), perceived usefulness (PU), attitude toward use (A) and behavioral intention to use (BI). The original TAM also measured other external variables such as actual use, however, the most significant constructs in explaining user acceptance of technology were perceived ease of use and perceived usefulness [10].

Venkatesh and Davis [11] proposed a revised TAM that maintained the core model of TAM, however, excluded user attitude. Social influences such as subjective norms were explored, defined by Fishbein and Ajzen [13] as a "person's perception that most people who are important to him think he should or should not perform the behavior in question". However, research has yielded mixed results around the significant effects of subjective norm on intention to use. Internalisation is another social influence that describes one's readiness to accept information and thus whether they should use a system [11]. However, the effects of subjective norms are theorised to become less significant with increased use of technology. Testing around these produced consistent results due to direct experience of technology or a system rather than relying on opinions of others. Thus, the experience of directly using a system will inform a user's intention to continue using it in the future, with this effect gradually growing stronger than the subjective norms or opinions during first-time and early use.

There is a plethora of research that evaluate economic evaluation tools, however, there is a paucity of those that review economic evaluation tools that are specifically designed for WHS interventions. Furthermore, there is no recent research evaluating the usefulness of publicly available government designed WHS economic evaluation tools.

2   Methods

This study consisted of three data sources: (1) Online Return on Investment Calculator [14] results submitted by ROI calculator users who input financial information relating to the implementation of a workplace health and safety intervention they wanted to calculate a ROI estimate for, (2) Google Analytics data from the tracking and reporting of the ROI calculator website, and (3) Usability questionnaire responses from users of the ROI calculator.

2.1   Participant Recruitment

Participant recruitment was only required for the usability questionnaire as Google Analytics and the ROI calculator data were secondary data sources. From the subset of the users who completed the online ROI calculator (n = 548), users who volunteered their email addresses (n = 59) on the online ROI calculator were contacted to seek voluntary participation in the online anonymous usability survey.

2.2   Usability Questionnaire Development

The usability questionnaire (Appendix 1) aims to evaluate the usability and practicality of the online work health and safety ROI calculator by exploring three (3) factors: perceived usefulness, ease of use, and self-predicted future use. This was delivered as an online anonymous questionnaire using Checkbox™.

The usability questionnaire was modelled on Babar's [15] software architecture evaluation process to measure 'perceived usefulness' and 'ease of use', which itself was 'based on the adapted Davis' [10] Technology Acceptance Model (TAM), a widely used general-purpose instrument for measuring users' attitude towards a particular technology" [11]. The subject was replaced with "ROI calculator", while the description of the task was replaced with "make an informed decision about work health and safety".

2.3   Data Collection and Analysis

Google Analytics results and ROI calculator results (as secondary data) were obtained via permissions-granted access to relevant Queensland Government databases. Google Analytic results were obtained from the period October 2016 to February 2018 and results from the online return on investment calculator database were from the period October 2016 to March 2018. As previously mentioned, users who provided a valid email address (n = 59) on the online ROI calculator site were contacted to seek voluntary participation in completing the online usability questionnaire. These results were valid as at 1 January 2018. All data were initially exported as Microsoft Excel spreadsheets and then descriptive statistics and initial reliability estimates were computed to determine correlation reliability.

3   Results

3.1   Google Analytics

Google Analytics show that whilst there were 18,633 sessions to the ROI calculator, the number of new users totalled 12,803, indicating that a proportion of these sessions were repeat visitors to the tool to account for the additional 31.29% sessions (Table 1). New users are classified as those with different IP addresses, and this does not differentiate between those with multiple devices accessing the site at multiple locations.

A bounce rate of 86.24% was seen, as the percentage of sessions with a single pageview. This is a high percentage, meaning users are not progressing to subsequent pages of the online tool, i.e. 86.24% of visit users are exiting the tool after viewing the first page. The average number of pages viewed per session was 1.12. This is a top-level metric for user engagement showing the average number of pageviews in each session. Supporting the high bounce rate, a low number of pageviews (11 pages in total) show that users are not progressing past the first two pages of the tool, where the first page is a disclaimer and the second page asks for demographics and basic information about the user and the intervention they intend to evaluate.

The average session duration recorded was 3 min 40 s. This provides a top-level view of how long users are spending on the website. Google Analytics, however, does not record the time spent for the last page viewed during a session. This means that average session duration will tend to be skewed lower than the actual amount of time people are spending on the online tool.

Table 1. Google Analytics measures (18 October 2016 to 12 February 2018)

Measure	
Sessions	18,633
New users	12,803
Pages viewed per session	1.12
Average session duration	3 min 40 s
Bounce rate	86.24%

3.2   Return on Investment Database

A total of 548 observations were recorded to the online ROI database, as of 1 March 2018 (Fig. 1). Of the 548 observations, 91 observations did not complete any section of the online ROI calculator (all fields were either 'zero' or blank) and 67 observations were case study results by the Principal Investigator, resulting in 390 observations remaining to be further analyzed. In other words, 71.17% of total recorded users (as recorded by the online database) entered some sort of information into the ROI calculator. Of the 390 observations that had entered data into the ROI calculator, 74 observations had ROI results that were non-blank. Therefore, 13.50% of all users that had visited the ROI calculator's hosting website reached the final step of the ROI calculator.

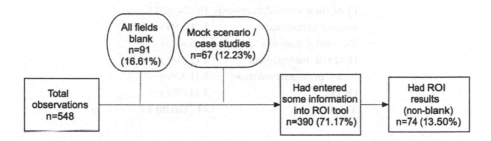

Fig. 1. ROI calculator completion conversion rates, as per online database results

The industries with the highest number of users accessing the ROI calculator website were from Manufacturing and Construction, 21.16% and 16.40%, respectively. Of the users who completed the ROI calculator, the industry with the greatest number of users was still the manufacturing industry (21.67%). There was an almost even split between small to medium enterprises (SME) and large businesses in terms of users that completed the ROI tool (48.65% and 51.35% respectively) (Table 2). This is in contrast to all visitors to the ROI calculator website, where 48.54% of users were from a SME and 31.75% indicated they worked for a large sized organization.

Table 2. Number of users by industry size (as defined by the Australian Bureau of Statistics)

	Users that completed the ROI tool	From all visitors to the website
Small (≤ 19)	9 (12.16%)	70 (12.77%)
Medium (20–199)	27 (36.49%)	196 (35.77%)
Large (≥ 200)	38 (51.35%)	174 (31.75%)
Blank	0 (0.00%)	108 (19.71%)
Total	**74 (100.00%)**	**548 (100.00%)**

Interventions were classified in a general manner (i.e. not solely by hazard or hierarchy of control) due to the nature of answers provided by users. The most common interventions evaluated were related to safety management systems (SMS), policies and procedures (25.68%) and new or modifications of plant and equipment (24.32%) (Table 3).

Table 3. Types of interventions evaluated by users of ROI calculator

Intervention type	
Manual handling	9 (12.16%)
SMS, policies, procedures	19 (25.68%)
WHS Training	5 (6.76%)
Wellness/Wellbeing program	12 (16.22%)
Plant (new or modifications)	18 (24.32%)
Fatigue management	1 (1.35%)
New employee/staff	4 (5.41%)
Heat risk management	2 (2.70%)
Injury management/rehab	1 (1.35%)
Other	3 (4.05%)
Total	**74 (100.00%)**

3.3 Usability Questionnaire

The online usability questionnaire was distributed to users of the ROI calculator who volunteered their email address on the website. Of the 59 emails contacted, 18 participants (30.51%) completed the online questionnaire, with 1 electing 'Do not agree to participate', thus giving a final response rate of 28.81%.

The results of the usability questionnaire are presented in Table 4, where there was a slightly positive attitude of perceived ease of use and perceived usefulness result by users, on a Likert scale of 1 to 7, where 1 represents Extremely difficult and 7 represents Extremely easy. Users scored *ease of use* a mean of 4.82, *clarity and understanding* a mean of 4.82, *ease of becoming skilful* a mean of 4.41 and *ease of remembering how to use*, a mean of 4.88. Similarly, users scored the ROI calculator a mean of 4.82; overall, they found the ROI calculator somewhat useful in helping them make an evaluation decision about their chosen WHS investment.

Table 4. Perceived ease of use, perceived usefulness by users from the return on investment calculator database.

	Mean
Perceived Ease of Use	
1. Having now used the ROI calculator	
(a) I found it easy to use the Return on Investment calculator	4.82 (±1.47)
(b) I found my interaction with the Return on Investment calculator was clear and understandable	4.82 (±1.33)
(c) I feel it would be easy to become skilful in using the Return on Investment calculator	4.41 (±1.70)
(d) I feel it would be easy to remember how to use the Return on Investment calculator again	4.88 (±1.58)
Perceived Usefulness	
1. I feel the WHSQ Return on Investment Calculator	
(a) made it easier to make an informed decision	4.65 (±1.00)
(b) improved my ability to make an informed decision	4.53 (±1.18)
(c) helped me make an informed decision more quickly	4.59 (±1.33)
(d) enhanced my effectiveness in making an informed decision	4.65 (±1.58)
2. Overall, I found the Return on Investment Calculator useful in making an informed decision	4.82 (±1.38)

The eigenvalue of the first factor was 6.44, representing 71.59% of the total variability. A reliability analysis was then run, resulting in a value of 0.947 for Cronbach's alpha, which indicates a high amount of internal consistency between questionnaire variables. This suggests high reliability, although, since the value is very high (very close to 0.95), it may also suggest that the questionnaire or the items it contains may be redundant.

Self-predicted Future Use

Whilst a majority of users (58.82%) indicated they would use the ROI calculator again (Table 5), a greater sample size is required to more accurately determine if this is representative of the larger population. Of the users who indicated that they would use the tool again, a mean score of 2.6 (±0.97) indicates that users would use the calculator to help them make a decision on their WHS investment between 40% to 60% of the time.

Table 5. Self-predicted future use by users from the return on investment calculator database.

1. Would you use the Return on Investment calculator in the future?	
No, I anticipate that I will not use it again	7 (41.18%)
Yes, I anticipate that I will use it again	10 (58.82%)
Total	**17 (100.00%)**

Demographics

Users who completed the usability survey worked primarily in the education and training (17.65%) and manufacturing industries (17.65%) (Table 6).

Table 6. Industries that users who completed the usability questionnaire worked in.

Industry	
Agriculture, forestry and fishing	1 (5.88%)
Arts and recreation services	1 (5.88%)
Education and training	3 (17.65%)
Manufacturing	3 (17.65%)
Retail Trade	1 (5.88%)
Other services	2 (11.76%)
Did not answer	6 (35.29%)
Total	**17 (100.00%)**

Users reported that they held the following positions: Health, Safety and Environment Advisor, Associate Director Facilities Management, WHS Business Partner, Human Services and Quality Systems Manager, Occupational Health and Safety Officer, WHS and Compliance/Biosafety, WHS and Rehabilitation and Return to Work, and General Manager Operations.

There was a variety of interventions implemented, including: Monorail lifting gantry, Competency training, injury prevention, large animal handling injuries, minimising direct interaction between man and beast, upgrade of safety management system, health and wellbeing and ageing workforce.

4 Discussion

The online ROI calculator recorded a good amount of site traffic (18,633 sessions and 12,803 new users), however, experienced low user completion of the online tool. The low conversion rate of 13.50% where 548 total observations to 74 that fully completed the ROI calculator was reflected in the Google Analytics results, which offer greater insight with more objective measures into user site behavior. The variance of observations captured by the ROI calculator database and the number of sessions measured by Google Analytics (18,633) may be due to the high number of people exiting the ROI tool at the first or second page. Page exits at the first and second pages would not be recorded as an observation in the ROI calculator database as the second page is where the tool begins collecting user-entered information. This reasoning for the variance in Google Analytics and ROI calculator information is supported by the high bounce rate of 86.24% and the average of 1.12 pages viewed per session. As the second page is where data collection begins, a large drop-out number of users is expected where the input of information has been shown to be a point of disengagement for many users of websites who are purely seeking information and do not want to give out their own data [16].

The low conversion rate from the beginning to completion could be due to a number of reasons. Users may have different expectations of the online ROI calculator compared to the intention of the developer, for example, the time commitment required, or the level of financial data required to properly complete the tool. This may be addressed in the future by surveying users on their motivations behind using ROI tools and their expectations of such tools and then using this data to inform how to ensure this information is communicated to users upfront. Secondly, users may have found the calculator too complicated, or too generic, for their purpose, which may also be resolved by setting clearer expectations of the ROI tool. The conversion rate could be skewed low as initial users seek to learn about and 'have a play' with the tool to determine whether they deem it useful to their situation. For example, users could open the ROI calculator webpage, click through the first few pages, exit it, then return to the calculator at a later date and thus, start a new session.

In addition, the low conversion rate could be due to poor website design; an unwieldy ROI calculator site could cause users to exit the site early. This is reflected with the preliminary results indicating that the perceived ease of use was average to somewhat easy to use (a score between 4 and 5 on a 7-point Likert scale), suggesting that issues such as poor design, unclear instructions or confusing wording may be apparent. The tool may not be as intuitive or have enough information around the use of it, which may affect the conversion rate as users may feel overwhelmed by the calculator itself or may not understand how to interpret and use the resulting outcomes. Providing training for the ROI tool, such as information webinars, video guides, tutorials or workshops may increase the value of the ROI calculator to users. With the above potential remedies, the tool may experience less web traffic overall, however, have more 'valuable' users, which will improve the conversion rate.

The greatest number of users reported that they were from the manufacturing and construction industries (21.16% and 16.40% respectively), which may be due to the promotion methods and industries targeted by the developer. Another possible explanation is that the construction industry is the third largest industry employer (by number of workers) in the state of Queensland, Australia, where the majority of promotions occurred. Additionally, small to medium sized businesses in the manufacturing and construction industry may be the most interested in using free ROI or cost benefit tools. Future research assessing usability of the online calculator categorizing by industry, business size and intervention type, such as hierarchy of controls or other classifications, may be useful in understanding and thus providing greater value of cost benefit tools targeted to users who are WHS professionals in various industries and businesses.

5   Conclusion

The results indicate users perceive the online ROI calculator to be a tool that is average to slightly positive in terms of its ease of use, usefulness and potential future use. Therefore, further research with current and potential users to improve the ROI calculator's usability would help improve the value to WHS professionals regarding the use of technology that aims to financially evaluate workplace health and safety interventions. The site, either due to effective promotion methods or by topic of interest, garners sufficient site traffic, however, faces a roadblock in seeing users complete the tool. Therefore, future research into measuring the usefulness of financial evaluations in justifying WHS costs and the awareness of available tools could provide important insight into the future of cost benefit analysis for WHS interventions. With greater understanding behind the decision making of whether to invest in a WHS control, future development of financial evaluation technology can then be better geared to truly aid WHS professionals in influencing WHS outcomes. Current studies are being undertaken, such as scenario walkthrough focus groups and a questionnaire on the awareness of available cost benefit tools and the necessity of financial evaluations in WHS, to achieve a better understanding of the low conversion rate and examine features of improvement to the online ROI calculator.

5.1   Limitations

There is a lack of clear definitions to measure site traffic, where the metrics are based on hits, impressions or pageviews [7]. Drawing conclusions about the performance of webpages based on ambiguous definitions of these metrics is therefore difficult. These metrics are heavily used by website owners to determine performance, however, attracting a large number of views and site traffic does not necessarily mean that there is good conversion. Finally, there was a modest survey response rate of the usability survey (n = 17), therefore, with a greater sample size, reliability analyses can be more confidently run and interpreted.

Acknowledgements. I would like to thank The Office of Industrial Relations, Queensland Government, Australia for funding this PhD research project, with special thanks due in particular to Sebastian Bielen and Nita Maynard.

Appendix

Appendix 1: Usability questionnaire

THE UNIVERSITY
OF QUEENSLAND
AUSTRALIA

Usability of the Return on Investment Calculator Questionnaire

The following questions ask you about your perceptions when using the Return on Investment Calculator for workplace health and safety (WHS) interventions. Please return this completed form along with the ROI Diary and your results from the ROI Calculator to Olivia Yu at o.yu@uq.edu.au.

Perceived ease of use

Having now used the ROI calculator:

a) I found it *easy* to use the Return on Investment calculator.

Extremely difficult						Extremely easy
1	2	3	4	5	6	7
○	○	○	○	○	○	○

b) I found my interaction with the Return on Investment calculator tool was *clear and understandable*.

Extremely confusing						Extremely clear and understandable
1	2	3	4	5	6	7
○	○	○	○	○	○	○

c) I feel it would be *easy to become skilful in using* the Return on Investment calculator.

Extremely difficult						Extremely easy
1	2	3	4	5	6	7
○	○	○	○	○	○	○

d) I feel it would be *easy to remember* how to use the Return on Investment calculator again.

Extremely difficult						Extremely easy
1	2	3	4	5	6	7
○	○	○	○	○	○	○

Perceived usefulness

1. ROI tools should facilitate making an informed decision about investing in workplace health and safety.
I feel the WHSQ Return on Investment Calculator:

a) *made it easier* to make an informed decision

Made it extremely complicated						Made it extremely easy
1	2	3	4	5	6	7
○	○	○	○	○	○	○

I feel the WHSQ Return on Investment Calculator:

b) *improved my ability* to make an informed decision

Did not improve   Improved my
my ability at all   ability immensely
1   2   3   4   5   6   7
○   ○   ○   ○   ○   ○   ○

c) helped me make an informed decision *more quickly*

There was no   Improved my
change in decision-   decision-making
making time   time immensely
1   2   3   4   5   6   7
○   ○   ○   ○   ○   ○   ○

d) *enhanced my effectiveness* in making an informed decision

Did not enhance   Enhanced my
my effectiveness   effectiveness
at all   immensely
1   2   3   4   5   6   7
○   ○   ○   ○   ○   ○   ○

2. Overall, I found the Return on Investment Calculator *useful* in making an informed decision.

Not useful   Extremely
at all   useful
1   2   3   4   5   6   7
○   ○   ○   ○   ○   ○   ○

Self-predicted future use

1. Would you use the Return on Investment calculator in the future?

   ● Yes, I anticipate that I will use it again.

   ○ No, I anticipate that I will not use it again.

2. If yes, please indicate how often you would use the Return on Investment Calculator to help you decide whether to invest in a WHS intervention?

20% or less   40%   60%   80%   100%
of the time   of the time of the time of the time of the time
○   ○   ○   ○   ○

References

1. Larson R, Berliner L (1983) On evaluating evaluations. Policy Sci 16:147–163. https://doi.org/10.1007/bf00138348
2. Blagec K, Romagnoli K, Boyce R, Samwald M (2016) Examining perceptions of the usefulness and usability of a mobile-based system for pharmacogenomics clinical decision support: a mixed methods study. PeerJ 4:e1671. https://doi.org/10.7717/peerj.1671
3. Mohamed K, Hassan A (2015) Evaluating federated search tools: usability and retrievability framework. Electron Libr 33:1079–1099. https://doi.org/10.1108/el-12-2013-0211

4. Nielsen J (2000) Designing web usability. New Riders, Indianapolis
5. Wang J, Senecal S (2007) Measuring perceived website usability. J Internet Commer 6:97–112. https://doi.org/10.1080/15332860802086318
6. Laitenberger O, Dreyer H (1998) Evaluating the usefulness and the ease of use of a web-based inspection data collection tool. In: Software metrics symposium fifth international
7. Carroll M (2003) Usability and web analytics: ROI justification for an internet strategy. Interact Mark 4:223–234. https://doi.org/10.1057/palgrave.im.4340184
8. Benbunan-Fich R (2001) Using protocol analysis to evaluate the usability of a commercial web site. Inf Manag 39:151–163. https://doi.org/10.1016/s0378-7206(01)00085-4
9. Barnard P, Hammond N, Morton J, Long J, Clark I (1981) Consistency and compatibility in human-computer dialogue. Int J Man-Mach Stud 15:87–134. https://doi.org/10.1016/s0020-7373(81)80024-7
10. Davis F (1989) Perceived usefulness, perceived ease of use, and user acceptance of information technology. MIS Q 13:319. https://doi.org/10.2307/249008
11. Venkatesh V, Davis F (2000) A theoretical extension of the technology acceptance model: four longitudinal field studies. Manag Sci 46:186–204. https://doi.org/10.1287/mnsc.46.2.186.11926
12. Turner M, Kitchenham B, Brereton P, Charters S, Budgen D (2010) Does the technology acceptance model predict actual use? A systematic literature review. Inf Softw Technol 52:463–479. https://doi.org/10.1016/j.infsof.2009.11.005
13. Fishbein M, Ajzen I (1975) Belief, attitude, intention, and behavior. Addison-Wesley Publishing Co., Reading
14. Work Health and Safety eTools (2018). https://worksafe.qld.gov.au/etools/. Accessed 1 Nov 2016
15. Ali Babar M (2010) A framework for groupware-supported software architecture evaluation process in global software development. J Softw Evol Process 24:207–229. https://doi.org/10.1002/smr.478
16. Benford S, Giannachi G, Koleva B, Rodden T (2018) From interaction to trajectories: designing coherent journeys through user experiences. In: SIGCHI conference on human factors in computing systems

What Can an Ergonomist Do to Manage Dangerous Substances Better?

Gyula Szabó[1,2](✉)

[1] Donát Bánki Faculty of Mechanical and Safety Engineering,
Óbuda University, Népszinház utca 8, Budapest, Hungary
szabo.gyula@bgk.uni-obuda.hu
[2] Federation of European Ergonomics Societies (FEES), Dortmund, Germany

Abstract. The objective of the European Month of Ergonomics (EME) of the Federation of Europe-an Ergonomics Societies (FEES) is to promote the ergonomics profession in Europe. The first theme "Know your ergonomics" gave an introduction in 2009. The following EMEs became specific, and the topics are dedicated to the actual campaigns of the European Agency for Safety and Health at Work (EU-OSHA). The EME goal is not only to relay the EU-OSHA message but to highlight the role of the ergonomics profession, to show how an ergonomist can contribute to a healthier and safer workplace, to improve performance and increase workers' satisfaction.

According to the three main ergonomics application fields defined by the International Ergonomics Association (IEA) the EME 2018/19 "Ergonomist to manage dangerous sub-stances better" covers the applicable physical, cognitive and organizational interventions related both to handling materials themselves and handling the relevant information.

Traditionally FEES prepares a slideshow to help federated societies and every interested professional to spread the EME message. This new slideshow contains a brief introduction on EME and EU-OSHA campaign, some definitions, the expectable benefits of the use ergonomics expertise and references to further information.

To make the EME more understandable and attractive several examples are presented.

Physically handling dangerous substances means e.g. material handling, lifting, moving, carrying which requires proper package shape and size, tools, process, task and organizational design. Cognitive ergonomics regarding dangerous substances e.g. can improve the visual appearance with colour and shape coding, symbols and pictograms, environmental design to easy recognition. Organisational ergonomics can help to eliminate modality changes, decrease exposures, define better processes.

We hope this EME will prove again that ergonomics can be a competitive factor in healthier and safer workplaces.

Keywords: Month of ergonomics · Hazardous materials

© Springer Nature Switzerland AG 2019
S. Bagnara et al. (Eds.): IEA 2018, AISC 819, pp. 438–443, 2019.
https://doi.org/10.1007/978-3-319-96089-0_47

1 Introduction

The Federation of European Ergonomics Societies has been organising the European Month of Ergonomics since 2009. The campaign was originally intended to achieve one of the main objectives of the FEES, namely the general promotion of ergonomics in the European area. The title of the first campaign was "Know your ergonomist", which focused solely on what ergonomics is, what is the practical use and results of ergonomics, and that high-level ergonomic solutions require skilled and well-trained professionals. The "Know your ergonomist" campaign [1] has shown that well-trained ergonomics professionals are around and also ergonomics professional organizations are present in many European countries. October, the month of ergonomics is the main period when campaign activity is the most vivid, but the message is relevant at all ergonomics related forums and channels and the communication period does not need to be limited to October.

1.1 The General Structure of the European Month of Ergonomics

Following the first campaign, the Communication and Promotion Committee of FEES decided to use the ergonomics campaign to support the current campaign of the European Agency for Safety and Health at Work (EU-OSHA), while keeping the previous objectives. In this system, the message of the EU-OSHA Healthy Workplaces Campaigns and the promotion of the ergonomic profession merged and present the ergonomics profession and approach, and highlight the added value or ergonomics to the development of health and safety at work in accordance with the objectives of the current campaign.

Each EU-OSHA campaign has ergonomic content, but the relevance varies. From an ergonomic point of view, it was particularly important the 2000 year "Turn your back on musculoskeletal disorders" and 2007 year "Lighten the load" campaign, addressing major workplace ergonomic risks, and focusing on the prevention of musculoskeletal disorders and lower back pain resulting from e.g. manual material handling. Among the new campaigns, the "Healthy Workplaces for All Ages" 2016–2017 was of high ergonomics value, as it encouraged the implementation workplaces that would enable workers to work from the first employment to retirement. This objective can obviously only be achieved taking full account of the worker's characteristics, that is, the high level of ergonomics application in the workplace. The FEES run that time "Ergonomics for all ages" EME [2].

The EU-OSHA "Healthy Workplaces Manage Stress" campaign 2014–2015 has been designed to take into account the psychological and social characteristics of workers and to design jobs according to individual differences. The "Ergonomics for managing work-related stress" promoted many ergonomic applications ranging from the creation of jobs and work activities to organizational ergonomic intervention [3]. The "Safe Maintenance" 2010–2011 EU-OSHA campaign required a widespread application of ergonomics, and FEES response was the European Month of Ergonomics 2010 - Ergonomics is a Key to Safe Maintenance [4] including physical abilities, support for practice, and prevention of defect with built-in security, just as designing and utilizing the design of work tools, tools, hand tools, and design.

All the EME materials mentioned above was prepared by the Communication and Promotion Committee of FEES, chaired by Martti Launis.

1.2   The Content of the Current EME

EU-OSHA launched the "Manage Dangerous Substances" campaign on April 2018. As usual, FEES joined in and volunteered to mediate the agency's message and to demonstrate the added value of the ergonomic profession in regard of the "Manage Dangerous Substances" campaign topic. The European Month of Ergonomics in 2018–19 is devoted to identify, elaborate and demonstrate how ergonomic profession can contribute to the healthy and safe handling of hazardous substances.

On one hand FEES as an European organisation of National Ergonomics organisations, is primarily responsible for preparing and organizing the EME in EU level, on the other hand the implementation is achieved through national ergonomic organizations that work closely with national EU-OSHA focal points.

The Communication and Promotion Committee prepared an initial EME communication package, and the first presentation held on May 24, 2018, in Poznan at the Polish Ergonomics Society seminar [5], marked the opening of the 2018–19 campaign activity of FEES. This EME presentation reiterates the ergonomic aspects of the agency's campaign material, explains the general ergonomic approaches, concepts and considerations [6] that are essential to the ergonomics promotion and provides opportunities for ergonomic handling of hazardous substances.

2   Ergonomic Consideration for Managing Hazardous Substances

In everyday vocabulary workplace ergonomics are often associated with work related musculoskeletal disorders (wMSDs), therefore it seems beneficial to apply this belief as a starting point. The practical section of the EME campaign begins with the manual handling of materials, and highlights how dangerous substances increase the risk, e.g. according to the EN 1005-2 standard [7] the risk of manual handling of hazardous material higher, due to the possibly more difficult to grasp, the less stable internal composition, or the restricted movement of the operator caused by the necessary personal protective equipment.

In the approach of the IEA and the FEES [6], ergonomics are far more than just preventing wMSDs in workplaces. The application of ergonomics for managing hazardous substances involves also the consideration of the characteristics of human information processing in the design of work processes. This includes taking human abilities, capabilities and limitations into account when designing products, such as packaging hazardous materials or tools used to handle them.

When dealing with hazardous materials the information provision is of high importance. The required deep understanding of human factors regarding e.g. the storing and displaying hard copies of material safety data sheets, the design of printed or electronic SDSs, the organisation of the downstream and feedback information flow and the provision of information to workers.

The particularity of ergonomics is its system approach which is sufficiently covers also the material properties, allowing achieving the optimum in terms of performance, well-being and satisfaction. The EME initial presentation also mentions the necessity to meet the requirements arising from individual human differences or special needs when designing management of hazardous substances.

The FEES EME material follows the STOP principle proposed by EU-OSHA [8], meaning the prevention actions in priority order are:

- Substitution (safe or less harmful alternatives),
- Technological measures (e.g. closed system, local exhaust ventilation)
- Organizational measures (e.g. limiting the number of exposed workers or the exposure time),
- Personal protection (wearing PPE).

So long the EME material does not include examples of ergonomic design implementing the first principle, the substitution of hazardous substances or replacing them with less hazardous substances.

In the elaboration of human factors for technological solutions, the FEES EME initial presentation builds on the Ramsey Accident Model [9]. Accordingly, prevention options include:

- avoid contact with hazardous material,
- sensory detection of hazard signals related to hazardous materials - better illumination, sound signals, tactile markings,
- cognition of danger - identify/recognize the materials,
- decision on how to avoid danger - responsible attitude and behaviour,
- ability to make a decision - anthropometrics, bio-mechanics, motion capabilities.

Several organizational-ergonomic methods are available to develop organizational measures to implement the 3rd step of the STOP principle.

Regarding PPE, ergonomics is the first essential health and safety design requirements in the PPE EU regulation [10], which is specified by a number of harmonized standards according to the ergonomics principles in PPE design standard [11].

2.1 Activities of FEES and Member Organizations

The European Month of Ergonomics offers the opportunity to broadcast a main ergonomic message and at the same time to relay the message of the Healthy Workplace campaign. In practice the EME utilise the snowball effect, which is

- the initial presentation material is used in different events, triggering some discussions and leading to conclusions,
- these results are always incorporated into the material, and
- this advanced material becomes the campaign material and used in the future.

With this solution, organizers of events around Europe can occasionally start from a well-prepared presentation, focusing on a professional issue, and then share their experience with the entire European community. The snowball continues to grow, and the great number of minimal individual effort will generate a big final superposed added value.

The expected activity of national ergonomic societies, companies, ergonomists or other interested parties may be:

- Organizing events or sections in this subject for a shorter or longer duration,
- Communicating EU-OSHA and FEES campaigns messages on existing channels (newsletter, web site, press conference),
- Displaying messages, values, and knowledge in education at different levels, including university lectures and courses,
- Initiate and conduct research and development programs, in particular student subject research, master thesis, PhD work,
- Participation in the EU-OSHA best practice competition with own application or jointly with other applicants,
- Translating/adapting the EME presentation material into national language. The material can be used freely while keeping the values, and indicating the source,
- Feedback the results on activities to FEES Communication and Promotion Committee, and sharing the generated material.

2.2 Challenging Examples

The first version of the EME material contains some obvious solutions, but the aim of this set of options is primary to trigger further discussions and activate ergonomist to elaborate solutions.

- Design process to Eliminate unnecessary contact by task design
- Install closed material transport and storage in different circumstances
- Eliminate contact by isolation of operators (e.g. telepresence)
- Prevention of inadvertent contact by packaging, locking, storage unit design
- Access control by design, e.g. special caps, zips
- Marking process, and product identification by design or using smart glasses
- Need to do – need to access policy, e.g. using biometric ID Design the environment of the hazardous substance to allow detection, e.g. be clear around the hazardous material
- Under the threshold delectability (visible, audible, odorous, tactile) with added odours, testing bitter, the touch should be disadvantageous (e.g. cause itching)
- Appearance design (size, graphic elements, colour, pitch), layout (visibility, position), intensity (repetition, volume, marking substance concentration).
- Designing dangers, prevention, treatment, compensation etc. markings for special user groups
- Substance documentation usability, e.g. structure, content, wording, accessibility
- Reorganise to eliminate modality changes, decrease exposures, define better processes
- Improve information management, e.g. the storing and displaying hard copies of material safety data sheets, the design of printed or electronic SDSs, the organisation of the downstream and feedback information flow and the provision of information to workers

– Regular cleaning, collection of residues and cleaning of cleaning tools and packaging.

3 Conclusion

At the beginning of the 18 months period it's not clear what can an Ergonomist do to manage dangerous substances better. However it's proved that that ergonomics has contributed countless times to technical progress and to the better world of people's living. The European Month of Ergonomics will provide the opportunity for the next one and a half years to develop and demonstrate that physical, cognitive and organizational ergonomics makes a useful contribution to health and safety at work and improves the handling of hazardous substances.

References

1. Federation of European Ergonomics Societies. European month of ergonomics 2009 - know your ergonomist. http://ergonomics-fees.eu/node/71. Accessed 26 May 2018
2. Federation of European Ergonomics Societies. European month of ergonomics - ergonomics for all ages. http://ergonomics-fees.eu/node/198. Accessed 26 May 2018
3. Federation of European Ergonomics Societies. European month of ergonomics - ergonomics for managing work-related stress. http://ergonomics-fees.eu/node/150. Accessed 26 May 2018
4. Federation of European Ergonomics Societies. European month of ergonomics 2010 - ergonomics is a key to safe maintenance. http://ergonomics-fees.eu/node/70. Accessed 26 May 2018
5. Szabó G (2018) Added value of ergonomics to manage hazardous materials. In: International seminar of ergonomics, Poznan, 23–25 May 2018
6. ISO 26800:2011 Ergonomics – General approach, principles and concepts
7. EN 1005-2:2003+A1:2008 Safety of machinery - Human physical performance - Part 2: Manual handling of machinery and component parts of machinery
8. EU-OSHA info sheet substitution of dangerous substances in the workplace. https://healthy-workplaces.eu/sites/default/files/publications/documents/WEB-info-sheet-substitution-of-dangerous-substances-HWC-2018-19.pdf. Accessed 26 May 2018
9. Jerry D (1985) Ramsey: ergonomic factors in task analysis for consumer product safety. J Occup Accid 7(2):113–123
10. Regulation (EU) 2016/425 of the European Parliament and of the Council of 9 March 2016 on Personal Protective Equipment and Repealing Council Directive 89/686/EEC ANNEX II
11. EN 13921:2007 Personal protective equipment. Ergonomic principles

Hospital Autopsy for Prevention of Sudden Cardiac Death

S. D'Errico$^{(\boxtimes)}$, M. Martelloni, S. Niballi, and D. Bonuccelli

Department of Legal Medicine, Azienda USL Toscana Nordovest Lucca,
Pisa, Italy
stefano.derrico@uslnordovest.toscana.it

Abstract. In the past 20 years, cardiovascular mortality has decreased in high-income countries in response to the adoption of preventive measures to reduce the burden of coronary artery disease and heart failure. Despite these encouraging results, cardiovascular diseases are responsible for approximately 17 million deaths every year in the world, approximately 25% of which are sudden cardiac death. The risk of sudden cardiac death is higher in men than in women, and it increases with age due to the higher prevalence of coronary artery disease in older age. Accordingly, the sudden cardiac death rate is estimated to range from 1.40 per 100 000 person-years in women to 6.68 per 100 000 person-years in men. Sudden cardiac death in younger individuals has an estimated incidence of 0.46–3.7 events per 100 000 person-years, corresponding to a rough estimate of 1100–9000 deaths in Europe and 800–6200 deaths in the USA every year. Cardiac diseases associated with sudden cardiac death differ in young vs. older individuals. In the young there is a predominance of channelopathies and cardiomyopathies, myocarditis and substance abuse, while in older populations, chronic degenerative diseases predominate. In younger persons, the cause of sudden cardiac death may be elusive even after autopsy, because conditions such as inherited channelopathies or drug-induced arrhythmias that are devoid of structural abnormalities are epidemiologically relevant in this age group. Identification of the cause of an unexpected death provides the family with partial understanding and rationalization of the unexpected tragedy, which facilitates the coping process and allows an understanding of whether the risk of sudden death may extend to family members. Accordingly, author present their experience with autopsies of unexplained sudden death young victims in which a cardiac origin was suspected and the relevance of a standardized protocol for heart examination and histological sampling, as well as for toxicology and molecular investigation.

Keywords: Sudden cardiac death · Hospital autopsy · Prevention

1 Introduction

Sudden death is defined as non-traumatic, unexpected fatal event occurring within 1 h of the onset of symptoms in an apparently healthy subject (unexplained). If death is not witnessed, the definition applies when the victim was in good health 24 h before the event [1]. Sudden cardiac death is a tragic complication of a number of cardiovascular

© Springer Nature Switzerland AG 2019
S. Bagnara et al. (Eds.): IEA 2018, AISC 819, pp. 444–454, 2019.
https://doi.org/10.1007/978-3-319-96089-0_48

diseases. More precisely, the term "sudden cardiac death" is used when a congenital, or acquired, potentially fatal cardiac condition was known to be present during life, or autopsy has identified a cardiac or vascular anomaly as the probable cause of the event, or no obvious extra-cardiac causes have been identified by post-mortem examination and therefore an arrhythmic event is a likely cause of death. The prevalence of sudden cardiac death is significant, with at least 3 million people worldwide dying suddenly each year. In the United States, sudden cardiac death occurs in up to 350,000 people each year, translating to 950 deaths per day, or 1 death every 1.5 min. The death can occur at all ages but is significantly more common in older age groups, and the incidence in young people aged less than 40 years is generally low. Although sudden unexplained death (SUD) in the young is statistically uncommon, affecting <5 per 100.000 persons per year, the sudden and unanticipated loss of an apparently healthy child or young adult remains a great tragedy with disastrous implications for the surviving family and the society. The public health burden of premature death for men and women is greater for sudden cardiac death than for all individual cancers and most other leading causes of death [2, 3]. The best estimate of the incidence of sudden cardiac death in the general population aged 20–75 years is 1 in every 1000 individuals, accounting for 18.5% of all deaths [4]. Estimates in studies of the incidence of sudden cardiac death in the young vary widely owing to differences in the age range of the study populations, sample size and study designs. Is reported that in the 1–40 age group, the incidence is up to 8.5 per 100,000 person years, including competitive athletes.

1.1   Causes of Sudden Cardiac Death

The causes of SCD can be broadly divided into structural and arrhythmogenic etiologies. Over the last two decades, our understanding of the cardiac causes of sudden unexplained death has increased importantly by the discovery of a genetic basis for a number of various cardiac diseases and the identification of the same genetic substrates in a significant portion of sudden cardiac death cases, including at least one-third of cases with structural cardiac changes at autopsy and up to one-third of cases where no cause of death is identified at post-mortem, so-called sudden arrhythmia death syndrome (SADS). The most common cause of sudden cardiac death is coronary artery disease in the elderly. In subjects over the age of 40 years, coronary artery disease and acute myocardial infarction account for over 90% of sudden cardiac death cases [5, 6]. In younger persons aged 1 to 40 years, sudden cardiac death is more likely to be caused by a variety of inherited heart diseases and cardiac arhythmia syndromes, and most are autosomal dominant disorder [7, 8]. **Of note, surviving family members with the same genetic substrate as their deceased relative may be at increased risk.** Structural causes of SCD in the young include inherited cardiomyopathies, such as hypertrophic cardiomyopathy (HCM), dilated and restrictive cardiomyopathies, arhythmogenic right ventricular cardiomyopathy (ARVC), and left ventricular noncompaction. HCM is the most common inherited heart disease with a prevalence of 1 in every 200 people. HCM remains the most common structural cause of SCD in the young, including competitive athletes. Other structural causes of SCD in the young include myocarditis, aortic dissection, congenital heart diseases and coronary artery disease. The main inherited arrhythmogenic causes of SCD in the young are arrhythmogenic disorders include familial long QT syndrome (LQTS), catecholaminergic polymorphic ventricular tachycardia (CPVT), Brugada syndrome (BrS), idiopathic ventricular fibrillation, early repolarization syndromes, and short QT syndrome. These disorders rarely cause any structural changes to the heart; often are associated with sudden cardiac death and a structurally normal heart. The underlying mechanism of these cardiac arrhythmia syndromes is dysfunction of the cardiac sodium, potassium and calcium ion-channel subunits and accessory proteins, which is not evident at the macroscopic or microscopic level. Therefore, arrhythmogenic causes of sudden cardiac death are difficult to identify at autopsy investigation and the cause of death is often unascertained. Unexplained sudden cardiac death is therefore a diagnosis of exclusion of all other possible causes of death, but is often presumed to have an arrhythmogenic basis.

1.2 Role of Post Mortem in the Investigation of Sudden Cardiac Death in the Young

Autopsy is a crucial step in defining the causes of sudden cardiac death, which are broadly categorized as having either a structural or arrhythmogenic basis [9, 10]. According to the 2015 ESC guidelines, identification of the cause of an unexpected death provides the family **with partial understanding and rationalization of the unexpected tragedy, which facilitates the coping process and allows an understanding of whether the risk of sudden death may extend to family members**. Accordingly, it appears reasonable that all unexplained sudden death victims undergo post-mortem expert examination to investigate whether a cardiac origin should be suspected. Overall, a properly conducted autopsy should provide answers to the following issues: (I) whether the death is attributable to a cardiac disease, (II) the nature of the cardiac disease (if present), (III) whether the mechanism of death was arrhythmic, (IV) whether there is evidence of a cardiac disease that may be inherited and thus requires screening and counselling of relatives and (V) the possibility of toxic or illicit drug use or other causes of unnatural deaths. Collectively, in all structural causes of SCD, often associated with severe atherosclerosis, chamber dilatation or left ventricular hypertrophy, the post mortem examination has a high probability of identifying the cause of death. In a three-year prospective study of sudden cardiac death in persons aged 1 to 35 years, a structural abnormality of the heart was found in 60% of cases [11]. However, the death remains unexplained in up to 40% of cases. Up to 12% of sudden cardiac death in the young have autopsy findings of uncertain significance, such as minor coronary artery disease, mild cardiac hypertrophy or minor myocardial fibrosis, which pose a challenge for the interpretation of the autopsy investigation [12, 13]. Such pathological findings may lead to the death being attributed to structural disease in some cases. Infact, clinical screening of the first-degree relatives of patients with inconclusive autopsy findings revealed an inherited arrhythmia syndrome in almost half of the families. A further challenge to the interpretation of the autopsy investigation is that sudden cardiac death due to structural causes can occur without characteristic pathology [14, 15]. In particular, hypertrophic cardiomyopathy caused by pathogenic variants in troponin T may have a high risk of sudden death with only mild hypertrophy and fibrosis. Furthermore, non-cardiac conditions can cause unexplained sudden death that is indistinguishable from unexplained sudden cardiac death. For example, patients with epilepsy have a higher mortality rate than patients without epilepsy and the leading cause of epilepsy related death is sudden unexpected death in epilepsy, so called SUDEP. This is of particular concern to the surviving family who do not know why their relative died or the risk of sudden death in other family members. It is important to define the precise cause of death, as this is relevant to the risk of sudden cardiac death and timely implementation of potential life-saving interventions in the relatives of the deceased. Important aspects of the postmortem include the proper and detailed conduct of the postmortem examination it self, collection of appropriate samples for subsequent analysis (including DNA analysis), careful evaluation of the findings, and a precise and accurate final conclusion are important aspects of the role of the postmortem. A comprehensive and detailed postmortem process can not only define the exact cause of death in a young sudden cardiac death case, but the results also have farreaching effects in identifying at risk relatives of the decedent. This information can thereby provide a therapeutic window for disease and sudden death prevention in at risk relatives.

1.3 Molecular Autopsy

Genetic studies over the last 20 years have shown that many inherited cardiac arrhythmia syndromes are caused by defective cardiac ion-channel subunits that regulate the cardiac action potential. Long QT syndrome is characterised by QT prolongation, T wave abnormalities, torsades de points, and a predisposition to syncope and sudden cardiac death. Approximately 75% of long QT syndrome is caused by pathogenic variants in the potassium ion channel subunits KCNQ1 (LQT1, 35%) and KCNH2 (LQT2, 30%), and the sodium ion channel subunit SCN5A (LQT3, 10%). An additional 5% of patients have a copy number variant in the KCNQ1 and KCNH2 genes. Genetic testing of a further 10 genes implicated in long QT syndrome increases the diagnostic yield by 5%. Brugada syndrome is associated with conduction delays, potentially lethal arrhythmias and a family history of sudden cardiac death. In a substantial portion of patients, Brugada syndrome is sporadic, but in some families follows an autosomal dominant inheritance pattern. At least eight genes are implicated in Brugada syndrome, and although pathogenic variants in SCN5A account for over 75% of genetically proven cases, they account for 20–25% of patients. Therefore, genetic testing of SCN5A in Brugada syndrome is useful, but of low yield. Catecholaminergic polymorphic ventricular tachycardia is a rare, adrenergically stimulated ventricular arrhythmia syndrome associated with syncope and sudden cardiac death, typically occurring during physical exertion or emotional stress. A highly penetrant autosomal dominant form accounts for 65% of cases and is caused by pathogenic variants in the very large calcium release channel, cardiac ryanodine receptor, encoded by RYR2. A rare recessive form of catecholaminergic polymorphic ventricular tachycardia is caused by pathogenic variants in CASQ2, encoding cardiac calsequestrin; an accessory protein of the ryanodine receptor. This genetic testing process has been termed the "molecular autopsy", and involves DNA extraction from postmortem blood, followed by DNA analysis of selected candidate genes responsible for the main inherited arrhythmogenic diseases [16, 17]. Guidelines for post-mortem evaluation of sudden cardiac death in the young recommend collecting blood as a source of DNA for the molecular autopsy. The major problem of genetic testing and the molecular autopsy is establishing genetic causality or pathogenicity. Determining the pathogenicity of the variants identified in the post mortem setting is complicated by the absence of a true "phenotype" in the deceased. Many factors need to be considered in determining pathogenicity including the type of mutation, the frequency of the variation in genetic population databases, the type of aminoacid change and its conservation, the predicted damaging effect using in silico tools, supportive functional data, and evidence of cosegregation of the variant within a family [18, 19]. Taking such factors in to account, the genetic variant is classified as pathogenic (disease-causing), benign, or variant of uncertain significance (VUS). The key consideration for the clinician seeing the surviving family of the decedent is that cardiac genetic results are "probabilistic".

2   A Post Mortem Protocol for Sudden Cardiac Death Prevention in Relatives: Lucca's Triennal Experience

According to specific guidelines for autopsy investigation of sudden unexpected death published by several scientific societies we developed a methodological approach in order to fulfil the basic requirements for an appropriate post-mortem assessment and to ensure standardization of the autopsy practice, including adequate ancillary testing and collection of suitable materials (i.e., blood and/or frozen sections of highly cellular tissues) for DNA extraction and genetic testing in the decedent if inheritable structural or non-structural cardiac disease are suspected.

2.1   Materials and Methods

A comprehensive postmortem study by an experienced team of pathologists was performed in all sudden cardiac cases in the young (0–40 years). A complete premorbid medical history was collected, including a history of syncopal episodes, exertional symptoms, intercurrent illnesses, recent pharmacological therapies, previous ECGs, and other relevant studies. The investigation was focused to detect any family history of cardiac disease, premature sudden death, or suspicious deaths (e.g. SIDS cases or drowning). Other relevant family history information included family members with epilepsy, identifying "fainters", and any other unusual symptoms or clinical presentations. Circumstances of sudden cardiac death were investigated in all cases when possible, including activity at the time of death, the level of physical activity, and the symptoms immediately preceding the death obtaining information from available ambulance and police reports, as well as talking to witnesses or those who found the deceased. Postmortem examination included a detailed macroscopic and histological evaluation of all organs with the purpose of identifying any non-cardiac causes of death, before focusing on specific cardiac pathologies. Heart was fixed in 10% buffered formalin and MNR was performed in all cases before macroscopic and histological examination. A standard histological examination of the heart includes mapped labelled blocks of myocardium from representative transverse slices of both ventricles. A 5–

10 mL blood sample was collected for subsequent toxicological analysis and DNA extraction and analysis. In addition, frozen sections of brain, liver, spleen, which are highly cellular and therefore rich in DNA, were collected and stored at a temperature of −20 °C. If a structural or non-structural genetic cardiac disease is rendered likely, blood or tissue samples suitable for future DNA extraction and genetic testing are procured and specialized analysis by an expert cardiac pathologist is requested. Cases in which this process results in a definite clinical diagnosis of structural genetic disease (e.g., HCM, ARVC, or DCM) were referred to a specialized Inheritable Cardiac Diseases Clinic (ICDC) for genetic testing using a candidate gene approach (i.e., phenotype-driven sequencing of a gene or a panel of genes previously associated with a certain genetic cardiac disease). In the cases in which no cause of death was found after thorough post-mortem analysis, genetic testing of the decedent's blood or tissue was performed by ICDC to identify a genetic cause for the sudden cardiac death. After establishing a clinical and/or genetic diagnosis of an inheritable cardiac disease as the underlying cause of sudden cardiac death, necessary actions to initiate the screening of the family members of the decedent were considered.

2.2 Results

The Azienda Toscana Nord Ovest includes a population of over one million two hundred thousand inhabitants with thirteen hospitals and twelve territorial areas. Since 2015, Department of Legal Medicine of Lucca performed 190 hospital autopsies in cases of sudden unexpected death. 20 cases occurred in young between 1–40 years (Table 1).

Table 1. Demographic characteristics of hospital autopsies performed on sudden unexpected deaths

Years	Hospital autopsies (n)	Sex (M/F)	Age (1–40 y.o.)
2015	34	24/10	1
2016	43	30/13	6
2017	76	54/22	7
2018 (Jan–May)	37	20/17	6
Tot.	190	128/62	20

In 75% of cases death was untestified and the deceased was found lifeless at bed, in the morning by relatives. In 25% of cases a sudden collapse occured in apparent well being. In three cases toxicological investigations revealed acute intoxication of drugs (neuropleptics) or cocaine and further analysis were excluded. In four cases coronary artery disease was indicated as the cause of death because of stenosis of one or more coronary arteries. Non cardiac pathologies (ab ingestis pneumonia, cerebral haemorrhage, post traumatic haemorrhagic shock) were diagnosed as relevant for death and excluded from molecular analysis as well as cases in which acute myocarditis and LAD artery myocardial bridging were observed. Dilatative cardiomypathy (1) and hypertrophic (1) cardiomyopathy were diagnosed at gross examination of the heart and

Table 2. Cases of sudden unexpected death in young between 1–40 year old.

Years	Sex	Age	Circumstances	Medical history	Toxicoloy	Cause of death after hospital autopsy	Molecular autopsy
2015	♂	38	Untestified	Negative	cocaine	Cocaine abuse in CAD	No
2016	♂	36	Untestified	Negative	Neuroleptics	Drug abuse	No
	♂	33	Sudden collapse	Past myocarditis	Negative	Acute myocarditis	No
	♀	35	Untestified	Negative	Negative	Dilatative cardiomyopathy	Ongoing on relatives
	♂	39	Untestified	Negative	Negative	Hypertrophic cardiomyopathy	Ongoing on relatives
	♀	38	Untestified	Mental retard	Negative	Ab ingestis pneumonia	No
	♂	32	Untestified	Negative	Negative	LAD artery myocardial bridging	No
2017	♂	34	Sudden collapse	Hypertension	Negative	Cerebral haemorrhage	No
	♂	39	Sudden collapse	Diabetes	Negative	Unexplained	Brugada Sdr (VUS)
							Ongoing on relatives
	♂	25	Untestified	Negative	Negative	Unexplained	Brugada Sdr (VUS)
							Ongoing on relatives
	♀	38	Sudden collapse	Recidivant syncopal episodes	Negative	Unexplained	Hypertrophic cardiomyopathy
							Ongoing on relatives
	♂	37	Untestified	Negative	Cocaine	Cocaine abuse	No
	♂	40	Untestified	Negative	Negative	CAD	No
	♂	40	Untestified	OSAS	Negative	Unexplained	Ongoing
2018	♂	40	Untestified	Negative	Negative	CAD	No
	♂	28	Untestified	Negative	Negative	Post traumatic haemorrhagic shock	No
	♂	39	Sudden collapse	Negative	Negative	CAD	No
	♂	40	Untestified	Negative	Negative	Unexplained	Ongoing
	♀	31	Untestified	Negative	Negative	Unexplained	Ongoing
	♀	32	Untestified	Negative	Negative	Unexplained	Ongoing

histopathological investigation; in these two cases relatives allowed for genetic test to verify inheritable cardiac structural disease. In 35% of cases (7/20) at both autopsy and toxicology investigations were inconclusive and frozen blood samples were referred to a specialized Inheritable Cardiac Diseases Clinic (ICDC) for genetic testing. In one case, mutation on LMNA gene "likely pathogenic" for dilated cardiomyopathy was detected and relatives (mother, father and 2 y.o. daughter) were immediately involved in genetic test to exclude inheritable cardiac structural disease. In two cases genetic variant for Brugada syndrome of uncertain significance was detected and relatives were invited to ICDC for explanation more explanation about risks. In last four cases, genetic test are still ongoing (Table 2).

3 Discussion

The term "sudden cardiac death" is used when a congenital, or acquired, potentially fatal cardiac condition was known to be present during life, or autopsy has identified a cardiac or vascular anomaly as the probable cause of the event, or no obvious extra-cardiac causes have been identified by post-mortem examination and therefore an arrhythmic event is a likely cause of death. A recent three-year prospective study of sudden cardiac death across Australia and New Zealand recorded an annual incidence of 1.3 cases per 100,000 persons aged 1 to 35 years [7]. A review of death certificates in England and Wales placed the annual sudden cardiac death incidence at 1.8 per 100,000 persons aged 1 to 34 years [20]. A retrospective study of sudden cardiac death in persons 1 to 35 years of age in Denmark showed a higher incidence of 2.8 per 100,000 person-years, or 1.9 per 100,000 person-years when only autopsied cases were considered [8]. These unselected, nationwide, patient cohorts likely provide the most accurate estimates of the incidence of sudden cardiac death in the young. The true overall incidence of sudden cardiac death is likely to be an underestimate, since primary arrhythmogenic disorders can predispose people to more overt causes of death, such as drowning and motor vehicle accidents and a complete methodological post mortem study is still far to be performed in all cases in which cause of death remains unknown. Autopsy is a crucial step in defining the causes of sudden cardiac death, which are broadly categorized as having either a structural or arrhythmogenic basis [21]. Identification of the cause of an unexpected death provides the family **with partial understanding and rationalization of the unexpected tragedy, which facilitates the coping process and allows an understanding of whether the risk of sudden death may extend to family members.** According to the 2015 ESC guidelines all unexplained sudden death victims undergo post-mortem expert examination to investigate whether a cardiac origin should be suspected. Up to 12% of sudden cardiac death in the young have autopsy findings of uncertain significance, such as minor coronary artery disease, mild cardiac hypertrophy or minor myocardial fibrosis, which pose a challenge for the interpretation of the autopsy investigation. Guidelines for post-mortem evaluation of sudden cardiac death in the young recommend collecting blood as a source of DNA for the molecular autopsy to investigate inherited cardiac arrhythmia syndromes. The major problem of genetic testing and the molecular autopsy is establishing genetic causality or pathogenicity. Determining the pathogenicity of the variants identified in the post mortem setting is complicated by the absence of a true "phenotype" in the deceased. This is an important consideration in family management and highlights the need to consider the genetic findings of the molecular autopsy with caution, and in conjunction with clinical findings derived from screening family relatives. These complexities highlight both the need for ongoing collaborative efforts to improve the ways we determine pathogenicity and the key role of the specialized multidisciplinary model of care for SCD and families with genetic heart diseases. This is especially important and relevant in the setting of SADS, where there is no phenotype in the decedent in up to 40% of cases. Finding a relevant clinical cardiac phenotype in family relatives of the decedent may significantly help in the determination of pathogenicity of genetic variants identified in young SADS cases. Given the possibility of an inherited cardiac disease as a cause of sudden cardiac death, appropriate evaluation and management of the surviving family is essential. The care of families in which sudden cardiac death has occurred aims to establish the cause of death in the victim and to clinically screen the surviving family members. In fact, a clinical diagnosis of an inherited cardiac disease can be made in up to half of families. Therefore, first-degree relatives and symptomatic relatives should have a comprehensive medical and family history, physical examination, resting and exercise ECGs and a standard transthoracic echocardiogram. Depending on the clinical situation, further second tier investigations may include CMR imaging, 24-h ECG monitoring and signal averaged ECG, and pharmacological challenge tests. Clinical evaluation alone in families with a sudden unexplained death may identify an underlying cause in up to 50% of selected and comprehensively evaluated families in tertiary centers [22, 23].

A common challenge with clinical evaluation is that not every at-risk family member will be revealed since arrhythmia syndromes display incomplete penetrance and variable expressivity. For example, incomplete penetrance is common in long QT syndrome with up to 40% of gene carriers having a normal QT interval. Therefore, asymptomatic relatives are generally screened up to age 40 years, as many genetic heart diseases most commonly manifest clinical disease in the second decade of life. Furthermore, given the challenges, genetic counselling as part of a multidisciplinary clinic is an important part of family management. Identification of a clearly pathogenic (disease causing) variant in a previously defined unexplained sudden cardiac death has high diagnostic value with at least two major clinical implications. First, the genetic confirmation of a cause of death has a major influence in families in coming to terms as to why their child or spouse died suddenly and brings some level of closure in this respect. Second, the pathogenic variant provides the family with a diagnostic test for screening other at-risk family members, in conjunction with clinical screening approaches. Offering cascade genetic testing to asymptomatic relatives should always be performed in conjunction with clinical evaluation, and only alongside comprehensive pre- and post-test genetic counseling. Defining the precise cause of death allows accurate identification of at-risk relatives and prudent initiation of therapeutic strategies, such as lifestyle modification, beta-blocker therapy and implantation of a cardioverter-defibrillator. Family management in the setting of sudden cardiac death of a young person is complex and ideally suited for a multidisciplinary specialized approach [24, 25].

References

1. Priori SG (2015) The Task force for the Management of patients with ventricular arrhythmias and the prevention of Sudden cardiac death of the European Society of Cardiology: 2015 ESC guidelines for the management of patients with ventricular arrhythmias and the prevention of sudden cardiac death. Eur Heart J 36:2793–2867
2. Ackerman M, Atkins DL, Triedman JK (2016) Sudden cardiac death in the young. Circulation 133:1006–1026
3. Stecker EC, Reinier K, Marijon E et al (2014) Public health burden of sudden cardiac death in the United States. Circ Arrhythm Electrophysiol 7:212–217
4. Semsarian C, Sweeting J, Ackerman MJ (2004) Sudden death in young adults: a 25-year review of autopsies in military recruits. Ann Intern Med 141:829–834
5. Zipes DP, Wellens HJ (1998) Sudden cardiac death. Circulation 98:2334–2351
6. Eckart RE, Shry EA, Burke AP et al (2011) Sudden death in young adults: an autopsy-based series of a population undergoing active surveillance. J Am Coll Cardiol 58:1254–1261
7. Bagnall RD, Weintraub RG, Ingles J et al (2016) A prospective study of sudden cardiac death among children and young adults. N Engl J Med 374(25):2441–2452
8. Winkel BG, Holst AG, Theilade J et al (2011) Nation wide study of sudden cardiac death in persons aged 1–35 years. Eur Heart J 32(8):983–990
9. Doolan A, Langlois N, Semsarian C (2004) Causes of sudden cardiac death in young Australians. Med J Aust 180(3):110–112
10. Puranik R, Chow CK, Duflou JA, Kilborn MJ et al (2005) Sudden death in the young. Heart Rythm 2(12):1277–1281
11. Bagnall RD, Weintraub RG, Ingles J et al (2011) A prospective study of sudden cardiac death among children and young adults. N Engl J Med 374(25):2441–2452
12. de Nronha SV, Behr ER, Papadakis M et al (2014) The importance of specialist cardiac histopathological examination in the investigation of young sudden cardiac death. Europace 16(6):899–907
13. Papadakis M, Raju H, Behr ER et al (2013) Sudden cardiac death with autopsy findings of uncertain significance: potential for erroneous interpretation. Circ Arrhytm Electrophysiol 6 (3):588–596
14. McKenna WJ, Stewart JT, Nihoyannopoulos P et al (1990) Hypertrophic cardiomyopathy without hypertrophy: two families with myocardial disarray in the absence of increased myocardial mass. Br Heart H 63(5):287–290
15. Varnava AM, Elliott PM, Baboonian C et al (2001) Hypertrophic cardiomyopathy: hitopathological features of sudden death in cardiac troponin T disease. Circulation 104 (12):1380–1384
16. Semsarian C, Hamilton RM (2012) The key role of the molecular autopsy in sudden unexpected death. Heart Rhythm 9:145–150
17. Semsarian C, Ingles J, Wilde AA (2015) Sudden cardiac death in the young: the molecular autopsy and a practical approach to surviving relatives. Eur J Heart 36:1290–1296
18. Ingles J, Semsarian C (2014) Conveying a probabilistic genetic test result to families with an inherited heart disease. Heart Rhythm 11:1073–1078
19. Ingles J, Semsarian C (2014) The value of cardiac genetic testing. Trends Cardiovasc 24:217–224
20. Papadakis M, Sharma S, Cox S, Sheppard MN, Panoulas VF, Behr ER (2009) The magnitude of sudden cardiac death in the young: a death certificate-based review in England and Wales. Europace 11(10):1353–1358

21. Bagnall RD, Semsarian C (2015) Role of the molecular autopsy in the investigation of sudden cardiac death. Progr Ped Cardiol 45:17–23

22. Behr ER, Dalageorgou C, Christiansen M et al (2008) Sudden arrhythmic death syndrome: familial evaluation identifies inheritable heart disease in the majority of families. Eur Heart J 28:1670–1680

23. Behr E, Wood DA, Wright M et al (2003) Cardiological assessment of first degree relatives in sudden arrhythmic death syndrome. Lancet 362:1457–1459

24. Amin AS, Wilde AM (2017) The future of sudden cardiac death research. Prog Ped Cardiol 45:49–54

25. Lahrouchi N et al (2017) Utility of post mortem genetic testing in cases of sudden arrhythmic death syndrome. J Am Coll Cardiol 69(17):2134–2145

Have a Healthy Lifestyle or Organize Work – Creating Healthy Shipboard Work Environments

Gesa Praetorius[✉], Cecilia Österman, and Carl Hult

Linnaeus University, 39182 Kalmar, Sweden
{gesa.praetorius, cecilia.osterman, carl.hult}@lnu.se

Abstract. This paper presents findings from a study concerning the work environment on board Swedish passenger vessels. The study explored work-related experiences of personnel in the service department (hotel, restaurant, catering, shops) based on individual and group interviews, observations, survey data and social insurance statistics concerning sick leave longer than 60 days. The results of this paper are based on ten semi-structured individual and group interviews with 16 respondents. The respondents were HR personnel from six shipping companies and crewmembers working onboard.

The results show that in the HR personnel's perception, healthy work environments are often associated to individual personal health activities, such as access to a gym or healthcare, lectures or other measures directed towards the individual seafarer. Aspects of the organizational and social work environments were barely mentioned as stressors or as contributing factors to an increasing number of sick absences. The interviewed crewmembers, however, highlighted the need for both organizational and social measures to foster healthy work environments. The need for employee participation within the organizational design and decision-making processes, including methods on how to conduct risk assessments prior to physical and organizational changes and follow up their consequences were emphasized. Thus, to create safe and sustainable work environments on board, more attention needs to be directed towards including shipboard personnel in the physical and organizational design of their own work environment rather than promoting a healthy lifestyle through measures directed towards the individual worker.

Keywords: Participatory ergonomics · Occupational safety and health Employee participation

1 Introduction

The maritime domain is an inherently international business with a long history of moving people and goods across oceans, rivers and across the borders of countries. The owner, or owners, of a ship need not to be from the same organization, or even country, as those operating it, which in turn may imply differences in technical and staff management. Often the crew on board can be of various nationalities and cultural background, not necessarily from the same country where the ship is registered. This

© Springer Nature Switzerland AG 2019
S. Bagnara et al. (Eds.): IEA 2018, AISC 819, pp. 455–464, 2019.
https://doi.org/10.1007/978-3-319-96089-0_49

complex nature of the global maritime industry means that the ship and its crew must adhere to several set of national, regional and global rules and regulations that are enforced by different Authorities and Recognized Organizations.

The modern ship is still a risky and hazardous workplace, despite significant changes due to increased automation and more sedentary and supervisory work tasks [1, 2]. Seafarers are exposed to constant background noise and whole-body vibrations from propulsion and auxiliary systems, and many arduous manual tasks undeniably remain. With changes in the organization of work, comes less people on board, and seafarers of today are to a large extent knowledge workers, in a slim and highly professional work system. This has led to a change, also in symptoms and origins of seafarer morbidity with a greater risk of psychosomatic disorders [3].

Personnel that work in the service department on ships, especially in the hotel, restaurant and shop departments on board cruise ships, report a greater degree of perceived exhaustion, job stress and fatigue than do other positions on board [4].

A high incidence of work related ill-health in this group suggests that many individuals suffer from pain and sometimes lifelong disability and relegation from the labor market. It may also affect the core business in the cruise industry; service quality and guest experience, as well as unnecessary societal costs. In the light of alarming news concerning seafarers' mental health and suicide rates [5], shipping companies are exposed to more public pressure to take responsibility for the creation of healthy and socially sustainable work environments on board. Furthermore, the introduction of the Maritime Labor Convention [6] in 2013 and its guidelines on the implementation of Occupational Safety and Health (OSH) management, has led to an increased pressure on ship owners and ship operators to promote safety and health as an integral part of the day to day management of work aboard.

While there is an increasing number of studies addressing human operators in sociotechnical ship systems, most of the studies addressing work environment aspects have focused on cargo vessels, and primarily on the bridge or engine room personnel [e.g. 7, 8]. In addition, the need for improved design of human—machine interaction, user-centered design of workspaces and technology, as well as end-user involvement has been the focus of several recent publications [e.g. 9–11]. Publications with a specific focus on the service department's work environment are rare and mostly focus on cruise liners from a more business-oriented perspective, addressing the impact of employee satisfaction and motivation on customer satisfaction [12, 13]. Other research in the cruise liner domain has addressed occupational communities [14], work perceptions [15] and working and living conditions onboard [16]. Two recent studies address the service department onboard of Swedish passenger vessels. Forsell and Eriksson [17], conducted a study on work environment for seafarers on board of Swedish vessels and found indications for work environment problems associated with noise exposure and heavy workload on neck, back and arms. Ljung and Oudhuis [18], primarily focused on the safety perceptions of service department employees, but also discussed the increased need for flexibility among the service crew as consequence of a constantly decreasing number of crew, temporary work contracts and financial pressures.

The present paper reports parts of the findings from a larger research project conducted during 2015–2017 within the Baltic ferry segment. Baltic ferries are ferries operated around the Baltic Sea and the Skagerak which offer full hotel, restaurant and entertainment services coupled with roll-on roll-off cargo transport service (RoPax) [19]. The overall aim of the project was to analyze work-related experiences based on interviews, observations, survey data and social insurance statistics concerning sick absences longer than 60 days. The project was a cooperation between researchers at Linnaeus University and the Swedish Social Insurance Agency for Seafarers (Försäkringskassan Sjöfart) funded by the Swedish Mercantile Marine Foundation.

This paper particularly focuses on contrasting the perceptions of HR personnel and personnel working in the service department on board passenger ships, concerning what constitutes a healthy work environment. The research questions that guided this partial study were the following:

1. How does HR personnel define and work for creating a healthy work environment on board?
2. What are the problems and concerns experienced by the personnel in the service departments onboard?
3. How can the work environment in the service department be improved to promote employee well-being and counterbalance the current trend of ill-health among employees?

2 Data Collection and Analysis

Ten semi-structured (n = 7) and group (n = 3) interviews were conducted with personnel (n = 11) and HR-professionals (n = 5) representing several different shipping companies within the Swedish passenger vessel segment. Five of the interviews were conducted with HR representatives for shipping companies representing a total 64% of the Swedish flagged merchant fleet, while the rest of the interviews were conducted with personnel from the catering department on board passenger vessels. All interviews with HR personnel were semi-structured and followed an interview guide. Beside demographic questions, the interview guide was centered on the role of HR in creating and promoting a healthy work environment in general, and for the service personnel specifically. Questions related both to general rules of conduct, *i.e.* policies for HR work with regards to health promotion, as well as to concrete measure the companies take to address increasing numbers of sick absences, especially long-term sick absences within the service department. Four of the five interviews were conducted via the telephone.

To contrast the HR department's perspective on the promotion of health and well-being at work, five interviews with representatives (n = 11) for the service departments were conducted on board two different passenger ships. Due to the organization of work and availability of respondents during their time on board, all interviews with personnel from the service departments but one was conducted as group interviews with two or more participants at a time.

All interviews were transcribed verbatim. Based on the transcriptions, codes were developed in an interactive inductive process. In a first step, the coding focused on identifying the role and assignment of the HR department in relation to the work environment on board. In a second step, aspects related to the physical, organizational and social work environment were identified in the data. These were then further divided into sub-codes addressing potential reasons and counter-measures for ill-health in the work environment. As an example, the code *organizational work environment* contains the sub-codes employment type, work environment responsibility, organization of work, leadership, education, ill-health and sick absences, status, symptoms for stress and ill-health, while the code *HR* contained the sub-codes assignment and responsibility, cooperation within and outside the shipping company.

3   Analysis and Results

3.1   The HR - Perspective

Healthy Work Environment. All the interviews were initiated by asking the respondents to give their view on the concept of health, what it means to them, and what characterizes a healthy workplace. To this set of questions, the respondents all identified both physical and psychological well-being as essential characteristics for a healthy work environment. While physical well-being is somewhat a precondition for overall well-being, it encompasses much more than the physical aspects of a workplace: *"It is essentially about being well and feeling good more than being* [physically] *healthy. Even if physical well-being contributes to it"* [HR respondent].

As all participants highlighted, aspects of psychological well-being are another important aspect and normally a healthy work environment offers both, the opportunity to take care of oneself and to be and stay physically healthy.

Further, as emphasized in the quote below, while acknowledging the responsibility of the company, and especially the HR department, the interviewed HR personnel also pointed to the responsibility of the individual employee: *"A health promoting work environment is characterized by having the opportunity to take care of yourself. You can never force people to do that, but you can provide them with an opportunity to do so."* [HR respondent]. Here, the respondents clearly stressed the importance of responsibility; a healthy work environment needs to be created and a company can only offer opportunities, but the willingness of the employees is needed to participate in, and at least partially take charge of their own well-being. All respondents stated that their respective companies have developed both group and individual measures to support a healthy work environment.

Health Promoting Measures. The HR personnel was further asked to elaborate on what could be done in general to promote healthy work places and specifically what they already had done or were planning to do within their own organizations. As illustrated in the quote below, the respondents gave many examples of various activities that, however, mostly focused on physical fitness. *"I can start some time back, or many years back in time, as we have worked with health promotion for very long, for example activities and other things such as participate in the Gothenburg half marathon or other competitions"* [HR respondent].

The examples of measures highlighted by the respondents encompassed lectures, sometimes streamed to the onboard personnel, a healthcare center and gym, individual support for running competitions and a health check and health profile for the individual employee. However, economic pressures as well as the working situation of onboard personnel set limits towards the overall use of the measures offered by the company.

In general, all five companies offer services to their employees, but some of the services might be specific to a healthcare center in a certain port, as told by one of the respondents. However, as emphasized already in the definition of health promotion and a healthy work environment, the respondents associated many of the measures to be the responsibility of the individual to take an active step towards a healthier lifestyle, rather than the company's responsibility for the aspects of the work environment stretching beyond the physical well-being and physical health were barely addressed in the responses. *"Well, I mean we are still fighting the basic problems, such as some people not using ear protection, despite the fact that those have been specially molded for them and everything, but they have never used those and will never do it"* [HR respondent].

One explanation, as highlighted in the quote above, might be that the physical work environment on board is very demanding and thus improvements that decrease the burden of work are at the core of the HR department's work. However, it was also emphasized that the provision of risk reducing measure needs to be accepted by the employees and that the offer alone does not solve the problem; the individual must accept the use of the measure, such as specific equipment. The respondents associated this to experiencing a certain "culture" which might hinder onboard personnel to agree to certain measures, such as wearing personal protective equipment (PPE), or participating in activities for a healthier lifestyle.

Slimmed Down Organization. Three of the five HR representatives used the term *"slimmed down organizations"* [translated from Swedish by the authors] when they discussed potential underlying causes for the high level of long sick leave, and the challenge of balancing work-life demands and resources, especially for the service personnel. Another respondent described the same phenomena, but in other words: *"In some way, we work harder and harder, to maximize earnings with as few people as possible."* The only respondent who did not talk about the economics of the work environment in terms like this, spoke of a long reorganization and change fatigue among the crew, adding that *"now we hope the organization is in place so that we can meet the demands with appropriate resources."*

3.2 Personnel's Perspective

Design of the Physical and Organizational Work Environment. Most of the respondents from the personnel described concerns of the limited leverage they have on the design of the physical and organizational work environment. They felt that company decisions are seldom communicated in advance to give employees and their safety representatives enough time to reflect before decisions are implemented. Nor did the respondents experience that employee participation was considered when changes are undertaken, which they are then forced to adapt to. *It often comes from the shore-side and then it is just "poof" and we must carry it out. It doesn't feel like that there is any participation at all. Instead of asking us "Do you have any clever ideas?" and "what would you think about this?"* [Respondent, restaurant].

Several examples of interventions that didn't have the desired effect on working conditions were raised by the participants. Without employee participation in the decision-making processes of purchases of new equipment, and the design and re-design of workspaces, some of the actions taken by the employer that were supposed to improve the work environment, rather generated new work tasks and increased burdens.

One example was the investment of a new (rather expensive) special carpet to decrease the strain on back and joints by dampening vibrations from the vessel's hull and machinery. The carpet was laid out surrounding a bar area. While the crew welcomed the effort by the company, it was emphasized that the desired effect was not achieved – the carpet was very heavy, difficult to move for cleaning, and was not very well adapted to the area covered. As a result, being able to clean the bar now requires several crewmembers to move the carpet. Therefore, an additional work task with increased weight on back and joints was created, which could have been avoided if personnel had been included when the carpet was bought. *"And especially if there is any change, something is re-built, or new things are ordered and so on, why not listening to those that have to work with it and are supposed to get the job done"* [Respondent, bar].

Further, the work environment in the service department is often characterized by narrow workspaces in which the use of supporting equipment such as trolleys might not be possible. Thus, supporting the personnel in their tasks and including them in decisions about their work equipment becomes particularly important.

Job Security, Shift Work and Working Hours. Instability in the work force was also highlighted by the respondents as influencing the personnel's health. Insecure employment contracts, sometimes week-to-week employments, have created a demanding situation with a high turnover rate of new employees who need to be trained and integrated into the workforce. As mentioned by a restaurant manager on board there have been new employees every week: *They must do that, every week there are two new ones. In general, every week that has passed, for half a year* [head chef]. This creates an additional burden on those with permanent positions, as new employees need to receive a work place introduction and training, which needs to be conducted during a normal shift increasing the stress experienced by some of the personnel. As one respondent expresses it, *"And then one already feels bad when I see that I have to work with three new people and I don't have a clue on how they work, what they are capable to do and what not"*.

Further, the short-term contracts, on which new employees often are hired on, create not only uncertainty about the capabilities of one's co-workers, but also con-tribute to the experienced psychological pressure of not knowing whether colleagues will suddenly disappear and not return to the vessel upon the next assignment. *"I don't know who I will work with next week when I come back, but suddenly three new names appear where I am working"* [Respondent, restaurant]. Many of the respondents experienced this as particularly impacting on their psychological well-being as a constant employee turnover would leave them reluctant to even try to get to know their new colleagues.

A normal working day is scheduled for at least 10 effective working hours, in which many of the employees change between various stations in the service department, *i.e.* after finishing a shift in the restaurant, a waiter might move to a bar and continue his/her work there. Many of the respondents thus felt that there is no time for a short recovery break during their working hours, let alone to socialize. While this used to be possible some years ago, according to the more experienced respondents, the slimming of the organization aboard and high employee turnover with a constant influx of new employees has cut down those possibilities.

Perception of Health Promoting Measures. Concerning health-promoting measures, the respondents acknowledged the companies' efforts with offering lectures, gym access and other health promoting measure, but stated that there are many other aspects in the organization of work, workspace design and access to equipment that should be prioritized. One of those examples were slip resistant work shoes that dampen the effects of vibrations. While these shoes are part of the PPE in some departments onboard, employees in other departments need to pay for those themselves. The personnel felt that the proper work shoes would be beneficial for waiters and all kitchen staff as well and should therefore not be paid for by the individual worker. Several other examples were given by the respondents were they felt that the shore-based organization has demonstrated a limited understanding of the settings of the working and living environment onboard and had made decisions that primarily affected the personnel's spare-time onboard, thus what they considered to be private and not a company matter.

Raised by almost all respondents was the need for more understanding of the shore-organization towards the physical job demands in the service department. Carrying heavy loads for long hours might leave the personnel reluctant to participate in physical exercises or make use of the gym to increase the physical health.

4   Discussion

The focus of this paper was the contrasting perceptions of HR personnel and personnel working in the service department on board passenger ships, concerning what constitutes a healthy work environment.

4.1   How Does HR Personnel Define and Work for Creating a Healthy Work Environment Onboard?

The HR respondents were found to focus strongly on individual measures for health - despite a clearly demonstrated awareness during the interviews, of what organizational and social factors are important for a healthy workplace. This indicates a lack of appropriate tools and possibly a limited understanding of the working and living conditions for service department employees onboard. The respondents know how it *should be* but not how to get there – at least not within their own limits of economic aspects and decision making as working in the HR department. The primary focus within the shore-based organizations seems to be vessel and passenger safety, *e.g.* fire safety, life-saving equipment, rather than the day-to-day occupational health and safety management within the various departments onboard. Furthermore, the HR respondents also assigned a large responsibility for the individual workers well-being, overlooking that ill-health can have underlying causes related to aspects such as workforce turnover, job insecurity, or lack of social support on and off working hours on board.

4.2   What Are the Problems and Concerns Experienced by the Personnel in the Service Departments Onboard?

The personnel in the service department raised concerns and problems with both the physical workspace design and the overall organization of work and work tasks. As shown above, not having access to supporting equipment such as trolleys in the restaurant, or proper work shoes adds to the experienced burden, despite measures being well-known and available. Furthermore, a lack of involvement in the decision-making processes about equipment purchases and design of workspaces has created frustration when well-intended interventions implemented by the company did not reach the desired positive outcome.

The job insecurity and high employee turnover was especially highlighted as something having a severe impact on the physical and psychological well-being. It is creating a work organization in which training and workplace introductions are recurring additional tasks during shift work for those already working long hours. Furthermore, the slimming of the organization and employee turnover also limited the personnel's ability to socialize during work hours and the time they spent off work.

Lastly, the respondents attributed many of the raised concerns to a limited understanding of the shore-based organization for their work and living situation.

4.3   How Can the Work Environment in the Service Department Be Improved to Promote Employee Well-Being and Counterbalance the Current Trend of Ill-Health Among Employees?

In 2010, the World Health Organization (WHO) published the WHO healthy workplace model [20]. It is presented as a comprehensive way of thinking and acting to create, promote and foster occupational health, safety, well-being and sustainability across industries. The model offers concrete guidance on how to improve the physical and psychosocial work environment and addresses work-related physical and psychosocial risks, the promotion and support of healthy behaviours, as well as broader social and environmental determinants. In the report, a healthy workplace is defined as one in which *"workers and managers collaborate to use a continual improvement process to protect and promote the health, safety and wellbeing of all workers and the sustainability of the workplace"*. The need for collaboration between employer and employees is especially highlighted and involved workers are seen as one of the key elements to healthy workplaces. While the personnel onboard perceived the efforts from the shore organization as positive, there was an obvious lack of interaction, cooperation and participation in the design and re-design of workspace, which the workers experienced as frustrating. Even if measures were intended to have a positive impact, some of them, such as the carpet mentioned by one of the respondents, rather increased the burden of work and created new tasks in an already very stressful environment.

Further, due to the high employee turnover, an uncertainty is created in the work environment, which often leads to yet another reorganization of the work, deviating from how it was planned, which is perceived as emotionally exhaustive. With a constant influx of new employees, the overall organization of work needs to change and acknowledge that a novel worker needs to be properly introduced and trained in the specific work environment. For that to be successful, it is necessary to set time aside so that social work experience, in terms of techniques and organization, can be transferred between the co-workers, and for the new employees to be properly integrated into the existing workforce. This also requires resources for socializing both during and after a shift.

Health promoting measures should take their beginning in the concerns of the employees, acknowledging the special needs that come with working onboard. While the HR respondents acknowledged that to a certain degree, there needs to be an increased understanding for the physical demands of work onboard recognizing the responsibility of the employer to create the preconditions for a sustainable work-life balance and a healthier lifestyle, without forcing the employees into certain choices.

5 Conclusion

The service department represents the largest group of employees on a passenger vessel. Thus, increasing numbers of sick absences create direct and indirect costs to both employers and the society. This paper has shown that there are many initiatives taken by the companies to promote a healthy work environment, but that they might fail to acknowledge the concerns and problems experienced by those working onboard. To close the gap between ship and shore, a better structure for employee participation in design, re-design and purchase decisions is suggested, as well as an improved communication plan for how personnel can raise their concerns towards those working to create a healthier work environment for them. Lastly, HR departments need to acknowledge that individual lifestyle choices and physical well-being are only one part of what shapes the preconditions for healthy employees – the ability to socialize with co-workers, build strong teams over time and have recreational facilities adapted to their needs are equally important.

References

1. Ellis N, Sampson H, Wadsworth E (2011) Fatalities at sea. In: Seafarers international research centre symposium proceedings 2011. Seafarers International Research Centre (SIRC). Cardiff University, Cardiff, pp 46–65
2. Oldenburg M, Baur X, Schlaich C (2010) Occupational risks and challenges of seafaring. J Occup Health 52:249–256
3. Oldenburg M, Hogan B, Jensen H-J (2013) Systematic review of maritime field studies about stress and strain in seafaring. Int Arch Occup Environ Health 86(1):1–15
4. Österman C, Hult C (2016) Administrative burdens and over-exertion in Swedish short sea shipping. Marit Policy Manag 43(5):569–579
5. UK Chamber of Shipping. Breaking the taboo of seafarer mental health. https://www.ukchamberofshipping.com/latest/breaking-taboo-seafarer-mental-health/
6. ILO (2006) Maritime Labour Convention. International Labour Organization, Geneva
7. Hetherington C, Flin R, Mearns K (2006) Safety in shipping: the human element. J Saf Res 37(4):401–411
8. Ljung M (2010) Function based manning and aspects of flexibility. WMU J Marit Aff 9 (1):121–133
9. Kataria A et al (2015) Exploring Bridge-Engine Control Room Collaborative Team Communication
10. Österman C, Berlin C, Bligård L-O (2016) Involving users in a ship bridge re-design process using scenarios and mock-up models. Int J Ind Ergon 53:236–244
11. Costa NA, Holder E, MacKinnon S (2017) Implementing human centred design in the context of a graphical user interface redesign for ship manoeuvring. Int J Hum Comput Stud 100:55–65
12. Sehkaran SN, Sevcikova D (2011) 'All aboard': motivating service employees on cruise ships. J Hosp Tour Manag 18(1):70–78
13. Larsen S, Marnburg E, Øgaard T (2012) Working onboard – job perception, organizational commitment and job satisfaction in the cruise sector. Tour Manag 33(3):592–597
14. Lee-Ross D (2006) Cruise tourism and organizational culture: the case for occupational communities. Cruise ship tourism, pp 41–50
15. Dennett A et al (2014) An investigation into hospitality cruise ship work through the exploration of metaphors. Empl Relat 36(5):480–495
16. Gibson P (2008) Cruising in the 21st century: who works while others play? Int J Hosp Manag 27(1):42–52
17. Forsell K et al (2017) Work environment and safety climate in the Swedish merchant fleet. Int Arch Occup Environ Health 90(2):161–168
18. Ljung M, Oudhuis M (2016) Safety on passenger ferries from catering staff's perspective. Soc Sci 5(3):38
19. Stopford M (2009) Maritime economics. Routledge, New York
20. WHO (2010) Healthy workplaces: a model for action: for employers, workers, policy-makers and practitioners. World Health Organization

Toward a Better Assessment of Occupational Exposure to Nanoparticles Taking into Account Work Activities

Louis Galey[1(✉)], Sabyne Audignon[1], Olivier Witschger[2],
Aude Lacourt[1], and Alain Garrigou[1]

[1] University of Bordeaux, Inserm, Bordeaux Population Health Research Center,
Team EPICENE, UMR 1219, 146 rue Léo Saignat, 33000 Bordeaux, France
louis.galey@u-bordeaux.fr
[2] Institut National de Recherche et de Sécurité (INRS), Rue du Morvan,
CS 60027, 54500 Vandoeuvre Les Nancy, France

Abstract. Numerous industrial sectors and processes may cause worker exposure to ultrafine particles or engineered nanoparticles (NPs). These exposures may affect workers' health if control measures (organisational, engineering, PPE…) are not properly defined, used and maintained. Today the available exposure data are insufficient for epidemiological studies. Part of the reason for this lack of data may be found in the variety of approaches used to assess worker exposure.

In this work, we propose an approach integrating aerosol measurement and work activity analysis to enhance exposure assessment, based on present recommendations to assess occupational exposure to NPs. This multi-disciplinary approach combines information gathering, real-time measurements and aerosol sampling (for physicochemical analyses), contextual information and work activity observation, video exposure monitoring, interviews with stakeholders, and development of appropriate safety measures.

Real exposure situations characterized as typical exposure situations, permitted to gather information on exposing work activities that can occur on workplaces thus making the determinants of exposure more visible. Work activity explains some variations on exposures objectified by video and measurements. Exposures become visible and documented, which triggers debate, discussions and even in some cases opens up new debate spaces. This method aims to help address the concerns of companies introducing new technology that is conducive to important work transformations. In this way, companies break through technological barriers, transform work practices, which makes an upgrade of prevention practices necessary. The approach also aims at setting a landmark method for future epidemiological studies.

Keywords: Exposure · Measurement · Typical exposure situation

1 Introduction

Almost all industrial sectors include workplace environments where worker exposure to airborne nanoparticles may occur. Nanoparticles (NPs) have a nominal diameter smaller than about 100 nm, or clump together as aggregates or agglomerates ranging in size from a few tens of nanometers up to several micrometers. These airborne NPs can be released as an unintentional by-product of a process, in which case they are generally designated as "ultrafine particles". Alternatively, they may be released during the production or handling of manufactured nanomaterials; called "engineered nanoparticles". Occupational exposure to ultrafine particles can occur where high-energy (temperature, speed or combustion) processes take place. These processes may be innovative (thermal spraying, additive manufacturing, etc.) or more common (welding, thermoplastics processing, etc.).

In many toxicological studies, the health effects caused by the inhalation of NPs are underlined: oxidative stress, genotoxicity and inflammatory responses (Bakand and Hayes 2016). It has also been shown that the biological responses differ from those observed from bulk material of the same chemical composition for the same mass concentration (Oberdörster et al. 2005). The size of these particles induces more intense biological responses and physiological barriers become permeable compared to bulk materials of the same chemical composition and mass. Such evidence has questioned the mass paradigm and has highlighted that other toxicity determinants (Bakand and Hayes 2016) such as particle number, particle surface area and reactivity, particle size and shape should be measured.

In this context of uncertainty regarding exposure and its associated human health effects, exposure assessment to NPs in occupational settings is a challenge. Nowadays, several strategies have been proposed without reaching a consensus and the ordinary practices of industrial hygienists and preventionists must change. Measurement currently plays a key role, when worker exposure assessment is involved and when addressing the management logics that underlie risk governance (Dagiral et al. 2016). However, ergonomic work analysis remains necessary and complementary in relation to highlighting exposures and discussing their determining factors (Garrigou et al. 2011). The work activity is still insufficiently considered in current strategies. Associating measurements and videos of work situations provide a medium for bringing together objective exposure data (Judon et al. 2015) and what is happening in the work situation.

The purpose of this paper is to contribute to developing a method – articulating measurements and work activity analysis – that could be used for the detection of typical exposure situations and thus to improve prevention in companies and develop knowledge on exposures.

2   Theoretical Frame

2.1   Occupational Exposure Assessment to Nanoparticles

Existing strategies to assess exposure to NPs are complex to follow because they differ from one another and require a lot of expertise. The diversity of objectives, nanoparticles, available measurement instruments and actors proposing these strategies also explain this diversity of current strategies. To characterize the most appropriate characteristics of NPs, several instruments must therefore be used in the field despite the disadvantages of this duplication (technical, cost, reliability, need for portability, etc.).

The importance to integrate contextual information that could influence exposure has been underlined (Woskie et al. 2010) to finally be integrated and described for some of them in the latter strategies (Bekker et al. 2015; CEN 2018; Eastlake et al. 2016; OECD 2015). These most recent documents examine how strategies can be made more operational and used by different practitioners who may not always be experts. Conversely, we observe that integrating a description of task or activity is becoming an issue of interest.

Other practices of industrial hygiene include a closer look at work activity. Task-based measurement enables to characterize more precisely some exposure situations (Brouwer et al. 2012). In other cases, video exposure monitoring consisting in synchronizing video and real time measurements shows greater consideration of the work actually carried out (Rosén et al. 2005; Beurskens-Comuth et al. 2011).

This possibility for a better characterization of exposure situation is expanded by new real time instruments (Asbach et al. 2017). The peculiarity of nanoparticles aerosol behavior also leads to taking into account work activity and contextual information to understand exposure.

2.2 Lessons from the Humanities

Several fields of humanities underline methodologies taking into account human work as it is really done. These practices enrich a functional approach of safety through a cultural approach, filling some gaps in the understanding of human work.

Typical action situation (TAS) from design processes in French ergonomist approach was prompted by the lack of consideration of human work in design (Jeffroy 1987). These TASs based on the current work must permit to build requirements for the stakeholders involved in the process or work design project and help identify critical points (Barcellini et al. 2014; Garrigou et al. 1995; Dutier et al. 2015). These authors underline that a global understanding of work will contribute to setting technical, organisational and human requirements. In this way, taking into account TASs can be more important than some technical criteria that can be too remote from work needs. For other authors, TASs will contribute to identifying strategies of operators and future skills needs by future operators (Grosjean and Neboit 2000).

For Daniellou (2004), reference situations from existing work practices have to be identified in contexts similar to future work so as to bring concrete and practical elements to the process that has to be designed. From these reference situations, TASs will be described. All the TASs constitute a library of TASs that will support the design project. These TASs gather variabilities that could be met in the future situations taking into account degraded and normal modes, transition and maintenance, events and incidents, etc. Among these variabilities, exposure to chemicals is rarely considered. Garrigou et al. (2001) noticed the importance of taking into account the potential presence of dusts, or revealed exposure to exhaust fumes from a truck (Garrigou et al. 1995) as health and safety topics discussed in working group sessions (Garrigou 1992). However, it remains difficult to be exhaustive in identifying and inventorying all possible TASs (Garrigou 1992).

Commonly (Daniellou and Garrigou 1992; Daniellou 1992; Garrigou 1992; Garrigou et al. 1995; Daniellou 2004; Duarte and Lima 2012), TASs are defined by: the objectives to realize (the task); the criteria; the stakeholders involved in them; the sources of information; the required means and tools; the constraints of accomplishing these objectives; the external or internal factors that can influence the realization of the objectives.

Once specified, TASs inform design project in different ways depending on the design stage. At the beginning of the project, TASs can enrich and detail project specifications. Then, TASs are gathered in scenarii used to support simulations of future work (Barcellini et al. 2014; Daniellou 2004; Garrigou 1992). The formalization of the TASs can be achieved by the means of prototypes, models, suggestions of work organization or technical suggestions for instance (Garrigou et al. 2001). Furthermore, using TASs (during simulation, in working groups, from a library of TASs…) will contribute to setting requirements integrated by designers. These requirements can be "descriptive" by formulating needs based on TASs and the corresponding variability, "prescriptive" when they are based upon recognized knowledge (anthropometry, toxicology (Garrigou et al. 2001)…) or "procedural" when dealing with the method to follow (stages, stakeholders and tools to be integrated during the design process) (Daniellou 2004). TASs can be described with varying degrees of details depending on what's is needed at this moment in the design process. In addition, TASs can have continuity over time or overlap.

The participation and co construction of the TASs is a key point of the process requesting a strong involvement of the company stakeholders and multidisciplinarity (Garrigou 1992; Garrigou et al. 2001). In this way, TASs are taking the form of intermediary objects (Jeantet 1998; Vinck 2009) as design stakeholders gradually own and shape them depending on the objectives of the project (Duarte and Lima 2012; Barcellini et al. 2014). This representation of work must foster the participation of people that are not always involved (workers for instance) and develop a certain amount of "control" (Garrigou et al. 2001). In some cases, the development of shared representations between workers and designers has been reported (Duarte and Lima 2012), which leads to a common reference frame. These discussions areas (Garrigou et al. 2001) created through then involvement of company stakeholders are a basis for reflection (Béguin 1998). They combine several rationales (Garrigou et al. 2001; Daniellou 2004) and consider the diverse constraints of participants (Daniellou 2004).

Stakes raised by the use of TASs in design ergonomics is to open "space to shape potential future activities" (Garrigou et al. 1995; Daniellou 2004) taking into account human work and its requirements.

Resulting from this literature review, the basic unit of work activity suggested by TASs appears to us relevant to integrate activity in exposure assessment to nanoparticles in an operational way.

3 Method

In this context where the actual work in presence of nanoparticles which has to be transformed, the aim of the method is to describe typical exposure situations (TESs).

The first stage of the study was to develop an operational method articulating the measurements of nanoparticles and the work activity analyses. A working group of experts from disciplines such as metrology of aerosols, industrial hygiene, epidemiology and ergonomics was created. Some of the participants were potential future users of the method. Measurement instruments were selected and tested for their relevance, describing the actual toxicity determinants of nanoparticles (number and mass concentration, size and chemical composition for instance). The intervention team received training to use the measurement instruments and record information on the work situations.

The second stage was to implement the method in two companies presented in Table 1.

Table 1. Main characteristics of companies

Field	Operation/process	Nature of particles
Aeronautics	-Cleaning of the thermal spraying (HVOF) booth	Tungsten carbide, cobalt, chromium
	-Production operation (preparation of the piece to be process, recovery of the piece after spraying…)	
Rubber industry	-Weighing of powders consisting in using bag (25 kg) or big bag to weigh dust in new bag on a scale.	Carbon black, titanium dioxide, zinc oxide, cellulose, calcium carbonate…
	-Mixing mill where worker has to channel the mixed rubber with powders previously weighed between cylinders	

When implementing this methodology (see Fig. 1), special attention was given to 'constructing' the intervention (Garrigou et al. 1995; Garrigou et al. 1998) for each company. Even though preventionists were the main discussion partners, the project was presented to health and safety committees at its start and end. Preliminary interviews with workers, management staff and preventionists were carried out to gather safety documents, discuss actual exposure situation representations and to know how safety rules were established. Then, three main steps were implemented:

1. This jointly performed work to construct the intervention prompted the selection of work situations. In-depth characterization of TESs combining video recordings, sampling and real-time measurements were done. A video exposure monitoring software (CAPTIV®), initially developed by INRS (France's national institute for research and safety), was used to synchronize videos and real time measurements.
2. We then processed the data gathered with the software in the following way:

   – spotting "exposure peaks" (see Fig. 2) in terms of either breathing/heart rate or pollutant concentration and identifying the activity corresponding to these peaks. Simply clicking on the peak displays footage of the activity performed at the time as it has been synchronized with the real time measurement data (see Fig. 2). This is particularly effective when conducting both hazard analyses (frequent in prevention) and activity analyses.

- post-coding of work activity parameters (see Fig. 2) allows time-related statistics (frequency and parameter intensity) to be generated and exposure conditions to be subsequently documented. Crossed statistics on cumulated time are generated when different parameters have significant values or when threshold concentrations are exceeded.

3. The resulting TES from implementation of the method were presented during confrontation interviews (Mollo and Falzon 2004; Clot 2008). These interviews were conducted with working groups comprising workers, management staff (production manager, manager, team leader), preventionists and occupational doctors. These interviews involved developing reflections on the activity (Schön 1983) by confronting one or more operators with traces of the activity (videos, photographs, etc.). The interview was based on video extracts of work activities footage and real time measurements. The exposure peaks recorded by the researchers and the corresponding work activity displayed by the video exposure monitoring program (see Fig. 2) supported the discussion. The objective was to focus discussion on the work activity content and practices actually implemented by the operators.

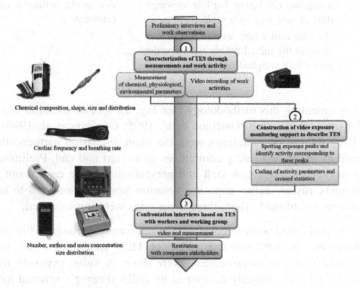

Fig. 1. Study design in companies to describe TES

4 Results

All situations that can expose the worker to NPs have been studied and looked into so as to define what could be called typical exposure situations, typical in the sense that they are representative of the repetition of the same determinants leading to exposure be it once a year or several times a day.

Figure 2 describes an increase of exposure to carbon black micro- and nanoparticles recorded when a weighed bag is closed by the worker. This kind of work activity also leads to vigorous physical strain due to bag handling. A coding of activity leads to correlate the level of exposure to the worker's actions.

From step 2 of the method implementation, several TESs have been objectified through the synchronization of video of work activity and measurements. In the rubber industry, exposure situations occurred when operators empty the bag for weighing, close the weighed bag, empty the bucket or the scoop used to weigh powders, load the bucket and clean the workplace with a broom. In aeronautics, cleaning of the top of the booth with a broom was the most exposing situation. Booth door opening and collection of the manufactured piece also led to residual exposure.

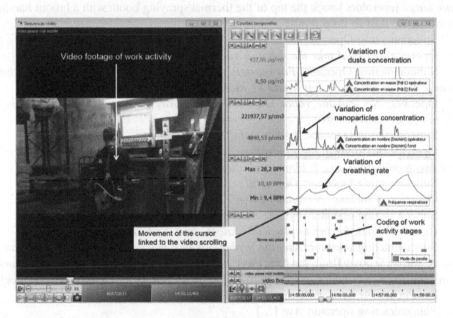

Fig. 2. Software interface to identify TESs articulating video footage of work activities and measurements of nanoparticles in the rubber industry

Step 3 of our method consisting in collective confrontations with workers and management contributed to developing an understanding of exposure situations (i.e. in aeronautics):

"Subcontracting operator: […] *No, it was on my arms. Yeah, on my arms* [dusts come through the (inadequate) suit and deposit on the arms]. *Because then we had the shitty suits? Right? If you don't have recycled air, I see that sometimes, after 15 min I'm taking it* [the respiratory protective equipment], *the thing is soaking wet.*

Researcher: *And do you think that the mask protects you?*

Subcontracting operator: *For me yes. So from time to time our mouth still feels* [mimes a hairy tongue] *Is it because we're constantly wearing the mask or because some dust is coming through, I wouldn't know.*

Researcher: *So what are you feeling then?*

Subcontracting operator: *Err, hairy tongue... tastes funny from time to time.*

Researcher: *What kind of taste?*

Subcontracting operator: *Can't explain it, really... I dunno, disgusting, like. It's odd, my mouth doesn't normally taste like that.*

Researcher: *And do you know whether there are any other control measures on that machine? Any meetings?*

Operator: *On that machine? Meetings, no. Or at least not with me.*

Researcher: *And do you think it could be done differently to avoid that kind of... knocking?* [operators knock the top of the thermal spraying booth with a broom handle to cause the dust to drop down]

Subcontracting Operator: *Well when...*

Operator: *It would take the robot, a vacuum, or get some partitions...*

Subcontracting Operator: *For starters it's odd that there isn't any big fan that would blow it all away, that would be nice. Then again, does it even exist?*

Operator: *No, but the robot has some... The way we do it, for all the bench cleaning, we use the robot.*

Subcontracting Operator: *Is that right???*

Operator: *All steps of the programme, it blows. We use the robot air because it's got two terminal air jets. It blows and after vacuuming* [it collects dusts]...

Subcontracting Operator: *Oh, right!*

Operator: *Knocking, personally I don't think it's great.*

Subcontracting operator: *No, me neither.*

Operator: *First, having to physically knock on it, and then on top of that...*

Subcontracting operator: *Aye, the problem is that you can't walk away. You can't knock and walk away* [from the gangway], *because there is so little space, so you knock and then you need to duck or hide to avoid, to avoid getting all the dust in your face!!*

Researcher: *At least, turn your face away...*

Subcontracting operator: *Aye* [...]

Researcher: *And do you feel hotter with the new suit then?*

Subcontracting operator: *Aye, a little bit. They're thicker so they feel a bit warmer.*

Operator: *Still, we're better off like this I think.*

Subcontracting operator: *You're right.*

Operator: *Look, you're closing your eyes!* [...]

Researcher: *We can see some variations* [in the data from the measurement instruments] *linked with.*

Subcontracting operator: *Linked with what I'm doing.*

Researcher: *So it can't be pollen either* [...]

Operator: *So in fact, the heart beat is when you're moving!*

Subcontracting operator: *Aye, that's when it's higher.*

Researcher: *Is that why you were talking about the robot? It would avoid movements?*

Subcontracting operator: *Arm movement and all the rest* [...]

Researcher: *Is that something you had considered? Perhaps not to that extent?*
Subcontracting operator: *Nope.*
Operator: *I thought we weren't really exposed, but seeing the footage!* [...]
Researcher: *So do you wear your own clothes?*
Subcontracting operator: *Well Yeah. This is mine, this is mine* [points to clothes on the screen]
Researcher: *So this means that you clean the booth up with you own clothes underneath? And?*
Subcontracting operator: *And then I put them in the washer. But I sort them. I don't clean my work clothes with other* [personal] *clothes.*»

Restitution of results based on TESs with management in aeronautics led to changing the cleaning procedure: another operator with appropriate protective equipment and training will ensure the cleaning. Furthermore, changes in programming the robot to automatically blow the dust are currently under consideration.

Putting step 3 into practice is at the heart of our methodology since it aims to enable the different stakeholders to discuss and debate their work activity. This step is a key point for a collective analysis of TESs. This collective analysis develops the identification of a TES (step 1) into a detailed specification of a TES (step 3). Simultaneous viewing of activity footage and synchronized exposure data during confrontation interviews makes it possible to understand how exposure occurs. This in turn helps workers discuss and question what they do, how they do it and consider exposure levels. Relevance or efficiency of implemented prevention practices are collectively debated in regard to the real exposure situation. The work as it is actually done and the situations where exposure occurs described by TESs are envisioned and considered by company stakeholders.

5 Discussion

While the first step of our methodology consists in establishing hypotheses on the exposure situations, the joint use of measurements and work activity analyses permit to support the objectification of exposure situations. Nevertheless, work observation is already a way to identify TES, as the first step of the method implementation has shown us. In some cases, measurements can make visible an exposure that would otherwise remain imperceptible by sight or smell. Moreover, measurements specify the nature of nanoparticles or the situation in which workers are exposed. A contribution of TESs is to integrate representations, beliefs, fears and concerns of workers in situations considered as exposing. Taking into account this subjectivity is necessary to improve safety practices. Depending on the objectives, several levels of TES description can me made (i.e. cleaning of the thermal spraying booth, or cleaning while the worker knocks with a broom the top of the thermal spraying booth from the gangway). Obviously, a higher level of details will be useful to identify determinants to change but may not be relevant for epidemiological studies.

Thus, TESs are a useful tool to build a more operational safety system and bring information in the design stage. The basic unit of real exposure situation analysis can be used by a diversity of stakeholders (Buchmann and Landry 2010) internal or external: industrial hygienists, preventionists, epidemiologists, ergonomists, occupational doctors… As an object for mediating discussion between different stakeholders and occupational health disciplines, TES is reminiscent of object-boundary selection in participative ergonomics processes. These objects are essential to encouraging stakeholder participation and involvement, hence a cooperative design. They are needed to create a common language that allows representation of the knowledge and ideas of stakeholders (Broberg et al. 2011). As it creates knowledge on exposures, epidemiologists can also integrate these factual situations in studies. Basic unit of TESs can be assembled in a scenario which is a valuable basis from which to transform actual work situations or enrich the design process as it can be done for TASs (Garrigou et al. 1995; Barcellini et al. 2014).

The social construction (Garrigou et al. 2001; Barcellini et al. 2014) work around the implementation of the method is important to enable company stakeholders to understand the observation data generated and their usefulness, while developing a climate of confidence and a dynamic around the project. In this regard, all the social construction spaces play a structuring role in relation to project performance. It is the ergonomist's role to develop social construction.

Involving operators in the process of exposure characterisation is meant to refine the comprehension of the proposed exposure situations and their determinants. Indeed, although situations may be approached thanks to the methods of ergonomics, understanding the causes of possible health effects remains a riddle (Garrigou et al. 2004). Solving this riddle requires investigating in the company, calling on all the stakeholders and using relevant exposure analysis techniques.

Actors involved in these working groups to build TESs can be occupational doctors, management, preventionists, workers, R&D representatives, engineers in charge of the processes or designers… Participation of stakeholders contributes to precise new TESs that were not initially identified or occurred during step 1 of exposure assessment. The expertise and experience of company stakeholders ensures TESs are accurate and detailed.

A significant advantage as opposed to design processes is that the situations already exist. They just need to be changed and those changes are precisely what the method aims to identify. Without this need of transposition that is a characteristic of TAS describing task, TESs are based on work activities. Sometimes, discussion aiming to transform actual situation can lead to project the future activity from TESs as it is done with TASs in a design project. New work situations implementing transformations and new preventive measures are discussed by participants of working groups. How big the change is determined by a better understanding of the risk, and of the exposure in particular. Figure 3 introduces the TES components, through technical, organisational and human determinants.

The issues raised by the development of nanotechnologies stress the need to integrate human work for efficiency and a safe development of innovation. TESs offer a possibility to question and inform choices in innovating process design, by exposure assessment on the field. Results of our study underlined the need to develop field intervention, and possibilities to implement field studies. In this way, setting up requirements for a safe development of nanotechnologies could become possible. TESs may offer the opportunity to integrate safety at the earliest stage of innovation.

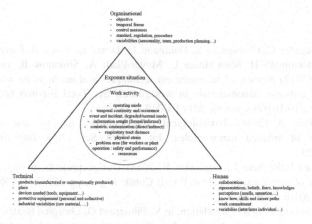

Fig. 3. Typical exposure situation components

6 Conclusion

Work activity analysis combined with nanoparticles measurements provides a better understanding of the exposure determinants, describing real-life exposure scenarii, and results in recommendations for tailored prevention. The implementation of this method reveals how workers and companies deal with risks resulting from nanoparticles in workplaces. We noticed different safety climates between aeronautics companies and rubber plant. High levels of nanoparticles exposure were recorded in rubber industry while laboratory and aeronautics companies wondered how to assess exposure generated by innovative processes to develop suitable prevention practices.

The tested method is operational and thus could be transferred and used by stakeholders involved in ultrafine and manufactured nanoparticles risk mitigation. Results contribute to enrich reflection for standardization. Relevant nanoparticles occupational exposure data could thus be produced and recorded in specific databases such as COLCHIC or Ev@lutil. This will contribute to knowledge improvement of NP exposure levels and associated health effects by public health professionals such as epidemiologists. The proposed approach takes into account limitations of the prevailing paradigm of risk prevention that hides some safety aspects, as human work, offering new perspectives to act on prevention. Using TESs offer the prospect to shift from technology, nanoparticles, and safety devices to future work activity. This critical point of view on the work exposure situations questions the functional approach of risk management.

Acknowledgments. The authors thank Rémy Anselm for his careful reading of the manuscript.

References

Asbach C, Alexander C, Clavaguera S, Dahmann D, Dozol H, Faure B, Fierz M, Fontana L, Iavicolif I, Kaminski H, MacCalman L, Meyer-Plath A, Simonow B, van Tongeren M, Todea AM (2017) Review of measurement techniques and methods for assessing personal exposure to airborne nanomaterials in workplaces. Sci Total Environ 603–604:793–806. https://doi.org/10.1016/j.scitotenv.2017.03.049

Bakand S, Hayes A (2016) Toxicological considerations, toxicity assessment, and risk management of inhaled nanoparticles. Int J Mol Sci 17(6). https://doi.org/10.3390/ijms17060929

Barcellini F, Van Belleghem L, Daniellou F (2014) Design projects as opportunities for the development of activities. In: Falzon P (ed) Constr Ergon, pp 187–204. CRC Press. https://doi.org/10.1201/b17456-16

Béguin P (1998) Participation et simulation, in V. Pilnière et O. Lhospital (textes rassemblés par), Participation, représentation, décisions dans l'intervention ergonomique, Actes des Journées de Bordeaux sur la pratique de l'ergonomie, Bordeaux, Éditions du Laboratoire d'ergonomie des systèmes complexes. Université Victor-Segalen, Bordeaux 2, pp 123–131

Bekker C, Kuijpers E, Brouwer DH, Vermeulen R, Fransman W (2015) Occupational exposure to nano-objects and their agglomerates and aggregates across various life cycle stages; a broad-scale exposure study. Ann Occup Hyg, mev023. https://doi.org/10.1093/annhyg/mev023

Beurskens-Comuth PAWV, Verbist K, Brouwer D (2011) Video exposure monitoring as part of a strategy to assess exposure to nanoparticles. Ann Occup Hyg 55(8):937–945. https://doi.org/10.1093/annhyg/mer060

Broberg O, Andersen V, Seim R (2011) Participatory ergonomics in design processes: the role of boundary objects. Appl Ergon 42(3):464–472. https://doi.org/10.1016/j.apergo.2010.09.006

Brouwer D, Berges M, Virji MA, Fransman W, Bello D, Hodson L, Gabriel S, Tielemans E (2012) Harmonization of measurement strategies for exposure to manufactured nano-objects; report of a workshop. Ann Occup Hyg 56(1):1–9. https://doi.org/10.1093/annhyg/mer099

Buchmann W, Landry A (2010) Intervenir sur les TMS. Un modèle des Troubles Musculo-squelettiques comme objet intermédiaire entre ergonomes et acteurs de l'entreprise ? Activités 07(7-2). https://doi.org/10.4000/activites.2418

CEN (2018) pr EN 17058 - Workplace exposure - Assessment of inhalation exposure to nano-objects and their agglomerates and aggregates. CEN

Clot Y (2008) Travail et pouvoir d'agir. Presses Universitaires de France

Dagiral É, Jouzel J-N, Mias A, Peerbaye A (2016) Mesurer pour prévenir? Terrains & travaux 28:5–20

Daniellou F (1992) Le statut de la pratique et des connaissances dans l'intervention ergonomique de conception (Habilitation à diriger des recherches). Université Victor Segalen-Bordeaux 2-ISPED, Laboratoire d'ergonomie des systèmes complexes

Daniellou F (2004) 21. L'ergonomie dans la conduite de projets de conception de systèmes de travail. In: Falzon P Ergonomie (1r éd, p 359). Presses Universitaires de France. https://doi.org/10.3917/puf.falzo.2004.01.0359

Daniellou F, Garrigou A (1992) La mise en œuvre des situations passées et des situations futures dans la participation des opérateurs à la conception. In Représentations pour l'action (Octarès éd). Annie Weill-Fassina, Pierre Rabardel, Danièle Dubois

Duarte F, Lima F (2012) Anticiper l'activité par les configurations d'usage : proposition méthodologique pour conduite de projet. Activités 09(2). https://doi.org/10.4000/activites.314

Dutier J, Guennoc F, Escouteloup J (2015) The ergonomist: a full design actor example of an ergonomic action. Procedia Manuf 3:5830–5837. https://doi.org/10.1016/j.promfg.2015.07. 837

Eastlake AC, Beaucham C, Martinez KF, Dahm MM, Sparks C, Hodson LL, Geraci CL (2016) Refinement of the Nanoparticle Emission Assessment Technique into the Nanomaterial Exposure Assessment Technique (NEAT 2.0). J Occup Environ Hyg. https://doi.org/10.1080/ 15459624.2016.1167278

Garrigou A, Daniellou F, Carballeda G, Ruaud S (1995) Activity analysis in participatory design and analysis of participatory design activity. Int J Ind Ergon 15(5):311–327. https://doi.org/ 10.1016/0169-8141(94)00079-I

Garrigou A (1992 janvier 1). Les apports des confrontations d'orientation socio-cognitives au sein de processus de conception participatifs : le rôle de l'ergonomie. CNAM, Paris. http:// www.theses.fr/1993CNAM0160

Garrigou A, Baldi I, Le Frious P, Anselm R, Vallier M (2011) Ergonomics contribution to chemical risks prevention: an ergotoxicological investigation of the effectiveness of coverall against plant pest risk in viticulture. Appl Ergon 42(2):321–330. https://doi.org/10.1016/j. apergo.2010.08.001

Garrigou A, Carballeda G, Daniellou F (1998) The role of 'know-how' in maintenance activities and reliability in a high-risk process control plant. Appl Ergon 29(2):127–131. https://doi.org/ 10.1016/S0003-6870(96)00060-9

Garrigou A, Peeters S, Jackson M, Sagory P, Carballeda, G (2004) 30. Apports de l'ergonomie à la prévention des risques professionnels. In: Falzon P (ed) Ergonomie (1r éd, p 497). Presses Universitaires de France. http://www.cairn.info/ergonomie–9782130514046-page-497.htm

Garrigou A, Thibault J-F, Jackson M, Mascia F (2001) Contributions et démarche de l'ergonomie dans les processus de conception. Perspectives interdisciplinaires sur le travail et la santé (3-2). https://doi.org/10.4000/pistes.3725

Grosjean J, Neboit M (2000) Ergonomie et prévention en conception des situations de travail. Hygiène et sécurité du travail, 179 (ND 2127-179-00)

Jeantet A (1998) Les objets intermédiaires dans la conception. Éléments pour une sociologie des processus de conception. Sociologie du travail 40(3):291–316. https://doi.org/10.3406/sotra. 1998.1333

Jeffroy F (1987) Maîtrise de l'exploitation d'un système micro-informatique par des utilisateurs non-informaticiens: analyse ergonomique et processus cognitif. Paris 13. http://www.theses. fr/1987PA131020

Judon N, Hella F, Pasquereau P, Garrigou A (2015) Towards an integrated prevention of chemical risk of dermal exposure to bitumen in road workers: development of a methodology based on ergotoxicology. Perspectives interdisciplinaires sur le travail et la santé, (17-2). https://doi.org/10.4000/pistes.4586

Mollo V, Falzon P (2004) Auto- and allo-confrontation as tools for reflective activities. Appl Ergon 35(6):531–540. https://doi.org/10.1016/j.apergo.2004.06.003

Oberdörster G, Oberdörster E, Oberdörster J (2005) Nanotoxicology: an emerging discipline evolving from studies of ultrafine particles. Environ Health Perspect 113(7):823–839. https:// doi.org/10.1289/ehp.7339

OECD (2015) Harmonized tiered approach to measure ad assess the potential exposure to airborne emissions of engineered nano-objects and their agglomerates and aggregates at workplaces. ENV/JM/MONO(2015)19, (No. 55). http://citeseerx.ist.psu.edu/viewdoc/ download?doi=10.1.1.696.3255&rep=rep1&type=pdf

Rosén G, Andersson I-M, Walsh PT, Clark RDR, Säämänen A, Heinonen K, Riipinen H, Pääkkönen R (2005) A review of video exposure monitoring as an occupational hygiene tool. Ann Occup Hyg 49(3):201–217. https://doi.org/10.1093/annhyg/meh110

478   L. Galey et al.

Schön DA (1983) The reflective practitioner: how professionals think in action. Basic Books, New York

Vinck D (2009) From intermediary object towards boundary-object. Revue d'anthropologie des connaissances 3(1):51–72

Woskie SR, Bello D, Virji MA, Stefaniak AB (2010) Understanding workplace processes and factors that influence exposures to engineered nanomaterials. Int J Occup Environ Health 16 (4):365–377. https://doi.org/10.1179/107735210799159950

The Assessment of Work-Related Stress in a Large Sample of Bank Employees

Giulio Arcangeli[1](✉), Gabriele Giorgi[2], Manfredi Montalti[1], and Francesco Sderci[1]

[1] Department of Experimental and Clinical Medicine, University of Florence, Florence, Italy
giulio.arcangeli@unifi.it
[2] Department of Human Sciences, European University, Rome, Italy

Abstract. Over the past decade there has been a marked increase in the amount of interest in issues involving work-related stress in Italy. Firstly, with the promulgation of the fundamental law for the protection of health and safety in the workplace, stress has been included as an element to be considered in the mandatory risk assessment. According to the European Foundation for the Improvement of Living and Working Conditions, research in Italy demonstrated high levels of stress in the banking sector. Moreover, since 2008, a deep financial crisis has widely spread around the world. Scientists expressed their worry about this crisis by pointing out that potential negative health effects can be created by collective fear and panic. Economic stress appeared consequently as a new important aspect of mental health. With this in mind, a study was conducted to evaluate peculiarities of hindrance and challenge stressors and its links with recovery in Italian banking population.

We contacted HR and H&S managers of a major Italian banking group and invited them to participate in a stress assessment. The questionnaires were administered online through the intranet company portal. Anonymity and confidentiality in the responses were, however, fully assured. Work-related stress was measured with the Stress Questionnaire (SQ) which assesses several psychosocial working variables.

Results of over 6,000 respondents demonstrated that female respondents lack more of job control and colleagues' support as compared to male respondents. Employees over 50 years old lack of supervisor' support. Employees with the shortest job seniority experienced greatest role ambiguity, lack of job control and colleagues support. Results of hierarchical regression analysis demonstrated that lack of colleagues' and supervisor's support as well as job demands and job control contributed in explaining recovery. The greatest contribution to the explained variance was of job demands and lack of colleagues' support.

Keywords: Work-related stress · Stress questionnaire · Banking sector

1 Introduction

Since 2008, a deep financial crisis, which started in the United States, has widely spread around the world [1]. In times of crisis, job loss occurs more frequently and finding a new employment becomes more difficult as unemployment rates rise [2].

© Springer Nature Switzerland AG 2019
S. Bagnara et al. (Eds.): IEA 2018, AISC 819, pp. 479–485, 2019.
https://doi.org/10.1007/978-3-319-96089-0_51

Literature findings suggest that crisis and recession are accompanied by an increase in mortality and suicide rates [3, 4].

Additionally, both a reduction in wages that an increase in unemployment; these phenomena are risk factors for the development of mental health problems were observed; consequently there was an increase in the use of mental health services by the population [5, 6]. Indeed, the perceived financial security is an important component of psychological subjective wellbeing and can result in psychological responses that lead to long-term physical and mental health consequences [7] like distress, depression, anxiety, psychosomatic symptoms and low self-esteem, as suggested by a Paul and Moser's meta-analysis of the mental health consequences of unemployment [6]. Moreover, the uncertainty of the future for employees working in organizations, as well as of the need for adaptation to many potential changes of work aspects could contribute, along with unemployment, to develop depression and mental health diseases [1, 3]. Lastly, during austere times, organizations could focus on survival rather than promoting a healthy work environment [8].

1.1 Fear of the Crisis

Most of the research on economic crisis and subsequent health effects has focused on the role of involuntary job loss and unemployment, and a few have looked at the effects of contracting economic markets on employees [8]. Although employed people might not be severely and adversely influenced by the financial crisis as unemployed people, nevertheless they might alter their attitudes towards work as a result of the perceived negative financial situation that the company or the market is facing [9, 10].

Employed people might be more cautious in investing emotionally in the company. Firstly, individuals reduce their commitment at work because of greater financial uncertainties and a reduction in the number of opportunities to receive benefits or advance their careers [11, 12]. On the other and we cannot forget that major restructuring, organizational failures, breakdowns and bankruptcy increase the feelings of job insecurity, which relates to the employees' perception about potential involuntary job loss [13]. This could lead to a higher unemployment risk and to potential stress [8].

A proper organizational support such as informing, communicating and backing employees to cope with the magnitude of the problems that are related to the crisis appear of outmost importance [14]. In fact, employees' negative attributions of crisis controllability have emotional and behavioral consequences for organizations. Moreover, managers might be reluctant to clearly communicate crisis antecedents and outcomes, which in turn develop in their subordinates a sense of insecurity and stimulate confusing attribution processes [15].

There is evidence that poor employability has general negative effects on workers, such as lower levels of organizational commitment, poorer performance, higher job dissatisfaction, or decreased health and well-being [16].

In addition, the gradual creation of a collective climate of fear and panic can lead to relevant effects of a social nature, particularly evident in the workplace [17]. Indeed, negative emotional responses may be enhanced within social networks of people who have similar concerns: individuals might easily empathize with those who are experiencing similar financial hardships [11] and consequently, emotion would be easily transmitted.

Therefore, fear of the crisis might stimulate a negative spiral in which people develop their own thoughts, feelings and behavior by influencing and being influenced by the economic environment [9, 10].

Besides, research on organizational membership behaviors suggests that stressful situations affect individuals' pro-social behaviors [18]. Both colleagues and supervisors may be less likely to be pro-social as they may be more preoccupied with their financial situation and their ability to hold on to their jobs and their benefits. In such a situation, social support among the employees might be lower which significantly impacts their stress [19, 20].

Finally, workers who experienced the crisis might feel that they did not accomplish anything through their jobs, because the extrinsic rewards might be reduced.

The workers might adopt an ambiguous view of their role, or perceive that hard work is more stressful because it did not allow them to reap beneficial rewards [11].

2 Methods

We contacted HR and H&S managers of a major Italian banking group and invited them to participate in a stress assessment. The questionnaires were administered online through the intranet company portal. Anonymity and confidentiality in the responses were, however, fully assured. Work-related stress was measured with the Stress Questionnaire (SQ) which assesses several psychosocial working variables [21]. One of the most important aspects of the questionnaire is the psychosocial risk scale; it is based on five main psychosocial risks which might lead to stress-related negative outcomes. Recovery was measured using one scale (3 items) from the SQ. Demographic aspects were detected by some questions, in which information was requested regarding the gender, age, territorial area, job seniority and job position of responders. The analysis and tabulation of results was performed using the statistical analysis package the Statistical Package for Social Sciences (SPSS). The following statistical analyses were performed: Descriptive Statistics, Correlations, One-Way Analysis of Variance (ANOVA) and hierarchical regression.

2.1 The Stress Questionnaire

The Stress Questionnaire (SQ) assesses several psychosocial working variables related to work-related stress. The first versions of the SQ were based on Karasek's (1979) demand, control and support model and the Health and Safety Executive's Management Standards work-related stress. Based on a review of the literature concerning work-related stress, the SQ was further developed adding emerging stress determinants such as the fear of crisis and non-employability [31]. Furthermore, the critical issues experienced in the job task were considered as well as in the physical and psychological context.

Literature [22, 23] has pointed out that one of the most important aspects of the questionnaire is the psychosocial risk scale; it is based on five main psychosocial risks which might lead to stress-related negative outcomes. The scale consists of 25 items and 5 subscales: job demand, job control, role, supervisors support, and colleagues support. Research showed that high job demands and low job control (i.e., how much discretion people have in their work), lack of colleague and supervisor support (i.e., the encouragement, the sponsorship and the resources provided by colleagues and the supervisor), role ambiguity and role conflict (i.e., whether people understand their role within the organization without having conflicting or ambiguous roles) may play a crucial role in work-related stress. Indeed, all the factors mentioned above are usually known as job stressors in the different models that have been proposed to explain the relationship between working conditions and the negative consequences of stress.

The scale is easily intelligible by occupational physicians and quickly administered; the results can be readily included in the preparation of the risk assessment document provided by the Italian Legislative Decree no. 81/2008, and subsequent amendments. In fact in Italy, the main operational guidelines on this issue also recommend performing a subjective assessment of work-related stress [24, 25].

3   Results

Results of over 6,000 respondents demonstrated that female respondents lack more of job control and colleagues' support as compared to male respondents. Employees over 50 years old lack of supervisor' support. Employees with the shortest job seniority experienced greatest role ambiguity, lack of job control and colleagues support. Results of hierarchical regression analysis demonstrated that lack of colleagues' and supervisor's support as well as job demands and job control contributed in explaining recovery. The greatest contribution to the explained variance was of job demands and lack of colleagues' support.

4   Discussion and Conclusion

Fear of the crisis in this study appeared to be an important construct for organizational and social wellbeing. In addition, this study highlighted the importance of employability during austere times. Employability might help employees to feel more secure during the crisis because people who have many alternatives in the labor market may be less affected by job insecurity than those with fewer alternatives [26].

Indeed, the current study researched the mediating role played by poor social support and job stress in the relationship between fear of the crisis and health.

It was found that both social support and job stress fully mediate the relationship of fear of the crisis with health, with all fit indices meeting their respective criteria, and with all path coefficients being significant.

The fact that the alternative models with its assumption, that instead of a total mediation, poor social support and job stress would play partial mediating roles, showed a poorer fit to the data, which suggested that this alternative relationships are less likely.

According to our model, in supposed and unstable crises, managers in particular need to support their subordinates. Whether opportunities and resources are perceived as scarce, people might not offer social support and they might be inclined to compete.

We hypothesized that if the subordinates overestimate the crisis effects managers have to argue that there is no 'real' crisis or that a crisis is not as bad as people think. In doing so, the harmful effects of the fear of the crisis might be reduced. On the other hand, if the fear of the crisis is more realistic, to change perceptions of organizations in crisis, leaders, especially managers, have to improve their relationship with their followers.

For example, encouraging subordinates or reminding them that the organization is supportive and trustful. Being aggressive and hard with employees is not the best strategy. Employees might think that the crisis must be worse than they thought, if the organization is responding aggressively [9, 10].

In addition, the full mediation of job stress on the relationship between social support and health means that when employees work with poor social support, they may perceive high job demand and role ambiguity.

This could facilitate the development of mental health problems.

A possible explanation may be that worried and unsupported employees, who were less proud of their achievements, felt that their work is more ambiguous and heavy. In short, fear of the crisis decreases social support and this in turn tends to increase stress. Then stress is associated with mental health problems. Possibly, the consequences will be more severe for some categories of workers. Our findings showed that fear of the crisis might develop within particular departments or among some categories of employees [9, 10].

In this research, women, full-time employees and temporary workers seemed particularly at risk. We consider that fear of the crisis might be higher for categories of workers that seem weaker in Italy, such as women and temporary workers.

4.1   Model Development

Following the above discussion, we developed the following theoretical model (Fig. 1). A development of fear of the crisis seems interesting and innovative in the literature. Firstly, in light of evidence from the literature we considered weak employability and fear of the company crisis as the determinants of the innovative construct fear of the crisis.

We hypothesized that individuals who doubt their employability and the company solidity will perceive the crisis to be more threatening and destabilizing. In addition, the model not only aims to evaluate determinants of fear of the crisis, but also the consequences. Indeed, we hypothesized that fear of the crisis is connected to mental health. In our model, we position social support and stress as mediators.

In particular, we suppose a total mediation of social support, rather than a partial one. Indeed, fear of the crisis might not be necessarily associated with high job demands, which is a highly important stress factor. As noted earlier, employees who are afraid of the crisis might perceive less work rather than overload. As far as the stress mediation is concerned, we hypothesized a total mediation. In time of crisis individuals might be less cooperative and consequently, the expectations of support by other employees might be

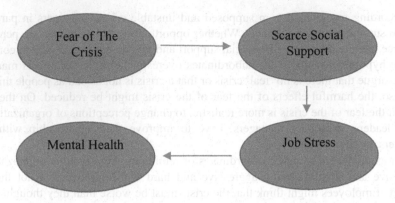

Fig. 1. The proposed theoretical model.

lower. Subsequently, the scarce support received by colleagues and supervisors might have an indirect effect on their mental health, especially in the short term.

To summarize, according to the literature review, fear of the crisis might influence social relationships, which in turn affect the job stress, particularly in terms of job demand and role ambiguity. Finally, job stress might be associated with mental health problems.

References

1. Spence AM (2009) The financial and economic crisis and the developing world. J Policy Model 31(4):502–508
2. Buss PM (2009) Public health and world economic crisis. J Epidemiol Community Health 63(6):417
3. Chang SS, Gunnell D, Sterne JA, Lu TH, Cheng AT (2009) Was the economic crisis 1997–1998 responsible for rising suicide rates in East/Southeast Asia? A time-trend analysis for Japan, Hong Kong, South Korea, Taiwan, Singapore and Thailand. Soc Sci Med 68 (7):1322–1331
4. Falagas ME, Vouloumanou EK, Mavros MN, Karageorgopoulos DE (2009) Economic crises and mortality: a review of the literature. Int J Clin Pract 63(8):1128–1135
5. Jenkins R, Bhugra D, Bebbington P, Brugha T, Farrell M, Coid J, Fryers T et al (2008) Debt, income and mental disorder in the general population. Psychol Med 38(10):1485–1493
6. Paul K, Moser K (2009) Unemployment impairs mental health: meta-analyses. J Vocat Behav 74(3):264–282
7. Price RH, Choi JN, Vinokur AD (2002) Links in the chain of adversity following job loss: how financial strain and loss of personal control lead to depression, impaired functioning, and poor health. J Occup Health Psychol 7(4):302–312
8. Houdmont J, Kerr R, Addley K (2012) Psychosocial factors and economic recession: the Stormont study. Occup Med (Lond) 62(2):98–104
9. Mucci N, Giorgi G, Roncaioli M, Fiz Perez J, Arcangeli G (2016) The correlation between stress and economic crisis: a systematic review. Neuropsychiatr Dis Treat 12:983–993. https://doi.org/10.2147/NDT.S98525

10. Giorgi G, Arcangeli G, Mucci N, Cupelli V (2015) Economic stress in workplace: the impact of fear the crisis on mental health. Work 51(1):135–142. https://doi.org/10.3233/WOR-141844

11. SooHoon L, Lim VKG (2001) Attitudes towards money and work – implications for Asian management style following the economic crisis. J Manag Psy 16(2):159–173

12. Padula RS, Chiavegato LD, Cabral CM, Almeid T, Ortiz T, Carregaro RL (2012) Is occupational stress associated with work engagement? Work 41(Suppl 1):2963–2965

13. DeWitte H (1999) Job insecurity and psychological well-being: review of the literature and exploration of some unresolved issues. Eur J Work Org Psy 8(2):155–177

14. Hansson RO, Robson SM, Limas MJ (2001) Stress and coping among older workers. Work 17(3):247–256

15. Coombs WT (2007) Ongoing crisis communication: planning, managing, and responding, 2nd edn. Sage Publications, Thousand Oaks

16. Wittekind A, Raeder S, Grote G (2010) A longitudinal study of determinants of perceived employability. J Organ Behav 31(4):566–586

17. Sperling W, Bleich S, Reulbach U (2008) Black Monday on stock markets throughout the world – a new phenomenon of collective panic disorder? A psychiatric approach. Med Hypotheses 71(6):972–974

18. Brief AP, Motowidlo SJ (1986) Prosocial organizational behaviors. Acad Manag Rev 11 (4):710–725

19. Arcangeli G, Giorgi G, Ferrero C, Mucci N, Cupelli V (2014) Prevalenza del fenomeno mobbing in una popolazione di infermieri di tre aziende ospedaliere italiane. G Ital Med Lav Ergon 36(3):181–185

20. Arenas A, Giorgi G, Montani F, Mancuso S, Perez JF, Mucci N, Arcangeli G (2015) Workplace bullying in a sample of Italian and Spanish employees and its relationship with job satisfaction, and psychological well-being. Front Psychol 15(6):1912. https://doi.org/10.3389/fpsyg.2015.01912

21. Mucci N, Giorgi G, Cupelli V, Gioffrè PA, Rosati MV, Tomei F, Tomei G, Breso-Esteve E, Arcangeli G (2015) Work-related stress assessment in a population of Italian workers. The stress questionnaire. Sci Total Environ 502:673–679. https://doi.org/10.1016/j.scitotenv.2014.09.069

22. Giorgi G, Perminienė M, Montani F, Fiz-Perez FJ, Mucci N, Arcangeli G (2016) Detrimental effects of workplace bullying: impediment of self-management competence via psychological distress. Front Psychol 7:60. https://doi.org/10.3389/fpsyg.2016.00060

23. Giorgi G, Montani F, Fiz-Perez J, Arcangeli G, Mucci N (2016) Expatriates' multiple fears, from terrorism to working conditions: development of a model. Front Psychol 7:1571. https://doi.org/10.3389/fpsyg.2016.01571

24. Giorgi G, Arcangeli G, Perminiene M, Lorini C, Ariza-Montes A, Fiz-Perez J, Di Fabio A, Mucci N (2017) Work-related stress in the banking sector: a review of incidence, correlated factors, and major consequences. Front Psychol 8:2166. https://doi.org/10.3389/fpsyg.2017.02166

25. Mucci N, Giorgi G, De Pasquale CS, Fiz-Pérez J, Mucci F, Arcangeli G (2016) Anxiety, stress-related factors, and blood pressure in young adults. Front Psychol 7:1682. https://doi.org/10.3389/fpsyg.2016.01682

26. Giorgi G, Leon-Perez JM, Cupelli V, Mucci N, Arcangeli G (2014) Do I just look stressed or am I stressed? Work-related stress in a sample of Italian employees. Ind Health 52(1):43–53. https://doi.org/10.2486/indhealth.2012-0164

Applying Ergonomics to Bathing Safety: Including Adoption of Unorthodox Practices for Slip-Resistant Underfoot Surfaces of Bathtubs Plus Showers and Provision of Effective Points of Control

Jake L. Pauls[1(✉)] and Daniel A. Johnson[2]

[1] Jake Pauls Consulting Services, Toronto, Canada
bldguse@aol.com
[2] Daniel A. Johnson, Inc., Olympia, WA, USA
dajincl@mac.com

Abstract. Of four key factors in bathing safety, related to fall prevention and mitigation, slip resistance is relatively easily addressed. However, in the field of underfoot slip resistance, water is not typically considered as a "friend." Surprisingly, when applied to an underfoot terry cloth towel, water dramatically improves slip resistance in otherwise slippery dedicated showers and in bathtubs used for standup showers. This is now empirically established but, as it defies convention and occurs in a relative research vacuum, there is reticence about adopting the practice of simply having a wet—*or even better*—damp towel underfoot while entering, using and exiting a shower facility. This paper presents what is known and what remains to be learned about the extent to which—*plus why*—water and terry cloth are a potent slip-resistance combination that, qualitatively, solves one-quarter of the shower safety problem with conventional bathtubs and even more with dedicated showers. Regarding potency, we have learned that thin towels are more effective than thick ones and contaminants such as shampoo appear not to affect slip resistance. Combined with code-required improvements to another factor—*effective points of control through the addition of grab bars or stanchions,*—a new, cost-effective practice can greatly reduce the large injury toll.

Keywords: Bathing safety · Slipping · Points of control

1 Introduction to Epidemiology, Etiology and Economics of the Problem

US injury statistics, including national annual estimates of hospital emergency department (ED) visits, from the US Consumer Product Safety Commission (CPSC) National Electronic Injury Surveillance System (NEISS) provide a useful picture of the size and nature of the injury problem with bathtubs, bathtub-shower combinations and showers. For example, using US CPSC NEISS data for 2010 (a year for which there are other useful data, noted below), an estimated 263,000 ED visits occurred due to bathtub

© Springer Nature Switzerland AG 2019
S. Bagnara et al. (Eds.): IEA 2018, AISC 819, pp. 486–500, 2019.
https://doi.org/10.1007/978-3-319-96089-0_52

and shower-related injuries. Note that NEISS deals separately with injuries associated with hot water scalds. (For convenience, the generic terms "bathing" and "bather" are used in this paper in relation to use of these facilities generally.) Examination of a convenience sample of short narratives available for the over 7,500 documented ED visits, on which such national estimates are based, revealed that falls are a typical mechanism leading to injuries, many of which occurred with bather movement before, during and after bathing when combinations of four key dangers are present:

1. Geometry of the impediments over which one must transfer (e.g., bathtub walls and high sills for dedicated showers)
2. Hard, unforgiving surfaces
3. Insufficient, effective points of control
4. Slippery underfoot surfaces.

Most dangerous are activities entailing transfers, both ambulatory—stepping in and out of a bathtub or dedicated shower—and stand-to-sit plus sit-to-stand transfers when horizontal forces underfoot increase relative to vertical ones.

Indeed, if bathing-related falls are compared to falls occurring on stairs, the exposure time-corrected risk of falls associated with bathing exceeds such risk for stairs. In other words, a single step into or out of a bathtub imposes a higher risk of a misstep and fall than occurs in a person's typical single step on stair flight—which entails moving ones foot the height of two risers. Each entails traversing about 400 mm vertically on a typical home stair.

Stairs accounted for the largest number of ED-treated injuries associated with consumer products typically found in buildings, according to the US CPSC NEISS data, with most occurring in homes. For 2010 in the US, stairs accounted for an estimated 1,232,000 visits to hospital EDs. Such stairs would likely be used for on the order of 100 steps per day per person, on average, versus on the order of one transfer step per person daily, on average, associated with bathing.

The relative growth of bathing-related falls versus those associated with stairs is also notable. Bath and shower-related injuries in the US grew in the two decades between 1991 and 2010 by a factor of two for those resulting in an ED visit and by a factor of three for those resulting in hospital admission after first going to the ED. For 2010, in the USA, there were about 263,000 ED-treated injuries associated with bathtubs and showers and about one million treated by medical personnel in all settings. Generally for all ages, stair-related injuries grew by about 65 percent over all ages for hospitalized cases between 1991 and 2010. Although outside the scope of this paper, toilet use involves some similar transfer issues to bathing with comparable mitigation measures, namely improving points of control. The vulnerability of older adults and their inability to forego toilet use, contrasted with bathtub and shower use, leads to larger proportions of older person injuries from toilet use. Thus dual use—for both bathing and toileting—of some points of control discussed below, is very important.

Most of the foregoing epidemiological and etiological data come from the author's own analyses using, the readily accessible US CPSC NEISS Web site (https://www. cpsc.gov/cgibin/NEISSQuery/ last accessed 2018/05/27) and his own survey of hotel bathing facilities and bathing experiences during extensive, world-wide travels.

There also are more-formal literature resources. A listing of 50 documents is available directly from the first author; these were examined in the course of preparing a bathroom usability and safety policy statement for the American Public Health Association in 2016.

A smaller set of references was examined in 2015 for a set of proposals to the National Fire Protection Association (NFPA) for new requirements covering all new bathing facilities within the scope of NFPA's model building code, NFPA 5000 and NFPA's *Life Safety Code*, NFPA 101.

US government epidemiological analyses are somewhat more dated than those presented here; e.g., US Centers for Disease Control and Prevention (2011). The best etiological studies are also dated; e.g., Aminzadeh *et al.* (2000); Kira (1966); Stone et al. (1975).

Especially recommended are synopses focused on both epidemiology and economics: e.g., Lawrence et al. (2015); Miller (2016). Miller (2016) provides some of the basic information needed to make the cost-benefit case for improved bathroom safety. In a later presentation, Miller (2017) provides more detail on cost-benefit analysis which was used by Pauls (2017) in a presentation video, focused on home bathroom safety. The economic bottom line of this is a close match in the annual societal cost per household, of bathing and toileting-related fall injuries in the US, and the cost of installing points of control, such as grab bars and, as a cost-effective more versatile innovation, stanchions. The latter, especially, could serve users of both the home bathing and toileting facilities that, in many small bathrooms, are often adjacent (as shown in Fig. 1).

2   Practice Innovations Addressing 3 of the 4 Types of Dangers

2.1   Points of Control to Mitigate Transfers over Impediments

See Fig. 1 for a two-option, demonstration installation in a traditional, small (2.1-m by 1.5-m) bathroom with bathtub-shower combination, a toilet and a lavatory (off the right side of the photo).

Unlike the conventional grab bars, wall-to-wall horizontal and tub-to-ceiling vertical stanchions do not require screws into the structure behind the ceiling, tiled walls or tub wall. They are not cantilevered out from walls but are held *between* wall, ceiling and other surfaces including, as in Fig. 1, the bathtub rim. Stanchions are typically seen on buses and trains, especially those used for commuting, and are typically within easy reach of most passengers, whether seated, standing or moving from one position to the other. The innovation here is to use stanchions in bathrooms.

In a typical small bathroom, the stanchion (the vertical one, with its bottom fixing on the bathtub rim, in Fig. 1) serves toilet plus shower and immersion bathing users. Either of options would meet the new NFPA requirements for locations of grab bar for new bathtub-shower combinations.

Although included as an option, as effective as conventional grab bars, stanchions are not required by the new NFPA rules where, in the 2018 edition of two codes, they

Fig. 1. Two options, one with stanchions (informally referred to as "poles," usually straight lengths of graspable tubing) and the other employing conventional grab bars providing two points of control.

are referred to as "poles" (NFPA 2018a, b). Stanchion-type bars are a relatively straightforward, cost-effective, NFPA-complying installation on existing bathtub-shower combinations, even in rental apartments. They can be easily installed and removed without wall, ceiling or tub damage as they can be held in place by special (automotive-grade) adhesives rather than screws into walls. The structural criterion applied is performance-based—a load of at least 250 lb must be sustained through the life of the installation. Similar requirements are currently being processed through two other national model building code organizations in both the US and Canada.

The costs of installing the two points of control (horizontal or diagonal and vertical) are comparable to the average USD280 societal cost of bathing and toileting-related injuries—expressed on an average, per-household basis—over a one-year period. Even a relatively expensive installation cost, say double the USD280 (e.g., for retrofitting conventional grab bars), would still make the cost-benefit very acceptable given the years of service possible with the two points of control.

The foregoing information deals with the points of control involving ones hands although the vertical pole can also provide support at ones back, for example when standing on one leg and drying the other leg with a towel. Feet typically provide one or two points of control. Table 1 shows options for points of control dealing with gravity and lateral loads. For bathing, we need two or three points of control for comfort and safety; hence the need for proper handholds for transfers.

Table 1. Minimum number of points of control currently provided with typical practices or imposed design rules.

Number of points of control	≤ 1	1	2	3	3–4
Standard walker for older adult					√
Occupational settings with risk of worker falls from heights				√	
Stairs			√		
Bathtubs with slip resistant underfoot surfaces when wet		√			
Bathtubs with slippery underfoot surfaces when wet	√				

2.2   Hard, Unforgiving Surfaces, Including Those of Impediments

The first two of the key dangers listed above are relatively difficult to prevent and, thus, mitigation approaches are needed. These dangers are geometry of the impediments one must traverse by stepping over (e.g., bathtub walls and high sills for shower enclosures) and hard, unforgiving surfaces (e.g., enamel surfaces of rigid tub walls, ceramic tiles on walls and floors, and metal water controls plus spouts). An additional complication affecting step-over of the tub wall is the difference between floor and bathtub bottom elevations; this can be as much as 100 mm. Retrofit cutouts of tub walls are one partial solution and another expensive solution is a purpose-designed, walk-in tub. These pose operational complications and do not entirely eliminate elevation differences and step-over dangers. Hence points of control are still important.

With the best dedicated showers, the step-over danger can be eliminated through design and construction of a water-draining shower pan and attention to confining water spray, including fixed enclosures (e.g., use of safety glass). Such showers offer additional benefits to users requiring a roll-in design. Some hotels have begun retrofit programs, replacing conventional combination shower-bathtubs with walk-in/ roll-in showers.

Many newer showers are delivered in one or a few pieces with resilient material for walls (instead of hard tiles) providing some cushioning in case of impact as well as controls and water spouts that would not be contacted so injuriously in a loss of balance or fall. However, even with newer materials, such showers share the danger posed by underfoot surfaces.

Showers require careful attention to underfoot slip resistance that is often inherent in wet conditions, even with certain tiles and surface roughness treatments underfoot. Adequacy of slip resistance should be confirmed, by competent expertise, in design and operation of such facilities. Unfortunately, for conventional bathtubs with their smooth surfaces, another approach to slip resistance is needed and this is the largest focus of this paper, especially as the recommended intervention is somewhat unorthodox, even heretical to some objecting to a virtually no-cost, simple solution to a complex problem.

3 Provision of Effective Underfoot Slip Resistance

3.1 Recent and Current Safety Standard Situation

Efforts to deal with slippery underfoot surfaces of bathtubs with manufactured surface treatments have not been successful (in the view of many safety professionals). An early effort to specify a minimum slip resistance criterion for new bathtub surfaces has floundered for multiple reasons including (as only several of many problems):

- Highly questionable minimum slip resistance (SR), also called static coefficient of friction (COF) of 0.04 originally specified in a ASTM standard, F462—79 (ASTM 1979) using a NIST-Brungraber portable slip-resistance tester.
- Application only to *new* bathing surfaces, for the duration of the warranty.
- Discontinued production of the single specified tester.
- Withdrawal, without replacement, of ASTM F462.
- The test foot material, used for original testing with the tester to simulate a bare foot, no longer being available.

As recently as early 2017, the situation was confused among bathtub manufacturing representatives and slip resistance experts (at the meeting of ASTM F15.03 committee responsible for safety standards for bathtub and shower structures). It was noted that the original decision on the 0.04 COF or SR threshold for "safe traction" was based on twice the traction from the best tested, untextured (wet) surface—which was also twice the traction of the worst textured surface. The threshold was *not* based on reasonable safety as determined with human subject testing, for example, to determine actually utilized (needed) slip resistance when stepping in and out of a bathtub (as discussed below).

The foregoing points and others, such as the summation that "F462 is only barely better than nothing," were presented within an ASTM's F15.03 Committee on January 31, 2017. One authority on the topic stressed what a replacement for F462 should be based on and how it should be employed to improve bathing safety from slip-related falls.

3.2 Study of Utilized Slip Resistance

Siegmund et al. (2010) performed a study to "quantify the friction used by barefoot subjects entering and exiting a typical bathtub/shower enclosure under dry and wet conditions." The study involved 60 subjects, equally divided by gender, from three age ranges, younger, middle age and older adults. Force plates with slip-resistant surfaces (with SR of 0.55 to 0.90) provided data to derive utilized friction which was lower, by about 0.06, in wet conditions than in dry conditions. Older subjects used less friction than young subjects ($p = 0.01$ to $p = 0.001$ depending on complexity of the required movement). The range of utilized friction was 0.102 to 0.442, with a median of 0.23. Between the highest and lowest mean plus/minus one standard deviation, the range of utilized friction was 0.13 to 0.36. Utilized friction was lower, by 0.058 ± 0.04, for wet conditions than for dry conditions. There were differences among age groups only for exiting the bathtub when young subjects used more friction. These results were significant at $p = 0.001$ and 0.006. The only grab bar provided was on the back wall and subjects were instructed to only use it in case of a slip or loss of balance (which did not occur); so, generally, it was not reported as a factor in stepping in and out of the bathtub during the testing.

Selected 'take home messages' from the paper's conclusions (with the following quotes taken out of order): "Overall, these findings suggest that subjects entering and exiting a bathtub adjust their utilized friction based on a perception of surface slipperiness." Referring to the outdated standard (described in the prior section) governing bathtub stipulating a "minimum friction of 0.04" (ASTM 1979), this is much "less than the friction subjects used here when entering a dry or wet bathtub." The authors also noted "the importance of using mats, textured or etched surfaces to increase the friction of smooth bathtubs."

3.3   Unreliable, yet Widely Recommended Slip Resistance Interventions

The advice given in the quote above, regarding "mats, textured or etched surfaces" is questionable based on the authors' personal experience with many bathtubs and showers worldwide. Some of the consumer-purchased mats, including those relying on suction cups, are relatively unreliable (compared to wet terry cloth towels addressed in this section), even if installed and used in accordance with product instructions. See Fig. 2.

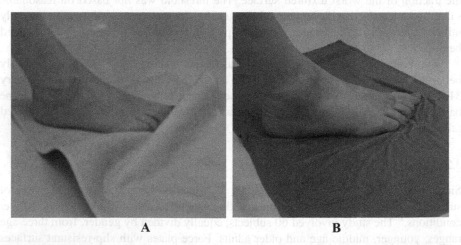

Fig. 2. Failure of (A) properly installed rubber suction cup mat, compared to a (B) wet terry cloth towel, during entry to wet bathtub with smooth enamel surface with relatively similar foot forces exerted.

Unreliability of both bathtub surface treatments and consumer-installed stick-on friction strips or rubber or vinyl mats relying on suction cups has been the expectation of the lead author for many years, buttressed by his many experiences with on the order of a hundred different bathtubs, combination bathtub-showers, and dedicated showers annually in the course of extensive international travel. Moreover, no relationship (except perhaps an inverse one) has been noted between price or opulence of a hotel room and the safety of its bathing facilities. This included one serious, injurious fall, he suffered decades ago, when attempting to get out of a bathtub, in a posh hotel room bathroom.

Only rarely—on the order of a few out of a hundred times—has a sufficiently slip-resistant, underfoot, bathtub surface been encountered in the daily showering experience, most notably with typical bathtub-shower combinations. When wet, etched surfaces and textures (which can be compromised with abrasive cleaners) almost always provide no better perceived slip resistance than does smooth enamel; i.e., with provided slip resistance less than a tenth of what is required for utilization (SR or COF in the 0.1 to 0.44 range) as discussed above in the subsection on utilized slip resistance.

Other solutions to achieving slip resistance, not relying on ineffective industry treatments of underfoot surfaces, were clearly needed (at least in the authors' view).

3.4   Underfoot—Dry, Damp and Sopping Wet—Terry Cloth Towels

The idea about use of wet terry cloth towels for bathtub mats likely came to the lead author from the usual hotel room mat for use on the floor outside a shower or bathtub. On smooth ceramic or stone flooring, such mats—often of somewhat thicker terry cloth construction, usually slide easily underfoot except when, if some water was splashed outside the tub or shower, the mat was thrown or placed on top of the wet floor. Result: the terry cloth mat no longer slid on the smooth floor. As soon as ones weight began to be applied to the mat, it stayed in place, *very reliably!* The water film under the mat had an adhesion effect. Soon the practice moved to insuring that there was some water on the floor before such a mat was set in place.

The full *Eureka* moment came when a standard hand-size, terry cloth towel was thrown into the wet, smooth enamel-finish bathtub being used for a shower. It was easy to position and smooth out the sopping wet towel with a light touch from one foot. As soon as (half) body weight was applied underfoot, there was no movement of the towel under wet conditions even when a horizontal force was exerted at about a 60-degree angle of one leg and the underfoot surface.

Experimenting with combinations of horizontal and vertical forces with one foot revealed what happens in the fraction of a second when the bare foot compresses the terry cloth towel. This expels excess water from underneath the ball and heel of the foot (as shown in Fig. 3B) so that the foot moves horizontally 1 to 3 cm before locking in place with no slipping between foot and towel as well as between towel and smooth bathtub surface. A similar effect is shown in Fig. 4 where the vertical force from the bather's body weight is applied to a curved portion of the bathtub so that there are both tangential and normal forces on the surface which varies in slope, underfoot, from 5 to 45° from horizontal (equivlent, as the angle tangent, to SR of 0.09 to 1.0).

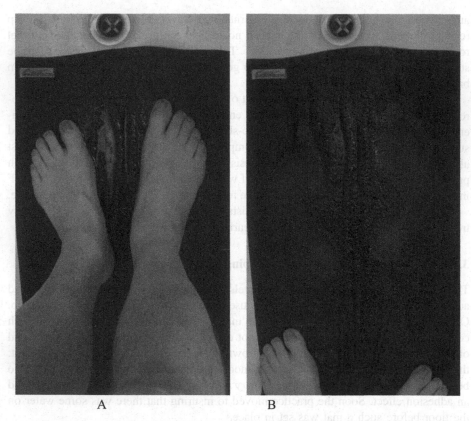

Fig. 3. Sequential views of (A) feet compressing the sopping wet terry cloth towel and (B) immediately after the feet are repositioned, the compressed towel—while briefly only damp —has visible impressions of the feet.

Testing with Variably-Inclined Surfaces. Subsequent, systematic testing of wet towels on a polished glass ramp surface of variable inclination confirmed that the wet towel, underfoot, is stable to nearly a 30-degree inclination. (See Fig. 5) At such high slopes the towel moves slowly, in a creep fashion (at about one cm per second), not precipitously. Note, unlike in the European Ramp Test (discussed by Di Pilla 2003), subjects here do not walk on the sloping surfaces; they stand in place. Effective SR up to about 0.5 can be tested with the adjustable slope available to the lead author.

Further testing, with an adjustable glass surface, as well as a polished surface comparable with plastic tub construction, is contemplated to provide more quantitative data, with multiple subjects, and to video record (in high-definition) the underside of the clear glass ramp to understand better the mechanism based on intimate interface of smooth glass, a film of water and the compressed, highly fibrous towel under bare foot pressure. But first, it is important to describe findings by the second author to complement the first author's empirical findings reported here.

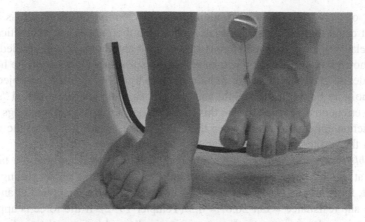

Fig. 4. Foot carrying all of bather's weight on the curved portion of the bathtub, where the slope ranges from 5 to 45° with a damp terry cloth towel interface. Note that the towel has only shifted less than a centimeter.

Fig. 5. 27.5-degree (tan 0.52) slope glass ramp surface with damp terry cloth towel showing slow creep just beginning.

Testing Slip Resistance of Terry Cloth Towels with a Tribometer. The second author of this paper, who is certified in the use of a tribometer (the Variable Incident Tribometer, VIT) has, independently been testing comparable terry cloth towel samples with a smooth granite surface as well as a calibrated test tile of known slip resistance (SR) comparable to what a glazed enamel tub provides under dry, damp and sopping wet conditions. As demonstrated in the following findings, these standard measurements corroborate what has been found in field settings with actual bathtub and shower surfaces and standard, hotel-supplied terry cloth towels (that are similar to towels used very widely in other residential settings, among others). The tribometer studies also considered the effect, to slip resistance, of various thicknesses, measured as weight per unit area, of terry cloth towels in addition to fiber composition, which is typically cotton but could include other materials such as bamboo and some synthetics.

One relatively early tribometer consisted of a weighted object being pulled across a dry, flat, horizontal surface. Slip resistance (SR) was simply the ratio of the horizontal force needed to start the object in motion divided by the weight of the object. But if there is moisture between the weight and the surface, then adhesion, called "stiction", can cause erroneous readings so that this method does not give valid readings (English 1996). Stiction appears to result when there is some residence time after the weight is placed on the surface and the horizontal force is applied.

Variable Incident Tribometer. An alternate method is to determine the tangent of the angle at which a force is applied to an object to just start the object sliding over the surface. If, for example, the object starts to slide when a force is applied at an angle of 26.6°, the slip resistance = tan 26.6° = 0.5 (Templer 1992). If the force is applied to a wet surface in such a manner so that no stiction is allowed to form, the readings of slip resistance are consistent with human experience (English 1996).

The English XL, commonly referred to as the Variable Incident Tribometer (VIT) (see Excel Tribometers), overcomes this problem and was used in subsequent tests reported below.

Slip Resistance Criteria. Slip resistance of a walkway can range from a theoretical zero up to and sometimes greater than 1.0. A slip resistant surface for those who do not expect to be walking on a slippery surface is generally considered to be one that provides a slip resistance of 0.5 or higher (Ekkebus and Killey 1973; Sacher 1993; Nemire et al. 2016). But one can walk without falling on a surface with a somewhat lower slip resistance, especially if one knows or expects the surface to be slippery.

Most dry surfaces provide adequate slip resistance while some of those same surfaces are dangerously slippery when wet (Templer 1992).

Initial Test of Concept with Smooth Granite Surface. A smooth granite counter top was initially tested both in a dry and wet condition, and with and without a towel between the VIT test foot and the granite. A VIT, recently calibrated by the manufacturer, was used for all tests.

Not surprisingly the wet granite exhibited a low SR (0.188) compared to the dry granite (SR = 0.864). The difference was highly significant (t test, $p < .01$). This result is consistent with reports that smooth hard surfaces, such as granite and marble, when wet, are slippery and associated with falls (e.g., Di Pilla, 2003, p.128).

The corner of a thin towel was inserted between the VIT foot and the granite. When a slip occurred, it was noted that the towel slipped on the granite rather than the test foot slipping on the surface of the towel. Distilled water was applied to the surface of the granite, or the top of the towel. The SR of the towel on the granite under dry and wet conditions is recorded in Table 2.

The difference between the towel on the dry granite vs towel on the wet granite was highly significant: t test: $p = 8.13$ E$-.06$. The difference between the Wet condition, with and without towel, was also highly significant: t test: $p = 6.12$ E$-.07$. The relatively higher SR on the wet vs the dry surface confirmed what Pauls had demonstrated; the foot did not readily slide when it was on a wet towel on a smooth hard surface.

Table 2. Effect of towel, wet or dry, on slip resistance of smooth granite counter top

Surfaces	No towel	No towel	With towel	With towel
Condition	Dry	Wet	Dry	Wet
	0.86	0.2	0.2	0.35
	0.86	0.16	0.24	0.41
	0.9	0.18	0.25	0.41
	0.85	0.21	0.26	0.4
	0.85			0.41
				0.4
Means	**0.864**	**0.188**	**0.238**	**0.387**

SR Testing on a Calibrated Test Tile. Further testing on towel density as well as when there were likely contaminates was conducted. A recently calibrated Certified Test Foot Calibration Tile from (Excel Tribometers, Tile #219) was used as a base upon which a portion of each towel was placed during testing. When tested wet without any towel the tile exhibited a SR of 0.188, a value within the range of 0.17 ($\pm$0.03) that was consistent with the range reported by Excel Tribometers for that calibrated tile.

With Soap and Shampoo Contaminants. Though expecting some effect on SR when the towels were contaminated with soap and shampoo, it was found that there was no significant effect on SR when testing towels that had been placed on a tub surface following bathing and the bather used either soap (Ivory® hand soap) or hair shampoo (Suave®). The testing was not done in that shower but, instead, on the calibrated test tile.

Towel Thickness. According to the Turkish Towel Company (2016) plushness is associated with its density, its weight and size. A towel's grams per square meter (GSM, or g/m^2) represents how many grams the dry towel weighs per square meter:
"The higher the grams per square meter (GSM), the denser the fibers of the towel are, making it softer and more absorbent. A GSM between 300 and 400 makes for a thinner towel, whereas a GSM of 450 to 600 is plush. A GSM of 700 or higher is considered luxury hotel quality." (Turkish Towel Company 2016)

The granite counter top was then tested with a thin hand towel (GSM = 406 g/m^2). SR with the dry towel (0.238) was relatively slippery but when distilled water was applied to the towel a SR of (0.397) was recorded. The difference was highly significant (t test, $p < .01$). It was decided to evaluate the concept using towels with characteristics, similar to those found in hotels and other commercial establishments, and to study any effect that contaminants (i.e., soap or shampoo) or how the use of thicker towels might affect the results.

Two towels of similar size ($\sim$410 mm by 460 mm) were measured with a tape measure and weighed on a digital scale so that the plushness of each towel could be calculated. Both towels had a GSM > 600 and so are considered "plush".

The SR of both towels on the certified test tile was ascertained under dry, damp and sopping wet conditions. The "damp" condition occurred when distilled water was applied to the towel but the water did not appear to pool above the base of the threads.

Table 3. Slip resistance of thin vs plush towels when dry, damp, and sopping wet on test tile

	A	B	C	D	E	F
GSM	643	683	643	683	643	683
Conditions	Dry	Dry	Damp	Damp	Sopping	Sopping
Thickness	Thin	Thick	Thin	Thick	Thin	Thick
	0.18	0.21	0.36	0.28	0.27	0.21
	0.17	0.2	0.33	0.26	0.26	0.23
	0.17	0.2	0.32	0.28	0.28	0.24
	0.19	0.19	0.31	0.31	0.29	0.24
				0.34		
				0.31		
Means	0.1775	0.200	0.33	0.297	0.275	0.23

Significance:
A vs B Thin vs Thick - Dry: significant i.e., $p < .05$ (t test,
$p = 0.0117$)
C vs D Thin vs Thick - Damp: not significant (t test, $p = 0.085$)
E vs F Thin vs Thick - Sopping: significant (t test, $p = 0.003$)
Dry vs Wet Towels Data pooled: Thin vs Thick (Damp &
Sopping Pooled)
C + D vs E + F not significant (t test, $p = 0.092$)
Wet vs Dry highly significant (t test $p \ll .001$)

The "sopping wet" condition was when the water was visible above the base of the fibers at the start of the test (see Table 3).

Tests in Porcelain Tub. Damp and sopping wet towels did not exhibit a statistically significant difference ($p = 0.26$) when tested in a nine-year-old bath tub that had been used almost daily. When data were pooled the damp and sopping wet towels had a higher SR (Mean = 0.296) than the wet tub with no towel underfoot (SR mean = 0.24). The difference was statistically significant ($p < 0.004$). The 95% Confidence Interval for the porcelain tub with a wet towel (damp or sopping wet); SR = 0.26 < 0.3 < 0.34.

4   Conclusions

Generally, the practice of using ordinary terry cloth towels to solve one of the main problems with bathing safety, along with installation of effective points of control—for example, using stanchions that integrate well with bathroom décor at low cost—should make bathing a less dangerous activity, at modest cost and low installation complexity in both new bathrooms and existing ones.

One bottom line is somewhat unorthodox, even heretical. Whereas in much of the work on slip resistance, water is considered an "enemy, it turns out that for slip resistance of smooth, wet surfaces typically found underfoot in a bathtub or shower, the combination of ordinary terry cloth towels and water is your "friend." Towels are ubiquitous and readily available in bathrooms in most homes and hotels. If people were advised to place damp towels on the floor of tubs and showers, as well as on the floor outside of the tub in a location where they would expect to be stepping when entering or exiting the tub, the probability of slips and related missteps could be greatly reduced.

Solutions to the slipping and other problems for bathing—especially showering— can be elegant, counterintuitive, inexpensive and immediately at hand (or should we say also "at foot") in every bathroom. Such solutions are addressed in freely accessible videos and, increasingly, those requiring structurally adequate installation of points of control are being enshrined in North American safety standards and building codes. Thus improved bathing safety could be a success story in applying ergonomics to heretofore inadequately addressed public health problems.

References

Aminzadeh F, Edwards N, Lockett D, Nair RC (2000) Utilization of bathroom safety devices, patterns of bathing and toileting, and bathroom falls in a sample of community living older adults. Technol Disabil 13:95–103

ASTM (1979) Standard consumer safety specifications for slip resistant bathing facilities, F462-79. Reapproved 1999, 2007. ASTM International, West Conshohocken

Bakken GM (2002) Human factors and legal aspects of fall injuries. In: Hyde A, Bakken G, Abele J, Cohen H, LaRue C (eds) Falls and related injuries: slips, trips, missteps and their consequences. Lawyers and Judges Publishing, Tucson, pp. 229–393

Di Pilla S (2003) Slip and fall prevention: a practical handbook. Lewis Publishers, Boca Raton, pp 163–165

English W (1996) Pedestrian slip resistance: how to measure it and how to improve it. William English, Alva

Excel Tribometers LLC. http://www.exceltribometer.com. Accessed 27 May 2018

Ekkebus CF, Killey W (1973) Measurement of safe walkway surfaces. J Soap/Cosmet./Chem. Spec.

Kira A (1966) The Bathroom. Viking Compass, New York

Lawrence B, Spicer R, Miller T (2015) A fresh look at the costs of non-fatal consumer product injuries. Inj Prev 21:23–29

Miller T (2016) Falls in home stairways and bathrooms: how many? how costly? 40-minute video of presentation at creating and implementing an evidence base for improved built environment standards and codes. Annual conference of the Canadian Public Health Association, Toronto, June 2016. https://vimeo.com/175101448. Accessed 27 May 2018

Miller T (2017) Will code changes save more than they cost? Presentation in meeting on the impact of building codes and standards in public health and safety, Melboune, AU. https://vimeo.com/channels/866600/239276202. Accessed 27 May 2018

Nemire K, Johnson DA, Vidal K (2016) The science behind codes and standards for safe walkways: changes in level, stairways, stair handrails and slip resistance. Appl Ergon 52:309–316

NFPA (2018a) Life safety code. National Fire Protection Association, Qunicy

NFPA (2018b) Building construction and safety code. National Fire Protection Association, Qunicy

Pauls J (2016) Bathrooms: obstructions, slipping and points of control. 22-minute video of presentation at slips, trips and falls 2016, London. https://vimeo.com/193507768. Accessed 27 May 2018

Pauls J (2017) For bathing safety, water can be your friend. Presentation at slips, trips and falls conference, Toronto. https://vimeo.com/channels/866600. Accessed 26 May 2018

Sacher A (1993) Slip resistance and the James Machine 0.5 static coefficient of friction – sine qua non. ASTM Stand. News

Siegmund GP, Flynn J, Mang DW, Chimich DD, Gardiner JC (2010) Utilized friction when entering and exiting a dry and wet bathtub. Gait & Posture 31:473–478

Stone RF, Blackwell D, Burton DJ (1975) A systematic program to reduce the incidence and severity of bathtub and shower area injuries. US Consumer Product Safety Commission Contract No. CPSC-P-74-334, report. Abt Associates, Cambridge

Templer J (1992) The staircase: studies of hazards, falls and safer design. MIT Press, Cambridge

Turkish Towel Company (2016). http://turkishtowelcompany.com/towel-guide-what-is-gsm/. Accessed 27 May 2018

US Centers for Disease Control and Prevention (2011) Nonfatal bathroom injuries among persons aged ≥ 15 years United States, 2008. MMWR 60:729–733

US Consumer Product Safety Commission, National Electronic Injury Surveillance System. https://www.cpsc.gov/cgibin/NEISSQuery/. Accessed 27 May 2018

VIT (2005). Slipmeter Manufacturing and Safety Engineering Consulting. http://www.exceltribometers.com. Accessed 27 May 2018

Crisis Management and Simulation Training: Analysis of Crisis Managers' Behavior Using Activity Logs

Sylvie Vandestrate$^{(\boxtimes)}$, Laurie-Anna Dubois, and Agnès Van Daele

University of Mons (UMONS), 18 Place du Parc, 7000 Mons, Belgium
sylvie.vandestrate@umons.ac.be

Abstract. Crises are key moments in companies' lifetimes, especially for high risks industrial systems. The Expert'Crise project trains managers to manage crises through theoretical training and accident simulation exercises. The trainees' performance during the exercise is a key factor in their learning process, which is completed by a debriefing. However, a detailed analysis is needed to understand what exactly happens between the managers inside the crisis room during the exercises. We have, therefore, developed a methodology, based on observations, in order to give feedback to managers and to suggest recommendations for improving emergency planning.

The purpose of this paper is to present this methodological approach, based on the construction of trainee activity logs. During the exercise, observational data collection is carried out through observation grids and camera recordings. After the exercise, the observers' notes and camera recordings are collated and incorporated into a global database, compiling key information about the trainees' actions and communications. Thereafter, a set of issues encountered by the crisis unit is identified from the database. Several descriptors of the level of completion of these issues are then defined.

Once the database has been sorted, according to these issues and their levels of completion, the log of the trainees' activity is complete. Each issue is now ready to be analyzed through some strategic crisis topics, such as communication, leadership inside the crisis unit, accordance with crisis roles set in the emergency instructions, difficulties and deviations encountered for each issue, and achieving the learning targets.

Keywords: Simulation exercise · Activity analysis · Crisis management

1 Introduction

1.1 Background: Crisis Management and Simulation Training

Crises are key moments in companies' lifetimes, especially for high risks industrial systems, and cover a large spectrum of realities (e.g. sanitary problems, natural disasters, extended power and water outage, etc.) [1]. In connection with the underlying project presented below, crises are considered, in this paper, as unexpected and disruptive situations following a severe accident, which are possibly not listed in the companies' emergency planning documents. The consequences of such events may

© Springer Nature Switzerland AG 2019
S. Bagnara et al. (Eds.): IEA 2018, AISC 819, pp. 501–508, 2019.
https://doi.org/10.1007/978-3-319-96089-0_53

threaten human lives both on and off the site, the environment, critical infrastructure, and sometimes the very survival of the company. These situations, therefore, require an urgent response provided by several stakeholders, including the internal crisis unit, which is often composed of the company's managers and decision makers who have to use a specific skillset. Indeed, these exceptional situations involve a particular and paradoxical reaction: while managers are not used to dealing with complex and rare accidental events, they still have to react immediately in an efficient and inventive way [2]. Thus, managers need to be trained to display specific behaviors and skills related to crisis management, such as decision-making and exchanging information [3].

From this perspective, simulation training – which is frequently used and studied in medical training – can be of interest, regarding learning targets. During a simulation, trainees can practice within a risk-free environment, and can, therefore, make errors without impacting anyone's health or safety [4]. Moreover, simulations enable trainees to be exposed to rare and complex conditions, which is very important for crisis management training [4]. Furthermore, simulation-based learning is experiential and reflexive [5]. Most of the time, this "learning by doing" process concludes with a debriefing providing feedback, also allowing trainees to adopt a reflexive overview of their own in-simulation activity [6].

1.2 Context: The Expert'Crise Project

Since 2015, the Expert'Crise project – funded by the European Social Fund – has aimed to develop crisis management training programs incorporating two components, depending on the target audience: emergency services, and high risks industrial systems (e.g. Seveso companies). This paper focuses on the latter. Managers and decision-makers are trained to manage crises through theoretical training and on-site accident simulation exercises (mainly simulated in chemical companies) [7].

Each exercise has three steps: briefing, simulation session, and two debriefings (the first at the end of the session, the second after in-depth analysis). The simulation scenario is based on a detailed analysis of the company's emergency management system. Hence, the scenario is highly bound by legal and internal procedures. The exercise is built in close collaboration with the company, but not with the trainees themselves, which leads to a credible scenario for the trainees.

This preparation process results in a partial exercise (i.e. not full-scale), which is close to a real world scenario, functionally speaking [8]. Some crisis management training makes use of virtual reality [9], but the Expert'Crise project employs learning role playing games [10] where the trainees play their own roles in the crisis management system, and the trainers play predefined fictional roles, acting as members of the emergency services, media and official authorities.

2   Methods

2.1   Framework

If we adhere to the professional didactic framework inspired by the ergonomic trends in French-speaking countries, the trainees' activity during the simulation sessions consists of two main parts: constructive (linked to skills) and productive (linked to performance) activity [11]. The trainees' activity during the exercise is, therefore, a key factor in their learning process [12], as is the debriefing which follows the exercise [13]. While the scientific literature on simulation training gives debriefing a key place in the learning process, concerning the reflexive approach in particular [14], a significant part of learning still relies on the activity performed by trainees during the simulation itself [15]. The experiential approach of simulation-based learning puts trainees' and trainers', activity at heart of research and training programs. This highlights the need for tools to analyze activity, especially when it is carried out in such a complex and uncertain context, such as a crisis, which places both the trainees and trainers into trouble.

Consequently, in addition to the first debriefing, a detailed analysis is needed to understand precisely what happened during the Expert'Crise simulation, specifically inside the crisis room, between the crisis resolution agents (i.e. decisions, communication and relational processes). We, therefore, developed a methodology based on observations in order to give feedback to managers and to suggest recommendations for improving emergency planning, both on material and organizational levels.

Up to this day, the Expert'Crise project has organized simulation exercises at eight companies[1]. Six exercises (Table 1) were analyzed through the methodology presented below.

Table 1. Expert'Crise exercises to date

Exercise	Companies	Exercise and environment	Type of exercise [16]
1[a]	A, B, C	Coordinated exercise between three companies on multi-operator site	Full-scale (A), functional (B), NA (C)
2	D	Single-site exercise on multi-operator site	Functional
3	E	Single-site exercise, isolated site	Functional
4	F	Single-site exercise, isolated site	Functional
5	C[b]	Single-site exercise on multi-operator site	Functional
6	G	Single-site exercise, isolated site	Functional
7	H	Single-site exercise, isolated site	Functional

[a]Exercise that was carried out before the completion of the methodology presented in this paper.
[b]Second exercise on the C-site.

---

[1] One of these eight companies (company C) was involved in two different exercises, but the first exercise (carried out with companies A and B) was not analyzed with the presented methodology.

2.2 Data Processing

The observational data collection was done through camera recordings, and observation grids (Table 2). During the exercise, the cameras and observers were located in strategic positions on the site, for example in the crisis room, control room, and disaster site.

Table 2. Excerpt of the observation grid (single line)

Time	Actions	Communication			
		Agents		Message content	
		Sender	Receiver	Sender's words	Receiver's words
		☐Person 1 ☐Person 2 ☐…	☐Person 1 ☐Person 2 ☐…		
		Vector			
		☐In person ☐Mobile ☐Computer	☐Radio ☐Landline ☐Other:		

After the exercise, all of the observers' grids were collated, in order to avoid data redundancy if several observers took notes of the same event, and then integrated into a single database (Table 3). The database was then completed with information from the camera recordings if grey areas remained. At this stage, the data were not completely processed as they still had to be sorted. In order to do so, a set of issues encountered by the crisis unit was identified from the database and the learning targets (i.e. situations-tasks) determined during the exercise preparation (e.g. warning level management, crisis unit mobilization, accident identification and management, media and authorities contact) were considered. In some cases, a single observation line could match several issues (e.g. several topics raised in one conversation), hence the observation lines needed to be split until they matched only one issue. Conversely, it was also possible that some observation lines matched no issue (e.g. missed call, conversation not related to the exercise), and were therefore considered extraneous to crisis management lead by trainees.

Several descriptors for the level of completion of these issues were then defined. The issue is raised (level 1, red flagged) when encountered for the first time by the crisis unit, or when mentioned without being resolved or questioned. The issue is being resolved (level 2, yellow flagged) if mentioned at least for the second time and is the subject of information, actions, questions or discussions by the crisis unit. At last, the issue is resolved (level 3, green flagged) when crisis unit members implement actions to resolve the issue, or receive confirmation that the problem is over.

Finally, this database includes key information about trainees' actions and communications: each action and communication (i.e. each processed observation line) relevant to crisis management is linked to an issue and a level of completion.

Table 3. Excerpt of the fifth exercise database (simplified)

Time	Actions	Message content		Issue	Level of completion
		Sender	Receiver		
10h04		Guard: (1) Accident in column 4, there's ammoniac smoke... (3) There's one person injured	First aid: (2) I will put on my equipment and alert my colleagues. Any other information? (4) Trigger the alarm	Alert	1
10h05	Alarm goes off			Alert	2
(...)					
10h17		Crisis unit chief: (1) Did you call for first aid? (3) Is someone injured? (5) Do we know who?	Guard: (2) All I know is that there's ammoniac smoke and we need a fire truck operator (4) Yes (6) No, they didn't tell me... He isn't moving anymore...	Crisis unit intervention	2
10h18		Evacuation manager: (1) John and Peter aren't here, so I'll do the evacuation count	Guard: (2) OK, let me give you the list	Evacuation count	1

2.3 Data Analysis

Once the database is sorted according to these issues and their levels of completion, the log of the trainees' activity is complete. This way, a table summarizing issue emergence and resolution can be achieved by means of a timeline taking the levels of completion (in the lines), and the issues concerned (in the columns) into consideration (Table 4). Each issue is now ready to be analyzed through strategic crisis management topics. These topics include communication, leadership inside the crisis unit, accordance with crisis roles set in the emergency instructions, issue resolution, difficulties and deviations encountered during each issue resolution, and achieving the learning targets.

From a global perspective, analysis includes crisis phase identification (i.e. warning chain, crisis management, and recovery process), communication flows within the crisis location, the means of communication and related difficulties (e.g. defecting mobile network or landline, means of communication in insufficient number or variety).

The next step was to count the interventions performed by each crisis unit member, first overall, to identify the one who was leading the crisis management team, and then issue by issue, to check the accordance with crisis roles set by the emergency instructions. For instance, it is obvious that the evacuation manager is expected to be involved in the evacuation and count processes, but it might not be the case in the simulation exercise, and the reason why then needs to be identified.

Other interesting elements in analyzing crisis management are the differences between the trainees' actions and the emergency procedures, which are sometimes necessary when responding to an unexpected situation (i.e. outside the emergency planning) [17]. Any differences may show that procedures are inappropriate, inefficient, contradictory, incomplete or simply not understood. Therefore, these procedures have to be adjusted, or the team members have to be trained. Besides any identified differences, it is necessary to pinpoint and explain difficulties encountered by the crisis unit during simulation, such as lost documents, undefined crisis roles or mission, and poorly prepared crisis management.

Table 4. Sixth exercise timeline

Time	Alert	Emergency services contact	Evacuation	Injured person care	Fire fighting	Pollution restricting	Message to media	Missing person care
09h30	1							
09h35	2	1	1					
09h40	2	2	2	1				
09h45	3			2	1			
09h50			3	2	2	1		
09h55		2		2	2	2		
10h00		2		2	2	2	1	
10h05		2			2			
10h10								1
10h15		2						

The analysis ends with a short crisis summary. This last point checks if the simulation exercise was a success concerning the learning targets (e.g. defining roles and missions, defining strategical priorities, cooperating, acquiring/collecting available information, coordinating used resources) and the situation-tasks (e.g. identifying the accidental situation, overcoming communication problems with field staff, coordinating with emergency services) prepared in advance of the simulation exercise [11].

3   Conclusion

In the end, the activity log is not as exhaustive as the initial database but it arranges raw data, classifies them into logical groups, and retains the crisis key crisis resolution elements in a simple and usable way. The observation grids are quite basic but give observers a wide degree of freedom to take notes, and they can be easily transferred from one exercise to another.

The purpose of this whole analysis process is to give concrete recommendations about crisis management resources and organization in order to help crisis units be better prepared to deal with the next crisis. For instance, while crisis management resources suggest changes focused on protective equipment, means of communication, and crisis room equipment, crisis organization adjustments are often related to the definition of crisis roles, training of operators and managers, information sharing, and crisis follow-up.

The major challenge leading to the methodology presented in this paper is to achieve cost effective trainee activity analysis. To do so, the Expert'Crise trainers chose to rely mainly on observational data from a third-person perspective without any complementary questionnaires or trainee interviews (besides the first debriefing, which is taken into account in the analysis process). Indeed, a literature review [18] legitimates the third-person perspective for analyzing activity by inferring "pragmatic judgments" from activity observation [19], which can also been expressed by trainees themselves during interviews.

Finally, the presented methodology is clearly not exhaustive and the analysis of trainee's actions in a crisis simulation exercise may include other factors, such as situation awareness, resource managing, planning or anticipation [4]. Therefore, self-confrontation interviews, for example, can complete this methodology to investigate these crisis management factors.

References

1. Lagadec P (1991) La gestion de crise - Outils de réflexion à l'usage des décideurs. MacGraw Hill, Paris
2. Lapierre D, Bony-Dandrieux A, Tena-Chollet F, Dusserre G, Tixier J, Weiss K (2015) Developing a tool to assess trainees during crisis management training for major risks. Saf Reliab Complex Eng Syst 26:195–202
3. Yee B, Naik V, Joo H, Savoldelli G, Chung D, Houston P, Karatzoglou B, Hamstra S (2005) Nontechnical skills in anesthesia crisis management with repeated exposure to simulation-based education. Anesthesiology 103:241–248
4. Jafferlot M, Boet S, Di Cioccio A, Michinov E, Chiniara G (2013) Simulation et gestion de crise. Réanimation 22:569–576
5. Haute Autorité de Santé: Guide de bonnes pratiques en matière de simulation en santé (2012). http://www.has-sante.fr/portail/upload/docs/application/pdf/2013-01/guide_bonnes_pratiques_simulation_sante_guide.pdf. Accessed 10 Apr 2018
6. Fanning R, Gaba D (2007) The role of debriefing in simulation-based learning. Summer 2 (2):115–125
7. Horcik Z, Savoldelli GL, Poizat G, Durand M (2014) A phenomenological approach to novice nurse anesthetists' experience during simulation-based training sessions. Simul Healthc 9(2):94–101
8. Duhamel P, Brohez S, Delvosalle C, Dubois L-A, Van Daele A, Vandestrate S (2017) Le projet Expert'Crise ou la formation à la gestion de crise en milieu industriel par des exercices de mise en situation: premiers résultats. Récents Progrès en Génie des Procédés, 110
9. Matsumoto E, Hamstra S, Radomski S, Cusimanon M (2002) The effect of bench model fidelity on endourological skills: a randomized controlled study. J Urol 167:1243–1247

10. Burkhardt J-M (2003) Réalité virtuelle et ergonomie: quelques apports réciproques. Le Trav Hum 66(1):65–91
11. Duhamel P, Brohez S (2018) Learning role-playing game scenario design for crisis management training: from pedagogical targets to action incentives. In: Communication presented at the 20th congress of the international ergonomics association, Florence
12. Samurçay R, Rabardel P (2004) Modèles pour l'analyse de l'activité et des compétences: propositions. In: Samurçay R, Pastré P (eds) Recherches en didactique professionnelle. Octarès, Toulouse, pp 163–180
13. Horcik Z (2014) Former par la simulation: de l'analyse de l'expérience des participants à la conception de formations par simulation. PhD thesis, University of Geneva, Geneva
14. Gardner R (2013) Introduction to debriefing. Semin Perinatol 37:166–174
15. Pastré P (2005) Apprendre par la résolution de problèmes: le rôle de la simulation. In: Pastré P, Rabardel P (eds) Appendre par la simulation: de l'analyse du travail aux apprentissages professionnels. Octarès, Toulouse, pp 17–40
16. Tena-Chollet F, Tixier J, Dandrieux A, Slangen P (2016) Training decision-makers: existing strategies for natural and technological crisis management and specifications of an improved simulation-based tool. Safety Science
17. Vidal-Gomel C (2016) Prévention des risques professionnels et formation: éléments de réflexion à partir de la didactique professionnelle et de l'ergonomie. Recherches en éducation, vol 26, pp 154–171
18. Filliettaz L (2014) L'interaction langagière: un objet et une méthode d'analyse en formation d'adultes. In: Friedrich J, Pita Castro JC (eds) Recherches en formation des adultes: un dialogue entre concepts et réalité. Editions Raisons et Passions, Dijon, pp 127–162
19. Pastré P (2011) La didactique professionnelle. Presses universitaires de France, Paris

In What Conditions Do People Adopt "Resilient" Behavior for Safety?

Experimental Study for Safety - II

Naoki Kubo[✉] and Miwa Nakanishi

Faculty of Science and Technology, Department of Administration Engineering,
Keio University, Yokohama, Kanagawa 223-8522, Japan
1687lnaoki@keio.jp

Abstract. People may follow a manual to achieve safety (Safety - I). On the other hand, it has been proposed in recent years that people sometimes respond flexibly to situations in order to achieve safety (Safety - II). Both modes (Safety - I and Safety - II) should be used mutually [1]; therefore, one strategy is to apply the Safety - I mode when the fluctuation of conditions is below a certain level and to apply the Safety - II mode when it is above that level. In this research, in order to clarify the level of situation fluctuation at which it is appropriate to switch modes from Safety - I to Safety - II, we conducted a simulator experiment involving a fire-extinguishing activity. As a result, it was revealed that people adopted resilient behavior when the situation fluctuation exceeded a certain level and resilient behavior led to success when the fluctuation was not too large or too small.

Keywords: Safety - I · Safety - II · Resilience engineering

1 Introduction

Under circumstances that change from moment to moment, Safety - I and Safety - II are seemingly contradictory concepts; however, they should be applied mutually. One way of thinking, to make the best use of both safety strategies, is that when the situation fluctuation is so large that it cannot be dealt with, it is considered effective to apply Safety - II. In that case, in order not to cause the situation to deteriorate, it is necessary to clarify how much the situation fluctuation is the boundary of the safety strategy to be applied.

The purpose of this study was to clarify whether a human being responds flexibly without using a manual when the situation fluctuation is large (resilient behavior) and whether some factors could prevent that resilient behavior from failing. This would be accomplished through a simulator experiment involving a fire-extinguishing activity.

© Springer Nature Switzerland AG 2019
S. Bagnara et al. (Eds.): IEA 2018, AISC 819, pp. 509–516, 2019.
https://doi.org/10.1007/978-3-319-96089-0_54

2   Method

2.1   Experimental Tasks and Conditions

We set up experimental tasks consisting of simulations of firefighting work on desktop screens. These were affected by weather conditions and resources, as in a task whose situation fluctuates from moment to moment. The experiment screen is shown in Fig. 1. The simulations of the firefighting work were that the tasks started, fire broke out from one place at the same time, and the experiment participants chose a resource (by clicking) and then a burning house.

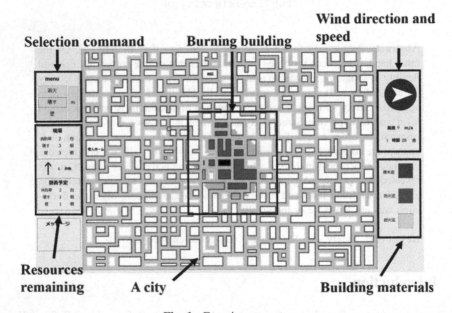

Fig. 1. Experiment screen.

Before the experiment, we read a manual that provided the standard guidelines for efficient firefighting work (created with reference to the firefighters' college of the Fire and Disaster Management Agency [2]). After the participants completely memorized the manual and were able to apply it well through practice, the experiment was conducted. The experimental conditions were the size of the situation fluctuation occurring during the task. We used the six scenarios (including one without any accident) shown in Table 1, each combined with an average wind speed (3, 7, or 11 m/s) to create 18 conditions in total. The participants conducted tasks lasting about seven minutes per condition after hearing an explanation of the experiment and being fitted with a measuring device. After completing one task, each participant performed the same task under all 18 experimental conditions. They could use three possible operations: applying a hose to a burning building, building a barrier to prevent the fire from spreading, .//or demolishing the building to prevent the fire from spreading. When a certain period of time had elapsed after the start of the task, one of the scenarios shown in Table 1 occurred. The screen showed a nursing home and a hospital, where people were had been left out in the street. But the other buildings were empty. Participants were taught that the purpose of the tasks was to minimize fire damage and to save the people in the nursing home and hospital (life priority). They were also taught to conduct firefighting work based on the manual. However, if they decided themselves that responding flexibly was a better option for achieving the above-mentioned purpose, we taught them that they could respond flexibly.

Table 1. Scenarios.

Scenario
(1) No fire trucks
(2) Fewer fire trucks
(3) Fewer resources to build walls and destroy buildings
(4) Construction of walls or destruction of buildings takes longer than usual
(5) Sudden reverse of wind direction
(6) Normal

2.2   Record Items

The fire-extinguishing activities of the experiment participants were recorded in an operations log. In order to examine the cognitive load of the experiment participants during the tasks, changes in cerebral blood volume were measured (OEG - 16, Spectratech). Furthermore, TIPI-J, a Japanese version of the Ten-Item Personality Inventory (TIPI), was used to examine the relationship between individual personalities ("Big Five" personality characteristics) and the judgment of each experiment participant.

2.3   Experiment Participants and Ethical Considerations

The experimental participants were 17 male university students aged 21 to 23. We explained the experiment to them orally and in writing and acquired signed consent forms. Furthermore, we analyzed the data after making personal identification impossible.

3   Results

3.1   Extraction of Resilience Behavior

During the tasks, attention was paid to the scene in which the experiment participants took actions (resilient behavior) for flexible extinguishing activities rather than extinguishing activities according to the manual. As an extraction criterion for resilient behavior, actions that deviated from the manual, but which could be regarded as acting to reduce the spread of fire or to save lives, were listed in advance. Examples (partial) of resilient behavior are shown in Table 2, below.

Table 2. Examples (partial) of resilient behavior.

Purpose/cause	Violation place of the manual
Protect the nursing home	Participant gave up the leeward fire and extinguished the windward fire
Minimize damage	Participant gave up the leeward fire and extinguished the windward fire
Minimize damage	Participant gave up the leeward fire and extinguished the windward fire
Minimize damage	Participant noticed that the wind direction was reversed but continue to extinguish the windward fire
Protect the hospital	Participant put off downwind fire extinguishing
Lack of resources	Participant put off downwind fire extinguishing
Wind was too strong	Participant extinguished the leeward fire after extinguishing the windward fire

3.2   Changes in Cerebral Blood Volume During Task Performance

In the brain region, it was hypothesized that the case of self-determined and resilient behavior would make (the dorsolateral prefrontal cortex (DLPFC), which is responsible for executing functions such as judgment and planning, more activate than the case of actions that followed the manual. Figure 2 shows the results of the oxygen–hemoglobin (Oxy–Hb) concentration variation (after normalization) in the DLPFC. Since resilient behavior appears when the situation fluctuation is large, the resilient behavior was observed against difficult tasks was high. Therefore, it was suggested that training to activate the DLPFC to enhance resilience may be effective. On the other hand, because resilient behavior tends to appear when a situation is difficult, we considered the possibility that the DLPFC might be activated simply because it was a difficult task. We divided the resource fluctuations into three groups (large, medium, and small) and conducted the same analysis. Figure 3 shows the results. There was no significant

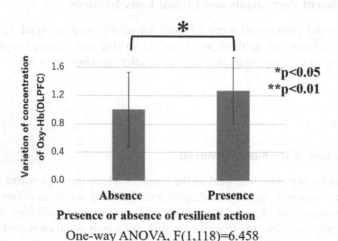

Fig. 2. Variation of concentration of Oxy-Hb in DLPFC.

difference in any of them, but the average value of the presence of resilient behavior was higher, and similar results could be obtained.

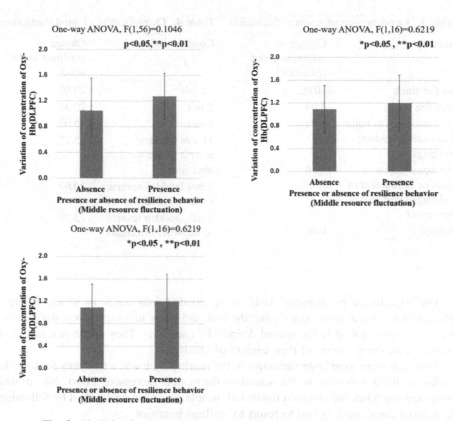

Fig. 3. Variation in concentration of Oxy-Hb in the DLPFC (three groups).

3.3   Situation Fluctuation and Resilient Behavior

The fluctuation caused by a lack of the resources in scenarios (1)–(4) was named "resource fluctuation," the fluctuation caused by a change in wind speed and direction was named "wind fluctuation," and that caused by the fire spread range was named "fire fluctuation." We indexed each as follows. First, the resource fluctuation was calculated by simulation of the average number of buildings burned out when resources fluctuated based on scenario (6) as the standard. Then, the wind fluctuation was calculated by simulation of the average number of buildings burned under the wind speed/wind direction condition based on the "no wind" condition as the standard. Finally, on the basis of the case in which the number of buildings included in the spread range is "1," the fire fluctuation was calculated by simulation based on the average number of buildings consumed per unit time. In order to index these fluctuations for each task, we decided to use the maximum value as the indicator of the fire fluctuation.

The index values of the resource and wind fluctuations for each experimental condition are shown Tables 3 and 4, respectively.

Table 3. Quantification of resource fluctuation

Sinario	Change of condition about resources
No fire track	10.00
Less fire tracks	5.00
Less resources to build walls and to destroy buildings	6.00
The construction of walls or destruction of buildings taked longer than usual	4.76
Normal	0.00

Table 4. Quantification of wind fluctuation

Condition of wind	Change of condition about wind
11 m/s	29.00
7 m/s	20.33
3 m/s	15.00
11 m/s (sudden reverse of the wind direction)	25.17
7 m/s (sudden reverse of the wind direction)	20.67
3 m/s (sudden reverse of the wind direction)	13.33

The experiment participants' tasks were divided into cases in which resilient behavior was seen at least once during the task and cases in which it was never seen. The cases were plotted in the spaced defined by each axis. They are shown in Fig. 4 together with the positions of their centers of gravity.

Although there were large variations in the results, there was a tendency for tasks to display resilient behavior as the situation fluctuation increased. From this, it was confirmed that when the situation fluctuated, people could not deal with it by following the manual alone, so they had to resort to resilient behavior.

In addition, we analyzed the tasks in order to clarify the relationship between successive patterns both with and without resilient behavior, and the success/failure of the results of each task based on those selections. In order to do this, we regarded a task as a failure if the fire spreads to more than 70% of the city, or if the hospital/nursing home burned. Figure 5 shows the percentage of cases in which resilient actions were involved and did not result in failure in the space shown in Fig. 4. In the same way, Fig. 6 shows the percentage of cases in which resilient actions were not involved and the task did not fail (the proportion is higher if the color is darker). From the results, there is a clear difference in both results, in the upper right area. Therefore, it is indicated that there are few failures in cases with resilient behavior when the situation fluctuation is large.

3.4   Relationship with Personality

We examined the correlation between the score of each of the Big Five personality characteristics for each experiment participant obtained by the TIPI-J [3] and the correlation between the number of tasks with resilient behavior that did not result in

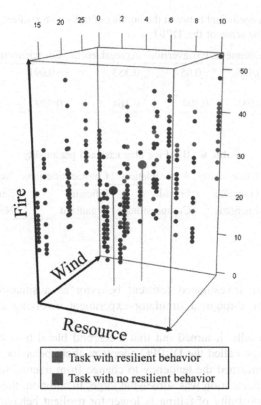

Fig. 4. Tasks divided into cases in which resilient behavior was seen at least once.

failure, and the results are shown in Table 5. From these results, a positive correlation was statistically significant at the 1-percent level in their openness scores. Openness is a characteristic that reflects the richness of imagination and curiosity, suggesting that this type of person tends not to fail with resilient behavior. We have concluded that it was easier for people with higher openness personalities to adopt resilient behaviors. This is because it was necessary to enrich imagination and curiosity in order to adopt those behaviors according to the situation and without relying on the manual (Table 6).

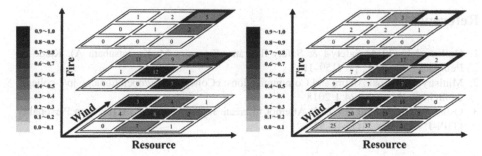

Fig. 5. Percentage of cases involving resilient behavior that did not result in failure.

Fig. 6. Percentage of cases without resilient behavior that did not result in failure.

Table 5. Correlation coefficient between the number of tasks with resilient behavior that did not result in failure and the score of the TIPI-J

	Openness	Extraversion	Agreeableness	Conscientiousness	Neuroticism
Correlation-coefficient	0.435	−0.05	0.353	−0.091	0.295
P-value	0.081	0.894	0.164	0.729	0.251

Table 6. Big Five character and personality

Openness	Extraversion	Agreeableness	Conscientiousness	Neuroticism
Inventive	Outgoing	Friendly	Efficient	Sensitive
Curious	Energetic	Compassionate	Organized	Nervous

4   Conclusion

In this research, we investigated resilient behavior as a situation fluctuated from moment to moment, through a simulator experiment involving a fire-extinguishing activity.

Based on the results, it turned out that a cerebral blood flow change is activated when the brain region called the DLPFC adopts resilient behavior.

First, people confirmed the tendency to change from manual to resilient behavior when the situation fluctuation was somewhat large. In addition, the experiment results showed that the probability of failing is lower for resilient behavior than for manual behavior, especially when the situation fluctuation is large.

Also, because the DLPFC, which is responsible for judgment and planning, is activated more for tasks involving resilient behavior than for other tasks, it was suggested that training that activates the DLPFC would be desirable to enhance resilience.

In addition, those participants with high openness scores tended not to fail during resilient behavior. From this, it was suggested that people with high openness scores can perform well during resilient behavior.

In future research, we want to clarify the boundaries of situation fluctuation in a form that is applicable beyond the fire extinguisher simulator.

References

1. Hollnagel E (2015) Safety-I & Safety-II (trans: Kitamura M, Komatsubara A). Kaido-do Publishing Co., Ltd., pp J159–J163
2. Ministry of Internal Affairs and Communications: eCollege. http://open.fdma.go.jp/e-college/danin.html. Accessed 24 1 2018
3. Oshio S, Abe S, Pino C (2012) Attempt to create Japanese version of Ten Item Personality (TIPI-J). Pers Res 21(1):40–52

Autopsy as on Outcome and Performance Measure: Three Years of Hospital Autopsy as an Instrument of Clinical Audit

S. D'Errico$^{(\boxtimes)}$, M. Martelloni, S. Niballi, and D. Bonuccelli

Department of Legal Medicine, Azienda USL Toscana Nordovest Lucca, Lucca,
Italy
stefano.derrico@uslnordovest.toscana.it

Abstract. An extensive literature documents a high prevalence of errors in clinical diagnosis discovered at autopsy. Multiple studies have suggested no significant decrease in these errors over time. Despite these findings, autopsies have dramatically decreased in frequency in the United States and many other countries. In 1994, the last year for which national U.S. data exist, the autopsy rate for all non-forensic deaths fell below 6%. The marked decline in autopsy rates from previous rates of 40–50% undoubtedly reflects various factors, including reimbursement issues, the attitudes of clinicians regarding the utility of autopsies in the setting of other diagnostic advances, and general unfamiliarity with the autopsy and techniques for requesting it, especially among physicians-in-training. The autopsy is valuable for its role in undergraduate and graduate medical education, the identification and characterization of new diseases, and contributions to the understanding of disease pathogenesis. Although extensive, these benefits are difficult to quantify. This review of the last three years of hospital autopsy in Lucca studied the more easily quantifiable benefits of the autopsy as a tool in performance measurement and improvement. Such benefits largely relate to the role of the autopsy in detecting errors in clinical diagnosis and unsuspected complications of treatment. It is hoped that characterizing the extent to which the autopsy provides data relevant to clinical performance measurement and improvement will help inform strategies for preserving the benefits of routinely obtained autopsies and for considering its wider use as an instrument for quality improvement.

Keywords: Hospital autopsy · Performance measure · Clinical audit

1 Introduction

There exists a general perception that necropsies are no longer necessary as antemortem diagnosis identifies the principal cause of death and other clinically significant diagnoses in the vast majority of cases. This perception has undoubtedly contributed to the progressive decline in necropsy rates over the past 30–40 years. In most jurisdictions, including the UK, Australia, France and US 10% or fewer of all natural deaths undergo necropsy [1–3]. These average rates reflect a wide range of institutional necropsy rates, with a small number continuing to performing relatively frequent necropsies but many

© Springer Nature Switzerland AG 2019
S. Bagnara et al. (Eds.): IEA 2018, AISC 819, pp. 517–521, 2019.
https://doi.org/10.1007/978-3-319-96089-0_55

performing almost none. Consequently, many clinicians, radiologists, and others involved in the antemortem diagnostic process never have the opportunity to learn of major missed diagnoses among patients who died under their care. Consequently, the number of important missed diagnoses among non necropsied deaths could approximate or even exceed the number observed at necropsy. In conditions where misdiagnosis confers substantial short term mortality, conventional estimates of diagnostic performance may substantially overstate diagnostic performance because they do not take into account the possibility of missed cases among non-necropsied deaths. The major reason for the decline in hospital post mortems is unclear. The rates of approach following death are not recorded and the willingness of medical staff to address postmortem examination may be affected by changing social perceptions. There may be confused thinking over gaining the trust and permission from families to agree to postmortem examination, following media attention surrounding inappropriate organ retention after death. Predictably, the need for post-mortem examination comes at a time when the clinician may be unsure of the reaction of the families at this juncture. Clinicians may be increasingly reluctant to discuss the subject of a post mortem. Identifying a missed diagnosis is a reason for physiscian's reluctance to seek post mortem, this is an obvious reason for declining rates. Only one study has addressed the issue of litigation following post mortem findings [4]. A review of 176 post mortems identified only one litigation, but the intent to proceed to litigation was present even before the patients's death. It is very clear that if current trends continue hospital clinical post mortems could be a rarity in the future.

2   Hospital Autopsy as an Instrument of Clinical Audit for Diagnostic Performance

Since 2015, 2900 hospital deaths occurred (including all age groups), post-mortem examinations were carried out in 83 patients with a **necropsy rate of 0.028%** (Table 1). In more than 50% autopsies were performed on sudden unexpected deaths occurred in Emergency Room to confirm clinical diagnostic suspects.

Indications for autopsy were identified and ante mortem and post mortem diagnosis in hospitalised medica patients compared. Results were shared with all physicians requesting hospital autopsy to discuss about concordance rate of clinical and post mortem diagnosis, causes of misdiagnosis. **Concordance between ante-mortem and post-mortem diagnoses was seen only in 45.7%**. The most common correct causes of death were cerebral haemorrhage, pneumonia, tumors and sepsis. Discordance between ante-mortem and post-mortem diagnoses was seen in patients who died in Emergency Room. Acute coronary syndrome, acute myocardial infarction, acute pulmonary embolism and aortic dissection were underestimated or undersuspected. Unsuspected medical conditions relevant to death were suggested by post mortem examination and were not recorded as known during life (i.e. structural cardiomyopathies) (Table 2).

Table 1. Demographic characteristics of hospital autopsies performed in 2015–2018

Years	Hospitalised medical patient (n)	Emergency room	Medical units	Surgical units
2015	7	7	-	-
2016	14	9	4 - Medicine (2) - Infective disease (1) - Intensive care (1)	1 - General surgery
2017	40	24	15 - Medicine (5) - Neurology (1) - Cardiology (1) - Pneumology (1) - Intensive care (7)	1 - ORL
2018 (jan–may)	22	12	8 - Medicine (1) - Cardiology (1) - Nephrology (2) - Intensive care (4)	2 - Orthopedics - Neurosurgery
	83	52	27	4

Table 2. Hospital autopsy and missing diagnosis

Post mortem diagnosis	Discordance rate (%)
ACS & AMI	62%
Pulmonary embolism	75%
Aortic dissection	50%
Pneumonia	20%
Tumor	-
Mechanical bowel obstruction	-
Digestive haemorrhage	-
Cerebral haemorrhage	-
Sepsis	12.5%
Thrombosis	-
Structural cardiomyopathies	100%

After auditing results, most of physicians asked to partecipate, in the future, to hospital autopsies and declared to be much more motivated to compare clinical diagnosis of death and post mortem diagnosis as an opportunity to enhance diagnostic accuracy, improve knowledge of disease and, finally, assist relatives with grieving and knowledge of potentially heritable diseases. They concluded that autopsies can be considered beneficial when findings are concordant by providing feedback on treatment decisions as well as educating young doctors but complains persist about the possibility of legal claims in case of misdiagnosis or diagnostic errors.

3 Discussion

The decline in the number of hospital autopsies ultimately reflects a simple fact: hospital autopsies as currently performed and used are for the most part viewed as less important than the other services pathologists provide, and just not worth their costs and effort. If the practice of autopsy is to remain (or perhaps become) relevant to the current practice of medicine, it must demonstrate that it clearly and objectively contributes to knowledge in ways that can inform and/or change medical practice. Discrepancies between ante-mortem and post mortem diagnosis are well known and have been documented repeatedly in the literature. Numerous studies document substantial rates of major clinically unsuspected diagnosis detected at necropsy including missed diagnoses that probably affected outcome [5–10]. The discrepancies are usually classified according to the Goldman criteria. A class I error is a major missed diagnosis, with potentially adverse impact on survival, that would have changed management. A class II error represents a missed major diagnosis, without potential impact on survival, that would not have changed therapy. The class III and IV errors are missed minor diagnoses not related to the cause of the main disease. Clinicians have generally attributed these persistent and roughly unchanged discrepancies between antemortem and postmortem diagnoses to selection bias, arguing that cases sent for necropsy are precisely those in which there is diagnostic uncertainty. Despite its plausibility, this view is not supported by the available evidence. If rates of clinically important diagnoses first detected at necropsy largely reflected case selection by clinicians, one would expect studies with high necropsy rates to report substantially lower error rates. A British study in which only 8% of decedents underwent necropsy reported clinically important missed diagnoses in 39% of cases [11]. Literature reported an inverse correlation between the post mortem rate and misdiagnosis rate (lower post mortem rates show higher error rates due to selection bias). One study suggested four kinds of errors that could lead to diagnostic inaccuracy, namely omission, premature losure, inadequate synthesis and wrong formulation. Omission and inadequate synthesis were negatively correlated with the degree of training of the treating physicians and led to false negative diagnoses. But premature closure was independent of clinical experience and correlated with overconfidence in findings. However despite the highest level of clinical skill and multiple diagnostic support, it would be unrealistic to expect no error in ante-mortem diagnoses. The aim of our study was to compare ante-mortem and post-mortem diagnoses in hospitalised medical patients. **Concordance between ante-mortem and post-mortem diagnoses was seen only in 45.7%**, which is consistent with other studies reported in the literature. Discordance between ante-mortem and post-mortem diagnoses was seen in patients who died suddenly in Emergency Room. Acute coronary syndrome, acute myocardial infarction, acute pulmonary embolism and aortic dissection were underestimated or undersuspected. Unsuspected medical conditions relevant to death were suggested by post mortem examination and were not recorded as known during life (i.e. structural cardiomyopathies). After auditing results, most of physicians asked to partecipate, in the future, to hospital autopsies and declared to be much more motivated to compare clinical diagnosis of death and post mortem diagnosis as an opportunity to enhance diagnostic accuracy, improve knowledge of disease and, finally, assist relatives with grieving and knowledge of potentially heritable diseases. They concluded that autopsies can be considered beneficial when findings are concordant by providing feedback on treatment decisions as well as educating young doctors but complains persist about the possibility of legal claims in case of misdiagnosis or diagnostic errors. We strongly suggest regular review of post mortem diagnosis as a new strategy of clinical audit to improve performance and outcome. We advocate that medical staff should consider the continuing practice of autopsy for select patients within their hospital communities as a valuable adjunct for clinical quality assurance and continuing medical education. Four conditions have been deemed to be essential for a post mortem to be a valid monitor of clinical performance: a high post mortem rate, standardised procedures during calculations of both sensitivity and specificity and an estimate of errors in post mortem diagnosis. The need for up to date facilities and the wider availability of diagnostic tools to bring post mortems to the same standard as clinical diagnostic has been recognised for a long time.

References

1. Burton JL, Underwood JC (2003) Necropsy practice after the "organ retention scandal": requests, performance, and tissue retention. J Clin Pathol 56:537–541
2. Royal College of Pathologists of Australasia Autopsy Working Party (2004) The decline of the hospital autopsy: a safety and quality issue for healthcare in Australia. Med J Aust 180:281–285
3. Chariot P, Witt K, Pautot V et al (2004) Declining autopsy rate in a French hospital: physicians' attitudes to the autopsy and use of autopsy material in research publications. Arch Pathol Lab Med 124:739–745
4. Nichols L, Aronica P, Babe C (1998) Are autopsies obsolete? Am J Clin Pathol 110:210–218
5. Shojania KG, Burton EC, McDonald KM, Goldman L (2005) Overestimation of clinical diagnostic performance caused by low necropsy rates. Qual Saf Health Care 14:408–413
6. Swaro A, Adhiyaman V (2010) Autopsy in older medical patients: concordance in ante- and post-mortem findings and changing trens. J R Coll Physicians Edinb 40:205–208
7. Saad R, Yamada AT, Pereira de Rosa FH et al (2007) Coronary artery disease. Comparison between clinical and autopsy diagnoses in a cardiology hospital. Heart 93:141–149
8. Gibson TN, Shirley SE, Escoffery CT et al (2004) Discrepancies between clinical and post mortem diagnoses in Jamaica: a study from the University Hospital of the West Indies. J Clin Pathol 57:980–985
9. Spiliopoulou C, Papadodima S, Kotakidis N et al (2005) Clinical diagnoses and autopsy findings. A retrospective analysis of 252 cases in Greece. Arch Pathol Lab Med 129:210–214
10. Sington JD, Cottrell BJ (2002) Analysis of the sensitivity of death certificates in 440 hospital deaths: a comparison with necropsy findings. J Clin Pathol 55:499–502
11. Perkins GD, McAuley DF, Davies S et al (2003) Discrepancies between clinical and postmortem diagnoses in critically ill patients: an observational study. Crit Care 7:129–132

Differential Effects of 8 and 12 Hour Non-rotating Shifts on Alertness, Sleep and Health of Public Safety Workers

Arijit Sengupta(✉), Zuleyha Aydin, and Samuel Lieber

New Jersey Institute of Technology, Newark, NJ 07079, USA
sengupta@njit.edu

Abstract. Shiftwork causes disruption in circadian and social rhythms of the shift workers. Extended hours shifts and non-rotating (permanent) shifts are increasingly being adopted in police agencies across United States. The aim of this study was to evaluate alertness, sleep, and wellness of workers in permanent shift systems in a public safety department. A self-reported questionnaire survey was administered to 39 police and security officers working in 8 and 12 h permanent shifts. When compared with the 8 h shift, 12 h shift work was associated with a significantly lower alertness level (p = 0.076), lesser sleep duration (p = 0.023), more perceived sleep insufficiency (p = 0.088), more perceived negative effect of shift type on sleep (p = 0.037), and higher frequency of back or lower back pain (p = 0.005). The results of this study are potentially useful when designing interventions to improve shift work experience.

Keywords: Shiftwork · Permanent shift · Alertness · Sleep · Wellness

1 Introduction

Shiftwork is designed to make use of all 24 h of a day and it is a common feature of today's work life across a broad range of occupations. Reasons for adopting shiftwork are varied. Manufacturing and engineering industries utilize shiftwork for economic reason. Equipment utilization increases when equipment is used 24 h per day. Other service sectors such as public safety, hospitals and air traffic control have to be available 24 h per day to meet the social need for these services. Whatever the reason for shiftwork may be, it is associated with increased health problems, including sleep disturbances, fatigue, productivity loss, digestive problems, emotional problems, and stress related illnesses [1]. One of the most common problems related to shift work is the decrease in quantity and quality of sleep which impacts alertness, fatigue level and health of the workers [2, 3]. As a consequence, these effects increase the risk of operator error, injuries and accidents [4].

Health studies on shift workers reported a series of physical, psychological and social problems due to the disruption of circadian and social rhythms [1]. Morning, afternoon and night shift work creates different kind of conflicts between the sleep and wakefulness hours and the internal timing mechanism [2]. Twelve or ten hour shift

© Springer Nature Switzerland AG 2019
S. Bagnara et al. (Eds.): IEA 2018, AISC 819, pp. 522–531, 2019.
https://doi.org/10.1007/978-3-319-96089-0_56

systems, as opposed to eight hours shift system, allow workers a compressed work week (CWW) of three to four days of work. The CWW shifts are gaining popularity because it requires fewer workdays and provides a larger block of time off, in a week [5]. It promotes fewer circadian changes, more free time for social life, reduced number of night shifts, and reduced commuting. However, the longer work shifts decrease the time available for sleep and may lead to increase of fatigue, reduced alertness, and increased safety risk during shift work [5].

In rotating shift systems workers take turns working on all shifts following a set schedule. The rotation or change of shifts adds more disruption for workers' routine of sleep and work [2]. Rotating shift system covers a wide variety of work schedules that varies in speed of rotation and direction of rotation. Speed of rotation is determined from the number of consecutive days a worker follows the same shift. The direction of rotation is determined from the chronological order of the change of shift. The combination of these two features of a shift schedule affects shift workers' physiological and psychological wellbeing [1]. To completely avoid the negative effects of rotating shifts, non-rotating or permanent shift systems were implemented [6].

Public safety departments operate 24 h to maintain security and safety services. An early study conducted in 1981 described the experimental implementation of non-rotating shift system in two Detroit Police Precincts for duration of one year [6]. Before and after survey showed favorable ratings for the non-rotating shifts. Several other indicators of health and quality of family life showed improvement. Later, in 1991, another study [7] in Lexington, Kentucky Police Department also noted improved sleep quality, sleep hygiene and reduction in absenteeism when changed from rotating to permanent shift system.

A recent study [8] investigating the effect of CWW on police personnel found negative effect on sleep, fatigue, and alertness, positive or no effect on quality of life, and insignificant effect on work performance, safety, and health [8]. Majority of other CWW studies came from health services or nuclear energy plants [5]. A review article on the effects of 12 h CWW shiftwork, summarized no great problems with sleep, health, safety, productivity or error rate when working 12 h compared to 8 h shifts [5]. All the above articles on CWW, employed rotating shift systems. No studies were found on CWW for non-rotating shift systems.

In recent times CWW as well as permanent shifts are increasingly being adopted in police departments [9]. A random telephone survey of 300 police agencies in United States revealed that between 2005 and 2009, the number of 8 h shift agencies dropped from 115 to 88, and the number of agencies with rotating shifts declined from 132 to 71 [9]. The aim of the present study was to evaluate alertness, sleep, and wellness of workers in permanent shift systems in a public safety department, and compare effects on 8 h and extended shift workers. A secondary objective was to determine the difference in impacts on rotating and non-rotating shift schedules in CWW workers.

2 Materials and Methods

The security staff and police officers from the Public Safety Department of New Jersey Institute of Technology participated in this study. The questionnaire survey and the study design were approved by both the Institutional Review Board (IRB) and the Chief of Public Safety Department. A detailed questionnaire survey was designed based on the standard shiftwork index [10]. The survey included 29 items covering alertness, sleep habits, and wellness factors. The printed copy of survey including a consent form was distributed to the shift workers on different days. Participation was voluntary, and all participants were guaranteed confidentiality and anonymity. All participants were adult, ranging age between 18 and 50, and English speaking.

The public safety department used three different permanent shift types as follows: 8 h (morning, afternoon, evening), 10 h (day, night) and 12 h (day, night). Job profiles of the shift workers participated in the survey are as follows: Sergeant, police officer, security staff and others (as dispatchers, lieutenant, chief of police and public safety officer) with a percentage of 16%, 22%, 43% and 19%, respectively of 78 total number of staff in the department. All participants were employed in shift system, and the broad task characteristics of their jobs were similar. The data collection continued for about two months and achieved a 50% response rate (39 participants out of 78). Participants were predominantly male (74%). They had an average (standard deviation) shift work experience of 5.7 (0.85) years, overtime per week 9.8 (9.91) h. The workload ratings were not significantly different between any of the groups studied.

The statistical significance of shift effects was determined by analyzing the survey data using a single factor ANOVA and Tukey's test with 90% confidence level. The effects on alertness, sleep and wellness were compared between groups of 8 h shift versus 12 h shift, 8 h morning shift versus 8 h afternoon shift, and 8 h morning shift versus 12 h day shift. Effects of 8 h or 12 h night shifts were not compared individually because of low participation. There were only 3 and 4 participants in 8 h and 12 h night shifts, respectively. Also, there were only two participants from 10 h shift, and as a result, data for 10 h shift system were excluded from the analysis.

3 Results and Analysis

3.1 Alertness

Participants rated their overall alertness level in the shift, and alertness levels during the Early-stage, Mid-stage and Late-stage of their shifts, in a scale of 1 to 5, where "1" is very alert and "5" is very sleepy. The mean rating of alertness levels of 12 h shift workers was lower than 8 h shift workers, for the overall shift and in every stage of the shift (Fig. 1). Also the alertness level for both shift types gradually decreased over the shift, which was expected. A significant decrease in mean alertness level was found for 12 h shift workers as compared to the 8 h shift workers for early-stage (p = 0.033), mid-stage (p = 0.052) and for the overall shift (p = 0.076).

To investigate the effect of the time of the day, alertness ratings of 8 h morning shift (from 07.00 a.m. to 03.00 p.m.) were compared with 8 h afternoon shift (from

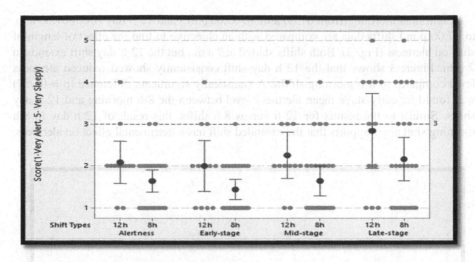

Fig. 1. Individual value plot of alertness ratings in the entire shift and, early, middle and late stage shifts for 8 h and 12 h shift workers with 95% confidence interval for the mean.

03.00 p.m. to 11.00 p.m.). Figure 2 illustrates that 8 h morning shift workers alertness level was almost unchanged over the entire shift, but the 8 h afternoon shift workers alertness level decreased gradually as the shift progressed with a large difference at the late-stage of shift (p = 0.088). The results support that working until late evening, 11 p.m., has a negative effect on alertness.

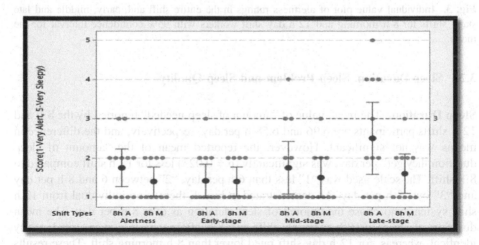

Fig. 2. Individual value plot of alertness ratings in the entire shift and, early, middle and late stage shifts for 8 h morning and 8 h afternoon shift workers with 95% confidence interval for the mean.

Eight hour morning (from 07.00 a.m. to 03.00 p.m.) and 12 h day (from 07.00 a.m. to 07.00 p.m.) shifts data are compared with an objective to find the effect of length of shift on alertness (Fig. 3). Both shifts started at 7 a.m., but the 12 h day shift extends to 7 p.m. Figure 3 shows that the 12 h day shift consistently showed reduced alertness level compared to 8 h morning shifht. A statistically significant difference (p = 0.037) was found for early-stage mean alertness level between the 8 h morning and 12 h day shifts. Similar to the results for 12 h versus 8 h shifts, this result of 12 h day to 8 h morning shift also supports that the extended shift has a detrimental effect on alertness.

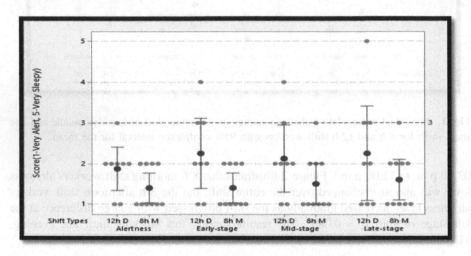

Fig. 3. Individual value plot of alertness ratings in the entire shift and, early, middle and late stage shifts for 8 h morning and 12 h day shift workers with 95% confidence interval for the mean.

3.2 Sleep Duration, Sleep Problem and Sleep Quality

Sleep Duration. The mean value of "amount of sleep needed" reported by the 8 h and 12 h shifts participants are 6.96 and 6.58 h per day, respectively, and the difference in means was not significant. However, the reported mean of the "amount of sleep duration they get" per day, was significantly (p = 0.023) less for 12 h shift compared to 8 h shift. The scale used was "1" less than 6 h per day, "2" between 6 and 8 h per day and "3" over 8 h per day. As demonstrated in Fig. 4, there is no individual from 12 h shift system who chose the amount of sleep duration as over 8 h per day. The mean duration of sleep for 8 h morning shift and 8 h afternoon shift, were rated nearly identical, whereas, for 12 h day shift rated lower than 8 h morning shift. These results indicate that sleep duration was affected by longer shift duration and effect of working in the afternoon and evening did not affect the sleep duration, even if they worked until 11 p.m.

Sleep Problems. Sleep problems included 11 questions and participants chose "1" (almost never), "2" (quite seldom), "3" (quite often) and "4" (almost always) to express

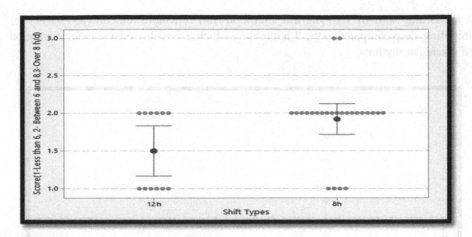

Fig. 4. Individual value plot of ratings of the amount of sleep duration with 95% confidence interval for the mean, for 8 h and 12 h shift workers.

how often they experience these sleep problems. Mean sleep problem indicators were compared for 8 h and 12 h shifts, and there was a statistically significant difference was observed for two sleep problems (see Fig. 5). Twelve hour shift workers scored significantly higher in sleep insufficiency ($p = 0.015$) and in the negative effect of shift type on sleep ($p = 0.082$) compared to normal 8 h shift. These results support the hypothesis that extended shift is associated with increased level of sleep problems.

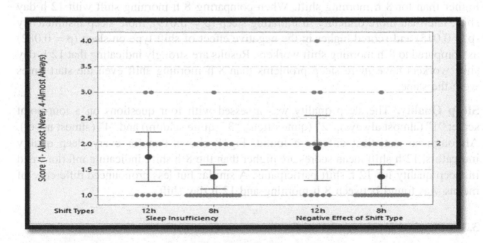

Fig. 5. Individual value plot ratings of sleep habit indicators with 95% confidence interval for the mean, for 8 h and 12 h shift workers.

Sleep problems scored by 8 h morning shift were almost identical to 8 h afternoon shift. Figure 6 shows that the mean score of 8 h afternoon shift of 1.8 is higher than the

8 h morning shift of 1.1. The 8 h afternoon shift felt more difficult (p = 0.017) for initiating sleep compared to the 8 h morning, which is possibly related to the disruption of circadian rhythms.

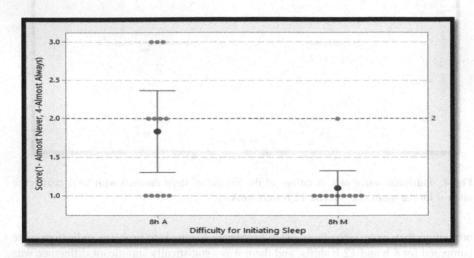

Fig. 6. Individual value plot of ratings of difficulty for initiating sleep for 8 h morning and 8 h afternoon shifts with 95% confidence interval for the mean.

Figure 7 shows that 12 h day shift mean scores for these three questions are much higher than for 8 h morning shift. When comparing 8 h morning shift with 12 h day shift, later had more difficulty in initiating sleep (p = 0.019), more sleep insufficiency (p = 0.015) and scored higher in the negative effect of shift type on sleep (p = 0.082) as compared to 8 h morning shift workers. Results are strongly indicating that 12 h day shift workers have more sleep problems than 8 h morning shift even the start times were the same.

Sleep Quality. The sleep quality was assessed with four questions on a four-point scale: "1" (almost always), "2" (quite often), "3" (quite seldom) and "4" (almost never). Although no statistical difference found, Fig. 8 shows that for every sleep quality indicators; 12 h shift mean scores are higher than the 8 h shift, indicating inferior trend in sleep quality for 12 h shift participants. A similar but less pronounced difference of means was found between 8 h morning and 12 h day shift.

3.3   Physical Health and Satisfaction

Six questions addressed the frequency of pain in various body locations, in a scale "1" (almost never), "2" (sometimes), "3" (usually), and "4" (almost always). Surprisingly, a significantly (p = 0.005) higher mean "back or lower back pain" frequency was reported by the 12 h group (mean 2.66) compared to the 8 h group (mean 1.78). A similar result was noted when comparing 8 h morning with 12 h day shift. 12 h day shift mean score of 2.37 is higher than 8 h morning shift 1.7, and the difference in mean

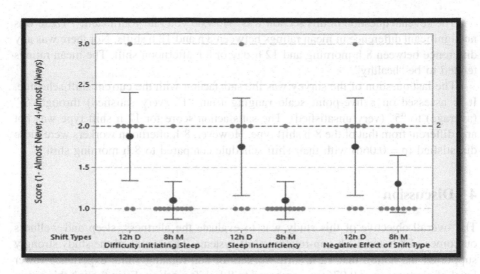

Fig. 7. Individual value plot of ratings of sleep habit indicators of 8 h morning and 12 h day shift workers with 95% confidence interval for the mean.

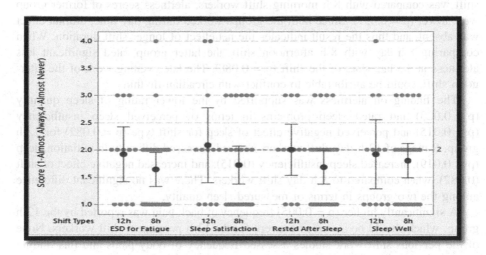

Fig. 8. Individual value plot of ratings of sleep quality indicators with 95% confidence interval for the mean for 8 h and 12 h shift workers.

was statistically significant (p = 0.019). These results support the notion that 12 h shift workers suffer from back/lower back pain more than 8 h shift workers.

A statistically significant difference (p = 0.043) for the frequency of "Pain in Arm/Wrist" was found between 8 h morning and 8 h afternoon shifts. The mean score for 8 h morning was 1.3 and for 8 h afternoon shift was 1.0. This difference was statistically significant but the difference is small, and the mean values are also small.

The seventh question in this section was "How do you rate your health". There was no significant difference in mean ratings between 8 h and 12 h shifts. Nor there was any difference between 8 h morning and 12 h day or 8 h afternoon shift. The mean ratings tended to be "healthy".

The last question of the survey was the satisfaction with the current shift schedule. It is assessed on a five-point scale ranging from "1" (very satisfied) through "3" (average) to "5" (very unsatisfied). The satisfaction score for 12 h shift type was not any different from that of the 8 h shift type. However, 8 h afternoon workers were more dissatisfied (p = 0.066) with their shift schedule compared to 8 h morning shift.

4 Discussion

The overall objective of this study was to evaluate the alertness, sleep and wellness outcomes of 8 h and 12 h non-rotating shift systems. The results of the study strongly supported the notion that 12 h shift workers of non-rotating shifts experience lower level of alertness (p = 0.076) as compared to 8 h shift workers. Even though this group receives four off days per week, they get less time to rest/sleep on the workdays, and longer work shift make them less alert and sleepy during the shift [5]. When 12 h day shift was compared with 8 h morning shift workers, alertness scores of former group were lower (p = 0.037). Since both the groups worked during day time, diurnal effect was absent, and thus the result indicates the net effect of longer shift duration. When comparing 8 h day with 8 h afternoon shift, the latter group rated significant less alertness at the late state of the shift (p = 0.088). The late evening work of the afternoon shift, could be attributable to conflict with circadian rhythm.

The finding on alertness was supported by the lower rating of sleep quantity (p = 0.023) and rated sleep problems in terms of perceived sleep insufficiency (p = 0.015) and perceived negative effect of sleep for shift type (p = 0.082) for 12 h group. Ratings of 12 h day shift workers showed increased difficulty in initiation sleep (p = 0.019), increased sleep insufficiency (0.015), and increased negative effect of shift (0.082), when compared to 8 h day shift workers. There was no significant difference among the two groups in terms of measured sleep quality.

A significantly higher (p = 0.005) frequency of back pain was reported by the 12 h group, which should be a serious concern for the 12 h permanent shift workers. None of the pervious shift work studies assessed frequency of body pains and this finding should be more carefully reviewed in future studies.

Interestingly, as opposed to the finding of a previous study of police work [8], the rating of satisfaction with current shift schedule was not any different for 8 h and 12 h shift workers. The higher satisfaction ratings after change from 8 h to CWW schedules were noted in several studies [5]. This may be attributable to the short term euphoria recoded in many studies [5]. This study results showed the long term effect of no difference in satisfaction.

It is also apparent that similar effects on alertness and sleep were also highlighted by the previous researchers on rotating 12 h shifts [11–15]. The results of this study are useful when designing interventions to improve shift work experience. One limitation of this study was lack of participation by the night shift workers, and, as a result, effects of non-rotating night shift for 8 or 12 h shifts could not be investigated.

References

1. Bambra CL, Whitehead MM, Sowden AJ, Akers J, Petticrew MP (2008) Shifting schedules: the health effects of reorganizing shift work. Am J Prev Med 34(5):427–434.e30. https://doi.org/10.1016/j.amepre.2007.12.023
2. Åkerstedt T (2003) Shift work and disturbed sleep/wakefulness. Occup Med 53(2):89–94. https://doi.org/10.1093/occmed/kqg046
3. Knutsson A (2003) Health disorders of shift workers. Occup Med 53(2):103–108. https://doi.org/10.1093/occmed/kqg048
4. Folkard S, Tucker P (2003) Shift work, safety and productivity. Occup Med 53(2):95–101. https://doi.org/10.1093/occmed/kqg047
5. Smith L, Folkard S, Tucker P, Macdonald I (1998) Work shift duration: a review comparing eight hour and 12 h shift systems. Occup Environ Med 55(4):217–229. https://doi.org/10.1136/oem.55.4.217
6. Owen J (1985) Changing from a rotating to a permanent shift system in the Detroit police department: effects on employee attitudes and behavior. Labor Law J 36:484–489
7. Phillips B, Magan L, Gerhardstein C, Cecil B (1991) Shift work, sleep quality, and worker health: a study of police officers. South Med J 84(10):1176–1184
8. Amendola K, Wyckoff L, Hamilton E, Jones G, Slipka M (2011) The impact of shift length in policing on performance, health, quality of life, sleep, fatigue, and extra-duty employment. Police foundation report. https://www.policefoundation.org/publication
9. Amendola K, Hamilton E, Wyckoff L (2011) Trends in shift length: results of a random national survey of police agencies. Police foundation report. https://www.policefoundation.org/publication
10. Barton J, Spelten E, Totterdell P, Smith L, Folkard S, Costa G (1995) The standard shiftwork index: a battery of questionnaires for assessing shiftwork-related problems. Work Stress 9 (1):4–30. https://doi.org/10.1080/02678379508251582
11. Di Milia L (1998) A longitudinal study of the compressed workweek: Comparing sleep on a weekly rotating 8 h system to a faster rotating 12 h system. Int J Ind Ergon 21(3–4):199–207. https://doi.org/10.1016/S0169-8141(97)00039-5
12. Kecklund G, Eriksen CA, Åkerstedt T (2008) Police officers attitude to different shift systems: association with age, present shift schedule, health and sleep/wake complaints. Appl Ergon 39(5):565–571. https://doi.org/10.1016/j.apergo.2008.01.002
13. Ong CN, Kogi K (1990) Shiftwork in developing countries: current issues and trends. Occup Med (Philadelphia, Pa.) 5(2):417–428
14. Paley MJ, Price JM, Tepas DI (1998) The impact of a change in rotating shift schedules: a comparison of the effects of 8, 10 and 14 h work shifts. Int J Ind Ergon 21(3–4):293–305. https://doi.org/10.1016/S0169-8141(97)00048-6
15. Schroeder DJ, Rosa RR, Witt LA (1998) Some effects of 8- vs. 10-h work schedules on the test performance/alertness of air traffic control specialists. Int J Ind Ergon 21(3–4):307–321. https://doi.org/10.1016/S0169-8141(97)00044-9

The Dialogue Workshop, a Method to Analyse the Coordination Needs Between Heterogenous Stakeholders for Risk Prevention in Micro and Small Enterprises (MSEs)

Sandrine Caroly[1(✉)], Deborah Gaudin[1,2], Marc Malenfer[3], and Patrick Laine[3]

[1] Pacte Laboratory, Grenoble Alps University,
IEP BP 48, 38040 Grenoble Cedex 09, France
sandrine.caroly@univ-grenoble-alpes.fr
[2] LIP, Grenoble Alps University, BSHM, 38040 Grenoble Cedex 09, France
[3] INRS, 65 Boulevard Richard Lenoir, 75011 Paris, France

Abstract. The aim is to describe the coordination conditions between OSH advisors and professional organisations in order to develop good practices in risk prevention in MSEs. Owner-managers of MSEs often have difficulties in applying OSH legislation and a lack of time and skills for doing it. The stakeholders' coordination develops a prevention program in two case studies conducted in both the road transport and construction sectors. Discussion groups were conducted during a one-day dialogue workshop with different stakeholders (OSH regulators, OSH advisors, professional organisations, worker and employer representations) of these programs. The cooperation between professional organisations and OSH advisors seems necessary to develop a proactive approach and to create a specific program, considering the particular needs of the professional branch.

Keywords: MSE · Risk prevention · Coordination

1 Objectives

The aim is to describe the coordination between occupational safety and health (OSH) advisors and professional organisations, in order to develop good practices in risk prevention in micro and small enterprises (MSEs), which they are often not demanding. We will explore a major factor: the role of the coordination of stakeholders in supporting the creation of prevention programs in MSEs.

In MSEs, the OSH barriers identified are [1]:

- OSH is not a priority: the business market is of higher importance for the owner-manager, who has rather a reactive attitude towards occupational risks,
- There is a gap between MSEs and prevention stakeholders: few companies are visited by the labor inspection, few occupational health services are consulted for OSH advices, support comes from private services,

© Springer Nature Switzerland AG 2019
S. Bagnara et al. (Eds.): IEA 2018, AISC 819, pp. 532–537, 2019.
https://doi.org/10.1007/978-3-319-96089-0_57

- There is a lack of resources: difficulties to applied the OSH legislation, no time to do it, nore internal skills,
- Few initiatives are realized to prevent psycho-social risks.

Our question is how to improve the capacity of owner-managers in MSEs to prevent occupational risks? The theoretical framework considered is:

- OSH practices according to the sector and its specific risks,
- the coordination of different stakeholders in a network, the creation of programmes taking into account the needs of the professional branch.

2 Methods

This study related to the French context is part of the Work Package 3 (WP3) of the SESAME research, which is financed by the European Agency for Safety and Health at Work (EU-OSHA) and involved nine different EU Member States: UK, Belgium, Denmark, Italy, Germany, France, Estonia, Romania, Sweden. During 3 years (2014–2017), this SESAME research aims at improving the OSH management in MSEs by providing evidence-based support, identifying good practices and increasing knowledge about the determinants of these identified good practices. This study is splitted into 4 work packages, focusing on different issues related to MSEs and risk prevention: a literature review (WP1), the existing good practices in OSH (WP2), the OSH programs, strategies, and policies targeting MSEs in each country (WP3), some recommendations (WP4).

In France, in the WP3 [2, 3], two networks were compared with their different OSH intermediaries that developed risk prevention actions in MSEs. Before choosing these two networks in road transport of goods and construction, we made an inventory and a selection among several good practices programs (4). For each of them, we had collective interviews (2) with the leader of each national program, completed by individual interviews with different stakeholders of the program (4). We obtained also data about the target (population, types of actions) and the assessment of the program.

In the construction sector, 4 MSEs' interviews (1 owner-manager and 1 employee) and a short visit of the companies were done. This data collection was obtained during the WP2.

For both the road transport and construction programmes, we wrote two case studies (2), linked with the themes of the SESAME project, which will be part of more global analyses and a comparison between different countries (WP4).

We also did some discussion groups during one dialogue workshop [4] bringing together different stakeholders of these programs. All identified stakeholders related to three defined sectors (road transport, construction and restauration), which include a large number of MSEs, were invited to a one-day workshop. OSH advisors and professional organisations had a strong interest to participate, while worker and employer representations were more difficult to convince, as well as some OSH regulators. The dialogue workshop design was a two-phased discussion groups: 1/by sector of activity about roles, practices, incentives and obstacles in OSH preventive activities in MSEs,

2/between peers about strategies and needs in OSH activities in MSEs for each type of stakeholder (regulator organisations, various external non-institutional stakeholders, that we called intermediaries, like professional organisations, chambers of commerce and industry, social partners).

In each group discussion of the dialogue workshop, notes were taken by a facilitator and a moderator, who then shortly discussed the insights in order to highlight the main relevant findings. At the plenary discussion, each moderator presented to all participants the findings coming from each discussion group and they were allowed to comment the summaries. Following the workshop day, each facilitator made a final document about notes taken, which was verified and completed by moderators. Based on these data, the researchers wrote a final summary about the entire dialogue workshop (minutes of discussions in the dialogue workshop), which was commented by the partners of the French Research and Safety Institute (INRS), before using it for writing the final national report. The discussion groups, the plenary discussion, as well as the additional interviews were recorded.

3   Results

The French Government and the National Health Insurance Fund for Salaried Workers (CNAMTS) defined four priorities for OSH actions and these priorities were deployed in 13 programs (2014–2017). Among the latter, we chose two national programs, which aim at creating OSH tools related to the needs of MSEs, in both the road transport and the construction sectors. These following examples were selected because their actions were supporting companies in multiple ways and targeting several specific points related to the sector of activity. We also considered their innovative character, the possibility to obtain evaluations and have further details about them, the involvement of various prevention stakeholders and their relevance regarding the effectiveness of their prevention system.

3.1   The Example of Road Transport Sector Program

The road transport sector programme is characterized by a combination of methods. These various methods aim to expand the ways to reach MSEs, in order to improve prevention:

- the specific design of the Online interactive Risk Assessment tool (OiRA), an e-tool which propose free of charge sectoral risk assessment for MSEs. This OiRA tool was adapted to the French road transport sector by the INRS,
- the specific design of the French tool-kits 'Synergie', two tools ('Synergie pédagogie' and 'Synergie accueil') aiming at welcoming new workers in firms as well as learners in vocational training (during practices) and at the same time providing an introduction to OSH. The French Government, the INRS, the CNAMTS, the Regional Health Insurance Funds (CARSATs), were involved in this preventive project,

- ongoing trainings given to the OPCA Transport & Service (public non-for-profit organisation which collects financial contributions for training) advisors about prevention,
- a financial aid intended for companies that purchase specific equipment, for example an adapted truck trailer to avoid falling from a height,
- advices for the design work environment, given by the MavImplant tool and related to particular situations, for example loading/unloading docks,
- an awareness campaign among owner-managers about specific risks and their legal obligations to face them.

This program required the coordination of various type of partners to reach the targeted companies: a training organisation (OPCA T&S), national and regional OSH stakeholders (INRS, CNAMTS, CARSATs and occupational health services) and equipment manufacturers.

An interesting point is that tools (OiRA and Synergie) were adapted to the specific MSEs' needs in the road transport sector: focused on the specific risks related to their occupational activities, easy to use, free of charge and accessible online.

3.2 The Example of the Construction Sector Program

The construction sector program is characterized by a central platform for distribution of easy to use tools and information, designed for employers who have little time and resources for OSH. This plateform is the Institute for Research and Innovation in Health and Safety at Work (IRIS-ST), founded by two professional organisations (the Confederation of Crafts and Small Construction Companies, CAPEB, and the National Union of Craftsmen of Public Works and Landscape, CNATP) to reduce accidents and improve OSH conditions in construction.

The IRIS-ST collaborate with various prevention organisations: a professional organisation specialized in prevention (the French Professional Agency for Risk Prevention in Building and Civil Engineering, OPPBTP), two national prevention stakeholders (the Social Security and the National Institute for Research and Safety, INRS). Equipment manufacturers were also involved in the program.

The kind of support offered is:

- advices about equipment for safety at work,
- information about risk exposure by profession,
- tools to make a risk assessment and documents explaining legal obligations related to prevention,
- information about safety training with an online survey that assists each company in identifying mandatory training,
- a financial support for purchasing materials or equipment.

The main point is that all these actions were focused on the specific needs of the construction companies. The information, tools, training courses and financial support were promoted by the professional organization. The tools for risk assessment and prevention were adapted to the target group and the working conditions in construction. The information on the website/via the app (smartphone application) was simple, short and effective which is necessary considering that the work is mainly done at temporal workplaces with little time available for OSH. Finally, the information support was free of charge for companies.

3.3   Suggestions Produced by Stakeholders During the Dialogue Workshop

In the Road Transport
Trainings have to be further developed in several dimensions: initial, continuous and specialized. The message about risk prevention should reach the priority targets, like companies with a high number of accidents at work (temporary workers included), and convince owner-managers that risk prevention is a profitable investment (improving both performance and health in the company). The coordination between stakeholders has to be improved (harmonization of the agreements, more information about the actions made by each stakeholders). A risk prevention leader, attached to a unique prevention organisation, should be appointed for creating links between several MSEs and this central OSH stakeholder.

In the Construction
Trainings for future owner-managers and workers should be improved and a culture of risk prevention in the construction sector developed. The adaptation of tools according to the trades is relevant and has to be continued. Better informing new MSEs on the roles and functions of the various stakeholders is suggested. There is a need to write into regional contracts for resources (CPOM) the distribution of concerted actions between the stakeholders. Construction companies that are rigorously engaged in improving risk prevention should be encourage with award labels. Finally, the French codification of risks should be simplified (NAF, codes related to the French activities).

4   Discussion

The aim is to describe the coordination between occupational safety and health (OSH) advisors and professional organisations, in order to develop good practices in risk prevention in micro and small enterprises (MSEs), which they are often not demanding. We will explore a major factor: the role of the coordination of stakeholders in supporting the creation of prevention programs in MSEs.

In both sectors (construction and road transport), the networks of stakeholders and the various methods used in the programs allow the promotion of advices among MSEs, most of the time given by professional partners. This contribution of the intermediaries is a key factor to develop OSH practices in companies. When the MSE has a question about health and safety at work, it contacts the intermediaries to be helped or obtain advices.

The cooperation between professional organisations and OSH advisors is necessary to develop a proactive OSH approach. A better network between different stakeholders can improve the efficiency of prevention in MSEs. But, this collective work among the various stakeholders depends on the share of operative and common references about

OSH and MSEs. One result of the dialogue workshop is that: *"it's the first time everyone is together around the same table"*. Then, we must continue to promote the meetings between stakeholders, in order to coordinate their actions in MSEs.

Globally, the capacity of owner-managers to develop an OSH approach in their MSEs, relies on their awareness of risks, their experience of implementing OSH measures, the public policies (financial support to acquire equipment, obligation of risk assessment). Furthermore, the required written risk assessment document determines the dynamics of MSEs in risk prevention. But all stakeholders, both institutionals and intermediaries, agree that owner-managers often have difficulties in identifying the roles and functions of the stakeholders surrounding them. The diversity of prevention communication supports and the redundancy of same actions made by multiple stakeholders do not help to simplify the owner-managers' understanding of prevention devices, useful contacts and adapted tools to their needs. The studied networks and programs in both sectors (construction and road transport) show us that many stakeholders have several actions in OSH prevention according to their specific missions. But, these initiatives do not take collective strategies into account, with for example, a better coordination of actions with partnership agreement. These system remains complex for the MSE. Networks between institutional stakeholders and intermediaries are created to answer the specific needs of the MSEs, according to the professional branch and with appropriated tools for risk prevention. We need to be able to better identify, support and pilote them.

References

1. EU-OSHA (2016) Contexts and arrangements for occupational safety and health in micro and small enterprises in the EU-SESAME projects. European Agency for Safety and Health at Work, Bilbao
2. EU-OSHA (2017) Safety and health in micro and small enterprises in the EU: from policy to practice. European Agency for Safety and Health at Work, Bilbao
3. EU-OSHA (2017) Safety and health in micro and small enterprises in the EU: the view from the workplace. European Agency for Safety and Health at Work, Bilbao
4. Gustavsen B, Engelstad PH (1986) The design of conferences and the evolving role of democratic dialogue in changing working life. Hum Relat 39(2):101–116

Associated Health Behaviors and Beliefs from a Self-paid Colonoscopy Population at a Regional Hospital in Northern Taiwan

Wan-yu Wang[✉] and Eric Min-yang Wang

Mackay Memorial Hospital, National Tsing Hua University,
10F (Health Inspection Center), No. 690, Section 2, Guangfu Road, East District,
Hsinchu, Taiwan
m908@mmh.org.tw

Abstract. OBJECTIVE: Understand the situation at Self-paid Colonoscopy Population the Health Behaviors, Health Beliefs and colon cancer Screening Knowledge. Impact of colonoscopy results and basic demographics, health behavior, colonoscopy knowledge, health beliefs. METHODS: This study was conducted 201 voluntary Self-paid Colonoscopy patients. Research questionnaire containing demographic data, Health Behavior, Colorectal cancer screening knowledge, Colonoscopy examination Health Beliefs. The results of this questionnaire will be analyzed statistically. RESULTS: Subjects have poor screening knowledge and less healthy behaviors. Positive relationship between health beliefs and screening knowledge ($P = 0.006$). Health beliefs had significant effects on gender, marital status, body mass index, health behaviors, screening experience and family history ($P < 0.05$). The results of colonoscopy had significant effects on age ($P < 0.05$). CONCLUSIONS: This study found that there is a positive correlation between screening knowledge and Health Beliefs. People's poor knowledge of colorectal cancer lead to people do not trust colonoscopy, is the biggest obstacle to screening for colorectal cancer. The male sex and people over the age of 50 in the colonoscopy knowledge absorption worse than others, especially for the knowledge of colorectal polyps is the worst. However, people under the age of 50 do not care much about the government subsidy stool occult blood related information. The study also found that patients over the age of 60 compared with 30 to 39 years old subjects have a higher probability of predicting the risk of adenoma.

Keywords: Colonoscopy · Health belief model · Health behaviors

1 Introduction

According to the World Health Organization in 2014 represents worldwide number more than 2 million dead each year, of which 21–25% of breast and colon cancer. Taiwan of colorectal cancer incidence and mortality has increased steadily. According to the Ministry of Health and Welfare of National Health Cancer Registry statistics show that the death toll stood at 3,386 in 2002, continued to rise in 2014 to reach 5,603 people [1]. Early colorectal cancer and asymptomatic, but early detection of problems

© Springer Nature Switzerland AG 2019
S. Bagnara et al. (Eds.): IEA 2018, AISC 819, pp. 538–547, 2019.
https://doi.org/10.1007/978-3-319-96089-0_58

through good health screening and early treatment of disease, the cure rate is very high cancer.

The average age of colorectal cancer is slightly lower in Taiwan than in other countries. People in the 30s and 40s in the age group in Taiwan have a trend of increasing colorectal cancer [2]. Colorectal cancer is mainly caused by the large intestine adenomatous polyps change, if early detection and removal of adenomatous polyps can reduce the incidence of colorectal cancer [3]. Colonoscopy is the most direct and reliable method of screening, and it is also the most accurate method of screening, with an accuracy of 95% or more. It was found that colonoscopy examination and removal of precancerous polyp reduce the occurrence of colorectal cancer, prevent death from colorectal cancer, and allow patients with higher survival rates [4]. A study in the United States calculated that colonoscopy can reduce the risk of conversion from premalignant to colorectal cancer [5].

In this study, Health Belief Model (HBM) revised as the basis to understand the object of study at their own expense to participate in colonoscopy behavior and other factors. Want to know from studies of risk factors associated with colorectal polyps related to genetic factors. And further explore the colonoscopy results and health behaviors, the correlation between health beliefs. The HBM was developed by the social psychologists Hochbaum (1952), Kegeles (1960) and Rosenstock [9] and others. According to the social psychology expert Rosenstock [9], it consists of the main received susceptibility, perceived severity, perceived benefits, perceived barriers, Cues to Action and Other Variables structures. The HBM is mainly used to explain the factors affecting individuals to take preventive health behaviors or personal use of preventive medical services. Let the subject feel the impact of the disease on self and decide whether to seek behavioral factors in health care services [6].

The most commonly used theory for the study of cancer prevention is HBM. In a study of colorectal cancer screening knowledge, attitudes, and barriers in 14 countries in the Asia Pacific region, Taiwanese respondents had the lowest cognitive risk scores for colorectal cancer screening. But the highest score in the recognition of the severity of colorectal cancer. And compared to other Asian countries, respondents over 50 in Taiwan have poor knowledge of colorectal cancer symptoms and risk factors. Taiwan colorectal cancer screening participation rate lower than Australia and Japan [7]. In the Dutch study in 2015, the subjects felt that colon cancer was a threat, but lack of knowledge of colorectal cancer led to distrust of the examination by the examinee, which was the biggest obstacle to colorectal cancer screening [8]. Therefore, the health belief model can effectively evaluate the behavior of cancer screening, and it is the most effective tool for measuring cognitive impact on healthy behavior.

In this study, health belief model revised as the basis to understand the object of study at their own expense to participate in colonoscopy behavior and other factors. Want to know from studies of risk factors associated with colorectal polyps related to genetic factors. And further explore the colonoscopy results and health behaviors, the correlation between health beliefs.

2   Methodology

The methodology used for this study is detailed in the following.

2.1   The Research Conceptual Model

Figure 1 shows this research bases on the health belief model value expectation theory as the basic framework, uses motivation and cognitive factors to predict and explain health-related behaviors, and constructs this research conceptual model by referring to relevant literature and collating and analyzing.

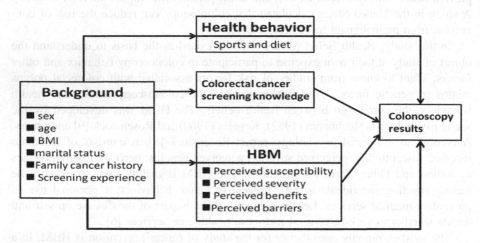

Fig. 1. The research conceptual model

2.2   Research Objects and Research Sites

This study was conducted between February 9 st, 2017 and April 19, 2017 for 201 voluntary Self-paid Colonoscopy patients. The research site is a health examination center within a regional teaching hospital. The investigators first explained the research purpose, importance, and confidentiality of the data to the subjects. After obtaining the subject's consent, the questionnaire was completed. A total of 201 questionnaires were returned after obtaining consent.

2.3   Research Tools

The contents of the questionnaire refer to relevant domestic literature, using a well-established, trustworthy, healthy belief model [6, 7, 9]. In addition, the "Colorectal Cancer Knowledge and Screening Behavior Knowledge Scale" has been revised and corrected through the measurement of expert validity. Experts include four senior specialists in gastroenterology, family medicine, and clinical psychologists. The questionnaire expert validity (Content Validity Index, CVI = 0.88) test. The compliance CVI value should be 0.8 or above, indicating that there is a good standard of expert validity [10].

Research questionnaire containing demographic data, Health Behavior (contains exercise habits, the amount of fruits and vegetables daily eat), Colorectal cancer screening knowledge (contains government grants screening, colorectal cancer and colonoscopy's knowledge), Colonoscopy examination Health Beliefs (contains perceived susceptibility, perceived severity, perceived benefits, perceived barriers).

2.4   Date Analyzing

The body mass index (BMI) is used in the questionnaire and is calculated as follows:

$$BMI = Weight(kg)/(height(m) \ x \ height(m))$$

Use SPSS for Window 19.0 statistical suite software. Anonymous data archiving and analysis, statistical method package inclusion Table, Percentage, Average, Standard Deviation t-test, chi-square, Pearson's correlation coefficient, Univariate logistic regression.

3   Results

The valid sample for the case is 201. Body mass index distribution: BMI posture was normal ($18.5 \leqq BMI < 24$) with 89 (44.3%), BMI was too light (BMI < 18.5) was 7 (3.5%), BMI was overweight ($24 \leqq BMI < 27$) There were 66 (32.8%) and 39 BMI (39.4%) with BMI body position. Gender distribution: Women studied 65 (32.3%) and males were 136 (67.7%). Mostly male. Age distribution: The study began to collect cases from the age of 21, with 4 (2.0%) between 21 and 29 years old, 52 (25.9%) between the ages of 30 and 39, and 102 (40. %), 34 (16.9%) aged 50–59, and 9 (4.5%) aged 60 and over. Distribution of Marital Status: Most of the participants in this study were 138 (68.7%) married and having children, 33 (16.4%) married but not having children, and 28 (13.9%) unmarried. There are 2 people (1.0%) divorced. The distribution of colorectal cancer among relatives was as follows: 165 (82.1%) of the relatives had no history of colorectal cancer and 36 (17.9%) had a history of familial colorectal cancer. Among them, only one relative in the family had a history of colorectal cancer which accounted for 31 (15.4%), and two or more relatives suffered from colorectal cancer which accounted for 5 (2.5%). Colorectal cancer screening experience: There were 117 people (58.2%) who had undergone colonoscopy or fecal occult blood tests. There were 84 (41.8%) patients who did not pass colorectal cancer screening or colonoscopy or fecal occult blood tests.

An average weekly exercise habit and daily intake of sufficient amount of fruits and vegetables for healthy behavior. 87.1% of the 201 subjects had poor health behaviors. There are 9 questions in screening knowledge of colorectal cancer, and the average correct score is 4.73 (SD = 1.66). Answers to the six questions or more accounted for only 33.8% of the 201 subjects.

The total scores of health behaviors, colorectal cancer screening knowledge and health beliefs were continuous variables. Logistic multiple regression was used to observe the correlation effect. The higher the score of health beliefs, the higher the health beliefs are. The independent variable is health behaviors and colorectal cancer screening knowledge. The results are shown in Table 1. The variance inflation factor (VIF) value is 1.007. It means there is no collinearity problem. Colorectal cancer screening knowledge and health beliefs had a significant impact ($P = 0.006$).

Table 1. Multivariate regression analysis of colorectal cancer screening knowledge and on health beliefs

Variables	Estimated B value	SE	Beta	T value	P value
Constant	85.690	1.969		43.525	0.000
Health behavior	−1.007	0.391	−0.178	−2.579	0.011
Screening knowledge	0.588	0.214	0.190	2.752	0.006**

**P < 0.01 significant correlation; VIF = 1.007

This study is based on the four variables of health beliefs. There are 4 problems in conscious susceptibility, 6 problems in conscious seriousness, 3 problems in conscious action benefits, and 9 problems in cognitive impairment. Using Likert Scale to calculate, 5 points are very consistent, 1 point is very non-conforming. The scores of the perceived barriers for a total of 9 questions and the score of the perceived susceptibility for a total of 1 question are inverse questions. Therefore, the reverse calculation process makes the higher scores of the four major health beliefs indicate that the health beliefs are more and more positive. Table 2 shows that among the four variables, the highest score is conscious action interests, and the standardization score is 95.47. The lowest score is the fourth, the standardized score is 57.30, and the conscious severity is standardized. It was 84.03, which was higher than the score of 70.11 for cognitive impairment.

Table 2. Health belief scale score analysis (n = 201)

Variable	Number	Score range	Total average	SD	z-score	Sequence
Susceptibility	4	4–20	11.47	2.79	57.35	4
Severity	6	6–30	25.21	3.68	84.03	2
Benefits	3	3–15	14.32	1.37	95.47	1
Barriers	9	9–45	31.55	6.10	70.11	3

There was a significant difference in gender with the perceived severity ($P = 0.047$). The results of a descriptive statistical analysis showed that the average number of women in the perceived severity was more than that of men, indicating that women had a higher perceived severity than men. Then we analyzed the gender differences in each of the six issues of perceived severity. We found that there were more women than men in the severity of the "psychological impact" of having colorectal cancer. There are more men than women in the severity of having a colorectal cancer that may increase family burdens (Table 3).

Table 3. Significant differences between gender and perceived severity

Health beliefs	F-test	P	T value	P
Psychological impact	1.367	0.244	2.333	0.021*
Physiological influence	7.870	0.006	2.511	0.013*

*P < 0.01 significant correlation

There was a significant difference between marital aspect and perceived benefits (P = 0.022). Table 4 presents a further post-hoc test and found that unmarried persons are significantly different from those who are married and have children (P = 0.027). Through descriptive analysis, people who are married and have children relative to those who are not married receive higher rates of perceived benefits.

Table 4. Post-hoc test result

Marital status	(I) status	(J) status	MD (I–J)	P
Tukey HSD	Unmarried	Married with no children	−0.21825	0.233
		Married and have children	−0.26173*	0.027*
		Divorce	0.22619	0.901

**P < 0.05 significant correlation

Table 5 shows that the body mass index (BMI) has a significant difference in the perceived susceptibility to "The opportunity now under state of health, suffering from colorectal cancer greatly" (P = 0.006). The body mass index BMI was significantly different for the perceived susceptibility "I feel embarrassed for colonoscopy" (P = 0.39). After the post-hoc test, Overweight had a significant difference for Obesity in "The opportunity now under state of health, suffering from colorectal visibly greatly" (P = 0.006). After descriptive analysis, Obesity felt that the chance of developing colorectal cancer under current health conditions was higher for overweight individuals (Table 6). However, Overweight performed significantly differently from those who had normal weight in regular "I feel embarrassed for colonoscopy" (P = 0.04). After descriptive analysis, the result was Overweight in "I feel embarrassed for colonoscopy" was higher in the perceived barriers than those in normal weight.

There was a significant difference between the experience of colorectal cancer examination and the perceived barriers (P = 0.037). There were significant differences between the relatives of the immediate family and the perceived severity (P = 0.022). Table 7 shows that A post-hoc test found that in "Psychological and physiological effects after suffering from colorectal cancer", two of the immediate family members with colorectal cancer had significant differences in the severity of those with no family cancer. Descriptive analysis showed that two people with a history of colorectal cancer in the immediate blood relatives had lower perceived severity than those without a family history.

Table 5. Significant differences in analysis of BMI and health beliefs

Health beliefs	F value	P value
Perceived susceptibility "The opportunity now under state of health, suffering from colorectal cancer greatly"	4.267	0.006*
Perceived barriers "I feel embarrassed for colonoscopy"	2.833	0.039*

Table 6. Post-hoc test results of BMI and health beliefs

Tukey HSD	(I) BMI	(J) BMI	MD (I–J)	SE	P
Perceived susceptibility	Overweight				
"The opportunity now under state of health, suffering from colorectal cancer greatly"		Normal	−0.220	0.167	0.554
		Underweight	−0.266	0.410	0.915
		Obesity	−0.693*	0.208	0.006*
Perceived barriers	Normal				
"I feel embarrassed for colonoscopy"		Underweight	0.579	0.588	−0.61
		Overweight	0.532*	0.028	0.04*
		Obesity	0.199	0.813	−0.38

*P < 0.05 significant correlation

Table 7. Family history of colon cancer versus perceived severity post-hoc test

Tukey HSD	(I) family history	(J) family history	MD (I–J)	SE	P
"I have psychological effects after suffering from colorectal cancer"	2 people	No	−0.891	0.361	0.038*
		1 people	−0.858	0.383	0.067
"I have physiological effects after suffering from colorectal cancer"	2 people	No	−1.230	0.369	0.003*
		1 people	−1.258	0.392	0.004*

*P < 0.05 significant correlation

In the study, 29.4% of subjects had found polyps after colonoscopy. Of these, 35.6% were adenomatous polyp, 1.7% carcinoma in situ. This section is mainly analyzed by univariate logistic regression. The results of the slice report were of a more malignant type of polyps, and the other classification was normal colonoscopy or the results of the slice report were less threatening. The results are shown in Table 8. The chi-square value of the overall pattern display test was 38.485 (P < 0.001). The Hosmer-Lemeshow test value was 0.144 (P > 0.05). For subjects over 60 years of age, the results of colonoscopy included 11.05 times (P = 0.026) ages of 30 to 39 years of age.

Table 8. Univariate logistic regression result (n = 201)

Variables	B	S.E.	Wald	P值	Exp(B)
Age					
21–29	−17.904	999	0.000	0.999	0.000
30–39 (reference)					
40–49	−1.348	0.762	3.126	0.077	0.260
50–59	0.406	0.808	0.253	0.615	1.501
60≧	2.403	1.081	4.942	0.026*	11.059

*P < 0.05 significant correlation

4 Conclusions

The results of this study found that the majority of men accepted colonoscopy, accounting for 67.66%, the same results as Lin and Kao in 2013. For the acceptance of colorectal cancer more men than women [11]. The related knowledge of colorectal cancer will affect the willingness of colonoscopy [12]. This result shows that the colorectal cancer screening knowledge and health beliefs have the same positive relationship. This study finds that those who are married and have children have a higher rate of the perceived benefits, and the results are the same as those of Australia's 2012 Health Belief Study, indicating that children who may have been married and have children are the mainstay of the family [13]. Therefore, there is a strong need to carry out colonoscopy at their own expense so that they can prevent colorectal cancer and maintain health, but also reduce the burden on the home after suffering from colorectal cancer. Studies have shown that 2 people of a family history of colorectal cancer have a higher degree of perceived susceptibility compared with those without a family history of colorectal cancer. In the study, it was also found that most of the people obtained the knowledge of colonoscopy through the media. However, people with a history of familial colorectal cancer gain knowledge of colonoscopy through their families. It is inferred that those who have familial colorectal cancer are more likely to get early screening for colorectal cancer and are more concerned with regular screening activities than those with no family history. Therefore, the perceived benefit rate for colorectal cancer is relatively low. Studies have shown that colonoscopy experience has a significant impact on the perceived barriers. Among them, colonoscopy experience is doubtful about the timing of the doctor's recommendations for regular examinations. Shows that the physician's explicit recommendation helps increase the rate of colorectal cancer screening [7]. This study found that BMI is significantly associated with "I feel embarrassed for colonoscopy" in the perceived barriers. And Overweight is higher than Obesity. In the process of colonoscopy, the subjects only wore thin trousers, and Overweight was more concerned with their size and less confident about their body shape. The BMI has significant correlation with "The opportunity now under state of health, suffering from colorectal cancer greatly" in the perceived susceptibility. Obesity feels that the chance of getting colorectal cancer is higher with his current health situation. Studies have pointed out that obesity is a risk factor for the continuous increase in the incidence of colorectal cancer [14]. At present, the Taiwan government has incorporated BMI into its cancer prevention program.

Now many people have more willingness to participate at Self-paid Colonoscopy, hoping to early detection of early signs of disease. Many scientists said that colonoscopy for the diagnosis of colorectal cancer the most effective examination and the highest accuracy diagnostic tools. Through this study, we found that there is a positive correlation between screening knowledge and Health Beliefs. Many studies show that the more positive the Health Beliefs, the more active the preventive screening activity. The study pointed out that screening knowledge is negatively correlated with behavioral impairment cognition, indicating that subjects feel colorectal cancer is threatening. People's poor knowledge of colorectal cancer lead to people do not trust colonoscopy, is the biggest obstacle to screening for colorectal cancer. The male sex and people over the age of 50 in the colonoscopy knowledge absorption worse than others, especially for the knowledge of colorectal polyps is the worst. However, people under the age of 50 do not care much about the government subsidy stool occult blood related information. The study also found that patients over the age of 60 compared with 30 to 39 years old subjects have a higher probability of predicting the risk of adenoma. It is recommended to increase regular health promotion activities to increase the awareness of colorectal cancer screening. And self-paying colonoscopy is worthy of promotion to check to achieve prevention rather than cure.

References

1. Health Promotion Administration, MOHW (2016) 2013 record annual cancer. http://www.hpa.gov.tw/Pages/List.aspx?nodeieid=269. Accessed 3 Mar 2017
2. Pallis AG, Papamichael D, Audisio R, Peeters M, Folprecht G, Lacombe D, Cutsem EV (2010) EORTC elderly task force experts' opinion for the treatment of colon cancer in older patients. Cancer Treat Rev 36:83–90
3. Urman J, Gomez M, Basterra M, Mercado MD, Montes M, Gómez Dorronsoro M, Garaigorta M, Fraile M, Rubio E, Aisa G, Galbete A (2016) Serrated polyps and their association with synchronous advanced colorectal neoplasia. Gastroenterol Hepatol 39 (9):574–583
4. Friedrich K, Grüter L, Gotthardt D, Eisenbach C, Stremmel W, Scholl SG, Rex DK, Sieg A (2015) Survival in patients with colorectal cancer diagnosed by screening colonoscopy. Gastrointest Endosc 82(1):133–137

5. Doubeni CA, Weinmann S, Adams K, Kamineni A, Buist DSM, Ash AS, Rutter CM, Doria-Rose VP, Corley DA, Greenlee RT, Chubak J, Williams A, Kroll-Desrosiers AR, Johnson E, Webster J, Richert-Boe K, Levin TR, Fletcher RH, Weiss NS (2013) The full report is titled "screening colonoscopy and risk for incident late-stage colorectal cancer diagnosis in average-risk adults. Ann Intern Med 158(5):1–48

6. Glanz K, Rimer BK, Viswanath K (2015) Health behavior: theory, research and practice, 5th edn. Jossey-Bass, San Francisco

7. Koo JH, Leong RWL, Ching J, Yeoh KG, Wu DC, Murdani A, Cai Q, Chiu HM, Chong VH, Rerknimitr R, Goh KL, Hilmi L, Byeon SJ, Niaz SK, Siddique A, Wu LC, Matsuda T, Makharia G, Sollano J, Lee SK, Sung JJY (2012) Knowledge of, attitudes toward, and barriers to participation of colorectal cancer screening tests in the Asia-Pacific region: a multicenter study. Gastrointest Endosc 76:126–135

8. Woudstra AJ, Dekker E, Essink L, Suurmond J (2015) Knowledge, attitudes and beliefs regarding colorectal cancer screening among ethnic minority groups in the Nether-lands – a qualitative study. Health Expect 19(6):1312–1323

9. Rosenstock IM (1974) Historical origins of the health belief model. Health Educ Mongraphs 2(4):328–335

10. Waltz CF, Strickland OL, Lenz ER (2010) Measurement in nursing and health research fourth edition, 2nd edn. Springer, New York

11. Lin YH, Kao CC (2013) Factors influencing colorectal cancer screening in rural southern Taiwan. Cancer Nurs 36(4):284–291

12. Tsai CF, Juan CW, Chiang MK, Chang HJ, Lin TY, Tsai CL, Lee SA (2011) The study of the colon cancer-screening test and associated health behaviors at a regional hospital in central Taiwan 4(2):39–49

13. Christou A, Thompson SC (2012) Colorectal cancer screening knowledge, attitudes and behavioural intention among Indigenous Western Australians. BMC Public Health 12:528

14. Chan AT, Giovannucci EL (2010) Primary prevention of colorectal cancer. Gastroenterology 138:2029–2043

How Nurses Perceive Organizational Climate Surrounding Patient Handoffs in Japanese Hospitals?

Xiuzhu Gu$^{(\boxtimes)}$ [iD] and Kenji Itoh [iD]

Tokyo Institute of Technology, 2-12-1 Oh-okayama, Meguro-ku,
Tokyo 152-8552, Japan
Xiuzhu.g.aa@m.titech.ac.jp

Abstract. The aims of the study were to extract organizational climate factors surrounding patient handoffs, and to capture their crucial characteristics in the current Japanese hospital context. A questionnaire survey was conducted between October and December 2017. A total of 5,117 valid responses were collected from nursing staff in 31 general hospitals with a response rate of 69%. The sample collected in 2011, which had 1,462 responses, was also used for comparison with the current data. Seven handoff factors were derived by applying principal component analysis to the 2017 sample with 44% of cumulative variance accounted for. Nursing staff perceived overall handoff adequacy and its elements moderately good. Significantly different views were observed between work units for all the factors. The most negative view was exhibited to information and responsibility continuity by respondents in ED. Regarding information transfer, compared with other intra-hospital handoff cases, information was transferred well in handoffs related to OR and ICU. There were also significant hospital differences in staff perceptions of patient handoff adequacy that the largest difference was identified in training and education. Comparing to six years ago, staff views became significantly more positive, and that the largest improvement was perceived for handoff process, training and education. Sufficiency of information transfer was also improved in the six-year interval. In conclusion, nursing staff perceptions of patient handoff practices and contributing factors have been improved in Japanese hospitals for the last six years, and currently viewed them moderately well. In addition, different views were extracted across work settings.

Keywords: Nursing handoff · Communication · Patient safety

1 Instruction

Patient handoff is a crucial process in health care. It can be defined as the transfer of professional responsibility and accountability for some or all aspects of care for a patient or groups of patients to another person or professional group on a temporary or permanent basis [1]. If handoff is conducted improperly, e.g., wrong or inadequate information is received and responsibility becomes unclear, patients may suffer serious harm. There are diverse types of patient handoff. Different factors, protocols, and mechanisms may be suitable for each, depending on work environment, specialties and systems. Intra-hospital patient handoff occurs frequently that it can be largely classified into two types: handoff between different departments/wards and that within the same work unit. Regarding the former type, there have been studies, for instance, on patients handed over from emergency department (ED) to general wards [2], from operating room (OR) to intensive care unit (ICU) [3], and from ICU to general wards [4]. A typical case of the latter type is the nurse-to-nurse shift handoff. This type of handoff occurs much more frequently than the other types, and has been examined extensively [5–7].

Regardless of the handoff type, poor communication is a main factor which leads to an adverse event in handoffs. In addition to communication, a number of factors have been reported in nursing handoffs, e.g., inadequate standardization, problems with equipment, busy wards, poor planning or use of time, complexity of cases or high caseloads, lack of training, interruptions, and fatigue [8]. These studies have been primarily conducted in North America and Europe. There are differences in healthcare systems and practices between Japan and Western countries as well as national cultures. Therefore, patient handoff in Japanese hospitals may be expected somewhat different from those in the Western countries, more specifically due to the different protocols, procedures, supporting aids, safety training, etc. However, only a few studies have been carried out on this topic in Japan.

There are several approaches frequently used in patient handoff studies, e.g., clinician's questionnaire-based self-assessment, interviews or focus groups, behavioral observation during handoff, and observational and experimental studies [9]. To avoid deriving results from isolated handoff episodes at a single organization, it is required to collect data from a variety of healthcare organizations. In addition, as the 'sharp end' of handoff practice [10], nursing staff knows well of how patient handoff is actually carried out in daily work. Thus, one of the most efficient and useful ways of getting an overview of nursing handoff is to obtain their views on this matter.

With such awareness, a self-administered questionnaire was developed and applied to various hospitals to investigate quality and safety of unit handoff and shift handoff by nurses in Japan [11]. However, there have been great changes and innovations not only in technologies but also management practices for patient handoffs in the last decade. With these changes as background, the present paper seeks to extract climate factors contributing to handoff quality and safety, and to capture their crucial characteristics in the current Japanese context. Changes in the climate surrounding patient handoffs for the last six years are also investigated and discussed.

2 Methods

2.1 Questionnaire

A questionnaire was developed by adapting from the former one with some revised, added and deleted question items [11]. Finally, 50 closed-ended question items about handoffs were included in the questionnaire, of which 31 were the same as the items used in the former survey. The questionnaire comprised two sections with an open-ended question and an additional demographic part. The questionnaire was constructed such that the respondents could remain anonymous. Section 1 had 26 unit handoff items to which a respondent rated his/her agreement level on a 5-point Likert-type scale from 1 (disagree strongly) to 5 (agree strongly). Three of these items were general questions: (1) "Patient safety has the highest priority for patient handoffs", (2) "Patient handoffs could be performed more efficiently," and (3) "Overall, patient handoffs between departments/wards are conducted well in my hospital". Section 1 included 8 more items asking about respondents' views of shift handoffs. One of these eight items was an overall evaluation of shift handoff adequacy as "Overall, shift handoff is conducted well in my department/ward". For details of the items, please refer to Appendix 1.

Section 2 dealt with insufficient information transfer as one of the most common failures in handoff incidents. Each respondent was asked to rate the frequency of handoffs with insufficient information in each of the 16 handoff cases on a five-point scale, from 1 (always) to 5 (never). The handoff cases included sending/receiving a patient to/from various departments/wards, organizations other than own hospitals, and between shifts, e.g., receiving patients from the emergency department (ED), operating room (OR), sending patients to the intensive care unit (ICU), outpatient department (OPD) and other hospitals.

2.2 Survey Sample

The survey was conducted from October to December 2017 after obtaining an approval from the Ethics Committee of the university to which both authors belonged. We sent invitation letters to 283 general hospitals having 300 beds or more. Of the 283 hospitals, 31 hospitals agreed to participate in the survey. We sent the questionnaires to a risk manager in each hospital by regular post mail. Administration staff assisted data collection in each hospital. The survey's anonymity and confidentiality were explained to each respondent. A questionnaire enclosed in an envelope was provided to each respondent who agreed to participate. When the respondent completed the questionnaire, he/she sealed the envelope containing his/her response and returned it to the administration office. Then all the responses collected in each hospital were sent to the authors by post mail (parcel) or courier. In total, 5,117 valid responses were collected, with a response rate of 69%. Respondents included 1,036 nursing staff working in internal medicine ward, 1,074 in surgical ward, 172 in ED, 413 in ICU, 255 in OR and 411 in OPD. For details of the survey sample, please refer to Table 1. In the former survey conducted in 2011, 1,462 valid responses were collected from six hospitals, with a response rate of 74%.

Table 1. Profile of the survey sample.

Attributes	N	%
Age		
<20	18	0%
20–29	2,058	40%
30–39	1,320	26%
40–49	1,038	20%
50–59	490	10%
>60	62	1%
NA	131	3%
Gender		
Male	330	6%
Female	4,654	91%
NA	133	3%
Working condition		
Full-time	4,703	92%
Part-time	214	4%
Others	44	1%
NA	156	3%
Professional group		
Nurse	4,739	93%
Nurse assistant	184	4%
Others	68	1%
NA	126	2%
Work unit		
Internal medicine ward	1,036	20%
Surgical ward	1,074	21%
Other inpatient wards	1,332	26%
Emergency department	172	3%
Intensive care unit	413	8%
Operating room	255	5%
Outpatient department	411	8%
Others	182	4%
NA	242	5%
Total	5117	

2.3 Statistical Analysis

We applied principal component analysis with Promax rotation to all the responses to 30 items in Sect. 1 (excluding 4 general question items) to elicit patient handoff factors. A difference in perception level for each handoff factor was investigated between seven work units, i.e., internal medicine ward (N = 1,036), surgical ward (N = 1,074), other inpatients wards (N = 1,332), ED (N = 172), ICU (N = 413), OR (N = 255) and OPD

(N = 411); and across hospitals by one-way analysis of variance (ANOVA), while the Kruskal–Wallis test was applied to each item of rank-based responses. The Kruskal–Wallis test was also applied to examine differences across hospitals for the evaluation of responses to Sect. 2. In addition, differences of insufficient information transfer frequency across handoff cases were explored using the Friedman test and the Wilcoxon signed rank test. Regarding changes in a six-year interval, i.e., differences between 2011 and 2017, the Mann-Whitney test was applied to responses from two surveys. All statistical analyses were performed using PASW statistics v.25 and AMOS v.25 (SPSS, Chicago, IL, USA).

3   Results

3.1   Patient Handoff Factors

The Kaiser–Meyer–Olkin measure of sampling adequacy was 0.89, and Bartlett's test of sphericity was significant at $p < 0.001$, indicating that the data was appropriate for factor analysis. Applying the principal component analysis, for eigenvalues > 1, six components were yielded with 44% of cumulative variance accounted for. The analysis result is summarized in Table 2 in terms of factor label, component items, their factor loadings, and Cronbach's alpha for each factor. Internal reliability, as assessed by Cronbach's alpha, was high enough at > 0.70 [12] or at an acceptable level as > 0.60 [13] for four out of six factors.

As for the first factor, some items highly loaded were questions about guidelines or procedures of either unit handoff or shift handoff, e.g., unit handoff process needs improvement, deviations from procedures or guidelines of both unit and shift handoff. The rest of highly loaded items were related to discontinuity of responsibility and information during unit handoffs, e.g., unclear responsibility after a unit handoff and insufficient information transfer in a unit handoff. Accordingly, we labelled this factor as "procedure and continuity during handoff." In this way, we interpreted all six factors as follows: (1) procedure and continuity during handoff, (2) unit handoff competence, (3) shift handoff information and environment, (4) communication between units, (5) training and education, and (6) unit handoff environment. As stated above, the first factor comprised a number of component items and was composed of two facets. Therefore, it was broken down to two sub-factors: (1-1) process and guidelines, and (1-2) information and responsibility continuity between units. Then, for understanding easily, the original factor structure is arranged and reordered as the following seven factors: (I) process and guidelines, (II) training and education, (III) information and responsibility continuity between units, (IV) unit handoff competence, (V) communication between units, (VI) unit handoff environment, and (VII) shift handoff information and environment. Hereafter, we designate the seven dimensions as *handoff factors*.

Table 2. Patient handoff factors elicited by principal component analysis.

Factors (variance [cumulative variance]) (Cronbach's alpha)	Component items	Loading
1. Procedure and continuity during handoff (19% [19%]) ($\alpha = 0.74$)		
1-1 Process and guidelines ($\alpha = 0.67$)	Q16. Patient handoff process needs improvement. (-)	0.685
	Q13. Current procedures, guidelines, or instructions need improvement. (-)	0.655
	Q25. Deviations from procedures or guidelines. (-)	0.645
	Q33. Deviations from procedures or guidelines (shift handoff). (-)	0.564
	Q27. Current procedures, guidelines, or instructions need improvement (shift handoff). (-)	0.519
1-2 Information and responsibility continuity between units ($\alpha = 0.55$)	Q14. Unclear responsibility after patient handoffs. (-)	0.489
	Q11. Receive too little information. (-)	0.476
	Q2. Views different from other departments/wards about required information. (-)	0.449
	Q22. No timely follow-up or support after patient handoffs. (-)	0.414
	Q23. The staff conducting a handoff with me is inexperienced or less capable. (-)	0.345
2. Unit handoff competence (8% [27%]) ($\alpha = 0.64$)	Q5. Receive accurate and updated information	0.736
	Q3. Current information technology supports well	0.691
	Q4. It is clear when responsibility is transferred	0.687
	Q6. I know what role is expected of me	0.385
3. Shift handoff information and environment (6% [32%]) ($\alpha = 0.60$)	Q29. Receive accurate and adequate information	0.693
	Q32. Sufficient time to make an adequate handoff	0.688
	Q28. Rarely interrupted	0.614
	Q31. Written message for important information	0.466
4. Communication between units (4% [36%]) ($\alpha = 0.65$)	Q21. Obtain information from patients/families	0.699
	Q19. Inform patients of their planned transfers	0.625
	Q17. Voice own concerns and express own opinions to share understanding	0.518

(*continued*)

Table 2. (*continued*)

Factors (variance [cumulative variance]) (Cronbach's alpha)	Component items	Loading
	Q9. Read back and ask questions to ensure that the information is correctly understood	0.480
	Q24. Understand patients'/families' complaints	0.412
	Q1. Find solutions and solve problems together with other departments/wards	0.343
5. Training and education (4% [41%]) ($\alpha = 0.53$)	Q7. New staff receive formal training	0.751
	Q12. Use patient handoffs as an opportunity for junior staff to learn good clinical practice	0.689
	Q30. New staff receive formal training (shift handoff)	0.514
6. Unit handoff environment (4% [44%]) ($\alpha = 0.49$)	Q20. Too busy to attend quickly to a newly received patient. (-)	0.708
	Q8. Sufficient time to make an adequate handoff	−0.560
	Q10. Rarely interrupted	−0.514

(-): Item having a negative meaning to the factor label.

3.2   Views on Handoffs Across Work Units

In general, about half of the respondents perceived that unit handoffs were conducted appropriately in their hospitals. Nursing staff working in inpatient wards, i.e., internal medicine ward (53% of agreement rate; Q25), surgical ward (49%), and other inpatient wards (50%), and ED (50%) exhibited more positive perception of the overall unit handoff adequacy than ICU (43%), OR (44%) and OPD (43%), having a significant difference (Kruskal–Wallis test; $p < 0.01$) between the seven work units. Significant differences were also observed between work units in their views of patient safety priority in unit handoff (80%, 77%, 82%, 77%, 82%, 81% and 74%; $p < 0.01$; Q15) and unit handoff efficiency ($p < 0.05$; Q18). A larger number of respondents in ED (42%; Q18) and ICU (40%) perceived that it could achieve higher efficiency, compared with internal medicine ward (33%), surgical ward (35%), other inpatient wards (35%), OR (34%) and OPD (34%). However, the differences of absolute positive percentages were not large. In addition, 50% of total respondents agreed that shift handoffs were conducted well in their own departments/wards (Q34) and there was no significant difference between the seven work units.

A mean score of each handoff factor was calculated over all of its component items for each respondent. Before the calculation, responses were reversed, e.g., from 1 to 5, and from 2 to 4, vice versa, for items having a negative meaning to the factor label. A positive respondent is determined for each factor as one had a mean score higher than 3.5 (score ranging from 1.0 to 5.0) and a negative respondent as lower than 2.5. Percentage of positive [negative] responses is referred to as that of positive [negative] respondents for each factor. Table 3 summarizes the comparative results in terms of percentage positive and negative responses based on the work unit groups and its significance level derived using one-way ANOVA for each handoff factor. Significant differences were identified in their views between the seven work units for all the handoff factors. In general, among the seven work units, nursing staff in inpatient wards exhibited high levels of positive perceptions to all the handoff factors except for unit handoff environment, and shift handoff information and environment (Factors VI and VII). For these two factors, ICU and OR respondents expressed their stronger perceptions than other units. However, OR staff demonstrated the most negative views of the current states in training and education of patient handoffs and communication between units. Nursing staff working in ED had relatively positive perceptions of training and education, and communication whereas their perceptions were lower than other unit groups for handoff process and guidelines, and information and responsibility continuity between units. In general, most respondents in every work unit showed high acknowledgement of adequate communication between staff members in different departments/wards and with patients in unit handoffs.

Table 3. Percentage positive and negative responses to handoff factors across work units.

Handoff factors	Work units							
	Internal medicine ward	Surgical ward	Other wards	ED	ICU	OR	OPD	p
I. Process and guidelines	29%	30%	34%	23%	25%	29%	27%	**
	9%	9%	8%	12%	10%	13%	9%	
II. Training and education	41%	44%	40%	47%	33%	22%	26%	***
	16%	13%	16%	19%	20%	32%	28%	
III. Information and responsibility continuity between units	24%	23%	23%	12%	21%	22%	21%	**
	12%	12%	12%	21%	15%	14%	16%	
IV. Unit handoff competence	26%	26%	26%	26%	23%	27%	21%	**
	9%	7%	8%	6%	9%	10%	14%	
V. Communication between units	48%	46%	48%	52%	46%	31%	44%	***
	2%	2%	2%	2%	3%	5%	3%	
VI. Unit handoff environment	23%	21%	25%	23%	39%	37%	19%	***
	23%	21%	20%	24%	10%	11%	27%	
VII. Shift handoff information and environment	24%	24%	21%	25%	28%	27%	20%	**
	15%	13%	17%	15%	10%	10%	12%	

ED: emergency department; ICU: intensive care unit; OR: operating room; OPD: outpatient department.
Upper row: % of positive responses; Lower row: % of negative responses.
**: $p < 0.01$; ***: $p < 0.001$ across the seven work units.

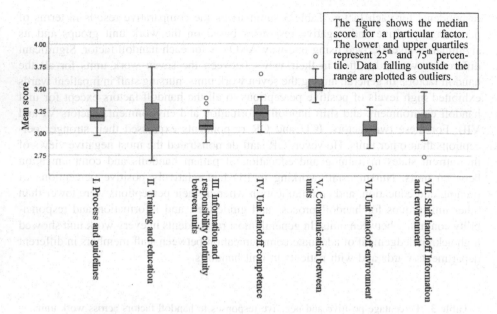

The figure shows the median score for a particular factor. The lower and upper quartiles represent 25th and 75th percentile. Data falling outside the range are plotted as outliers.

Fig. 1. Distribution of responded scores to each handoff factor for 31 hospitals.

3.3   Organizational Differences in Staff Views

Differences of staff views of general aspects in handoff and handoff factors were investigated across 31 hospitals. A significant difference (Kruskal–Wallis test; $p < 0.001$) was identified across hospitals for overall adequacy of unit handoff (Median = 3.40, SD = 0.15, Min = 3.20 and Max = 3.28; Q26). There was a significant difference in staff views of patient safety priority across all 31 hospitals (Median = 4.04, SD = 0.13, Min = 3.78 and Max = 4.28; $p < 0.001$; Q15), but no significant difference about possibility of increased efficiency (Median = 3.22, SD = 0.10, Min = 2.90 to Max = 3.37; $p > 0.05$; Q18). There was also a significant organizational difference in overall adequacy of shift handoff (Median = 3.41, SD = 0.17, Min = 3.14 and Max = 3.92; $p < 0.001$).

Similarly, significant differences were observed across hospitals to all the seven factors ($p < 0.001$; one-way ANOVA). The distribution of responded scores to each handoff factor across 31 hospitals was shown in Fig. 1 using a box plot chart. In this figure, it can be seen that respondents from all the hospitals rated relatively high scores to communication (Median = 3.51; SD = 0.09). In general, the differences of absolute scores between hospitals were not large for all the factors except for handoff training and education (Median = 3.15; SD = 0.25). The hospital having the highest percentage of positive responses to this factor was 63% whereas the lowest was 12%.

3.4   Changes of Staff Views in Six-Year Interval

Handoff Factors. Changes of staff views on patient handoffs in a six-year interval were investigated by comparing shared items in both surveys in 2017 and 2011. For two general items of unit handoffs, no significant difference was observed: patient safety priority (80% of agreement in 2017 versus 76% in 2011) and possibility of increased efficiency (35% in 2017 versus 35% in 2011). There were 16 shared items in the two surveys. For each item, a comparative result is shown in Table 4, including percentage of positive responses and significant difference level between two surveys derived by Mann-Whitney test. Except for a few items relevant to communication and shift handoff information and environment, there were significant differences for most items. In general, nursing staff surveyed in 2017 had significantly more positive views of almost all items than six years ago. In particular, large changes were captured in their perceptions of handoff process, training and education. For instance, the percentage of positive responses about formal training received by new staff members was increased from 27% to 55% for unit handoff (Q7; $p < 0.001$) and from 31% to 60% for shift handoff (Q30; $p < 0.001$). The only exception was about information transfer in shift handoffs. In 2011, 58% of respondents agreed that written message was used for important information transfer and the percentage of positive responses was decreased to 41% in 2017.

Frequency of Transfer with Sufficient Information. Regarding sufficiency of information at patient handoff, Table 5 summarizes comparative results of percentages of positive responses about information sufficiency for each patient handoff case in 2011 and 2017, i.e., percentage of respondents who selected the response option of "never or almost never" or "rarely" experienced lack of required patient care information. In general, nursing staff were likely to perceive lack of information less frequently when *sending patients* to than receiving patients from other departments/wards or other organizations. For instance, only 24% of respondents in 2017 expressed that they never or rarely experienced insufficient information when receiving patients from inpatient wards, whereas this percentage increased to 47% when they sent patients there (Wilcoxon signed rank; $p < 0.001$). Significant differences in the frequency of insufficient information transfer were also identified across handoff cases in receiving patients ($p < 0.001$) and in sending patients ($p < 0.001$). For instance, they perceived information particularly insufficient when receiving patients from inpatient wards, but transferred more sufficient information from OR, ICU and ambulance.

Changes of staff views on information sufficiency were also investigated in a six-year interval. Significant differences were observed for most of the handoff cases except for some cases when receiving patients from non-inpatient ward units, i.e., ED, OR and ICU. Nursing staff views of information sufficiency in handoffs became significantly more positive from 2011 to 2017 in most cases although absolute percentages had not large changes.

Table 4. Comparisons of item responses in handoff factors between 2017 and 2011.

Handoff factors	Items	2017	2011	p
I. Process and guidelines	Q16. Patient handoff process needs improvement. (-)	35%	21%	***
II. Training and education	Q7. New staff receive formal training	55%	27%	***
	Q12. Use patient handoffs as an opportunity for junior staff to learn good clinical practice	28%	16%	***
	Q30. New staff receive formal training (shift handoff)	60%	31%	***
III. Information and responsibility continuity between units	Q14. Unclear responsibility after a patient handoff. (-)	44%	39%	***
	Q11. Receive too little information. (-)	39%	39%	
	Q2. Views different from other departments/wards about required information. (-)	48%	35%	***
IV. Unit handoff competence	Q3. Current information technology supports well	27%	25%	**
	Q4. It is clear when responsibility is transferred	39%	34%	***
	Q6. I know what role is expected of me	64%	60%	**
V. Communication between units	Q21. Seek to obtain required information from patients/families	80%	75%	
	Q19. Inform patients planned transfers	59%	61%	
VI. Unit handoff environment	Q10. Rarely interrupted	61%	53%	***
VII. Shift handoff information and environment	Q29. Receive accurate and adequate information	52%	50%	
	Q28. Rarely interrupted	33%	33%	
	Q31. Written message for important information transfer	41%	58%	***

Figure: % of positive responses.
**: $p < 0.01$; ***: $p < 0.001$ between 2017 and 2011.

3.5 Organizational Differences in Sufficiency of Information Transfer

Differences in staff views of frequency of patient transfer with insufficient information were also investigated across 31 hospitals. Significant differences were observed across hospitals in all handoff cases ($p < 0.001$; Kruskal-Wallis test). A mean score of frequency of insufficient information transfer was calculated for each hospital. Higher score stands for a more positive view of the frequency of insufficient information transfer, i.e., lower frequency (ranging from 1.0 to 5.0). The distribution of responses to each handoff case across 31 hospitals is shown in Fig. 2 using a box plot chart. From the figure, the larger difference was seen when receiving patients rather than sending patients, excluding the shift handoff case. In particularly, a large difference across hospitals was seen in the case of receiving patients from ambulance (Median = 3.56;

Table 5. Positive responses to information sufficiency in patient handoffs.

Handoff case		2017	2011	p
Receive patients from	ED	32%	34%	
	OR	60%	61%	
	ICU	49%	49%	
	OPD	31%	–	
	Inpatient wards	24%	22%	***
	Clinic	33%	30%	
	Ambulance	55%	47%	**
	Other hospitals	32%	28%	**
	Shift handoff	32%	30%	**
Send patients to	OR	66%	62%	*
	ICU	56%	52%	*
	OPD	56%	–	
	Inpatient wards	47%	44%	**
	Clinic/homecare	68%	–	
	Other hospitals	69%	62%	**
	Shift handoff	42%	40%	*

Figure: % of respondents who "never or almost never" or "rarely" experienced insufficient information at patient transfer.

ED: emergency department; OR: operating room; ICU: intensive care unit; OPD: outpatient department.

*: $p < 0.05$; **: $p < 0.01$; ***: $p < 0.001$ between 2017 and 2011.

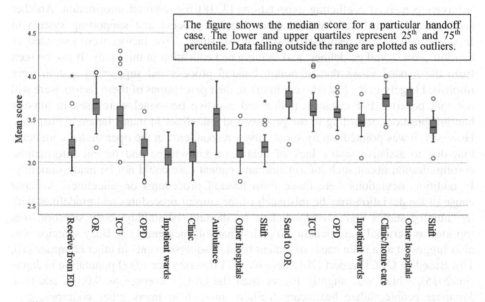

The figure shows the median score for a particular handoff case. The lower and upper quartiles represent 25th and 75th percentile. Data falling outside the range are plotted as outliers.

Fig. 2. Distribution of responses to sufficiency of information transfer for 31 hospitals.

SD = 0.20). In this case, the highest percentage of sufficient information transfer among 31 hospitals was 76% whereas the lowest was 30%.

4 Discussion

4.1 Characteristics of Patient Handoff Factors

It was suggested that nursing staff seemed to have moderately positive perceptions of overall adequacy of patient handoff in Japan. This result was different from a former study in Norway where 90% of nursing staff felt the overall handoff quality was high [14]. It was also reported that factors other than completeness in information transfer may affect nursing staff perceptions of handoff quality [14]. As such a factor, communication was elicited as a handoff factor from Japanese nurses' responses. Comparing to other handoff factors, Japanese nursing staff had relatively positive views of communication and there was no significant change in this factor for the last six years. Importance of communication in patient handoff has been also suggested in previous studies, for instance, communication during handoff affects handoff quality [9, 15] and its breakdown is a leading risk factor [16].

Transfer of information and responsibility are the most important tasks in patient handoff. However, the results of the present study suggested that many handoff cases were conducted with a lack of required patient information in Japan. Such insufficient information transfer was reported as one of the major handoff failures occurred frequently in Japanese hospitals [17]. There might be three major causes leading to insufficient information transfer in patient handoffs. One is heterogeneity of structural factors related to handoff across departments/wards: different clinical specialties and departments use different supporting systems and aids, and therefore senders and receivers perceived conflicting expectations [2, 18] for required information. Another cause may be attributed by inadequate handoff process and supporting system in Japanese hospitals as results suggested in this study. Two factors were extracted as handoff process and guidelines, and training and education in this study. It can be seen from the nurses' views that: although handoff process and supporting system were improved largely in the last six-year interval, their perceptions of these factors were still not that positive. For instance, unlicensed assistive personnel were used in nursing handoff practices according to the process and guidelines in many Japanese hospitals. However, it was pointed out by some survey respondents in the open-ended comments that due to assistive staff's lack of healthcare knowledge and patient information, communication about such information and patient care could not be made smoothly. In addition, deviations were made from handoff procedures or guidelines. A major cause of the deviation may be mismatch of the current procedures and guidelines with the nursing staff's high workload. Due to shortage of staffing, busy situation was reported as a crucial factor leading to patient handoff incidents [17]. Busy situation was also suggested as a main cause of patient handoff adverse events in other countries [10, 19]. Based on OECD report [20], there were 11.0 nurses per 1,000 population in Japan (in 2015), which was slightly higher than the OECD average as 9.0. In addition, Japanese people utilize healthcare facilities more than many other countries, e.g., higher healthcare expenditure per capita and average length of stay [20]. Therefore, Japanese nurses must provide healthcare service to three times of average hospital beds, i.e., 13.2 beds per 1,000 population in Japan, comparing to 4.7 beds as the OECD average [20]. Nurses' busy situation is a part of handoff environment factor as extracted in this study. Besides busy situation, a lack of time and interruptions during handoffs may be a cause leading to failures in not only information continuity but also responsibility continuity. For instance, sometimes staff could not attend quickly to the new received patient because in a busy situation as results in this study. Another cause of responsibility discontinuity may be the sender and receiver lacking awareness of mutual responsibility [10]. Survey results showed that sender's follow-up and support after handoffs as responsibility sharing were inadequate.

4.2 Factors Leading to Different Views

We explored several factors influencing nurse views of patient handoffs in this study. For instance, significant differences were seen across hospitals, and this may partly stem from different handoff processes and systems as well as organizational differences in safety culture. The largest difference across hospitals was detected in handoff training and education. Significant differences were also detected across work units and these differences may depend, in part, on different job requirements. ED, ICU and OR nurses, for instance, are typically engaged in more technical tasks than inpatient ward nurses, and their patients' critical levels [21] are different from those of the inpatient ward nurses, who tend to have longer and repetitive interactions with patients. Therefore, inpatient ward nurses had positive perceptions of handoff adequacy for most factors whereas for nurses in other work units, their perceptions were quite high for some factors but extremely low for the others. For instance, ED nurses exhibited the most positive perceptions of handoff training, education and communication, but their perceptions were the lowest to information and responsibility continuity among all work unit groups. One possible reason for the lowest perceptions by ED nurses, it may be rooted, in part, in conflicting expectations with regard to other departments/wards [22] and variations in the clinical information required [2]. ED staff typically view their roles as stabilization and disposition of patients, not as definitive patient care and management. Therefore, they are likely to sign out patients with pending data [21]. In contrast, based on the survey results, patient information was more sufficiently transferred in handoff cases involving ICU and OR than other work units. The sufficiency information transfer with ICU and OR, may be contributed by better handoff circumstances, i.e., sufficient time, rare interruptions and adequate staffing, considering ICU and OR nurses' positive perceptions of these factors.

Regarding information adequacy perceived more positively when sending versus receiving patients, the results obtained in this paper are similar to those of former studies that senders had more positive perceptions of information sufficiency than receivers [4, 14, 23]. One of the possible reasons may be the Dunning–Kruger effect [24], i.e., nursing staff are biased to overestimate their own performance. As another reason, it is plausible that even if a receiver perceives the information as insufficient, usually he/she tries to find it out by him/herself. To overcome perceived information asymmetry between senders and receivers, feedback and communication about handoff errors among providers have been suggested as a means for better information transfer [22].

4.3   Implications for More Effective and Safe Handoffs

Based on the survey results, there are mainly two suggestions for safer and more effective patient handoffs in Japanese hospitals. Firstly, it is suggested to improve handoff process, procedures and guidelines. A proper standardization of the handoff process is required under consideration of appropriate handoff staffing [25]. Secondly, effective use of unlicensed assistive staff should be more positively accepted for nursing handoffs like healthcare support workers. To improve current busy situations, it is required to share nurses' workload with them. However, adequate handoff training is essential to give unlicensed personnel for safe, effective and efficient handoffs, e.g., to let them know clearly their roles and handoff protocols, process and procedures. In addition, for smooth handoffs and error prevention, it is necessary to involve them more in handoffs. For instance, not to use them for patient movement only, but also share necessary patient information with them.

4.4   Limitations

A primary limitation of this study is a single methodology we used, i.e., a subjective survey approach. There is a possibility of respondents' over-estimation for their own practices as discussed. Therefore, we plan to collect objective data related to safety outcomes such as adverse event rate and safety behavior in a future study. Such safety performance data will be used to explore relationships with contributing factors.

5   Conclusions

This paper investigated organizational climate factors surrounding patient handoffs in Japanese hospitals, from the viewpoint of nursing staff perceptions through a questionnaire-based survey. Seven handoff factors were yielded from nursing staff responses to both unit and shift handoff items: (1) process and guidelines, (2) training and education; (3) information and responsibility continuity between units, (4) unit handoff competence, (5) communication between units, (6) unit handoff environment, and (7) shift handoff information and environment. Based on the elicited handoff factors, characteristics of nursing handoffs in Japanese hospitals are summarized as follows: (1) In general, nursing staff perceived patient safety priority highly positive, perceived overall adequacy of handoff moderately positive, and perceived the current handoff efficiency rather negatively. Significant differences were observed for all handoff factors between the seven work units, i.e., internal medicine ward, surgical ward, other impatient wards, ED, ICU, OR and OPD, and across hospitals. In general, inpatient ward nurses exhibited relatively positive views for all factors except for unit handoff environment and shift handoff information and environment. For these two factors, ICU and OR nurses perceived more positively than the other work units. There were significant hospital differences in nurse perceptions. In particular, the largest difference was obtained from views of handoff training and education. (2) In a six-year interval from 2011 to 2017, staff views became significantly more positive except for communication and shift handoff information. The largest improvement was extracted on handoff process, training and education. (3) Focusing on sufficient information transfer, nurses reported less frequently when receiving patients than when sending patients, especially when receiving patients from inpatient wards. Compared with other intra-hospital handoff cases, information was transferred more adequately in handoffs with the OR and ICU. (4) Differences of sufficient information transfer frequency were also observed across hospitals. Large differences across hospitals were seen in handoff cases when receiving patients from other work units or organizations rather than sending patients or in shift handoffs. In addition, frequency of sufficient information transfer was significantly increased from 2011 to 2017 in all the handoff cases except for the cases receiving patients from non-inpatient ward work units in own hospitals. Based on the results, for safer and more effective nursing handoffs, it is suggested to improve the current handoff process by proper standardizations and to effectively use non-licensed assistive staff.

Acknowledgments. This work was in part supported by Grant-in-Aid for Young Scientists (B) (No. 15K16291), Japan Society for the Promotion of Science. The authors thank to the risk management personnel of the hospitals and nurses who participated in the survey.

Appendix 1. Question Items in Section 1

Q1 to Q26 are about handoffs between your department/ward and other departments/wards.

- Q1. Regarding problems discovered during patient handoffs, my department/ward always finds solutions and solves them together with other relevant departments/wards.
- Q2. My department/ward often has views different from others about what information is required for patient handoffs.
- Q3. Current information technology supports patient handoffs well.
- Q4. It is clear to me when responsibility for a patient is transferred.
- Q5. We always receive accurate and updated information in patient handoffs from other departments/wards.
- Q6. I know what role is expected of me in a patient handoff.
- Q7. In my department/ward, new staff members always receive formal training for patient handoffs.
- Q8. We have sufficient time to make an adequate patient handoff.
- Q9. In my department/ward, it is a common practice that staff read back information they received and ask questions regarding unclear parts to ensure that the information is correctly understood.

Q10. In my department/ward, we are rarely interrupted during patient handoffs.

Q11. In my department/ward, we sometimes talk about having received too little information in patient handoffs.

Q12. My department/ward uses patient handoffs as an opportunity for junior staff to learn good clinical practice.

Q13. In my department/ward, current procedures, guidelines, or instructions for patient handoffs need improvement.

Q14. It is sometimes unclear who is responsible for a patient after a patient handoff.

Q15. In my department/ward, patient safety has the highest priority for patient handoffs.

Q16. In my hospital, the patient handoff process needs improvement.

Q17. In patient handoffs, we always voice our own concerns and express our own opinions to share understanding with other handoff staff.

Q18. In my hospital, patient handoffs could be performed more efficiently.

Q19. We always inform our patients of their planned transfers to other departments/wards whenever they are capable of understanding this.

Q20. In my department/ward, sometimes we are too busy to attend quickly to a newly received patient.

Q21. In my department/ward, we always seek to obtain required information from patients themselves or their families.

Q22. In my department/ward, we usually do not make timely follow-up and support for patient care after patient handoff.

Q23. Sometimes, the staff conducting a handoff with me is inexperienced or less capable.

Q24. In patient handoffs, it is a common practice that we discuss about communications with patients/families to understand their complaints.

Q25. In my department/ward, sometimes we make deviations from procedures or guidelines of patient handoffs.

Q26. Overall, patient handoffs between departments/wards are conducted well in my hospital.

Q27 to Q34 are about shift handoff

Q27. In my department/ward, current procedures, guidelines, or instructions for shift handoffs need improvement.

Q28. In my department/ward, we are rarely interrupted during a shift handoff.

Q29. In my department/ward, we always receive accurate and adequate information about patients from departing nurses in shift handoffs

Q30. In my department/ward, new staff members always receive formal training for shift handoffs.

Q31. In my department/ward, we use written message for transferring important information in shift handoffs.

Q32. In my department/ward, we have sufficient time to make an adequate shift handoff.

Q33. In my department/ward, sometimes we make deviations from procedures or guidelines of shift handoffs.

Q34. Overall, shift handoff is conducted well in my department/ward.

References

1. British Medical Association (2005) Safe handover: safe patients. Guidance on clinical handover for clinicians and managers. https://www.bma.org.uk/-/media/Files/PDFs/Practical %20advice%20at%20work/Contracts/safe%20handover%20safe%20patients.pdf. Accessed 23 May 2018
2. Lawrence S, Spencer LM, Sinnott M, Eley R (2015) It takes two to tango: Improving patient referrals from the emergency department to inpatient clinicians. Ochsner J 15(2):149–153
3. Gleicher Y, Mosko JD, McGhee I (2017) Improving cardiac operating room to intensive care unit handover using a standardised handover process. BMJ Open Qual 6(2):e000076
4. Stelfox HT, Leigh JP, Dodek PM, Turgeon AF, Forster AJ, Lamontagne F, Fowler RA, Soo A, Bagshaw SM (2017) A multi-center prospective cohort study of patient transfers from the intensive care unit to the hospital ward. Intensive Care Med 43(10):1485–1494
5. Holly C, Poletick EB (2014) A systematic review on the transfer of information during nurse transitions in care. J Clin Nurs 23(17–18):2387–2395
6. Johnson M, Sanchez P, Zheng C (2016) The impact of an integrated nursing handover system on nurses' satisfaction and work practices. J Clin Nurs 25(1–2):257–268
7. Robertson ER, Morgan L, Bird S, Catchpole K, McCulloch P (2014) Interventions employed to improve intrahospital handover: a systematic review. BMJ Qual Saf 23(7):600–607
8. Riesenberg LA, Leitzsch J, Cunningham JM (2010) Nursing handoffs: a systematic review of the literature. Am J Nurs 110(4):24–34 quiz 35-6
9. Manser T (2013) Fragmentation of patient safety research: a critical reflection of current human factors approaches to patient handover. J Public Health Res 2(3):e33
10. Ekstedt M, Odegard S (2015) Exploring gaps in cancer care using a systems safety perspective. Cogn Technol Work 17(1):5–13
11. Gu X, Andersen HB, Madsen MD, Itoh K, Siemsen IM (2012) Nurses' views of patient handoffs in Japanese hospitals. J Nurs Care Qual 27(4):372–380
12. Nunnally JC (1978) Psychometric theory, 2nd edn. McGraw-Hill, New York
13. Hair JF, Black WC, Babin BJ, Anderson RE, Tatham RL (2006) Multivariate data analysis, 6th edn. Pearson Prentice Hall, Upper Saddle River
14. Reine E, Raeder J, Manser T, Smastuen MC, Rustoen T (2018) Quality in postoperative patient handover: different perceptions of quality between transferring and receiving nurses. J Nurs Care Qual. [Epub ahead of print]
15. Richter JP, McAlearney AS, Pennell ML (2016) The influence of organizational factors on patient safety: Examining successful handoffs in health care. Health Care Manag Rev 41 (1):32–41
16. WHO Patient Safety Alliance (2007) Communication during patient hand-overs: Patient safety solutions, volume 1, solution 3. http://www.who.int/patientsafety/solutions/patientsafety/PS-Solution3.pdf. Accessed 23 May 2018
17. Gu X, Seki T, Itoh T (2017) Developing an error taxonomy system for patient handoff events. In: Proceedings of the IEEE international conference on industrial engineering and engineering management - IEEM 2017, Singapore
18. McElroy LM, Macapagal KR, Collins KM, Abecassis MM, Holl JL, Ladner DP, Gordon EJ (2015) Clinician perceptions of operating room to intensive care unit handoffs and implications for patient safety: a qualitative study. Am J Surg 210(4):629–635
19. Andersen HB, Siemsen IMD, Petersen LF, Nielsen J, Ostergaard D (2015) Development and validation of a taxonomy of adverse handover events in hospital settings. Cogn Technol Work 17(1):79–87

20. Organization for Economic Cooperaion and Development (2017) Health at a glance 2017: OECD indicators. https://read.oecd-ilibrary.org/social-issues-migration-health/health-at-a-glance-2017_health_glance-2017-en#page1. Accessed 23 May 2018
21. Horwitz LI, Meredith T, Schuur JD, Shah NR, Kulkarni RG, Jenq GY (2009) Dropping the baton: a qualitative analysis of failures during the transition from emergency department to inpatient care. Ann Emerg Med 53(6):701–710 e4
22. Lee SH, Phan PH, Dorman T, Weaver SJ, Pronovost PJ (2016) Handoffs, safety culture, and practices: evidence from the hospital survey on patient safety culture. BMC Health Serv Res 16:254
23. Gu X, Liu H, Itoh K (2017) Patient handoff quality and safety in China: health care providers' views. In: Cepin M, Briš R (eds) Proceedings of European safety and reliability conference - ESREL 2017, safety and reliability - theory and applications. Taylor & Francis Group, London, pp 1675–1682
24. Kruger J, Dunning D (1999) Unskilled and unaware of it: How difficulties in recognizing one's own incompetence lead to inflated self-assessments. J Pers Soc Psychol 77(6):1121–1134
25. Wears RL (2015) Standardisation and its discontents. Cogn Technol Work 17(1):89–94

Physical Fitness Comparison of Trained and Untrained Industrial Emergency Brigades

Esteban Oñate[✉] and Elías Apud

Unit of Ergonomics, University of Concepción, Barrio Universitario s/n,
Concepción, Chile
estebanonate@udec.cl

Abstract. Emergencies can occur at any time and may reach unpredictable magnitudes. Unfortunately, in many companies, emergency brigades are often organized with people who perform work of sedentary nature and they are not prepared to face sporadic tasks of high physical demands.

The objective of this study was to compare the physical fitness of workers with different degree of training. They worked for the same company, but in different industrial plants. In one of them they had a physical training program, guided 3 times per week during working hours by a physical educator. In the other plant there was no facilities for training.

Age, stature, body mass, body fat and aerobic capacity were evaluated with conventional methods in 57 physically trained and 21 untrained workers.

Results showed that aerobic capacity was 20.8% higher in the trained brigade. Body fat content was significantly higher in the untrained brigade. A further analysis showed that according to Chilean standards for emergency brigades, 58% of the untrained brigade members do not reach the recommended level, while only 19% of the trained workers are below the reference level.

As conclusion, the results only confirmed the importance of physical training to improve working capacity and this is particularly important for workers who perform light activities and in isolated occasions have to face high and dangerous demands. Therefore, the main recommendation is to stimulate training within working hours as part of the preparation of the brigade members.

Keywords: Physical training · Work capacity · Body composition

1 Introduction

From an ergonomic point of view, the work of emergency brigades is one of the most difficult to adapt to achieve efficiency without exposing brigade members to risks to their physical and mental health. This is because emergencies can occur at any time and reach very different magnitudes. Studying the physiological and psychological response during real emergencies would require a team of researchers dedicated to the continuous monitoring of the brigades, having available non-invasive techniques that allow these evaluations without altering the work. This has a high cost, demands a lot of time and it is a contradiction to expect a real emergency to occur to study its risks [1].

For this reason, it is necessary to apply anticipatory methods that allow looking for the most efficient and safest forms of organization to control the various emergencies

© Springer Nature Switzerland AG 2019
S. Bagnara et al. (Eds.): IEA 2018, AISC 819, pp. 567–572, 2019.
https://doi.org/10.1007/978-3-319-96089-0_60

that the brigades must face. Although technical training is of paramount importance, the selection criteria and the physical and psychological maintenance programs of the brigade members constitute a fundamental element to face emergencies, whose success depends to a large extent on the organization and efficiency of the crews [2]. It is important to mention that the Ergonomics Unit of the Universidad de Concepción in Chile has conducted studies in emergency brigades and has established minimum requirements based on four variables, of which the most important are aerobic capacity, expressed in milliliters of oxygen per kilogram of weight and body fat percentage [1]. The levels required for young workers are greater than for those responsible for directing the crews, as can be seen in the Table 1.

Table 1. Recommendations of aerobic capacity and body fat levels for brigade members.

Young brigade members	Senior brigade members
Aerobic capacity greater than 43.5 ml/min/kg body weight	Aerobic capacity greater than 37.5 ml/min/kg body weight
Fat mass not exceeding 15%	Fat mass not exceeding 20%

It is necessary to state that the above-mentioned values are optimal, but flexible criteria are required and also to consider that these indices are modified with training and with a sedentary lifestyle.

In the last decades, there have been important technological advances for the detection and control of emergencies. However, independent of the efficiency in the management of resources, it is necessary to consider that in every workplace there are human beings and their response is fundamental to promptly control the different types of emergencies [2]. From this point of view, there are many concerns about the limits of physiological demands of emergency brigades, which need to be studied systematically. Unfortunately, in many companies in Chile, emergency brigades are organized with people who perform different types of work, often of a sedentary nature, without any special preparation to comply with these dangerous tasks.

Consequently, the objective of this study was to compare the physical fitness, in terms of body composition and response to controlled exercise, of two groups of workers one with scheduled training and the other without any physical training at work.

2 Methodology

The subjects were 78 volunteers who carried out different activities in two different plants of a holding of wood industries. One of these plants had a training program for their volunteers, which was performed by a physical education teacher, for 45 min, three times a week, during work hours. In the plant that was used for comparison the volunteers that made up the brigades did not perform any special physical activity in their working time.

Before beginning the study, the volunteers were given an explanatory talk about the objectives of the study and also about the tests they would undergo. The subjects participating in the study did not carry any pathology that prevented them from being members of the brigades of the two companies.

The studies were carried out in the polyclinics of the companies, in a room with air conditioning that was maintained at 20 °C. Age was recorded and body mass and stature measured on a clinical beam scale (Precisión Hispana, Madrid). VO2 max was estimated from the technique of Maritz et al. [3]. The subjects exercised on a mechanically braked cycle ergometer (Body Guard, Norway), at three loads, so as to obtain the VO2 max from extrapolation to a predicted maximal cardiac frequency. Body composition, in terms of fat mass and fat free mass was estimated from the technique of Durnin and Womersley after measuring biceps, triceps, subscapular and suprailiac skinfolds [4]. These technique was demonstrated valid for the evaluation of Chilean workers [5]. The skinfolds were measured with a Holtain/Tanner caliper (Holtain, Crymmych, U.K.) which exerts a constant pressure at all openings and can be read to the nearest 0.1 mm.

3 Results and Discussion

Table 2 summarizes the set of variables measured in the members of the trained and untrained emergency brigades.

Table 2. Summary of variables measured in trained and untrained emergency brigades.

Variable	Trained brigade			Untrained brigade		
	n	Average	St. dev.	n	Average	St. dev.
Age (years)	57	37.5	8.1	21	44.0	10.0
Weight (Kg)	57	74.2	8.2	21	81.4	11.7
Height (m)	57	1.70	0.1	21	1.70	6.1
Aerobic capacity (l/min)	57	3.1	0.6	21	2.7	0.5
Aerobic capacity (ml/min/kg)	57	41.4	7.0	21	33.0	6.0
Fat mass (%)	57	22.3	3.6	21	26.2	3.3
Fat mass (Kg)	57	16.6	3.6	21	21.6	5.2
Fat free mass (MLG) (Kg)	57	57.6	5.9	21	59.9	7.5
Body mass index (BMI) (Kg/m^2)	57	26.1	2.7	21	28.0	2.6
MLG/height (Kg/m)	57	34.1	3.0	21	35.1	3.2

As can be seen in Table 2, untrained brigade members have lower average aerobic capacities and a higher percentage fat mass than those trained. The analysis summarized in Table 3 reveals that these differences are statistically significant.

Table 3. Statistical analysis of the differences between trained and untrained brigades.

Variable	Trained brigade average	Untrained brigade average	t-value	df	p
Aerobic capacity (ml/min/kg)	41.4	33.0	4.89871	76	0.000005
Fat mass (%)	22.3	26.2	−4.41321	76	0.000033

One aspect that needs to be highlighted is that the untrained brigade was made up of workers whose average age was higher than that of the brigade with training. Therefore, it is important to consider this variable in the analysis, which can be seen in Table 4.

Table 4. Comparison of aerobic capacity and percentage of fat mass of trained and untrained brigades classified by age groups.

Brigade	Age	n	Aerobic capacity (ml/min/kg) average	Aerobic capacity (ml/min/kg) st. dev.	Fat mass (%) average	Fat mass (%) st. dev.
Untrained	20–29	0	-	-	-	-
Untrained	30–39	7	36.6	5.9	23.2	2.0
Untrained	40–49	8	33.0	5.0	27.5	3.5
Untrained	≥ 50	6	28.6	5.3	28.1	1.3
Trained	20–29	9	44.4	5.3	18.9	3.1
Trained	30–39	24	42.8	7.1	21.4	2.9
Trained	40–49	19	39.1	6.5	24.4	3.4
Trained	>50	5	38.2	8.7	24.4	2.0

As shown in Table 4, the differences in aerobic capacity and body composition clearly reveal that trained workers, at equivalent ages, have a better physical condition.

Looking in more detail, in Fig. 1 it is possible to see the distribution of the aerobic capacity of the trained and untrained brigade.

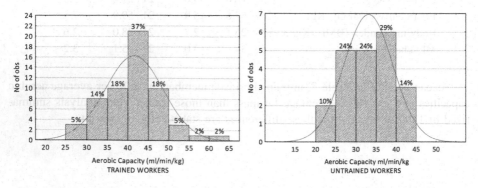

Fig. 1. Aerobic capacity distribution of trained and untrained brigade's members

Looking at Fig. 1 and taking the reference level for aerobic capacity of 37.5 ml/min/kg body weight, indicated as a reference for senior brigade members, 58% of the untrained do not reach that level, while only 19% of trained workers do not reach this figure.

On the other hand, body composition in terms of percentage of fat mass indicates better levels in the trained group since, as can be seen in Fig. 2, 58% of the untrained workers are at a level higher than 25% and there are no workers with less than 20% fat mass. In other words, everyone has some degree of overweight. In the case of the trained brigade, the number of workers with more than 25% fat mass is reduced to 27% and there are 22% of brigade members with less than 20% body fat.

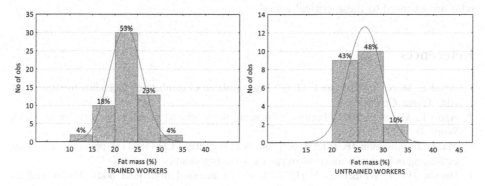

Fig. 2. Percentage body fat distribution of trained and untrained brigade members.

The results are not surprising since there is a significant amount of information that reveals that programmed physical activity improves physical fitness. In Chile has been observed that volunteers from a steel company and an oil refinery, in both cases without training, had low aerobic capacities [1], even lower than those found in this study. Moreover, in forest fire brigades integrated by seasonal workers, the groups subjected to systematic training significantly improved their response to effort and body composition, while those who did not participate in any programmed physical activity tended to worsen. Therefore, it is important to consider the need for emergency brigades to be in optimal conditions, particularly when they have sedentary work and they have to face very heavy activities during emergencies. The important thing is that a low physical fitness can be reversed with training and adequate feeding in reasonable periods of time. Although the word reasonable appears as very vague, the truth is that the changes depend on the intensity and frequency of training and also the willingness of people to regulate their diet.

Finally, it is important to characterize the members of the brigades, but to refine the criteria of selection and control of the physical aptitude it would be necessary to evaluate simulations close to the real situations to have an estimation of the demands of the work.

4 Conclusions

The study reveals that, both in terms of aerobic capacity and body composition, the trained brigade has a better physical fitness than the untrained brigade.

The results allow to recommend that the brigades participate in regular training programs run by qualified professionals.

It would also be advisable to strengthen education for good nutrition, since the tendency to overweight and obesity is a problem that affects the response to high physical demands.

Finally, it is essential to conduct studies in different simulations to determine the demands of each of them and improve the selection and control criteria of the people who are exposed to these critical tasks.

References

1. Apud E, Meyer F, Maureira F (2002) Ergonomía en el combate de incendios forestales, 1st edn. Trama, Concepción
2. Apud E, Meyer F (2011) Factors influencing the workload of forest fire-fighters in Chile. Work 38(3):203–209
3. Maritz JS, Morrison JF, Peter J, Strydom NB, Wyndham CH (1961) A practical method of estimating individual's maximal oxygen uptake. Ergonomics 4(1–4):97–122
4. Durnin JVGA, Womersley R (1974) Body fat assessed from total body density and its estimation from skinfolds thickness: measurements on 481 men and woman aged from 16 to 72 years. Br J Nutr 32(1):77–97
5. Apud E, Jones P (1980) Validez de la medición de los pliegues de grasa subcutánea en estudios de composición corporal. Rev Méd Chile 101(1):661–673

A Preliminary Study Examining the Feasibility of Mini-Pig Assisted Activity for Elderly People in Nursing Homes in Japan

Himena Mano and Iiji Ogawa[✉]

Teikyo University of Science,
2525 Yatsusawa, Uenohara, Yamanashi-Ken 409-0193, Japan
iiji@ntu.ac.jp

Abstract. This study examined the effects of introducing a mini-pig in an Animal Assisted Intervention (AAI) to increase the quality of life for elderly people in Japan. Two intensive-care nursing homes and a day-care center located in Tokyo were selected for this study. In one of the intensive-care nursing homes and the day-care center, a mini-pig AAI was carried out twice in a two month period with more than one week between each activity. In the other nursing home, the AAI was carried out once. In each facility, 10 to 25 clients, who were between 70 and 95 years of age, participated in the activity. Each session was 30 to 40 min long. A total of 68 participants participated in the study: 55 older adults participated in the mini-pig AAI and 13 caregivers provided feedback. There were seven older adults without dementia, and their relaxation levels were measured before and after the mini-pig AAI using the Visual Analogue Scale (VAS). The caregivers filled out a questionnaire after the activity in the three facilities. Two clients showed improvement on their level of relaxation after the mini-pig AAI. The caregivers' evaluations elicited the following answers frequently regarding the responses of the older participants: 'exhibited expression never seen in daily life', 'became more expressive and animated', and 'looked forward to participating in the next activity', In conclusion, the mini-pig AAI was found to be effective similarly to companion dogs and, in some cases, even better as it was novel to see and touch a mini-pig.

Keywords: Elderly people · Animal assisted intervention · Mini-pig

1 Introduction

The world has paid attention to the growing number of elderly people in Japan. It is estimated by 2045 on average 36.8% of the population throughout Japan will be 65 years or older, with Tokyo having the least at 30.7% and Akita having the most at 50.1% [1]. The advent of such an extremely aged society has been increasing the need for pleasure and recreational activities in nursing home facilities in Japan. There is a growing interest in using Animal Assisted Intervention (AAI) in nursing home facilities for therapeutic and recreational purposes; i.e., rehabilitation and mental support. These programs, primarily using dogs, have become increasingly popular in Japan.

© Springer Nature Switzerland AG 2019
S. Bagnara et al. (Eds.): IEA 2018, AISC 819, pp. 573–577, 2019.
https://doi.org/10.1007/978-3-319-96089-0_61

Previous studies have found that dogs introduced into the workplace not only provide an enjoyable and relaxing atmosphere but also improve workers' performances [2–5]. In the field of Ergonomics, these findings are noteworthy. A recent study examined the effects of introducing a cat and a dog on a creative performance task resulting in five out of the nine participants (55.6%) showing an improvement in their performance [6].

In recent years, Animal Assisted Therapy (AAT) has introduced other animals besides cats and dogs. The San Francisco Society for the Prevention of Cruelty to Animals (SPCA) has introduced AAT dogs into San Francisco International Airport terminals for travelers' enjoyment since 2013. In 2016, a certified mini-pig, LiLou, joined the team and roams the terminals entertaining travelers [7].

Similarly, mini-pigs have been gaining popularity as a non-traditional home pet. However, there seems to be no current research investigating the effects and benefits of utilizing mini-pigs in AAI programs in nursing homes. Therefore, a challenge was undertaken to investigate the effects of introducing a mini-pig AAI to increase the quality of life for elderly people in Japan.

2   Method

2.1   Participants

Three nursing homes located in Tokyo were selected for this study. The three facilities differ in terms of the number of caregivers and clients with dementia as shown in Table 1. One of the facilities is a day-care center, and the other two are intensive-care nursing homes. In one of the intensive-care nursing homes and the day-care center, the mini-pig AAI was carried out twice in October and November, 2017 with more than one week given between the sessions. In the other nursing home, the AAI was carried out once. Each session took 30 to 40 min beginning at 2 p.m. and was carried out either indoors or outdoors. In each facility, 10 to 25 older adults, who were between 70 and 95 years of age, participated in the mini-pig AAI. In addition to the older participants, caregivers participated in the study by answering a questionnaire. The Institutional Review Boards (IRB) of Teikyo University of Science approved all procedures and methods. All protocols for this study and consent procedures were also approved by the IRB.

2.2   Companion Mini-Pig

The mini-pig is a 3 year old female, weighing 45 kg. She has been living with a family, is well-trained, and loves to be touched by people. This study was conducted under the permission of the animal welfare boards of Teikyo University of Science. All procedures and methods for the study were approved.

Table 1. Background information of the three facilities.

Facilities	Intensive-care nursing home A	Intensive-care nursing home B	Day-care center
Capacity	100	60	10
Total employees	61	37	8
Participants in the mini-pig AAI	20	25	10
Clients classification			
Self-reliant	0	2	0
Support level*			
Level 1	6	10	0
Level 2	3	7	0
Care level**			
Level 1	9	12	9
Level 2	10	9	6
Level 3	19	6	8
Level 4	23	4	4
Level 5	13	2	2

*The level of support needed: classified into two levels. These clients do not need nursing care but rather need assistance at times. Support level 2 signifies the higher level of support needed.
**The level of care needed: classified into five levels in accordance with the conditions of the client. Care level 5 signifies the highest level of nursing care needed and care level 1 the lowest.

2.3   Measurements

Older adults who did not have dementia were asked to fill out a basic questionnaire that includes favorite animals, pet breeding experience, mini-pig petting experience etc. They were also asked to rate the level of relaxation before and after the session using the Visual Analogue Scale (VAS). In addition to the older adults, caregivers were given a questionnaire after each session.

The mini-pig's stress was monitored throughout the AAI to protect animal welfare. The activity measurement unit, Actiwatch, was attached to the neck of the mini-pig for a month to measure baseline activity level both daytime and nighttime. The mini-pig also wore the Actiwatch throughout the AAI, and the results were compared to the baseline activity level.

3   Results

3.1   Participants

A total of 68 human participants participated in the study: 55 older adults and 13 caregivers. The average age of the older study participants was 85.1 and 72.7% were women. Seven older participants (3 in the nursing homes, and 4 in the day-care center), ranging in age from 80 to 92, who did not have dementia, completed the study measurements. Thirteen caregivers (6 in the nursing home A, 2 in the nursing home B, and 5 in the day-care center) completed the caregiver questionnaire. The average age of the caregivers was 38.8 and 23.1% were women.

3.2 Visual Analogue Scale (VAS)

It was found that two of the seven clients improved their level of relaxation after the mini-pig AAI. One of the participants, who showed improvement, previously reported in the questionnaire that she disliked animals. Four participants did not show any change and one participant was unable to provide any feedback.

3.3 Caregivers' Evaluation

The results of the 13 caregivers' evaluation after the mini-pig AAI are shown in Fig. 1. Most frequent answers were 'exhibited expression never seen in daily life,' 'became more expressive and animated', and 'looked forward to participating in the next activity'. One of the caregivers in the day-care center reported noticeable behavior changes in an older client with dementia, who never spoke in daily life, during the mini-pig AAI. As soon as she saw the mini-pig, she started to talk and pet the mini-pig. The five caregivers in the day-care center answered 'no change' after the AAI.

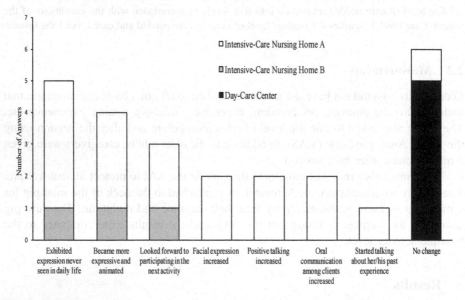

Fig. 1. Results of caregivers' evaluation after the mini-pig AAI (Multiple answers permitted) (N = 13).

3.4 Mini-Pig Activity Level

The comparison between the baseline activity levels of daytime and nighttime indicated that whenever the mini-pig participated in the AAI in the daytime, she had less activity at night compared to a more typical day.

4 Discussion

Under the limited conditions, the following conclusion can be made:

(1) One client who did not like any animals in the day-care center showed improvement on relaxation after the mini-pig AAI. The mini-pig may be more appropriate for people who do not like animals.
(2) One of the most surprising results was that a client with dementia in the day-care center, who never spoke in daily life, even though she loves animals and children, started to talk and touch the mini-pig as soon as she saw it. A similar response has been reported during equine assisted activity.
(3) The mini-pig had more than usual activity levels on the days that she participated in AAI compared to her more typical days. It is possible that she was under greater stress which resulted in less activity during night.

In conclusion, the mini-pig AAI was found to be effective similarly to companion dogs and, in some cases, even better since it was novel to see and touch a mini-pig in Japan. However, continuous research is necessary and further considerations are needed for animal welfare and zoonotic diseases.

References

1. National Institute of Population and Social Security Research. http://www.ipss.go.jp/. Accessed 22 Apr 2018
2. Ogawa I, Sakamoto T (2003) What is the most beneficial type of recreation for computer operators? In: Proceedings of human-computer interaction international 2003, vol 3, pp 118–122
3. Ogawa I (2006) The effect of companion animal dogs on office workers to relieve job stress - an experimental study using the Profile of Mood States (POMS)-. Health Sci 22(2):227–239
4. Ogawa I (2006) Does playing with a companion animal during a break period reduce job stress and improve work performance for computer operators? In: Proceedings of human-computer interaction international 2006, vol 3, pp 118–122
5. Takeshima R, Ogawa I (2011) Is the presence of a companion animal dog beneficial for computer operators? In: Robertson MM (ed) Ergonomics and Health Aspects of Work with Computers. Springer, Heidelberg, pp 78–87
6. Ogawa I (2015) Is the presence of a cat/dog beneficial for a creative task performance? Presented at poster no P109, 19th triennial congress of the international ergonomics association congress, Melbourne, Australia
7. Meet the Wag Brigade/San Francisco International Airport. https://www.flysfo.com/services-amenities/wag-brigade. Accessed 22 Mar 2018

Subjective Evaluation of the Physical Work Environment and the Influence of Personality

Knut Inge Fostervold[(✉)] and Anne-Marie Halberg

Department of Psychology, University of Oslo, Oslo, Norway
k.i.fostervold@psykologi.uio.no

Abstract. Subjective evaluations of physical working conditions are common in the field of human factors and ergonomics (HFE). The aim of this study was to investigate the effect of personality factors on judgements of important physical factors in the work environment. The sample consisted of 80 informants working in a white-collar organization. User evaluations of Lighting, Indoor air, Privacy, Negative affect (NA) and Positive affect (PA) were collected twice, over two consecutive years. The results confirm that personality characteristics do affect appraisals of physical conditions, although only for Negative affect. The results also revealed high stability in subjective judgements across time, both with regard to trait-like emotions (NA and PA) and with regard to appraisal of environmental conditions.

Keywords: User evaluations · Physical work environment
Personality characteristics

1 Introduction

User evaluations of physical conditions in the work-environment are common in research and practice in the field of human factors and ergonomics (HFE). The method is simple, cost effective, and enjoy high face validity. User evaluations repeatedly reveal high degree of heterogeneity as different evaluators often rate the same physical setting quite differently. Users differ, for example, substantially in the preferred light setting. This variability is more than a methodological problem. Research shows that light settings adjusted in accordance with personal preferences influence feelings of well-being and satisfaction positively [1].

Personality characteristics may account for some of this variability. Dispositional tendencies to experience negative emotions coined negative affect (NA), is perhaps one of the most interesting aspects. Such trait-like pervasive feelings of negative emotions have repeatedly been discussed as an important factor in the appraisal process of environmental conditions [2–4]. Dispositional trends toward positive emotions or positive affect (PA) have not been studied to the same extent. Nevertheless, studies indicate that also PA are prone to affect subjective appraisal of environmental conditions [5, 6].

Research into judgement and decision-making point to that the ease of an evaluation task is dependent on the "evaluability" of relevant attributes [7]. Following this line of thinking, environmental conditions containing salient objective attributes should

© Springer Nature Switzerland AG 2019
S. Bagnara et al. (Eds.): IEA 2018, AISC 819, pp. 578–582, 2019.
https://doi.org/10.1007/978-3-319-96089-0_62

be easier to evaluate and consequently should be less influenced by personality characteristics.

The aim of the present study was to investigate the relationship between both negative and positive affect and important physical factors in the work-environment.

2 Methods

The study was conducted as a part of a general work environment survey in a white-collar organization in Norway. The sample consisted of 39 women and 41 males (80 in total). The mean age was 41 years (SD = 10.6 years). Data were collected twice, over two consecutive years, making it possible to study both short and long-term effects.

2.1 Measures

Negative and positive affect were measured by the short-form of the dispositional version of the Positive and Negative Affect Schedule (PANAS) [8]. The scale consists of 10 words, describing negative and positive emotions. The informant is asked to rate each word on a numeric 5-step interval scale, measuring to what extent the individual experience the emotion "in general".

Items developed in house measured subjective evaluations of physical conditions. The questionnaire was summarized into three indices; Lighting quality was measured by seven items, Indoor air quality by six items, and satisfaction with Privacy by three items. Each item was assessed by a numeric 7-step interval scale, with semantic descriptors at both ends and in the middle.

2.2 Statistics

Structural equation modelling (SEM) was used as primary analytical tool. The goodness of fit measures included the model Chi-square (χ^2_M), including its degrees of freedom and p value, the Comparative fit index (CFI), and the Standardized Root Mean Square Residual (SRMR) [9]. To limit the number of estimated parameters, each construct was modelled as a single-indicator latent factor. The error variance for each factor $(Var(\epsilon_i))$ was set according to the formula [10]:

$$[(1 - \hat{\rho}_{y_iy_i})var(y_i)]$$

For the sake of consistency, the error variances were calculated using reliability estimates obtained in the present sample.

3   Results

Three models were fitted, one model for each environmental condition. Each model contained environmental condition, PA, and NA, at time1 and time2. Hypothesized paths were drawn from PA and NA at time1 pointing towards environmental condition at time1 and environmental condition at time2. Paths were also added between PA-time1 and PA-time2, and between PA-time2 and environmental condition at time2. Similar paths were added for NA. Finally, paths representing inter-correlations between PA and NA were added at time1 as well as time2.

The goodness of fit measures, for all three models, are shown in Table 1. The Chi-square was non-significant for indoor air and privacy. Lighting was bordering significant. Further, the results showed a CFI above the criteria of .95 [11] for indoor air and privacy. The CFI for lighting was slightly below this criteria. The SRMR were below the criteria of .08 [12] for all three models. Thus, the measures indicate good or satisfactory fit with the empirical data for all three models.

Table 1. Goodness of fit measures for all three models.

Condition	$\chi2$	Df	p	CFI	SRMR
Lighting	9.501	4	.050	.928	.0651
Indoor air	4.896	4	.298	.988	.0444
Privacy	7.475	4	.113	.956	.0538

The structural relationships between each environmental condition and NA and PA at time1 (T1) and time2 (T2) are shown in Fig. 1. All parameters shown in the figure were significant except for the relation shown for NA on Lighting.

The results showed strong stability across time (T1–T2) both for evaluations of physical condition and personality characteristics (NA and PA). In all three relationships higher path coefficient were observed for NA compared to PA at time1. However, the relationship was not uniform across environmental conditions. The strongest influence of NA seems to be present for Privacy (β = .43). The influence of NA was somewhat lower for Indoor air (β = .37), while the result for Lighting (β = .11) was non-significant.

4   Discussion

The main results showed high stability in subjective judgements across time, both with regard to trait-like emotions (NA and PA) and with regard to appraisal of environmental conditions. NA and PA measured by means of the "general time frame" instruction have previously shown high long-term stability [13, 14]. The results obtained in the present study consequently seems to confirm this finding.

Stability measures of environmental conditions seem to be sparsely reported in the literature. Investigating the stability of physical properties in the work environment may be difficult for many reasons. An obvious problem is that physical conditions tend

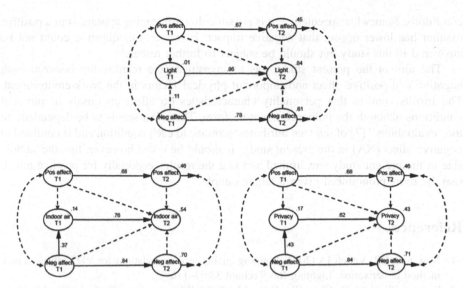

Fig. 1. Path diagrams depicting the structural relationship between environmental condition, negative and positive affect at time1 and time2. Numbers are shown as standardized estimates. Empirical indicators and error terms are omitted to enhance readability. Non-significant paths are displayed as dashed lines.

to change both seasonally and as results of changes being made at the workplace. Thus, a reasonable expectation should be lower stability estimates than what seems to be present. However, according to Vischer [15], physical properties in the work environment are best understood as a type of hygiene factors. Inferior physical conditions seem to generate strong negative impact on feelings of well-being and self-perceived productivity. Conditions considered as being satisfactory on the other hand, seem to be neutral and generate no specific effect, either positive or negative. This tendency towards a dichotomization may affect appraisals of physical properties in the work-environment directly, making it difficult to discern between quality differences in physical conditions that are satisfactory.

Another main finding in this study was that the influence of NA seems to vary with environmental condition. This finding accord with the hypothesis derived from research on judgement and decision-making. Lighting, Indoor air, and Privacy were presumed to comprise differing degree of saliency. Lighting, being a clearly visible factor in the work environment, was considered to retain the highest evaluability and should consequently be lesser affected by trait-like emotions. Privacy was presumed to have lowest evaluability and should thereby be expected to be influenced the most by personality characteristics.

The effect of personality on user evaluations of physical properties in the work-environment were found only for negative affect (NA). The results for positive affect (PA) were smaller and non-significant. Nevertheless, it could be noted that a weak trend towards the same pattern was observed also for PA. One explanation could be the tendency towards a dichotomization between non-satisfactory and satisfactory conditions. Somewhat speculative, it is possible that influencing appraisals in a positive manner has lower impact that negative impact. However, this question could not be answered in this study but should be subject to further research.

The aim of the present study was to investigate the relationship between both negative and positive affect and important physical factors in the work-environment. The results confirm that personality characteristics do affect appraisals of physical conditions although the influence is not uniform. The effect seems to be dependent on the "evaluability" [7] of relevant attributes germane to each condition and is confined to negative affect (NA) in the present study. It should be noted however, that the sample size in the present study was limited and that the results, especially for positive affect, may be more pronounced given a larger sample.

References

1. Newsham GR, Veitch JA (2001) Lighting quality recommendations for VDT offices: a new method of derivation. Lighting Res Technol 33:97–116
2. Spector PE, Zapf D, Chen PY, Frese M (2000) Why negative affectivity should not be controlled in job stress research: don't throw out the baby with the bath water. J Organ Behav 21:79–95
3. Girardi D, Falco A, Dal Corso L, Kravina L, De Carlo A (2011) Interpersonal conflict and perceived work stress: the role of negative affectivity. TPM 18:257–273
4. Campbell JM (1983) Ambient stressors. Environ Behav 15:355–380
5. Baker J, Cameron M (1996) The effects of the service environment on affect and consumer perception of waiting time: an integrative review and research propositions. J Acad Mark Sci 24:338–349
6. Zhang X, Zuo B, Erskine K, Hu T (2016) Feeling light or dark? Emotions affect perception of brightness. J Environ Psychol 47:107–111
7. Hsee CK, Lowenstein GF, Blount S, Bazerman MH (1999) Preference reversals between joint and separate evaluations of options: a review and theoretical analysis. Psychol Bull 125:576–590
8. Thompson ER (2007) Development and validation of an internationally reliable short-form of the positive and negative affect schedule (PANAS). J Cross Cult Psychol 38:227–242
9. Kline RB (2016) Principles and practice of structural equation modeling. The Guilford Press, New York
10. Bollen KA (1989) Structural equations with latent variables. Wiley, New York
11. McDonald RP, Ho MHR (2002) Principles and practice in reporting structural equation analyses. Psychol Methods 7:64–82
12. Hu LT, Bentler PM (1999) Cutoff criteria for fit indexes in covariance structure analysis: conventional criteria versus new alternatives. Struct Equ Model: Multidisciplin J 6:1–55
13. Watson D, Walker LM (1996) The long-term stability and predictive validity of trait measures of affect. J Pers Soc Psychol 70:567–577
14. Leue A, Lange S (2011) Reliability generalization: an examination of the positive affect and negative affect schedule. Assessment 18:487–501
15. Vischer JC (2008) Towards an environmental psychology of workspace: how people are affected by environments for work. Archit Sci Rev 51:97–108

Influence of Select Modes of Load Carriage on Movement Biomechanics of Industrial Workers

Rauf Iqbal[✉] and Arundhati Guha Thakurta

Ergonomics and Human Factors Engineering, National Institute of Industrial
Engineering (NITIE), Mumbai, India
raufiqbal@nitie.ac.in, arundhati18@gmail.com

Abstract. A study was conducted with an aim to determine the changes of
human lower limb biomechanics biomechanics while carrying loads in selective
modes among industrial workers of India and to suggest the biomechanically
efficient mode of load carriage. Industrial workers (n = 20) in the age range of
22 to 58 years were selected and asked to carry loads on head, across one
shoulder and in hands. Qualisys Motion Capture System (Sweden), Kistler
Force Plate (Switzerland) and Polar S810i HR monitor, Finland were used to
record data. Heart rate was recorded at rest and during select modes of load
carriage with 40% of individual body weight i.e. loads on head, shoulder and
hand. Walking speed for head load, shoulder load and hand load were
4.19 ± 0.55 km.hr^{-1}, 4.09 ± 0.82 km.hr^{-1} and 3.94 ± 0.84 km.hr^{-1} respectively. Double limb support time while carrying loads has been found lowest in
case of head mode (0.65 ± 0.46 s) followed by hand mode (0.67 ± 0.14 s) and
shoulder mode (0.68 ± 0.06 s) of load carriage. There is a gradual increase in
hip flexon from head to shoulder and hand load condition and gradual decrease
in ground reaction force from head to shoulder and hand load conditions.
Considering the changes in the physiological parameters along with the
biomechanical parameters, head load was found to be more productive in
comparison to shoulder and hand load modes. Results demonstrated a significant
changes in kinetic and kinematic parameters of workers at selective load conditions. Findings would provide substantial input for designing work and work
rest cycle for different industrial manual workforce.

Keywords: Load carriage · Lower limb biomechanics · Stride length
Double limb support · Energy cost · Gait · Heart rate

1 Introduction

Human motion analysis is an important basis for designing of work in industries and
other occupations. The main three functions of lower limbs are to bear weight, provide
means for locomotion and maintain equilibrium. In developing countries of the world,
a large population is employed in carrying and transporting loads manually as their full
time work or on contractual basis [3, 12]. Human movement while load handling, and
the corresponding resultant energy expenditure play a significant role for maintaining

© Springer Nature Switzerland AG 2019
S. Bagnara et al. (Eds.): IEA 2018, AISC 819, pp. 583–589, 2019.
https://doi.org/10.1007/978-3-319-96089-0_63

safety, health, and efficiency of the workforce in different industries. There are many different ways in which loads may be carried, and the mode of load carriage adopted can be determined by certain factors such as weight, shape and size of the load, the duration for which it has to be carried, the terrain, physical characteristics and condition of the individual and the clothing as well as personal preference [5]. It is evident previous research surveys that biomechanics of human movement changes with modes of load carriage in any situation of developing countries like India, where manual work encompasses a major population. Further, biomechanical load coupled with energy cost of work provides a significant input for design of a work system. Energy cost of work is derived from the rate of oxygen consumption directly or from the heart rate responses at work indirectly. Heart rate response has been found to be an easy assessment technique in industrial situation, since heart rate can be easily measured without using any sophisticated equipment. Physiological Cost Index (PCI) also known as Energy Expenditure Index (EEI) is used for measuring the energy cost of walking indirectly [11]. EEI is measured as the difference of walking heart rate and resting heart rate divided by walking speed. There have been various studies where the Energy Expenditure Index were used to measure energy cost [1, 6, 7, 9]. It is evident from previous studies that along with the biomechanical changes of human movement (lower limb), a notable change in EEI pattern has been observed during load carriage tasks. So, there is a need to understand how the selective modes of load carriage changes the temporal-spatial, angular and force parameters of the human lower limb movement analysis (Gait-the pattern of walking) among industrial workers in India doing manual handling of loads. Consequently, the purpose of this study was to (i) study the changes in the lower limb biomechanics (gait parameters analysis) and physiological parameters with selective modes of load carriage among industrial workers and (ii) to suggest biomechanically efficient mode of load carriage for the workers on the basis of the changes in EEI and gait parameters.

2   Materials and Methods

2.1   Participants

Volunteers who participated in this study were working as manual handlers in various small and medium scale industries in Mumbai, the western region of India. The volunteers were selected by using purposive sampling method. While selecting volunteers, care was taken that none of them had existent history of severe musculoskeletal disorders as such or restriction of movements. The sample size of the study was 20 industrial workers of the age range of 22 to 58 years.

2.2   Instrumentation

Biomechanics was studied by using Qualisys Motion Capture System (Sweden). For tracing the subject's movements, an array of six high speed cameras (type: 'Oqus'500) from 'Qualisys Motion Capture System' was used. Kistler (Switzerland) force plates (2 Nos) were used along with the Qualisys system. This system altogether captures motion data along with the force data. Weighing scale, flexible measuring tape and anthropometric kit were used to measure the physical dimensions of the volunteers. Polar S810i Heart Rate Monitor (Finland) was used to measure heart rate during carrying of load in select modes.

Written consent from the volunteers was obtained and then a screening examination was conducted to ensure each volunteer was free of lower extremity musculoskeletal problems.

2.3 Data Collection Procedure

Thirty six reflective markers were placed on specific locations of each volunteer's legs and pelvic region. Six high speed cameras (type OQUS) by Qualisys Motion Capture System, Sweden were used to film the volunteers walking along a 7 m walkway. Each volunteer first walked along the walkway at their self-selected speed without any load. This was used to establish a baseline and then, they walked along the walkway while carrying 40% of respective body weight (sand box of 40 cm x 40 cm x 16 cm) [4] load on head, across one shoulder and hand (frontal plane). Each volunteer followed the same protocol for the testing conditions. The volunteers were allowed to walk on both the force plates (Kistler, Switzerland). Force plate-1 and force plate-2 were stepped by left foot and right foot respectively. The volunteers were asked to walk for 15 min, considering the time consumed to complete one whole cycle of load carriage task for the selected industrial situation during which 4 trials were recorded by the OQUS infrared digital cameras and was considered for analysis. Resting and peak heart rates were also noted for each activity.

2.4 Data-Processing

To record the motion and force plate data the software Qualisys Track Manager was used and was analyzed by Qualisys Visual 3D software (professional version). The lower limb biomechanical parameters (gait parameters) and Energy Expenditure Index (EEI) were generated for each of the volunteers and comparison was made with different modes of load carriage.

2.5 Analysis

To ensure that the cameras and force plates recorded properly from both sides of the y-axis (defined as the direction of walking), the test subject entered the analyze-space from positive y-direction half of the trials, while the other half from the negative y-direction. The reflective markers were used to define the different anatomical points on the test subject's body. It was ensured that the cameras and force plates recorded properly all the three coordinate axis. In order to test the comparability of the force plates, half the trails were done stepping with the subject's left foot on force plate-1 and second half of the trails the subject stepped with his right foot on force plate-1 while traversing the same pathway in reverse direction.

3 Results and Discussions

The volunteers consisted of 20 healthy industrial workers in the age range of 22–58 years. None of the subjects had any history of neurological or orthopedic disorders likely to affect their mobility.

Table 1. Demographic parameters of the subjects

Groups	Age (years)	Height (cm)	Weight (Kg)	BMI (Kg/m²)
Mean ± SD	38.1 ± 12.99	167.4 ± 7.45	63.9 ± 7.02	22.8 ± 2.42
Range	22–58	157–182	54–77	17.2 ± 26.8

In Table 1 the demographic details of the workers have been presented. The mean age was 38.1 ± 12.99 years. The mean height and weight of the test volunteers were 167.4 ± 7.45 cm and 63.9 ± 7.02 kg respectively. The Body Mass Index of was 22.8 ± 2.42 kg/m² and 74.0 ± 12.45 beats/min respectively.

In Table 2, changes in the basic gait parameters like walking speed, stride length and stride width have been shown in the three selective load carriage conditions. Here, the considered parameters were measured on the basis of time and space and hence known as temporal and spatial parameters of gait which are the vital signs of human walking pattern.

Table 2. Biomechanical parameters (time-space) with modes of load carriage

Parameter	Stride length (meter)	Speed (Km/hr)	Stride width (meter)	Double limb support time (sec)
Head load	1.21 ± 0.45	4.19 ± 0.55	0.25 ± 0.20	0.65 ± 0.46
Shoulder load	1.12 ± 0.04	4.09 ± 0.82	0.26 ± 0.05	0.68 ± 0.06
Hand load	1.19 ± 0.01	3.94 ± 0.84	0.29 ± 0.06	0.67 ± 0.14

Larger stride length was recorded during head mode of load carriage in comparison to shoulder and hand mode. Walking speed of the workers was maximum while carrying load on head and minimum speed while carrying load in hand. Comparison of stride width shows that it increases from head load to shoulder load to hand load condition. Since, load on hand creates a moment arm anterior to spine and shifts the Centre of Gravity (CG) outside the body, hence, to counteract shifting of CG, there is increase of Base of Support (stride width). Decrease in speed and increase in stride width is a sustained effort to maintain the stability while carrying load [6]. Wide-based gait is considered indicative of imbalance [2]. So, hand load caused maximum imbalance as speed decreased and stride width increased which indicated as adaptability for the workers to carry on their work. Though the difference is insignificant but double limb support time was found minimum in case of head load followed by hand and shoulder load. Double limb support time being inversely proportional to movement stability, it is indicated that head load could be more stable than hand and shoulder load [12].

In Table 3, changes in the angular gait parameters like knee and hip angles, and force parameter i.e. ground reaction force have been shown in different load conditions. These are the essential parameters of gait, the changes in any of the parameter might change the pattern of walking of an individual as well along with the temporal-spatial parameters of gait.

Table 3. Biomechanical parameters (angular and force) with modes of load carriage

Parameter	Hip flexion (degree)	Knee flexion (degree)	Ground reaction force (F1) heel strike (% of body weight)	Ground reaction force (F2) toe off (% of body weight)
Head load	56.051 ± 3.41	59.95 ± 3.46	164.25 ± 12.52	149.7 ± 12.23
Shoulder load	62.4 ± 3.25	62.38 ± 3.43	147.33 ± 15.11	143.93 ± 1539
Hand load	67.88 ± 4.12	66.78 ± 4.53	156.08 ± 15.43	148.63 ± 14.05

The functional movement of hip and knee have been found to be increased from head load to shoulder and hand load despite the age variations [10]. The increased flexion angle may be due to an effort to counteract the shifting of Centre of Gravity and maintaining balance while carrying load (Table 3). There is a tendency towards a greater impact during heel strike and comparatively less forceful push-off in all the three modes of load carriage. Ground reaction force is higher in case of head load in comparison to shoulder and hand load. Higher flexion angle and consequent lower ground reaction force is an indicative of an effort for maintaining stability during load carriage.

In Table 4, the heart rate (pre-work and post-work), heart rate cost, Energy Expenditure Index in different load conditions has been presented. There is a gradual increase in heart rate cost and EEI from head load to shoulder load and hand load, the reason for which may be attributed to increased activity of muscle groups to maintain posture and stability during movement with load.

Table 4. Heart rate responses with modes of load carriage

Parameter	Pre-load heart rate (beats/min)	Post-load heart rate (beats/min)	HR cost (beats/min)	EEI (beats/meter)
Head load	74.0 ± 12.45	124.0 ± 15.78	50.0 ± 10.54	0.74 ± 0.23
Shoulder load	79.05 ± 10.17	132.0 ± 10.96	52.95 ± 10.88	0.82 ± 0.28
Hand load	80.15 ± 9.11	137.6 ± 11.39	57.45 ± 10.04	0.93 ± 0.93

As a consequence, there is also a gradual increase in oxygen consumption and energy cost of work from head load to shoulder load and hand load (Table 5).

Table 5. Oxygen consumption and energy cost with modes of load carriage

Parameter	VO_2 (Lit/min)	Energy cost work (KJ/min)
Head load	1.98 ± 0.44	41.62 ± 9.28
Shoulder load	2.21 ± 0.31	46.33 ± 6.45
Hand load	2.36 ± 0.32	49.61 ± 6.69

Cardio-vascular responses like, EEI and energy cost of work were used to compare the economy of walking at various load conditions by the workers. In order to note the magnitude of the changes in the physiological variables of the individuals, the oxygen consumption and physiological cost index were taken into account in previous studies as well [8]. The changes in those physiological parameters signify the impact of different modes of load carriage on industrial workers while carrying out their daily job involving manual handling of loads.

4   Conclusions

The findings of the study would be beneficial to provide objective information for evaluating the influence of gait function with increase in different loaded gait intervention. Carrying load on head could be inferred as the most efficient way of manual handling of loads in the present condition in which the volunteers were given a load of 40% of body weight to carry in head, shoulder and hand modes of load carriage. Minimizing all those biomechanical changes seems a rational design goal for the industrial workers doing manual handling of load. Findings would provide substantial input for designing work and work rest cycle for industrial workforce.

References

1. Butler P, Engelbrecht M, Major RE, Tait JH, Stallard J, Patrick JH (1984) Physiological cost index of walking for normal children and its use as an indicator of physical handicap. Dev Med Child Neurol 26(5):607–612 ISSN 0012-1622
2. Fredericson M, Yoon K (2006) Physical examination and patellofemoral pain syndrome. Am J Phys Med Rehabil 85(3):234–243 ISSN 0894-9115
3. Guha Thakurta A, Iqbal R, De A (2017) The influence of three different load carrying methods on gait parameters of Indian construction workers. MOJ Anat Physiol 3(4):00098
4. Kinoshita H (1985) Effects of different loads and carrying systems on selected biomechanics parameters describing walking gait. Ergonomics 28(9):1347–1362 ISSN 0014-0139
5. Legg SJ (1985) Comparison of different modes of load carriage. Ergonomics 28(1):197–202
6. Maki BE (1997) Gait changes in older adults: predictors of falls or indicators of fear? J. Am Geriatr Soc 45(3):313–320 ISSN 0002-8614

7. Mossberg KA, Linton KA, Friske K (1990) Ankle-foot orthoses: effect on energy expenditure of gait in spastic diplegic children. Arch Phys Med Rehabil 71(7):490–494 ISSN 0003-9993
8. Mukherjee G, Samanta A (2001) Physiological response to the ambulatory performance of hand-rim and arm-crank propulsion systems. J Rehabil Res Dev 38(4):391–399 ISSN 0748-7711
9. Rose J, Gamble JG, Burgos A, Medeiros J, Haskell WL (1990) Energy expenditure index of walking for normal children and for children with cerebral palsy. Dev Med Child Neurol 32 (4):333–340 ISSN 0012-1622
10. Shkuratova N, Morris ME, Huxham F (2004) Effects of age on balance control during walking. Arch Phys Med Rehabil 85(4):582–588 ISSN 0003-9993
11. Suzuki T, Sonoda S, Saitoh E, Murata M, Uno A, Shimizu Y, Kotake T (2005) Development of a novel type of shoe to improve the efficiency of knee-ankle-foot orthoses with a medial single hip joint (Primewalk orthoses): a novel type of shoe for Primewalk orthosis. Prosthet Orthot Int 29(3):303–311 ISSN 0309-3646
12. Thakurta AG, Iqbal R, Bhasin HV, De A, Khanzode V, Maulik S (2015) Characteristics of gait variability among healthy Indian construction workers during different load carrying modes. In: Ergon in caring for people. Proceedings of the international conference on humanizing work and work environment 2015, pp 9–16 ISBN 978-981-10-4979-8

Development an Office Ergonomic Risk Checklist: Composite Office Ergonomic Risk Assessment (CERA Office)

Gyula Szabó[1,2] and Edit Németh[3(✉)]

[1] Donát Bánki Faculty of Mechanical and Safety Engineering,
Óbuda University, Budapest, Hungary
szabo.gyula@bgk.uni-obuda.hu
[2] Federation of European Ergonomics Societies (FEES), Budapest, Hungary
[3] Institute of Business Economics, Eötvös Loránd University,
Budapest, Hungary
nemethe@gti.elte.hu

Abstract. Despite the continuous efforts the burden of work-related Musculoskeletal Disorders (wMSDs) hasn't been eliminated. Ergonomics risk assessment tools had been often developed on causes of different occupational diseases related to different body parts, and implemented in various forms ranging from paper-pencil worksheets to immersed virtual reality simulations.

The Composite Ergonomic Risk Assessment (CERA) was developed to assess ergonomic risks and determine intervention points by Hungarian experts for industrial and office workplaces, according to the EN 1005 Safety of machinery - Human physical performance standard.

Professionals with different background e.g. OSH specialists, occupational doctors, ergonomist participated in the development and evaluation process of CERA Office when various risk assessment methods had been tested.

The objectives of the new tool was to provide a reliable but quick health risks assessment of computer workplaces and define correction measures and improvement action plans according to different risk levels.

In the pilot period 20 different office tasks were assessed by the new risk assessment method as well as with three already widely accepted and used risk assessment tools. The control tools was Cornell University's Cornell Musculoskeletal Discomfort Questionnaires (CMDQ), the Rapid Upper Limb Assessment (RULA), the Rapid Office Strain Assessment (ROSA).

The statistical analysis highlighted that CERA Office tends to result on higher risk than the traditional tools. This alteration is due to the higher attention demand required by CERA Office on evaluators to analyse the movements, postures, activities more precise. CERA Office discovered weaknesses of workstations which was not identified by traditional methods used by not-proficient evaluators.

The added value of CERA Office is the collection of well defied work situations requiring intervention. CERA Office is an example to efficient and reliable ways to identify the ergonomic risks of office workplaces.

Keywords: Composite Ergonomic Risk Assessment (CERA)
Office ergonomics · wMSD · Checklists

© Springer Nature Switzerland AG 2019
S. Bagnara et al. (Eds.): IEA 2018, AISC 819, pp. 590–597, 2019.
https://doi.org/10.1007/978-3-319-96089-0_64

1 Introduction

In the 1980s the new information technologies exploded in daily life. With their appearance it could be expected a large-scale substitution of manual assembly work by automation, and also elimination of exacting office work resulting in a lower rate of work related musculoskeletal disorders and better possibilities to prevent work-related illnesses.

However, with the spread of personal computers, new risks have emerged and it has become clear that computer work itself poses significant risks, mainly due to the awkward postures required by the use of devices and the very high repetition of movements.

Various types of computer work - initially, data entry - are characterized by a motionless static body position, and a high repetition rate of upper limb, while simultaneously loading the shoulder and neck. The visual work of seeing screen devices have often led to visual problems.

Not surprisingly, the emerging ergonomic risk has provoked considerable interest. On the one hand, self-evaluation tools that take into account the specificities of computer work was developed, and on the other hand, regulations have been set up to meet the requirements of computer-based workplaces specifying the workplace and the computer work organisation. For example, the EU VDU directive [1] sets visual requirements related to screen visibility, monitor setting, mandatory eye ophthalmic check, defines posture requirements, such as table adjustability, leg support requirements, high-quality chair seating, and in addition, the correct design and setting of input devices.

The technology advance produced solution in workstation design e.g. cathode-ray tube displays has been overtaken by the ever-increasing quality of screens, or ergonomic designed keyboards and electric height-adjustable tables has been widespread.

Good technical solutions became less and less expensive in office job design, and the focus shifted towards the proper use of tools. Supportive services were provided to train workers the correct use of the tools in the work setting, such bringing the ability to sit correctly and follow proper working methods to the daily life. The modern approach, as a stand-alone ergonomic function, prevents musculoskeletal disorders in the office during work, provides good access to equipment and includes preventive movements, such as workplace gymnastics for office workers or back massage.

2 The Composite Ergonomics Risk Assessment for Office Work (CERA Office)

The use of ergonomic risk assessment tools for office work has also become popular. Some of them are available on the web and allow workers self-assessment and help to improve working conditions immediately. Ergonomic risk assessment methods were initially based on an analysis of the causes of various musculoskeletal disorders and often determine the importance and urgency of the require measure based on an evaluation of the worker's movement, effort and body position, i.e. by assessing the direct risk factors.

Workplace safety assessment often involves assessing working conditions, such as assessing working height, noise or lighting, and taking into account the indirect ergonomic risk factors that can even be measured. Screening, ergonomic problem-spotting methods are primarily based on the evaluation of direct risk factors, while indirect factors are evaluated by checklists when engineers assesses the fulfilment of the health and safety design requirements.

To meet the Hungarian professionals' need for comprehensive ergonomic risk assessment tool the Composite Ergonomics Risk Assessment (CERA) has been developed [2]. CERA implements the requirements of the EN 1005 standard series as a paper pencil tool, thus assessing the risk of body positions, manual material handling, effort and repetitive movements. The CERA has spread to the Hungarian community as expected, and it has been used as a part of the work safety risk assessment, for the preparation of ergonomic development programs, and beyond it has became part of ergonomic and OSH education thanks to its didactical structure.

2.1   The CERA Office

The CERA can be further developed by means of computerization, the development of an interactive evaluation sheet and the automated processing of evaluations. A PDF and an Excel table format was developed [3], and there was a trial with the ErgoCapture application utilising Kinect based motion capturing [4].

The next direction for the further development of CERA is the preparation of new individual alternates for different type work. Specific CERA versions are provided for welding, manual waste sorting [5] and this paper presents the CERA version for computer workplaces. In the course of new tools development it became clear that risk factors absent at the given work situation should be left out from the general CERA assessment, for example for the computer workplace evaluation there's no reason to consider risk factors related to manual handling. On the other hand, the inclusion for evaluation of indirect risk factors is unavoidable for the specific methods, such as working height (Fig. 1).

The CERA office includes all of the direct ergonomic risk factors potentially present during screen work, thus assess posture, especially the position of the hand according to EN 1005-4 [6]. On the other hand, it includes an assessment of the workplace elements, such as the position of the monitor, seat and keyboard. The result of the evaluation of the indirect factors depends on the setting of these elements. The assessment can be done according to the real, state of the art situation or can also be done with optimum utilization of the workplace setting options, so with an ideal user behaviour. This distinction shows that the CERA office, like other ergonomic self-assessment tool, is capable of improving user/employee behaviour, that is to help workplace health promotion or work safety training programs. The advantage of CERA Office in the office environment compared with the use of the original CERA is that it includes all the risk factors associated with all typical computer workplaces and is substantially shorter than the full CERA assessment, as it ignores irrelevant factors.

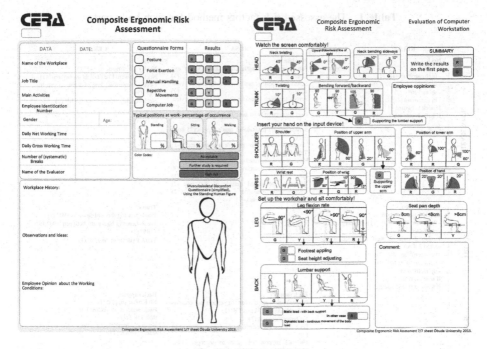

Fig. 1. CERA office

Laptops, which are not fit to work without docking and secondary keypad, tablets, smart phones and other additional new device and applications appear in workplaces by the development of information technologies. These also pose challenges that we did not take into account when assessing the risks at the yet traditional office work.

3 Methods

The original version of the CERA implemented the EN 1005 standard series, and was validated by evaluations of industrial workplaces simultaneously with other widely accepted ergonomic evaluation methods. Similarly the validation and the usability assessment of CERA Office was made with several office workstation evaluation together with Cornell University's Cornell Musculoskeletal Discomfort Questionnaires (CMDQ) [7], the Rapid Upper Limb Assessment (RULA) [8], the Rapid Office Strain Assessment (ROSA) [9].

The assessed risk factors for each method are shown in Table 1 for each body part. RULA and CERA risk assessment methods use direct risk factors, and evaluate the body postures and movements, while ROSA uses indirect risk factors, and relies on the layout of work equipment and furniture. Note, that CERA-Office only records non-compliance and RULA scores a good posture by 1 point, while ROSA scores bad posture only.)

Table 1. The assessed risk factors method for each body part

	RULA		ROSA	CERA
HEAD, NECK	**Neck** Locate Neck position Positon: - between 0° to +10° (1) - between +10° to 20° (2) - up to +20° + (3) - in extension (4) If neck is twisted +1 If neck is side-bending +1	position Positon:	**Monitor** Screen at arm's length/screen positioned at eye level (1) Screen too low (causing neck flexion to view screen) (2) Screen too high (causing neck extension to view screen) (3) User required to twist neck in order to view screen (+1) Screen too far (outside of arm's length (75 cm)) (+1) Document holder not present and required (+1) **Telephone** Headset used/one hand on telephone and neck in a neutral posture, telephone positioned within 300 mm (1) Telephone positioned outside of 300 mm (2) Neck and shoulder hold used (+2) No hands free options (+1)	**Neck** Neck - Twisting up to +45° (1) Neck - Upward line of sight up up to 0° (1) Neck - Downward line of sight up up to +45° (1) Neck bending sideway up to +10° (1)
TRUNK	Trunk position Locate Trunk Position - between 0° to +10° (1) standing erect, seated -20° 1 also if trunk is well supported while seated 2 if not - between 0° to +20° (2) - between 20° to 60° (3) - up to 40° + (4) If trunk is twisted +1 If trunk is side-bending +1			**Trunk** Trunk bending forward down to +90° (1) Trunk bending backward between +90° to +105° (1) Trunk twisting up to +10° (1)
BACK			**Back support** Proper back support a lumbar support and chair is reclined between 95 and 110 (1) No lumbar support (2) Back support is reclined too far (greater than 110) (2) No back support (i.e., stool or improper sitting posture) (2) Back support is non-adjustable (+1)	**Back support** No lumbar support (1) Back support is reclined too far (greater than 110) (1) No back support (i.e., stool or improper sitting posture) (1)
ARM	**Upper arm position** Locate Upper Arm Positon: - between -20° to +20° (1) - down to >-20° (2) - between +20° to +45° (2) - between +45° to +90° (3) - up to 90°+ (4) If shoulder is raised: +1 If upper arm is abducted: +1 If arm is supported or person is leaning: -1 **Lower arm position** Locate LowerArm Position: - between -60° to +100° (1) - between 0° to +60° (2) - up to 100°+ (2) - across (4) If arm is working across midline of the bodi +1 I farm out to side of body +1		**Armrest** Elbows are supported at 90 , shoulders are relaxed (1) Armrests are too high (shoulders are shrugged) (2) Armrests are too low (elbows are not supported) (2) Armrests are too wide (elbows are not supported, or arms are abducted while using the armrests (+1). The armrests have a hard or damaged surface e creating a pressure point on the forearm (+1) Armrests or arm support is non-adjustable (+1)	**Shoulder** Shoulder is raised (1) **Upper arm position** Position of upper arm: - down to >-20° (1) - between +20° to +60° without support (1) - up to 90°+ (1) **Lower arm position** Position of lower arm: - between 0° to +60° (1) - up to < +100° (1)
WRIST	**Wrist** Locate Wrist Position: - it is 0° (1) - between 0° to +15° (2) - up to +15°+ (3) - between 0° to -15° (2) - up to -15°+ (3) - across (2) If wrist is bent from the midline +1 Wrist Twist If wrist is twisted mainly in mid-range =1 If wrist at or near end of twisting range =2		**Mouse** Mouse in line with the shoulder (1) Reach to mouse/mouse not in line with the shoulder (2) Pinch grip required to use mouse/mouse too small (+1) Mouse/keyboard on different surfaces (+2) Hard palm rest/pressure point while mousing (+1) **Keyboard** Wrists are straight, shoulders are relaxed (1) Wrists are extended beyond 15 of extension (2) Wrists are deviated while typing (+1) Keyboard too high e shoulders are shrugged (+1) Keyboard platform is non-adjustable (+1)	**Wrist** Wrist without rest (1) Position of wrist - between 0° to -90° (1) - between +15° to +90° (1) Wrist twist Wrist is twisted mainly in mid-range up to +/- 20° (1)
LEG	**Legs** If legs & feet supported and balanced +1 If not +2		**Chair** Knees bent to approximately 90° (1) Seat too low a knee angle less than 90° (2) Seat too high a knee angle greater than 90° (2) No foot contact with ground (3) Insufficient space for legs beneath the desk surface (+1) Seat pan height is non-adjustable (+1) Pan depth (3) Approximately 7.5 cm of space between the edge of the chair and the back of the knee (1). Seat pan length too long, less than 7.5 cm of space between the edge of chair and the back of the knee (2). Seat pan too short, more than 7.5 cm of space between the edge of chair and the back of the knee (2). Seat pan depth is non-adjustable (+1).	**Leg** Leg flexion rate Seat too low a knee angle less than 90° (1) Seat too high a knee angle greater than 90° (1) No foot contact with ground (1) Seat pan depth (3)

3.1   Data Collection

In the pilot a total of 20 workstation was assessed at Hungarian companies operating in different industries, including secretary and IT workstations of the Chamber's College and of an Educational Institution, and an onsite office of a construction company.

The workplaces was selected to represent the diversity both jobs and work activities and worker characteristics, therefore the sample consisted of 9 women and 11 men. Based on the work content the following three jobs categories was created: administrative (8 person), clerk (5 person), middle manager (7 person).

The precondition of the 4 ergonomic assessment tools was met for all the 20 workstation. For example CERA Office is applicable to 4 to 8 h computer work which involves repetitive movements and visual screen and data input device is used. At every workstation in the study data input devices were separate from the screen and the data input devices were located at different distances and in the working height.

The pilot was part of real ergonomic workplace evaluations, therefore the assessment was done according to the real, state of the art situation without warning workers in advance to adjust workstations to their best knowledge.

3.2   Results

The number of observations completed were 20 workstation. The participants used for statistical analyses for each ergonomics assessment tool. However, a smaller sample size n = 20 was used when testing the relationship between the ergonomics assessment tools and the musculoskeletal outcome variable matched. Participants' job type demographics consisted of: administrative (40%), clerk (25%), and middle manager (35%). The ratio of the sitting-standing (walking) postures of the work activity was measured which influence the direct risk factors and also modifies the duration of static body positions and work involving communication and telephony (Table 2).

Table 2. The research sample

Average time in shift	Sit at the desk	Stand next to desk	Walk	Telephony
Administrative (8 person)	90%	4%	5%	1%
Clerk (5 person)	73%	3%	22%	2%
Middle manager (7 person)	63%	2%	34%	1%

Table 3 compares the results of the ergonomics assessment tools. Testing of the within-subject effect was conducted using the non-parametric measure comparison; results are shown in Table 3. Non-parametric tests showed a difference within using ergonomics risk method (with a significance of $p < 0.05$) but no difference was appointed in between assessed by risk assessment methods expect where RULA and CERA was compared. The result is a moderately strong, significant ($p < 0.02$) correlation ($r = 0.644$) between RULA and CERA assessment tools.

The body postures of ergonomic assessment tools were compared with each other. Table 4 shows the results of trunk showed very strong correlation between two

Table 3. Ergonomic assessment tools compared (N = 20)

		CMDQ	RULA	ROSA	CERA
CMDQ	Correlation Coefficient	1.000	.158	.278	.046
	Sig. (2-tailed)	.	.507	.235	.846
RULA	Correlation Coefficient		1.000	−.198	.644**
	Sig. (2-tailed)		.	.402	.002
ROSA	Correlation Coefficient			1.000	.044
	Sig. (2-tailed)			.	.852
CERA	Correlation Coefficient				1.000
	Sig. (2-tailed)				.

**. Correlation is significant at the 0.01 level (2-tailed).
*. Correlation is significant at the 0.05 level (2-tailed).

methods (RULA and CERA Office) (r = 0.768, p < 0.000). The results of arm, shoulder (r = 0.473, p < 0.035) and the wrist (r = 0.733, p < 0.000) showed strong correlation. The results of our studies carried out on a small sample indicate correspondence between the two factors.

Table 4. The body postures of ergonomic assessment tools compared (N = 20)

		CMDQ	RULA	ROSA
NECK	Correlation Coefficient	−.405	−.049	.079
	Sig. (2-tailed)	.076	0.838	.739
TRUNK	Correlation Coefficient	.128	.768**	.402
	Sig. (2-tailed)	.590	.000	.079
LEG	Correlation Coefficient	.064	.266	.651**
	Sig. (2-tailed)	.788	.258	.002
LUMBAR	Correlation Coefficient	.380	.	.361
	Sig. (2-tailed)	.099	.	.118
ARM, SHOULER	Correlation Coefficient	.117	.473*	.356
	Sig. (2-tailed)	.623	.035	.124
WRIST	Correlation Coefficient	.157	.733**	−.352
	Sig. (2-tailed)	.510	.000	.128

4 Conclusion

In everyday practice, there is a need for methods that evaluate the ergonomic risk of different work activities in a complex manner and also enable self-assessment of the work for the workers. CERA office is such a method to evaluate the ergonomic risks of computer workplaces. In CERA office on one hand, the direct factors are based primarily on the body position based on the EN 1005-4 standard and the indirect risk factors according to the requirements of the screen directive and computer workplace requirements.

CERA office and all the similar ergonomics methods are well used and necessary to evaluate working conditions, monitor and determine appropriate measures, as well as to improve user behaviour.

References

1. Directive 90/270/EEC - display screen equipment of 29 May 1990 on the minimum safety and health requirements for work with display screen equipment (fifth individual Directive within the meaning of Article 16 (1) of Directive 89/391/EEC)
2. Szabó, Gy (2012) Evaluation and prevention of work-related musculoskeletal disorders in Hungary. In: Ahran TZ, Waldemar K (eds) Advances in physical ergonomics and safety. CRC Press - Taylor and Francis Group, Orlando, pp 195–202
3. Szabó Gy, Mischinger G (2013) Just an other ergonomic tool: the 'Composite Ergonomic Risk Assessment'. In: Mijović B, Čubrić IS, Čubrić G, Sušić A (eds) Ergonomics 2013: 5th international ergonomics conference. Croatian Academy of Sciences and Arts, Zadar, pp 169–174
4. Szabó Gy (2014) ErgoCapture – a motion capture based ergonomics risk assessment tool. In: Abraham T, Karwowsky W, Marek T (szerk.) Proceedings of the 5th international conference on applied human factors and ergonomics 2014 and the affiliated conferences, Krakow, pp 8023–8031
5. Szabó Gy (2017) Composite ergonomic risk assessment of municipal manual waste sorting. In: 8th international conference on applied human factors and ergonomics, Los Angeles, California, USA (2017)
6. EN 1005-4:2005 + A1:2008 Safety of machinery - Human physical performance - Part 4: Evaluation of working postures and movements in relation to machinery
7. Erdinc O, Hot K, Özkaya M (2008) Cross-cultural adaptation, validity and reliability of Cornell Musculoskeletal Discomfort Questionnaire (CMDQ), Research report
8. Hignett S, McAtamney L (2000) Rapid entire body assessment: REBA. Appl Ergon 31:201–205
9. Sonne MWL, Villalta DL, Andrews D (2012) Development and evaluation of an office ergonomic risk checklist: the rapid office strain assessment (ROSA). Appl Ergon 43(1):98–108

The Development of Resilience Management Guidelines to Protect Critical Infrastructures in Europe

Luca Save[1,5(✉)], Matthieu Branlat[2,5], William Hynes[3,5],
Emanuele Bellini[4,5], Pedro Ferreira[4,5], Jan Paul Lauteritz[5,6],
and Jose J. Gonzalez[5,7]

[1] Deep Blue srl, Piazza Buenos Aires 20, 00198 Rome, Italy
luca.save@dblue.it
[2] SINTEF, Strindveien 4, 7034 Trondheim, Norway
[3] Future Analytics Consulting, 23 Fitzwilliam Square (South), Dublin 2, Ireland
[4] University of Florence, Via S. Marta 3, Florence, Italy
[5] IST-CENTEC, University of Lisbon, Avenida Rovisco Pais, 1049-001 Lisbon, Portugal
[6] Fraunhofer IAO, Nobelstraße 12, 70569 Stuttgart, Germany
[7] CIEM, University of Agder, Serviceboks 509, 4898 Grimstad, Norway

Abstract. The capability to be resilient in the face of crises and disasters is a topic of highest political concern in Europe especially as far as critical infrastructures and urban environments are concerned. Critical infrastructures are systems or part of systems essential for the maintenance of vital societal functions, the disruption or destruction of which would have a significant impact on the well-being of people. Examples of them are transportation services, energy infrastructures, water and wastewater systems, health and emergency services, financial services, communication infrastructures, etc. The symposium focuses on the experience of four different projects funded under the Horizon 2020 Programme: DARWIN, RESILIENS, RESOLUTE, SMR. The projects are all dealing with the application of resilience engineering, community resilience and urban resilience concepts to concrete examples of crises and situations of emergency. Such principles are translated into guidelines covering different resilience abilities that the organizations managing critical infrastructure should possess.

Keywords: Resilience engineering · Critical infrastructures · Urban resilience

1 Topics of the Symposium

The capability to be resilient in the face of crises and disasters is a topic of highest political concern worldwide especially as far as critical infrastructures and urban environments are concerned. Critical infrastructures are systems or part of systems essential for the maintenance of vital societal functions, the disruption or destruction of which would have a significant impact on the well-being of people, with major concerns for political institutions at national, supranational or local level. Examples of

© Springer Nature Switzerland AG 2019
S. Bagnara et al. (Eds.): IEA 2018, AISC 819, pp. 598–606, 2019.
https://doi.org/10.1007/978-3-319-96089-0_65

them are transportation services, energy infrastructures, water and wastewater systems, health and emergency services, financial services, communication infrastructures, etc. Such infrastructures are endangered by both natural hazards (e.g. floods, storms, earthquakes, volcanoes and tsunamis) and man-made threats, with the second category including events caused by a deliberate action of the human (e.g. terrorist attacks) or by unwanted accidents (e.g. organizational accidents triggered by technical failures and human errors). In recent years, the topic of improving resilience capabilities has acquired increasing importance in the research agenda of the EU and of its Member States and Associated Countries, also as a consequence of lessons learnt from major crises and disasters. As a confirmation of such increasing interest, the present Symposium has been organized to compare the experience of four different research projects dealing with the development of Resilience Management Guidelines for Critical Infrastructures, under the Horizon 2020 Programme:

- DARWIN (Expect the unexpected and know how to respond).
- RESILIENS (Realising European Resilience Critical for Infrastructure)
- RESOLUTE (RESilience management guidelines and Operationalization applied to Urban Transport Environment)
- SMR (Smart Mature Resilience).

The experiences of the different projects are dealing with the application of resilience engineering, community resilience and urban resilience concepts to concrete examples of crises and situations of emergency. Such principles are translated into guidelines covering different resilience abilities that the organizations managing critical infrastructure should possess. These includes the abilities to: (a) anticipate a crisis or emergency, (b) monitor the factors that could trigger a crisis, (c) respond and adapt to the consequences of a crisis, (d) learn from a crisis which has already occurred.

2 The Development of Resilience Management Guidelines in the DARWIN Project

2.1 Background

H2020 project DARWIN aims to build resilience management guidelines to support organisations in developing and enhancing their resilience in the context of crisis management. During the first 6 months of the project, a vast review of associated literature, standards and operational documentation, as well as interviews of practitioners, was undertaken. A significant number of requirements were identified to guide the subsequent development of the DARWIN guidelines. Those requirements included especially conceptual requirements that captured resilience management capabilities the guidelines are addressing. The presentation will describe the nature of the guidelines, the process of their development and its results, achievements and limitations.

2.2   Nature of the DARWIN Resilience Management Guidelines (DRMG)

The context is that of organizations that already have a number of processes and tools in place to support their management of crises (e.g., preparation activities, contingency plans, procedures, learning activities). As a result, the guidelines are positioned at a meta-level: rather than replacing them, they provide an approach to adopt a critical view and revise or develop such processes and tools. Such view is based on research and practice on resilience management inspired by the fields of Resilience Engineering and Community Resilience.

2.3   DRMG Development

The guidelines' development follows a 4-step process established to be collaborative and iterative, and to include operational input early and as often as possible. The process changed and solidified during the course of the project as a result of the evolving understanding of what type and content of guidelines would be useful to develop and of how to produce such guidelines while fulfilling the various objectives of the project.

The following points are essential aspects of the development of the guidelines:

- The building blocks are individual pages that address specific topics about resilience management and propose interventions in order to develop and enhance the corresponding capabilities captured in the conceptual requirements. They are at the heart of the guidelines development process described in this document. As a whole, the guidelines build on the topics by organising and relating them, because the capabilities they refer to are not independent.
- Operational perspectives are incorporated through the participation of end-user practitioners and experts from Air Traffic Management, healthcare and other Critical Infrastructures as co-creators throughout the development life-cycle.
- A Knowledge Management platform facilitates the development, management and use of the guidelines. The platform offers opportunities to reconsider common views on the nature of guidelines, their necessary evolution and their multi-faceted, multi-purpose content.

2.4   Acknowledgments

The development of the DARWIN guidelines is a collaboration between SINTEF (Norway), Deep Blue (Italy), FOI (Swedish Defence Research Agency), Ben-Gurion University (Israel), ENAV (Italian air navigation service provider), ISS (Italian National Institute of Health) and KMC (Centre for Teaching & Research in Disaster Medicine and Traumatology, Sweden). It has benefited from the involvement of experts from the DARWIN Community of Practitioners (DCoP). The research leading to the results received funding from the European Union's Horizon 2020 research and innovation programme under grant agreement number 653289. Opinions expressed in this publication reflect only the author's view. The European Commission is not liable for any use that may be made of the information contained in this paper.

Project website: https://h2020darwin.eu/.

3 The Pilot Evaluation of Resilience Management Guidelines in the DARWIN Project

3.1 Background

H2020 project DARWIN aims to build resilience management guidelines to support organizations managing critical infrastructures to develop and enhance their resilience in the context of crisis management. The project focuses on the Healthcare (HC) and Air Traffic Management (ATM), although other sectors are involved via a large DARWIN Community of Practitioners (the DCoP) and by taking into account the infrastructures that may suffer, as a cascading effect, the consequences of a crisis or emergency occurring in the HC and ATM domain before the end of the project and which actions were recommended to implement them.

3.2 The Guideline Evaluation Process

The DARWN Resilience Management Guidelines (DRMGs) can be accessed via a WIKI platform and are composed of so-called Concept Cards (CC), each one linked to a resilience management principle. The CCs suggest a number of actions that an organization managing a critical infrastructure should perform to revise or integrate its own polices and guidelines, in order to increase its level of resilience. The evaluation of the CCs included a variety of exercises, with the core part consisting of four pilot exercises conducted in Italy and Sweden, in collaboration with three end-user organizations participating to the DARWIN consortium. The organizations were ENAV (the Italian air navigation service provider), ISS (Italian National Institute of Health) and KMC (Centre for Teaching & Research in Disaster Medicine and Traumatology, Sweden). The feedback collected during the Pilot Exercises was the main input of a Summative and Formative Evaluation process used to determine which CCs required more improvements before the end of the project and which actions were recommended to implement them.

3.3 The Pilot Exercises

Three pilot exercises were organized in Rome (Italy) and one in Linköping (Sweden). Each of them was associated to a reference scenario, representing a situation of crisis or emergency:

- Pilot 1 "Aircraft crashing in urban area close to Major Italian airport shortly after taking off"
- Pilot 2 "Total Loss of Radar Information at Rome Area Control Centre"
- Pilot 3 "Disease Outbreak During an Incoming Flight"
- Pilot 4 "Collision between Oil Tanker and Passenger Ferry leaving Gotland islands in severe weather conditions".

The scenarios were intended as representative situations to reflect on how to apply the principles and actions indicated in a selected number of CCs. However, the resilience practitioners involved in the exercises - and representing a large variety of organizations - were free to propose also different scenarios, based on their individual experiences. The exercises were organized in different evaluation sessions, depending on the specific CC the attendants were asked to use and to the method adopted to get prepared to the management of the crisis. Some of these consisted of operative techniques, such as Command Post Exercises and Table Top Exercises. Others were more speculative in nature, such as checklist-based evaluations, used in the context of dedicated workshops.

3.4 Acknowledgments

The DARWIN partners taking part in the evaluation of guidelines were the following: Deep Blue (Italy), ENAV (Italian air navigation service provider), FOI (Swedish Defence Research Agency), ISS (Italian National Institute of Health) and KMC (Centre for Teaching & Research in Disaster Medicine and Traumatology, Sweden), TUBS (Braunschweig University of Technology, Sweden). The evaluations has benefited from the involvement of experts from the DARWIN Community of Practitioners (DCoP). The research leading to the results received funding from the European Union's Horizon 2020 research and innovation programme under grant agreement number 653289. Opinions expressed in this publication reflect only the author's view. The European Commission is not liable for any use that may be made of the information contained in this paper.

Project website: https://h2020darwin.eu/.

4   The Development and Application of a European Resilience Management Guideline in the RESILENS Project

4.1   Background

Critical infrastructure (CI) provides the essential functions and services that support European societal, economic and environmental systems. As both natural and man-made disaster and crises situations become more common place, the need to ensure the resilience of CI so that it is capable of withstanding, adapting and recovering from adverse events is paramount. Moving resilience from a conceptual understanding to applied, operational measures that integrate best practice from the related realm of risk management and vulnerability assessment is the focus of the RESILENS project. RESILENS (Realising European ReSILiencE for CritIcaL INfraStructure) developed a European Resilience Management Guideline (ERMG) to support the practical application of resilience to all CI sectors. Accompanying the ERMG is a Resilience Management Matrix and Audit Toolkit which enables a resilience score to be attached to an individual CI, organisation (e.g. CI provider) and at different spatial scales (urban, regional, national and transboundary) which can then be iteratively used to direct users to resilience measures that will increase their benchmarked future score.

4.2 Nature of the RESILENS Resilience Management Guidelines (ERMG)

The ERMG and resilience management methods was tested and validated through stakeholder engagement, table-top exercises and three large-scale pilots (transport CI, electricity CI and water CI). The ERMG and accompanying resilience methods are hosted on an interactive web-based platform, the RESILENS Decision Support Platform (RES-DSP). The RES-DSP also hosts an e-learning hub that will provide further guidance and training on CI resilience. Overall, RESILENS aims to increase and optimize the uptake of resilience measures by CI providers and guardians, first responders, civil protection personnel and wider societal stakeholders of Member States and Associated Countries.

4.3 ERMG Development

The first Draft ERMG was based on an integral master document composed of the following chapters: Introduction and instructions on how to use the document; Professional terminology; Generic chapters dealing with resilience management for CI: requirements; Specific chapters on resilience management by case sectors. E.g. power utilities, energy (Gas & Oil), transportation (surface transport, aviation, maritime), telecommunication, water and sewage; Finally instructions for the measurement, evaluation and audit of resilience of CI.

The ERMG provides standardized terminology recommendations within and across sectors and a methodological approach to the development of an integration matrix, derived and taken from risk assessment and resilience management across various CI sectors, rescuers and other end users. Key to success is that the final, validated and harmonised ERMG addresses the needs and requirements of end-users (CI providers, rescuers and the wider society), and can be operationalised in a variety of sectors.

Project Website: http://resilens.eu/.

5 The RESOLUTE Project: Guidelines for Resilience Control Room Design in the Smart Cities

5.1 Background

RESOLUTE is based on the vision of achieving higher sustainability of operations in European UTS. The project recognises foremost the ongoing profound transformation of urban environments in view of ecological, human and overall safety and security needs, as well as the growing importance of mobility within every human activity.

RESOLUTE considers resilience as a useful management paradigm, within which adaptability capacities are considered paramount. This led to the creation of the European Resilience Management Guidelines (ERMG), aiming to support cities in developing the right capacities and monitoring the resource availability to sustain its adaptive capacity. To this end, the ERMG development has adopted a system's perspective, by applying the Functional Resonance Analysis Method (FRAM) to model a generic CI and to identify which are the desired functions and the related interdependencies that should be managed towards the improvement of resilience in a CI.

5.2 The ERMG Operationalization for Resilience Control Room

The ERMG were operationalized exploiting big data generated by Smart City features. The main outcome is represented by the Collaborative Resilience Assessment and Management Support System (CRAMSS). The CRAMSS aims at improving the communication among involved actors, including organizational decision makers and citizens.

The growing complexity of socio-technical systems requires a shift from centralized decision making to distributed and collaborative decision making. To this end, cooperation and coordination mechanisms become fundamental, namely between multiple operational control centers of CI stakeholders. Control room systems and processes design becomes fundamental to ensure accurate and context related flows of information (both pull and push).

5.3 The Resilience Control Room Design Recommendation

The development of CRAMSS and all the related RESOLUTE outputs were steered by a set of information and control room design principles. These can be summarized as follows:

- Suitability for the task's nature and performance conditions
- Self-descriptiveness (being self-explanatory)
- Controllability
- Conformity with user expectations
- Error tolerance (ability to reverse and correct decisions)
- Suitability for customization
- Suitability for learning

These systems requirements were called on throughout project development and they reflect on project achievements. From a resilience perspective, human centred design principles are fundamental to ensure that human operators and, in the case here addressed, decision makers at control rooms, are able to cope with complexity and uncertainty towards enhanced adaptive capacities within the operation of CI.

5.4 Acknowledgments

This work has been supported by the RESOLUTE project (www.RESOLUTE-eu.org) and has been funded within the European Commission's H2020 Programme under contract number 653460. Opinions expressed in this publication reflect only the authors' view. The European Commission is not liable for any use that may be made of the information contained in this paper.

Project Website: http://www.resolute-eu.org.

6 Contribution of the SMR (Smart Mature Resilience) Project Towards the European Resilience Management Guideline

6.1 Background

The project H2020 SMR – Smart Mature Resilience, along with four other H2020 projects on resilience funded by the topic DRS-7-2014: Crisis management topic 7: Crises and disaster resilience – operationalizing resilience concepts, is a contributor to the set-up of the European Resilience Management Guidelines.

6.2 Consortium

The SMR consortium consists of four universities (Agder, Norway; Linköping, Sweden; Strathclyde, UK; Tecnun, Spain), seven cities (Bristol, Glasgow, Kristiansand, Riga, Rome, San Sebastian, and Vejle), the German Institute for Standardization– DIN, and ICLEI – Local Governments for Sustainability. The fact that four of the cities (Bristol, Glasgow, Rome and Vejle) belong also to the 100 Resilient Cities initiative, by the Rockefeller Foundation, created synergy and added value to SMR.

6.3 Development of Tools

One of the outstanding results of SMR is the development of specific tools for the assessment and enhancement of Urban Resilience. The tools have been tested and validated in the cities participating in the project.

The tools developed during the project are:

- The Risk Systemicity Questionnaire (RSQ) is an advanced participatory instrument aimed at promoting discussion and raise awareness on risks and resilience challenges cities are facing.
- The Maturity Model (MM) is a framework model adapt to define, on the basis of measurable and qualitative parameters, the resilience level of a city, along with policies to advance the resilience level.
- The City Resilience Dynamics Model is a training tool that helps cities explore different strategies regarding the implementation of resilience policies, simulate the results of each strategy and learn about the resilience building process.
- The Resilience Building Policies tool complements the SMR Maturity Model, the Risk Systemicity Questionnaire and the City Resilience Dynamics tool by providing a collection of case studies as reference to cities for further information.
- The SMR Resilience Information Portal provides a collaborative environment in order to facilitate awareness and engagement among key partners in resilience building.

In addition, work was carried out in standardization of resilience concepts, to formally define the terms and indicators to be utilized in this new science. Resilience – and urban resilience, in particular – is a sector nowadays under development, and definition of its standards is an unquestionable need.

6.4   Acknowledgments

Smart Mature Resilience has received funding from the European Union's Horizon 2020 research and innovation programme under grant agreement number 653569.

Project Website: http://smr-project.eu/home/.

Averting Inadequate Formulations During Cause Analysis of Unwanted Events

Jean-François Vautier[1], Guillaume Hernandez[1(✉)],
Catherine Sylvestre[2], Isabelle Barnabé[3], Stéphanie Dutillieu[4],
Michèle Tosello[4], Cécile Lipart[1], Virginie Barrière[1],
Jean-Marc Jullien[5], Christophe Dufour[1],
and Diana Paola Moreno Alarcon[6]

[1] CEA/Paris-Saclay, Paris, France
guillaume.hernandez@cea.fr
[2] CEA/Valduc, Salives, France
[3] CEA/CESTA, Le Barp, France
[4] CEA/Cadarache, Cadarache, France
[5] CEA/Marcoule, Chusclan, France
[6] MINES ParisTech, Paris, France

Abstract. This paper provides feedback of inadequate ways of formulating the causes of an unwanted event in the high risk industry field. The elements we studied originated from unwanted events analyses reports from the CEA (French Alternative Energies and Atomic Energy Commission) and from worldwide accident analyses exercises that have been carried out by students. This communication details two ways of inadequate formulations: the first expressed as a "lack of" a safety provision that was not requisite throughout the duration of the unwanted event; and the second expressed as "comments". This article provides tips to identify each type of inadequate formulation and proposes methods to prevent the occurrence and reoccurrence of such formulations.

Finally, the discussion addresses how an adequate formulation of "the causes" allows the analyst to focus on the real causes of the unwanted event thus not only improving the quality of the investigation but also promoting a more in-depth analysis. More specifically, we emphasize the importance of distinguishing between the search of the possible causes from the identification of the possible sanctions of an unwanted event.

Keywords: Cause analysis · Unwanted event · Safety

1 Introduction

The purpose of this paper is to present feedback on inadequate ways of formulating the causes of an unwanted event in the high risk industry field. The formulations of causes discussed in this document, were chosen as they do not conform to the specifications or requirements provided by industrial guidelines [1–3] - guidelines which detail appropriate ways to formulate the causes of unwanted events. The non-compliant elements of these cause analyses originated from both unwanted events analyses from the CEA

© Springer Nature Switzerland AG 2019
S. Bagnara et al. (Eds.): IEA 2018, AISC 819, pp. 607–612, 2019.
https://doi.org/10.1007/978-3-319-96089-0_66

(French Alternative Energies and Atomic Energy Commission) as well as from worldwide accident analyses exercises that have been carried out by students.

INRS [3] indicates that a cause is a fact i.e. information, state or action expressed in a concise manner. It is something that occurs with certainty and whose existence can be verified. The document further specifies two inadequate ways of formulating a cause:

- By detailing the lack of something, a negative expression "not ...". In other words it is something that did not occur;
- By detailing an opinion, an interpretation or a value judgement of something that occurred. In this case, it is the expression of the cause from the perspective of one analyst. In other words, it is a comment on a cause, and hence likely to vary from one analyst to another.

This paper builds on this pre-existing framework and aims to further clarify and illustrate the aforementioned two ways of inadequately formulating causes of unwanted events. The first part of this feedback analysis provides examples of inadequate formulation of causes resulting from a "lack of" and proposes methods to prevent its occurrence. The second part discusses inadequate formulation of cause through the use of "comments" and proposes methods to counteract this. Finally a discussion is presented.

2   Inadequate Way of Formulating a Cause by Expressing the "Lack of" a Safety Provision that Was not Required at the Moment of the Unwanted Event

2.1   Context

The first inadequate way of formulating a cause is connected to the forms of reasoning emphasizing the lack of safety provisions.

In this particular case, the analyst misidentifies a lack of means at the moment of the event for a cause. Particularly, as the problem or unwanted event may have been adverted if the means had been implemented. An example is the lack of maintenance of a machine containing broken critical pieces and thus resulting in the total breakdown of the machine, particularly as the maintenance would have identified the broken pieces. Nevertheless, this lack of maintenance cannot be considered as a cause if no maintenance of the machine's critical pieces had been required until the moment of the unwanted event occurred. Particularly as we previously saw that a cause is something, which exists at the moment of the unwanted event.

Moreover, in addition to the lack of maintenance, another analyst could have conceived failure to replace the machine as a cause as, from his point of view, the machine was far too old and not replaced as a prevention with a newer one (even if this preventive measure was not required at the moment of the unwanted event).

In the same manner, the machine's brake-down could have been ascribed to an inadequate design or production of the critical pieces, despite the designers meticulously execution of the required methods during the design process.

2.2   Tips to Identify This Type of Inadequate Formulation

The lack of a safety provision may be expressed with a conditional form of the verb. For example, "the machine broke down because it should have been properly maintained and this maintenance was not carried out". In this case, we do not know if this safety provision (in this example the maintenance) was required or not.

The lack of a safety provision may also be explicitly expressed by "Lack of…" "Non…" "… not…". However, using this type of language fails to provide information regarding the actions or the state of parameters that existed at the moment of the unwanted event. Similarly, indicating that a worker suffered a hand injury since, he did not wear protecting gloves fails to provide information whether wearing the gloves was a requirement or not.

2.3   Propositions to Prevent the Occurrence of This Inadequate Formulation

1. When the conditional is used in a causal analysis it is important to question if the safety provisions was required or not. If yes, the preterit form of the verb should be employed.
2. Use forms that indicate "omitting something" rather than "not carrying out something". In the same way employ the terms "performing only operation A and B" or "focusing only on something" rather than using "non-compliance of a safety rule". The objective is to propose active forms that describe what happened rather than what did not happened and was expected.

In all cases, always question the conditions that took place when the event occurred (and not what could have taken place at the moment of the unwanted event) [4]. One objective of this is to avoid the retrospective bias. It is a cognitive error of judgment that indicates the tendency of people to retrospectively overestimate the fact that events could have been anticipated with more foresight. The most common example is the estimation ex-post-facto that specific maintenance requirements should have been incorporated in a safety rule set, despite the designers' unawareness of the need to incorporate these maintenance procedures into the rule set at the time they were written.

Therefore, it is necessary to verify if the safety provision was required at the moment of the unwanted event. This is important information as we may only consider this lack of a safety provision to be a cause, if the safety provision was required at the moment of an unwanted event. For example, not wearing a required helmet may be considered as a cause of an injury if it was required (at the time of the unwanted event) and if the helmet could have prevented the injury.

3   Inadequate Way of Formulating a Cause Related to the "Comments"

3.1   Context

The second inadequate way of formulating a cause is connected to the expression of comments surrounding the cause; comments that may be mistaken as a cause by some analysts [5].

More precisely, several types of comments may be considered:

- an expression using a portmanteau word,
- an expression using a word related to an emotion,
- an expression using a legal term or word.

3.2   Tips to Identify This Type of Inadequate Formulation

A comment surrounding the cause may be expressed using a portmanteau word. It concerns some words such as "climate of safety" or other expressions such as "his mind was elsewhere". They are at times considered as a unique explanation of the problem that occurred, thereby often preventing analysts from examining the root causes, since these comments resemble sentences or conclusions.

Another kind of comment may be expressed with words related to an emotion. It concerns the use of certain verbs that amplify an idea or a state of something. For example, employing verbs such as "the earthquake devastated the entire plant" rather than stating that "the plant was destroyed by the earthquake".

Finally, a third kind of comment may be articulated using legal terms, such as expressing a cause using the judge's terminology. For example, directly transposing the judge's terminology by employing legal terms such as the notion of "liability" or the "fault of someone" to describe the cause.

3.3   Propositions to Prevent the Occurrence of This Inadequate Formulation

1. Verify that the cause is not formulated with portmanteau words, but particularly that this "cause" is not the last in a chain of causes. Particularly, as such words have an "ability" to impede the "movement" of examining further in-depth organizational causes.
2. Delete all the words within the lexical field of empathy or affect, especially as these emotions inhibit the analysis of all the determinants of the activity. For example, in the case of "the operator was in distress and could not see the red alarm light up", the term "in distress" is an interpretation and not a fact nor an analysis.
3. Continuously search to explain the facts that occurred rather than finding liability of someone. Particularly, a judge never convicts a tool for the occurrence of a breakdown but someone may be convicted for that, whereas in a real causal analysis the characteristics of the tool will be considered as technical causes of the occurrence of the event.

In all cases, it is essential to build a team of analysts and to manage the interaction between all team members in order ensure that all members' viewpoints are considered, thus preventing the projection of only a single member's vision.

4 Discussion and Conclusion

This part of the paper discusses how adequate formulation of causes may improve the quality of analyses by focusing on the real causes of an unwanted event and promoting an in depth analysis [6].

During the analysis of an unwanted event, it is important to first focus on explaining the occurrence of an event (i.e. on the conditions that induce this event) before contemplating the actions that would have improved the situation or averted the event. For this reason, we highlight the shortfall of formulating a cause by expressing the "lack of" safety provision not required at the moment of the unwanted event.

Next, it is important to distinguish between the search for the possible causes from the identification of the possible sanctions of an unwanted event. Thus, the analyst has to refrain from using legal lexical field terms (fault, sanction, guilty, responsibility…). Particularly, as the concept of "fault" in law makes reference to sanctions attributed to the non-compliance of prescribed rules. It is therefore essential to differentiate between terminologies of different domains: the technical/human factors domain and the judicial field [7]. Moreover, the legal logic of repairing the damage caused by a work accident may lead company actors to consider the individual as responsible of negligence or lack of vigilance [8]. Therefore, this difference in approach also requires a great deal of pedagogical effort, as operators often experience feelings of guilt and a fear of sanctions following an event.

Finally, further research on the following research topics would be interesting:

- The significance of causal links (between two causes that are well defined).
- The benefits of using methods that position different formulations of a "comment" and compare them to the real causes. One such method is structural analysis, which studies the relationships between different variables [9, 10]. Certain elements (words) have a strong influence on certain variables (e.g. expression with portmanteau words), while other elements may have no influence at all.

References

1. Monteil B, Périgord M, Raveleau G (1985) Les outils des cercles et de l'amélioration de la qualité. Les Editions d'Organisation, Paris
2. Villemeur A (1988) Sûreté de fonctionnement des systèmes industriels. Editions Eyrolles, Paris
3. INRS (2013) L'analyse de l'accident du travail – La méthode de l'arbre des causes. ED 6163. INRS
4. Noyé D, Ravenne C (1987) Animer un cercle de qualité – Une équipe, un projet, une méthode. INSEP Editions, Paris

5. Galley M (2007) Six Common Errors when Solving Problems. ThinkReliability
6. Dien Y, Dechy N, Guillaume E (2012) Accident investigation: from searching direct causes to finding in-depth causes – problem of analysis or/and of analyst? Saf Sci 50(6):1398–1407
7. Bourcier D, De Bonis M (1999) Les paradoxes de l'expertise – Savoir ou juger? Institut Synthélabo pour le progrès et la connaissance, Le Plessis-Robinson
8. Garrigou A, Peeters S, Jackson M, Sagory P, Carballeda G (2004) Apports de l'ergonomie à la prévention des risques professionnels. In: Ergonomie. PUF, Paris, pp 497–514
9. Tenière-Buchot P-F (1989) L'ABC du pouvoir – Agir, Bâtir, Conquérir... (et sourire). Les Editions d'Organisation, Paris
10. Godet M (1985) Prospective et planification stratégique. Ed. Economica, Paris

Safety for Industry, Threat for Drivers? Insights into the Current Utility of Heath Assessments for Rail

Janine Chapman[1] , Joshua Trigg[2], and Anjum Naweed[2(✉)]

[1] Flinders Centre for Innovation in Cancer, Flinders University,
Bedford Park, Australia
[2] Appleton Institute for Behavioural Science, Central Queensland University,
44 Greenhill Road, Wayville, SA 5034, Australia
anjum.naweed@cqu.edu.au

Abstract. The rail driver workplace is full of challenges for effective health management, posing a significant threat to the sustainability of the industry. In Australia, train drivers undergo periodic health assessment as part of a nationally standardised approach to reducing sudden incapacitation risk; however, studies suggest that the current assessment protocol is not operating as effectively as they might. To improve this, there is a need to understand the experiences of drivers undergoing workplace health assessments, and how they engage with them.

Drawing on research of known barriers and enablers of positive health status, this study sought to examine train drivers' perceptions and experiences of recurring organisational health assessments and how they subjectively engage with this process. Five focus groups with train drivers ($n = 29$) were held across four Australian rail organisations, seeking to gain their understanding of the National Standard and their attitudes towards health assessments. Transcript data were subjected to thematic analysis.

Preliminary findings identified four primary factors: drivers' unmet information needs, low perceived assessment reliability and validity, need for psychological assessment and support, and the use of maladaptive assessment-threat avoidance strategies. This paper presents an overview of these preliminary findings and suggests that driver engagement with health assessment may be improved by proactively addressing these factors in occupational health initiatives and preventative interventions to tackle the growing problem of train driver health impairment.

Keywords: Rail industry · Risk and safety management · Occupational health

1 Introduction

The importance of train driver health for safety is evident from rail incidents associated with health-related factors. In Australia, the 2003 Waterfall train derailment occurred after the driver collapsed from a cardiac event at the helm, leading to many deaths and injuries [1]. This accident identified a requirement for better occupational health

© Springer Nature Switzerland AG 2019
S. Bagnara et al. (Eds.): IEA 2018, AISC 819, pp. 613–621, 2019.
https://doi.org/10.1007/978-3-319-96089-0_67

assessment processes recognizing the specific risks in this sector. A year later, Australia's National Standard for Health Assessment of Rail Safety Workers [2, 3], was introduced to obtain periodic individual health assessment of rail workers.

Research evaluating driver health status since the introduction of the health assessment has garnered mixed results. On the one hand, it reports reduced blood pressure, lowered cholesterol markers and significant decreases in self-reported smoking in train drivers, but on the other, no improvement in diabetes, and other cardiac risk factors [4]. Furthermore, it has shown an increase in body mass indicative of obesity [4, 5], and greater prevalence of obstructive sleep apnea [6]. Australian train drivers' obesity rates, and morbidity and mortality risk, has now exceeded that of the Australian general population [7].

Recent studies have identified numerous barriers impeding train driver health [8, 9]. Drawn from the driver's perspective, the studies demonstrate various individual-level health barriers, but also highlight a "systems" problem, meaning that significant change at the organisational and the job-design levels is required in order to successfully promote autonomous health management (see Fig. 1).

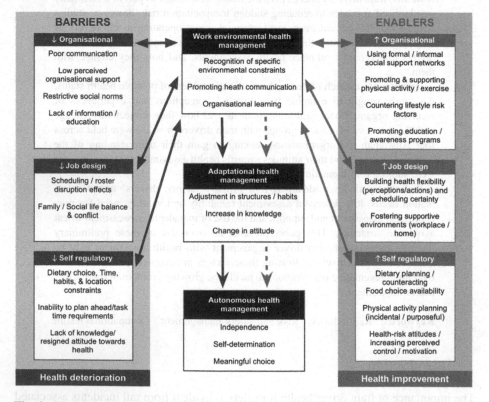

Fig. 1. Model showing the influence of barriers and enablers of positive health status on overall health and wellbeing in train drivers. The influence of organisational- and job-design levels point to the requirement for a system-level change [8].

These findings, along with the mixed success of the Standard and contradictions within the literature concerning train driver health status, [4–6] have served to place a spotlight on improving the existing health assessment. To inform policy and provide the best outcomes for rail safety, the Standard and assessment parameters are currently undergoing consultation with the medical community, consumer health groups, transport departments, industry groups and associations, and unions and regulators. This has included rail transport operators and their employees; however, little consultation has been undertaken with the rail workers themselves.

Inadequate input into the Standard and/or consideration of how system actors could come to view and engage with the health assessment may influence behaviour in ways that are complex and unpredictable. Whilst the importance of health assessments for reducing cardiac risk and sudden driver-incapacitation is clear, train drivers' own experiences of engaging with them is not. Train drivers may, for example, view and interact with the assessment process in meaningfully different ways than the designers of the Standard expect or intend. This has important implications, as train drivers are not only the focus of the Standard, but the object of the corresponding assessment. Gaining a better understanding of how train drivers currently engage with the assessment process will shed light on how the issue may be best tackled at a systems level.

1.1 Research Aims

The aim of this study was to examine Australian train drivers' perceptions and experiences of their health/medical assessments as featured within the Standard with the aim to understand how these experiences influenced their engagement with the process.

2 Methods

2.1 Study Design

This study adopted a qualitative exploratory research design, an approach that has been used to good effect in context of substantive transport research [e.g., 10–13].

2.2 Participants and Recruitment

Experienced Australian train drivers ($n = 29$, 26 (89%) male, $M_{age} = 46.5$ years, SD = 11.75) were recruited to participate. Four rail organizations supported the study and a total of five focus groups were held across Australia in New South Wales (2), Victoria (2), and South Australia (1). Participants were employed as train drivers in a full-time ($n = 19$) or part-time ($n = 10$) role.

2.3 Procedure

After consent, two moderators conducted each focus group with the stated purpose of capturing their 'understanding of the role of periodic health assessments in the rail context'. Discussion length varied based on size of the group and ranged from 80 min to over 2 h. Following ice-breaker questions, demographics and employment history, focal interview topics included experiences with the Standard and associated assessments (awareness and understanding; experience of assessments; attitudes towards assessments); and perceptions of health and wellbeing (lifestyle risk factors/diet/exercise; personal approach to health and wellbeing). Both open-ended and closed-format questions were used to capture individual perspectives.

After discussion of personal approaches to health management and health-risk mitigation, and how these approaches related to management of health and wellbeing within the broader context, participants discussed what they felt could be modified to benefit themselves and their organisations. The interview guide was refined to account for new information provided in each successive focus group. Focus groups proceeded until no new conceptual information was elicited (i.e. saturation) [14, 15]. Institutional ethical clearance was obtained from the human research ethics committee of Central Queensland University (Approval number: *H15/03-039*).

2.4 Data Analysis

Voice recordings were transcribed. Thematic analysis was used to inductively identify themes and subthemes [see 16, 17], with coding performed using NVivo (v. 10, QSR International) as follows: (i) transcripts were read closely; (ii) primary themes were coded if recurring across participants and focus groups; (iii) subthemes were iteratively coded into meaningful units; (iv) data, codes, and conceptual themes were constantly compared in an increasingly hierarchical manner to clearly describe concepts [e.g., 18]; (v) (sub)themes were discussed between researchers for consistency; and (vi) then distilled into overarching themes. As part of this last stage, each theme was embedded into a thematic network to usefully aid analysis by synthesising connections and capture an overarching theme for the findings [19].

3   Results

Preliminary findings from analysis identified four themes: (1) unmet information needs of drivers regarding the Standard and personal health; (2) drivers' perceptions of the reliability and validity of assessments; (3) need for expanded psychological wellbeing consideration within health assessments; and (4) maladaptive strategies used to manage feelings of personal threat engendered by assessments. Each theme is depicted in relation to the health assessments within the Standard and illustrated with sample quotes.

3.1   Unmet Information Needs

Nearly every participant in the study was unclear about the reasons for the Standard and why it was introduced. While they knew that an assessment was required, they expressed confusion relating to the purpose, focus, and practical application of the process, meaning that they had a limited understanding of what the assessment was looking for and what was expected of them. In the absence of such information, participants questioned the validity of the health assessment criteria and outcomes:

> You look at some blokes, and you can't figure out how they get through medicals. [They] can't fit through the back door of the cab, so how are they passing their medicals? [F2-P5]

> You think, that bloke can't be too healthy, but they get through their medicals. [There's] still an issue waiting to happen there for a few of them. [F4-P6].

While participants were aware that medical assessments focused on, and indeed, were able to pick up indicators related to cardiac issues and sudden incapacitation, some felt that they did not measure certain chronic health-risk indicators that they perceived to be of equal or greater concern:

> What do they test for? Glucose, cholesterol, and that's all. They don't test us for any signs of lung cancer or anything like that [and should] because we're near the diesel all the time. [F3-P8].

This reflected a potential misunderstanding about the purpose of the assessment, which is not clearly communicated to drivers.

3.2 Perceived Low Reliability and Validity of the Assessment Process

Attitudes towards the medical assessments were significantly shaped by drivers perceived low reliability and validity of the assessment process. Many participants, for example, thought that the BMI metric was imprecise and offered this as a reason for discounting the validity of medical assessments more generally:

> Features of [BMI] give you an idea of where you're at, but they are flawed ... it's very impersonal, it's generalised across the entire population where everyone's different. [F1-P2].

The value of workplace health assessments was also questioned, for example, in contrast to private healthcare consultation:

> [For] a health professional you go to your local doctor... You'd be in a lot of trouble if you relied on your work medical as your only guide to [your health]. [F3-P4].

Some drivers reported that external healthcare sources had provided conflicting information to that received at work, which had the effect of eroding the perceived credibility of workplace assessments:

> At my last [assessment] [the health assessor] said 'oh your cholesterol's 5.8'. I said ah 3 days ago I had a full blood test done, said I was 4.9. Isn't that strange how it can change in 3 days?' [The health assessor] goes 'oh don't worry about that one'. He wasn't concerned about a proper blood test that was done under proper fasting. [F3-P8].

3.3 Need for Psychological Wellbeing Assessment and Support

Participants felt that psychological wellbeing should be a greater priority in their occupation, such that the need for psychological wellbeing assessment and support within health assessments was a key theme, for example:

I just find it personally very demanding … [as] there's a constant, maybe low-level, anxiety attached to the job. [F3-P2].

The interconnection of physical and mental wellbeing was considered an important part of a holistic approach to health and a contributor to safety outcomes (T3, Q4). Participants had access to some support services, such as Employee Assistance Programs (EAPs), which were implemented to assist employees with personal and/or work-related problems. Participants felt that EAPs were useful, with many aware of how the program worked and the frequency of sessions per year. The general view, however, was that wellbeing was not being proactively supported by the organisation:

We've got a health and wellbeing officer just been appointed around 6 months ago and still nobody knows what [the officer] does. [F4-P3].

Drivers also did not feel that psychological health was featured proactively enough within the organisations, and thought more recognition should be made about the link between mind-body health in assessments:

[Assessments should] also focus on the mental side of things … [F5-P4].

It's your physical and mental state … it's a combination of both. [F2-P2].

3.4   Maladaptive Assessment-Threat Avoidance

There was a tendency to view assessments as a punitive monitoring device with potential to trigger adverse organisational responses. Assessments were perceived as a threat, which formed a recurrent theme. The perceived threat was to job security (T4, Q1), and was tied in with a lack of personal context for health assessors:

You do find the medical a bit of a threat, because if you do have a problem … basically your job's put on hold … [and] it could mean that you're no longer driving. So, it's sort of held against you… [F3-P5].

[Health assessors] don't know how we live our lifestyles, not like your [own] doctor does…[So] there's the threat that you could lose your job if they make the wrong decision. That's the only thing that I think about. [F3-P8].

The Authorised Health Professionals performing the assessments and fulfilling the assessor role were generally distrusted, and their relationship with train drivers was considered combative. The self-reported health information collected in the assessments was viewed by participants as something that could and would be used against them, rather than to support them to make healthy changes. Accurate health and lifestyle information was consequently omitted, or disclosed selectively:

You don't mention you're a smoker, when they ask you, you lie through your teeth and say 'no I'm not a smoker.' I haven't had an issue in a medical since. [F1P4].

You wouldn't go in [the medical] and offer information. [F3-P4].

4 Discussion

The aim of this study was to examine Australian train drivers' perceptions and experiences of their occupational health assessments to understand how these they influenced their engagement with the process. Figure 2 presents a diagram summarizing the themes from preliminary findings, illustrating how the participants' experiences influenced their perception of health assessments.

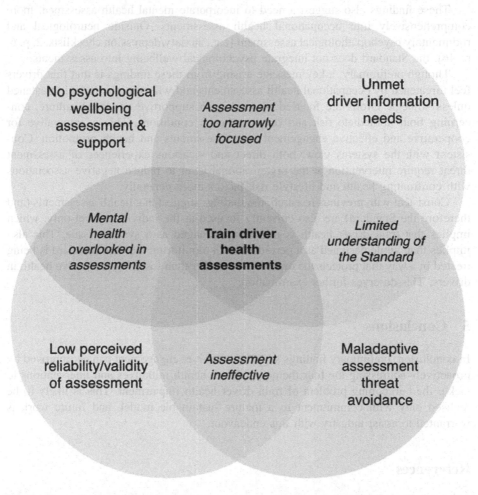

Fig. 2. Venn diagram showing preliminary conceptualization of study findings. Outer circles depict themes, overlapping circles depict influence on train driver perspectives on the health assessment.

Impact evaluations of the Standard have identified a number of objective effects of its assessments [4, 5], but other areas show mixed and somewhat counterintuitive outcomes. The findings of this study explain some of the findings of this work, for example, purported decreases in self-reported smoking cessation may in fact be a reflection maladaptive assessment threat avoidance in health assessments, which suggests that the assessment outcomes may not be entirely accurate. From a systems thinking point of view, this lends support for normalisation of deviancy [20], which in the absence of catastrophic results, suggests that maladaptive behaviour has become the social norm. The findings of the study suggest therefore that train drivers have not been consulted and/or consulted adequately in the development of the Standard, and consider the process to be punitive rather than supportive.

These findings also suggest a need to incorporate mental health assessment more comprehensively into occupational health assessments. Outside neurological and rudimentary psychopathological assessment [e.g., anxiety/depression checklists, 2, p. 6, p. 46], the Standard does not integrate psychological wellbeing into assessments.

Though preliminary, a key message arising from these findings is that that drivers feel threatened by occupational health assessments and will continue to feel threatened unless strategic efforts are focused on creating a supportive and open culture, concerning both immediate risk and ongoing health conditions. This is imperative for cooperative and effective engagement with assessments and health promotion. Consistent with the systems view, both direct and vicarious experiences of assessment threat require intervention at the organisational level to reduce negative associations with confronting health and lifestyle risk factors more generally.

Consistent with previous research, the findings suggest that health assessments (and therefore the Standard) are also currently focused at the individual-level only, which implies that train driver health is not being managed as a systems issue. This also implies that, while designed as a periodic health monitoring tool, the Standard is being treated in a way that protects the organisation, rather than fostering preventive health in drivers. This deserves further examination.

5 Conclusions

In conclusion, preliminary findings suggest that driver engagement can be improved by proactively addressing the four themes in future health initiatives and interventions to tackle the burgeoning problem of train driver health impairment. This is likely to be realised only with commitment to a mature sustainable model, and future work is warranted to assist industry with this endeavour.

References

1. Hocking B (2006) The inquiry into the waterfall train crash: implications for medical examinations of safety-critical workers. Occup Health Saf 184(3):126–128
2. National Transport Commission (2012) National Standard for Health Assessment of Rail Safety Workers. Author, Melbourne
3. National Transport Commission (2017) National Standard for Health Assessment of Rail Safety Workers. Author, Melbourne

4. Mina R, Casolin A (2012) The Australian National Standard for rail workers five years on. Occup Med 62:642–647
5. Mina R, Casolin A (2007) National standard for health assessment of rail safety workers: the first year. Med J Aust 187(7):394–397
6. Colquhoun CP, Casolin A (2015) Impact of rail medical standard on obstructive sleep apnoea prevalence. Occup Med 66(1):62–68
7. Chapman J, Naweed A (2015) Health initiatives to target obesity in surface transport industries: review and implications for action. Evid Base 2:1–32
8. Naweed A, Trigg J, Allan M, Chapman J (2017) Working around it: rail drivers' views on the barriers and enablers to managing workplace health. Int J Workplace Health Manag 10 (6):475–490
9. Naweed A, Chapman J, Allan M, Trigg J (2017) It comes with the job: work organizational, job design, and self-regulatory barriers to improving the health status of train drivers. J Occup Environ Med 59(3):264–273
10. Lo SH, van Breukelen GJ, Peters G-JY, Kok G (2013) Proenvironmental travel behavior among office workers: a qualitative study of individual and organizational determinants. Transp Res Part A: Policy Pract 56:11–22
11. Nielsen JR, Hovmøller H, Blyth P-L, Sovacool BK (2015) Of "white crows" and "cash savers:" a qualitative study of travel behavior and perceptions of ridesharing in Denmark. Transp Res Part A: Policy Pract 78:113–123
12. Vaezipour A, Rakotonirainy A, Haworth N, Delhomme P (2017) Enhancing eco-safe driving behaviour through the use of in-vehicle human-machine interface: a qualitative study. Transp Res Part A: Policy Pract 100:247–263
13. Hafner RJ, Walker I, Verplanken B (2017) Image, not environmentalism: a qualitative exploration of factors influencing vehicle purchasing decisions. Transp Res Part A: Policy Pract 97:89–105
14. Guest G, Bunce A, Johnson L (2006) How many interviews are enough? An experiment with data saturation and variability. Field Methods 18(1):59–82
15. Morgan DL (1997) Focus groups as qualitative research. Sage, Thousand Oaks
16. Saldana J (2009) The coding manual for qualitative researchers. SAGE Publications, Los Angeles
17. Braun V, Clarke V (2006) Using thematic analysis in psychology. Qual Res Psychol 3 (2):77–101
18. Wasserman JA, Clair JM, Wilson KL (2009) Problematics of grounded theory: innovations for developing an increasingly rigorous qualitative method. Qual Res 9(3):355–381
19. Attride-Stirling J (2001) Thematic networks: an analytic tool for qualitative research. Qual Res 1(3):385–405
20. Vaughan D (1997) The challenger launch decision: risky technology, culture, and deviance at NASA. University of Chicago Press

Protection of Pregnant Women at Work in Switzerland: Implementation and Experiences of Maternity Protection Legislation

Alessia Zellweger[1,2](✉) ⓘ, Peggy Krief[2] ⓘ,
Maria-Pia Politis Mercier[1] ⓘ, Brigitta Danuser[2] ⓘ, Pascal Wild[2,3] ⓘ,
Michela Zenoni[2] ⓘ, and Isabelle Probst[1] ⓘ

[1] School of Health Sciences (HESAV), University of Applied Sciences and Arts
Western Switzerland (HES-SO), Avenue de Beaumont 21, 1011 Lausanne,
Switzerland
alessia.zellweger@hesav.ch
[2] Institute for Work and Health (IST), Universities of Lausanne and Geneva,
Route de la Corniche 2, 1066 Epalinges, Switzerland
[3] INRS Scientific Management, Nancy, France

Abstract. *Objectives.* Like most industrialized countries, Switzerland has introduced legislation to protect the health of pregnant workers and their unborn children from workplace hazards. This study aims to assess legislation's degree of implementation in the French-speaking part of Switzerland and understand the barriers to and resources supporting its implementation.

Methods. Data were collected using mixed methods: (1) an online questionnaire send to 333 gynecologist-obstetricians (GOs) and 637 midwives; (2) exploratory semi-structured interviews with 5 workers who had had a pregnancy in the last 5 years.

Results. Questionnaire response rates were 32% for GOs and 54% for midwives. Data showed that several aspects of the implementation of maternity protection policies could be improved. Where patients encounter workplace hazards, GOs and midwives estimated that they only received a risk assessment from the employer in about 5% and 2% of cases, respectively. Preventive leave is underprescribed: 32% of GOs reported that they "often" or "always" prescribed preventive leave in cases involving occupational hazards; 58% of GOs reported that they "often" or "always" prescribed sick leave instead.

Interviews with workers identified several barriers to the implementation of protective policies in workplaces: a lack of information about protective measures and pregnancy rights; organizational problems triggered by job and schedule adjustments; and discrepancies between some safety measures and their personal needs.

Conclusions. Results demonstrate the need to improve the implementation and appropriateness of maternity protection legislation in Switzerland. More research is required to identify the factors affecting its implementation.

Keywords: Pregnancy · Occupational exposure
Maternity protection legislation

© Springer Nature Switzerland AG 2019
S. Bagnara et al. (Eds.): IEA 2018, AISC 819, pp. 622–633, 2019.
https://doi.org/10.1007/978-3-319-96089-0_68

1 Introduction

1.1 Background

Overall, the international medical literature shows that work in itself does not pose a risk to pregnancy [1, 2]. Nevertheless, chemical, physical, or biological exposure may affect pregnancy outcomes (miscarriage, preterm birth, small for gestational age) and child development (malformation, cognitive faculties) [2–5]. Even though a recent meta-analysis suggested that their impact may be weaker [6], certain work activities are also suspected of representing a risk to pregnancy and the unborn child, such as poor posture, lifting, work schedules, or psychological stress [4, 7, 8].

Most industrialized countries have introduced legal provisions for the protection of maternity at work [9]. This *maternity protection legislation* (hereafter MPL) requires that occupational risks to pregnancy be assessed and measures be taken to avoid the exposure of pregnant workers. This should primarily be done by eliminating those risks or adapting working conditions. If those options prove infeasible, employees should be transferred to another post or, as a last resort, granted paid leave.

In Switzerland, the Labor Law, its ordinances, and the Ordinance on Maternity Protection at Work (OProMa) set out which types of jobs are considered dangerous or arduous, the processes to be put in place to counter the risks, and the responsibilities of all the actors involved. If a company carries out specific activities which might be dangerous or arduous in case of pregnancy, then it must call in an officially authorized specialist to carry out a risk analysis. The gynecologist-obstetricians (GOs) must verify whether their patients are exposed to any professional activities banned under the OProMa. If they are, the GOs must ask employers for their risk analyses and decide on whether expectant mothers can safely continue employment at their workstations. In the absence of a risk analysis or workplace accommodations, but in the presence of presumed dangers, the GOs will prescribe preventive leave according to the precautionary principle. Employers finance preventive leave directly. Preventive leave is different from sick leave, which is financed either directly by the employer for short periods or by the employer's loss-of-income insurance for longer periods. Although midwives can and do monitor pregnancies, they have no legally defined role in or authority over maternity protection in the workplace: their role is both informal yet very important.

The authors have carried out an international literature review [10] which points out shortcomings in the implementation of MPL in several states - e.g., Belgium, Quebec, UK [11–15] - in addition to Switzerland [16, 17]. One initial finding concerned the use of sick leave certificates instead of preventive leave certificates [18, 19]. Furthermore, in several contexts, workers are put on either preventive or sick leave rather than retained at work following ergonomic adaptations to their workstations or a transfer to another position [20]. A second finding was the complexity of the factors which either facilitate or hinder the implementation of MPL. At the individual level, the representations which stakeholders give to pregnancy at work and its risks play a crucial role in the implementation of MPL. Indeed, workers' individual attitudes can lead certain women to choose to continue working despite medically identified risks to their health [15, 21]. Workers often considered other "risks", such as the deterioration of relationships with colleagues and impacts on career paths or job retention. At the organizational level, the needs for adjustments to workstations, job reassignments, or maternity-leave cover makes companies see pregnancy as a complex problem [12, 13]. Moreover, the multiplicity of different stakeholders creates the requirement for a single, clear definition of the occupational risks faced by employees [22]. At the macrosocial level, some of the problems and solutions may reside in national-level incentives and policies to implement the law [23]. Another issue lies in social and cultural representations, which, for example, make women's occupational health less visible [24].

1.2   Objectives

This article presents the first results from a wider study [25]. It focuses on: (1) the degree to which MPL are implemented, as reported by GOs and midwives, and (2) the barriers and resources identified by women workers in exploratory interviews.

1.3   Theoretical Approach

Considering that MPL are complex intervention, this study is inspired by realist approaches [26]. A realist approach attempts to reveal the circumstances in which interventions are implemented or not, which mechanisms are at work in each context, and which effects are produced, expected, or not expected.

2   Methods

This paper draws on a mixed methodology combining quantitative and qualitative data sets.

GOs and midwives in the French-speaking part of Switzerland filled in an ad hoc online questionnaire about their experiences, practice, and difficulties in implementing MPL. The main themes concerned the frequency with which they received or asked for risk assessments, contacted employers, and prescribed sick leave or preventive leave[1], and their sensitivity to occupational health.

The qualitative investigation consisted of 5 exploratory, semi-structured interviews with workers (nurses), who had had a pregnancy in the last 5 years. Interviews focused on workers' perceptions of workplace hazards during their pregnancy, their experiences of protective measures, and their perceptions of barriers to and resources helping maternity protection measures. Records were anonymized, and transcripts were analyzed thematically.

---

[1] Because midwives cannot legally sign a mother off for preventive leave, we investigated how frequently they referred their patients to GOs who could do so.

3 Results

3.1 Degree of Implementation of MPL (Quantitative Section)

The questionnaires were sent online to 333 gynecologist-obstetricians (GOs) and 637 midwives, in April 2017. The response rates were 32% (n = 105) for GOs and 54% (n = 356) for midwives. The question *"Do pregnancy consultations form a part of your professional activity?"* was used to filter these populations. Those answering "Yes" included 93 GOs and 205 midwives. Table 1 displays simple descriptive statistics of the questionnaire responses.

Table 1. Simple descriptive statistics

		GOs (n = 93)	Midwives (n = 205)
Estimated percentage of risk analyses received for patients facing an occupational risk: mean (sd)		5.4 (15.9)	2.1 (5.8)
		% (n)	% (n)
Frequency at which professionals asked for an occupational risk analysis	Never/rarely	35 (30)	79 (159)
	Sometimes	37 (32)	15 (31)
	Often	13 (11)	3 (6)
	Always/nearly always	15 (13)	2 (5)
Contact with the employer of a patient whose work poses a risk to pregnancy		58 (50)	9 (19)
Difficulties implementing OProMa with the employer		70 (35)	53 (10)
Reasons explaining the difficulties in implementing OProMa with the employer	The employer asked for sick leave to be granted	97 (34)	32 (6)
	Absence of any risk analysis	66 (23)	32 (6)
	Lack of knowledge about employers' obligations	60 (21)	21 (4)
Frequency of prescription of preventive leave during normal pregnancies (or midwife refers patient to a GO for prescription of preventive leave)	Never/rarely	36 (31)	30 (60)
	Sometimes	32 (28)	27 (54)
	Often	20 (17)	23 (46)
	Always/nearly always	12 (10)	19 (37)
Frequency of prescription of sick leave during normal pregnancies (or midwife refers patient to a GO for prescription of sick leave)	Never/rarely	15 (13)	13 (26)
	Sometimes	28 (24)	30 (60)
	Often	40 (34)	31 (62)
	Always/nearly always	17 (15)	25 (49)
Reasons why professionals "always" or "nearly always" prescribe sick leave instead of preventive leave	A request by the patient	60 (51)	45 (55)
	Habit	34 (29)	28 (25)
	A lack of competency	34 (29)	59 (59)

Risk Analysis. On average, GOs and midwives estimated that they received employers' risk analyses in only about 5% and 2%, respectively, of cases where their patients had a job involving a maternity protection risk. Furthermore, 35% of GOs and 79% of midwives declared that they rarely or never asked for a risk analysis when consulting a patient whose job entailed a risk to her pregnancy.

Contact with the Employer and Difficulties Implementing the OProMa. Some 58% of GOs and 9% of midwives stated that they contacted the employers of patients whose work posed a risk to their pregnancies. These professionals also said that they encountered complications when attempting to implement the OProMa with employers (70% of GOs; 53% of midwives). The main reasons mentioned as the causes of these difficulties were, for both types of professionals, the absence of any risk analysis (66% of GOs; 32% of midwives) and above all that employers asked for their employees to be put on sick leave (97% of GOs; 32% of midwives).

Preventive Leave or Sick Leave. In cases involving a normal pregnancy and a confirmed risk, only 32% of GOs "often" or "always" prescribed preventive leave, whereas 57% of GOs declared that they "often" or "always" prescribed sick leave. In the same situations, 42% of midwives stated that they "often" or "always" referred their patients to a GO for the prescription of preventive leave, and 56% of them "often" or "always" prescribed sick leave or asked a GO to do so. The reasons why GOs would "often" or "always" prescribe sick leave rather than preventive leave were: a request by the patient (60%), habit (34%), or a perceived lack of competency specifically in the domain of occupational risk (34%). The main reasons declared by midwives were: a perceived lack of competency (59%), a request by the patient (45%), and habit (28%).

3.2 Workers' Qualitative Experiences

Five exploratory interviews were organized with nurses who had had a pregnancy in the last 5 years. Table 2 shows participants' principal characteristics.

Workers pointed out several hindrances to the implementation of protective policies in workplaces.

Table 2. Characteristics of participants

Pseudonym	Workplace when pregnant	Activity rate when pregnant
Marie	Cantonal hospital, Surgery Department	100%
Diane	University hospital, Adult Intensive Care Unit	80%
Léa	Cantonal hospital, Oncological Outpatients Unit	80%
Amanda	1st pregnancy: University hospital, Pediatric Intensive Care Unit 2nd pregnancy: University hospital, Neonatology Department	100% 70%
Julia	University hospital, Pediatric Intensive Care Department	1st pregnancy: 80% 2nd pregnancy: 60%

Lack of Information About Protective Measures and Pregnancy Rights. Three women stated that they had no prior knowledge about MPL when they announced their pregnancies to their management. Diane and Julia, on the other hand, thought that they had good knowledge about these measures when they announced their pregnancies. None of the nurses interviewed knew about the possibility of preventive leave in cases where the work is arduous or dangerous, nor did they know of the existence of any risk analysis in their departments.

The majority of the nurses felt that they received little or no information from their supervisors about their rights as a pregnant employee. Notably, related Amanda, the little information which she received from her superiors seemed to focus exclusively on questions surrounding maternity leave. Julia and Diane, on the other hand, not only stated that they knew about the protection measures in place (see paragraph above) but they also felt that they had been well informed by their supervisors: *"We had this lady in HR who was super, who really explained things to us well. [...] every time that a woman announced that she was pregnant, she got a meeting with her, and we were very, very, very well informed."* (Julia).

Organizational Problems Triggered by Job and Schedule Adjustments. The principal difficulty, mentioned by all the participants, was the perception of a certain incompatibility between their state of pregnancy and the work asked of them in their hospital departments. The physical arduousness of the work, ergonomic difficulties, contact with and manipulation of dangerous substances, and the feeling of never being suitable for the tasks at hand, were all factors making nurses believe that reconciling work and pregnancy was very complicated. Marie, for example, admitted that the difficulties which she encountered during her pregnancy were not linked to any medical condition, but rather to ergonomic constraints: *"[...] my pregnancy was totally normal, I had no problems at all, but it was really in relation to the space there was between the bedside table and the bed. I had become so voluminous that in the end, bending down was a bit of a problem. We have to be able to look after people."*

Work in healthcare is perceived to be very arduous, yet three nurses described how their hospital proposed absolutely no adaptations for dealing with pregnant employees. *"No, so basically, either you continued working and you did the same things, or you were put on leave: there was no middle ground. They couldn't put us in an office, or anything like that. No."* (Julia).

Colleagues and Supervisors as Resources. Contrastingly, colleagues and a positive reception to the announcement of the pregnancy by their supervisors were key aids in achieving balance between work and pregnancy. Indeed, all the study participants mentioned that support from their colleagues was a precious resource: *"And then there was the kindness of all my colleagues, [...] they were thoughtful with regard to all the physical handling, they did it to relieve me, which meant that I was able to work a little longer."* (Diane).

Perceived Discrepancy Between Some Protective Measures and Workers' Needs. Most of the employees raised the issue of the mismatch between certain characteristics of the MPL and their implementation in the reality of the workplace. *"Once it has been announced, they don't have the right to make you do overtime. So, all the extra hours* *that you do can't be counted, because they'll be breaking the law. And you do those extra hours! There is no other way [...] and then at 16h00 I have to leave, I have to hand back my patients, my colleagues can't take them on because they've got something else, and then you stay one hour, you stay two hours more... that was really very complicated. It's clear, there's no choice! I can't say, 'Right. See you. I'm leaving!' Really, that's just not possible. There are patients involved."* (Julia)

Two nurses also had the perception that the protective measures suggested for their benefit in fact represented extra work for them. This was notably the case when they were meant to cut their working hours by 50%. *"It is always difficult to manage a 50% part-time role in these departments, [...] I had another child at home, so I had to get up early just as often as usual to work during the busiest time of the day, the morning [...]. So, in the end, I was tired and stressed out because I had to have finished my work before 11h30, and it's just impossible."* (Marie).

The inappropriateness of the measures put in place vis-à-vis the real needs of employees generated several negative effects felt by the participants, such as feelings of being a burden on their colleagues if no replacement staff had been foreseen or feelings of frustration or a drop in motivation linked to their jobs. *"I did lots of little supporting jobs, actually. So, I mean that when I arrived at 10h30, I put everybody on their drips Because I didn't feel really... I was doing the support jobs, it was like I was doing the dirty work, and I wasn't responsible for any of the patients."* (Amanda).

Personal Strategies. Workers sometimes put in place unexpected strategies to deal with the perceived discrepancy between protective measures with their needs. For example, three nurses decided to delay the announcement of their pregnancies, for different reasons. Although Amanda decided not to reveal her pregnancy immediately because *"during the first three months, you don't announce it because you're always scared about miscarriage"*, the others delayed announcing their pregnancies so as to avoid any adaptations being set up that they would have considered inadequate anyway. *"My strategy was to announce my pregnancy as late as possible, because as soon as I announced it to my boss, in fact, she'd have put me on 8-hour shifts when I had been working 12-hour shifts. And that would have raised the number of days that I worked per month."* (Marie).

4 Discussion/Conclusions

This study is limited to the French-speaking part of Switzerland, and the exploratory interviews include only one profession (nurses) with a very small sample, so that results cannot be generalized to the overall situation in Switzerland. However, the present study's results highlight the need to improve the implementation of MPL in Switzerland and to adapt its provisions.

Quantitative results revealed three difficult aspects in applying MPL.

(1) The absence of risk analyses, or the fact that employers fail to provide them to healthcare staff, represents a failure of companies to apply MPL. Yet only a minority of GOs and midwives asked for analyses when consulting for a patient whose job involved a risk to her pregnancy.

(2) The majority of the professionals questioned claimed to have encountered difficulties implementing the OProMa with employers. The principal reasons for this were that employers asked for their employees to be prescribed sick leave, even when there was no medical diagnosis to justify this.

(3) Indeed, professionals themselves under-prescribed preventive leave and over-prescribed sick leave. The two professional groups cited different reasons, however. Although the majority of GOs mentioned that they usually prescribed sick leave at the patient's request, midwives instead mentioned problems linked to their perceived lack of competency in this area. It should be noted here that the OProMa gives midwives no official authority in the prescription of preventive leave, which may well explain that perception.

GOs seem to have difficulty taking up the essential role which Swiss legislation has conferred upon them, namely the prescription of preventive leave when pregnant employees face an occupational danger. Statements in the literature concerning Sweden [27] revealed that GOs considered themselves to be *"bad judges"* when it came to evaluating the arduousness of their patients' working conditions. Likewise, in the USA, Stotland et al. [28] showed that GOs felt that they did not know enough about occupational health to be able to answer the patients' questions about issues of occupational exposure. These realizations raise questions about whether GOs have the necessary skills to deal with occupational risks. One solution might be to redefine roles, associating the occupational health physician more closely with decisions about preventive leave. Swiss legislation should also make the role of midwives more explicit. Notably, the essential tasks which should be given to midwives include informing patients about the law, identifying activities involving risks, referring women to the appropriate occupational healthcare specialist, and collaborating with the GOs. Another interesting element was GOs' prescription of sick leave in place of preventive leave. Several publications in the international literature have shown increasing rates of sick leave among pregnant employees [27, 29]. These increases involve many factors, however, consistent evidence indicates that occupational exposure and arduous working conditions are leading to higher rates of sick leave during pregnancy [29–32]. Granting sick leave thus seems to be one means adopted by GOs, or requested by pregnant employees themselves or by their employers, to react to a potential danger in the workplace.

The results from the qualitative section of our study corroborate the existence of deficient application of MPL within some healthcare institutions in Switzerland. Analysis of our interviews showed furthermore that the types of adaptations implemented by employers were not necessarily in line with the needs of their pregnant employees. Indeed, interview participants perceived some of the adaptations proposed by their employers to be additional burdens on them. The perceived inappropriateness of the protection measures can also engender a lack of employee motivation and reduce their enjoyment at work. The finding that certain workplace adaptations were perceived to have been poorly implemented was also revealed by Fanello et al. [19] in France and Malenfant et al. [13] in Quebec.

In a recent qualitative study in ten hospitals in Quebec Gravel et al. [33] showed that the majority of pregnant nurses are now maintained at the workplace longer thanks to job adjustments or reassignment. Our data bring a different situation to light. Indeed most of the interviewees felt that adaptations were difficult to reach in their hospital: *"basically, either you continued working and you did the same things, or you were put on leave"* (Julia).

The studies included in our literature review [10] often demonstrated a deterioration in professional relationships (negative comments from colleagues and supervisors) following pregnancy announcements or the implementation of MPL [15, 20, 34]. Our interviews failed to duplicate this finding, maybe because nursing is a heavily female profession and lots of colleagues have been through this experience. Nevertheless, it seems clear that colleagues and supervisors play a vital role in reconciling pregnancy and work.

Delaying the announcement of their pregnancy was one of the employees' strategies for combatting the perceived inappropriateness of some of the workplace adaptations put in place for them (for example, in order to continue to work 12-h shifts, something which Swiss legislation prohibits). These personal strategies echoed a key concept in ergonomics and occupational health: the concept of *margin for manoeuvre*, which is a testament to the wide range of personal strategies implemented by all workers in order to continue doing their jobs while being able to adjust their professional tasks to their personal state of health [35, 36].

Without wishing to have pregnancy considered an illness, the condition's extraordinary nature cannot be denied and nor can the fact that it can put limits on an employee's ability to carry out certain professional tasks and activities. Thus, it seems pertinent to take margin for manoeuvre into consideration when looking at the personal strategies put in place by pregnant workers. For example, if in Gravel et al. [33] the interviewed pregnant nurses display an important margin for manoeuvre, notably by challenging not only working conditions but also reassignments that they judge inappropriate for their health or their career, in our interviews some nurses choose instead to acting upon the announcement's conditions of their pregnancy. However, this choice is not a trivial one. A pregnancy has to be announced before a woman can benefit from MPL. Recent studies have shown that the period presenting the greatest risks of fetal malformation or miscarriage is between the 3rd and 8th weeks of gestation [37]. Employees might be encouraged to announce their pregnancies to their supervisors earlier if the adaptations put in place for them were closer to their perceived needs and by establishing some sort of participative management for the implementation of those adaptations.

More research is required to evaluate the degree of implementation of this legislation in workplaces. Qualitative interviews with others relevant stakeholders from various workplaces would be useful in order to more accurately understand the barriers to maternity protection's implementation and establish ways to overcome them.

Acknowledgments. This work is supported by the Swiss National Science Foundation (grant number 162713), by Canton Vaud's Public Health Service and by a research fund belonging to the University of Applied Sciences and Arts Western Switzerland (HES-SO).

References

1. Casas M et al (2015) Maternal occupation during pregnancy, birth weight, and length of gestation: combined analysis of 13 European birth cohorts. Scand J Work Environ Health 41:384–396. https://doi.org/10.5271/sjweh.3500
2. Fowler JR, Culpepper L (2018) Working during pregnancy. UpToDate
3. Figà-Talamanca I (2006) Occupational risk factors and reproductive health of women. Occup Med 56:521–531. https://doi.org/10.1093/occmed/kql114
4. Lafon D (2010) Grossesse et travail: quels sont les risques pour l'enfant à naître? EDP Sciences edn. Institut National de recherche et de Sécurité (INRS), Les Ulis
5. Goldman RH, Wylie JB (2017) Overview of occupational and environmental risks to reproduction in females. UpToDate
6. Palmer KT, Bonzini M, Bonde J-PE (2013) Pregnancy: occupational aspects of management: concise guidance. Clin Med 13:75–79. https://doi.org/10.7861/clinmedicine.13-1-75
7. Larsen AD (2015) The effect of maternal exposure to psychosocial job strain on pregnancy outcomes and child development. Dan Med J 62:B5015
8. Glover V (2014) Maternal depression, anxiety and stress during pregnancy and child outcome; what needs to be done. Best Pract Res Clin Obstet Gynaecol 28:25–35. https://doi.org/10.1016/j.bpobgyn.2013.08.017
9. International Labour Organization (2010) Maternity at work: a review of national legislation: findings from the ILO database of conditions of work and employment laws, Geneva
10. Probst I, Zellweger A, Politis Mercier M-P, Danuser B, Krief P (Under review) Implementation, mechanisms, and effects of maternity protection legislation: a realist narrative review of the literature
11. Lembrechts L, Valgaeren E (2010) Grossesse au travail. Le vécu et les obstacles rencontrés par les travailleuses en Belgique. Etude quantitative et qualitative. Institut pour l'égalité des femmes et des hommes, Bruxelles
12. Lippel K (1998) Preventive reassignment of pregnant or breast-feeding workers: the Quebec model. New Solut J Environ Occup Health Policy 8:267–280
13. Malenfant R, Gravel A-R, Laplante N, Plante R (2011) Grossesse et travail: au-delà des facteurs de risques pour la santé. Revue Multidisciplinaire sur L'emploi, le Syndicalisme et le Travail 6:50–72
14. Adams L et al (2016) Pregnancy and maternity-related discrimination and disadvantage: experiences of employers. Department for Business, Innovation and Skills and the Equality and Human Rights Commission, London
15. Adams L et al (2016) Pregnancy and maternity-related discrimination and disadvantage: experiences of mothers. Department for Business, Innovation and Skills, Equality and Human Rights Commission, London
16. Grolimund-Berset D, Krief P, Praz-Christinaz S (2011) Difficultés pratiques de la mise en application de l'Ordonnance sur la protection maternité (Oproma) en Suisse à la lumière de deux cas cliniques. Archives des Maladies Professionnelles et de L'environnement 73:524–525
17. Aellen G, Nicollier L, Outdili Z, Ribeiro K, Stritt K (2013) Application de l'Ordonnance sur la protection de la maternité chez les femmes médecins. Revue médicale suisse, pp 1433–1434
18. Makowiec-Dabrowska T, Hanke W, Radwan-Wlodarczyk Z, Koszada-Wlodarczyk W, Sobala W (2003) Working condition of pregnant women. Departures from regulation on occupations especially noxious or hazardous to women. Med Pr 54:33–43

19. Fanello C, Ripault B, Drüker S, Moisan S, Parot E, Fontbonne D (2005) Déroulement des grossesses du personnel d'un établissement hospitalier. Evolution en vingt ans. Archives des Maladies Professionnelles et de l'Environnement 66:244–251. https://doi.org/10.1016/S1775-8785(05)79089-5

20. Malenfant R, De Koninck M (2002) Production and reproduction: the issues involved in reconciling work and pregnancy. New Solut J Environ Occup Health Policy 12:61–77

21. Legrand E (2015) Santé reproductive et travail: la prévention des risques reprotoxiques. Rapport final dans le cadre du Programme national de recherche Environnement-Santé Travail. Université du Havre, Le Havre

22. Aviles-Palacios C, Lopez-Quero M, Garcia-Lopez M-J (2013) Gender and maternity considerations and techniques in occupational health services: the Spanish case. Saf Sci :27–31 https://doi.org/10.1016/j.ssci.2013.03.011

23. Tarchi M, Bartoli D, Demi A, Dini F, Farina GA, Sannino G (2007) Emerging problems in enforcement of safe maternity and feeding protection at work: a public prevention service experience. Giornale Italiano di Medicina del Lavoro ed Ergonomia 29:385–386

24. Messing K, Boutin S (1997) Les conditions difficiles dans les emplois des femmes et les instances gouvernementales en santé et en sécurité du travail. Ind Relat 52:333–363. https://doi.org/10.7202/051169ar

25. Krief P, Zellweger A, Politis Mercier M-P, Danuser B, Wild P, Zenoni M, Probst I (2018) Protection of pregnant women at work in Switzerland: practices, obstacles and resources. A mixed-methods study protocol. BMJ Open

26. Pawson R, Greenhalgh T, Harvey G, Walshe K (2005) Realist review–a new method of systematic review designed for complex policy interventions. J Health Serv Res Policy 10 (Suppl 1):21–34. https://doi.org/10.1258/1355819054308530

27. Larsson C, Sydsjo A, Alexanderson K, Sydsjo G (2006) Obstetricians' attitudes and opinions on sickness absence and benefits during pregnancy. Acta Obstet Gynecol Scand 85:165–170

28. Stotland NE et al (2014) Counseling patients on preventing prenatal environmental exposures–a mixed-methods study of obstetricians. PLoS ONE 9:e98771. https://doi.org/10.1371/journal.pone.0098771

29. Sydsjo G, Sydsjo A (2002) Newly delivered women's evaluation of personal health status and attitudes towards sickness absence and social benefits. Acta Obstet Gynecol Scand 81:104–111. https://doi.org/10.1034/j.1600-0412.2002.810203.x

30. Hansen M, Thulstrup A, Juhl M, Kristensen J, Ramlau-Hansen C (2015) Occupational exposures and sick leave during pregnancy: results from a Danish cohort study. Scandin 41

31. Henrotin JB, Vaissiere M, Etaix M, Dziurla M, Malard S, Lafon D (2017) Exposure to occupational hazards for pregnancy and sick leave in pregnant workers: a cross-sectional study. Ann Occup Environ Me 29 https://doi.org/10.1186/s40557-017-0170-3

32. Kaerlev L, Jacobsen LB, Olsen J, Bonde JP (2004) Long-term sick leave and its risk factors during pregnancy among Danish hospital employees. Scand J Public Health 32:111–117. https://doi.org/10.1080/14034940310017517

33. Gravel AR, Riel J, Messing K (2017) Protecting pregnant workers while fighting sexism: work-pregnancy balance and pregnant nurses' resistance in Quebec hospitals. New Solut J Environ Occup Health Policy 27:424–437. https://doi.org/10.1177/1048291117724847

34. Bretin H, De Koninck M, Saurel-Cubizolles M-J (2004) Conciliation travail/famille: quel prix pour l'emploi et le travail des femmes? À propos de la protection de la grossesse et de la maternité en France et au Québec. Santé, Société et Solidarité 3:149–160. https://doi.org/10.3406/oss.2004.1006

35. Durand MJ, Vezina N, Baril R, Loisel P, Richard MC, Ngomo S (2009) Margin of manoeuvre indicators in the workplace during the rehabilitation process: a qualitative analysis. J Occup Rehabil 19:194–202. https://doi.org/10.1007/s10926-009-9173-4

36. Durand MJ, Vezina N, Baril R, Loisel P, Richard MC, Ngomo S (2011) Relationship between the margin of manoeuvre and the return to work after a long-term absence due to a musculoskeletal disorder: an exploratory study. Disabil Rehabil 33:1245–1252. https://doi.org/10.3109/09638288.2010.526164

37. Romano D, Moreno N (2010) Barriers for the prevention of chemical exposures in pregnant and breast-feeding workers? J Epidemiol Community Health 64:193. https://doi.org/10.1136/jech.2008.085282

Identifying Ambulation-Related Missteps and Falls with Ergonomically Descriptive Terminology, Not as "Slips, Trips and Falls"

Jake L. Pauls[✉]

Jake Pauls Consulting Services, Toronto, Canada
bldguse@aol.com

Abstract. Several generic missteps—*departures from normal human gait in bipedal ambulation*—were systematically identified in 2003 by the HFES Forensics Professional Group, e.g., an air step, heel scuff, overstep, slip, stumble, trip, understep and unstable footing. Yet overwhelmingly, missteps and falls are commonly referred to as "slip, trip and fall", leading to incident misidentification in epidemiology, etiology, forensics, etc. The need for change and international efforts on this, begun in 2007, are discussed. This includes efforts to change the name of IEA's relevant Technical Committee, up to now called the "Slips, Trips and Falls Committee". Opposition by fellow ergonomists has been muted but effective in preventing a name change. It is especially ironic that the field of ergonomics—*which should lead in this matter*—appears content to misrepresent the ergonomic aspects of missteps and falls. This only continues a flawed tradition and misleads other disciplines about the underlying facts about how and why falls occur and what should be done by way of prevention and mitigation. A change now could help save many people from predicable and preventable falls, a leading cause of injuries in most countries.

Keywords: Missteps · Falls · Semantics

1 Disruptions in Normal Human Bipedal Gait Resulting in Falls

1.1 Historical Introduction

Human bipedal gait—a defining human characteristic—has intrigued us for a long time. Through photographic documentation of such gait in the 1880s, by Muybridge (1955), we began to understand our gait with the insights from many tight sequences of multiple, stop-motion photographs, a techniques Muybridge had first used to document the gait of a running horse. Today, bipedal gait can be studied by even more sophisticated documentation methods and other instrumentation. We are thus now capable of understanding and describing more-accurately what constitutes normal bipedal gait and what happens when gait is disrupted and a misstep occurs, often leading to a fall to the walking surface or worse, resulting in injuries.

Muybridge's work also encompassed stairs, so even a long time ago, we were able to recognize that gait on stairs differs from gait on a planar walkway. Moreover we

S. Bagnara et al. (Eds.): IEA 2018, AISC 819, pp. 634–638, 2019.
https://doi.org/10.1007/978-3-319-96089-0_69

were able to distinguish between ascent gait and descent gait, the latter being more dangerous and important in design, construction and their regulation (Templer 1992).

1.2 Classification and Consequences of Missteps and Falls

Certain missteps can be so disruptive and jarring that a person can be injured from the misstep alone, even in the absence of a complete fall or before other injuries are sustained from an ensuing fall. An "air step" can be such an injurious event, resulting to injuries to ankles, knees, hips and/or back. As it is the most easily understood, and long remembered of several types of misstep, it is appropriate that it also is first in the following alphabetically ordered listing with the eight definitions based on the best treatment of missteps and falls terminology (Thompson *et al.* 2005) with some minor editing for form by this paper's author.

Air Step. A misstep consisting of "stepping onto an unexpected depression, [slope] or step down".

Heel Scuff. "A misstep occurring in stair descent when the rear of a person's foot or shoe rubs or scrapes the riser or nosing of a stair step".

Overstep. "A misstep occurring when a pedestrian descending stairs places a foot on a stair tread or nosing of a landing such that the weight on the ball of the foot is not sufficiently supported by the tread or nosing, and the ball of the foot slips or rotates over the nose [nosing] of the tread or landing".

Slip. A misstep "occurring when insufficient friction exists between a pedestrian's foot and a supporting surface to prevent the foot from sliding".

Stumble. A misstep occurring "when a person trips or missteps, and then takes one or more rapid steps forward in an attempt to recover balance".

Trip. A misstep "occurring when the toe or some other portion of the foot in the swing phase of the walk cycle contacts a surface, an obstruction or a protrusion. The swinging foot may fail to adequately clear the ground or it may encounter an obstacle or an irregularity on the ground surface. May also occur when the bottom of shoe locks on the surface with excessive friction. May involve unstable or erratic foot trajectories on uneven or discontinuous surfaces, or obstacles on the floor".

Understep. A misstep "occurring when, in ascending stairs, a foot is not adequately placed on the nosing of a tread or landing to support the weight of the pedestrian".

Unstable Footing. A misstep occurring on a "walkway surface that is not sufficiently even, or which deforms under a pedestrian's foot or footwear pressure".

Related Terms. Thompson *et al.* (2005) also provided the following definitions or commentary. Together, the excerpts provided here for missteps and falls are but a small part of an extensive, instructive glossary of terms and related commentaries.

Misstep. "An unintentional departure from pedestrian gait appropriate for the walkway surface. If the misstep results in a loss of balance and the pedestrian does not recover that balance, a fall may result".

Fall. "An unintended impact between a person and a walkway or other surface caused by a loss of balance or support, misstep, or other physical or physiological condition [such as syncope or loss of balance, for example with significant postural hypotension]. Also, an undesirable descent due to the force of gravity, usually from a standing posture or during ambulation, to a lower level, usually the ground or floor".

Stair. "A change in elevation consisting of one or more risers".

Stairway. "One or more flights of stairs, either exterior or interior, with the necessary landings and platforms connecting them, to form a continuous and uninterrupted passage from one level [or floor] to another".

2   Do the Misstep Types Each Get a Fair Share of Attention?

This question was asked and partly answered in the author's paper (Pauls 2007) given at the International Conference on Slips, Trips, and Falls 2007: From Research to Practice, held at the Liberty Mutual Research Institute for Safety, Hopkinton, MA in the USA. Indeed the conference title suggests a succinct answer—*"No"*—and the paper set out some of the problems created due to the semantic limitation to "Slips" and "Trips." The paper's abstract noted:

> Some types of missteps are not covered by the conventional label, "slip, trip and fall". This is not a trivial omission. Air steps, for example, are a major source of pedestrian injuries, which can be serious, even in cases where there is not an ensuing fall to the walking surface. One consequence of the latter is that such pedestrian problems get relatively little attention at the "research" end of the spectrum addressed in "Slips, Trips and Falls" conferences held to date....
> At the "practice" end there are also shortcomings in standards, codes and regulations as well as guides and other practice-oriented materials that could do much more to prevent or mitigate the problems.

Today, over a decade later, these statements from 2007 still hold but there is more evidence now of the problems caused by the semantic choices made in the falls field that, more pointedly than ever before, are on the agenda for the IEA2018 Congress, both in a session on forensics and ergonomics (in the case of this paper) and within the program of the IEA Slips, Trips and Falls Committee.

One aspect shared in each of these IEA sessions is the effect the semantic issue has within the legal system. Far too often when there is fall-related, civil litigation, the case Complaint refers to the incident (leading to the injury) as either a "slip" or a "trip". Complaints, the legal instruments used widely within the civil litigation field, are typically drawn up prior to a proper, detailed, and unbiased investigation of the evidence, thus it is dangerous to prejudge causation at the Complaint stage of a case even if such causation is based on the recall of the injured person. People often refer to their fall as, for example, "I slipped" (on a stairway or sidewalk or some other pedestrian feature). The term "slipped" is used widely by non-experts to mean, "I lost my footing" or something like that. Ergonomics experts in the falls field should interpret such statements as, "there was a departure from normal gait and the consequence was a subsequent fall and injury". Then the evidence-based conclusion, after proper professional investigation of the site and the circumstances, should be able to include one or more of the several misstep types listed above.

For example, over a span of several decades, the author's investigations of falls in a variety of settings in two countries (USA and Canada), have typically been classified, by the client attorneys (and more generally their professional field), as "slip, trip and fall cases". Yet almost no incident was due to a true "trip" or "slip". This was partly because the author tends to be requested to examine many stair-related falls and injuries for which true slips and true trips are relatively rare as an *initiating* misstep. For example, an overstepping misstep of an undersized stair tread—very common in home settings—might have a slip or trip as a secondary or subsequent misstep after an overstepping misstep. Ones foot, if (necessarily) located too far forward on an undersized tread will rotate around the step nosing (the leading edge); that foot will often end up—uncontrollably—on a lower tread. That could be because, *secondarily,* the foot rotated on the nosing and dropped or there could have been an element of a slip. No amount of anti-slip treatment at the nosing will change the outcome. One's foot will lose secure placement on the tread and, with loss of balance, a fall will ensue. This should be classified, and addressed forensically, as an overstepping misstep with secondary missteps (all departures from normal gait) resulting in a fall that injures a person due to uncontrollable trauma due to impacts with the stair, enclosing walls or railings, and the floor below the stair.

In short, missteps and falls can involve much more than either a slip or a trip and experts in ergonomics working in a forensics context should not be satisfied with an overly simplistic diagnosis of the event. Moreover, they also should not effectively prejudge the event or incident as an "accident". If an expert's client, especially an attorney representing an injured party, describes an incident as an "accident," they should be gently but effectively cautioned that this can prejudge a case as the term "accident" is best reserved for an event that is both unpredictable and unpreventable. This distinction is addressed in a paper on the history of the terms "accident" and "act of God" (Loimer and Guarnieri 1996).

Finally, addressing directly the question posed in the heading to this section, the answer is "No". The eight misstep types described above, as identified over a decade ago, do not each get a fair share of attention. This is clear in conferences on falls, especially those in which there is a role by IEA's Slips, Trips and Falls Committee. It is also true in organizations such as ASTM International where slipping dominates most of the discussion on falls and related standards/guides preparation by its committees, particularly, Committee F13 on Pedestrian/Walkway Safety and Footwear. Such domination might be forgivable if two conditions were present:

1. Most falls were due to slipping.
2. Interventions to control slipping were effectively implemented.

Neither of these conditions appears to apply—in this author's view.

3 Conclusions

This paper is a brief reminder of a largely unheeded suggestion directed to an important field of professional activity within ergonomics generally and within IEA in particular. If we are really going to be serious about falls investigation plus prevention or mitigation—*for example, through the development and application of environmental safety standards and codes*—addressing the huge injury toll from falls, we need to begin using language that has long been identified by ergonomists in relation to forensics issues generally and falls in particular. Reasonably safe, productive bipedal ambulation is at stake as the societal costs of falls keep rising rapidly, especially as human population characteristics change worldwide.

References

Loimer H, Guarnieri M (1996) Accidents and acts of god: a history of the terms. Am J Public Health 86(1):101–107

Muybridge E (1955) The human figure in motion. Dover Publications, Inc., New York. (Based on 11-volume original book, Animal Locomotion, published in 1887)

Pauls J (2007) Predictable and preventable missteps that are not "slips, trips and falls", but result in serious injuries. In: Proceedings of international conference on slips, trips, and falls 2007: from research to practice. Liberty Mutual Research Institute for Safety, Hopkinton, pp 15–19

Templer JA (1992) The Staircase (Volume 2): Studies of hazards, falls and safer design. MIT Press, Cambridge

Thompson DA, et al (2005) A guide to forensic human factors terminology. In: Noy YI, Karwowski W (eds) Handbook of human factors in litigation. CRC Press, New York, pp 38-1–38-46

The Impact of Physical Activity Enjoyment on Motor Ability

Akari Kamimura[1,2]($\boxtimes$), Yujiro Kawata[3,4], Shino Izutsu[2,5],
Nobuto Shibata[3,4], and Masataka Hirosawa[3,4]

[1] Department of Child Development and Education, School of Humanities,
Wayo Women's University, 2-3-1, Konodai, Ichikawa-City,
Chiba 272-8533, Japan
kamimura@wayo.ac.jp
[2] Institute of Health and Sports Science and Medicine, Juntendo University, 1-1,
Hiraga-gakuendai, Inzai-shi, Chiba 270-1606, Japan
[3] School of Health and Sports Science, Juntendo University, 1-1,
Hiraga-gakuendai, Inzai-shi, Chiba 270-1606, Japan
[4] Graduate School of Health and Sports Science, Juntendo University, 1-1,
Hiraga-gakuendai, Inzai-shi, Chiba 270-1606, Japan
[5] Faculty of Sports and Health Sciences, Japan Women's College of Physical
Education, 8-19-1, Kitakarasuyama, Setagaya-ku, Tokyo 157-8565, Japan

Abstract. This study aimed to examine the impact of physical activity enjoyment on the motor ability of school-age children. We collected data from 351 elementary school students (180 boys and 171 girls; Mage = 8.78, standard deviation = 1.85) from public elementary schools in Japan. We investigated individual profiles (sex, age, birth date, and school year), children's physical size (height, weight, and Rohrer index), children's motor ability (performance during a 50-m sprint, standing broad jump, and throwing a soft ball), and enjoyment of physical activity using the Physical Activity Enjoyment Scale (PACES). We assessed the effect of physical activity enjoyment (independent variable) on motor ability (dependent variable) using binomial logistic regression analysis. An examination of confounders (chi-square test) indicated that weight and Rohrer index were the influential factors in the 50-m sprint (ps < .05). Therefore, the adjusted ORs for these factors were calculated using logistic regression analysis. The adjusted analysis results indicated that enjoyment (PACES score $\geq$ 4.00) is a promoting factor for motor ability.

Keywords: Motor ability · Physical activity enjoyment · Children

1 Introduction

The decrease in children's motor ability has been regarded as an important social issue in many developed countries [1–6], because it could lead to health problems during their (children) lifetime. Over the past few decades, a considerable number of studies have been conducted on chronic childhood diseases and the amount of physical exertion [7, 8]. However, little attention has been paid to the association between the psychological aspects of physical activity and children's motor ability. It is imperative

© Springer Nature Switzerland AG 2019
S. Bagnara et al. (Eds.): IEA 2018, AISC 819, pp. 639–645, 2019.
https://doi.org/10.1007/978-3-319-96089-0_70

for researchers to gain the knowledge to improve positive awareness of physical activity among children in order to increase their motor ability.

Kamimura and colleagues have been investigating the relationship between relative age, physical size, motor ability, and psychological aspects of physical activity (e.g., physical competence and positive awareness) among children. Among children aged 4–5 years, Kamimura et al. [9] indicated that relative age could influence motor abilities or a teacher's evaluation, mediating in terms of physical size. However, in 5-year-old children, motor ability was higher among those who demonstrated positive awareness (high enjoyment, strong liking, and high confidence) as compared with loss of enjoyment, dislike, and negative awareness [10]. Among children aged 6–12 years, Kawata et al. [11] discovered that relative age could influence the psychological aspects of sports participation (physical activity enjoyment and competence) in children. In addition, it was indicated that the frequency of exercise outside of school was associated with physical competence [12].

Physical activity enjoyment has been reported to be associated with various physical activity factors [13], such as self-determination [14] and perceived competence [15]. Moreover, DiLorenzo et al. examined several psychological and environmental variables in relationship to physical activity in elementary school children and found that physical activity enjoyment was the only consistent predictor of physical activity levels [16]. Therefore, it is possible that physical activity enjoyment is also a predictor of motor ability in elementary school children.

Despite the growing interest in physical activity enjoyment and the importance of promoting children's physical activity, current research lacks findings on the impact of physical activity enjoyment on motor ability. Therefore, we examined the impact of physical activity enjoyment on the motor ability of school-age children.

2   Methods

2.1   Participants

We collected data from 351 elementary school students (180 boys and 171 girls; Mage = 8.78, standard deviation = 1.85) from public elementary schools in Tokyo, Japan.

2.2   Measurements

We investigated individual profiles (sex, age, birth date, and school year), children's physical size (height, weight, and Rohrer index), children's motor ability (performance during a 50-m sprint, standing broad jump, and throwing a soft ball), and physical activity enjoyment using the Physical Activity Enjoyment Scale (PACES).

Demographic Data. We used a questionnaire to obtain information from all participants about their sex, age, birth date, and school year.

Children's Physical Size. Children's physical size was measured using body height and weight. Rohrer index was calculated using body height and weight data.

Motor Abilities. Children's motor ability was assessed using a 50-m sprint, standing broad jump, and softball throwing. For the 50-m sprint, we set a 50-m straight alley and created start and finish lines. A measurer stood at the finish line and recorded each child's time from the start cue ("set and go") to the moment the child crossed the 50-m finish line. For the standing broad jump, we set a balk line in front of the outside sandbox. We instructed the children to jump as far as they could from the balk line, using both their right and left feet at the same time. We recorded their distance. For the softball throwing, we used an official softball (No. 1 size for 1st and 2nd grade students and No. 2 size for 3rd to 6th grade students). We instructed the children to throw the ball as far as possible, using their dominant hand. We recorded the distance.

Enjoyment of Physical Activity. Children's physical activity enjoyment was measured using PACES [17]. This scale has a one-factor structure and 16 items. Respondents were requested to answer each item with a 5-point Likert-type scale (1 = strongly disagree; 2 = disagree; 3 = no opinion; 4 = agree; 5 = strongly agree). The score of PACES was calculated from the mean of the 16 items.

2.3 Data Analysis

We assessed the effect of physical activity enjoyment (independent variable) on motor ability (dependent variable) using binomial logistic regression analysis. Prior to analysis, we examined the demographics that might influence motor ability (confounding factors) using a chi-square test and calculated adjusted odds ratios (ORs). Statistical significance was set at $p < .05$.

Because logistic regression analysis requires categorical variables, we converted all scores into categorical variables. Since the Japanese school year runs from April 1 to March 31, children's birth date was converted into the following four groups based on their birth month: Group A (April 2 through June), Group B (July through September), Group C (October through December), and Group D (January through April 1). The Rohrer index was converted into categorical variables based on the popular indicators for defining childhood obesity (into groups of "Underweight: ≤ 114 points," "Normal: 115–144 points," and "Overweight: ≥ 145 points"). Motor ability scores and physical size were converted into categorical variables based on the national average of each age (into groups of "low" and "high"). PACES score was converted into categorical variables based on ≥ 4.00 (into groups of "low-middle" and "high").

2.4 Ethical Consideration

We obtained ethical approval from the Research Ethics Committee of Juntendo University. Prior to the study, we obtained permission from the school principals and board of education. Informed consent was obtained from the parents of the participants.

3 Results

3.1 Correlations Between Awareness of Physical Activity and Motor Ability

Table 1 shows the results of the examination of confounding factors using the chi-square test. It was confirmed that weight (χ^2 = 4.61, df = 1.00, p < 0.05) and Rohrer index (χ^2 = 20.11, df = 2.00, p < 0.001) were the influential factors in the 50-m sprint. Therefore, adjusted odds ratio for sex was calculated using logistic regression analysis (LRA).

Table 1. Examination of confounding factors (χ^2 test)

		Motor ability								
		50-m sprint			Standing broad jump			Softball throwing		
Attribution	Categories	<cut off	≥cut off	χ^2	<cut off	≥cut off	χ^2	<cut off	≥cut off	χ^2
Birth month	April 2–June	42	53	n.s.	67	28	n.s.	64	31	n.s.
	July–September	25	54		51	28		57	22	
	October–December	26	58		59	25		48	36	
	January–April 1	39	54		67	26		62	31	
Height	<National average	75	105	n.s.	130	50	n.s.	124	56	n.s.
	≥National average	57	114		114	57		107	64	
Weight	<National average	80	157	*	166	71	n.s.	158	79	n.s.
	≥National average	52	62		78	36		73	41	
Rohrer index	Underweight: ≥114	32	76	***	74	34	n.s.	73	35	n.s.
	Normal range: 115–144	80	137		148	69		137	80	
	Overweight: ≥145	20	6		22	4		21	5	

Note. Cut off is the national average by age.
n.s.: not significant, *p < 0.05, **p < 0.01, ***p < 0.001

3.2 Classification of Children's Awareness of Physical Activity

Table 2 shows the results of the logistic regression analysis of the motor ability and physical activity enjoyment scores (LRA).

Table 2. Logistic regression analysis of the motor ability and physical activity enjoyment scores (LRA)

Motor ability									
	50-m sprint			Standing broad jump			Softball throwing		
Variables	A-OR	95% CI	P	OR	95% CI	P	OR	95% CI	P
PACES ≥4.00	2.516	1.565–4.045	***	1.883	1.122–3.102	***	2.225	1.337–3.702	***

***p < 0.001

The adjusted analysis results indicated that physical activity enjoyment (PACES score ≥ 4.00) is a promoting factor for motor ability (50-m sprint: adjusted OR 2.52, 95% confidence interval [CI]: 1.157–4.045, p < 0.001; standing broad jump: OR 1.88, 95% CI: 1.12–3.10, p < 0.05; throwing a soft ball: OR 2.23, 95% CI: 1.337–3.702, p < 0.01).

4 Discussion

This study primarily aimed to examine the effects of physical activity enjoyment on the motor ability of school-age children. Our study has two major findings.

First, the results of the examination of confounding factors confirm that weight and Rohrer index were the influential factors in the 50-m sprint. These results support that the ability of moving the center of gravity is lower among children with high Rohrer index (or heavy body weight) compared to those with standard Rohrer index (or standard body weight) [18].

Second, the results of the examination of the impact of physical activity enjoyment on motor ability suggest that high physical activity enjoyment (PACES score ≥ 4.00) is an influential factor for motor abilities (sprint, jump, and throw). These results support that physical activity enjoyment is a consistent predictor of physical activity levels [16], and that there is a possibility that children's motor ability improves through experiences of any physical activity.

5 Conclusions

We conclude that physical activity enjoyment has a significant influence on motor ability. Furthermore, children who enjoy physical activity are 1.8–2.5 times more likely to be above the national average of motor ability compared with children who do not enjoy it. Therefore, increasing and maintaining high physical activity enjoyment is important to improve the motor ability of school-age children.

Acknowledgement. This work was supported by JSPS KAKENHI Grant Number JP18K13123 (PI: Akari Kamimura). We are extremely grateful to 351 children who participated in the study and provided invaluable data. We gratefully acknowledge the work of all staffs of our research group.

References

1. Nakamura K (1999) Transfiguration of children's play. J Health Phys Educ Rec 49:25–27 (in Japanese)
2. Montgomery C, Reilly JJ, Jackson DM, Kelly LA, Slater C, Paton JY, Grant S (2004) Relation between physical activity and energy expenditure in a representative sample of young children. Am J Clin Nutr 80:591–596
3. Reilly JJ, Jackson DM, Montgomery C, Kelly LA, Slater C, Grant S, Paton JY (2004) Total energy expenditure and physical activity in young Scottish children: mixed longitudinal study. Lancet 363:211–212
4. Finn K, Johannsen N, Specker B (2002) Factors associated with physical activity in preschool children. J Pediatr 140:81–85
5. Pate RR, Pfeiffer KA, Trost SG, Ziegler P, Dowda M (2004) Physical activity among children attending preschools. Pediatrics 114:1258–1263
6. Williams HG, Pfeiffer KA, O'neill JR, Dowda M, McIver KL, Brown WH, Pate RR (2008) Motor skill performance and physical activity in preschool children. Obesity 16(6):1421–1426
7. Ogden CL, Carroll MD, Curtin LR, McDowell MA, Tabak CJ, Flegal KM (2006) Prevalence of overweight and obesity in the United States, 1999–2004. JAMA 295:1549–1555
8. Ogden CL, Troiano RP, Briefel RR, Kuczmarski RJ, Flegal KM, Johnson CL (1997) Prevalence of overweight among preschool children in the United States, 1971–1994. Pediatrics 99:E1
9. Kamimura A, Kawata Y, Hirosawa M (2016) The relationship between birth month, physical size, motor ability and physical activity evaluated by kindergarten teachers among Japanese young children. Juntendo Med J 62(Suppl. 1):104–108
10. Kamimura A, Kawata Y, Izutsu S, Hirosawa M (2017) The effect of awareness of physical activity on the characteristics of motor ability among five-year-old children. In: International conference on applied human factors and ergonomics. Springer, Cham, pp 100–107
11. Kawata Y, Kamimura A, Oki K, Yamada K, Hirosawa M (2016) Relative age effect on psychological factors related to sports participation among Japanese elementary school children. In: Salmon P, Macquet A-C (eds) Advances in human factors in sports and outdoor recreation. Springer, Cham, pp 199–211
12. Kawata Y, Kamimura A, Hirosawa M (2016) Relationship between physical competence and frequency of exercise outside of school among elementary school students. J Phys Fit Sports Med 5(6):563
13. Rovniak L, Anderson E, Winett R, Stephens R (2002) Social cognitive determinants of physical activity in young adults: a prospective structural equation analysis. Ann Behav Med 24(2):149–156
14. Ntoumanis N (2002) Motivational clusters in a sample of British physical education classes. Psychol Sport Exerc 3(3):177–194
15. Boyd MP, Yin Z (1996) Cognitive-affective sources of sport enjoyment in adolescent sport participants. Adolescence 31(122):383–396

16. DiLorenzo TM, Stucky-Ropp RC, Vander Wal JS, Gotham HJ (1998) Determinants of exercise among children. II. A longitudinal analysis. Prev Med 27(3):470–477
17. Moore JB, Yin Z, Hanes J, Duda J, Gutin B, Barbeau P (2009) Measuring enjoyment of physical activity in children: validation of the physical activity enjoyment scale. J App Sport Psychol 21(S1):116–129
18. Beunen G, Malina RM, Ostyn M, Renson R, Simons J, Van Gerven D (1983) Fatness, growth and motor fitness of Belgian boys 12 through 20 years of age. Hum Biol 599–613

Effect of the Fragrance on Concentration

Yuka Saeki[✉]

Ehime University, Toon, Ehime 791-0295, Japan
yukas@m.ehime-u.ac.jp

Abstract. This study was designed to clarify whether the fragrance of peppermint oil has the effect of enhancing the concentration by using Psychomotor Vigilance Test (PVT). PVT was applied to the participants (n = 16, female) for 10 min in the room with peppermint oil and without the odor (control). The fragrance of peppermint was allowed to fill the room 10 min before the start of the measurement. The participants were asked to press a response button, located on the right side of the device, as soon as the visual stimuli appeared at random from 2 to 10 s in PVT. Visual Analogue Scale (VAS) was also used for subjective sensation (0, no concentrate at all; 100, concentrate very much). Reaction time (RT) to visual stimulation was measured for evaluation of concentration using PVT. Median of RT was significantly smaller in peppermint oil than that of control. Subjective sensation by VAS also showed significantly high concentration in peppermint oil compared with control. These results indicate that the concentration must increase by smelling the fragrance of peppermint oil. When people get tired during work, it suggests that concentration might improve by smelling the fragrance of peppermint followed by the prevention of human errors.

Keywords: Peppermint oil · Psychomotor Vigilance Test · Concentration

1 Introduction

Essential oils have been widely applied for getting relaxation or comfort [1], or antimicrobial effect [2]. Some essential oils have the effect of refreshing or increased concentration [3, 4]. Furthermore, the previous study reported that the inhalation of valerian caused a prolonging in sleep latency and prolonging in total sleep, whereas significant prolonging in sleep latency was observed with lemon inhalation in rats [5]. Thus, the odor may influence the cognitive function and concentration.

The Psychomotor Vigilance Test can measure the ability to remain alert, aware, and vigilant through responses to stimuli coming from a visual signal and this device has been used in many studies [6–8].

The purpose of this study was to clarify whether the fragrance of peppermint oil has the effect of enhancing the concentration by using PVT.

2   Methods

2.1   Subjects

Sixteen female participated in this study (21.3 ± 0.8 yrs.). They had no history of disease, drinking, smoking, or uses of any medications that could affect olfactory sensation and nervous system. The nature, purpose and risks of the present study were explained to each subject before written informed consent was obtained.

2.2   Measurements

Psychomotor Vigilance Test (PVT) for 10 min was used to measure sustained attention (PVT-192, Ambulatory Monitoring Inc., NY, USA). The device combines a four digit red LED screen, an alpha numeric information display screen and two response push buttons. The interstimulus interval, defined as the period between the last response and the appearance of the next stimulus, varied randomly from 2 to 10 s. The protocol adopted used only visual response test [9]. Vigilance was assessed by reaction time (RT) and number of lapses of attention. Lapses were defined as RT > 500 ms and false starts were defined as responses that occurred before the stimulus was presented or with a RT < 100 ms [10, 11]. The values obtained were analyzed using the software React (Ambulatory Monitoring Inc., Ardsley, NY).

Visual Analogue Scale (VAS) was also used for subjective sensation (0, no concentrate at all; 100, concentrate very much).

2.3   Study Protocols

Experiments were performed at a quiet air-conditioned room (room temperature 25.2 ± 0.6°C, humidity 30.2 ± 2.3%). PVT was applied to the participants for 10 min in the room with peppermint oil and without the odor (control). These experiments were performed in different room and the order of measurement was at random. Some drops of peppermint oils were dripped on the thick filter paper in the diffuser and the oil was warmed by a miniature bulb under the filter paper. The fragrance of peppermint was allowed to fill the room at 10 min before the start of the measurement. The participants were asked to press a response button, located on the right side of the device, as soon as the visual stimuli appeared at random from 2 to 10 s interval. Following the PVT in control and peppermint condition, VAS was measured as subjective sensation.

2.4   Statistical Analysis

The Kolmogorov-Smirnov test was used to confirm that the data obtained from each group showed a normal distribution. After confirming that the data showed a normal distribution, the significance of difference between control and peppermint was evaluated by applying Student's t-test and were considered statistical at p < 0.05.

This study was submitted to and approved by Ethics Committee of Ehime University Graduate School of Medicine under protocol no. 28-12.

3 Results

Median of RT was 253.2 ± 34.9 ms or 240.3 ± 33.8 ms in control or in peppermint oil, respectively, and the median RT of peppermint oil was significantly smaller than that of control (Fig. 1). Subjective sensation by VAS showed 46.1 ± 17.8 for control and 69.0 ± 21.5 for peppermint, there was a significant difference between two groups (Fig. 2). There was no significant correlation between RT and subjective sensation in both groups. Furthermore, there was no difference in lapse which means more than 500 ms of reaction time, between two groups.

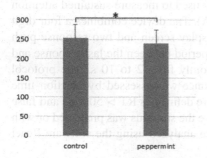

Fig. 1. Reaction time (msec) in PVT in 2groups, *; p < 0.05

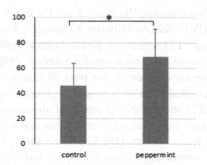

Fig. 2. VAS in 2groups, *; p < 0.05

4 Discussion

Peppermint oil is one of essential oils which is thought to be refreshing and increased concentration [3]. In this study, arousal effect of peppermint was investigated using the PVT.

The PVT is widely used for the research concerning sleep [7, 9], fatigue [10], the effects of shift work on the psychomotor performance of night workers [8] and these results have been evaluated. Therefore, it must be reasonable to use this device as an indicator of concentration in this study.

Since the reaction time to visual stimuli was faster in peppermint than control without odor, inhaling peppermint oil might have the effect of enhancing the concentration. The previous study shows that peppermint oil can reduce sleepiness when pupillay fatigue oscillation was measured as the indicator of sleepiness [3], consistent with this study.

Interestingly, there was no significant correlation between RT and subjective sensation, consistent with the previous study [3]. This result might suggest that there were some participants who despite saying that they could not concentrate at all under the scent of peppermint, RT for peppermint was faster than control. These results indicate that the concentration must increase by smelling the fragrance of peppermint oil apart from the subjective sensation.

In conclusion, peppermint oil has the effect to enhance concentration. When people get tired during work, it suggests that concentration might improve by smelling the fragrance of peppermint followed by the prevention of human errors.

References

1. Lehrner J, Marwinski G, Lehr S, Johren P, Deecke L (2005) Ambient odors of orange and lavender reduce anxiety and improve mood in a dental office. Physiol Behav 86:92–95
2. Seow YX, Yeo CR, Chung HL, Yuk HG (2014) Plant essential oils as active antimicrobial agents. Crit Rev Food Sci Nutr 54:625–644. https://doi.org/10.1080/10408398.2011.599504
3. Norrish MI, Dwyer KL (2005) Preliminary investigation of the effect of peppermint oil on an objective measure of daytime sleepiness. Int J Psychophysiol 55:291–298
4. Sakai N, Yoshimatsu H, Ikenishi T, Niikura Y, Kondo N, Sako N (2011) The promoting effect of mint odor on recovery from mental stress evoked by simple summation task. Tohoku Psychol Flolia 70:26–35
5. Komori T, Matsynoto T, Motomura E, Shiroyama T (2006) The sleep-enhancinf effect of valerian inhalation and slee-shortening effect of lemon inhalation. Chem Senses 31:731–737
6. Yamamoto I, Haseba T, Ohno Y, Nakagawa K, Ohira H, Yamada Y (2010) Evaluation of drunkenness by measuring psychomotor vigilance tasks (PVT). Bull Kanazawa Dent Col 39:138–140
7. Basner M, Mcguire S, Goel N, Rao H, Dinges D (2015) A new likelihood ratio metric for the psychomotor vigilance test and its sensitivity to sleep loss. J Sleep Res 24:702–713
8. Narciso FV, Barela JA, Aguiar SA, Carvalho ANS, Tufik S, de Mello MT (2016) Effects of shift work on the postural and psychomotor performance on night workers. PLOS ONE 11: e0151609. https://doi.org/10.1371/journal.pone.0151609 Accessed 5 Nov 2016
9. Van Dongen HP, Dinges DF (2005) Sleep, circadian rhythms, and psychomotor vigilance. Cli Sports Med 24:237–249. https://doi.org/10.1016/j.csm.2004.12.007 vii–viii, Epub 2005/05/17. S0278-5919 (04)00143-7. PMID: 15892921
10. Matthews RW, Ferguson SA, Sargent C, Zhou X, Kosmadopiulos A, Roach GD (2015) Using interstimulus interval to maximize sensitivity of the Psychomotor Vigilance Test to fatigue. Accid Anal Prev 99:406–410. https://doi.org/10.1016/j.aap.2015.10.013 Epub 2015 Nov 10. PMID: 26563739
11. Basner M, Dinges DF (2011) Maximizing sensitivity of the Psychomotor Vifilance Test (PVT) to sleep loss. Sleep 34:581–591

OSH Implementation in SMEs in Malaysia: The Role of Management Practices and Legislation

Lilis Surienty[✉]

School of Management, Universiti Sains Malaysia, 11800 Minden, Pulau Pinang,
Malaysia
lilis@usm.my

Abstract. In an operational analysis report of the Occupational Safety and
Health Regulations (Penggunaan dan Standard Pendedahan Bahan Kimia Ber-
bahaya Kepada Kesihatan) 2000 has found that 80% of more than 800 workplaces investigated failed to adhere fully to regulations. In addition, the Small
Medium Enterprises (SMEs) have been reported to contribute some 30 to 50%
of industrial accidents. Since SMEs constitute almost 99% of the Malaysian
business establishments, and contributing about 36% of the country's GDP,
improvements to its workers' safety and health issues should go hand in hand
with Malaysia economic development. Malaysia faces greater challenge to
monitor OSH requirements are adhered to in spite of trying to stay competitive
and survive with its limited capital or financial resources. Reviews have shown
that the involvement of various parties including the public in policy making,
the development of appropriate infrastructure and human resources, enforcement
autonomy, focused job scope within department, appropriate language usage,
training inclusive of all forms of diversity at work and appropriate penalty are
key success factors at reducing death rates, accidents and lost work days at the
workplace for these two countries. Through questionnaire distributed to 150
SMEs in Malaysia, results were analysed and findings with regard to the
implementation of OSH in workplace is discussed. The results showed that
safety rewards and employee participation in management practices are significant in OSH implementation in the SMEs. Practical implications are discussed.
Recommendations include open management and safety behaviour acknowledgement as reward.

Keywords: OSH implementation · Safety management practices
Legislation

1 Introduction

According to the most recent figures published by National SME Development Council
(2012), small medium size enterprises (SMEs) with less than 200 employees made up
98.5% of all firms up to 1st quarter of 2018 and accounted for an average of 65.5% of
jobs in all sectors and contributing 36.6% to total GDP (as compared to 55% in Japan
and rising to 41% by 2020 (Department of Statistics 2016; Lee 2018; Malaysian

© Springer Nature Switzerland AG 2019
S. Bagnara et al. (Eds.): IEA 2018, AISC 819, pp. 650–671, 2019.
https://doi.org/10.1007/978-3-319-96089-0_72

Productivity Report, 2017; SME Annual Report, 2015/16). However, despite their economic importance and the number of employees dependent on them, small firms have received very little attention from occupational health and safety (OHS) researchers in Malaysia. It is difficult to quantify OHS problems in small firms. For example, neither Department of Occupational Safety and Health (DOSH) nor National Institute of Occupational Safety and Health (NIOSH), two institutions responsible for occupational safety in Malaysia published risk indicators by firm size, and many people believe that small firms are more likely than larger firms to fail to declare injuries. In addition, accidents may be relatively rare in these firms due to their smaller workforce, and this may be a factor in the lack of attention given to the subject.

In addition, Lim (2002) reported that the number of reported accidents at the workplace especially in SMEs was worrying and statistics showed that SMEs in Malaysia had the workplace accident rate of 30% to 50% higher than big companies. Specifically, they are the major contributors to 80% of total accidents in Malaysia (Abdul-Aziz et al. 2015). Results from audits conducted by DOSH for two thousand and six hundred SMEs in 2002 showed poor compliance (Yahaya, 2002). Thus, although SMEs are important to the country, their contribution to workplace accident statistics is equally substantial.

Generally speaking, all employers and employees have the same OHS rights and obligations. The provisions of Malaysian's Occupational Health and Safety laws concerning prevention programs and joint health and safety committees, however, apply differently to small firms with 40 employees or less. Furthermore, a unit contribution rate applies uniformly to all firms, thus depriving them of an economic incentive to manage occupational health and safety. The models developed for larger corporations have proved to be ineffective and the difficulty of contacting smaller firms, their geographical dispersal and their short life spans, have all helped ensure that they have been left more or less to their own devices in terms of occupational safety (Lahm 1997; McKinney, 2002).

SMEs have significantly different characteristics from large businesses in terms of their financial, expertise and staffing capabilities. These issues affect the performance of SMEs in terms of compliance with safety regulations and have generated substantial on-going debate between practitioners about designing regulatory and enforcement strategies that optimize compliance levels. However, the way in which these factors affect SME compliance in the past has been difficult to assess due to the problems in gaining published references to empirical research with regard to safety in Malaysian SMEs. This article builds upon works in this field by exploring the results of empirical research undertaken within the area of occupational safety around the world on SMEs. This research provides a significant contribution to the existing field of knowledge because it evaluates the way that factors perceived to affect SME compliance actually impact upon SME behaviour and compliance levels within individual employees. The issues found to affect SME behaviour and attitudes towards occupational safety have important implications for both occupational safety and environmental policies in terms of regulatory and enforcement strategies.

In Malaysia, SMEs is categorized based on either their annual sales turnover or the number of full-time employees as shown in the table below:

	Micro enterprise	Small enterprise	Medium enterprise
Manufacturing Sector	Sales turnover of less than RM300,000 or full time employees less than 5	Sales turnover between RM300,000 and less than RM15 million or full time employees between 10 and 75	Sales turnover between RM15 to less or equal to RM50 million or full time employees between 75 and 200
Services and other services	Sales turnover of less than RM300,000 or full time employees less than 5	Sales turnover between RM300,000 and less than RM3 million or full time employees between 5 and 30	Sales turnover between RM3 million and less or equal to RM20 million or full time employees between 30 and less or equal to 75

(*Source:* SME Corp, 2013)

SMEs in Malaysia continue to face many challenges- both traditionally and new ones. One of the many challenges that SMEs face is the high workplace accidents rate which may reflect negatively to workers' wellbeing and safety and eventually company's goodwill as a potential partner. Personal communication with vendors for big automakers in Malaysia, revealed that the main automakers monitor the accidents report incurred in each of the vendors closely and are very concerned over unsafe practices that may exist and the vendors believed that this could be a factor for a termination of a business contract. Researches have shown that accidents at workplace can be reduced if employees and employer are more sensitive or have good safety behaviour (Makin and Sutherland 1994; Christian et al. 2009). In addition, effective implementation of OSH will assist in the formation of good safety behaviour because OSH implementation requires employer to cater the safety needs of the employees and the employees to have some responsibility towards their own safety. Accidents at workplace involve monetary cost as companies paid out compensation to employees involved in the workplace accidents. The amount has been high in the last few years and this cost a lot of money to the economy which can otherwise be used for other productive purposes such as an investment in new technology to boost productions. Importantly, this accidents would also incur talents lost and would cost companies to manage such talents lost to accidents.

An effective implementation of the Occupational Safety and Health (OSH) practices could reduce accidents thus decrease compensation paid. Studies (Jaselskis and Suazo 1993; Teo and Phang 2005) have found that safety measures taken in the workplace can lead to better safety performance and Laukkanen (1999) also emphasised on safety as being part of a skilful job performance. In order to generate higher returns by reducing cost incurred to pay out to compensation, it is important for the SMEs to acknowledge the importance of OSHA. If safety problems and health risks in the work environment is reduced, the country's economic competitiveness will eventually improve.

Workplace safety and health in Malaysia is the responsibility of the Department of Occupational Safety and Health (DOSH) an agency under the Ministry of Human Resources. The National Institute of Occupational Safety and Health (NIOSH) which provides training courses, seminars and research in OSH assists DOSH in its works to improve companies implementation of OSH. In term of regulations, OSHA 1994 and Control of Industrial Major Accidents 1966 (CIMAH 1996) are the major laws that provide the provision on areas such as the responsibilities of employers to provide a safe working environment, training, and emergency preparedness (Shaluf and Fakhru'l-Razi 2003). These bodies oversee all workplace safety and health matters no matter what the business size is although for business operations, the country has set up a specific agency, i.e. SME Corp, to promote the SMEs activities.

SMEs and large multinational companies are very different in many aspects. These differences have huge impact when it comes to the implementation of OSH and this raises several issues. The size of the company plays a big part in the effectiveness of OSH implementation (Cook 2007; Hong 2011; Saksvik et al. 2003). Large multinational companies often have the financial muscle and structure to effectively implement a good occupational and safety (OSH) system. They have the financial capability which in most cases, lacking in SMEs in committing and developing a safety program inside their organisation. Large multinational companies also have the necessary size and structure to effectively implement and benefit from OSH. Implementing OSH can consume a huge amount of financial resources and it represents a considerable investment by the company. When such an amount of financial resources are invested, the company expects it to benefit every employee inside the organisation. In this case, large companies with bigger number of employees lead OSH program to benefit more people thus spreading the cost per person much thinner. In the case of SMEs, OSH is often perceived as irrelevant as they do not have a huge workforce and OSH implementation cannot be translated into direct monetary gain for the company and thus seems as unimportant for companies' survival (Lahm 1997; McKinney 2002). In conclusion, SMEs as important they are to the country's economic development and unique in their set up should be subjected a different OSH management than for those used to evaluate general establishments (such as MNCs and local business conglomerates) in order to ensure that SMEs do follow the health and safety requirements that could ensure that workers' safety is not compromised.

This paper investigates relationships between organizational characteristics (i.e., management practices and legislation) and the extent of implementation of OSH in 150 smaller businesses in Northern Region of Malaysia. We also examine whether legislation moderates the relationships between management practices and OSH Implementation in smaller enterprises.

2 Literature Review

2.1 OSH Implementation

'Implementation' can be defined as a course of action taken to put into use an idea, decision, procedure or program (Klein and Sorra 1996). It is the move to adopt the safety practices different from compliance which is in relation to a certain requirement. Implementation does not necessarily referring to following the OSH Act as an example but more of capturing the act of adopting an extent of the safety practices known to employer to ensure safety practices exist. Then, the immediate outcome of interest is initial use or early use of this new idea. OSH implementation that is measured here would be implementation that is following at least the minimum requirements outlined by the combination of the local Occupational Safety and Health Act (OSHA) 1994 and ILO OSHA requirements. This activity is an organisational level construct as implementation OSH must be interdependent and coordinated among the many functional, departments, work shifts and locations. It is organisation wide initiatives and not an individual basis. In addition, the outcome should be a collective benefit, such as to workforce health, improve productivity and lower health care costs.

Occupational Safety and Health Act (OSHA) 1994 is the guiding act that governs occupational safety practices in the country. It became the main legislation in OSH regulation in Malaysia, replacing all other legislations before it. These regulations apply to all establishments in the country and do not differentiate between SMEs and other types of company such as MNCs. As such, the regulations may pose a burden to SMEs as they have different capabilities compared to other large establishments and the regulations fail to address the differences. The idea of OSHA 1994 is more of a self-regulation of occupational safety and health (DOSH 2006). This means that both employer and employee share the responsibilities in the implementation of OSH at the workplace. However, Section 15 of the Act places that it is the general duty of an employer to ensure the safety, health and welfare at work of all employees. Thus, employer including those of SMEs has a higher burden to ensure implementation of OSH effectively following Section 15 of the Act.

Implementation of OSH measured in this study covers aspects such as assessing the availability of an organisation's OSH policy, employer's safety leadership, provision and easy access to safety training, documentation system and management, emergency preparedness and response arrangements, the appointment or the availability of a safety officer, risk assessment record and provision of the conducive work environment to develop work safety at the workplace.

2.2   Safety Management Practices

Ruth (2004) has found that construction site injuries often result from managerial issues, rather than engine defects. Management practices are important in helping to reduce accidents in the workplace. Effective management practices are found to lead to positive organizational outcomes (Becker and Gerhart 1996) such as turnover (Huselid 1995) and productivity (Katz et al. 1987). Specifically, management practices are generally categorized into the following: hiring and selection practices, incentives and rewards, safety training, communication and feedback, worker participation, management commitment, and performance evaluation. These are reviewed below.

Management Commitment. Management commitment is defined as the management's involvement and engagement in actions towards achieving a goal (Cooper 2006). Gilkey et al. (2003) found that management support is important in the implementation of OSH. Management commitment is the most important factor to a successful preventive programme of safety (Champoux and Brun 2002). It is important for managers to be actively involved to be able to identify and be made responsible for the allocation of resources for a successful implementation of OSH. In addition, initiatives taken by management are considered fundamentals in ensuring a successful implementation (Saksvik et al. 2003). Ashill, Carruthers and Krisjanous (2006) found that management commitment is manifested through various ways such as having safety education and training, giving rewards, and empowerment of employees to make decisions. Investment in safety education and training will allow employees to gain the necessary safety knowledge and help them to work safely. Furthermore, giving rewards to employees that report unsafe behaviours of co-workers during working is also an important aspect of OSH successful implementation. All the measures mentioned require a considerable amount of financial investment. For example, the company needs to pay for the safety training of the employees. The management must be committed to invest money into those activities in order to effectively implement OSH. On the other hand, management commitment can also be viewed from their commitment in penalizing employees who do not follow safety measures such as the use of personal protective equipments (Harper, 1998; Holmes 1999). This view is supported by Dejoy (1985) as well as Lin and Mills (2001) with the former stressing on a two way communication between employees and managers to facilitate the effective implementation of OSH. It is a fact that an SME organization structure usually conform of a simple structure that include a managing director who sometimes acts as the financial manager and also as the human resource manager. Because of the small size of the organization, every additional head count means cost to the organization and people are usually assigned more than just one job responsibility. And, management of SMEs spends little time on safety issue- a phenomenon that is also prevalent in the United Kingdom and Spain (Vassie et al. 2000). Thus, it is hypothesized that,

H_1: Management commitment is positively significant towards OSH implementation.

Hiring Practices. Hiring and selection are the first group of HR practices adopted during the recruitment process. Terpstra and Rozell (1993) found that well-designed selection procedures improve overall organizational performance. During selection, the applicant's personality and qualifications are checked. Employees' personality and safety behaviours are found to be correlated (Keehn 1961; Lai, Liu and Ling 2011). For example, employees' characteristics such as extroversion (Powell et al. 1971), aggression (Hansen 1989) and intelligence (Tiffin and McCormick 1962) have been found to affect safety outcomes. Physical ability is sometimes considered during selection of employees. Every individual has different physical abilities and capacity to carry out the work (Hale and Hale 1972). It is important to select individuals who are able to work safely within the organization's environment (Letho and Salvendy 1995). Personal characteristics and physical attributes such as age, stature and body weight have been found to be significantly correlated with safety management outcomes (Lai, Liu and Ling 2011) in US as well as Singapore.

Selecting individuals with relevant working experience has been found to produce better safety outcomes (Powell et al. 1971). Hansen's (1989) causal model of accidents shows that inadequate job experience contributes to accidents. Furthermore, Mohamed (2003) has found that workers' perception of risk is influenced by their work experience. This suggests that working experience is one of the important factors to be considered during recruitment. Moreover, some studies believe that much consideration has to be put on the employees' alcohol and drug consumption habits during the hiring process in order to improve safety outcomes. Spicer et al. (2003) found that alcohol usage leads to injuries at the workplace. Other studies had found that alcohol and drug abuse affects workers' ability to work safely (Normand et al. 1990). Thus, it is hypothesised that,

H_2: Hiring practices are positively significant toward OSH implementation.

Safety Training. Tam and Fung (1998) found that safety training is important to decrease accident rate in the workplace. Safety training that includes more level of employees would lead to more reduction in accident rates because it is the most direct way to increase workers' technical skills and awareness of safety. According to Vinodkumar and Bhasi (2010), safety training plays a significant role in accident prevention programme and in any occupational safety and health programme. Safety training should include making workers more familiar with the tasks on site, and teaching them how to perceive potential hazards and consequences (Lingard 2002; Ricci, Chiesi, Bisio, Panari and Pelosi 2016). Gillen and others (2004) have found that safety training is an important measure used to improve safety performance. Construction sites managers who were interviewed place safety training high on their priority as they believed it will raise employees' awareness on workplace hazards and other safety issues in general. Eventually, they will display safe behaviour because those safety issues affect their own safety at work. Cheng, Li, Fang and Xie (2004) who studied the construction sites in China have found that site managers and safety officers strongly advocate the use of safety training to train their employees on safety issues. Finally, Burke et al. (2006) who had done a meta-analysis on 95 studies have found that safety training significantly improves employees' safety behaviour. Specifically, engaging training method is more effective at improving safety behaviour. Therefore, it is hypothesised that,

H_3: Safety training is positively significant towards OSH implementation.

Employee Participation. Walters and Nichols (2006) have found that joint arrangement that involves consultation with employees on safety and health issues lead to improved employees' awareness on safety and health issues and consequently, safety performance. Moreover, study by Dillard (1997) confirmed that employee participation leads to better safety performance and the reduction of worker compensation cost as results of decreased accident cases. Latham and Yukl (1975) also found that if employees are allowed to participate in setting safety goal, they exhibit greater performance and improves safety participation in OSHMS (Khor and Surienty 2017). This suggests that safety performance through safety implementation can be improved if employees participate in the decisions-making and goal-setting of health and safety issues. Finally, Fuller (1999) had found that the creation and implementation of a consensus program leads to the improvement of workplace safety. Specifically, it is suggested that management and employees works together in the assessment of safety issues and identification of strategies to solve safety problems faced by them. Thus, it is hypothesised that,

H_4: Employee participation is positively significant towards OSH implementation.

Safety Incentive. Study conducted by Saari and Latham (1982) on mountain beaver trappers found that monetary strengthens the relationship between hourly wage and employee performance. The mountain beaver trappers were already performing (catching beavers) when they are given a fixed hourly wage. Then, monetary incentive (bonus) was given in addition to the hourly wage to the mountain beaver trappers. Their performance improved significantly (as measured by the number of beavers caught) when the incentive is given. Therefore, this finding suggested that incentive can act as moderator that strengthens the relationship between pay and performance and also with safety performance. In addition, a study conducted by Shaw, Gupta and Delery (2002) in the trucking industry confirmed that incentive can act as moderator that influences the relationship between pay dispersion and safety performance of employees (measured by frequency of accidents). Truckers in the companies participating in the study receive different level of pay according to their experience and seniority. In addition, individual incentive that is tied to individual truckers' performance is given together with the basic pay. The study found that when the level of difference in pay is high, employees will experience fewer accidents when there are plenty of individual incentives given to them. On the other hand, common punishments such as fines and suspension from work are used as deterrents against violating safety rules. Previous studies by Aksorn and Hadikusumo (2008) and Lai, Liu and Ling (2011) have identified that punishment as one of the critical success factors in influencing safety program performance on sites.

Past studies have found a significant positive relationship between giving out monetary rewards and frequency of 'other accidents' (Lai, Liu and Ling, 2011). The more monetary rewards are given out to workers when there are fewer accidents, the frequent 'other accident' happened on sites. On the other hand, Haines and others (2001), Hong, Surienty, and Daisy (2011), Khor and Surienty (2017), and Vredenburgh's (2002) studies in Canada, Malaysia and the US discovered that monetary reward has no effect in reducing injuries at workplace or encouraging safety performance. Teo and others (2005) have found that monetary rewards lead to better firm's safety performance. This suggests that monetary reward is not a stable incentive for safety management outcomes. Therefore, it is hypothesised that,

H_5: Safety incentive is positively significant towards OSH implementation.

Legislation Moderates the Relationship Between Management Practices and OSH Implementation. According to The Reinforcement Theory, negative reinforcement is theorised to reduce unwanted behaviours. Legislation which is conceptualised as guiding rules that could act as a negative reinforce to reduce unsafe practices (behaviour). Singapore, which has one of the lowest workplace accident rates in the region is subjected to a strict enforcement of safety standard, training of workers and safe work practice through its Employment and The Factories Acts that aim to protect the safety and health of employees (Koh and Jeyaratnam 1998). Similarly, in Australia, employers have a duty of care to provide for their workers with a safe working environment. Studies have shown that effective enforcement of the legislation influences observance of rules and regulations as shown in Norwegian companies that showed increasing adoption safety practices (Saksvik et al. 2003). It is found that legislation that in place encourages implementation and adoptions among companies and even showed improvements over a three-year assessment. In this case, it is the health and safety requirements at the workplace (Cooke & Gautschi III, 1981; McQuiston, Zakocs and Loomis, 1998). Moreover, Cooke and Gautschi III (1981) found that the inspections on workplaces carried out by the authority managed to reduce workplace accidents. If rules and regulations are followed to the letter then it is concluded that OSH is implemented. Furthermore, Eakin and Weir (1995) have found that SMEs lack both the resources and motivation to deal with OSH issues. And with lack of enforcement from the legislative bodies overseeing the implementation of OSH Act, SMEs are not driven to fulfil their role as implementation would also means additional cost. Enforcement is able because legislation written supports the action that follows. Thus, it is hypothesized that,

H_{6a}: Legislation moderates the positive relationship of management commitment, hiring practices, employee involvement, safety reward and safety training with general safety compliance, safety documentation, safety procedure and suppliers' OSH such that the relationships will be stronger when legislation is high.

H_{6b}: Legislation moderates the negative relationship of management commitment, hiring practices, employee involvement, safety reward and safety training with safety resources and risk assessment such that the relationships will be weaker when legislation is high.

3 Methodology

Sample and Procedure. This study aims to measure the OSH implementation status as well as the management practices that contribute to OSH implementation in SMEs. However, this study is carried out in the Northern Corridor of Economic Region (NCER) - an economic development area designated by the Malaysian covering the state of Penang, Perak, Perlis and the northern area of Perak. Thus, the sampling frame would be all SME operating in all four states of NCER.

Listing of SMEs operating in NCER was obtained from the directory provided by SMECorp (formerly known as SMIDEC) through their website. Firstly, the SME was divided into 4 strata following the 4 states in NCER. After that, random stratified sampling method was used to select the SMEs for this study. SMEs in each stratum were numbered prior to the selection and a web-based number generating software was used to generate a random set of number representing the SMEs from each of the stratum. Then, based on the information provided in the SMECorp listing, the researcher contacted the SMEs and explained the study to them as well as to solicit their cooperation to participate in this study. Companies that declined to participate or those that are out of operation would be replaced by another one using the same selection procedure mentioned. In the end, a total of 222 SMEs agreed to participate in this study- 168 from Penang, 46 from Kedah, 6 from Perak and 2 from Perlis. The questionnaire was posted to the safety and health officer (SHO) or personnel involved in OSH management of the SMEs together with a cover letter and self-addressed and postage paid envelope.

Questionnaire Design. The questionnaire for this study is divided into 5 sections. Section 1 contains items that measure the implementation status of OSH in the organisation. There is a total of 89 items for this section. The items were adapted from an OSH compliance study done for Proton vendors based upon Part 1:2005 OSH MS (Occupational Safety and Health Management Systems – Requirements) and ILO-OSH-2001 Guidelines (ILO 2001). An example of the item includes: "There is an OSH policy in the organization."

Section 2 contains 26 items that measure the 6 management practices in this study. There are 7 items measuring worker participation, 4 items measuring safety training, 3 items measuring hiring practices, 3 items measuring reward system, and 5 items measuring management commitment. Items in worker participation were adapted from Vredenburgh (2002) and Dillard (1997). Items were adapted from these 2 studies because they reflect worker participation in an organisation clearly. An example includes: "Management solicits opinions from employees before making final decisions on safety issues."

Items in safety training were adapted from Vredenburgh (2002) and Worsfold and Griffith (2003). Items in these 2 studies were chosen because they reflect safety training in a direct and clear manner. For example, "On-the job training is given to all new starters" would instantly give the respondent a clear statement on safety training.

Items in hiring practice were adapted from Vredenburgh (2002). There are not many studies that examine hiring practices in relation to organisational safety. Study done by Vredenburgh (2002) is one of the few that provide a clear measurement to measure the management's hiring practices to ensure that potential workers for the organisation have good prior safety records. Therefore, items from Vredenburgh (2002) were used for this variable. An example includes: "Employees are hired based on a good safety record in their previous positions."

Items in reward system were adapted from Vredenburgh (2002). Items from this study were chosen because it clearly defines the reward mechanism used in an organisation to reward employees on their safety performance. An example of the item includes: "Employees are rewarded for reporting a safety hazard (e.g. thanked, recognized in organisation's bulletin board, give cash or other reward)."

Items in management commitment were adapted from Vredenburgh (2002). The items from this study were used because they define clearly the way management show their commitment and concern for the safety of employees inside the organisation. An example of the item includes: "Management is concerned about general health and safety of employees."

Section 3 contains items on the legislation or control factor of OSH that affects the implementation of OSH in organisation. Information about legislation was obtained from McQuiston, Zakocs, and Loomis (1998) and Fairman and Yapp (2005). Elements of legislation were extracted from these studies and adapted to become the survey items. An example includes: "Legislation requiring consultation with workers regarding safety issues will lead to the organization complying with this rule."

Section 4 contains items on demographic variable. Personal information such as age, gender, and education level of the respondents are asked. Besides, information on the participating companies such as type of business and years of establishment was also gathered. Response scale was a seven-point Likert scale (ranging from 1 = "Strongly Disagree" to 7 = "Strongly Agree").

The respondents are company employees who are responsible for OSH issues at their workplace. It is believed that the knowledge on OSH possessed by these individuals would enable them to answer the survey because they are involved in the day-to-day operation of OSH in their company. Therefore, for this study, the survey is targeted at employees who are in charge of OSH issues at their workplace. One hundred and fifty two usable questionnaires were collected and analysed using SPSS. These 150 respondents represents 35 companies in the manufacturing, packaging, maintenance, services and construction. The majority of the employees occupied the junior management position but there are also from the middle and the top management. The mean age was 38.5 years, ranging from 23 to 70 years. Of the 150 respondents, 99 (66%) were male and 51 (34%) were female.

4 Results

The descriptive statistics (means and standard deviations) and the correlations between OSH implementation, management practices, and legislation are presented in Table 1.

Management commitment predicted general safety compliance ($\beta = 0.65$, $p < 0.05$), safety resources ($\beta = -0.45$, $p < 0.05$), and risk assessments ($\beta = -0.52$, $p < 0.05$), components of OSH implementations providing partial supports for Hypotheses 1 as management commitment is negatively predicting safety resources and risk assessment. Hiring practices predicted safety documentation ($\beta = 0.38$, $p < 0.05$), safety procedures ($\beta = 0.33$, $p < 0.05$), safety resources ($\beta = 0.19$, $p < 0.05$) and suppliers' OSH ($\beta = 0.22$, $p < 0.05$) providing only partial support for Hypothesis 2. Safety training predicted safety documentation ($\beta = 0.18$, $p < 0.05$), safety procedure ($\beta = 0.34$, $p < 0.05$), and suppliers' OSH ($\beta = 0.27$, $p < 0.05$), components of OSH implementations providing partial supports for Hypotheses 3. It is because safety training does not significantly relate to general safety compliance and risk assessment. Employee participation predicted general safety compliance ($\beta = 0.25$, $p < 0.05$), safety documentation ($\beta = 0.21$, $p < 0.05$), and safety procedure ($\beta = 0.24$, $p < 0.05$), components of OSH implementations providing partial supports for Hypotheses 4. Safety reward predicted general safety compliance ($\beta = 0.10$, $p < 0.05$) and safety procedure ($\beta = -0.26$, $p < 0.05$), 2 out of 5 of the components of OSH implementations providing partial supports for Hypotheses 5 (see Table 2).

Table 1. Correlation matrix, mean and standard deviation of study variables.

Variables	1	2	3	4	5	6	7	8	9	10	11
1. General compliance	**.96**										
2. Safety procedures	.41**	**.97**									
3. Safety documentation	.57**	.61**	**.92**								
4. Safety resources	−.51**	.02	−.17*	**.75**							
5. Supplier's OSH	.44**	.56**	.45**	−.14	**.77**						
6. Risk assessment	−.47**	−.15	−.21**	.53**	−.22	**.61**					
7. Management commitment	.87**	.24**	.37**	−.50**	.40**	−.46**	**.93**				
8. Hiring practices	.10	.48**	.46**	.20*	.33**	.07	.02	**.82**			
9. Employee participation	.73**	.30**	.42**	−.36**	.35**	−.26**	.65**	.08	**.79**		
10. Safety reward	.43**	−.09	.24**	−.25**	.13	−.05	.34**	−.01	.42**	**.76**	
11. Safety training	.40**	.53**	.45**	−.06	.46**	−.09	.34**	.37**	.34**	.15	**.48**
M	5.70	4.44	4.97	2.75	4.18	2.81	5.76	2.99	5.03	4.57	4.67
SD	0.82	1.31	1.12	1.06	1.23	1.05	0.79	1.37	1.11	1.27	1.01

Note: ** $p \leq 0.01$, * $p \leq 0.05$. Number in diagonal refer *to Cronbach alpha.*

Legislation has a positive significant relationship with only general safety compliance of the OSH implementation ($\beta = 0.12$, $p < 0.05$). Specifically, the interaction of management commitment and legislation emerged as a significant predictor for safety documentation ($\beta = 1.48$, $p < 0.05$); safety procedures ($\beta = 1.63$, $p < 0.05$), safety resources ($\beta = -2.38$, $p < 0.05$), and suppliers OSH ($\beta = 2.15$, $p < 0.05$). Hiring practices and legislation emerged as a significant predictor of general safety compliance ($\beta = -0.61$, $p < 0.05$) and risk assessment ($\beta = 1.55$, $p < 0.05$). Safety reward and legislation is a significant predictor of general safety compliance ($\beta = -0.70$, $p < 0.05$), safety documentation ($\beta = -1.72$, $p < 0.05$), safety resources ($\beta = 2.69$, $p < 0.05$), suppliers OSH ($\beta = -1.59$, $p < 0.05$), and risk assessment ($\beta = 1.17$, $p < 0.05$). Safety training and legislation is a significant predictor of only suppliers OSH ($\beta = -0.66$, $p < 0.05$). Legislation did not moderate any of the employee participation and OSH implementation relationships. Thus, initial results showed that Hypotheses 6a and 6b are partially supported.

To further explore the interactions, the simple slopes for management commitment, hiring practices, safety reward and safety training at high and low levels (1 SD above and below the mean) of OSH implementation were plotted. As predicted high levels of legislation enhanced the effect of management commitment on safety documentation, safety procedure and suppliers' OSH (Figs. 1, 2 and 3). In addition, high levels of legislation weakened the negative effects of management commitment on safety resources and risk assessment. These findings partially support Hypothesis 6a and 6b suggesting that safety resources and risk assessment would be improved when there is high levels of legislation and management commitment was highest, as shown in Figs. 4 and 5.

Table 2. Results of hierarchical regression analysis for osh implementations, management practices, and legislation, their interactions with the moderator

OSH implementations:	General compliance		Safety documentation		Safety procedure		Safety resources		Suppliers OSH		Risk assessment	
	β	R^2	β	R^2	β	R^2	β	R^2	β	R^2	β	R^2
Management practices												
Management commitment	.65		–		–		–.45		.26		–.52	
Hiring practices	–		.38		.33		.19		.22		–	
Employee participation	.25		.21		.24		–		–		–	
Safety reward	.10		–		–.26		–		–		–	
Safety training	–	.81^c	.18	.42^c	.34	.45^c	–	.30^c	.27	.32^c	–	.23^c
Moderator												
Legislation	.12	.82^c	–	.42	–	.45	–	.31	–	.32	–	.23
2-way interactions												
Management commitment X legislation	–		1.48		1.63		–2.38		2.15		–2.61	
Hiring practices X legislation	–.61		–		–		–		–		1.55	
Employee participation X legislation	–		–		–		–		–		–	
Safety reward X legislation	–.70		–1.72		–		2.69		–1.59		1.17	
Safety training X legislation	–	.84^b	–	.50^c	–	.48	–	.42^c	–.66	.39^a	–	.33^c

Note. $N = 149$. Nonsignificant coefficients are not included. ^{a} $p < .05$. ^{b} $p < .01$. ^{c} $p < .001$.

Figure 6 shows that hiring practices-general safety compliance relationship was positive for legislation and was significant for the positive slope for low levels of legislation. Thus, this result does not support Hypothesis 6a suggesting that general safety compliance increases as hiring practices that is related to safety becomes higher when legislation is low. However, contrary to our expectations, when organisations are high in their hiring practices, risk assessment would be higher when legislation is high. In addition, risk assessment is low if legislation is high when hiring practices is also low (see Fig. 7).

Figures 8, 9, 10, 11 and 12 show the safety reward-safety implementation relationships. Specifically, Fig. 8 shows that the relationship between safety reward and general safety compliance was positive for low level of legislation. Thus, our result does not support Hypothesis 6a suggesting that general safety compliance increases as safety reward increases only when legislation is low, as shown in Fig. 8. Similarly, Fig. 9 shows the safety reward-safety documentation relationship. Safety documentation s higher as safety reward increases for low level of legislation. But, safety documentation would be higher when safety reward is low but legislation is high. Thus, our result does not support Hypothesis 6a suggesting that safety documentation increases as safety reward increases when legislation is low. However, safety documentation would be lowest when organisation's low on safety reward and legislation is low, as shown in Fig. 9.

Figure 10 shows the safety reward-safety resources negative relationship with legislation low to moderates the relationship. Safety resources is reduced as safety reward increases for low level of legislation. Thus, our result supports Hypothesis 6b suggesting that safety resource reduces as safety reward increases when legislation is low. However, safety resources would be highest when organisation's low on safety reward and legislation is low, as shown in Fig. 10.

On the other hand, Fig. 11 shows the significant positive safety reward- suppliers OSH relationship. Suppliers' OSH increases as safety reward increases for low level of legislation. Thus, our result partially support Hypothesis 6a. However, Suppliers' OSH would be lowest when organisation's low on safety reward and legislation is still low, as shown in Fig. 11.

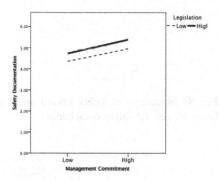

Fig. 1. Interaction of management commitment and legislation on safety documentation.

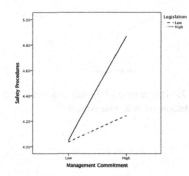

Fig. 2. Interaction of management commitment and legislation on safety procedures.

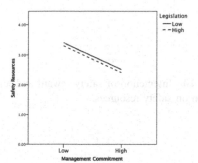

Fig. 3. Interaction of management commitment and legislation on safety resources.

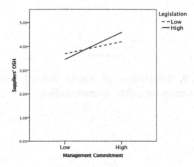

Fig. 4. Interaction of management commitment and legislation on suppliers' OSH.

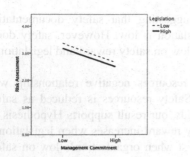

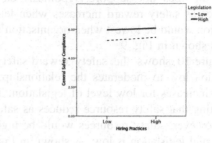

Fig. 5. Interaction of management commitment and legislation on risk assessment.

Fig. 6. Interaction of hiring practices X legislation on general safety compliance

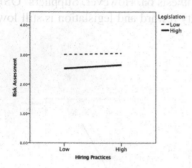

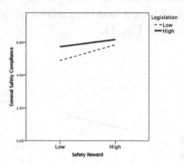

Fig. 7. Interaction of hiring practices and legislation on risk assessment.

Fig. 8. Interaction of safety reward and legislation on general safety compliance.

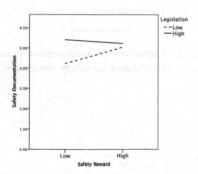

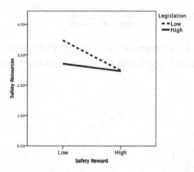

Fig. 9. Interaction of safety reward and legislation on safety documentation.

Fig. 10. Interaction of safety reward and legislation on safety resources.

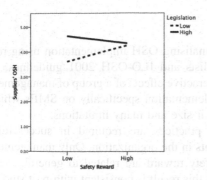

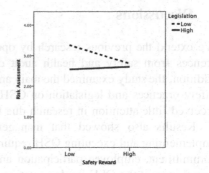

Fig. 11. Interaction of safety reward and legislation on suppliers' OSH

Fig. 12. Interaction of safety reward and legislation on risk assessment

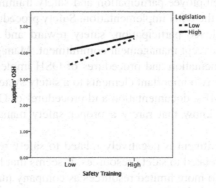

Fig. 13. Interaction of safety training and legislation on suppliers' OSH

Similarly, Fig. 12 shows the safety reward-risk assessment negative relationship when legislation is low. Safety resources is reduced as safety reward increases for low level of legislation. Thus, our result partially support Hypothesis 6b. However, safety resources would be highest when organisation's low on safety reward as well and legislation is low, as shown in Fig. 12.

Finally, Fig. 13 illustrates the relationship of safety training and suppliers' OSH. As safety training increases suppliers' OSH implementation when legislation is low and when safety training is low, suppliers' OSH is also reduced for low level of legislation. Thus, our result does not support Hypothesis 6a suggesting that suppliers' OSH is highest when safety reward is high and legislation is low instead of high, as shown in Fig. 13.

5 Discussions

We extend the previous research by operationalising OSH implementation using references from safety and health audit checklists and ILO-OSH 2001 guidelines. In addition, the study examined the main and interactive effects of a group of management safety practices and legislation on OSH implementation specifically on SMEs which received little attention in research due to their size and many limitations.

Results also showed that management practices are required in successfully implementing and executing OSH requirements in the organization. Only management commitment, employee participation and safety reward are related to general safety compliance of the OSH implementation. And this result is consistent with past studies where management initiatives such as providing appropriate reward, training, and investing in the environment are important in the successful implementation of OSH (Ashill, Carruthers, and Krisjanous 2006; Gilkey et al. 2003; Hong 2011). Furthermore, only hiring practices, employee participation and safety training are positively related to safety procedures of the OSH implementation. Safety procedures are affected by all 4 (hiring practices, employee participation, safety reward and safety training) of the management practices except management commitment. Management commitment do not affect safety documentation and procedures of OSH implementation. Documentation and procedures are two important elements to a safety system (Yu and Hunt 2004). It is possible that in SMEs, documentation and procedures of OSH implementation are not relevant since we know that rarely a proper safety management system is also present in SMEs.

Management commitment is negatively related to safety resources whereas hiring practices is positively related to safety resources. It seems that hiring practices exercise would end up leading to more limited resources as company may think that since there are safety competent personnel that has been recruited being able to spearhead safety issues. Only three of the 5 management practices encourage Suppliers' OSH of the OSH implementation. They are management commitment, hiring practices and safety training. As this involves an external party, management commitment is important to lead the initiative and good hiring practices and safety training would allow the company to find the right match to ensure the Suppliers' OSH practices are implemented. Consistent with findings from Mullen and others (2017), employers who are perceived to have fulfilled safety-related obligations, i.e. management commitment towards safety practices, employees tend to reciprocate with positive safety performance behaviours.

As for risk assessment implementation, only management commitment is negatively related to it. Specifically, management commitment would lessen no practices for risk assessment. Since safety resources and risk assessment involves allocation of budget, only management commitment in SMEs would be able to move those aspects of OSH implementation. And, this situation may be a specific SMEs condition due to its single ownership situation and owner action as the only top management which is unlike a big company.

The positive effects of legislated employer safety obligations appear to be stronger when management within the organization are perceived to promote safety and engage in effective commitment for OSH implementation. Our data suggest that the positive relationship between management commitment and OSH implementation is strengthen with lots of legislations. Thus, is important for organizations to recognize that the success of safety programs and policies depend on effective commitment from the management. Our results are consistent with previous research demonstrating that workplace safety may be shaped by various organizational safety influences including management's concern for safety (Griffin and Neal 2013), leader safety practices (Zohar and Polachek 2014; Mullen et al. 2017). Researchers have also shown that leader safety practices buffer the negative effects of organizational and individual influences on safety performance (Kao et al. 2015).

The other interactive effects showed that low level of legislation instead of high level of legislation that either enhances or weaken the effect to OSH implementation. As explained, as SMEs have many limitations attached to them that too much legislation would only break them instead of strengthening their position (Hong, Surienty and Daisy 2011). It is a known fact that SMEs have low awareness with regard to OSH requirements and that is the reason why SMEs became the target community for OSH compliance for DOSH Malaysia Plan until 2015. In addition, it is possible that legislation role is over-shadowed by the lack of legislation awareness among SMEs and thus, they are not aware of legislation requirements. Thus, this factor would need to be considered as a controlling factor instead of a direct influencing factor of SMEs OSH implementation.

One obvious finding would be that legislation did not moderate the relationship between employee participation and OSH implementation. Obviously, when there is employee participation, legislation becomes irrelevant to safety implementation. Consistently, Khor and Surienty (2017) have found that employee involvement are found to have a direct significant relationship to safety participation in OSHMS. As suggested, specifically, safety training encourages safety participation in OSHMS. Similarly, when an employee is involved in the operation decision making process, it helps safety participation in OSHMS thus future study may want to test the moderating effect of safety training on the above direct relationships.

Saksvik et al. (2003) study of OSH implementation in Norwegian companies was the nearest to the intention of this study but the implementation measures used for their study were simple and simply evaluating respondents' opinion of the implementation progress in one word and not descriptive enough to describe either the level of implementation or aspects of safety implementation. In this study, the implementation measures capture the OSH components that are important for a wholesome organisation safety practices that proves to be richer and comprehensive.

This study acknowledges some limitation can be attributed to the number of factors being studied which may not reflect the overall picture of factors that could associate SMEs OSH implementation. For instance, Saksvik et al. (2003) has suggested that allocated time is important to achieve an effective implementation of OSH. Strict and realistic time allocated is important in managing implementation of OSH for Norway. Future research may want to examine the role of time in managing effective implementation of OSH.

Sample selection is also limited to the Northern Region of Malaysia so the findings need to be addressed with caution. Future study may want to replicate the study to get a better view of the situation for SMEs nationwide.

6 Conclusion

Existing SMEs-OSH issues posed a need to understand the influence of SMEs capabilities such as the management safety practices, legislative role, and demographic profiling, on current OSH implementation status. Further empirical work is needed to further understand how these factors and other possible factors impacted OSH implementation in SMEs. This study did not make any direct association with the extent of OSH implementation at SMEs. In addition, SMEs selected must represent the true population of SMEs in Malaysia so that results could be generalized to the true population.

References

Aksorn T, Hadikusumo BHW (2008) Critical success factors influencing safety program performance in Thai construction projects. Saf Sci 46(4):709–727

Ashill NJ, Carruthers J, Krisjanous J (2006) The effect of management commitment to service quality on frontline employees' affective and performance outcome: an empirical investigation of the New Zealand public healthcare sector. Int J Nonprofit Volunt Sector Mark 11:271–287

Abdul-Aziz A, Baruji ME, Nooh MN, Shah SNMMI, Mohammad-Yusof N, Nik-Him NF (2015) A preliminary study on accident rate in the workplace through occupational safety and health management in electricity service. J Res Bus Manag 2(12):9–15

Bibbings R (2007) Simplifying standards. RoSPA Occup Saf Health J 37:46–47

Bradshaw LM, Curran AD, Eskin F, Fishwick D (2001) Provision and perception of occupational health in small and medium-sized enterprises in Sheffield UK. Occup Med 51:39–44

Champoux D, Brun JP (2001) Occupational health and safety management in small size enterprises: an overview of the situation and avenues for intervention and research. Saf Sci 41:301–318

Charles MB, Furneaux C, Pillay J, Thorpe D, Castillo CP, Brown K (2007) Uptake of an OHS code of practice by australian construction firms. World Building Congress

Christian MS, Wallace JC, Bradley JC, Burke MJ (2009) Workplace safety: a meta-analysis of the roles of person and situation factors. J Appl Psychol 94:1103–1127

Cooper D (2006) Exploratory analyses of the effects of managerial support and feedback consequences. J Organ Behav Manag 26:41–82

Definition of SMEs by size (2009). http://www.smidec.gov.my/detailpage.jsp?page=defsme. Accessed 21 Aug 2009

Dejoy D (1985) Attributional process and hazard control management in industry. J Saf Res 16:61–71

Department of Safety and Health (DOSH) (2006) Guidelines on Occupational Safety and Health Act 1994 (Act 514). http://dosh.mohr.gov.my/. Accessed 26 Oct 2010

Department of Statistics (2016) Press Release Small Medium Enterprises Gross Domestic Product (SMEs GDP). https://www.dosm.gov.my/v1/index.php?r=column/pdfPrev&id=YzI2NWE2U0tXS1VEdnFsWHpqM1Fudz09. Accessed 1 Jan 2018

De Valk MMA, Oostrom C, Schrijvers AJP (2006) An assessment of occupational health care in the Netherlands (1996–2005). Occup Med 56:475–479

DOSH annual report (2008) Department of Occupational Safety and Health, Putrajaya

Dugdill L, Kavanagh C, Barlow J, Nevin I, Platt G (2000) The development and uptake of health and safety interventions – aimed at small businesses. Health Educ J 59:157–165

Eakin JM, Weir N (1995) Canadian approaches to the promotion of health in small business. Can J Public Health 86:109–122

EASHW (2004) Promoting Health and Safety in small and medium sized enterprises SMEs. European Agency for Safety and Health at Work

European Agency for Safety and Health at Work (2003) Improving occupational safety and health in SMEs: examples of effective assistance. Office for Official Publications of the European Communities, Luxembourg

Fairman R, Yapp C (2005) Making an impact on SME compliance behaviour: an evaluation of the effect of interventions upon compliance with health and safety legislation in small and medium sized enterprises. Research report 366. Kings Centre for Risk Management, Kings College London

Flagstad K (1995) The functioning of the internal control reform. Dissertation, NTNU, Trondheim

Gilkey DP, Keefe TJ, Hautaluoma JE, Bigelow PL, Herron RE, Stanley SA (2003) Management commitment to safety and health in residential construction: homesafe spending trends 1991–1999. Work 20:35–44

Griffin MA, Hu X (2013) How leaders differentially motivate safety compliance and safety participation: the role of monitoring, inspiring, and learning. Saf Sci 60:196–202

Hale A, Baram M (1998) Safety management. Pergamon Books, Amsterdam

Haines VY III, Merrheim G, Roy M (2001) Understanding reactions to safety incentives. J Saf Res 32(1):17–30

Hasle P, Limborg HJ (1997) A method for introduction of preventive working environment activities in small enterprises. Centre for Alternative Social Analysis

Holmes N (1999) An exploratory study of meanings of risk control for long term and acute occupational health and safety risk in small business construction firms. J Saf Res 30:61–71

Hong KT (2011) Safety management practices and safety behaviour: a study of SME in NCER, Malaysia. Unpublished Thesis, Universiti Sains Malaysia, Georgetown

Hong KT, Surienty L, Hung DKM (2011) Occupational safety and health in malaysian small medium enterprise (SME) and effective safety management practices. Int J Entrep Technopreneurship 1(2):321–338

ILO-OSH 2001 (2001) Guidelines of occupational safety and health management systems. International Labour Office, Geneva

Jaselskis EJ, Suazo G (1993) A survey of construction site safety in Honduras. Constr Manag Econ 12:245–255

Jaselskis E (1996) Strategies for achieving excellence in construction safety performance. J Constr Eng Manag 122:61–70

Kao K, Spitzmueller C, Cigularov K, Wu H (2015) Linking insomnia to workplace injuries: a moderated mediation model of supervisor safety priority and safety behavior. J Occup Health Psychol. https://doi.org/10.1037/a0039144

Khor LK, Surienty L (2017, in print) The strategic role of safety advice towards safety participation in OHSMS in malaysian manufacturing firms. J Occup Saf Health

Klein KJ, Sorra JS (1996) The challenge of implementation. Acad Manag Rev 21:1055–1080

Koh D, Jeyaratnam J (1998) Occupational health in singapore. Int Arch Occup Environ Health 71:295–301

Kongtip P, Yoosook W, Chantanakul S (2008) Occupational health and safety management in small and medium-sized enterprises: an overview of the situation in Thailand. Saf Sci 46:1356–1368

Labour and Human Resources Statistics (2008). http://www.mohr.gov.my/Statistik_Perburuhan_2008.pdf. Accessed 2 Jan 2010

Lai DNC, Liu M, Ling FYY (2011) A comparative study on adopting human resource practices for safety management on construction projects in the united states and singapore. Int J Proj Manag 29:1018–1032

Lamm F (1997) Small business and occupational health and safety advisors. Saf Sci 25:153–161

Laukkanen T (1999) Construction work and education: occupational health and safety reviewed. Constr Manag Econ 17:53–62

Lee J (2018) SMEs Need to Rise to the Challenge. The Star Online: StarBiz. https://www.pressreader.com/malaysia/the-star-malaysia-starbiz/20180101/281603830846187. Accessed 1 Mar 2018

Lin J, Mills A (2001) Measuring the occupational health and safety performance of construction companies in Australia. Facilities 19:131–138

Lindoe PH, Olsen OE (2004) Implementing quality and health/safety systems in the hospitality industry. a comparison with the aluminium industry in Norway. Scand J Hosp Tour 4:154–172

Lingard H, Rowlinson S (1994) Construction site safety in Hong Kong. Constr Manag Econ 12:501–511

Lingard H, Rowlinson S (2005) Occupational health and safety in construction project management. Spon Press, Abingdon

Makin PJ, Sutherland VJ (1994) Reducing accidents using a behavioural approach. Leadersh Organ Dev J 15:5–10

McKinney P (2002). Expanding HSE's ability to communicate with small firms: a targeted approach (No. 420/2002). Prepared by AEA Technology plc for the Health and Safety Executive

Mizoue T, Huuskonen MS, Muto T, Koskinen K, Husman K, Bergstrom M (1999) Analysis of Japanese occupational health services for small- and medium-scale enterprises in comparison with the finnish system. J Occup Health 41:115–120

Mohamed S (2003) Scorecard approach to benchmarking organizational safety culture in construction. J Constr Eng Manag 129(1):80–88

Mullen J, Kelloway EK, Teed M (2017) Employer safety obligations, transformational leadership, and their interactive effects on employee safety performance. Saf Sci 91:405–412

Neville H (1998) Workplace accidents: the cost more than you might think. Ind Manag 40:7–10

Ng WK (2003). OSH profile in the service sector in particular the small and medium sized enterprises. http://www.mtuc.org.my/osh_profile_service.htm. Accessed 16 Jan 2010

Prahbu VB, Robson A (2000) Impact of leadership and senior management commitment on business excellence: an empirical study in the North East of England. Total Qual Manag 11:399–409

Ricci F, Chiesi A, Bisio C, Panari C, Pelosi A (2016) Effectiveness of occupational health and safety training: a systematic review with meta- analysis. J Workplace Learn 28(6):355–377

Saksvik PO, Torvatn H, Nytro K (2003) Systematic occupational health and safety work in Norway: a decade of implementation. Saf Sci 41:721–738

Seneviratne M, Phoon WO (2006) Exposure assessment in SMEs: a low-cost approach to bring OHS services to small-scale enterprises. Ind Health 44:27–30

Shaluf I, Fakhru'l-Razi A (2003) Major hazard control: the Malaysian experience. Disaster Prev Manag 12:420–427

SME Performance Report 2008 (2008). http://www.smecorp.gov.my/sites/default/files/SME%20AR08%20Eng%20Text.pdf. Accessed 2 Jan 2010

SME Annual Report 2010/2011 (2012) Leveraging Opportunities, Realising Growth. National SME Development Council, Kuala Lumpur

Storey DJ, Westhead P (1994) Management training and small firm performance: a critical review. Working paper No. 18. Warwick University: Small and Medium Enterprise Centre

Tang S (1997) Safety cost optimization for building projects in Hong Kong. Constr Manag Econ 15:177–186

Teo AL, Phang TW (2005) Singapore's contractors' attitudes towards safety culture. J Constr Res 6:157–178

Teo EAL, Ling FYY, Ong DSY (2005) Fostering safe work behaviour in workers at construction sites. Eng Const Archit Manag 12(4):410–422

Torner M, Cagner M, Nilsson B, Nordling P (2000) Occupational injury in Swedish fishery: promoting implementation of safety measures. Occup Ergon 2:91–104

Vassie L, Tomas JM, Oliver A (2000) Health and safety management in UK and Spanish SMEs: a comparative study. J Saf Res 31:35–43

Vinodkumar MN, Bhasi M (2010) Safety management practices and safety behaviour: assessing the mediating role of safety knowledge and motivation. Accid Anal Prev 42(6):2082–2093

Vredenburgh AG (2002) Organizational safety: which management practices are most effective in reducing employee injury rates? J Saf Res 33(2):259–276

Wilson J, Joe M, Koehn E (2000) Safety management: problems encountered and recommended solutions. J Constr Eng Manag 126:77–80

Yapp C, Fairman R (2004) Factors affecting food safety compliance within small and medium-sized enterprises: implications for regulatory and enforcement strategies. Food Control 17:42–51

Yu CK, Hunt B (2004) A fresh approach to safety management systems in Hong Kong. TQM Mag 16:210–215

Zohar D, Polachek T (2014) Discourse-based intervention for modifying supervisory communication as leverage for safety climate and performance improvement: a randomized field study. J Appl Psychol 99(1):113–124. https://doi.org/10.1037/a0034096

Role of Dispositional Affect on Coping Strategies of Turkish Drivers

Burcu Arslan[(⊠)] and Bahar Öz

Middle East Technical University, Ankara, Turkey
barslan.psy@gmail.com

Abstract. Driving is a complex and demanding activity, thus, is mostly accompanied with stress, which is found to influence psychological and physiological health of driver, resulting with increased probability of fatal accidents. Matthews developed first driver stress model, which is a demonstration of interactions between environmental stressors, driver cognitive stress processes, personality traits related with stress vulnerability and stress outcome experiencing by driver. Dispositional affect characteristics are one of the most notable elements of coping strategies. Many studies prove that affects can predict coping strategies. This study is conducted to understand the relationship between dispositional affect characteristics of Turkish drivers and their coping strategies. Results indicate that dispositional affect of drivers are able to predict their coping strategies. Possible implications of these findings are discussed.

Keywords: Driver stress · Coping with stress · Dispositional affect

1 Introduction

Most used definition of stress is the cognitive process which includes interactional relationship between person and environment, including appraisal of whether situation exceeds or taxes individual resources and whether individual wellbeing is endangered [1]. Cognitive theory of stress includes mediator processes of stress and their outcomes, which are cognitive appraisal and coping. Cognitive appraisal process begins with individual's primary appraisal of situation as whether is harmful, threating or challenging, or whether is familiar or novel [2]. Secondary appraisal is related with individual's resources (what can be done?). Along with primary and secondary appraisals of situation, individuals choose a strategy to cope with situational demands. With this point of view, individuals have two coping options, whether they regulate their emotions in face of threating or challenging stimulus (emotion-focus coping), or they approach or avoid that stimulus (problem-focus coping). These two distinction options dominate most of stress literature. However, they are not sufficient to explain driver stress concepts.

Stress often accompanies driving which is a complex and demanding activity within itself. Psychological and physiological health of driver are found to be heavily affected by this stress which, in turn, increases probability of fatal accidents [3]. First stress model in the literature is developed by Matthews [4]. This transactional model is a demonstration of interactions between environmental stressors, driver cognitive stress

© Springer Nature Switzerland AG 2019
S. Bagnara et al. (Eds.): IEA 2018, AISC 819, pp. 672–678, 2019.
https://doi.org/10.1007/978-3-319-96089-0_73

processes, personality traits related with stress vulnerability and stress outcome experiencing by driver. Matthews et al. [5] define five driver specific coping strategies, which are confrontive (e.g. Relieved my feelings by taking risks or driving fast), emotion focused (e.g. Blamed myself for getting too emotional or upset), task-focused (e.g. Tried to watch my speed carefully), reappraisal (e.g. Felt I was becoming more experienced driver) and avoidance coping strategies (e.g. Stayed detached or distanced from situation). Confrontive coping strategy is found to be related with risk-taking behaviors [6], higher mistakes and violations [7]. Likely, emotion-focused coping strategy is found to be related with higher mistakes [7] and higher injury rates [8]. On the other hand, task-focused coping strategy is found to be an important factor for explaining driver's performance and safety behaviors [9]. Coping strategies are related not only with driver behaviors, but also driver health. According to a study by Machin and Hoare [9], confrontive coping is related with need for recovery and physical symptoms of bus drivers. Bus drivers who prefer confrontive coping strategies, experience more severe symptoms that warrant medical attention, became more fatigued and reported more negative affect. On the other hand, appraisal coping strategy is found to be related with more positive affect at the end of workday.

Dispositional affect characteristics are one of the most notable elements of coping strategies. Studies prove that affect can predict coping strategies. In detail, confrontive coping has aggression and hostility in its nature, as expected, and it is found to be related with negative affectivity [9]. On the other hand, positive affect is found to be related with adaptive and, task focus coping strategies, in which drivers mainly focus on safe operation of driving [10]. Although these results are mostly expected, some contradicting findings are also presented in literature. Rhodes and Pivik report that positive affect may be associated with risky driving, however only in high risk drivers, such as young male drivers [11]. The reasoning could be that young drivers are more prone to overrate their skills and underestimate their likelihood of being involved in an accident. This finding may point out that, for high risk drivers, confrontive coping strategy could be related with positive affect. Lawton et al. [12] state another contradicting result, which is that negative affect is found to be related with speeding behavior and likelihood of losing control over vehicle, and it is discussed as to reduce driver's likelihood of risky driving. Therefore, negative affect may be related with increase in task-focus coping strategy, which can also be concluded in light of this result.

This study is conducted to understand the relationship between dispositional affect characteristics of Turkish drivers and their coping strategies.

2 Method

2.1 Participants

The study included 360 Turkish drivers with age range of 19–60 (M = 24.10, SD = 6.06). 46.3% of drivers were females and 53.7% were males. Education level of drivers were ranged between elementary school to postgraduate level. Mean of the year of license ownership was 4.86. Average mileage of drivers was 55172 km (SD =

135623.23). 33.8% of participants never had an accident. 98.1% of accidents reported by drivers did not include any injury or fatality.

2.2 Instruments

Demographic Information Form. Demographic information form was developed to collect information about driver's characteristics and driving history. This form included questions about age, gender, education level, license year, and mileage during last year, total mileage, amount of involved accidents, amount of drivers' active accidents (participant driver being responsible for an accident), and amount of accidents that caused an injury or fatality.

Driver Coping Questionnaire (DCQ). Driver Coping Questionnaire was developed to measure driver's style of coping with stress in traffic [5], and to assess cognitive reactions of drivers to stressful situation in traffic environment. Drivers were asked how they react in face of stressful situations while driving. They were instructed to remember a stressful situation in traffic environment, for example, a near accident, traffic jam, poor visibility etc. Then, they were asked to rate how often they engage in each reaction in on a scale of 1 to 5. The questionnaire had 35 reactions composing 5 coping strategies, namely Confrontive (e.g. Flashed my car lights or used horn in anger), Task-focus (e.g. Made sure I kept my distance from the car in front), Emotion-focus (e.g. Criticized myself for not driving better), Reappraisal (e.g. Tried to gain something worthwhile from the drive) and Avoidance (e.g. Though about good times I'd had). Cronbach alpha of each factor was satisfied; .76, .78, .76, .70, .53, respectively.

Positive and Negative Affect Scale (PANAS). The Positive and Negative Affect scale was developed to be a brief measurement to assess individual's affective states or traits, by Watson and Clark [13]. The scale consisted of 20 adjectives composing 2 factors. Positive Affect (PA) was assessed by participants' ratings of 10 adjectives; alert, attentive, active, determined, enthusiastic, excited, interested, proud and strong. Negative Affect (NA) was also assessed by 10 adjectives; afraid, ashamed, distressed, guilty, hostile, irritable, jittery, nervous, scared and upset. High score of PA indicated high pleasure and engagement toward environment, whereas low score of PA indicated low vigor and tendency to depression. High scores of NA meant aversive mood and distress, while low scores of NA meant calm and relaxed state. Participants were asked to rate how often they experience such mood states in general frame of time.

Turkish adaptation of the scale was developed by Gençöz [14]. The original study of scale development and the study of adaptation to Turkish used a 5 point Likert type scale. Şimşek [15] utilized a 7 point Likert type scale in her study with very satisfactory psychometric properties. In the current study, 7 point Likert type scale (1 = never to 7 = always) was used. Likewise, internal consistency coefficients for PA and NA were found as .78 and .84, respectively.

2.3 Procedure

Participants were recruited via social media channels and Sona (data collection system) application of Middle East Technical University. Participants joined by Sona system were received extra points for their courses. Before receiving questionnaires, participants read an informed consent and signed as agreed. Informed consent included information about researchers, explanations about confidentiality and anonymity of responses. Completing all questionnaires approximately took 15 min.

3 Results

Table 1 demonstrates means and standard deviations of demographic information, subscales of DCQ, and Positive and Negative Scale. Mean score of license year refers to that participants have had their driving license for 4.86 years on average. The participants reported more Positive Affect ($M = 50.4$, $SD = 7.19$) than Negative Affect ($M = 32.01$, $SD = 8.29$). Maximum and Minimum scores for each of PANAS subscales was 70 and 10, respectively. Looking at subscales of DCQ, drivers were reported Task-focus coping at most ($M = 73.53$, $SD = 16.17$) and Avoidance at least ($M = 40.35$, $SD = 14.65$).

Table 1. Descriptive statistics for demographic information and subscales of DCQ and PANAS

Scale	Mean	Std. deviation
Age	24.1083	6.07309
Gender	1.54	.499
Education	3.97	1.079
License year	4.86	5.111
Total mileage (km)	83107.28	544196.384
Faulty accident	.71	.944
Active accident	.73	.982
Injury accident	.03	.240
Ticket	1.227778	2.5241575
Positive affect	50.4167	7.19941
Negative affect	32.0167	8.29671
Confrontive	40.4921	19.09874
Emotion focus	46.0952	19.21642
Reappraisal	62.1111	16.33976
Avoidance	40.3571	14.65218
Task focus	73.5397	16.17612

Table 2 indicates Pearson r inter correlations between demographic variables, subscales of PANAS and subscales of DCQ. Confrontive coping is positively correlated with reappraisal coping ($r = .206$, $p < .01$). This is conceptually unexpected. On the other hand, reappraisal is positively correlated with emotion-focus coping

($r = .266, p < .01$). Task-focus coping is negatively correlated with confrontive coping ($r = -.163, p < .01$), whereas negatively correlated with other coping strategies that are emotion-focus, reappraisal and avoidance ($r = .381, p < .01; r = .499, p < .01; r = .248, p < .01$, respectively). Negative Affect is found positively correlated with faulty accidents ($r = .124, p < .01$) which is in parallel with literature. Drivers who report more negative affect are also reported more faulty accidents. Meanwhile, drivers who report more negative affect are also reported less reappraisal coping, which is known as adaptive coping strategies ($r = -.108, p < .01$). Positive affect is found to be positively correlated with total mileage ($r = .121, p < .05$). It was very surprising that Positive Affect is found to be positively related with Confrontive coping ($r = .228, p < .01$). Further analyses will be likely to reveal more detailed results.

A simple linear regression is conducted to predict DCQ factors by affective trait of drivers. Analyses are calculated based on relationships that are found in correlations. First simple linear regression analysis is done to predict confrontive coping by positive affect trait of drivers. Results show that positive affect scores predicted confrontive coping strategies of drivers, $\beta = .228, CI\ 95\%\ [.336, .873], p < .001$, with a R^2 of .052. The direction of relationship is found as positive. Hence, it may be concluded that drivers with more pleasure and engagement with environment, tend to report more aggressive, risky coping strategies.

Table 2. Linear regression analysis on confrontive coping and positive affect

	R^2	Adj R^2	F	β	p
Confrontive coping	.052	.049	19.589	.228	.000

Second simple linear regression is calculated for Reappraisal coping scores by Negative Affect scores. It is found that there is a significant relationship between Negative Affect and Reappraisal coping strategy, $F\ (1,\ 358) = 4.26, p < .04$ with $R^2 = .012$. Reappraisal coping strategy is predicted significantly by drivers affective traits, $\beta = -.108, CI\ 95\%\ [-.471, -.010], p = .040$. Direction of relationship is found as negative, meaning that people with more aversive mood and negative emotions likely to report less effort to reappraise or looked at bright side of the situation (Table 3).

Table 3. Linear regression analysis on reappraisal coping and negative affect

	R^2	Adj R^2	F	β	p
Confrontive coping	.012	.009	4.257	-.108	.040

4 Discussion

Main focus of this study was to determine in which extent drivers' dispositional affect predict their coping strategies. Results of the present study supported this claim, as dispositional affect of drivers are correlated with the ability to predict drivers' coping strategies. Especially, two of coping strategies (confrontive coping and reappraisal coping) were found to be related with drivers' dispositional affect. The relationship that was found between positive affect and confrontive coping strategy was unexpected. Drivers with high positive affect were found to report more confrontive coping strategies. Some assumptions may be done to explain this relationship. Positive affect may lead drivers to appraise the stressful situation challenging, more than threating. Folkman state that as a primary appraisal of stress inducing situations, challenge is related with positive emotions, such as excitement and eagerness [2]. Also, challenge is an opportunity to mastery over environment [2]. Confrontive coping is also about mastery and approaching to problem [5]. Positive affect drivers can see stressful situations as challenging and try to obtain mastery over them. This tendency may lead them to confront with stress-inducing situation by taking risks (e.g. speeding) or by behaviorally (e.g. flashing car lights or using horn).

Another important result of the study was the relationship between reappraisal coping and negative affect. Drivers with negative affect were found to be reported less reappraisal coping, which were in line with literature. Drivers with negative affect were found to experience aversive mood and distress as personality traits. As in Matthews's definition, reappraisal is a positive evaluation of a situation [5]. In stressful situations, this aversive mood and distressfulness traits may enhance stress levels of drivers, and interfere with reappraisal of situation.

There are several implications of the study. Primary appraisal of situation is as important as dispositional affect of drivers; and, indeed, it may even be more important in certain cases. Stress-related accidents are increasing with each passing year. To prevent this, drivers can be educated on how to appraise and approach to stressful situations. Organizations should add interventions programs aimed at driver appraisals in face of stress. Challenge appraisals should also be inhibited or directed towards more task-focus coping. Negative affect is seen as inhibitor of important adaptive coping, reappraisal. Besides education for appraising of situation, organizations should create more positive affect enhancing environments to drivers. Lowering negative affect of drivers is not only benefit their wellbeing, but also enhance productivity, lessen sickness absences and need for recovery after workday [9]. Decreasing time pressure, workload and increasing safety and assurance of drivers could be first and most important step toward this fundamental problem, which is as known as driver stress, and its undesirable outcomes.

References

1. Lazarus RS, Folkman S (1984) Stress, appraisal and coping. Springer, New York
2. Folkman S (1984) Personal control and stress and coping processes: a theoretical analysis. J Pers Soc Psychol 46(4):839

3. Hennessy DA, Wiesenthal DL, Kohn PM (2000) The influence of traffic congestion, daily hassles, and trait stress susceptibility on state driver stress: an interactive perspective. J Appl Biobehav Res 5(2):162–179

4. Matthews G (2002) Towards a transactional ergonomics for driver stress and fatigue. Theor Issues Ergon Sci 3(2):195–211

5. Matthews G, Desmond PA, Joyner L, Carcary B, Gilliland K (1996) Validation of the driver stress inventory and driver coping questionnaire. In: International conference on traffic and transport psychology, Valencia, Spain, pp 1–27

6. Emo AK, Matthews G, Funke GJ (2016) The slow and the furious: anger, stress and risky passing in simulated traffic congestion. Transp Res Part F: Traffic Psychol Behav 42:1–14

7. Kontogiannis T (2006) Patterns of driver stress and coping strategies in a Greek sample and their relationship to aberrant behaviors and traffic accidents. Accid Anal Prev 38(5):913–924

8. Shamoa-Nir L, Koslowsky M (2010) Aggression on the road as a function of stress, coping strategies and driver style. Psychology 1(01):35

9. Machin MA, Hoare PN (2008) The role of workload and driver coping styles in predicting bus drivers' need for recovery, positive and negative affect, and physical symptoms. Anxiety Stress Coping 21(4):359–375

10. Desmond PA (1997) Fatigue and stress in driving performance. Doctoral dissertation, University of Dundee

11. Rhodes N, Pivik K (2011) Age and gender differences in risky driving: the roles of positive affect and risk perception. Accid Anal Prev 43(3):923–931

12. Lawton R, Parker D, Manstead AS, Stradling SG (1997) The role of affect in predicting social behaviors: the case of road traffic violations. J Appl Soc Psychol 27(14):1258–1276

13. Watson D, Clark LA, Tellegen A (1988) Development and validation of brief measures of positive and negative affect: the PANAS scales. J Pers Soc Psychol 54(6):1063

14. Gençöz T (2000) Positive and negative affect schedule: a study of validity and reliability. Türk Psikoloji Dergisi 15(46):19–28

15. Şimşek ÖF (2005) Paths from fear of death to subjective well-being: a study of structural equation modeling based on the terror management theory perspective. Unpublished Master thesis, METU Graduate School of Social Sciences, Ankara

Social Networks Applied to Zika and H1N1 Epidemics: A Systematic Review

Diná Herdi Medeiros de Araujo, Elaine Alves de Carvalho,
Claudia Lage Rebello da Motta, Marcos Roberto da Silva Borges,
José Orlando Gomes$^{(\boxtimes)}$, and Paulo Victor Rodrigues de Carvalho

Programa de Pós Graduação em Informática, Universidade Federal do Rio de
Janeiro PPGI/UFRJ, Rio de Janeiro, RJ, Brazil
herdidina@gmail.com, nane.alves@gmail.com,
joseorlando@nce.ufrj.br

Abstract. Background: Health crises and health emergencies occur regionally and globally. In this context, online social networks are technical resources widely used to share large amounts of information with increasing reach and speed. This capacity of dissemination of information on epidemics through social media interactions creates the possibility of collaboration between population and health professionals or agencies. This article intends to present, through results obtained in a systematic review, examples on how social interactions enable collaboration in health surveillance to treat the epidemic situations of Zika and H1N1.

Methodology: The methodology applied in this article was the systematic review, to answer how social networks are being applied in a collaborative context to assist in the identification, monitoring and generation of information on Zika Virus and H1N1 epidemics

Results: The works presented demonstrated that social media interactions are important tools for the rapid dissemination of information at low cost, reaching audiences who need clarification, and also favoring the formation of specific collaborative networks among researchers, in the monitoring and generation of epidemiological information for making decision.

Conclusion: Collaboration through social networks as pointed out in the papers is an essential aspect to connect individuals with common interests, generating information that allows the monitoring and control of epidemics, as well as creating networks of research for the development of science in the service of life.

Keywords: Social media · Collaboration in health · Health surveillance
Zika · H1N1 · Twitter · Facebook · Instagram

1 Introduction

In recent decades, many public health crises have occurred because of epidemics such as H1N1 in 2009, Ebola in 2013 and 2016, Zika in 2015. These epidemics have reached many people, some on a global scale, causing mass deaths. The Ebola virus, killed about 11.000 people, and the H1N1 flu led to the death of approximately 8.000 to

© Springer Nature Switzerland AG 2019
S. Bagnara et al. (Eds.): IEA 2018, AISC 819, pp. 679–692, 2019.
https://doi.org/10.1007/978-3-319-96089-0_74

18.000 people. In these periods, there was considerable ambiguity with regard to information on diseases, their geographical location, risks and ways of prevention [21]. As these health crises do not occur as frequently as floods and hurricanes, information on health ends up becoming partial and conflicting, mainly in social media. This has consequences for society because people need to make decisions about travel, treatment and prevention using, many times, this kind of information ambiguous [41].

Social media today represents a change in how the information is accessed and shared, allowing the sharing of opinions, texts, and videos. In addition, it is possible to use resources from these means of communication to sort, give opinions and recommendations. Thus, social networks allow people with characteristics or common goals to connect and collaborate on common interests. This dynamic creates value for healthcare organizations because directly affects the population. In the age of social networking, the sharing of public health information and local events on the theme are increasingly appearing in literature [30]. Social media information can be used to understand how local population view health problems of or even predict and identify more quickly those outbreaks [1]. The adoption of *online communication*, particularly through social networks like Facebook, Instagram, and Twitter, has created new forms of collaboration between people for sharing information about emergencies, especially in health [7].

Different researches have shown a correlation between information from social media and events in the real world. It was identified that 66% of the scientific papers resulted in a positive relationship between the data from the traditional epidemiological surveillance system and the results of social media monitoring [24].

Thus, in cases of more urgent action, as in epidemics, governments can use the monitoring of social media as a complementary tool in the identification and geographical location of problems [13].

Public health professionals can also use health data obtained from social media to detect potential risks of epidemics. When performing qualitative and quantitative analysis of data from social media it is possible to draw a picture of the behavior and the opinion of the population [13]. Therefore, with the information available, they can establish a cycle of monitoring and public response *online*, identifying emergencies and acting on the efficacy of communication strategies for adaptation of educational campaigns [13].

With this review article, we are interested to show how social media use the collaborative participation of its members, researchers and health professionals to generate relevant information and expand communication on epidemiological health emergencies. Our research question was how social networks was used to help on the identification, monitoring and generation of useful information during the H1N1 and Zika virus epidemics. The H1N1 and Zika diseases were selected due their high dissemination capacity and potential for major global health damage. The result of this review showing the contributions of social networks to aid health surveillance, can be used to evaluate opportunities of technological improvement for a better use of social media information for collaboration.

2 Background

Social networks is define as a social structure formed by people and their relationships directly and indirectly. In this way, all people who have had relationships are part of the social network [18]. In modern society, the social networks share information, knowledge, interests and generate ways to achieve common goals [40]. They have been tools of great importance also for the health, connecting people in search of information [32]. Professionals and researchers in the field of public health also have used as a tool for information, education, and training on health issues, mainly for emergency problems, such as epidemics [35].

The social networks provide access to millions of records, which provides an unprecedented opportunity for inferences and spread of information in large scale, covering the general population. This allows the development of new strategies for monitoring and reporting for application in health [11], mainly in surveillance health.

In 1963, Alexander Langmuir, conceptualized health surveillance as the continuous observation of the distribution and trends of the incidence of diseases through the systematic collection, consolidation, and evaluation of reports of morbidity and mortality, as well as other relevant data, and to the regular dissemination of such information to all who need to know it [7]. In epidemiological context, the selected works in systematic review deal with Influenza A or H1N1 and Zika virus.

The human influenza is an infectious viral disease highly communicable, global distribution and which features seasonal epidemics. Small changes antigens common (*drifts*) make individuals infected before partially protected, which facilitates the circulation of the virus annually. The influenza type A (H1N1), an integral part of the study, caused by changes in major antigenic *shifts called*. The symptoms are similar to those caused by viruses or other influenzas [15]. The H1N1 flu known when affected much of the world's population between 2009 and 2010. This was thus the first influenza pandemic to happen in a globalized world, with a rapid and intense movement of people and goods, which has great influence on the spread of the virus among countries. This epidemic, in globalized, promoted the sharing and transmission of information in real time, which introduced new tools for your follow-up and monitoring in an environment of collaboration [3].

Another disease is called Zika, which is a flavivirus transmitted by the bite of the *Aedes aegypti mosquito*. This disease produces moderate fever and skin rashes during a period of 2 to 7 days. However, it is estimated that 80% of people infected show no symptoms [28]. Despite the symptoms have a short duration, this disease has a high potential for dissemination and ability to generate significant complications for people's health. These complications are related to the syndrome of showed Guillain-Barre, which is an autoimmune disease capable of reaching the peripheral nervous system causing weakness, numbness and even paralysis. In addition, to reach people of any age, this disease does not have a cure identified [28]. Another possible complication of the Zika virus is the microcephaly. This disease causes malformation of the brain of the fetus, making it less than expected for their age. Occurs when the infection from their mothers by the Zika virus, during pregnancy, reaches the fetuses, which can cause brain damage, severe and irreversible, causing birth defects or even death [36].

3   Methodology

The methodology of the systematic review is shown in Fig. 1.

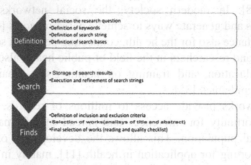

Fig. 1. Steps of review

The databases used were Science Direct (www.sciencedirect.com); Springer (www. springerlink.com); Scielo (www.scielo.org); IEEE Xplorer (www.ieeexplore.ieee.org), and the key words used: Zika, H1N1, influenza, social media and health. Table 1 presents a summary of the search terms and respective variations derived from the research question. We considered these databases appropriate because of the quantity of indexed journals and coverage of relevant disciplines such as health sciences and computer sciences [27]. We have used free search terms with no controlled descriptors in order to have a broader search.

Table 1. Summary of the search terms.

Search string	Search results, N
(H1N1 or Zika) and social media	1641
(H1N1 or influenza) and zika and social media	113
(H1N1 or influenza) and zika and social media and health	44
H1N1 and influenza and zika and social media and health	20

The selection criteria for publications were:

1. Works published and available in full on a scientific basis, in Portuguese and English.
2. Only works with the theme of the health area or computing/information technology, with a focus on the application of social networks and cases of collaboration for health surveillance.
3. Works published since the year 2009.
4. Exclusion of works that treat of a generic form the collaboration on health surveillance.
5. Exclusion of works that do not address the theme H1N1, Zika and Social Media.

4   Results

Only **20** papers were selected because they met the inclusion/exclusion criteria. After reading the titles/abstracts and texts, only **5** of them actually answered the research question, and considered in the discussion of the proposed theme (Table 2). The Chart 1 shows the papers analyzed per search base, and Chart 2 the illnesses find per bases.

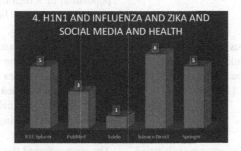

Chart 1. Distribution of works selected by basis. Source: Authors

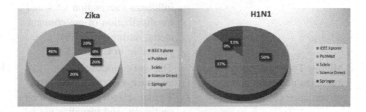

Chart 2. Distribution of bases for illnesses Zika and H1N1. Source: Authors

Table 2. Sumary of selected papers.

Research base	Title	Abstract	Status
Scielo	Visual archives concerning Zika virus: images on Instagram as part of the constitution of an epidemic memory [4]	Visual files posted on the Instagram network, concerning the Zika virus epidemic were studied by Antunes et al. (2016), helped to build a visual memory of the epidemic and concluded that the feeling of Internet users can provide solutions for care and prevention	Included

(continued)

Table 2. (*continued*)

Research base	Title	Abstract	Status
PubMed	Detecting influenza epidemics using search engine query data" [19]	This article presents a method of analysis of a large number of queries from Google searches that estimated with precision the influenza epidemic in a population of 9 regions of public health in the United States	Included
PubMed	Pandemics in the age of twitter: content analysis of tweets during the 2009 H1N1 outbreak [13]	Via *Twitter*, understand public perceptions in relation to the pandemic of influenza H1N1. The authors concluded that *Twitter* is potentially suitable for text mining and longitudinal analysis	Included
PubMed	Health Tweets: an exploration of health promotion on Twitter [16]	The study showed the Twitter was used as a communication modality to update the public about waiting times at influenza vaccination sites and for distribution of government alerts, during the H1N1 pandemic in Canada	Included
IEEE Xplore	Mining Twitter data for influenza detection and surveillance [8]	The work proposes an approach to extract data from Twitter, classify based on their sentiment, and visualize via a real-time interactive map	Excluded
IEEE Xplore	Predicting flu trends using Twitter data [2]	Which monitors messages posted on Twitter with a mention of flu indicators to track and predict the emergence and spread of an influenza epidemic in a population	Excluded
IEEE Xplore	Forecasting influenza levels using real-time social media streams [31]	The authors presented a model that predicts the weekly percentage of U.S. population with Influenza-Like Illness, using multilayer perceptron with backpropagation algorithm on a large-scale social media stream	Excluded

(*continued*)

Table 2. (*continued*)

Research base	Title	Abstract	Status
IEEE Xplore	Distance-based outliers method for detecting disease outbreaks using social media [14]	This research used Twitter data to detect influenza outbreak, with distance-based outliers method to transform the noisy Twitter data into regions. The simulations show a good accuracy and robustness	Excluded
IEEE Xplore	Immediate and long-term effects of 2016 Zika outbreak: a Twitter-based study [29]	The article explored discussions in twitter about Zika virus. Were performed volumetric and text mining analysis, to explore the different underlying themes in the Twitter	Excluded
Science Direct	Effective surveillance and predictive mapping of mosquito borne diseases using social media [26]	The article showed proposed framework focused outbreak management using spatial and temporal information which help in identification, characterization and modeling of user behavioral patterns on the web	Excluded
Science Direct	Zika virus pandemic—analysis of Facebook as a social media health information platform [40]	Study about information source for the Zika virus pandemic in the United States, found that the misleading posts were far more popular than the real ones, dispersing accurate and relevant public health	Included
Science Direct	Toward a collaborative model of pandemic preparedness and response: Taiwan's changing approach to pandemics [38]	Taiwan Case Study article provides evidence that by implementing an approach where all society is prepared to respond to pandemics	Excluded
Science Direct	Using online social networks to track a pandemic: a systematic review [3]	The article conducted an electronic literature search for articles about online social networks (OSNs) published between 2004 and 2015, for studying and understanding social interaction and communication among populations in pandemics	Excluded

(*continued*)

Table 2. (*continued*)

Research base	Title	Abstract	Status
Science Direct	Public information officers' social media monitoring during the Zika virus crisis, a global health threat surrounded by public uncertainty [6]	Monitoring trends are explored about Zika Virus by survey was conducted with public information officers (PIOs) at local public health departments across the United States to reveal how they utilize and listen to social media during public health crises	Excluded
Science Direct	Disaster management, crowdsourced R&D and probabilistic innovation theory: Toward real time disaster response capability [9]	This paper argues the developing field of probabilistic innovation, and its methodologies, such as those drawing from crowdsourced R&D and social media, may offer useful insights into enabling real-time research capabilities, with important implications for disaster and crisis management	Excluded
Springer	Ebola, Zika and the International Health Regulations: implications for Port Health Preparedness [20]	This work reviews the legislative framework and actions taken improving Port Health preparedness of Ireland, in response to the declaration of Public Health Emergency of International Concern for Ebola Virus disease in August 2014	Excluded
Springer	Survey of the perceptions of key stakeholders on the attributes of the South African Notifiable Diseases Surveillance System Benson et al. [6]	Study of perceptions of a sample of the key stakeholder of South African Notifiable Diseases Surveillance System, and the study findings inform the revitalisation need of health notification system in South Africa	Excluded
Springer	Modernising epidemic science: enabling patient-centred research during epidemics [37]	The article proposes a framework that includes an integrate patient-centered developing methods and tools that address the very real epidemiology and contextual challenges of emerging and epidemic	Excluded

(*continued*)

Table 2. (*continued*)

Research base	Title	Abstract	Status
Springer	The Ebola outbreak: catalyzing a "shift" in global health governance? [33]	This article examined reports issued by four high-level commissions/panels convened to specifically assess WHO's performance post-Ebola	Excluded
Springer	Guillain–Barré syndrome in Colombia: where do we stand now? [34]	The aimed of article was to assess the Guillain–Barré syndrome (GBS) in Colombia, through a systematic literature review. Is a onset muscle weakness disease currently preceded by Zika virus (ZIKV) infection	Excluded

4.1 Work Related to H1N1 Influenza

According to the literature, all the people of the world have already had contact with the virus H1N1 1918, either by the third or fourth generation. Swapping and changing their genes, the H1N1 has managed to stay for more than a century circulating in the population [22], having its apex in 2009, with the global outbreak that, according to the World Health Organization, caused the death of more than 18,000 people. In the latter case, the globalized world has used the approaches of collaborative communication through social networks, disseminating and sharing information about the H1N1. As result of this event, numerous academic publications produced, confirming the potential contribution of social networks in public health.

In the last outbreak of H1N1 virus, the social network that showed greater collaboration in academic papers was the *Twitter* social network and server to *microblogging*, in texts of up to 140 characters), a result of intense overall utilization, with the purpose of measuring the perceptions of the population in emergency situations and interaction between official agencies and health professionals.

The Influenza A virus H1N1 emerged in Mexico, in early April 2009. As soon arose and spread rapidly around the world, generating concerns and putting on alert to Global Public Health. In the second half of April of that same year, the World Health Organization (WHO) has issued an alert about the new epidemic [15]. In this context, the use of social networking tools like Twitter helped to predict the behavior of the population, signaling actions of health surveillance.

As influenza is a disease seasonal and that cause yearly respiratory diseases in thousands of people around the world, it becomes a concern for the health surveillance. The study *"Detecting influenza epidemics using search engine query data"* [19] selected in the review, presents a method of analysis of a large number of queries from Google searches that estimated with precision the influenza epidemic in a population of 9 regions of public health in the United States. In the selection process of consultations was automated used linear regression, cross-validation, a percentage of ILI per region and the transformation of Fisher Z. with updated data health professionals respond better to seasonal epidemics.

In the study *"Pandemics in the age of Twitter*: the *content analysis of tweets during the 2009 H1N1 outbreak"*, Eysenbach and Chew (2010) [13] describe how they, via *Twitter,* understand public perceptions in relation to the pandemic of influenza, the H1N1. The authors concluded that *Twitter* is potentially suitable for text mining and longitudinal analysis. With its 140 characters, the sharing of users with the followers of your thoughts, feelings, and activities contain a wealth of data on public opinion and behavior. The authors demonstrated in this study that the sharp increase of messages on *Twitter* with the term H1N1 related to the volume of news on the subject.

Another example of the use of social media in health was presented in a qualitative study *"Health Tweets: An Exploration of Health Promotion on Twitter"* Donelle and Booth (2012) [16]. The study demonstrated the efficiency of the tool Twitter, used during an influenza pandemic (H1N1 in 2009, in Canada. This network was used as a form of communication to update the public on waiting times at places of influenza vaccination and for distribution of government alerts. Results of the present study demonstrated that the issues that determine the "health" are widely discussed by users of technology of social networks like Twitter, and can play a fundamental role in the dissemination of information.

For public health educators, Twitter and other social networks can play an important role in disseminating information only if strategic communication processes are reliable (for example, a dense network of followers, consistent and real-time contacts) [43].

4.2   Work Related to Zika Virus

The Zika virus was detect in Brazil in 2015 May, and in November of the same year, the Ministry of Health confirmed the probable relationship between Zika and microcephaly. As already explained above, the microcephaly is a neurological disturb. By January 2016, the Ministry of Health investigated 3.448 cases of microcephaly in the country and its possible relation with the contagion by the Zika virus [30]. Faced with these facts, the World Health Organization has called the Emergency Committee to Regulate International Health to analyze the Zika virus and its dissemination, as well as the consequences for public health in the global context. As a result, were defined strategies for research to clarify aspects of the association between Zika virus and microcephaly [42]. Another organization, the Pan-American Health Organization (PAHO), which is considered a regional structure of the World Health Organization (WHO), provides that only in the American continent must have occurred between three and four million cases of Zika in 2016 [42].

In this epidemiological panorama, is the importance of the work of the health surveillance. In this context, there are social networks to generate collaboration and information sharing in mass. In spite of rumors on social networks does not replace the traditional research in the health area, they can used as complementary tools to identify and monitor population reports regard to health [30]. This becomes an important instrument for health surveillance in the epidemiological context.

It is in this context of collaboration between community and organization/health professionals, through social networks, that the work selected in the systematic review about social network Fiocruz for Zika arises. It becomes an institutional tool capable of assisting the public health organization and its professionals in the surveillance of the Zika virus epidemic in Brazil. The main social network tool adopted by Fiocruz for the case of the Zika virus epidemic is Facebook [17]. Facebook's Fiocruz network, applied for monitoring and disseminating information about Zika virus, was widely used by the institution, especially in the period of the epidemic to promote media campaigns. Many visuals and standardization of records were employed to ensure that the general public identified the information as coming from a secure source. Figure 2 illustrates this form of communication on the network.

Fig. 2. Zika virus campaign in the media - Facebook. Source: Highlights Fiocruz Network 2016 [25]

Two works on the Zika virus selected in the review examined the effective use of social media Facebook and Instagram with positive and negative outcomes. According to Antunes et al. [4] the analysis of images on the epidemic and its debate on social networks, specifically of Instagram, with cut in the months of November and December 2015, images arranged in a mosaic, which collective construction became a representative sample, based in linear viewing from the quantities of "tanned" (or "likes") in the images referring to the hashtag "#Zikavirus", forming an "image cloud" (or "cloud") according to their importance and redundancy. These images have been analyzed by the software ImageCloud (application developed by the laboratory studies on image and Cyberculture (Labic), of the Universidade Federal, do Espírito Santo - Brazil) helped to materialize senses about the disease, and the society has information about the Zika virus. And The study that examined the Facebook (Facebook Inc, Menlo Park, CA) [39] as an information source for the Zika virus pandemic in the United States, found that the misleading posts were far more popular than the real ones, dispersing accurate and relevant public health.

5   Discussion

The systematic review demonstrated that social media are important tools for rapid dissemination of information at low cost, reaching an audience in need of clarification, especially in matters of emergency in health. In addition, these networks are very useful to professionals and health departments in monitoring and generation of information on epidemics.

In many cases, the social networks can become platforms for promoting campaigns in health education for the public, both general and specific.

This use of social media in health arises, among other reasons, the ability to have many close contacts, social influence and access to resources to bind people. With this, these media become important in disease prevention, detection, and treatment of epidemics, as well as dissemination of information on earth [24].

Although many are using social networks to communicate health problems, there are still few studies on how to best use social networks in the theme.

In spite of its importance in health, the social networks treated in this study present points of attention in relation to aspects such as reliability, buoyancy, dynamism, objectivity and relevance. Among these points, one can highlight the rumors of gravity. It is considered a real disservice to the population, and can induce people to make wrong decisions, even causing panic. The greatest hindrance to this type of situation is the inability to verify the source of person-to-person information [23]. This is still an issue that must be improved in technological tools applied to social networks.

Another issue to be improved in the use of social networks for health issues is completeness and clarity of information. Since the same widespread message on the network may have multiple posts, chances are ambiguity. Thus, the more appropriate the subject is the published message, the greater the chances of adequate understanding and clarity of the data by the population [23].

The dynamic aspect of networks is another challenge for epidemiological scenarios. Constant changes occur at all times in the field of health. Therefore, postings should be being monitored frequently and responses should be posted in a short enough time to generate the desired mass effects, and be used for health surveillance.

6 Conclusions

The new media, especially social media, bring possibilities of collaboration never experienced, eliminating physical barriers, storms and providing space for new forms of social mobilization in health. An instrument, so that you can ensure a greater range of essential information to society with respect to your health, as policies of prevention, vaccination campaigns, among others. Thus, social networks are more than a technology, but also a means of communication, interaction and social organization [5].

In terms of collaboration, the works analyzed, it becomes clear that it is an aspect present and essential in social networks applied to health to generate the dynamics of communication expected population-health professionals-population. However, it also concludes that there are still opportunities for improvement, especially in technology, regarding *frameworks*, mechanisms, and algorithms of treatment and search for sensitive data to the context, monitoring information and evaluation of relationships and feelings between members of the network. Thus, problems related to objectivity, dynamism, rumor and other aspects related to the quality of information could be diminish or at least improved. In this field of activity are advancement opportunities for future work.

References

1. Abbdullah S, Wu X (2011) An epidemic model for news spreading on twitter. In: 2011 International conference on tools with artificial intelligence, Proceedings, Vermont. IEEE Xplore, pp 163–169
2. Achrekar H, Gandhe A, Lazarus R, Yu SH, Liu B (2011) Predicting flu trends using twitter data. In: IEEE conference on computer communications workshops, INFOCOM WKSHPS, pp 702–707
3. Al-garadi MA, Khan MS, Varathan KD, Mujtaba G, Al-Kabsi AM (2016) Using online social networks to track a pandemic: a systematic review. J Biomed Inf 62:1–11
4. Antunes MN, et al (2016) Visual archives concerning Zika virus: images on Instagram as part of the constitution of an epidemic memory. e-ISSN 1981-6278
5. Avery EJ (2017) Public information officers' social media monitoring during the Zika virus crisis, a global health threat surrounded by public uncertainty. Public Relat Rev 43(3):468–476
6. Benson FG, Musekiwa A, Blumberg L, Rispel LC (2016) Survey of the perceptions of key stakeholders on the attributes of the South African notifiable diseases surveillance system. BMC Public Health 16(1):1–9
7. Brazil. The Ministry of Health. The Secretary of Health Surveillance (2005) Basic course of epidemiological surveillance. Ministry of Health, Brasilia
8. Byrd K, Mansurov A, Baysal O (2016) Mining twitter data for influenza detection and surveillance. In: Proceedings of the international workshop on software engineering in healthcare systems - SEHS 2016, pp 43–49
9. Callaghan CW (2016) Disaster management, crowdsourced R&D and probabilistic innovation theory: toward real time disaster response capability
10. Castells M (2005) The Network Society. vol 1 Trad. Roneide Venancio Majer in collaboration with Klauss Brandini Gerhardt. 8 ed. Peace and land, São Paulo
11. Centola D (2013) Social media and the science of health behavior. Circulation 127 (21):2135–2144
12. Chauhan A (2014) Online media use and adoption by hurricane sandy affected fire and police departments. All Graduate Theses and Dissertations, Paper 3895
13. Chew C, Eysenbach G (2010) Pandemics in the age of Twitter: the content analysis of tweets during the 2009 H1N1 outbreak. PLoS ONE 5(11):1–12
14. Dai X, Bikdash M (2016) Distance-based outliers method for detecting disease outbreaks using social media. In: Conference proceedings - IEEE SoutheastCon (2016)
15. Development and history of Influenza A H1N1 in Brazil. http://www.efdeportes.com. Accessed 28 May 2017
16. Donelle L, Booth RG (2013) Health tweets: an exploration of health promotion on twitter. Online J Issues Nurs 17(3)
17. Fiocruz (2012) Manual of Social Media Fundação Oswaldo Cruz. Coordination of Social Communication/President of Fiocruz
18. Fuks H, Pimentel M (2011) Groupware systems. Elsevier, Rio de Janeiro
19. Ginsberg J, Mohebbi MH, Patel RS, Smolin MS, Brilliant L (2009) Detecting influenza epidemics using search engine query data. Nat Int Wkly J Sci 457:1012
20. Glynn RW, Boland M (2016) Ebola, Zika and the international health regulations - implications for port health preparedness. Globalization and Health 12(1):1–5
21. Gui X, et al (2017) Managing uncertainty: using social media for risk assessment during a public health crisis. In: Proceedings of the 2017 CHI conference on human factors in computing systems. ACM, pp 4520–4533

22. H1N1. http://scienceblogs.com.br/rainha/2009/08/h1n1_mais_de_90_anos_entre_nos/. Accessed 30 May 2017
23. Halse SE, et al (2016) Tweet factors influencing trust and usefulness during both man-made and natural disasters. In: ISCRAM
24. Hempel M (2014) The use of social media in environmental health research and communication: an evidence review. Environmental Public Health, Vancouver
25. Highlights The Fiocruz (2016) Presentation of the results of the network
26. Jain V, Kumar S (2017) Effective surveillance and predictive mapping of mosquito-borne diseases using social media. In: IEEE 19th international conference on e-Health networking
27. Jatoba A, Burns CM, Vidal MCR, Carvalho PVR (2016) Designing for risk assessment systems for patient triage in primary health care: a literature review. JMIR Hum Factors 3(2): e21
28. Kadri SM, Trapp-Petty M (2016) The Zika virus fight on social media. Arch Clin Microbiol 7:3
29. Khatua A, Khatua A (2016) Immediate and long-term effects of 2016 Zika outbreak: a twitter-based study. In: 2016 IEEE 18th international conference on e-Health networking, applications and services, Healthcom (CDC)
30. Klein GH, Pedro GN, Rafael T (2017) Big data and social media: monitoring of networks as a management tool. Health Soc 26(1):208–217
31. Lee K, Agrawal A, Choudhary A (2017) Forecasting influenza levels using real-time social media streams. In: Proceedings - 2017 IEEE international conference on healthcare informatics. ICHI, pp 409–414
32. Lima SGP, et al (2015) The use of social networks in the digital area of health: a systematic review of health and research, vol 8, Special Edition, pp 79–91
33. Mackey TK (2016) The Ebola outbreak: catalyzing a "shift" in global health governance? BMC Infect Dis 16(1):1–12
34. Mahecha MP, Ojeda E, Vega DA, Sarmiento-Monroy JC, Anaya JM (2017) Guillain-Barré syndrome in Colombia: where do we stand now? Immunol Res 65(1):72–81
35. Medronho RA (1995) Geoprocessing and health: a new approach to space in health-illness process. ISTC/Nect/Fiocruz, Rio de Janeiro
36. Morens DM, Fauci AS (2016) Zika Virus Infection and Zika- Associated With Fetal/Microcephaly: General Update to Chapter 233: Arthropod-Borne and Rodent-Borne Virus Infections
37. Rojek AM, Horby PW (2016) Modernising epidemic science: enabling patient-centred research during epidemics. BMC Med 14(1):1–7
38. Schwartz J, Yen MY (2017) Toward a collaborative model of pandemic preparedness and response: Taiwan's changing approach to pandemics. J Microbiol Immunol Infect 50 (2):125–132
39. Sharma M, Yadav K, Yadav N, Ferdinand KC (2017) Zika virus pandemic - analysis of Facebook as a social media health information platform. Am J Infect Control 45(3):301–302
40. Social Network. https://pt.wikipedia.org/wiki/Rede_social. Accessed 25 May 2017
41. Villa R (2016) Zika, or the burden of uncertainty. La Clin Ther 167(1):7–9
42. WHO - World Health Organization. Zika virus. http://www.who.int/topics/Zika/en/. Accessed 03 May 2017
43. Winston ER, Medlin BD, Romaniello BA (2012) An e-patient's end-user community (EUCY): the value added of social network applications. Comput Hum Behav 28(3):951–957

Ergonomic Safety and Health Activities to Support Commercial Nuclear Power Plant Control Room Modernization in the United States

Jeffrey C. Joe[1]([✉]) [iD], Casey Kovesdi[1] [iD], Jacques Hugo[1] [iD], and Gordon Clefton[2]

[1] Idaho National Laboratory, Idaho Falls, ID 83415, USA
{Jeffrey.Joe,Casey.Kovesdi}@inl.gov,
Jacques.Hugo0l@gmail.com
[2] Clefton Enterprises, Swanton, MD 21561, USA
gclefton68@gmail.com

Abstract. Affordable, abundant, and reliable electricity generation is essential to fueling a nation's robust and globally competitive economy. In the United States (U.S.), commercial nuclear power plants (NPPs) account for approximately 19% of reliable and cost-competitive base load electricity generation. Other technologies that reduce reliance on fossil fuels and provide base load electricity cost-competitively at a national scale are still under development. Thus, without suitable replacements for nuclear power, the generating capacity of nuclear energy in the U.S. must be continued through the safe and efficient operation of commercial NPPs. The U.S. Department of Energy's (DOE) Light Water Reactor Sustainability (LWRS) research and development (R&D) program provides the technical bases for the long-term, safe, and economical operation of NPPs. One area in the LWRS program is the Plant Modernization pathway, which includes human factors R&D, human factors engineering (HFE), and ergonomics to enable the modernization of the instrumentation and control (I&C) technologies in NPP main control rooms. DOE researchers, including ergonomics specialists at Idaho National Laboratory (INL), have collaborated with numerous commercial NPP utilities over the last few years on control room modernization. This paper summarizes recent ergonomics safety and health R&D and HFE performed in collaboration with a U.S. commercial utility to modernize their NPP control rooms.

Keywords: Ergonomic safety & health · Nuclear power plants
Control room modernization · Digital instrumentation & control systems
Light water reactor sustainability program · Human factors engineering

© Springer Nature Switzerland AG 2019
S. Bagnara et al. (Eds.): IEA 2018, AISC 819, pp. 693–700, 2019.
https://doi.org/10.1007/978-3-319-96089-0_75

1 Introduction: Why Modernize Nuclear Power Plant Control Rooms?

Existing commercial nuclear power plants (NPPs) are valuable assets in the infrastructure portfolio of the United States (U.S.). They safely and reliably provide approximately 19% of all the electricity generated. While other non-carbon emitting renewable electrical generation technologies are making advances, their intermittency poses challenges to their large-scale integration into the grid. Indeed, historical analyses [1] have shown that previous energy transitions have taken decades, and so NPPs are still needed for the foreseeable future to generate base-load electricity.

The instrumentation and control (I&C) systems in commercial NPPs are the 'eyes and ears' of the operator, allowing them to control important primary and ancillary systems [2]. That is, NPP I&C systems are important because they are the way in which the operators maintain good situation awareness, thereby allowing the plant to operate safely and efficiently for all phases of operation.

Given this need, upgrading the existing I&C in commercial NPPs is very important because they potentially limit the operation of NPPs into their second license extension. Current NPP control room I&C is comprised mostly of analog technologies, but in other energy sectors analog I&C has largely been replaced with digital technologies. Thus, without the manufacturing and product support base for these legacy analog systems, spare and replacement parts are becoming increasingly scarce. As a result, analog systems, although still reliable, have reached the end of their useful service life.

Given the importance of I&C systems in NPP main control rooms and the need to upgrade them, the U.S. Department of Energy (DOE) sponsors the Light Water Reactor Sustainability (LWRS) Program, which facilitates and encourages cost-shared collaborations with utilities to safely extend the life of currently operating commercial NPPs. The LWRS Program has a broad scope and includes human factors research and development (R&D), human factors engineering (HFE), and ergonomics to support plant modernization, and in particular, control room I&C modernization. The remainder of this paper describes the ergonomic safety and health R&D and HFE performed in collaboration with multiple U.S. commercial utilities to modernize their NPP control rooms.

2 Guidance for the Ergonomic and Human Factors Engineering Approach

The U.S. Nuclear Regulatory Commission (NRC) closely regulates the U.S. civilian nuclear industry to protect people and the environment. The HFE and the ergonomics of the control room are within the purview of the NRC's oversight, and so utilities leverage NRC guidance on human factors such as NUREG-0711 [3] and NUREG-0700 [4], as well as industry guidance developed by the Electric Power Research Institute (EPRI) [5] and Idaho National Laboratory (INL) [6] when engaging in control room modernization. Having a common regulatory HFE program review model and established industry guidance allows for a level of standardization in the approach taken to conduct HFE R&D for control room modernization.

3   Recent Ergonomic Activities for Control Room Modernization

In 2016, a collaboration started that involved performing cost-shared R&D activities with a U.S. nuclear utility on control room modernization at four of its commercial NPPs. The purpose of this project was to conduct R&D to ensure these NPPs are able to operate for beyond their original licensing period with new and/or revitalized technologies. Additionally, this R&D project needed to consider the differences that must be addressed in fleet settings and in unregulated (i.e., deregulated or merchant) markets where different decision factors weigh on investment decisions and to demonstrate methods and techniques for modernization and investment in NPPs in these settings.

The utility partner was in the process of upgrading the non-safety related (NSR) nuclear steam supply systems (NSSS) and balance of plant (BOP) systems for four commercial NPP units. Although the utility had properly maintained and upgraded the NSR NSSS and BOP systems with like-for-like replacement parts, the systems had been in service for more than 30 years, and upgrading them to a new digital I&C system presented an opportunity to improve equipment reliability, reduce the likelihood of plant transients, and in general improve safety margins. For this upgrade, the changes to the control room included: (1) the addition of soft controls and digital control system (DCS) based alarm points on video display units added to the control boards, (2) the deletion of a number of analog controls and indicators, and (3) changes to procedures, the conduct of operations, and training. Given these changes, this work focused on the effects a hybrid human system interface (HSI) could have on the safety and health ergonomics of the control room, including issues such as: changes and inconsistencies in HSI design and operation, and impacts on operator workload, situation awareness, and the conduct of operations.

At the beginning of the project, a number of technical HFE R&D activities for control room modernization were performed. The focus was on performing ergonomic and other HFE technical analyses of digital I&C system hardware that was going to be installed, and in particular, the DCS's HSI.

Figure 1 shows the results of ergonomic and HFE analyses of the I&C hardware and evaluations of the DCS's HSI. Using state-of-the-art, three-dimensional modeling, researchers were able to identify that physical placement of touch screen monitors on the control boards was beyond the reach of some operators and that other aspects of their design (e.g., font size) and placement (e.g., viewing angle) affected screen legibility because they were not designed in a manner that is consistent with HFE design recommendations.

Fig. 1. Ergonomic and HFE evaluations using three-dimensional modeling help identify and prevent the introduction of new human error traps when performing digital I&C upgrades.

4   Ergonomic and Human Factors Simulator-Based R&D Activities

The latter stages of the project involved conducting operator-in-the-loop studies at the utility's control room simulators. The purpose of the workshops was to conduct HFE validation assessments of the planned control room I&C upgrades for the four NPP units, and had the objective of identifying potential HFE issues with the upgrades prior to installation in the main control room.

The operator-in-the-loop workshops entailed direct observations and assessments of key operator interactions with the existing and new HSIs across a number of normal, abnormal, and emergency scenarios. The scenarios were designed to evaluate the functional ergonomic and human factors aspects of the existing and upgraded I&C systems, with particular emphasis on the ability of the new I&C system to support operators' cognitive processes and their ability to facilitate the operators' ability to perform the correct control actions. The general workshop flow is depicted in Fig. 2.

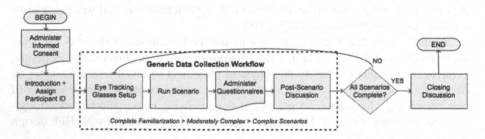

Fig. 2. Workshop execution workflow.

The assessment strategy focused on assessing plant safety and overall human-system performance. As such, measures of usability, workload, and situation awareness were collected during select scenarios to evaluate the suitability of the new HSIs in an operational context. During each scenario, the evaluation team also identified key information that determined the success or failure of the new HSIs as installed.

Human factors researchers and operationally experienced monitors observed the operators who performed actions in the scenarios using relevant procedures (see Figs. 3 and 4). Observations focused on operator interactions that were affected by the upgraded controls and indicators. Changes in time available for operator action, as well as information availability, were identified by observers during the scenarios.

Fig. 3. Ergonomics researchers observing NPP operators.

4.1   Workshop Results

If any potential performance deviations or difficulties were identified, then the LWRS HFE project team held a post-scenario discussion with the operators to identify potential contributors (e.g., HMI design issue(s), procedure issue(s), training/experience, simulator artifacts, etc.) so that appropriate actions could be taken. This assessment, attribution, and disposition process is shown in Fig. 5.

The results of the operator-in-the-loop workshop showed that the upgraded HSIs support operators in emergency operation tasks in that they did not adversely affect operator performance in highly complex scenarios where the operators' responses to transients and casualties that challenge the safety of the plant are important. Operators were also able to complete tasks without losing global situation awareness. This result indicated that the upgrades do not adversely affect the operators' mental models of the plant, particularly with respect to their ability to perform their critical safety-related actions.

Fig. 4. NPP operator interacting with the new DCS HSI.

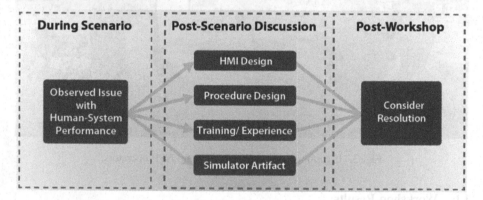

Fig. 5. Ergonomics and HFE Assessment Strategy.

For normal and abnormal operating conditions, the operators were able to successfully complete the tasks correctly, completely, and without confusion or misunderstanding using the upgraded HSIs. The operators were sufficiently alerted, provided with usable controls, and received adequate feedback from the HSIs.

The results also showed a number of operational improvements that will be realized with the successful installation of the new DCS. The most apparent example of this is in how the DCS automatically swaps controlling channels through a median select control function instead of manual channel selector switches, which therefore eliminates any potential operator errors in failing to correctly diagnose that a transmitter or sensor has failed. In addition, the new HSIs were well designed and presented more critical plant parameters in a user-friendly manner to the operators, which facilitated their ability to get a big picture understanding of the plant's state. This was most apparent in scenarios where operators had to detect, diagnose, and mitigate small leaks in the reactor coolant system. Observations and analyses of the scenarios identified a few medium and low level human factors issues, but the performance issues observed can and will be addressed satisfactorily with training and revised procedures. Overall, in the context of evaluating human-system performance, the control room operators appear to be sufficiently supported by upgraded HSIs, operating procedures, and their training to safely control the plant with the upgraded DCS.

5 Additional Work Performed

LWRS program researchers also developed frameworks for an HFE program management plan and a business case for fleet-based control room modernization. This work was done because it is recognized that control room modernization is the convergence of scientific, engineering, and economic considerations [7]. The HFE program plan provides structured and systematic guidance on an upgrade process that goes beyond obsolescence management. It defines an endpoint design concept for a fully modernized control room and a migration strategy for describing the phasing and ordering of I&C and control board changes. By considering main control room improvements that produce harvestable cost savings, the business case analysis allows utilities to move beyond performing like-for-like replacements that historically have not reduced operations and maintenance costs. The development of an HFE program plan and business case are additional activities that help utilities to achieve their long-term sustainability goals.

6 Conclusion

In summary, LWRS program researchers conducted cutting-edge R&D that furthered DOE's objective to support the long-term sustainability of the light water reactor fleet by ensuring the human factors aspects of control room upgrades are addressed. By performing R&D that addresses reliability and obsolescence issues of legacy analog control rooms, and by demonstrating and documenting the human factors processes that utilities should undertake to perform control room modernization for a fleet of NPPs operating in a merchant market, this work provides the technical bases that help reduce the uncertainty and risk of modernizing control rooms, thereby helping provide the incentives for the industry to make investments required for nuclear power operation periods to 60 years and beyond.

Disclaimer. INL is a multi-program laboratory operated by Battelle Energy Alliance LLC, for the United States Department of Energy under Contract DE-AC07-05ID14517. This work of authorship was prepared as an account of work sponsored by an agency of the United States Government. Neither the United States Government, nor any agency thereof, nor any of their employees makes any warranty, express or implied, or assumes any legal liability or responsibility for the accuracy, completeness, or usefulness of any information, apparatus, product, or process disclosed, or represents that its use would not infringe privately-owned rights. The United States Government retains, and the publisher, by accepting the article for publication, acknowledges that the United States Government retains a nonexclusive, paid-up, irrevocable, world-wide license to publish or reproduce the published form of this manuscript, or allow others to do so, for United States Government purposes. The views and opinions of authors expressed herein do not necessarily state or reflect those of the United States government or any agency thereof. The INL issued document number for this paper is: INL/CON-18-45643.

References

1. Smil V (2010) Energy transitions: history, requirements, prospects. ABC-CLIO, LLC, Santa Barbara
2. International Atomic Energy Agency (1999) Modern Instrumentation and Control for Nuclear Power Plants: A Guidebook, IAEA Technical report Series #387, Vienna, Austria
3. U.S. Nuclear Regulatory Commission (2012) Human Factors Engineering Program Review Model, NUREG-0711 (Rev. 3), Washington, DC
4. U.S. Nuclear Regulatory Commission (2002) Human-system interface design review guidelines, NUREG-0700 (Rev. 2), Washington, DC
5. Electric Power Research Institute (2015) Human Factors Guidance for Control Room and Digital Human-System Interface Design and Modification: Guidelines for Planning, Specification, Design, Licensing, Implementation, Training, Operation, and Maintenance for Operating Plants and New Builds. EPRI TR 3002004310, Palo Alto
6. Boring RL, Ulrich T, Joe JC, Lew R (2015) Guideline for operational nuclear usability and knowledge elicitation (GONUKE). Procedia Manuf 3:1327–1334
7. Joe JC (2017) A Human Factors Meta Model for U.S. Nuclear Power Plant Control Room Modernization. Proceedings of the 10th International Topical Meeting on Nuclear Plant Instrumentation, Control, and Human-Machine Interface Technologies (NPIC & HMIT 2017), San Francisco, pp 1833–1842

Slips, Trips and Falls

Effect of Heel Area on Utilized Coefficient of Friction During High-Heeled Walking

Sumin Park⬤ and Jaeheung Park(✉)⬤

Department of Transdisciplinary Studies, Graduate School of Convergence
Science and Technology, Seoul National University, Suwon 16229,
Republic of Korea
park73@snu.ac.kr

Abstract. The purpose of this study is to investigate the effect of heel area on
the utilized coefficient of friction (uCOF) during high-heeled walking and to
explain the change of the uCOF by understanding the gait patterns. Four high
heels with different heel areas (Narrow: 0.9 * 0.9 cm, Moderate: 1.5 * 1.7 cm,
Wide: 2.8 * 2.9 cm, Wedges: one piece of the sole and heel) were manufactured
for walking experiments. Ten females walked at 1.25 m/s on a treadmill with
two force plates inserted, and the walking data were collected by motion capture
system. The uCOF during stance phase was calculated from ground reaction
forces (GRFs). The peak uCOF and the peak GRFs during loading response
period was identified according to the heel area. One-way repeated measure
analysis of variance and correlation analysis were performed for the peak uCOF
and the peak GRFs. As the heel area became narrower, the peak uCOF increased
significantly. However, there was no significant difference in the peak uCOF
between the wide heels and wedge heels. The changed peak uCOF according to
the heel area correlated with the change in timing of the peak uCOF. The peak
uCOF and the peak anterior-posterior GRF occurred in early loading response
period when wearing the narrow heels than the wide heels, although the magnitudes of the peak GRFs were consistent. The potential for slipping during
high-heeled walking can be increased at narrow heel area since the horizontal
shear force reaches the peak quickly before the vertical force is sufficiently large
with weight acceptance.

Keywords: Utilized coefficient of friction · High-heeled walking
Heel area

1 Introduction

Slips occur during walking when the adhesive friction between the foot and the floor is
lost. Unexpected slips cause muscle sprains and strains, and even more severe musculoskeletal injuries if slips lead to falls [1]. 40–50% of slips contribute to falls [2], and
slips and falls account for about 20% of all occupational accidents [1].

In industry, it is important to understand slip-resistance of the shoes and the
workplace floor in order to reduce the occupational accidents caused by slips and falls.
The slip-resistance is calculated as the available coefficient of friction (aCOF), which is
either the static or dynamic coefficient of friction (COF) between the outsole of the shoe

© Springer Nature Switzerland AG 2019
S. Bagnara et al. (Eds.): IEA 2018, AISC 819, pp. 703–709, 2019.
https://doi.org/10.1007/978-3-319-96089-0_76

and the floor surface. Researchers investigated the aCOF in workplace floors covered with various materials, because the slip-resistance changes when the floor is contaminated. A study reported that slips did not happen during walking across a clay-covered beam with static COF of 0.41 in shoes with urethane sole, but did occur during walking across an oil-covered beam with static COF of 0.20 [3]. Perkins and Wilson [4] mentioned that dynamic COF varies depending on how the shoe moves on the surface under different contaminant conditions and slips can occur in a wet vinyl tile floor with dynamic COF of 0.30 if the heel strikes the floor at a landing speed of more than 2 cm/s. Miller [5] suggested aCOF of 0.50 as a minimum safety standard of the shoe-floor interface for the workplace surface.

To predict the potential for slipping, the utilized coefficient of friction (uCOF) was also investigated as well as the aCOF [6–8]. The aCOF is usually measured by tribometers to illustrate the slip-resistance of the shoes and the floor [6], while the uCOF means the friction required to continue the movement and is calculated by force plates as the ratio of the resultant shear ground reaction force to the vertical ground reaction force during walking [9]. The uCOF is affected by gait biomechanics, and thus it depends on age, sex, walking speed, etc. [9–11]. If the peak uCOF exceeds aCOF of the shoe-floor interface, the foot slides due to the loss of traction with the floor [12]. It is reported that wearing high heels increases the peak uCOF during walking and the risk of slipping is higher during high-heeled walking than normal walking due to the increased peak uCOF [12].

The high risk of slipping during high heeled walking can increase more depending on the heel area of high heels even with the identical heel height due to changed uCOF. It is hypothesized that the peak uCOF increases as the heel area becomes narrower, although the heel height is identical. Therefore, the purpose of this study is to investigate the effect of heel area on uCOF during high-heeled walking with the identical heel height and to explain the change of uCOF by understanding the gait patterns according to the heel area.

2   Materials and Methods

2.1   Subjects

Ten females with a shoe size of 235 mm, who are experienced in wearing high heels, participated in walking experiments. Their average height, weight, and age are 159.3 ± 3.02 cm, 50.5 ± 4.25 kg, and 24 ± 2.72 years, respectively. The participants read and signed a written informed consent document, which is approved by the Institutional Review Board of Seoul National University, before the experiments. All participants reported no musculoskeletal disorders.

2.2   Instrumentation

Four shoes with different heel areas having the identical heel height of 9 cm were made of the identical material from a manufacturer. Figure 1 shows the manufactured shoes (Narrow heels: 0.9 * 0.9 cm, Moderate heels: 1.5 * 1.7 cm, Wide heels: 2.8 * 2.9 cm,

Wedge heels: one piece of the sole and the heel). Two force plates (AMTI, MA, USA), which are inserted in a treadmill, were used to obtain ground reaction forces (GRFs) during walking. The treadmill's floor is made of two separate rubber belts for each of the right foot and left foot. The reflective markers attached to the toe and heel of the shoes were used to calculate stride length and stride time. Six cameras (Motion Analysis, CA, USA) were used to capture the markers by motion capture system.

Fig. 1. The manufactured shoes (Narrow heels: 0.9 * 0.9 cm, Moderate heels: 1.5 * 1.7 cm, Wide heels: 2.8 * 2.9 cm, Wedge heels: one piece of the sole and the heel)

2.3 Experimental Procedures

It was checked whether the shoes of 235 mm are fit with the subjects' feet or not. In case there is a small gap between the shoes and feet of the subjects, thin soles were inserted inside of the shoes. The participants walked on the treadmill at a speed of 1.25 m/s for 30 s. Force data and marker data were recorded at 800 Hz and 200 Hz, respectively. Since comfortable walking speed for women was reported as approximately 1.272 m/s [13], the speed of 1.25 m/s was chosen for the experiments. Walking trials according to the shoe condition were performed in a randomized order with a 3 min break time after each trial.

2.4 Data Process

The force data and marker data were filtered by a 5th-order low-pass Butterworth filter with a cutoff frequency of 30 Hz and 10 Hz, respectively. Anterior-Posterior GRF (GRF_{AP}), Medial-Lateral GRF (GRF_{ML}), and Vertical GRF (GRF_{V}) of five cycles were extracted for each of the left foot and right foot. The timing of heel-strike and toe-off were determined by GRF_{V} to distinguish stance phase, swing phase, and a cycle of walking. The averaged GRF_{AP}, GRF_{ML}, and GRF_{V} of each subject according to the heel area were calculated and normalized by the body mass of each subject. The utilized coefficient of frictions (uCOF) of each subject during stance phase was calculated as the ratio of the averaged resultant shear GRF (GRF_{R}) to the averaged GRF_{V}.

$$uCOF = \frac{Resultant\ shear\ GRF}{Vertical\ GRF} = \frac{\sqrt{GRF_{AP}^2 + GRF_{ML}^2}}{GRF_{V}}$$

The peak uCOF was defined as the maximum uCOF during weight acceptance since most of the slips occur in the forward direction during the period [4]. Stride lengths of five cycles for each of the left foot and right foot were determined based on the midpoint between the toe and heel markers at the heel strike. The averaged stride length, stride time, stance ratio, and swing ratio were calculated according to the heel area. The stance ratio and swing ratio indicate the percentages of stance time and swing time per stride time respectively during a cycle of walking. Statistical analysis was performed using one-way repeated measure analysis of variance (RM-ANOVA) with the Huynh-Feldt correction and the LSD post-hoc test. Correlation analysis was also performed to understand the relationship between the changes in uCOF and GRFs. The significance level was set to less than 0.05. Data process and statistical analysis were conducted by MATLAB (MathWorks, MA, USA) and SPSS (IBM, NY, USA) software respectively.

3   Results

The magnitude of the peak uCOF increased significantly as the heel area of the high heels became narrower (Table 1(A)).

The peak uCOF when wearing the narrow heels and the moderate heels were 1.08 times (post-hoc test, p = 0.003) and 1.06 times (post-hoc test, p = 0.011) higher respectively than the peak uCOF when wearing the wide heels (Table 1(A)). There was no significant difference in the peak uCOF between the wedge heels and the wide heels (post-hoc test, p = 0.138).

The changed magnitude of the peak uCOF according to the heel area correlated with the change in timing of the peak uCOF (correlation analysis, $r = -0.674$, $p < 0.001$). The peak uCOF occurred in the early loading response period when wearing the narrow heels (post-hoc test, $p = 0.003$) and the moderate heels (post-hoc test, $p = 0.001$) than the wide heels (Table 1(A)). There was also a relationship between the timing of the peak uCOF and the timing of the peak GRF_{AP} (correlation analysis, $r = 0.889$, $p < 0.001$). Like the peak uCOF, the peak GRF_{AP} occurred in the early loading response period when wearing the narrow heels than the wide heels, while the timing of the peak GRF_{ML} and the peak GRF_V did not differ significantly (Table 1(B)). The magnitudes of the peak GRF_{AP}, the peak GRF_{ML}, and the peak GRF_V were consistent regardless of the heel area (Table 1(B)).

According to the results of the magnitudes of the GRFs at the peak uCOF (Table 1 (C)), both the GRF_{ML} and GRF_V were smaller when wearing the narrow heels than the wide heels due to the changed timing of the peak uCOF. Meanwhile, the GRF_{AP} and GRF_R at the peak uCOF were not statistically significant depending on the heel area.

The stride length and stride time did not show any difference according to the heel area during walking at 1.25 m/s. However, the stance ratio and swing ratio when wearing the narrow heels decreased and increased respectively than the wide heels and the wedge heels (Table 2). The changed stance or swing ratio also correlated with the change in the timing of the peak GRF_{AP} (correlation analysis, $r = 0.650$, $p < 0.001$).

Table 1. Change of uCOF and GRFs according to heel area during high-heeled walking

	Heel area				p-value
	Narrow (0.9 * 0.9 cm)	Moderate (1.5 * 1.7 cm)	Wide (2.8 * 2.9 cm)	Wedge (One piece)	
(A) uCOF in loading response					
Peak uCOF (unitless)	0.225[a]	0.222[a]	0.209	0.216	0.003*
Timing of the peak uCOF (%)	15.8[a,b]	16.2[a,b]	18.7	18.5	<0.001*
(B) GRFs in loading response					
Peak GRF_{AP} (N/Kg)	2.406	2.391	2.340	2.421	0.405
Timing of the peak GRF_{AP} (%)	17.2[a]	17.7[a,b]	19.0	18.7	0.022*
Peak GRF_{ML} (N/Kg)	1.090	1.049	1.095	1.112	0.191
Timing of the peak GRF_{ML} (%)	29.6	28.9	30.4	29.9	0.123
Peak GRF_V (N/Kg)	12.405	12.316	12.570	12.426	0.165
Timing of the peak GRF_V (%)	23.4	24.3	23.9	24.9	0.508
(C) GRFs at the peak uCOF					
GRF_{AP} (N/Kg)	2.308	2.292	2.297	2.385	0.286
GRF_{ML} (N/Kg)	0.411[b]	0.395[a,b]	0.537	0.578	0.007*
GRF_V (N/Kg)	10.543[a,b]	10.603[a,b]	11.544	11.512	<0.001*
GRF_R (N/Kg)	2.357	2.339	2.382	2.476	0.080

NOTE: *means there is a significant difference by RM-ANOVA. [a]indicates a significant difference in comparison with the wide heels by post-hoc test. [b]indicates a significant difference in comparison with the wedge heels by post-hoc test. The timing of the uCOF or GRFs is presented as a percentage of stance phase.

4 Discussion

It has been reported that the peak uCOF during high-heeled walking increases due to the increased GRF_R and decreased GRF_V as the heel height of the shoe increases, and it leads to the high risk of slipping [12]. The results of this study show that the potential

Table 2. Change of gait parameters according to heel area during high-heeled walking

	Heel area				p-value
	Narrow (0.9 * 0.9 cm)	Moderate (1.5 * 1.7 cm)	Wide (2.8 * 2.9 cm)	Wedge (One piece)	
Stride length (m)	1.15	1.16	1.16	1.16	0.122
Stride time (s)	0.92	0.93	0.92	0.93	0.100
Stance ratio (%)	62.98[a,b]	63.39[a,b]	63.92	63.88	0.011*
Swing ratio (%)	37.02[a,b]	36.61[a,b]	36.08	36.12	0.011*

NOTE: *means there is a significant difference by RM-ANOVA. [a]indicates a significant difference in comparison with wide heels by post-hoc test. [b]indicates a significant difference in comparison with wedge heels by post-hoc test. The stance ratio and swing ratio indicate the percentages of stance time and swing time per stride time respectively.

for slipping during high-heeled walking can vary not only by heel height but also by heel area. In case of wearing high heels with the narrow heel area, the peak GRF_{AP} occurred in early loading response period than wearing high heels with the wide heel area (Table 1(B)). The changed timing of the peak GRF_{AP} can be explained by the fact that the center of pressure of the foot moves quickly from the rearfoot to the forefoot in the forward direction due to the narrow heel area, and it resulted in the changed timing of the peak uCOF and the increased peak uCOF by the reduced GRF_V (Table 1(A) and (C)). In summary, the resultant shear force reaches the peak quickly by the fast forward roll-over of the foot before the vertical force is sufficiently large with weight acceptance, and this changed gait patterns increase the peak uCOF during high-heeled walking with narrow heel area. The reduced stance ratio (Table 2) is also related to the fast forward roll-over of the foot (correlation analysis, r = 0.650, p < 0.001) since the fast foot roll-over brings quick toe-off and induces short stance time.

The peak uCOF when wearing the wide heels was 0.209 and the peak uCOF increased to 0.225 by about 0.02 when wearing the narrow heels (Table 1(A)). Beschorner et al. have reported that a small difference in uCOF can increase the probability of a slip dramatically, showing a slipping odds ratio of 1.7 for an increase of 0.01 in uCOF [8]. According to the logistic regression result in the study of Beschorner et al., the uCOF of 0.2 shows 25% of probability for slipping during walking on a floor having aCOF of 0.165, while the uCOF of 0.22 shows 50%. It implies that wearing narrow high heels can induce two times more slipping than wearing wide high heels during walking on the floor having aCOF of 0.165 due to the increase in the peak uCOF from 0.209 to 0.225. Wearing narrow high heels should be carefully concerned during walking on surfaces having low aCOF since it can increase the risk of slipping.

In this study, the walking experiments were conducted on the treadmill at the specified speed. The effect of the heel area on uCOF at the self-selected speed may be different from the results of this study if the walking speed is significantly affected by the heel area.

5   Conclusion

This study presented that the heel area of high heels influences on the uCOF during walking at an identical walking speed even with the same heel height. Wearing high heels with a narrow heel area increases the peak uCOF since the peak uCOF occurs in early loading response period and the vertical GRF is not sufficiently large at the timing of the peak uCOF. Wearing high heels with a wide heel area is recommended for women to reduce the peak uCOF and the risk of slipping.

References

1. Buck PC, Coleman VP (1985) Slipping, tripping and falling accidents at work: a national picture. Ergonomics 28(7):949–958
2. Courtney TK, Sorock GS, Manning DP, Collins JW, Holbein-Jenny MA (2001) Occupational slip, trip, and fall-related injuries – can the contribution of slipperiness be isolated? Ergonomics 44(13):1118–1137
3. Swensen EE, Purswell JL, Schlegel RE, Stanevich RL (1992) Coefficient of friction and subjective assessment of slippery work surfaces. Hum Factors 34(1):67–77
4. Perkins PJ, Wilson MP (1983) Slip resistance testing of shoes – new developments. Ergonomics 26(1):73–82
5. Miller JM (1983) "Slippery" work surfaces: towards a performance definition and quantitative coefficient of friction criteria. J Saf Res 14(4):145–158
6. Burnfield JM, Powers CM (2006) Prediction of slips: an evaluation of utilized coefficient of friction and available slip resistance. Ergonomics 49(10):982–995
7. Hanson JP, Redfern MS, Mazumdar M (1999) Predicting slips and falls considering required and available friction. Ergonomics 42(12):1619–1633
8. Beschorner KE, Albert DL, Redfern MS (2016) Required coefficient of friction during level walking is predictive of slipping. Gait Posture 48:256–260
9. Kleiner AFR, Galli M, do Carmo AA, Barros RM (2015) Effects of flooring on required coefficient of friction: elderly adult vs. middle-aged adult barefoot gait. Appl Ergon 50:147–152
10. Burnfield JM, Powers CM (2003) Influence of age and gender on utilized coefficient of friction during walking at different speeds. In: Metrology of pedestrian locomotion and slip resistance. ASTM International
11. Fendley AE, Medoff HP (1996) Required coefficient of friction versus top-piece/outsole hardness and walking speed: significance of correlations. J Forensic Sci 41(5):763–769
12. Blanchette MG, Brault JR, Powers CM (2011) The influence of heel height on utilized coefficient of friction during walking. Gait Posture 34(1):107–110
13. Bohannon RW (1997) Comfortable and maximum walking speed of adults aged 20–79 years: reference values and determinants. Age Ageing 26(1):15–19

Static Postural Stability on Narrow Platforms to Prevent Occupational Stepladder Falls

Atsushi Sugama[1(✉)] and Akihiko Seo[2]

[1] National Institute of Occupational Safety and Health, Japan, Kiyose, Japan
sugama@s.jniosh.johas.go.jp
[2] Faculty of System Design, Tokyo Metropolitan University, Hino, Japan

Abstract. Falls from heights are the most common causes of occupational fatal accidents in many countries; using agents, such as stepladders or scaffolds, is one of the main causal factor. This study aims to evaluate the postural stability of static standing on narrow platforms. Eleven male participants stood on five platforms that had anterior/posterior widths ranging from 6 to 25 cm and maintained their position for 50 s. The coordinates and velocities of center of mass (CoM) and center of pressure (CoP) were calculated from kinematic data of human body and foot reaction forces. The results showed that the relative position of CoP to the platform width and the translational velocity non-linearly increased with shortened platform width and more significantly changed than the relative position of CoM, while there was no significant difference between the 15-cm and 25-cm platforms. The regression lines of the relative position and the velocity of CoP were approximated as a function of the inverse of the platform width or the square, respectively. Shortened platforms make the postural balance of static standing non-linearly unstable, whereas platforms that are 15 cm or wider stabilize the postural perturbation comparable to that achieved on the ground. Therefore, the equipment with a platform or rungs at least 15 cm or wider should be recommended for tasks at elevated places.

Keywords: Postural stability · Falls from height · Occupational safety
Stepladder

1 Introduction

Ladders are common equipment to access elevated locations and are used for various tasks in occupational situations. Falls from height, however, are one of the leading cause of occupational fatalities [1], and using agents, such as ladders and scaffolds, are one of the leading cause of occupational falls [2, 3].

Ladder falls occur due to loss of balance of either human or ladder [4, 5]. Therefore, previous studies have examined postural balance during ladder tasks as well as the structural stability of ladders [6–9]. Static postural stability is generally defined as the ability of a body to restore its original static equilibrium after it has been slightly displaced [10]. Since decreasing the area of a platform corresponds to decreasing the base of support (BoS) for a worker, a wider BoS and a state where the center of mass (CoM) of the body is located close to the center of the BoS are preferable for maintaining static postural stability [11, 12]. However, few previous studies have investigated the postural balance on narrow platforms that are narrower than one foot in length.

The purpose of this study was to evaluate the postural stability on narrow platforms based on the theory of postural control for prevention of falls from height. In this study, we analyzed postural perturbation based on kinematical indices of the human body. This paper includes a part of contents that were reported in our previous study [13].

2 Methods

2.1 Participants

Eleven men aged 22–27 years participated in the study. The means and standard deviations of age, height, and weight were 23.4 ± 1.5 years, 169.3 ± 4.9 cm, and 61.2 ± 7.9 kg, respectively. For safety, all participants wore a safety harness and the same model of safety shoes (MZ010J; Midori Anzen Co. Ltd., Japan). The average participant foot length [14] was 24.7 ± 1.1 cm, and shoe sizes ranged from 24.5 to 27.0 cm. The safety shoes had non-flat soles and there was a bump beneath the arch of the foot.

An oral overview of the experiment was presented to each participant to obtain their informed consent. This experiment was conducted with approval from the ethics committee of Hino campus of Tokyo Metropolitan University.

2.2 Experimental Procedure

A square column-shaped platform was placed on a force plate (9280AA; Kistler Instrument Corp., Amherst, NY, USA). The height and lateral (left–right) length of the platform were 8 and 48 cm, respectively. Five platform widths (anterior–posterior) of 6, 8, 10, 15, and 25 cm were used in the experiments.

The participant stood on the platform in a comfortable stance with their feet shoulder-width apart and maintained this position for 50 s as shown in Fig. 1. Participants were required to look the point of gaze that was located anterior to the eye level.

Experimental orders were randomized, and measurements were repeated two times for each condition.

2.3 Measurements and Analysis

Horizontal Displacements and Velocities of CoM and CoP. Winter [15] described postural sway on the sagittal plane during static standing by the equation about the positions of the center of pressure (CoP) and CoM. The anterior–posterior position of CoM (X_{CoM}) was calculated as follows:

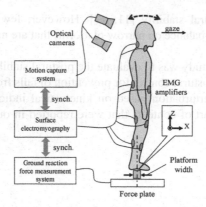

Fig. 1. Diagram of experiment and measurement system [13].

$$X_{\text{CoM}} = \frac{1}{M} \sum_{i=1}^{N} m_i x_i \tag{1}$$

where M is the mass of the total body; N is the number of segments; m_i and x_i are the mass and anterior–posterior positions of the ith segments, respectively. The length and mass of each segment were determined as described by previous studies [16, 17].

In this study, body posture was measured with an optical motion capture system (OptiTrack; Natural Point, Inc., USA) and its software (Motive 1.8.0; Natural Point, Inc., USA). The coordinate system was a right-handed z-up world, and the origin was a point on the platform's surface. Three-dimensional coordinates of reflective markers placed on the participants' bodies were recorded at a sampling frequency of 100 Hz. The coordinate data were assigned to the body posture of a multi-segment model.

The signals from the force plate were recorded at a sampling frequency of 500 Hz and were amplified and used to calculate the anterior–posterior coordinate of CoP X_{CoP} as follows:

$$X_{\text{CoP}} = \frac{F_x \cdot (z_0 + z_s) + M_y}{F_z} \tag{2}$$

where F_x and F_z are the measured reaction forces in the X and Z directions, respectively. M_y is the moment of force in the sagittal direction; z_0 and z_s are the vertical distances from the detection axis of the force plate to i^{th} surface and the height of the platform, respectively.

Next, relative values of X_{CoM} and X_{CoP} to the platform width w (R_{CoM}, R_{CoP}) were calculated as follows:

$$R = \frac{2X}{w} * 100 \tag{3}$$

In addition, mean velocities of CoM or CoP (V_{CoM}, V_{CoP}) for the anterior–posterior direction were calculated as shown in Eq. 4. Here, n is the number of data points, and t is the measurement time.

$$V = \frac{\sum_{j=1}^{n-1} \sqrt{[X(j+1) - X(j)]^2}}{t} \qquad (4)$$

Statistical Analysis. To compare the effects of platform width on the measured indices, an analysis of variance (ANOVA) with a randomized block design for participants and a post hoc Tukey's test were performed. Levels of statistical significance of tests were set at 0.05.

3 Results

3.1 Anterior–Posterior Positions of CoM and CoP

Figure 2 shows the maximum and minimum values of R_{CoM} and R_{CoP}, and these varied non-linearly. Maximum values occurred at a platform width of 6 cm for both R_{CoM} and R_{CoP}, where the maximum value of R_{CoM} was approximately 40% and that of R_{CoP} was approximately 70%.

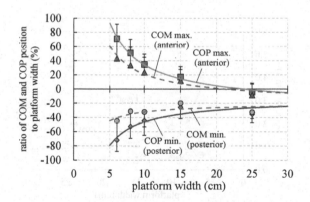

Fig. 2. Ratio of anterior/posterior position of CoM and CoP to platform width [13].

Figure 3 shows the maximum, mean, and minimum values of differences between X_{CoM} and X_{CoP}. ANOVA analysis showed that there were significant differences between 6 cm and 15 cm conditions for the maximum and minimum values.

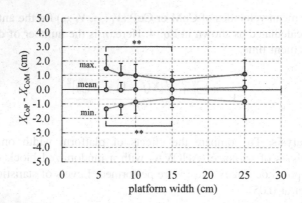

Fig. 3. Maximum differences of anterior/posterior position of CoM and CoP (X_{CoP}–X_{CoM}).

3.2   Translational Velocity of CoM and CoP

Translational velocities of CoM (V_{CoM}) and of CoP (V_{CoP}) are shown in Fig. 4. These results demonstrate non-linear relationships in both V_{CoM} and V_{CoP}, and V_{CoP} were markedly higher than V_{CoM} in all conditions. The values of V_{CoP} ranged from 1.0 to 3.8 cm/s, while the values of V_{CoM} were approximately 0.5 cm/s in all conditions. Moreover, ANOVA analysis showed significant differences for all pairs excluding between platforms of 15 cm and 25 cm (p < 0.001 in all cases).

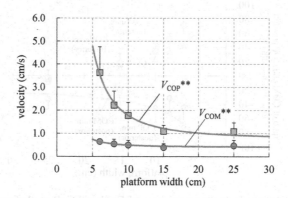

Fig. 4. Anterior/posterior velocity of CoM and CoP [13].

4   Discussion

4.1   Approximate Expressions of R_{CoM} and R_{CoP}

Evaluation functions, of which the platform width w was set as an explanatory variable, were produced to assess postural stability on narrow platforms. The objective variables were chosen from the measured indices: motion ranges of CoM and CoP relative to platform widths (R_{CoM}, R_{CoP}) and translational velocities (V_{CoM}, V_{CoP}).

Both R_{CoM} and R_{CoP} were approximated as functions of w^{-1} with following model equation: $y = a/x + b$. V_{CoM} and V_{CoP} were considered as derivative functions of R_{CoM} or R_{CoP}, respectively, and approximated as functions of w^{-2} with following model equation: $y = a/x^{\wedge}2 + b$.

The results of fitted curves were shown in Figs. 2 and 4 and the coefficients of correlation, r, were ≥ 0.9 for all indices excluding the minimum value of R_{CoM} ($r = 0.68$). These results show that body motions relative to the platform and subjective senses of instability can be modeled with high accuracy.

4.2 Characteristics of Postural Stability on Narrow Platforms

The anterior–posterior position of CoM and CoP relative to platform width significantly increase with decreasing of platform width, and these values are proportional to the reciprocal of the platform width. These are due to decreasing of BoS for the person who stands on the narrow platform. Since decreasing of BoS causes the decreasing in the contact area between the soles of the feet and the platform, that might decrease the accuracy of postural control. Previous studies have suggested that input signals acquired through the visual, vestibular, and somatosensory systems are organized in the central nervous system (CNS) for postural control [10]. In this study, the weight of input information from somatosensory system is varied by decreasing of platform width. As the results, the value of X_{CoP}–X_{CoM} increased in a condition with platform of 6 cm compared with other conditions.

In addition, velocities of CoM and CoP approximate to functions of the reciprocal of the square of platform width. One reason for these trends may be related to kinematical need as shown in Eq. 1. On the narrow platforms, theoretical controllable range of CoP, which is equal to BoS, in the anterior–posterior direction decreases. Therefore, subjects had to move CoP both further and faster with decreasing platform widths to produce a difference in horizontal displacement between CoM and CoP.

These results indicate that the changes in postural stability on narrow steps are not linear with respect to the width of the platform and grow rapidly worse when the width is below 15 cm.

5 Conclusion

This study aimed to evaluate the postural balance of static standing on narrow platforms. The anterior–posterior positions of CoM and CoP and the translational velocities were both non-linearly increased with decreasing of the platform width. The positions and velocities were approximated as the function of the reciprocal of platform width and the reciprocal of the square of platform width, respectively.

The data obtained and predictive models derived showed that there was little difference in the indices of postural stability (i.e. the positions and velocities of CoM and CoP) when the platform width was 15 cm or deeper. Therefore, agents with platforms of 15 cm or wider is recommended for occupational equipment such as ladders and stepladders to decrease the risk of loss of balance.

Acknowledgements. The authors would like to thank I. Moriya for technical assistance with the experiments. This work is supported by JSPS KAKENHI Grant Number JP17K12995.

References

1. Nadhim EA, Hon C, Xia B, Stewart I, Fang D (2016) Falls from height in the construction industry: a critical review of the scientific literature. Int J Environ Res Public Health 13:638–657. https://doi.org/10.3390/ijerph13070638
2. Simeonov P (2017) Ladder safety: research, control, and practice. In: Hsiao H (ed) Fall prevention and protection, principles, guidelines, and practices. CRC Press, New York, pp 241–269
3. Sugama A, Ohnishi A (2015) Occupational accidents due to stepladders in Japan: analysis of industry and injured characteristics. Procedia Manuf 3:6632–6638. https://doi.org/10.1016/j.promfg.2015.11.003
4. Björnstig U, Johnsson J (1992) Ladder injuries: mechanisms, injuries and consequences. J Saf Res 23:9–18. https://doi.org/10.1016/0022-4375(92)90035-8
5. Huang X, Hinze J (2003) Analysis of construction worker fall accidents. J Constr Eng Manag 129:262–271. https://doi.org/10.1061/(asce)0733-9364(2003)129:3(262)
6. Seluga KJ, Ojalvo IU, Obert RM (2007) Analysis and testing of a hidden stepladder hazard – excessive twist flexibility. Int J Contr Saf Promot 14:215–224. https://doi.org/10.1080/17457300701722601
7. Yang BS, Ashton-Miller JA (2005) Factors affecting stepladder stability during a lateral weight transfer: a study in healthy young adults. Appl Ergon 36:601–607. https://doi.org/10.1016/j.apergo.2005.01.012
8. Axelsson PO, Carter N (1995) Measures to prevent portable ladder accidents in the construction industry. Ergonomics 38:250–259. https://doi.org/10.1080/00140139508925102
9. Cohen HH, Lin LJ (1991) A scenario analysis of ladder fall accidents. J Saf Res 22:31–39. https://doi.org/10.1016/0022-4375(91)90011-J
10. Shumway-Cook A, Woollacott MH (2013) Motor control: translating research into clinical practice, 4th edn. Lippincott Williams & Wilkins, Philadelphia
11. DiDomenico A, Nussbaum MA (2005) Interactive effects of mental and postural demands on subjective assessment of mental workload and postural stability. Saf Sci 43:485–495. https://doi.org/10.1016/j.ssci.2005.08.010
12. Kirby RL, Price NA, MacLeod DA (1987) The influence of foot position on standing balance. J Biomech 20:423–427
13. Sugama A, Seo A (2017) Postural stability of static standing on narrow platform for stepladder safety. Jpn J Ergon 53:125–132 (in Japanese)
14. ISO 7250-1:2017. Basic human body measurements for technological design – Part 1: Body measurement definitions and landmarks
15. Winter DA, Patla AE, Prince F, Ishac M, Gielo-Perczak K (1998) Stiffness control of balance in quiet standing. J Neurophysiol 80:1211–1221. https://doi.org/10.1152/jn.1998.80.3.1211
16. Chaffin DB, Andersson GBJ, Martin BJ (2006) Occupational Biomechanics, 4th edn. Wiley-Interscience, pp 37–51
17. Michiyoshi A, Hai-peng T, Takashi Y (1992) Estimation of inertia properties of the body segments in Japanese athletes. J Soc Biomech 11:23–33. https://doi.org/10.3951/biomechanisms.11.23 (in Japanese)

Pushing Induced Sliding Perturbation Affects Postural Responses to Maintain Balance Standing

Yun-Ju Lee[1]([✉]), Bing Chen[2], Jing-Nong Liang[3],
and Alexander S. Aruin[4]

[1] Department of Industrial Engineering and Engineering Management,
National Tshing-Hua University, Hsinchu, Taiwan
yunjulee@ie.nthu.edu.tw
[2] Department of Neurological Surgery, The Miami Project to Cure Paralysis Lois
Pope Life Center, University of Miami, Miami, FL, USA
[3] Department of Physical Therapy, University of Nevada Las Vegas,
Las Vegas, NV, USA
[4] Department of Physical Therapy, University of Illinois at Chicago,
Chicago, IL, USA

Abstract. Pushing is considered a manual handling activity in industry and its functional role is also in our daily life, such as pushing moving strollers or grocery carts, which is under examined. In such situations, two perturbations need to be handled simultaneously, specifically, pushing (voluntary movement) and pushing-induced slipping when walking on a slip surface (sliding perturbation). Eight participants were instructed to push a handle while standing on a locked or unlocked movable board, which was placed on a force plate. Three accelerometers were attached to the handle to detect the moment of the handle moving away (T_{handle}), which denotes time zero, the pelvis to detect the moment of trunk movement, and the movable board to detect the moment of board movement. The onset time, magnitude of center of pressure (COP) at T_{handle} and maximum pushing force were calculated. The onsets of board movement, trunk movement, and COP were initiated prior to T_{handle}. Pushing while standing on the unlocked sliding board significantly affected the onset of COP but did not affect onset of trunk movement. In addition, the direction of acceleration on the board was corresponding to the reaction force of the pushing forces at hands in the backward direction. Studying the combined effects of varying movability of the support surface and pushing tasks may contribute to the development of new environment safety for workers and elderly.

Keywords: Pushing force · Sliding surface · Occupational biomechanics Balance

© Springer Nature Switzerland AG 2019
S. Bagnara et al. (Eds.): IEA 2018, AISC 819, pp. 717–724, 2019.
https://doi.org/10.1007/978-3-319-96089-0_78

1 Introduction

Pushing is a common manual material activity and one risk factor of working environment is the floor condition [1]. It has been suggested that working posture is extremely dependent upon the quality of the foot-to-floor interface [2]. Slipping occurs when the ratio of the friction force to the normal force at the feet exceeds the maximum coefficient of friction and the risk of slipping was associated with the maximum friction at the feet in pushing [3, 4]. When pushing on the difference surfaces, changes in posture of knee flexion and trunk extension has been observed [5]. In addition, changes in anticipatory and compensatory postural adjustments are observed in the center of pressure (COP) displacements when pushing in the anterior-posterior direction; the backward COP displacements were seen prior to the pushing followed by forward COP displacements after the movement [6]. In the medial-lateral direction, the COP displacement is affected by the placements of lower extremities on the floor [6, 7] and by the magnitude of the weight to be pushed: larger COP displacements were reported when pushing heavier weight [8]. Meanwhile, the orientations of the COP displacements and vertical torques were corresponding to the upper and lower extremities positions in pushing [8]. However, all previous studies only focused on pushing tasks itself and did not take subsequent perturbations to postural control and balance into considerations.

Performing a voluntary movement while pushing an object or responding to the movements of an unstable surface (that are considered as a primary task); these tasks are performed simultaneously with maintenance of vertical posture. The central nervous system (CNS) uses sensory information from the vestibular, visual, and somatosensory systems in control of upright posture as well as in dealing with the primary task. In addition, many daily life activities that are associated with carrying out motor tasks are performed simultaneously (e.g. holding a cup of tea or talking on the cell phone while walking). Another example is standing wearing rollerskates and performing voluntary movements. It was reported that people who use rollerskates and perform fast arm flexion movements showed larger shank muscle activities and muscle co-contraction patterns. In addition, the inverted direction of COP displacement in the anterior-posterior direction was recorded as compared to standing wearing regular shoes [9]. It has been shown in experiments involving both perturbations (pushing and externally induced forward and backward translations of the platform) that the onset time of the tibialis anterior muscle in such conditions depends on the direction of moving platform and is different compared to pushing only (while standing on a non-movable platform) [10]. Thus, dealing with two perturbations simultaneously modifies patterns of muscle activities and COP displacement as compared to experiencing a single perturbation [10, 11].

Pushing as a perturbation resulted in different performing strategies was observed in elderly compared to healthy young adults [12]. Aging population is growing unprecedentedly worldwide and it has been reported that a clear relationship between working capacity and workers' age [13]. A better understanding of how sliding surface affects pushing and balance performances could provide new information crucial in designing working environment. The objective of the present study was to investigate how a combination of voluntary pushing movement and translational perturbation affects pushing performance and balance. To do that the experimental paradigm involved two body perturbations: one was induced by the upper body performing the pushing an object while standing on a free to move surface and the other perturbation was induced when the sliding board moved. We hypothesized that onsets of the COP displacement would be affected by the presence of a secondary perturbation; we also expected to see a decrease in the magnitudes of the COP displacements due to the translation perturbation.

2 Methods

2.1 Participants

Eight young male volunteers (age = 26.50 ± 1.70 years, height = 1.68 ± 0.02 m, mass = 72.43 ± 2.49 kg) participated in the experiment. All participants were free from any musculoskeletal disorder and neurologic disease that could affect performing the experimental tasks. The project was approved by the University of Illinois at Chicago Institutional Review Board, and all participants provided written informed consent before taking part in the experimental procedures.

2.2 Instrumentation and Procedure

The customized sliding board was 0.06 m height and made of two layers connected by a linear bearing system. The sliding board had a lock mechanism allowing the top layer (lengths 0.50 m and width 0.45 m) to be free to slide in the anterior-posterior direction or remain stationary. The bottom layer was 0.60 m length and width, which was positioned on top of the force platform (model OR-5, AMTI, USA). Three accelerometers were used in the experiment. The moment of the handle moving away (T_{handle}) was detected and denoted time zero by an accelerometer (model 208CO3, PCB Piezotronics Inc, USA) attached to the handle. The moment of trunk movement (T_{trunk}) was detected by an accelerometer (model 1356a16, PCB Piezotronics Inc, USA) attached to the dorsal surface of the subject at the level of L5S1. The moment of the sliding board movement (T_{board}) was detected by an accelerometer (model 208CO3, PCB Piezotronics Inc, USA) attached underneath the top layer of the sliding board.

The participants were instructed to stand barefoot on the top layer of the sliding board with their upper arms at 90° of elbow flexion and wrist extension, and palms slightly contacting the load cell (model LSB300, Futek Inc, USA) extending from the wooden handle as well as their feet shoulder width apart and in parallel to push the horizontal flat wooden handle (62 × 9 × 2 cm) attached to an aluminum pendulum. The pendulum was affixed to the ceiling with additional loads of 30% individual body weight; the height of the wooden handle was adjusted to each individual's hand position accordingly. Feet position was marked on the sliding board and was reproduced across the trials. Two conditions were performed: the not moving sliding board would be referred to as the "locked condition", while the free-moving sliding board would be referred to as the "unlocked condition". Each participant was given two practice trials prior to data collection to allow familiarization with the task and awareness of the locked or unlocked experimental condition. Five trials were collected in each condition.

2.3   Data Processing

All data were processed offline using MATLAB software (MathWorks, Natick, MA, USA). The onsets of the accelerometer signals were detected T_{handle}, T_{trunk}, and T_{board} using the Teager-Kaiser onset time detection method [14]. The ground reaction forces and the moments of forces were filtered with a 20 Hz low-pass, 2nd order, zero-lag Butterworth filter. Time-varying COP traces in the anterior-posterior direction were calculated using the approximations described in the literature [15]. The onset of the COP moved away from the baseline was detected by the Teager-Kaiser method (COP_{onset}).

2.4   Statistics

The dependent t-tests were performed to evaluate effects of sliding perturbation on the onset of T_{trunk}, COP_{onset}, and maximal pushing force. Means and standard errors are presented in the results and figures. Significant difference was set at $p < 0.05$.

3   Results

Figure 1 shows a typical example of changes in the acceleration of the handle (T_{handle}), of the trunk (T_{trunk}), and the sliding board (T_{board}). The onset of T_{handle} was set as time 0 and the first changes in related to time 0 was observed in T_{board} and followed by T_{trunk}. Regardless of the sliding board being locked or unlocked, the participants performed pushing task with similar peak acceleration of T_{handle} (t(7) = 2.168, p = 0.067). The first changes in the acceleration of the unlocked board were seen at 142.03 ± 30.48 ms prior to the 0; the sliding board did not move in the locked condition. Additionally, the orientation of the accelerometer attached to the board indicated that the sliding perturbation (backward movement) was directed opposite to trunk movement (forward movement).

The dependent t-test revealed that the onset time of trunk movement were not significantly different between the locked and unlocked conditions (t(7) = 0.254, p = 0.807). The onsets of COP were initiated prior to T_{handle} and it was 279.10 ± 35.22 ms preceding T_{handle} in the locked condition and 189.93 ± 15.71 ms in the unlocked condition (Fig. 2). Pushing while standing on the unlocked sliding board significantly affected the onset of COP (t(7) = −2.379, p = 0.049). The magnitudes of the COP displacement (t(7) = 3.064, p = .018) were significantly smaller when standing on the unlocked board (0.05 ± 0.01 m) than the locked condition (0.03 ± 0.01 m). Furthermore, the maximum pushing force was significantly affected by the board mobility (t(7) = −2.386, p = 0.048) and it was 135.51 ± 18.79 N in the locked condition and 113.31 ± 17.49 N in the unlocked condition.

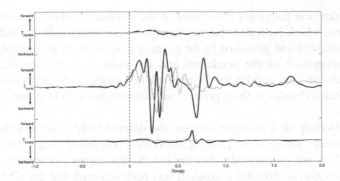

Fig. 1. A typical example of the changes in the signals obtained from the three accelerometers. T_{handle} is the acceleration of the handle, T_{trunk} is the acceleration of the trunk, and T_{board} is the acceleration of the sliding board. The vertical lines indicate the moment of the start of the handle moving away (0). Gray lines reflect changes in the acceleration while standing on the locked sliding board and black lines are for the unlocked sliding board.

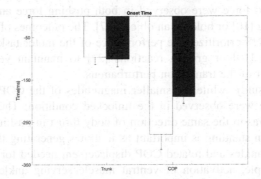

Fig. 2. The onset time of trunk movement and center of pressure (COP) in the locked (white boxes) condition and the unlocked (black boxes) conditions.

4 Discussions

The current study investigated effects of dual perturbations on pushing and balance performances when pushing while standing on the sliding surface. The later onset of COP was observed when pushing with translation perturbations (the unlocked condition). While in the unlocked condition, the magnitudes of the maximal exerted force and the COP displacement were significantly smaller compared to the locked condition. Thus, our hypotheses that the onset time and magnitudes of muscle activities and the COP displacement would be affect by the movability of the sliding board was supported.

The unlocked sliding board in the current study was similar with a skateboard. Pushing in the unlocked condition was comparable with that one person on a skateboard pushes another person on the other skateboard and results in both skateboarders are pushed away in the opposite directions. In the unlocked conditions, the sliding board was detected backward movement at the moment of starting to move by the accelerometer, which indicated that the translation perturbation in the unlocked condition was followed and provoked by the pushing movement. In the current study, both maximal acceleration of the pendulum and the trunk were not statistic significant different between two conditions. It implied that participants performed the same pushing task and changes in these postural adjustments due to in response to translation perturbation.

When standing on a movable surface, the delayed COP onset and the backward board movement induced by pushing movement suggested that CNS treated the pushing task as the primary perturbation and the translation perturbation as the collateral perturbation to postural control. It has been reported that the CNS prioritizes performance of the task when balance is not in imminent danger [25, 26], but the CNS reverses the priorities to maintain postural balance over motor tasks when one or two tasks involve postural control [23] or threat to the balance [24]. Thus, changes in the onset time of the tibialis anterior muscle were observed when that pushing an object while standing on a moving platform [10]. On the other hand, when simultaneously experiencing translation perturbations but imposed externally by a mechanical device increases in exerted force were observed in both pushing force and grip force when performing pushing [10] or holding an object [27]. The outcomes of these prior studies indicate that the CNS prioritizes the performance of the motor task (focal arm movements in pushing) [10] or griping reactions [27] to maintain vertical posture and overcome the effect of the translation perturbations.

In the current study, while the smaller magnitudes of the COP displacement and exerted hand force were observed in the unlocked condition. The backward sliding board movement was in the same direction of body thrust in pushing [10]. Activation of shank muscles in standing is important as it allows generating the ground reaction torque around the ankles and related COP displacement needed for control of vertical posture. For example, activation of ventral muscle serving ankle joint in standing allows moving COP backward and activation of dorsal muscle allows moving COP forward [16–18]. In order to assist trunk bending movement, anticipatory activation of either ventral muscles or inhibition of dorsal muscles could be used to shift COP backward [19, 20]. In addition, activation of dorsal muscles plays an important role for backward translation perturbation [21] and to shift COP forward. Hence, trunk inertia could be used a tool for increasing pushing force to reach the sufficient exerted force [22] and counteracting the destabilizing force of gravity for the smaller COP displacement in the unlocked condition compared to the locked condition.

Some limitations should be mentioned and noted. The current findings might not generate to older adults since older adults already showed muscle co-contraction and inefficient pushing strategies when performing pushing tasks while standing on a solid surface. In case, older adults utilize the same strategy of remaining too strong muscle co-contraction under the sliding surface and performing the pushing task. It has been reported that segments with stiffness affect postural adjustments corresponding to forthcoming perturbations [24], which might subsequently put elderly at higher risk of falling. However, the further study should be conducted with middle aged or aging workers to evaluate their pushing and balance responses when standing on the moving surface.

5 Conclusion

Exerted pushing force induced backward sliding board movement when standing on the unlocked surface. Standing on the unlocked sliding board resulted in the delay of the COP onset time and decreases in the magnitudes of the COP displacement. The pushing task was considered as the primary perturbation and the moving surface was the collateral perturbation to balance control. These results reveal that the CNS is capable of controlling the motor task and postural component when dealing with two self-initiated perturbations: primary (pushing) and subsequent (translational). In addition, the inertia of the trunk was used to assist in generation of the pushing force needed to suffice the pushing task supported a decreased magnitude of the pushing force in the unlocked condition. Meanwhile, body inertia allowed counteracting the destabilizing effect of force of gravity as smaller COP displacements were seen when the surface was movable. The findings also provide a basis for future studies focused on the development of potential working environment design for middle aged or aging workers by modifying movability of the standing platform.

References

1. Chaffin DB, Andres RO, Garg A (1983) Volitional postures during maximal push/pull exertions in the sagittal plane. Hum Factors 25(5):541–550
2. Lavender SA et al (1998) Trunk muscle use during pulling tasks: effects of a lifting belt and footing conditions. Hum Factors 40(1):159–172
3. Grieve DW (1983) Slipping due to manual exertion. Ergonomics 26(1):61–72
4. Hoozemans MJ et al (1998) Pushing and pulling in relation to musculoskeletal disorders: a review of risk factors. Ergonomics 41(6):757–781
5. Boocock MG et al (2006) Initial force and postural adaptations when pushing and pulling on floor surfaces with good and reduced resistance to slipping. Ergonomics 49(9):801–821
6. Lee YJ, Aruin AS (2013) Three components of postural control associated with pushing in symmetrical and asymmetrical stance. Exp Brain Res 228(3):341–351
7. Lee YJ, Aruin AS (2014) Isolated and combined effects of asymmetric stance and pushing movement on the anticipatory and compensatory postural control. Clin Neurophysiol 125 (4):768–776
8. Lee YJ, Aruin AS (2015) Effects of asymmetrical stance and movement on body rotation in pushing. J Biomech 48(2):283–289
9. Shiratori T, Latash M (2000) The roles of proximal and distal muscles in anticipatory postural adjustments under asymmetrical perturbations and during standing on rollerskates. Clin Neurophysiol 111(4):613–623
10. Dietz V et al (2000) Effects of changing stance conditions on anticipatory postural adjustment and reaction time to voluntary arm movement in humans. J Physiol 524(Pt 2):617–627
11. Chen B, Lee YJ, Aruin AS (2018) Standing on a sliding board affects generation of anticipatory and compensatory postural adjustments. J Electromyogr Kinesiol 38:168–174
12. Lee YJ, Chen B, Aruin AS (2015) Older adults utilize less efficient postural control when performing pushing task. J Electromyogr Kinesiol 25(6):966–972
13. Talini D et al (2017) Effects of the work limitations on the career path of a cohort of health workers. Med Lav 108(6):434–445

14. Li X, Zhou P, Aruin AS (2007) Teager-Kaiser energy operation of surface EMG improves muscle activity onset detection. Ann Biomed Eng 35(9):1532–1538
15. Winter DA et al (1996) Unified theory regarding A/P and M/L balance in quiet stance. J Neurophysiol 75(6):2334–2343
16. Muller ML, Redfern MS (2004) Correlation between EMG and COP onset latency in response to a horizontal platform translation. J Biomech 37(10):1573–1581
17. Nardone A, Schieppati M (1988) Postural adjustments associated with voluntary contraction of leg muscles in standing man. Exp Brain Res 69(3):469–480
18. Runge CF et al (1999) Ankle and hip postural strategies defined by joint torques. Gait Posture 10(2):161–170
19. Alexandrov AV, Frolov AA, Massion J (2001) Biomechanical analysis of movement strategies in human forward trunk bending. II. Experimental study. Biol Cybern 84(6):435–443
20. Oddsson LI (1990) Control of voluntary trunk movements in man. Mechanisms for postural equilibrium during standing. Acta Physiol Scand Suppl 595:1–60
21. Torres-Oviedo G, Ting LH (2007) Muscle synergies characterizing human postural responses. J Neurophysiol 98(4):2144–2156
22. Lee WA, Michaels CF, Pai YC (1990) The organization of torque and EMG activity during bilateral handle pulls by standing humans. Exp Brain Res 82(2):304–314
23. Muller ML, Redfern MS, Jennings JR (2007) Postural prioritization defines the interaction between a reaction time task and postural perturbations. Exp Brain Res 183(4):447–456
24. Shumway-Cook A, et al (1997) The effects of two types of cognitive tasks on postural stability in older adults with and without a history of falls. J Gerontol A Biol Sci Med Sci 52 (4):M232-40
25. Chen B, Lee YJ, Aruin AS (2016) Control of grip force and vertical posture while holding an object and being perturbed. Exp Brain Res 234(11):3193–3201
26. Mitra S (2004) Adaptive utilization of optical variables during postural and suprapostural dual-task performance: comment on Stoffregen, Smart, Bardy, and Pagulayan (1999). J Exp Psychol Hum Percept Perform 30(1):28–38
27. Bateni H et al (2004) Resolving conflicts in task demands during balance recovery: does holding an object inhibit compensatory grasping? Exp Brain Res 157(1):49–58

Ergonomic Analysis of Labor Applied to Scaffolders in a Shipyard in Brazil

Guilherme Deola Borges[✉], Angelica Mufato Reis,
and Antonio Renato Pereira Moro

Federal University of Santa Catarina, Florianopolis, SC, Brazil
guilhermedeola@gmail.com

Abstract. Modules of 6 oil and gas platforms are being built at a shipyard in Brazil. It was verified that, among the disciplines of construction and assembly (welding, structure, electrical, scaffolding, painting, instrumentation, commissioning, piping and load handling), scaffolders are the workers who are exposed to the greatest amount of unsafe conditions due to the risks inherent to the work at heights and the lack of safety behavior. This study aims to analyze the working conditions of scaffolders analyzing their job and applying a tool designed for the whole shipyard safety. It was used the direct observation of the activity during 8 months and the Ergonomic Analysis of Labor (EAL) were applied to ascertain the diagnosis of the context that covers the working routine of the scaffolders. ErgCAP – Ergonomics of Correction, Awareness and Participation was implemented as a tool in which the construction and assembly teams were responsible for identifying and registering the unsafe conditions. From the observation it was realized that the physical structure offered to the scaffolders is not ideal and the workers frequently go beyond some limits of the procedure to achieve production goals, putting themselves and the others at risk. ErgCAP registered 599 unsafe conditions in the area of scaffolding, among which the most recurrent are: opening in floor, scaffolding without skirting board, lack of inspection, activity being carried out without issuing a work permission, scaffolding material spread and seat belt not used properly. From the observations it is important to note that it is not always possible to attach the seat belt to a structure other than the scaffold being erected, which is unsafe and requires investment in the structure of the construction site. Another characteristic observed is that couplers, wooden planks and even scaffold tubes are often thrown at the ground. Training and awareness should be improved. The conclusion is that ErgCAP is an effective tool against one of the main causes of accident in the construction industry, fall from height accidents, because it transfers responsibilities to all workers and immediately acts on the perception of other risks. Not only who is doing the job is informed about the risk, but they also learn how to perceive it during their activities, as all the others become guardians of the safety in the construction site.

Keywords: Ergonomics · Construction · Scaffolder

© Springer Nature Switzerland AG 2019
S. Bagnara et al. (Eds.): IEA 2018, AISC 819, pp. 725–738, 2019.
https://doi.org/10.1007/978-3-319-96089-0_79

1 Introduction

The construction of modules is inserted in the context of the FPSO (Flotation, Production, Storage and Offloading) platforms to be operated in Brazil's pre-salt area. The total investment is around 12 billion dollars and it is relevant to achieve brazilian production targets. Each platform can produce 150,000 barrels of oil and 6 million cubic meters of gas per day.

Civil and naval construction are directly linked to the development and infrastructure of a country. To build a platform it is needed several disciplines that work together. The main ones are welding, electrical, instrumentation, structure, piping, painting, load handling, commissioning, inspection and scaffolding. According to OSHA (Occupational Safety and Health Administration), falls, being struck by an object, electrocutions and caught-in/between are known as "the fatal four". Within them, falls account for 36% of fatal accidents (OSHA 2016).

According to the International Labor Organization (ILO), occupational accident is an incident that occurs at work, and may or may not result in fatality. ILO estimates that among the 2.7 billion workers worldwide, more than 2 million deaths a year are due to occupational diseases (ILO 2013). The construction industry has one of the highest accident rates, with more than 60,000 fatal accidents per year. One of the primary causes of fatal construction accidents is falls from height (Lipscomb et al. 2014). According to López et al. (2008), the chances of an accident having serious consequences increase when scaffolds are involved. Even with the increasing attention of construction workers to the hazards and risks involved in working at height, falls still appear to be leading cause of fatal construction accidents (Goh and Sa'adon 2015).

The scaffolder is a professional who provides the necessary access to the other workers in a construction site. Its function is to assembly the scaffolding, stairs, handrails and skirting boards to guarantee the safe access of all, and after that disassemble them. The work can be performed in open or closed locations, exposed to the climatic conditions of heat, cold, winds, humidity and while exercising activities is subject to noises, varied work positions, physical effort, material movement, use of tools and work at height. Therefore, these aspects of work environment of the scaffolder have importance in the physical, cognitive and organizational field of the ergonomic analysis.

Among the techniques applied in the prevention of occupational hazards, Ergonomics is currently being used as a multidisciplinary tool dedicated to examining working conditions in order to achieve the best possible harmony between man and the work environment, also achieving the development of optimal conditions of comfort and productive efficiency (Guérin et al. 2001).

The problem of fall from height has been studied around the world with theoretical and applied studies, but the fall is still the biggest security problem in the construction industry, especially in the presence of temporary structures such as scaffolding.

This study aims to analyze the working conditions of scaffolders analyzing their job and applying a tool designed for the whole shipyard safety.

2 Methods

2.1 Case Study

Scaffolders were observed to determine the typical working conditions and their variables: types of scaffolding, handling tools, protective equipment and environmental conditions. From the direct observation of the activities, aiming at a better understanding of the work context, an ergonomic analysis of the work was performed. At the end some recommendations were made to mitigate the risks. ErgCAP was used for 8 months in order to involve the entire work force in eliminating unsafe conditions.

2.2 Ergonomic Analysis of Labor - EAL

The methodology used was to observe and record the activities throughout the day, during several stages of the work. The EAL approach is structured in the stages of demand analysis, task analysis, activity analysis, diagnosis and recommendations. All are linked in order to understand the work and to transform it (Guérin et al. 2001).

The data referring to the work and the workers were collected, as well as the risks involved. The documents and procedures for carrying out the activities were analyzed. The activity was understood through observation and by talking with scaffolders and supervisors. Their routines were recorded by pictures and notes. As a result, the question to be answered is: what are the strategies used by the scaffolders and which are not written or foreseen by the organization? ErgCAP was implemented for a period of 8 months to record the unsafe conditions.

2.3 Ergonomics of Correction, Awareness and Participation - ErgCAP

In Goh and Goh (2016), the effectiveness of the fall prevention plan was verified through a document that recorded information such as fall prevention policy, roles and responsibilities, fall risk assessment and emergency response. The analysis indicated that the fall prevention plan was effective because it requires a clear allocation of responsibilities, increases commitment, and makes explicit the competency requirements.

The ErgCAP is a risk management tool where everyone is invited to participate in the safety of the site. The goal is that everyone when perceiving an unsafe condition must act until resolved. After solving the problem must be recorded in a specific spreadsheet for further analysis of results.

Figure 1 shows how the unsafe conditions were systematically recorded in a spreadsheet and served as the basis for risk analysis.

ErgCAP – Ergonomics of Correction, Awareness and Participation					
Author	Date	Workplace	UnsafeCondition	Done	Subject

Fig. 1. Recording of unsafe conditions (Source: Authors)

2.4   Workplace

At the construction site there are approximately 1,000 employees, including 90 scaffolders. Being a function that requires a lot of physical effort, all of the scaffolders are male. The age ranges from 20 to 45 years working in the company from 1 to 4 years and total experience from 1 to 15 years. They receive specific training in their activities when starting in the company and there are monthly safety training. Table 1 shows the profile of scaffolders.

Table 1. Profile of scaffolders

Scaffolders	90 workers
Gender	100% male
Age	20–45 years
Scholarity	High school
Company time	1–4 years
Experience time	1–15 years

Figure 2 show the modules in this construction site and Fig. 3 a complete FPSO.

Fig. 2. Power generation modules (Source: Authors)

Fig. 3. Platform for operation in the pre-salt area (Source: http://www.petrobras.com.br/fatos-e-dados/p-66-deixa-estaleiro-rumo-ao-campo-de-lula.htm)

2.5 Characteristics of Scaffolder Work

According to Brazilian Regulatory Norm - NR 34, "the assembly and dismantling of scaffolding shall stop immediately in the event of insufficient light and adverse weather conditions, such as rain, winds, among others".

Work at height is performed since it is considered "any activity performed above 2.00 m (two meters) from the lower level, where there is a risk of falling". (NR 35 - Work at Height/Safety and Occupational Health).

3 Results

3.1 Ergonomic Analysis of Labor - EAL

EAL results is that the prescribed task of the scaffolders is: to assembly scaffolding, setting and fixing their pieces of wood or metal; installing the frame, securing the vertical modules and securing the horizontal crossbars, to form the platform support; mount the platforms, fixing their pieces of wood or metal on the crossbar; mounting accessories, installing handrails, ladders and other devices, to allow access and transit and to provide safety to workers; modifying scaffolding, to adapt them to the progression of tasks; dismantling the scaffolding, once the work is completed, to enable the reuse of these structures. The scaffolders are still responsible for verifying the quality of the equipment, checking the clamps or fittings necessary for each model, arranging the mooring of the structure of the building, reporting any problems observed in the assembly material to the direct supervisor and release the scaffolding together with the safety professional.

They carry out their work exposed to the diverse meteorological. While exercising his activities is subject to noise, physical effort, varied positions of work, movement of materials, tools and equipment and work at height. For these reasons, it is needed to use complete Personal Protective Equipment - PPE.

In analyzing the activity, the ergonomist should describe how the results are obtained and how the prescribed means are used by the worker. Part of this stage is the investigation of psychological phenomena that characterize the human being in the performance of his acts. It is also intended to understand, what are the strategies of the worker to meet the prescribed work (Laville 1977; Guérin et al. 2001). In the end, it is also the role of the ergonomist to present possible recommendations, countermeasures, for problems.

Every scaffolder, before beginning his activities at the company, receive 8 h of theoretical and practical training. The training contains the following contents: assembly and dismantling of scaffolding and work at height, materials used in scaffolding, techniques in assembling and dismantling, use of PPE, preliminary risk analysis, occupational risks, lessons related to accidents, basic concepts of firefighting, rescue and first aid.

The day begins with the Daily Safety Dialogue, Fig. 4, which is usually conducted by the supervisor or the safety professional. The conversation addresses the key security issues that can occur during the workday. For example on rainy days, attention is drawn to the wet floor, and when there is a special activity to be performed to pay attention to details.

Fig. 4. Daily Safety Dialogue (Source: Authors)

During the assembly and dismantling of scaffolding, it is mandatory to use PPE and tools in good conditions, such as Fig. 5: overalls, safety helmets with jugular, ear protectors, seatbelts with a girdle, gloves, safety boots, ratchet and hammer.

Fig. 5. PPE and work tools (Source: Authors)

In the reception of the structural elements that make up the scaffolding (pipes, wooden planks, clamps, stairs) it is verified if they are in perfect conditions of use, that is to say, without kneading, drilling, welding or corrosion.

The structural elements of the scaffold are stored in a specific location, the scaffolding storage in Fig. 6, allowing them to be removed following the intended use sequence. Tubes are stored by sizes on identified shelves.

Fig. 6. Scaffolding storage (Source: Authors)

The main elements of the scaffold assembly are: Rohr tube, which can be from 1 to 6 m in length and weighs 3.65 kg/m; The clamp weighs 1.2 kg each; and the wooden plank that measures $0.30 \times 0.04 \times (0.50$ to $4.00)$ m. This material is always transported by hand or with the help of a cart which is pushed on uneven floor, as shown in the Fig. 7.

Fig. 7. Moving material from the storage to the work front (Source: Authors)

Arriving at the place where the scaffolding will be assembled, the team prepares to divide tasks as shown in Fig. 8. Every tube need clamps, so this activity is anticipated before climbing. Such torque force is applied to the bolts that some tubes are oval shaped by kneading.

Fig. 8. Scaffolding assembly (Source: Authors)

Also according to the Brazilian Regulatory Norms (Brazil 2017a, b):

NR 34 – 34.11.25.1 – Scaffolders shall stop assembling or dismantling in case of insufficient light or bad weather, such as rain or strong wind;
NR 34 – 34.11.9 – The areas around scaffolding shall be signaled and protected against the impact of vehicles or mobile equipment.

The biggest risk of the scaffolder is the fact of working at height on a temporary structure and still being assembled. The combination of these factors puts the scaffolder in an unsafe condition much more frequently than any other activity.

This job requires attention, concentration, agility, balance, strength and something that is not required in job interviews, but it is fundamental according to the scaffolders themselves: "To be a scaffolder, you have to be brave".

Once the service is finished, all checklist items are checked to put the release board and the other disciplines can perform their activities (Fig. 9).

Fig. 9. Scaffolding released (Source: Authors)

3.2   Ergonomics of Correction, Awareness and Participation - ErgCAP

As every construction process involves risks, especially in large works there is great turnover of people and teams. Many of them coming from other citie. It is more difficult to create a safety culture where the work place is dynamic and teams are not fixed.

Ergonomics of awareness seeks to empower the workers themselves to identify and correct routine or emergency problems. And the ergonomics of participation seeks to involve the worker himself in the solution of problems (Iida 2005).

There were 599 potential hazards registered in the scaffolding area, mainly caused by the risk of falling scaffold material in an area close to assembly and dismantling (non-isolated area) or risk of scaffolder fall from height (although always secured by the seat belt).

The diagnoses of the situation investigated are constructed and it is an essential product of EAL. It was noticed that the scaffolding activity is the one that offers the most risks and the workers are exposed to unsafe conditions. This fact is mainly due to the great rigor of the safety procedures that must be followed since work at height is in itself a risk factor.

Among the records of unsafe conditions on scaffolding, Fig. 10, the main ones were categorized, among them are: opening in floor, scaffold without skirting board, lack of inspection and release plate, activity being performed without issuing work permit; scattered scaffold material, belt not used properly.

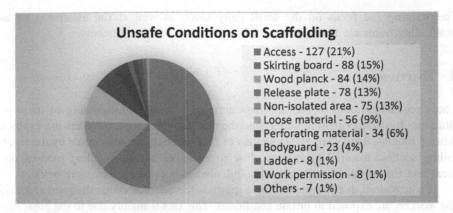

Fig. 10. Unsafe conditions on scaffolding (Source: Authors)

The conditions of the ErgCAP tool were classified according to the risks they represent, following examples recorded from the worksheet:

- Access - people are blocked, obstructed or insecure;
- Skirting board - there is no skirting board, not well fixed or do not dimensioned according to standards;
- Wood plank - there are holes in the wooden planks, they are not well fixed or they are deteriorated;
- Release plate - there is no release plate or it is not suitable;
- Non-isolated area - the work of the scaffolder is not well insulated or the material used for assembly and disassembly put at risks people on the same level;
- Loose material - scaffold material with risk of falling;
- Perforating material - usually nails used in the assembly of wooden planks that are exposed, offering risks to scaffold users;
- Bodyguard - there is no guardrail, it is not securely fixed or it is not dimensioned according to norm;
- Ladder - improperly mounted, putting users at risk;
- Work permission - not issued;
- Others - offer risks that are not included in other classifications;
- Clamping - scaffolding not locked for use;
- Safety belt - not coupled or improperly coupled.

There were 127 situations of risks caused by blocked, difficult or insecure access.

In addition, 337 unsafe conditions can be grouped together that can cause an accident due to falling object or presence of object in an inadequate place: lack of skirting (88), ranging between wooden planks (84), lack of insulation area (75), material loose on the scaffold platform (56) and perforating material left by the scaffolder (34).

Since it is the function of the scaffolder to give safe access to the other disciplines, it was to be expected that in this main issue there should not be so many deviations.

The training and focus on the small daily tasks, as well as the awareness of the scaffolding teams are fundamental for the reduction of these numbers.

4 Discussion

The direct observation of the activities aims to a better understanding of the work situation. It was important to understand the risks to which the scaffolders are exposed. After 8 months, there were 599 situations of potential risk, mainly caused by the risk of falling scaffold material in an area close to assembly and dismantling (non-isolated area) and risk of fall from height (although always trapped by the safety belt).

It was noticed that the scaffolding activity is the one that offers the most risks and the workers are exposed to unsafe conditions. This fact is mainly due to the great rigor of the safety procedures that must be followed since work at height is in itself a risk factor.

Firstly, it is not always possible to fasten the safety belt to a structure other than the scaffolding being assembled. Of course, this is a serious mistake as it puts at risk the safety efficiency of the belt in case of fall or imbalance of the scaffolder. Another characteristic observed is that they are often throwing clamps, wooden planks and even scaffold tubes to the floor. Even taking cake of who are below, this puts someone who is moving at risk, often the area is not isolated and the pitch may not hit the target. In both cases an extra risk of accident due to fall (of person or material) is added, caused partly by the imprudent assumption of the scaffolder and partly by requirements of productivity.

Some recommendations are applicable for this study:

- To construct fixed structures to tie the safety belt of the scaffolder, at a suitable height, never below the safety belt line;
- To provide a basket for placing the clamps, so that they do not fall from the tubes during disassembly and are not thrown by the scaffolder;
- To signal and to isolate the areas where the scaffolding is being assembled and dismantled, to avoid accidents in case of material fall;
- To guarantee the efficiency in the training and qualification of the professionals regarding security;
- Before starting the work, to remember the main aspects to be noticed in order to avoid errors during the execution of the procedures;
- To perform Daily Safety Dialogue focused on the activities of the day and remember common deviations in order to avoid them;
- To perform daily inspection of all tools and equipment to ensure they are in good conditions;
- To check the accesses to the work area, as well as the activities that are occurring simultaneously;
- To ensure that PPE are adequate to carry out the activity;
- To collect all scaffold material at the end of the day leaving the area clean, avoiding the risk of fall in the absence of the scaffolders;
- To Build rails at the construction site for cargo transportation.

If these recommendations are followed the number of unsafe conditions will decrease.

5   Conclusion

This work presented the characteristics of scaffolder, as well as the results of observations over 8 months. These results extracted from ErgCAP and the recommendations can be used in other works to assist in the safety of these professionals.

The EAL associated with the ErgCAP tool evidenced a large number of important risk situations to be addressed. This management tool proved to be efficient and easy to apply. It can be used in addition to construction sites, such as in factories or any environment where the number of incidents is relevant. The management expectation should be to register less and less risky situations, since it is understood that the environment is becoming more secure since the implementation of the tool.

The observation of the activity showed that the concern with production overtakes safety procedures. During the dismantling of scaffolding, it is frequent to not comply with the procedure that says that it is forbidden to throw scaffold material. This practice is motivated by the fulfillment of the deadlines of the production and it puts at risk the other scaffolders who are at the level below, but also the other workers, especially when the area is not well isolated.

The ideas and results presented here can help future research and guide construction workers in dealing with the problem of falling from height. We also emphasize that before using the best equipment, the most technological tools and the best written procedures, it is necessary to guarantee the effectiveness of the training and awareness of the workers, since the man is the basis of the safety in the work.

References

Brasil (2017a) Ministério do Trabalho e Emprego. Normas Regulamentadoras de Segurança e Medicina do Trabalho. NR 34 Condições e Meio Ambiente de Trabalho na Indústria da Construção, Reparação e Desmonte Naval. http://www.mte.gov.br. Accessed 10 July 2017

Brasil (2017b) Ministério do Trabalho e Emprego. Normas Regulamentadoras de Segurança e Medicina do Trabalho. NR 35 Trabalho em altura. http://www.mte.gov.br. Accessed 10 July 2017

Goh YM, Goh WM (2016) Investigating the effectiveness of fall prevention plan and success factors for program-based safety interventions. Saf Sci 87:186–194

Goh YM, Sa'adon NFB (2015) Cognitive factors influencing safety behavior at height: a multimethod exploratory study. J Constr Eng Manag 141(6):04015003

Guérin F, Laville A, Daniellou F, Duraffourg J, Kerguelen A (2001) Compreender o trabalho para transformá-lo: a prática da ergonomia. Edgard Blücher, São Paulo

Iida I (2005) Ergonomia. Projeto e Produção, 2nd edn. Edgard Blücher, São Paulo

ILO (2013) ILO calls for urgent global action to fight occupational diseases. http://www.ilo.org/hanoi/Informationresources/Publicinformation/Pressreleases/WCMS_211708/lang--en/index.htm. Accessed 01 Aug 2016

Laville A (1977) Ergonomia. EduspEpu, São Paulo

Lipscomb HJ, Schoenfisch AL, Cameron W, Kucera KL, Adams D, Silverstein BA (2014) How well are we controlling falls from height in construction? Experiences of union carpenters in Washington State, 1989–2008. Am J Ind Med 57(1):69–77

López MAC, Ritzel DO, Fontaneda I, Alcantara OJG (2008) Construction industry accidents in Spain. J Saf Res 39(5):497–507

OSHA (2016) Safety and Health Regulations for Construction: Fall Protection: Standard 29 CFR 1926, Subpart M. https://www.osha.gov/pls/oshaweb/owadisp.show_document?p_table= standards&p_id=10922. Accessed 01 Aug 2016

Characteristics of Surface EMG During Gait with and Without Power Assistance

Seiji Saito[1](✉) and Satoshi Muraki[2]

[1] Faculty of Computer Science and Systems Engineering, Okayama Prefectural University, 111 Kuboki, Soja-shi, Okayama 719-1197, Japan
s.saito@ss.oka-pu.ac.jp
[2] Faculty of Design, Kyushu University, 4-9-1 Shiobaru, Minami-ku, Fukuoka 815-8540, Japan

Abstract. Technology that assists and extends various functions of human beings will soon be available not only to medical and welfare but also to healthy individuals. This study aimed to characterize surface electromyography (EMG) signals in response to walking assistive equipment. Ten healthy male students walked on an 8-m uphill road (5.8% incline) using an assist walker (RT.2, RT.WORKS Co., Ltd) under assist and non-assist conditions. The EMG signals were recorded from four muscles (the rectus femoris [RF], biceps femoris, tibialis anterior, and lateral gastrocnemius). During loading response and terminal stance, the percent maximum voluntary isometric contraction (% MVC) peak value for RF was achieved more quickly in the assist condition than in the non-assist condition. However, during loading response and mid-swing, the %MVC peak value of RF was significantly lower in the assist condition than in the non-assist condition. These results indicate that humans alter muscle exertion patterns in specific muscles to adapt to walking assistance; such a change in the muscle exertion pattern may be adapted for smoother walking.

Keywords: Power assistance · Gait · EMG

1 Introduction

In the near future, technology that assists and extends various functions of human beings will be available not only to medical and welfare sites but also to healthy individuals. In other words, humans will enter into an era of coexistence with assistive technology. Recently, several reports regarding such devices have been reported [1–4]. For assistive devices to be successful, the user and device must appropriately cooperate. Therefore, it is believed that humans adapt physiological responses to assistive devices. Thus, the present study aimed to characterize surface electromyography (EMG) signals in response to walking assistive equipment.

© Springer Nature Switzerland AG 2019
S. Bagnara et al. (Eds.): IEA 2018, AISC 819, pp. 739–743, 2019.
https://doi.org/10.1007/978-3-319-96089-0_80

2 Methods

In this study, 10 healthy male students (age: 21.2 ± 1.9 years, height: 166.8 ± 2.2 cm, and weight: 62.8 ± 6.8 kg) participated. We excluded individuals who had injuries or surgical interventions to the lower extremities in the past year or who had any neurological problem. This study was conducted in accordance with the Declaration of Helsinki, and informed consent was obtained from all participants following the ethics committee of the Okayama Prefectural University.

The participants walked on an 8-m uphill road (5.8% incline) using an assist walker (RT.2, RT.WORKS Co., Ltd) under assist and non-assist conditions. To increase the assist effect, a 10-kg weight was placed on the assist walker. The participants walked under each condition thrice with the same stride (70 cm), walking speed (5 km/h), and cadence (120 steps/min).

The surface EMG signals were recorded from four muscles, namely, the rectus femoris (RF), biceps femoris (BF), tibialis anterior (TA), and lateral gastrocnemius (LG), using a surface EMG system (Surface EMG Sensor SX230, Biometrics Ltd, UK). The EMG electrodes were placed on the right side of the body on RF, BF, TA, and LG. The electrodes were placed at the midpoint of the line from the anterior spina iliaca superior to the superior part of the patella for RF; at the medial position on the line between the ischial tuberosity and the lateral epicondyle of the tibia for BF; at one-third the distance between the tip of the fibula and that of the medial malleolus for TA; and at one-third the distance on the line between the head of the fibula and the heel for LG.

Before initiating the experiment, the maximum voluntary isometric contraction (MVC) of the muscles was measured. The measurements were performed in the sit-ting position for RF and LG; in the standing position for TA; and in the prone position for BF. Each muscle was measured for 10 s in 2 trials, with a 2-min break between each trial. The MVC value was calculated based on the 5 s of data between 3–7 s of the 10-s trial. The EMG amplitude was processed using standard filtering and rectifying methods and was normalized using the 5-s MVC value. Walking cycle was deter-mined using foot switches in the right shoe (heel and toe), and the EMG data for three walking cycles were extracted. The normalized percent IEMG, percent MVC peak value (% MVCp), and the time of the appearance of the %MVC peak value (%MVCpt) were calculated for each gait cycle and compared under the assist condition. The EMG and footswitch signals were sampled at a 1-kHz rate and were stored in the data logger's memory.

All statistical data were analyzed using SPSS Statistics version 20.0 for Windows (IBM Corporation, Armonk, NY, USA), and the significance level α for all analyses was 0.05. To examine the general characteristics of the participants, means and standard deviations were calculated. To compare with assist and non-assist conditions, the data were analyzed by paired t-test.

3   Results

During a single gait cycle, no significant difference was observed between the conditions in terms of %MVCp, %MVCpt, and %IEMG for all muscles. Similarly, during each gait phase, no significant difference was observed between the conditions in terms of %MVCp, %MVCpt, and %IEMG of RF, TA, and LG (Fig. 1). The results for RF are shown in Table 1. During the loading response, the %MVCpt value of RF was significantly faster under the assist condition than that under the non-assist condition. In addition, the %MVCp and %IEMG values of RF were significantly lower under the assist condition than those under the non-assist condition during the loading response. During the terminal stance, the %MVCpt value of RF was significantly faster under the assist condition than that under the non-assist condition, but there was no significant difference between %MVCp and %IEMG. Lastly, during mid-swing, the %MVCp and %IEMG values of RF were significantly lower under the assist condition than those under the non-assist condition.

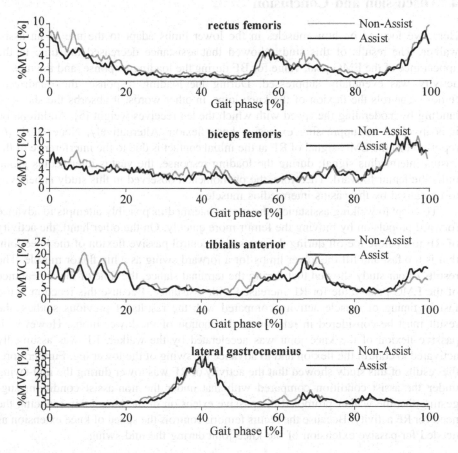

Fig. 1. Change in %MVC of rectus femoris, biceps femoris, tibialis anterior, and lateral gastrocnemius under the assist and non-assist conditions

Table 1. %MVCp, %MVCpt, and %IEMG of RF in each gait phase

	%MVCp (%)		%MVCpt (%)		%IEMG (%)	
	Non-assist	Assist	Non-assist	Assist	Non-assist	Assist
LR	15.2 ± 10.8	12.3 ± 9.4*	5.4 ± 2.7	4.0 ± 3.4*	77.1 ± 50.4	62.5 ± 35.0*
MSt	7.1 ± 6.4	5.4 ± 3.9	16.6 ± 4.7	18.5 ± 6.2	48.1 ± 45.7	33.0 ± 25.0
TSt	2.9 ± 2.9	2.3 ± 1.7	41.1 ± 6.2	38.5 ± 5.6*	22.3 ± 32.0	15.5 ± 20.4
PSw	7.6 ± 8.1	7.6 ± 8.5	59.2 ± 3.1	58.5 ± 3.1	37.8 ± 42.1	36.5 ± 42.6
ISw	5.4 ± 6.4	5.1 ± 4.3	66.2 ± 2.2	66.9 ± 3.1	41.7 ± 51.5	30.7 ± 37.9
MSw	3.3 ± 3.6	2.2 ± 2.5*	80.2 ± 4.6	81.6 ± 4.4	16.5 ± 22.6	11.1 ± 20.1*
TSw	11.1 ± 13.4	10.3 ± 11.9	98.4 ± 2.2	98.8 ± 1.5	56.9 ± 66.7	52.4 ± 60.2

LR, loading response; MSt, mid-stance; TSt, terminal stance; PSw, pre-swing; ISw, initial swing; MSw, mid-swing; TSw, terminal swing; %MVCp, normalized EMG peak value; %MVCpt, time to % MVC peak value; %IEMG, normalized IEMG. *$P < 0.05$

4 Discussion and Conclusion

Here, we focused on how muscles in the lower limbs adapt to the use of an assist walker. The results of this study showed that assistance decreased the time to the appearance of the EMG peak value for BF during the loading response, and the muscle activity was eventually suppressed. During the loading response, the quadriceps femoris controls the flexion of the knee joint. In other words, it absorbs the shock of landing by moderating the speed with which the leg receives weight [5]. Additionally, it is involved in suppressing excessive knee flexion. Alternatively, Nene et al. [6] reported that the EMG signal of RF at the initial contact is due to the interference of the vastus intermedius signal; during the loading response, the vastus intermedius group pulls the femur forward. Therefore, the phenomenon observed in this study is believed to be caused by the vastus intermedius muscle.

To adapt to walking assistance, the vastus intermedius possibly attempts to advance forward propulsion by moving the femur more quickly. On the other hand, the activity of RF just before toe-off during the pre-swing control passive flexion of the knee joint that is too fast and lift the lower limbs for a forward swing as a hip flexor muscle. The results of our study showed that during the terminal stance, the time of the appearance of the EMG peak value for RF increased with assistance. Because this reaction is too fast in timing of muscle activity compared with the results of previous studies, the result must be considered in relation to the motion of the lower limbs. However, if passive flexion of the knee joint was accelerated by the walker, RF was assumedly activated to control the flexion and to advance the swing of the lower leg. Furthermore, the results of this study showed that the activity of RF was lower during the mid-swing under the assist condition compared with that under the non-assist condition, suggesting that the walker promoted the passive extension of the knee joint, reducing the need for RF activity. Because the rectus femoris controls the speed of knee extension as needed for passive extension of the knee joint during the mid-swing.

Overall, our study findings indicate that humans alter the activation exertion patterns of specific muscles to adapt to walking assistance. Such a change in the muscle exertion pattern may be adapted for assistance for smoother walking.

Acknowledgements. This work was supported by JSPS KAKENHI Grant Number JP17H01454.

References

1. Kasai R, Takeda S (2016) The effect of a hybrid assistive limb® on sit-to-stand and standing patterns of stroke patients. J Phys Ther Sci 28(6):1786–1790
2. Chihara H, Takagi Y, Nishino K, Yoshida K, Arakawa Y, Kikuchi T, Takenobu Y, Miyamoto S (2016) Factors predicting the effects of hybrid assistive limb robot suit during the acute phase of central nervous system injury. Neurol Med Chir 56(1):33–37
3. Ogata T, Abe H, Samura K, Hamada O, Nonaka M, Iwaasa M, Higashi T, Fukuda H, Shiota E, Tsuboi Y, Inoue T (2015) Hybrid assistive limb (HAL) rehabilitation in patients with acute hemorrhagic stroke. Neurol Med Chir 55(12):901–906
4. Fukuda H, Samura K, Hamada O, Saita K, Ogata T, Shiota E, Sankai Y, Inoue T (2015) Effectiveness of acute phase hybrid assistive limb rehabilitation in stroke patients classified by paralysis severity. Neurol Med Chir 55(6):487–492
5. Lay AN, Hass CJ, Nichols TR, Gregor RJ (2006) The effects of sloped surfaces on locomotion: an electromyographic analysis. J Biomech 40(6):1276–1285
6. Nene A, Byrne C, Hermens H (2004) Is rectus femoris really a part of quadriceps? Assessment of rectus femoris function during gait in able-bodied adults. Gait Posture 20(1):1–13

Using OpenSim to Investigate the Effect of Active Muscles and Compliant Flooring on Head Injury Risk

Jonathan Mortensen[✉] and Andrew Merryweather

Department of Mechanical Engineering, University of Utah,
Salt Lake City, UT 84112, USA
Jon.Mortensen@utah.edu

Abstract. The force and acceleration of the head during an impact from a fall are shown to decrease on compliant flooring. Our understanding of the benefits of compliant flooring is limited, because studying head impacts resulting from falls is often not feasible for human subject studies. The OpenSim platform allows investigation of muscle activations in fall induced head impacts using dynamic simulation. The purpose of this study is to determine if simulations in OpenSim can provide a greater understanding of the effect of neck muscle activations during falls onto compliant flooring. This is accomplished by simulating a fall from bed height onto a compliant floor. Force profiles and head motions from simulations without active neck muscles are compared to values obtained from published literature to evaluate the suitability of OpenSim for simulating falls onto compliant flooring. Force profiles and head motions from simulations with active neck muscles are also investigated and compared to related literature. The results indicate that OpenSim is capable of providing realistic results and can be used for investigating the effect of active neck muscles during falls onto compliant flooring. The results also support that active muscles can substantially reduce brain injury risk from a fall. Future studies are aimed at investigating the role of active neck muscles in populations with greater fall risk to evaluate head injury avoidance strategies.

Keywords: Compliant flooring · OpenSim · Falls · Head injury

1 Introduction

Slips and falls are the cause of at least 60% of admissions to hospitals for traumatic brain injuries in adults age 65 and older [1]. The severity of injury from a fall is correlated with the stiffness of the floor [2]. Force and acceleration of the head have been experimentally shown to be significantly reduced on compliant flooring during impacts from falls [3]. Compliant flooring absorbs more energy than traditional flooring, and may be used effectively to reduce injury severity. However, the effect of compliant flooring is difficult to demonstrate experimentally with human subjects. This is particularly true for measuring impact forces of the head on the floor [4]. Understanding the ability of compliant flooring to reduce head forces and accelerations can be improved with musculoskeletal models if these models have the capability to

© Springer Nature Switzerland AG 2019
S. Bagnara et al. (Eds.): IEA 2018, AISC 819, pp. 744–751, 2019.
https://doi.org/10.1007/978-3-319-96089-0_81

realistically represent head motion, contact forces, and active muscle forces. OpenSim may be an effective modeling platform for this purpose [5].

OpenSim has provided insight in many fields of study. OpenSim has aided in understanding active muscles during gait, including subjects with impaired gait [5]. The effect of tendon transfer studies has also been simulated using OpenSim prior to effective surgeries [6]. OpenSim has also proved useful in sports analysis, prosthetic design, and metabolic analyses [7–9]. It is possible that OpenSim may be able to provide useful results in the study of compliant floors and head impacts.

Contact surfaces in OpenSim employ Hertz contact theory [10]. This theory determines forces and deformations based on linear elastic theory [10]. In addition to this theory, a dissipation model is also employed [10]. The resultant model requires several inputs, including a composite material stiffness value and a dissipation value [10].

A recently developed model of the head and neck is well suited for simulating impacts to the head with or without active neck muscles [11]. This model is capable of developing realistic strength in flexion, extension, lateral bending, and axial rotation without the need of reserve actuators.

The purpose of this study is to determine if simulations in OpenSim can provide a greater understanding of the effect of muscle activations during falls on compliant flooring by exploring the role of neck muscle strength on injury risk. This is accomplished by first comparing contact surface force profiles and linear head accelerations to values found in related literature with no activation present in the muscles. Second, the role of active neck muscles to prevent injury from a fall is explored and compared to results found in related literature.

2 Methods

2.1 The Model

The head and neck model used for this study is a progression of two previous OpenSim models [12, 13]. The current model's novel contribution is the inclusion of hyoid muscles, and is known as the HYOID model. The hyoid muscles help overcome a flexion strength limitation in previous models, and is capable of realistic strength in all directions. This improvement is essential to this study. Without realistic strength, reserve actuators are necessary for simulating dynamic movements. Reserve actuators supply forces that muscles cannot. A model that can provide realistic strengths without reserve actuators is able to more realistically simulate the effect of active neck muscles during impact scenarios.

2.2 The Fall

A backward fall was simulated in the OpenSim environment (Fig. 1). This was accomplished by applying gravity to the model in a supine position above a contact surface. The fall height was set to 62.5 cm to represent a fall out of bed [3]. This height is expected to result in a maximum skull velocity of 3.5 m/s. The simulated skull velocity was compared to this expected velocity.

Three contact surface materials were simulated: concrete, Sorbothane, and a composite of concrete and Sorbothane. Sorbothane is a visco-elastic polymer with excellent shock absorption and damping properties [14]. The composite concrete and Sorbothane surface is meant to represent either a thin layer of Sorbothane over concrete or Sorbothane combined with a more rigid material.

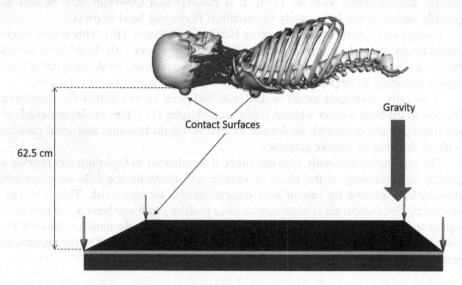

Fig. 1. OpenSim drop test from a height of 62.5 cm. Applied gravity causes the model to fall until the contact surfaces collide.

The contact pair between the skull and ground require a composite stiffness value derived from the stiffness of the two materials in contact. This is accomplished according to Eqs. (1) and (2) [10], where E_i^* is the plain strain modulus of material i, E_i is the Young's modulus of material i, v_i is the Poisson's ratio of material i, and E^* is the composite stiffness value used in OpenSim simulations. The stiffness values for the contact pair between the skull and ground are shown in Table 1. A dissipation value is also required, and was assumed to be 0.5 s $\cdot$ m^{-1}. This assumption is similar to values published by Lintern et al. [15].

$$E_i^* = \frac{E_i}{1 - v_i^2} \tag{1}$$

$$E^* = \left(\frac{E_1^{*\frac{2}{3}} E_2^{*\frac{2}{3}}}{E_1^{*\frac{2}{3}} + E_2^{*\frac{2}{3}}}\right)^{\frac{3}{2}} \tag{2}$$

Table 1. Composite stiffness values for contact surfaces between the skull and three materials.

	Contact stiffness with bone (GPa)
Concrete	5.57
Sorbothane	0.18
Concrete Sorbothane mix	2.87

As the motion of the torso was not a focus of this study, the contact between the torso and ground surfaces was simplified. This was done by using a value of 10 GPa for the composite stiffness and a dissipation value of $1000 \text{ s} \cdot \text{m}^{-1}$. This essentially treats the collision between the torso and ground as inelastic and causes the torso to stop very quickly upon contact with the floor.

2.3 Muscle Activations

To investigate the effect of active muscles, the flexion muscles in the model were activated upon the beginning of the fall of the model. Drops were simulated with the model's flexion muscles inactive, activated to 10% of their capacity, and activated to 50% of their capacity. The active muscles were allowed to bring the head into flexion as the model fell. Choi et al. have reported an experiment involving healthy adults intentionally falling with active neck muscles and allowing their skulls to contact extremely compliant flooring [4]. The resultant decrease of impact velocity due to active neck muscles is compared with the results published by Choi et al. [4].

2.4 Injury Risk

The resultant linear acceleration of the skull's center of gravity and contact force between the skull and ground were recorded in the OpenSim environment using the model setup shown in Fig. 1. Wright et al. reported maximum linear accelerations and contact forces of a surrogate human head for impact velocities of 1.5, 2.5, and 3.5 m/s on various surfaces [3]. The simulated results from this study were comparable to the

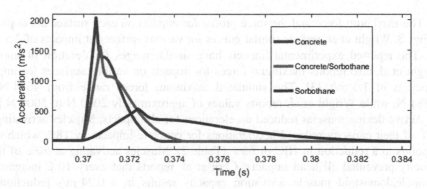

Fig. 2. Linear acceleration profile of the skull's center of mass during simulated falls onto each surface.

experimental results of Wright et al. The reduction of the Head Injury Criterion (HIC) [16] due to active neck muscles was also calculated. The HIC is used to quantify severity of head impacts, based on the linear acceleration of the center of mass of the head. A reduction in the HIC indicates a reduction in head injury risk. The reduction of HIC due to active neck muscles was calculated.

3   Results

Free fall from a height of 62.5 cm, would result in an impact velocity of 3.5 m/s. However, due to the torso hitting the ground before the skull the impact velocity of the skull in these simulations was only 1.625 m/s. The maximum linear acceleration of the skull during these impacts was 208 g for concrete, 160 g for a surface with stiffness halfway between concrete and Sorbothane, and 54 g for Sorbothane (Fig. 2). For an impact velocity of 1.5 m/s, Wright et al. reported maximum acceleration values ranging from 27 g to approximately 90 g [3].

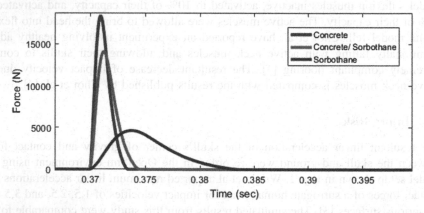

Fig. 3. Force applied to the skull's contact surface during a simulated fall onto each surface.

The maximum force and the force profile for impacts on each surface are depicted in Fig. 3. Wright et al. reports similar curves for various surfaces at impacts of 2.5 m/s [3]. The reported experimental impacts have similar ranges for duration of impact. Wright et al. also reports maximum forces for impacts on various surfaces for impact velocities of 1.5 m/s [3]. The simulated maximum forces range from 4624 N to 18384 N, while Wright et al. reports values of approximately 2000 N to 5000 N [3].

Active flexion muscles reduced accelerations for all impacts. Muscles activating at 10% of their capacity reduced accelerations for concrete impacts by 18% which corresponds to a reduction in HIC of 63% (Table 2). Muscles activating at 50% of their capacity prevented all head impacts. Choi et al. reports that every 10% increase of sternocleidomastoid muscle activation capacity results in a 0.24 m/s reduction of impact velocity [4]. The impact simulated impact velocity with all flexion muscles activated to 10% of their capacity resulted in a decrease of 0.361 m/s.

Table 2. HIC without active muscles and with 10% activation in flexion muscles.

	HIC	HIC with 10% flexion
Concrete	306	113
Concrete/Sorbothane	204	81
Sorbothane	36	17

4 Discussion

The accelerations and force profiles produced in the OpenSim environment are comparable to the values in the literature, indicating that OpenSim models can effectively simulate falls onto compliant flooring. Wright et al. reporting lower accelerations and peak forces than were simulated in this study may be due to a number of factors [3]. Wright et al. dropped a surrogate human head onto different surfaces than were simulated in this study [3]. Concrete is a more rigid material than any of the material's used in the study by Wright et al. [3]. It is also possible that the added mass of the neck, point of impact, or area of impact may have resulted in higher accelerations and forces in this simulation study. Experimental values for the dissipation constant were also not available. Obtaining these values for simulated materials may offer improved simulation results. The accelerations and peak forces of this study having similar magnitudes to the study by Wright et al. Combining the ability to model contact parameters with OpenSim's ability to simulate active muscle has the potential to advance research in fall prevention and injury risk reduction.

The results of this study support that a 50th percentile male can substantially reduce the risk of head injury from a fall out of bed, given awareness of the impending impact resulting in a muscle clenching response. Activating neck muscles to only 50% of their capability completely prevented contact of the skull with the ground. If contact with the ground cannot be avoided through active neck muscles, the results of this study indicate that even a minor amount of muscle activation can result in a substantial decrease in head injury risk. The decrease in impact velocity in this study due to active neck muscles are comparable to results published by Choi et al. [4]. The study conducted by Choi et al. has higher impact velocities without active neck muscles than are reported in this study [4], which likely contributes to the lower impact velocity due to the active neck muscles being larger in this study. Active neck muscles are more effective at reducing impact velocities when the impact velocities are initially smaller. The method used in this study could easily be adapted to explore the effect of active neck muscles in older populations, that have reduced neck strength, by decreasing the strength of each muscle in the model appropriately. Response time delays associated with age or pathological conditions could also be incorporated in similar future studies. Data from such studies would be useful in determining neck strength goals for populations at risk for falls.

5 Conclusion

OpenSim modeling can effectively be used to simulate falls onto compliant flooring. The results of this study support OpenSim as a promising tool for investigating the effects of active neck muscles during falls. Compliant flooring significantly reduces impact force and head injury criteria from a fall. The simulation parameters used in this study should be supported by additional material testing in future studies to improve simulation results. Active neck muscles have the capability to mitigate head injury risk during falls. Building on this work, future studies have the potential to provide insight on necessary muscle strength and reaction time for leveraging the capability of active neck muscles to reduce the prevalence of head injuries as a result of falls.

References

1. Schonnop R et al (2013) Prevalence of and factors associated with head impact during falls in older adults in long-term care. Can Med Assoc J 185(17):E803–E810
2. Lachance CC et al (2017) Compliant flooring to prevent fall-related injuries in older adults: a scoping review of biomechanical efficacy, clinical effectiveness, cost-effectiveness, and workplace safety. PLoS ONE 12(2):e0171652
3. Wright AD, Laing AC (2012) The influence of headform orientation and flooring systems on impact dynamics during simulated fall-related head impacts. Med Eng Phys 34(8):1071–1078
4. Choi W et al (2017) Effect of neck flexor muscle activation on impact velocity of the head during backward falls in young adults. Clin Biomech 49:28–33
5. Delp SL et al (2007) OpenSim: open-source software to create and analyze dynamic simulations of movement. IEEE Trans Biomed Eng 54(11):1940–1950
6. Holzbaur KR, Murray WM, Delp SL (2005) A model of the upper extremity for simulating musculoskeletal surgery and analyzing neuromuscular control. Ann Biomed Eng 33(6):829–840
7. Buffi JH et al (2015) Computing muscle, ligament, and osseous contributions to the elbow varus moment during baseball pitching. Ann Biomed Eng 43(2):404–415
8. Machado MF et al (2012) Challenges in using OpenSim as a multibody design tool to model, simulate, and analyze prosthetic devices: a knee joint case-study. In: EUROMECH Colloquium 524, Multibody System Modelling, Control and Simulation for Engineering Design
9. Baskar H, Nadaradjane SMR (2016) Minimization of metabolic cost of muscles based on human exoskeleton modeling: a simulation. Int J Biomed Eng Sci 3(4):9
10. Sherman MA, Seth A, Delp SL (2011) Simbody: multibody dynamics for biomedical research. Procedia Iutam 2:241–261
11. Mortensen J, Vasavada AN, Merryweather A (2018, under review) The inclusion of hyoid muscles improve moment generating capacity and dynamic simulations in musculoskeletal models of the head and neck. PLoS ONE
12. Cazzola D et al (2017) Cervical spine injuries: a whole-body musculoskeletal model for the analysis of spinal loading. PLoS ONE 12(1):e0169329
13. Vasavada AN, Li S, Delp SL (1998) Influence of muscle morphometry and moment arms on the moment-generating capacity of human neck muscles. Spine 23(4):412–422
14. Sorbothane. https://www.sorbothane.com/material-properties.aspx

15. Lintern T (2014) Modelling infant head kinematics in abusive head trauma. ResearchSpace@ Auckland
16. Hutchinson J, Kaiser MJ, Lankarani HM (1998) The head injury criterion (HIC) functional. Appl Math Comput 96(1):1–16

Slips, Trips and Falls in Crowds

Roger Haslam[1](✉) ⓘ and Victoria Filingeri[2] ⓘ

[1] Loughborough University, Loughborough, UK
r.a.haslam@lboro.ac.uk
[2] University of Derby, Derby, UK
v.filingeri@derby.ac.uk

Abstract. Crowd situations are commonplace and involve circumstances known to lead to slips, trips and falls (STF). Data from focus groups with crowd participants (5 groups, n = 35 individuals); observations of crowd situations (n = 55); and interviews with crowd organisers (n = 41) were analysed to examine understanding of and responses to the risk of STF in crowds. Although safety was a high priority for both crowd participants and organisers, explicit consideration of STF as a safety concern was low among both groups. Crowd observations found STF risks mitigated on some occasions and present on others, without any discernible pattern for the variation. A risk management framework for STF risk in crowds is proposed. It is concluded that improved understanding is needed of the nature and pattern of STF occurrence in crowds and the efficacy of measures for prevention.

Keywords: Crowd ergonomics · Crowd safety · Fall prevention

1 Introduction

Slips, trips and falls (STF) are a leading cause of unintentional injury for both occupational and leisure activities [1]. Falls involve a loss of balance due to some reason, which results in a person falling to the ground or other lower level. Aspects of the environment involved in falls are the foot-floor interface and the presence of trip hazards. The frictional characteristics of footwear and flooring materials affect the likelihood of slipping, with these influenced by their condition and maintenance and also the presence of contaminants (e.g. water, ice, litter). Obstacles in the walkway may lead to tripping if they go unnoticed by the pedestrian. Because the clearance between feet and the floor is small during normal gait, deviations in the walking surface of as little as 10 mm may be sufficient to cause a trip. Other perturbations of balance may result from contact with objects or people (e.g. bumping into or being pushed by something or someone). Although the multi-factorial causality of STF is reasonably well understood, efforts to reduce the incidence of injury, however, have had mixed success [2].

Gatherings of large numbers of people, crowds, are a ubiquitous part of our human experience. Crowds are encountered in circumstances such as using public transport, visiting retail environments or participating in entertainment events (e.g. sport matches, music festivals, amusement parks, museums etc.). Crowds occur in wide ranging

© Springer Nature Switzerland AG 2019
S. Bagnara et al. (Eds.): IEA 2018, AISC 819, pp. 752–758, 2019.
https://doi.org/10.1007/978-3-319-96089-0_82

environments e.g. streets, fields, sports stadia, concert halls, shopping malls, airports, railway stations etc. In some situations, the management and infrastructure are permanent; in others the conditions and oversight are transitory. The nature of a crowd can vary from good humoured, sedate and purposeful to angry, excited and in disarray.

Crowding can exacerbate risk factors for STF, with the proximity of large numbers of individuals leading to deterioration of the walking surface or obstructed vision, for example. Crowd movement can result in jostling or pushing. STF in crowd situations can have serious consequences, for both the faller and others involved in the gathering. An individual STF in a moving crowd can result in obstruction to the crowd flow and further multiple fall or crush injuries. Such crowd 'disasters', receive considerable attention from news media, with reporting of deaths and injuries at events such as pilgrimages and festivals regular occurrences. Data on the incidence, injuries and causes of STF in crowds, however, are not collected in a systematic manner and with significant underreporting. Data that are available suggest that slips are more frequent than trips and that risk increases with older age and when carrying items or in the presence of luggage. Consumption of alcohol is also implicated [3].

Filingeri et al. [4, 5] reported the findings of interview and observations studies of multiple crowd situations, which examined the factors affecting crowd participant experience. This paper presents further analysis of the data from these studies, focusing on risk and management of STF in crowd situations.

2   Method

The research employed a combination of focus groups, observations and interviews. Five focus groups (35 individuals, age range: 21–71 years) were conducted to collect in-depth accounts of the features of crowd situations important to crowd participants. Observations were then undertaken of 55 crowd situations, to identify aspects that contribute to positive and negative experience of crowds. The researcher observed crowds as a participant observer, using a standardised checklist as a prompt for recording audio notes, video and photographs, enabling crowd situations to be observed consistently and systematically. The crowds observed encompassed a wide variety of type, size, purpose and participant culture. Observations covered a range of seasons and weather conditions. Semi-structured interviews were conducted separately (n = 41) with organisers responsible for different aspects of the design, planning, management and operation of events and other crowd situations. The objective was to understand organisers' priorities, along with the consideration given to the experience of crowd participants.

3   Results

Safety was one of the most frequently mentioned topics in the focus groups. This was regarded as an important concern affecting crowd satisfaction across crowd participant demographics, with STF hazards, trampling risks and violence being discussed. In the words of one focus group member:

"But you just... you would just feel like you were being pushed along on a wave wouldn't you...? And if anything happened... if somebody stumbled over or something you'd... phh... not a hope have you...?"
(Healthy Adults focus group, female aged 47 years).

The crowd observations identified many instances where STF hazards had been identified and risks controlled (Table 1). There were also many examples seen where STF risk was present (Table 2). Figure 1 illustrates the litter problems that can occur with large scale events and the ensuing risk of slipping or tripping on the debris.

Table 1. Examples from observations of mitigated STF risks

Hazard	Observation notes
Wet floor surface	Retail events 1 - Christmas Market (city centre): Non-slip floor boarding in the cathedral square area. Important when the rain began to pour
Muddy floor surface	Music events 6 - Outdoor Music Festival: Straw and hay placed down over very muddy patches of ground, to prevent accidents
Severe weather conditions	Retail events 1 - Christmas Market (city centre): Event cancelled for the first time in many years, due to heavy snow and ice and risk of STF
Temporary cables	Conferences and exhibitions 26 - University careers fair: Cables were stuck to the ground with thick black tape to reduce the risk
Low level steps	Participatory events 36 - Fairground (town centre): Pavement slopes implemented to prevent tripping on curb
Stair safety	Theatre event 42 - Musical show (London): Handrails along stairs in-between the theatre seated tiered sections
Poor lighting	Music events 8 - Classical Concert (Vienna): Clear exit routes were lit up within the concert area

Interviews with organisers of crowd situations found that health and safety was a key consideration, receiving greater attention than other aspects such as crowd participants' goal achievement, comfort and satisfaction. The protection of crowd users, prevention of accidents, protecting venue reputation and legal obligations were discussed as reasons for devoting time and resources to health and safety during crowd events. STF related aspects, however, were only mentioned explicitly in 3 of the 41 organiser interviews. One interviewee highlighted the serious consequences that can ensue from falls in a crowd:

And then the doormen start coming back in... because it can be that dangerous. Because crowds surging... moving and so we're trained in specific ways, to see that people can get them out. And again, the worst thing is somebody falling over in a crowd..."
(Security officer, music events, interviewee 10)

An event planner involved in sporting events and a police interviewee discussed the problems arising from poor weather. The event planner, for example, stated:

"That's the trickiest thing in the bad weather. I mean, the main thing is having good trained marshals in place ... so if there are slippery banks or bottlenecks, then you try to clear them."
(Event planner, outdoor spectator events, interviewee 16)

Fig. 1. Litter posing risk of STF (Music events 4 – Victoria Park, London)

In both these extracts, the interviewees pointed to the importance of having trained personnel to identify and respond to problems.

Table 2. Examples from observations where STF risk was present

Hazard	Observation notes
Wet flooring	Music events 5 – Pub (city centre): Getting a number of drinks from the bar back to the other people in your group is difficult, causing spillages. Wet areas on the floor were dangerous when walking (especially when carrying drinks)
Wet flooring	Retail 47 – Shopping centre (city centre): Water from outside brought inside on shoes, causing a slip hazard
Litter	Music events 2 – Arena (city centre): Rubbish thrown on the floor – plastic cups/bottles. Did not see bins, or areas to dispose rubbish. No staff seen to be collecting litter. On exiting event – plastic cups lined the floor. All creating STF hazards
Litter	Participatory events 35 – Carnival (Brazil): Litter everywhere, food and litter lined the streets. No personnel to collect the litter throughout the event. STF hazards
Inadequate lighting	Theatre event 43 – Cinema (town centre): Dark as you enter and exit the cinema screen, the room is dark, and although the seat numbers are lit up, it provides a safety hazard

(continued)

Table 2. (*continued*)

Hazard	Observation notes
Restricted space for entry/egress	Sporting events 14 – Rugby match (rugby stadium): Standing on tiered seating section, having to climb across people to get out. Steep and restricted space. Some people do not stand to allow others to pass, and instead try to move their feet to one side allowing limited space
Escalator fall	Transport hub 31 – Underground (London): Two older women fell backwards at the base of the escalator causing a blockage. No one pressed the emergency stop button and so oncoming pedestrians had to try and jump over the women, as people at the bottom attempted to help them to stand. The emergency stop buttons were placed too far away from the point the fall occurred. The emergency stop buttons were also not close enough to other participants who were on the escalators who tried to help
Escalator fall	Transport hub 34 – Metro (Vienna): An older man (estimated over 70 years) fell backwards on the escalator. I pressed the emergency button from the bottom of the escalator and other passengers helped to get the man to the top of the escalator until assistance came
Encumbrances	Theatre event 44 – Arena (city centre): Little space available to store belonging like bags and coats. Cloakroom available but put off by long queues

4 Discussion

Our investigations found safety to be an important concern for both crowd organisers and crowd participants. Although STF are a prominent cause of injury, the relatively few references in the focus group and interview discussions to STF was not in line with the safety implications that these incidents pose. The STF mitigations seen in the observation study, however, indicated that STF risks on these occasions had been considered and addressed, to some extent at least.

Where STF risks were apparent in the crowd situations observed, these arose from shortcomings with the built environment and permanent infrastructure as well as more variable environmental and behavioural components and the organiser responses to these. Although it might be expected that STF safety of locations designed for hosting crowds would be to a high standard, previous studies have highlighted deficiencies with the design of walkways, steps, handrails etc. in major event venues. In a premier league football stadium, for example, problems were identified with the physical design and layout of seating areas and entrance and egress routes, especially the narrow clearway between seat rows, irregular and excessive step dimensions and unmarked step edges [6, 7]. Regarding escalators, the risk of falls posed by these is well known, with public education campaigns visible as a response [8].

The role of crowd organisers is pivotal in controlling STF risks, both to design out hazards but also in managing the complex individual and collective behaviour of people in crowds. This is especially the case for crowds occurring in temporary surroundings (e.g. street or field festivals). In a wider examination of the activities of crowd organisers, Filingeri et al. [5] found that organisers tended to rely on experience and judgement in approaching their planning and decisions, without accessing training or the guidance materials available. The organisers of infrequent or small-scale events can have the greatest knowledge and experience gap. It was suggested that the non-use

Table 3. Prevention measures for STF in crowds, adapted from [1, 2]

Primary prevention – eliminate STF hazards at source

- Provide slip resistant flooring
- Design facilities (e.g. catering, washrooms) to avoid presence of fall risks
(systems approach, with attention to environment, equipment, layout, tasks and people)
- Cover outside walkways to keep off rain, snow, ice, leaves
- Design walkways to exclude trip hazards
- Plan pedestrian routes to allow sufficient space between individuals
- Separate pedestrians from moving machinery and vehicles
- Provide sufficient, convenient space for onsite storage
- Avoid presence of low steps
- Install steps and stairs of appropriate dimensions
- Provide step edges with good contrast
- Install handrails
- Avoid visual distraction in step/stair locations
- Avoid need for walking/standing on surfaces that move unpredictably
- Install grab rails and hand holds
- Install adequate lighting
- Design and select environment features to facilitate cleaning and maintenance
- Design and select environment features for durability and resistance to damage

Risk reduction – reduce risk of hazards that continue to be present

- Educate and raise awareness of fall risks and fall consequences
- Perform fall risk assessments and implement controls
- Organise sustainable housekeeping procedures for inspection, cleaning and maintenance
- Manage fall risks introduced during installation, cleaning and maintenance
- Provide warning signs for slip hazards
- Mark trip hazards
- Provide durable marking of step edges
- Fit additional handrails
- Fit additional grab rails
- Fit barriers for edge protection
- Encourage use of lighting
- Discourage carrying of awkward, heavy items
- Avoid creating circumstances that encourage rushing
- Implement risk management protocols for inclement weather
- Implement risk management protocols for those at increased risk of falling
- Provide assistive mobility aids for those in need

Maximise capability – individual responses

- Encourage use of suitable footwear
- Encourage use of suitable clothing
- Facilitate moderation with alcohol consumption

of guidance is in part due to problems with the guidance currently available, both its content and its form. With regard to STF, key guidance available concerning safety at sporting and music events gives only cursory attention to how STF hazards can be addressed [9, 10].

Along with the lack of data and analysis on the extent and nature of STF in crowds, there is also a scarcity of research evidence on the effectiveness of different forms of control and intervention to prevent STF injury. In the absence of this, an adapted STF risk control framework for crowds is proposed (Table 3), based on the approach we have presented previously for generic and occupational circumstances [1, 2].

5   Conclusions

Crowd situations, combining a high density of people, high volumes of pedestrian movement, sometimes in unfamiliar, unpredictable and less than ideal locations, provide a combination of circumstances that can lead to STF. STF in crowds, however, are surprisingly under-researched. Better understanding is needed of the nature and pattern of STF occurrence in crowds and the efficacy of measures for prevention.

References

1. Haslam R, Stubbs D (2006) Understanding and preventing falls. CRC Press, Boca Raton
2. Chang W-R, Leclercq S, Lockhart TE, Haslam R (2016) State of science: occupational slips, trips and falls on the same level. Ergonomics 59:861–883
3. Kendrick VL (2013) The user experience of crowds. PhD thesis, Loughborough University. https://dspace.lboro.ac.uk/2134/13888
4. Filingeri V, Eason K, Waterson P, Haslam R (2017) Factors influencing experience in crowds – the participant perspective. Appl Ergon 59:431–441
5. Filingeri V, Eason K, Waterson P, Haslam R (2018) Factors influencing experience in crowds – the organiser perspective. Appl Ergon 68:18–27
6. Au SYZ, Gilroy J, Livingston AD, Haslam RA (2004) Assessing spectator safety in seated areas at a football stadium. In: McCabe PT (ed) Contemporary ergonomics. CRC Press, Boca Raton, pp 23–27
7. Haslam RA, Au SYZ, Gilroy J, Livingston AD (2004) Slip, trip, fall related hazards at a football stadium. In: McCabe PT (ed) Contemporary ergonomics. CRC Press, Boca Raton, pp 18–22
8. Waterson PE, Kendrick VL, Ryan B, Jun GT, Haslam RA (2016) Probing deeper into the risks of slips, trips and falls for an ageing rail passenger population. IET Intell Transp Syst 10:25–31
9. Health and Safety Executive (HSE) (2010) Managing crowds safely: a guide for organisers at events and venues. http://www.hse.gov.uk/pubns/priced/hsg154.pdf. Accessed 5 Feb 2017
10. Events Industry Forum (EIF) (2016) The purple guide to health, safety and welfare at music and other events. http://www.thepurpleguide.co.uk/. Accessed 5 Feb 2017

Improving Slip Resistance on Ice: Surface-Textured Composite Materials for Slip-Resistant Footwear

Z. S. Bagheri[1,2(✉)], A. A. Anwer[3], G. Fernie[1], H. E. Naguib[3], and T. Dutta[1]

[1] Toronto Rehabilitation Institute, University Health Network, Toronto, ON, Canada
shaghayegh.bagheri@uoit.ca

[2] Department of Automotive, Mechanical and Manufacturing Engineering, University of Ontario Institute of Technology (UOIT), Oshawa, Canada

[3] Department of Materials Science and Engineering, University of Toronto, Toronto, ON, Canada

Abstract. Falls present a massive health risk for older adults. Half of those over 80 will fall at least once a year with 1 in 5 suffering a serious injury. Footwear outsole material that provides good grip on ice and snow can potentially prevent many of these injuries. Our team has developed our own promising patent-pending composite outsole materials with a unique structure that consist of soft rubber compound with hard microscopic fibers protruding out from the surface. In this study, we attempt to optimize the ice friction performance of our composite for extended use. We investigate the effect of manufacturing and testing parameters using the Taguchi method for robust design. Our results on optimization of process parameters demonstrate that fiber content at 8% volume fraction with mold temperature sets at 120 °C lead to maximum ice friction properties before and after simulated wear. The optimized composite design showed a higher coefficient of friction (COF) on ice than any on the market now, which highlights the capability of the material to provide improved traction on icy surfaces and prevent fall-related injuries.

1 Introductions

One in three seniors over 65 and one in two over 80 will fall each year [1, 2]. Twenty percent of injury-related death in older adults can be traced back to falls. Slower reaction times prevent these individuals from keeping their balance after a slip [3]. The fear of falling can also lead to declines in health for older adults if it leads to individuals becoming sedentary. For instance, many older adults avoid going outside in the winter during slippery weather and can become isolated for extended periods as a result. These periods of low activity and isolation can trigger a downward spiral of negative health effects.

Poor footwear is often identified as one of the risk factors for falls in older adults [4]. Several features of footwear design, such as heel height, midsole cushion, slip resistance of the outer-sole, fixation method, heel collar height and midsole geometry,

© Springer Nature Switzerland AG 2019
S. Bagnara et al. (Eds.): IEA 2018, AISC 819, pp. 759–766, 2019.
https://doi.org/10.1007/978-3-319-96089-0_83

have been associated with maintaining balance in seniors and preventing falls [5]. In particular, the lack of a slip-resistant outer-sole has been identified as one of the most important characteristics related to the risk of slip-and-fall incidents in older adults [6].

Until very recently there was no objective information available to consumers to help them select footwear for use in icy weather. Our team at Toronto Rehabilitation Institute (TRI) has recently developed a method to evaluate the slip resistance of winter footwear by having participants walk up and down icy slopes [7]. This method is called the Maximum Achievable Angle (MAA) test and it measures the steepest icy incline that participants can walk up and down without experiencing a slip. This test is currently under review with the Canadian Standards Association and ASTM International for adoption as the standard for footwear slip resistance measurement on snow and ice. To-date, of the 200 footwear models that have been tested, only ~ 30 have been found to show reasonable performance on ice (the results of all of our testing can be found at www.ratemytreads.com.

We recently completed a real-world evaluation of some of the best performing boots we have identified in the MAA test. A real-world study we ran in January and February of 2018 showed that homecare workers given the best slip-resistant footwear on client visits report three times fewer slips and nearly five times fewer falls compared to workers wearing their own footwear over an eight week period. While it is clear that this footwear has benefits, the composite material technology they are based on have two limitations that may prevent wide-spread adoption:

- The composite materials are still fairly new which makes the best performing boots are quite expensive ($\sim \$200$) and are therefore likely to be unaffordable for many older adults.
- We have on-going work evaluating the wear resistance of these new materials that is showing that the slip resistance decreases over as little as 100,000 steps of use. (For context, it would not be unusual for healthy individuals to walk 10,000 to 20,000 steps in a single day).

To overcome these limitations of existing footwear outsole materials, we have developed novel textured composite materials with superior slip-resistance and abrasion-resistance properties [8–10]. While our previous work confirmed the superior properties of our initial formulations of this composite, the design parameters that result in optimum performance still need to be identified. This study optimizes the design of the composite by systematically varying a number of manufacturing and testing parameters using the Taguchi method for robust design to ensure the composite can retain its grip on ice for extended use.

2 Methods

2.1 Materials Selection

Our material is a hybrid high friction composite material that utilizes both soft fibers (Poly-Phenylene-Benzobisoxazole) and hard fibers (Carbon) to create a textured surface for use in footwear outsole applications. The hard microscopic fibers extending out can penetrate icy surfaces, while the softer component can sufficiently deform and mold against hard substrates, ensuring good grip on both slippery outdoor surfaces as well as hard indoor surfaces (Fig. 1). This means our textured composite materials can be worn in both indoor and outdoor environments safely.

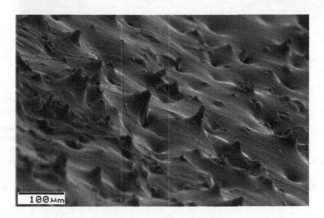

Fig. 1. The image of our fiber-reinforced composite taken using a scanning electron microscope.

2.2   Specimen Preparation

The manufacturing process involves the following steps: (a) melt blending of fibers (PBO, Carbon) and matrix (TPU) using a melt compounder, (b) injecting this uniform blended composite into a mold designed specifically to enhance fiber alignment during the injection process in footwear and, (c) cutting the end of the injection-molded surface using a sharp cutting tool as to expose fibrous phase on the composite surface. The overall orientation of the fiber is strongly enhanced during the injection process making it even more favorable for the protruding fibers to grip the ice surface (Fig. 2).

2.3   Ice Friction Testing

Taguchi Design of Experiment was used to identify a set of manufacturing parameters that result in the best ice friction properties before and after abrasion with the least sensitivity to the testing condition. Four factors, namely fiber content, mold temperature, sliding distance, and applied vertical load each at three levels (Table 1), was considered in this study in accordance with $L^{27}(313)$ orthogonal array design.

A full-factorial experiment with 4 factors each at 3 levels would require 34 = 81 separate experiments. Using Taguchi's factorial experiment approach, we are able to reduce the number of experiments to 27 for a given composite material per condition, which leads to 54 experiments in total. A small sample ($\sim 1\ cm^2$) of composite material was manufactured for each experimental run which consists of measuring the slip resistance on ice before and after applying a specific amount of simulated wear to each sample. Slip resistance on an ice surface was measured using the SATRA STM 603 Slip Resistance Testing machine and the samples was abraded using a DIN rotary

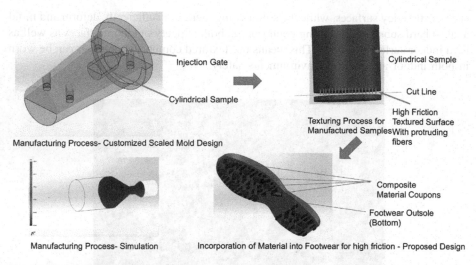

Fig. 2. Schematic of current manufacturing process and specific mold design to enhance fiber alignment

Table 1. Control factors and their level used for the experiments.

Control factor	Level			
	Low	Medium	High	Unit
A: Fiber content	1	4	8	%
B: Mold temperature	120	130	140	°C
C: Sliding velocity	0.3	0.4	0.5	m/s
D: Normal load	250	400	500	N

drum abrasion tester (Haida International, reference standard: DIN-53516, ISO-4649) to simulate approximately one year of real-life wear of outsoles to determine how well these materials retain their slip-resistance over extended use. SATRA STM 603 Slip Resistance Testing machine measures the Coefficient of Friction (COF) for different footwear outsole materials by applying a specified normal force pressing the test sample onto a test surface and then moving the test surface horizontally while recording the shear force. The ratio of the normal force and shear force is used to calculate the COF for a given material and surface combination.

The COF was transformed into signal-to-noise ratio as logarithmic transformation of the loss function using the "larger-is-better" performance characteristic as follow:

$$\frac{S}{N} = -10 * log\left(\frac{\sum\left(\frac{1}{y^2}\right)}{n}\right)$$

Where "n" is the number of observations and "y" is the observed data. This "larger-is-better" characteristic is suitable to determine the optimal manufacturing parameters leading to maximized COF (i.e. improved slip-resistance properties). Analysis of Variance (ANOVA) was performed in order to determine the factors that have the greatest impact on the slip-resistance performance of the textured composite material. With the results of S/N ratio and ANOVA, the optimal combination of manufacturing parameters was determined to an acceptable level of accuracy.

3 Results and Discussion

Our results on optimization of process parameters demonstrate that fiber content at 8% volume fraction with mold temperature set at 120 °C would lead to maximum ice friction properties before and after abrasion. A maximum of 8 vol% fibrous phase was used to retain compliancy of composite material. The minimum temperature of 120 °C was chosen such that the melted composite blend would have sufficient viscosity to fill the entire mold.

An ANOVA was performed for a level of significance of $p < 0.05$ to determine the factors and interaction that have significant effect on the output results. Table 2 shows the results of the ANOVA for the optimized composite design, with the last column of the table represents the percentage of contribution for each control factor and their interaction. These are employed to determine the corresponding effect on the final output. As it is evident from the table, the only contribution factor that affects the ice friction properties of the developed composite before abrasion is fiber content with a 40% contribution ($p = 0.024$). After abrasion, however, the ice frictional performance of our composite was independent of all control factors and their interactions, indicating the capability of our composite material to retain its enhanced grip even after extensive wear (COF = 0.61 ± 0.05).

Table 2. Analysis of variance for signal-to-noise ratios for ice slip-resistance properties of PBO-CF composite. Seq SS: sequential sum of squares, P: percentage of contribution

Source	CF-PBO-before abrasion				CF-PBO after abrasion			
	Seq SS	F test	P value	P (%)	Seq SS	F test	P value	P (%)
A: Fiber content	31.12	7.44	**0.024**	**40.08**	0.76	0.49	0.637	5.46
B: Mold temperature	1.61	0.39	0.696	2.07	0.07	0.04	0.956	0.50
C: Sliding velocity	12.08	2.89	0.132	15.58	1.94	1.24	0.355	13.78
D: Normal load	16.22	3.88	0.083	20.89	3.71	2.36	0.175	26.36
A * B	3.61	0.43	0.782	4.65	1.07	0.34	0.841	7.61
A * C	0.06	0.01	1	0.48	1.71	0.55	0.71	12.17
B * C	0.37	0.04	0.995	3.02	0.09	0.03	0.998	0.66
Error	12.54				4.71			
Total	77.63				14.08			

Our optimized composite materials have a higher COF on ice than any on the market now based on SATRA STM 603 Slip Resistance Testing machine. The optimized composite materials have resulted in COF values greater than 0.6. In contrast, the best footwear on the market today have COFs in the range of 0.1 as measured by the SATRA machine. The difference between a COF of 0.1 and 0.6 represents a large difference in predicted performance of our new materials compared to existing materials in our maximum achievable angle testing in WinterLab. We can estimate that a footwear outsole with COF 0.5 will be able to maintain grip on icy inclines of approximately 30° while outsoles with COF of 0.1 will likely only maintain grip on icy slopes of up to 6°. This conversion is done by taking the inverse tangent of the COF value since the incline angle of the ice surface represents the ratio of the normal force to the shear force generated when a participant walks up or down the icy slope. A larger MAA indicates that the footwear is safer on all surfaces. A user wearing footwear with higher MAA scores will be less likely to slip on an unexpected patch of ice on a level surface or on sloped surfaces in the built environment. In particular, it is important to note that snow and/or ice buildup can increase the effective slope in a particular area as shown in Fig. 3.

Fig. 3. Snow/ice buildup that increases the effective slope of a surface.

The fact that our materials maintain a high coefficient of friction on ice even after they have experienced simulated wear makes this material unique and will help to make the material commercially viable. In the scientific literature, the only existing uses of composite materials for modulating fiction either focus on reducing friction [11] or are not appropriate for use with footwear because of cost or complexity [12, 13]. Our team is aware of two commercially available footwear models that incorporate a glass-rubber composite in the outsole that have come onto the market very recently [14, 15]. Little is known about these propriety materials in terms of their composition or the methods used for manufacturing them. Our testing has demonstrated that at least one of these materials loses its slip-resistant properties very quickly with use and has a very limited useful life. It is possible that these and other companies have their own proprietary development and testing programs though few details have been published. An alternative method for providing traction on ice is with metallic cleats embedded into footwear outsoles to dig into snow and ice surfaces. However, a growing body of evidence on metallic cleats demonstrates that users often dislike them because they force changes in gait patterns and may be associated with lower extremity injury, including ankle and knee sprains. In addition, outsoles with built-in spikes and cleats become hazardous in the transition from outdoor to indoor environments and therefore may not be appropriate for older adults. Our testing shows that these hard cleats become very slippery on wet tile surfaces commonly found in entrance areas.

Slip-resistant footwear made possible through this work will provide good traction on ice without the downsides associated with metallic cleats. Since our new materials consist of mostly of a soft rubber compound, they can be expected to have little impact on the biomechanics of walking compared to metallic cleats. The dimension of the embedded fibrous phase is in the order of microns and therefore has no practical effect on contact area during different stages of gait cycle. Our testing also demonstrated our composites have good wear resistance and are therefore able to maintain their slip resistance after being exposed to considerable abrasion (simulated wear). Therefore, we believe our materials have good potential for being incorporated in footwear outsoles to create footwear that can outperform existing footwear to prevent winter slips and falls.

4 Conclusion

In this study we developed a novel hybrid high friction composite material that utilizes both soft fibers and hard fibers to create a textured surface for applications that requires high slip-resistant properties for extended use on icy surfaces. Using Taguchi experimental design analysis, we discovered that fiber content at 8% volume fraction with mold temperature sets at 120 °C would lead to maximum ice friction properties after simulated wear (COF = 0.61 ± 0.053), which highlights the point that our material can retain its ice friction properties even after extensive wear. Our results have demonstrated that a number of our optimized composites have superior wear-resistance and slip-resistance properties and therefore have good potential for being incorporated in commercial footwear and may prevent many winter slips and falls.

References

1. Prudham D, Evans JG (1981) Factors associated with falls in the elderly: a community study. Age Ageing 10(3):141–146
2. Campbell AJ et al (1981) Falls in old age: a study of frequency and related clinical factors. Age Ageing 10(4):264–270
3. Porciatti V et al (1999) The effects of ageing on reaction times to motion onset. Vis Res 39 (12):2157–2164
4. Berg WP et al (1997) Circumstances and consequences of falls in independent community-dwelling older adults. Age Ageing 26(4):261–268

5. Menant JC et al (2008) Optimizing footwear for older people at risk of falls. J Rehabil Res Dev 45(8):1167–1181
6. Jessup RL (2007) Foot pathology and inappropriate footwear as risk factors for falls in a subacute aged-care hospital. J Am Podiatr Med Assoc 97(3):213–217
7. Hsu J et al (2015) Assessing the performance of winter footwear using a new maximum achievable incline method. Appl Ergon 50:218–225
8. Rizvi R et al (2016) Multifunctional textured surfaces with enhanced friction and hydrophobic behaviors produced by fiber debonding and pullout. ACS Appl Mater Interfaces 8(43):29818–29826
9. Rizvi R et al (2015) High friction on ice provided by elastomeric fiber composites with textured surfaces, vol 106, p 111601
10. Anwer A et al (2017) Evolution of the coefficient of friction with surface wear for advanced surface textured composites. Adv Mater Interfaces 4(6):1600983
11. Zhang SW (1998) State-of-the-art of polymer tribology. Tribol Int 31(1):49–60
12. Dickrell PL et al (2005) Frictional anisotropy of oriented carbon nanotube surfaces. Tribol Lett 18(1):59–62
13. Wang H et al (2014) Anisotropy in tribological performances of long aligned carbon nanotubes/polymer composites. Carbon 67:38–47
14. Slip-resistant rubber composition, outsole using the rubber composition and method of manufacturing the outsole. 2013: United State no US 8,575,233 B2
15. Colonna M et al (2016) Ski boot soles based on a glass fiber/rubber composite with improved grip on icy surfaces. Procedia Eng 147:372–377

Impact Analysis on Human Body of Falling Events in Human-Exoskeleton System

Jing Qiu$^{(\boxtimes)}$, Ye Chen, Hong Cheng, and Lei Hou

Center for Robotics, School of Automation Engineering, University of Electronic
Science and Technology of China, Chengdu 611731, China
qiujing@uestc.edu.cn

Abstract. Powered lower extremity exoskeletons (LEE) can support and
ambulate individuals in an upright posture, and meanwhile bring rehabilitative
benefits. However, guaranteeing the safety of human-exoskeleton system is still
a big challenge. The type and extent of probable risks of these devices have not
yet been understood. Falls can result in serious injuries for LEE users and awful
consequences. In this paper, we focus on the process of human-exoskeleton
system falling and assess the injury of human. Herein, impacts of falling events
were assessed and analyzed based on finite element method in the current study.
In order to collect kinematics of the human-exoskeleton system for simulated
impact analysis, an experiment platform consisting of an air bend and a ceiling
rail was designed. Volunteers individually or wearing exoskeleton were asked to
lean body in different directions (e.g. forward, side and backward) until they
inevitably fell. The results of the experiment and simulation indicated that the
main parts injured were head, thorax, spine, arm and pelvis when human-
exoskeleton system fell. The maximum impact velocity of head can be 6.5 m/s,
if no buffer actions were taken, and that can cause traumatic brain injury (TBI).
No fractures occurred in other parts, but local squeezing and bruising could be
found in the simulation. It is anticipated that the study of human-exoskeleton
falling event will be useful in making safety regulations and safety exoskeleton
designing.

Keywords: Falling · Injury assessment · Lower limb exoskeleton

1 Introduction

Spinal cord injury (SCI) is a major cause of gait disability. Every year around the
world, between 250000 to 500000 people suffer from a SCI [1]. LEE has attracted
attention due to the rehabilitation and walking-assist functions especially for people
with SCI. In recent years, many companies developed LEE products successfully and
also entered the market [2–5]. ReWalk^{TM} personal exoskeleton was approved by the
Food and Drug Administration (FDA) of the United States as a classIImedical device in
2014. Indego^{TM} was the second exoskeleton gaining FDA approval in the US, then
Ekso GT™ exoskeleton and Hybrid Assistive Limb (HAL^{TM}) for medical use. Many
studies focus on the benefit of using LEE for a long time [6–8]. However, probable
risks, especially the problem of human-exoskeleton falling, still exist.

© Springer Nature Switzerland AG 2019
S. Bagnara et al. (Eds.): IEA 2018, AISC 819, pp. 767–776, 2019.
https://doi.org/10.1007/978-3-319-96089-0_84

This paper defines the falling of the human-exoskeleton system as follows: the sudden drop of the human-exoskeleton system to the ground or a lower plane in the case of non-external impact. The reason of falling can been classified into three types: human, exoskeleton and surroundings. The wearer misoperation, LEE system error or the rough ground can all cause falling accident. In order to avoid falling, many measures have been proposed [9–11], such as extra support frame or walking stick. But the extra support makes the system large and aren't suitable for outdoor personal using and using walking stick are still risky. The result of falling can be serious and falling is the second leading cause of accidental or unintentional injury deaths worldwide [12].

The injury of human body relates to the impact location and it depends on the direction and the posture of falling. Even some studies perceived the seriousness of human-exoskeleton falling [13] and proposed safe falling strategies for LEE [14], the mechanism of LEE falling is yet to be figured out. This paper focuses on the process and the final injury of human-exoskeleton falling. And that can be useful in LEE safety designing and developing relevant regulations.

Structures of this article are as follows: For purpose of analyzing the human-exoskeleton falling mechanism, we built a human-exoskeleton falling experimental platform. Volunteers were asked to fall with LEE under protective measures, and a motion capture system (VICON) was used to get the kinetic parameter. But the injury of human can't be obtained from experiments. Accordingly, we simulated the impact process of human body with ground based on the finite element method (FEM). Data from experiments were used to set initial conditions of falling simulation and contrastive analysis the accuracy of simulation results. Then we inspected the Von Mises nephogram and velocity changes of the impact parts, such as pelvis and head. Finally, we assessed the possible injury of human in the falling impact.

2 Methods

Different from free human falling, human-exoskeleton can be more complicated. The exoskeleton affects the posture and direction of falling, and the impact object is not only body to ground but also body to exoskeleton. Exoskeleton in parallel with lower extremities, the center of gravity is lower than human alone. Motors are set in the knee joint and hip joint. The joints are locked while the power was on and stood by or rotate control by the exoskeleton system. Human can't transform the posture independently in the falling accident caused by the disability of lower extremities or the resistance of reducer. Therefore, the posture of human-exoskeleton appear to be stiffer. In order to analyze the kinematics parameter and compare the distinction between human and human-exoskeleton in falling accidents, the falling experiment platform was set up. In the experiment, volunteers were asked to fall in three directions (sideward, forward and backward) with or without exoskeleton. Also we have some protective measures to decrease the impact velocity of human body. Hence, the data in the final impact moment can't be used. To study the human biomechanical response in the impact, we built a three-dimensional human-exoskeleton model to simulate the situation based on the finite element method.

2.1 Falling Experiment

Four healthy young volunteers participate in this experiment. They are all male, aged between 23 and 28, weighing 70 ± 7 kg and their overall height is 172 ± 4 mm. All the volunteers are graduate students, and had no training or experiment relating to falling previously. The ethics committee approved the falling experiment, and all the volunteers read and signed informed consent before the experiment.

The experimental platform is set up in an indoor environment, and the experimental area is approximately 6 m * 6 m. Flexible air cushion is placed in the center of the area, and its size is 2.0 * 1.8 * 0.2 m. A contour block is placed on one side of the air cushion for standing. To prevent the block from sliding during the falling, weight objects are set around the block to stabilize it. We also used ceiling-mounted dynamic Body-Weight Support (BWS) as a protective measure. The BWS can avoid human falling and is often used in gait training. It contains an actor which can adjust rope length, a proprietary Rail System set on the ceiling, and a strap. We have extended the rope tied to the human body, which can reduce or prevent the final impact and have no effect on the pre-stage of falling.

A motion capture system (VICON) was used for getting kinematic parameters. Around the experiment area, 8 high-speed cameras were placed to record the falling process. The cameras were placed towards the experiment area, circle placement, and the adjacent angle was about 45°. The VICON processor was connected to 8 high-speed cameras and computers via signal lines. The data collected by the high-speed camera was finally imported to the computer for modeling and analyzing (Fig. 1).

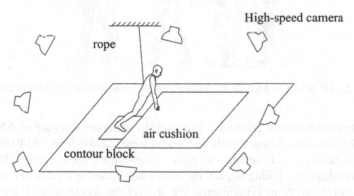

Fig. 1. Falling experiment platform

Experiments were divided into two categories: human falling and human-exoskeleton falling. Volunteers were instructed to fall from different directions including sideward, forward and backward. Given the back of the LEE could hurt the human spine in falling backward, we didn't carry out human-exoskeleton falling backward experiment. Moreover, given the change in human posture during falling, subjects were asked to fall with or without joint rotation in human free falling. However, the joint rotation in human-exoskeleton falling can't be achieved due to the resistance of exoskeleton. Subjects with or without exoskeletons stood on the contour block, 39 marks were affixed in the whole body. Marks were used for tracking body motions. Subjects were asked to lean their body until they inevitably fell. Data obtained from cameras were imported to the software named Nexus (provided by VICON) to establish human link model. Then we can study the process of falling and collect the velocity of human part.

2.2   Impact Simulation

The simulation method was used to analyze injuries to human body caused by falling impact.

Firstly, a LEE three-dimensional model was built. The dummy model used in this article is the M50-PS simplified pedestrian model, provided by Global Human Body Models Consortium (GHBMC). Then we combined the LEE with the dummy model, connecting them with two-dimensional belts which distributed in shoes, lower leg, thigh and waist (Fig. 2).

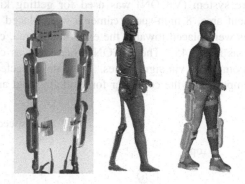

Fig. 2. 3D model of LEE (left), dummy (middle) and human-exoskeleton (right)

Next, hexahedral mesh grids were structured by using the software of ANSYS. The number of human-exoskeleton cells and nodes were 2236735 and 814028. We also checked the Jacobian and skewness of cells to improve the mesh quality. Then we set the initial conditions of falling impact, the model was leaned in a particular angle based on the data obtained from experiments. We also set the acceleration of gravity in the vertical direction. Simulation was divided into three directions: sideward, forward and backward (Fig. 3).

Finally, the results were calculated, and biomechanical response in important body parts was extracted.

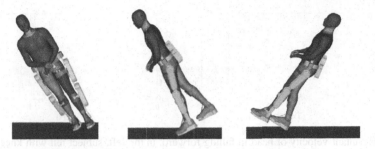

Fig. 3. Falling in different directions

3 Results

3.1 Human Free-Falling

According to the human falling posture, it can be divided into 5 situations in the human free-falling experiment: backward, forward, sideward, backward with waist bending and forward with knee bending. When the joint bends while falling, the impact part will change and head impact can be mitigated or avoided. This is also a falling protection strategy adopted by some humanoid biped robots [15]. Resultant velocity was used to denote the kinetics of human body, the definition is as follows:

$$V_R = \sqrt{V_x^2 + V_y^2 + V_z^2} \tag{1}$$

In Eq. (1), V_R is the resultant velocity, and V_x, V_y and V_z respectively represent velocity along the x-axis, y-axis and z-axis. We extracted the center of head mass velocity curve as a reference to analyze falling process.

Falling Forward. When the low limb hit an obstacle on the ground while walking, human body may fall forward. In the falling forward experiment, head velocity curve was affected by falling posture and subject height. The first impact with the ground took place on the knee and reduced kinetic energy while falling forward with knee bending. There were differences in every trial, so we selected a subject with height of 172 cm as a representative. The head velocity was showed in Fig. 4.

We can see the maximum velocity of falling straight is 4.8 m/s, higher than falling with knee bending which is 4.0 m/s. The speed changes slowly in the early falling process, and the speed rises quickly when the falling tendency is inevitable. Although the subject was asked not to take a protective action while falling, the arm will also support to the ground due to natural reactions.

Falling Backward. The human body tends to fall backwards when the sole slip. Usually, upper body will rise to reduce the injury caused by the impact on the head or spine. Subjects were asked to fall with or without waist bending. The velocity of head is shown in Fig. 5.

In Fig. 5, the maximum head speed in the left is 4.0 m/s, lower than the right which is 4.5 m/s. In the left trial, head impact occurred at the maximum head speed. However,

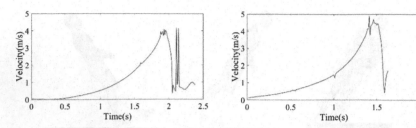

Fig. 4. Resultant velocity of head in falling forward. In the left, subject fell with knee bending, and in the right subject fell straight. The data after reaching the peak of curve can be ignored for that the impact occurred on the air cushion.

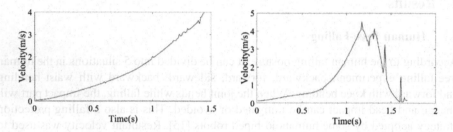

Fig. 5. The velocity of head in falling backward. In the left figure, subject fell backwards without waist bending. The data after head impact were missed because the marks slid from human body. In the right figure, subject fell backwards with waist bending.

in the right trial, head impact occurred at 1.4 s and the impact speed was 2.0 m/s. The speed fluctuations that occurred at from 1.2 s to 1.4 s were caused by the impact between knee and air cushion.

Falling Sideward. It is potentially fall sideward, when one lower limb support is insufficient. The velocity of head in the falling sideward trial is shown in Fig. 6. The maximum speed of head is 5.2 m/s, and this is also head impact speed.

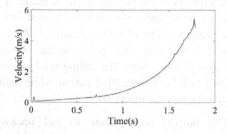

Fig. 6. Velocity of head in falling sideward

3.2   Human-Exoskeleton Falling

Different from human falling, the impact in the human-exoskeleton can be more complicated. The center of gravity will decline, but the force of impact will increase due to LEE parallel to the lower limbs. Human injury is related to various factors such as collision speed, impact location, and impact force. The velocity curve of human-exoskeleton falling is shown in Fig. 7. The characteristic of human-exoskeleton falling is stiff rotate like an inverted pendulum. The knee and waist is locked by the LEE, and can't rotate during falling. The maximum speed of head in the human-exoskeleton falling forward and sideward are 4.4 m/s and 3.6 m/s. In human-exoskeleton falling, the subjects' arms tried not to take protective measures as we required. In the experiment, human-exoskeleton falling backward was rejected due to safety issues.

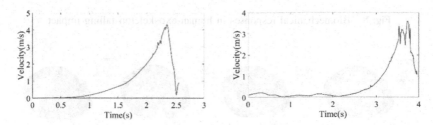

Fig. 7. Velocity of head in human-exoskeleton falling. In the left figure, the subject fell forward; in the right figure, the subject fell sideward.

3.3   Human-Exoskeleton Impact Simulation

During the falling, human could take some buffer actions to reduce injury. However, most LEE users are disabled or the aged and have difficulties to react. Therefore, in the simulation, human model fallings without autonomic responses and biomechanical responses are shown in Fig. 8.

If human-exoskeleton fell without buffer posture, the impact position can be the head and that can cause traumatic brain injury.

As is shown in Fig. 9, impact will cause large deformation of the head. No fracture feature was found in other parts by observing the vonMises of bones. Nevertheless, there were local scratches or extrusion distributed in thigh, lower leg and arm (Fig. 10).

4   Discussion

The purpose of this paper is to analysis and assess the injury of humans in human-exoskeleton falling accident. The falling process was studied by experiments, and the falling impact was analyzed by falling simulation. Results show that there were some differences between human-exoskeleton falling and human free falling. The falling postures were similar in sideward falling but different in backward and forward falling. In human free-falling, the joint will bend to reduce the impact energy while human-exoskeleton fell straight. We found that the duration from leaning inevitably falling

Fig. 8. Biomechanical responses in human-exoskeleton falling impact

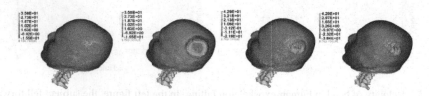

Fig. 9. Thickness reduction of head in backward falling.

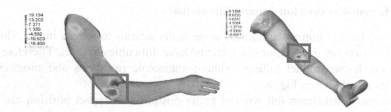

Fig. 10. Thickness reduction of arm and leg in human-exoskeleton falling impact.

angle to impact of human-exoskeleton is 0.60 ± 0.08 s, which is similar to the duration of the falling of elderly people in the real life which is 0.59 ± 0.25 s [16]. The maximum velocity of head in human free-falling experiments is 4.6 ± 0.6 m/s, and the data is 6.5 ± 0.5 m/s in human-exoskeleton falling simulation. The main reason for this difference is that human took buffer measures to reduce the impact speed. It is worth noting that the wearer of LEE is disabled or elderly, their time of reaction can be longer than the falling duration. Therefore, the human-exoskeleton falling simulation is similar to the real situation.

According to the results of falling simulation, we found that injuries occurred when human hit the ground and when human react with LEE. The impact parts were different when falling in different directions, but the head impact velocity was similar. The arm can be hurt in falling forward or sideward. Human body is easily squeezed in the human body and LEE contact parts when falling with LEE. The part most seriously injured was the head. If no protection is taken as in the simulation, the consequences of a falling will be particularly serious in the real life. That means we should take measures to avoid head impact in the human-exoskeleton falling accident or reduce the impact velocity.

It should be noted that this study has only examined human-exoskeleton fall individually and the ground leveling. In the real life, the situation can be more complicated, such as falling when supported by a crutch or with obstacles around. But the experiments and simulations in this paper represent a general situation of falling injuries without protection. In the experiments, the subjects are healthy young people that might be different from users of LEE. Due to long-term inactivity of lower limbs, they may be suffering from complications of osteoporosis. They are therefore more likely to have a fracture in a falling accident. We asked the subjects to lean until they inevitably fell and asked them to keep their joints fixed or rotated. There are differences between the falling experiment and the real-life falling. In the next step, we will design human-exoskeleton falling experiments in different situation and imitate the real-life falling. And the subjects will be people of different ages, heights, and genders. Also, we will improve the falling simulation. Natural reactions of human body will be added into the dummy, such as the arm support action and waist bending action.

5 Conclusions

This paper studies the process and the consequences of human-exoskeleton falling. Its motivation is to assess the risk of the LEE falling, thus we can improve the LEE safety design and reduce the injury of users. The LEE constrains the lower limbs joint, and that makes the falling motion rigid. When human-exoskeleton fall without protection and buffer actions, the head will hit the ground, causing traumatic brain injuries. In addition, falling with LEE can also cause local bruising of arms and legs.

References

1. World Health Organization. http://www.who.int/news-room/fact-sheets/detail/spinal-cord-injury. Accessed 25 Apr 2018
2. Esquenazi A, Talaty M, Packel A, Saulino M (2012) The ReWalk powered exoskeleton to restore ambulatory function to individuals with thoracic-level motor-complete spinal cord injury. Am J Phys Med Rehabil 91(11):911–921
3. Tsukahara A, Hasegawa Y, Eguchi K, Sankai Y (2015) Restoration of gait for spinal cord injury patients using HAL with intention estimator for preferable swing speed. IEEE Trans Neural Syst Rehabil Eng 23(2):308–318
4. Benson I, Hart K, Tussler D, van Middendorp JJ (2016) Lower-limb exoskeletons for individuals with chronic spinal cord injury: findings from a feasibility study. Clin Rehabil 30 (1):73–84
5. Kazerooni H, Steger R, Huang L (2006) Hybrid control of the Berkeley lower extremity exoskeleton (BLEEX). Int J Robot Res 25(5–6):561–573

6. Sale P, Franceschini M, Waldner A, Hesse S (2012) Use of the robot assisted gait therapy in rehabilitation of patients with stroke and spinal cord injury. Eur J Phys Rehabil Med 48 (1):111–121
7. Hidler J, Nichols D, Pelliccio M, Brady K, Campbell DD, Kahn JH, Hornby TG (2009) Multicenter randomized clinical trial evaluating the effectiveness of the Lokomat in subacute stroke. Neurorehabil Neural Repair 23(1):5–13
8. Lajeunesse V, Vincent C, Routhier F, Careau E, Michaud F (2016) Exoskeletons' design and usefulness evidence according to a systematic review of lower limb exoskeletons used for functional mobility by people with spinal cord injury. Disabil Rehabil: Assistive Technol 11 (7):535–547
9. Li L, Hoon KH, Tow A, Lim PH, Low KH (2015) Design and control of robotic exoskeleton with balance stabilizer mechanism. In: 2015 IEEE/RSJ International Conference on Intelligent Robots and Systems (IROS). IEEE, pp 3817–3823
10. Huynh V, Bidard C, Chevallereau C (2016) Balance control for an underactuated leg exoskeleton based on capture point concept and human balance strategies. In: 2016 IEEE-RAS 16th International Conference on Humanoid Robots (Humanoids). IEEE, pp 483–488
11. Goffer A (2014) U.S. Patent Application No 13/535,485
12. World Health Organization. http://www.who.int/news-room/fact-sheets/detail/falls. Accessed 25 Apr 2018
13. He Y, Eguren D, Luu TP, Contreras-Vidal JL (2017) Risk management and regulations for lower limb medical exoskeletons: a review. Med Devices (Auckland) 10:89
14. Khalili M, Borisoff JF, Van der Loos HM (2017) Developing safe fall strategies for lower limb exoskeletons. In: 2017 International Conference on Rehabilitation Robotics (ICORR). IEEE, pp 314–319
15. Moya J, Ruiz-del-Solar J, Orchard M, Parra-Tsunekawa I (2015) Fall detection and damage reduction in biped humanoid robots. Int J Humanoid Rob 12(01):1550001
16. Choi WJ, Wakeling JM, Robinovitch SN (2015) Kinematic analysis of video-captured falls experienced by older adults in long-term care. J Biomech 48(6):911–920

The Analysis of Foot-Eye Coordination Strategies Among Middle-Aged and Elderly Adults: An Example of Foot Positioning Tasks

Yi-Chen Wang and Jun-Ming Lu[⊠]

Department of Industrial Engineering and Engineering Management,
National Tsing Hua University,
No. 101, Section 2, Kuangfu Road, Hsinchu 30013, Taiwan
jmlu@ie.nthu.edu.tw

Abstract. Falls are the most important issue among elderly population. It is needed to develop early interventions for middle-aged and elderly people to reduce the falling risk. Therefore, this study aims to help prevent trips among middle-aged and elderly people by analyzing their foot-eye coordination strategies while performing foot positioning tasks under different conditions. 10 middle-aged and elderly males above 55 years old and 10 young males in the range of 20 to 30 years old were recruited. There were 3 levels of obstacle height (0 cm, 5 cm, 10 cm), whereas 2 levels (1step, 2steps) were set for the distance between the obstacle and the target. Each participant walked from the starting position and stepped over the obstacle while both feet onto specified targets positioning. The positioning error was defined as the performance measure of primary task. Besides, two different colors were displayed from toe-off to mid-swing and from mid-swing to heel-strike of the leading foot respectively. The participant had to recall the colors been seen as the performance measure of secondary task. Principal component analysis (PCA) was conducted with the gait data collected. 16 PCs accounted in total of 81.3% of variance were extracted. Cluster analysis was then used to determine four different strategies based on the PC scores. It was found that the foot-eye coordination strategies had significant effects on the performance of primary task ($p < .001$), as well as on secondary task ($p < .001$). Furthermore, the results showed that the strategies were related to age and environmental condition.

Keywords: Gaze behavior · Gait · Foot positioning
Principal component analysis · Clustering analysis

1 Introduction

Falls often result from a variety of factors associated with aging, medications, the environment, and so on. The factors can be mainly separated into internal and external contributors. The internal factors include muscle loss, reduced ability to control lower limbs, and decline in cognitive function. The external factors include shoe selection, inadequate lighting, and wet floor [3]. Chang et al. [4] indicated that reduced strength of the lower extremities, experience of falls, and reduced dynamic balance are the major

© Springer Nature Switzerland AG 2019
S. Bagnara et al. (Eds.): IEA 2018, AISC 819, pp. 777–785, 2019.
https://doi.org/10.1007/978-3-319-96089-0_85

causes of falls in the elderly. Besides, middle-aged people have shown degeneration in gait stability that is similar with elderly people. It is necessary to develop early interventions for both middle-aged and elderly adults to prevent from falls [5]. Fortunately, exercise programs, which have been considered the most cost-effective strategy for preventing falls, can improve the fall rate in community-dwelling older adults [6].

There are different types of exercise program to improve muscle strength and maintain balance, increase target stepping accuracy, and reduce risks of falls significantly [7]. From the perspective of the mechanisms associated with falls, Hollands, Hollands, and Rietdyk [8] highlighted the fact that people use vision to guide lower-limb movements to cope with environmental demands in daily lives. The visual cues provide the environmental information to program the foot placement and control stepping movement to cross an obstacle or step on a safe place. The relationship between gaze behavior that people sampled the visual information and lower-limb movements can be considered as "foot-eye coordination." There is much evidence showing that older adults at high risk of falling have different foot-eye coordination strategies from those at a lower risk of falling. According to Chapman and Hollands [9], high-falling-risk participants look away earlier from target before foot contacted with it, compared with low-falling-risk participants. Besides, the earlier gaze transfer performed by high-falling-risk one led to a poor stepping performance. In other words, high- and low-falling-risk groups might have different gaze behavior to control their locomotion, which results in different stepping accuracy. Other studies showed that if there are visual secondary tasks along with the stepping-target or crossing-obstacle tasks, it would change participants' gaze behavior. The increased head movement angle and the distracted attention would lead to a higher chance to contact an obstacle and higher falling risk [10, 11]. It is possible that exercises aimed at improving visually guided stepping actions will improve visuomotor and physical function while reducing fall risk [8].

Young and Hollands [12] measured the stepping performance of both control and intervention groups, in which only the intervention group was instructed to maintain gaze on stepping target until heel contact. After few days, all participants returned to perform the same tasks again. The results showed that the intervention group had a significant reduction in stepping errors on average. In other words, it may be a new way to reduce falling risk among elderly adults through adopting a proper foot-eye coordination strategy.

Therefore, it is worthwhile to investigate the relationship between lower-limb movements and gaze behaviors among middle-aged and elderly people, as well as identifying different foot-eye coordination strategies among them. In order to further compare different strategies, stepping accuracy can be considered as an indicator to reflect gait stability and falling-risk, whereas the responses to unexpected visual stimulus can be treated as a measure of residual visual attention.

2   Method

2.1   Participants

10 older males above 55 years old (age: 65.6 ± 3.8 years; height: 166.2 ± 3.0 cm; weight: 66.9 ± 11.7 kg) and 10 younger males ranging from 20 to 30 years old (age: 23.4 ± 0.8 years; height: 172.7 ± 4.4 cm; weight: 64.7 ± 7.6 kg) were recruited. In order to identify the differences in foot-eye coordination strategies among them, the younger participants were taken as the control group who are assumed to adopt better strategies. All participants had 20/40 vision or better, achieved a score higher than 41 on Berg Balance Scale (BBS), completed Timed Up & Go Test (TUG) in no more than 12 s, and self-reported no known musculoskeletal or neurological impairments.

2.2   Protocol

The experiment was a 3 × 2 factorial design. There were 3 levels of obstacle height (0 cm, 5 cm, 10 cm), whereas 2 levels (one-step, two-step) were set for the distance between the obstacle and the target. In every trial, each participant had to walk along a 1.2 m wide path and step over the obstacle to complete the primary task of foot positioning (see Fig. 1). If the participant's both feet are placed within the target without contacting the obstacle, the trial is defined as a success. Every of the six conditions needs to be completed with a success for five times, or the participant cannot proceed to the next condition. Therefore, every participant needs to succeed 30 times in total. Besides, as shown in Fig. 2, two different colors (out of red, orange, yellow, green, blue and purple) will be displayed through the monitor in front of the walkway from toe-off (A) to mid-swing (B) and from mid-swing to heel-strike (C) of the leading foot respectively as the visual secondary task. After the participant completed the foot positioning task, he had to recall the color(s) been seen as the performance measure of the secondary task. The order of six conditions and colors displayed in secondary task had been randomized to minimize learning and fatigue effects. Each participant was required to wear the eye-tracking glasses and motion capture suit, for collecting gaze and motion data.

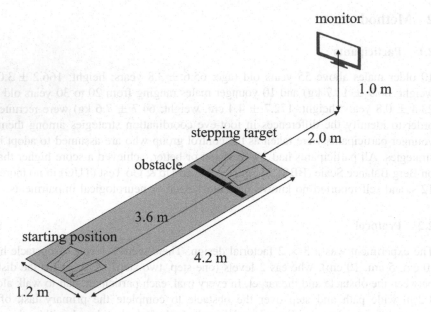

Fig. 1. Experimental setup.

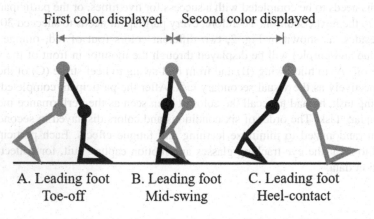

Fig. 2. Periods of displaying colors in the visual secondary task. (Color figure online)

2.3   Data Analysis

In order to identify the key frames that represent the characteristics of the lower-limb movement, the duration starting from the moment when leading foot stepping on the target to the moment when trailing foot stepping on the target was first considered. This last (n^{th}) step was then divided into 10 time-equivalent segments. The $(n-1)^{th}$, $(n-2)^{th}$, $(n-3)^{th}$ steps were also treated in the same way. Finally, 40 segments with 41 key frames were generated for each trial (see Fig. 3). There are 9 parameters obtained, including the flexion/extension angles of hips, knees, ankles, spine, and neck, as well as the vertical gaze angle. Principal component analysis (PCA) was first conducted to reduce data dimension. All PCs were extracted until they accounted in total of 80% of variance. Then, Ward's hierarchical clustering method was used to divide the samples into multiple groups based on principal component scores, in which one group was defined as one foot-eye coordination strategy. Subsequently, MANOVA was performed to test if the strategies affect the foot positioning errors significantly. The measure of foot placement errors was calculated with the percentage of the center position of target subtracted from the participant's actual foot placement, and then divided by the target size in anterior-posterior and medial-lateral direction. In addition, chi-square test was performed to find out if different strategies result in different performance of the secondary task, by comparing the percentages of correct responses. Similarly, chi-square was used to see the relationship between the strategies and ages, obstacle heights, distances between obstacle and target.

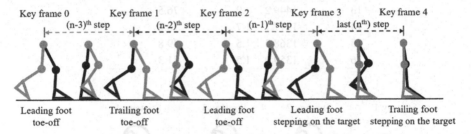

Key frame 0   Key frame 1   Key frame 2   Key frame 3   Key frame 4
$(n-3)^{th}$ step $(n-2)^{th}$ step $(n-1)^{th}$ step   last (n^{th}) step

Leading foot   Trailing foot   Leading foot   Leading foot   Trailing foot
toe-off   toe-off   toe-off   stepping on the target   stepping on the target

Fig. 3. Key frame definition (leg in orange means leading foot) (Color figure online)

3 Results and Discussion

According to the results of principal component analysis, 16 components could explain 81.3% of the total variance. Table 1 gives the amount of variance that was explained by each component. Subsequently, the cluster analysis divided the samples into four groups (i.e. strategies). As Fig. 4 shows, the differences among these four strategies were mainly found in the periods of the last (n^{th}) step (key frames 3–4) and the $(n-2)^{th}$ step (key frames 1–2). For those who adopt strategy 1, the spine was bent forward more while trying to place the trailing foot on the target. Strategy 2 is characterized by the more flexed knee of trailing foot while it was about to be placed on the target. As for strategy 3, the participant would place his trailing foot on the target more slowly, whereas it was lifted fast. While taking strategy 4, the hip of trailing foot for obstacle-crossing was more flexed in the $(n-2)^{th}$ step.

Table 1. Variance explained by components

Component	Total	% of Variance	Cumulative %
1	3084.1	12.9	12.9
2	2674.2	11.2	24.1
3	2562.0	10.7	34.8
4	2481.6	10.4	45.1
5	1612.9	6.7	51.9
6	1346.5	5.6	57.5
7	1103.8	4.6	62.1
8	1803.2	3.4	65.5
9	1658.9	2.8	68.2
10	1560.2	2.3	70.5
11	1507.7	2.1	72.7
12	1462.5	1.9	74.6
13	1449.2	1.9	76.5
14	1417.2	1.7	78.2
15	1368.3	1.5	79.8
16	1359.0	1.5	81.3

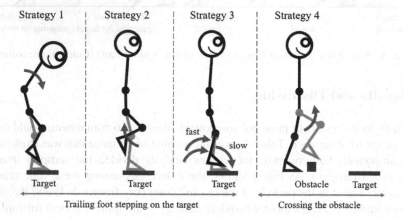

Fig. 4. Characteristics of the four foot-eye coordination strategies (leg in orange means leading foot for stepping on the target or crossing the obstacle) (Color figure online)

It was found that the foot-eye coordination strategies had significant effects on the performance of primary task ($p < .001$), and also on the secondary task ($p < .001$). In the foot positioning task, using strategy 4 resulted in the most accurate foot positioning among all strategies. Strategy 2 led to the largest leading foot placement errors, but has nearly the same accuracy as strategy 4 on trailing foot placement errors. On the contrary, strategies 1 and 3 were with the worse trailing foot positioning accuracy, but with better performance of leading foot positioning. In conclusion, strategy 4 contribute to best performance on primary task, so it is considered as the safest strategy (Fig. 5).

In the visual secondary task, as shown in Fig. 6, the percentages were calculated with the number of the four possible combinations in each strategy based on whether the first or the second color was seen. The trials that the participants adopted strategy 1 had 42.9% of both correct responses. Strategy 2 contributed to the highest 55.3% of both correct answers. As for strategy 3, none of the trials had both correct responses (0.0%), and the most participants could merely see the second color displayed by the monitor (43.8%) or none of the two colors (47.3%). While taking strategy 4, the most participants could see both colors (41.5%) or none of them (40.4%). Generally speaking, based on the percentages of correct responses, strategies 1 and 2 resulted in better performance of secondary task, which means it is possible to have residual visual attention. However, strategy 3 led to the worst performance.

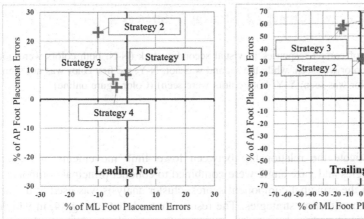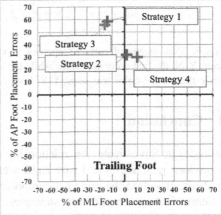

Fig. 5. Means of foot ML foot placement errors among the four strategies (the "+" markers means the average foot placement errors of each strategy)

In addition, it was found that the strategies were significantly related to age groups ($p < .001$) and distances between obstacle and target ($p < .001$). Most middle-aged and elderly adults adopted strategy 1 (48.8%) and strategy 3 (27.6%) in the experiment, whereas most young people used strategy 2 (46.3%) and strategy 4 (18.8%). Regarding the six environmental conditions, participants would adopt strategy 2 (48.3%) in one-step distance. Besides, strategy 1 (47.1%) and strategy 4 (27.1%) were more commonly seen in two-step distance. The results showed that strategy 4 contributed to the most accurate foot placement, and was adopted by young participants a little bit more (59.6%) than middle-aged and elderly ones. Considering the environmental factors, it was more frequently used at two-step distance.

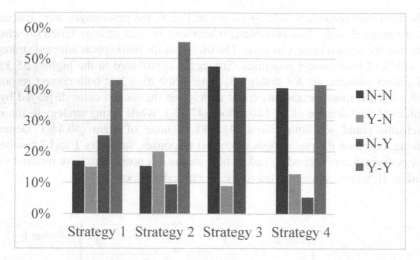

Fig. 6. Percentages of correct responses in visual secondary task (N-N: both colors were not seen; Y-N: the first color was seen, while the second was not; N-Y: the first color was not seen, while the second color was seen; Y-Y: both colors were seen) (Color figure online)

4 Conclusion

In order to investigate the relationship between lower-limb movements and gaze behaviors, joint angles and gaze angle were combined to conduct principal component analysis. Based on the principal component scores, cluster analysis was used to define different foot-eye coordination strategies. The results showed that strategy 4, in which the hip of trailing foot for obstacle-crossing was more flexed in the $(n-2)^{th}$ step, has the best performance of foot positioning, and it was adopted by young participants more and also more often in the two-step condition. In summary, an intervention or training program to guide middle-aged and elderly people to use a proper foot-eye coordination strategy may have the potential to help increase foot positioning accuracy while ensuring residual visual attention.

References

1. Health Promotion Administration, Ministry of Health and Welfare. https://www.hpa.gov.tw/Pages/Detail.aspx?nodeid=1253&pid=7664. Accessed 06 May 2018
2. Fuller GF (2000) Falls in the elderly. Am Fam Phys 61(7):2159–2168
3. Rogers ME, Rogers NL, Takeshima N, Islam MM (2003) Methods to assess and improve the physical parameters associated with fall risk in older adults. Prev Med 36(3):255–264
4. Chang JT, Morton SC, Rubenstein LZ, Mojica WA, Maglione M, Suttorp MJ, Roth EA, Shekelle PG (2004) Interventions for the prevention of falls in older adults: systematic review and meta-analysis of randomised clinical trials. BMJ 328(7441):680

5. Wu Y, Wang YZ, Xiao F, Gu DY (2014) Kinematic characteristics of gait in middle-aged adults during level walking. In: 2014 36th annual international conference of the IEEE engineering in medicine and biology society (EMBC), Chicago, IL, USA. IEEE, pp 6915–6918
6. Karlsson MK, Magnusson H, von Schewelov T, Rosengren BE (2013) Prevention of falls in the elderly - a review. Osteoporos Int 24(3):747–762
7. Bauman A, Merom D, Bull FC, Buchner DM, Fiatarone Singh, MA (2016) Updating the evidence for physical activity: summative reviews of the epidemiological evidence, prevalence, and interventions to promote "Active Aging". Gerontologist 56(Suppl 2): S268–S280
8. Hollands M, Hollands K, Rietdyk S (2017) Visual control of adaptive locomotion and changes due to natural ageing. In: Locomotion and posture in older adults. Springer, Cham, pp 55–72
9. Chapman GJ, Hollands MA (2007) Evidence that older adult fallers prioritise the planning of future stepping actions over the accurate execution of ongoing steps during complex locomotor tasks. Gait Posture 26(1):59–67
10. Menant JC, St George RJ, Fitzpatrick RC, Lord SR (2010) Impaired depth perception and restricted pitch head movement increase obstacle contacts when dual-tasking in older people. J Gerontol Ser A: Biomed Sci Med Sci 65(7):751–757
11. Serchi V, Cereatti A, Cinelli ME, Della Croce U (2015) Gaze strategies while negotiating obstacles in a virtual environment with distractors. Gait Posture 42:S3
12. Young WR, Hollands MA (2010) Can telling older adults where to look reduce falls? Evidence for a causal link between inappropriate visual sampling and suboptimal stepping performance. Exp Brain Res 204(1):103–113

The Influence of Information Acquisition Strategies on Foot Proprioception and Obstacle Avoidance Pattern in People with Low Vision

Tadashi Uno[1,2]([∆]), Ping Yeap Loh[3], and Satoshi Muraki[3]

[1] National Institute of Technology, Tokuyama College,
Yamaguchi 745-8585, Japan
t-uno@tokuyama.ac.jp
[2] Graduate School of Design, Kyushu University, Fukuoka 815-8540, Japan
[3] Faculty of Design, Kyushu University, Fukuoka 815-8540, Japan

Abstract. The purpose of this study was to understand the influence of various information acquisition strategies on foot proprioception and obstacle avoidance in people with low vision. Ten adult males (41.0 ± 7.1 years) with pigmentary retinal degeneration were recruited for this study. Participants acquired obstacle information (obstacle height: 4 cm and 15 cm) through three different strategies, namely, front (A), downward (B), and tactile (C). Subsequently, the participants performed two different tasks; Task 1: After identification of the obstacles, the participants reproduced the obstacle height by lifting their foot while standing still (10 times). Task 2: Following the acquisition of the obstacle information through conditions B and C, participants performed obstacle step-over from a standing position. In task 1, condition B showed significantly higher toe-rise and coefficient of variance in toe-rise ($p < 0.05$) than in conditions A and C, which both displayed similar toe-rise. Likewise, in task 2, the highest points of the leading and trailing feet while stepping over the obstacle were significantly higher ($p < 0.05$) in condition B than in condition C. Additionally, the coefficient of effort in condition B was significantly larger ($p < 0.05$) than that of condition C. These results suggest that differences in information acquisition strategies have an impact on the foot trajectory during obstacle step-over. Out of the three methods used in this study, information acquisition through the tactile sense may be the best obstacle avoidance feedback method for people with low vision.

Keywords: Low vision · Pigmentary retinal degeneration
Information acquisition strategies

1 Introduction

People with low vision face a major challenge of independent travel from place to place. Several reasons contribute to this mobility issue such as lack of information on the surrounding environment, orientation to the environment, and lack of information about barriers. However, the advancement in engineering technology such as GPS technology and wearable devices has enhanced the mobility of people with low vision

© Springer Nature Switzerland AG 2019
S. Bagnara et al. (Eds.): IEA 2018, AISC 819, pp. 786–790, 2019.
https://doi.org/10.1007/978-3-319-96089-0_86

[1–3]. In addition, when mobile, their remaining functional visual power does not allow acquisition of comprehensive information about the surrounding environment. Consequently, people with low vision incorporate low-vision aids for orientation and mobility to address the lack of visual information, as well as to optimize use of their remaining visual power.

Previous studies reported the importance of a multisensory approach in orientation and mobility training including auditory, tactual, olfactory, and kinesthetic senses [4–6]. Particularly, information acquisition by tactile and auditory senses may enhances the learning process of obstacle step-over among people with low vision. Obstacle step-over is an essential skill to prevent trip and fall which involves identification of steps and obstacles during walking. However, limited research has reported on the usefulness and efficacy of tactual training in obstacle step-over in people with low vision. Therefore, the purpose of this study was to investigate the influence of various information acquisition strategies on foot proprioception and obstacle avoidance in people with low vision due to pigmentary retinal degeneration.

2 Methods

This study is approved by was approved by the Ethics Committee of the Faculty of Design, Kyushu University. Ten adult males (41.0 ± 7.1 years) with pigmentary retinal degeneration were recruited for this study. All participants held grade 2 physical disability certificates issued by the Ministry of Health, Labor and Welfare of Japan (vision: 0.02–0.04, field of view: 10° and loss rate: 95%).

Participants acquired obstacle information (obstacle height: 4 cm and 15 cm) through three different strategies, namely, front: looking at the obstacle by standing directly in front of it (condition A), downward: looking down at the obstacle by standing at 20 cm (toe-obstacle; condition B), and tactile: identifying the obstacle (block) by hand exploration, with no time limit (condition C) (Fig. 1).

Subsequently, the participants performed two different tasks; Task 1: After identification of the obstacles, the participants reproduced the obstacle height by lifting their foot while standing still (10 times with each foot). Task 2: Following the acquisition of the obstacle information through conditions B and C, participants performed obstacle step-over from a standing position, in a single step. In both tasks 1 and 2, the lower limb movement for each condition was recorded using a high-speed camera at 250 frames per second. Gait parameters, such as the height of the lifted foot (measured at the toe), the step length, as well as characteristics of the foot trajectory, such as the highest points of the leading and trailing feet while stepping over the obstacle, were analyzed. The coefficient of effort (the ratio of the highest point of the leading foot to the height of the obstacle) was calculated.

One-way analysis of variance with repeated measures was used to analyze the differences between each condition. Post-hoc pairwise Bonferroni-corrected comparison was used to examine mean differences in each condition.

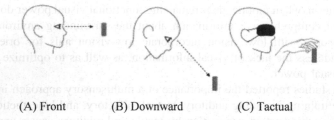

(A) Front   (B) Downward   (C) Tactual

Fig. 1. Acquisition strategy of obstacle information

3   Results

In task 1, condition B showed significantly higher toe-rise and coefficient of variance in toe-rise ($p < 0.05$) than in conditions A and C, which both displayed similar toe-rise (Fig. 2). Likewise, in task 2, the highest points of the leading and trailing feet while stepping over the obstacle were significantly higher ($p < 0.05$) in condition B than in condition C (Fig. 3). Additionally, the coefficient of effort in condition B was significantly larger ($p < 0.05$) than that of condition C.

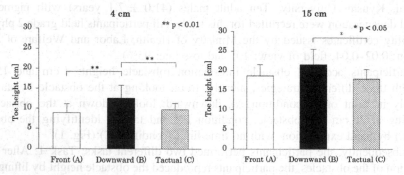

Fig. 2. The height of the toe of foot during elevation of lower extremity

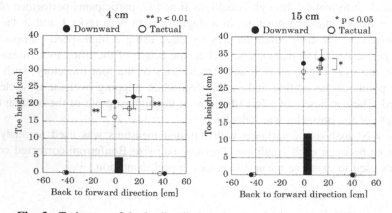

Fig. 3. Trajectory of the leading limb at 4 cm and 15 cm obstacle height

4 Discussion

Pigmentary retinal degeneration is a progressive disease that involves degeneration of rod photoreceptor cells in the eye with the main visual characteristic of decrease in peripheral vision field and reliance on central vision. Therefore, the participants showed greater ability to identify and reproduce the height of obstacle at a near distance to the obstacle compared to a far distance. Moreover, it was difficult and more challenging to identify the obstacle at a farther distance or when moving towards the obstacle. In task 1, condition B resulted in higher foot lifting which indicates acquisition of poor and divergent environmental information. The divergence of acquired information was reduced by tactile sensing in condition C (Fig. 2).

In task 2, participants performed obstacle step-over better with the tactual method (Fig. 3), as compared to the looking-downward method. A combination of various sensory information such as tactile and visual information is useful in obstacle identification, as integration of the acquired information establishes higher level perception [7]. Additionally, acquisition via multiple sensory input may improve the accuracy of perceived information [8, 9]. Our results indicated that the tactual method (condition C) was a useful way to improve the participants' lack of visual information on the obstacle. By the tactual method, the participants were able to acquire more information on the obstacle, such as dimension and shape. As a result, they could perform a more desirable and safer obstacle step-over motion.

Acknowledgement. This work is supported by JSPS KAKENHI Grant Numbers 15K16413.

References

1. Masaki T, Motohiro O (1995) A present situation and problems of independent travel of blind and visual impaired. Soc Instrum Control Eng 34(2):140–146. https://doi.org/10.11499/sicejl1962.34.140
2. Cardin S, Thalmann D, Vexo F (2007) A wearable system for mobility improvement of visually impaired people. Vis Comput 23(2):109–118
3. New Energy and Industrial Technology Development Organization. http://www.nedo.go.jp/activities/ZZ_00446.html. Accessed 23 Apr 2018
4. Zijlstra GR, Ballemans J, Kempen GI (2013) Orientation and mobility training for adults with low vision: a new standardized approach. Clin Rehabil 27(1):3–18. https://doi.org/10.1177/0269215512445395
5. Cappagli G, Finocchietti S, Baud-Bovy G, Cocchi E, Gori M (2017) Multisensory rehabilitation training improves spatial perception in totally but not partially visually deprived children. Front Integr Neurosci 11:1–11. https://doi.org/10.3389/fnint.2017.00029
6. Ishmael D (2015) The use of auditory, tactual, olfactory and kinaesthetic senses in developing orientation and mobility (O & M) skills to learners with congenital blindness (CB). IOSR J Humanit Soc Sci 20(2):34–44. https://doi.org/10.9790/0837-20213444
7. Newell FN (2004) Cross-modal object recognition. In: Galvert G, Spence C, Stein BE (eds) The hand-book of multisensory processes. Oxford University Press, Oxford, pp 123–139

8. Wada Y (2007) Vision influences on tactile discrimination of grating orientation. Jpn J Psychol 78(3):297–302
9. Ernst MO, Bülthoff HH (2004) Merging the senses into a robust percept. Trends Cogn Sci 8 (4):162–169. https://doi.org/10.1016/j.tics.2004.02.002

Overstep Slips on Stairway Treads During Descent

Rodney A. Hunter(✉) (iD)

Hunarch Consulting, Balwyn, VIC, Australia
info@hunarch.com.au

Abstract. A prototype tribometer, with a mechanical foot fitted with shoes, and that mimics the action of the leg and foot at ground contact to the moment of potential slip, was used to test the slip resistance of two stairway treads using four types of footwear, and at two ground reaction force angles. Based on slip distance, velocity and the location of the first metatarsal heads with respect to the nosing, it is concluded that there is greater overstep slip risk for shoes with arched outsoles than for shoes with continuous (non-arched) outsoles. It is also concluded that integrating slip resistive attributes near the horizontal tangency of the nosings might be beneficial, although possibly only marginally and that, in any case, it might be ineffective in offsetting the high slip velocities of feet as they slip off treads.

Keywords: Overstep · Slip · Stairway

1 Introduction

This study proceeds on the basis that many falls from oversteps during stairway descent can be attributable to "overstep slips"—occurring when the outsole at the foot's first metatarsal head (FMH) is too near or anterior to the horizontal tangency point (HTP) of the nosing[1]. Key issues are: (a) a slip that is inconsequential if it is further away from the HTP facilitates an overstep slip if it is too close; (b) the slope of arched midsoles[2] (between the foresole and heel) facilitates a slip if it is the part of the foot that first contacts the tread (at the nosing)[3]; (c) the metatarsal heads, particularly the first, is critical in slip dynamics on stair treads.

The primary goal of this study is to investigate the effect on overstep slips of the location of the FMH relative to the nosing at initial foot contact on the tread. The effect of outsole convexity and ground reaction force angle is also considered.

The study continues from previous investigations by Hunter into the slip-resisting role of tread nosings [1–4].

---

[1] The "nosing" is the upper-front quadrant of the tread front.

[2] "Midsole" denotes the part of the outsole between the forefoot (anterior to the metatarsal heads) and the heel (corresponding with the calcaneus). An "arched midsole" (or "arched outsole" or "continuous outsole") is where the outsole has a protruberant heel.

[3] The slope is also attributable to the normal dorsiflexion of the forefoot during stair descent.

© Springer Nature Switzerland AG 2019
S. Bagnara et al. (Eds.): IEA 2018, AISC 819, pp. 791–799, 2019.
https://doi.org/10.1007/978-3-319-96089-0_87

2 Apparatus

2.1 Tribometer

Tests were conducted with a prototype tribometer that mimics the key dynamics of the foot at ground contact (up to the moment of potential slip), including on stair treads during descent[4]. Load is applied by an inclinable strut that is double-articulated to enable its lower section to slip and, for stair descent, to decrease in its vertical angle after tread contact[5]. The strut's foot is articulated at its "ankle", "metatarsal heads" and "proximal-phalangeal heads" and can be fitted with typical footwear (see Fig. 2). The foot's sole is partly padded with neoprene foam to simulate the impact-attenuation of real feet. The tribometer's fundamental attribute is its simulation of dorsiflexion of the foot at its ankle and of the forefoot at the metatarsal heads during stair descent[6] (see Fig. 1).

Fig. 1. Tribometer test foot showing articulation between the "metatarsal heads" and "proximal-phalangeal heads"; the metatarsal heads and the foot's main body; and at the ankle (indicated by broken lines). Electronic goniometers record dorsiflexion and plantarflexion.

For more-biofidelic loading, adjustable friction systems control the rate of declination of the foot and, for stair descent, the posterioral rotation of the strut and declination of the metatarsal heads. Load is applied by friction-limited release of suspended weights, the combined mass of which is adjustable, including up to typical body mass. Vertical separation between foot and ground is adjustable (but minimal separation is considered necessary for realistic loading).

---

[4] The tribometer was originally inspired by the Brungraber Inclinable Articulated Strut Slip Tester (Marks 2 and 3) and is a developed version of one that was previously reported by Hunter [4].

[5] The strut represents the angle between the body's centre of mass and the ground contact, that is, the angle of the ground reaction force.

[6] And plantarflexion during level walking.

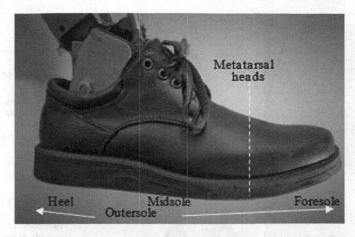

Fig. 2. View of the tribometer's test foot inside one of the test shoes, with identification of the terms used in this report.

Strut angle is digitally displayed, and slip distance and velocity, and declination of the ankle and metatarsal heads are electronically recorded.

Preliminary validation of the tribometer with a three-axis force platform indicated good loading biofidelity.

2.2 Tread Samples

Two treads of dense high-grade Australian hardwood were used. Each one was smoothly dressed, and one was also thickly coated with polyurethane; both treads were smoothed with a 1200-grit sheet (see Fig. 3). The treads had rounded nosings of 5 mm radius. They were mounted on a horizontally-adjustable platform to allow a range of shoe-to-nosing distances.

2.3 Footwear

Three "Oxford" style and one elevated high-heeled shoes were used, the outersoles of three of which were of Neolite and the fourth of smooth leather, each smoothly finished with 1200-grit. The outersoles of three shoes were arched as shown in Fig. 4.

3 Method

Two hundred and thirty-five tests were performed with the two tread samples and four shoe types[7]. However, most of the tests were used to validate and calibrate the tribometer; ultimately, analyses of the results of 95 tests were used for this study.

---

[7] A small number of tests were also conducted with other shoe types for comparative purposes.

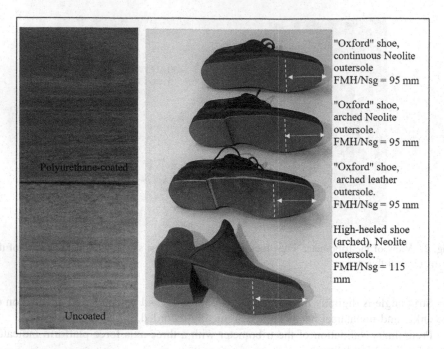

Fig. 3. View of the tread samples (on the left) and shoe samples (on the right), showing also the terms used in this report.

Outersoles and coated treads were lubricated with a water-glycerol solution, but not the leather outersole and uncoated tread. The second, uncoated, tread was chosen so that the leather outersole shoe could be tested without liquid. All tread and outersole combinations constituted highly slippery conditions.

After trial-and-error, initial strut angles of 73° and 82° to the horizontal, and foot angles of 29.5° and 32° respectively were employed, except that the foot angle for the high-heel shoe was 5°. The 82° initial strut angle was adopted as being close to typical, but the 73° was an artificial angle adopted to increase the probability of slips.

The anterior or posterior distance between the front of each shoe and the vertical tangent to the tread nosing was measured before each test and automatically adjusted with the previously determined position of the metatarsal head in each shoe: the recorded slip distances were therefore with respect to the distance between the FMH and the tread nosing (FMH/Nsg distance). Vertical distances of 1 mm, 2 mm and 3 mm between the outersoles and treads were employed. FMH/Nsg starting distances in the tests varied between 130 mm posteriorly and 35 mm anteriorly for non-arched outersoles, and between 70 mm posteriorly and 5 mm anteriorly for arched outersoles.

Several tests were video-recorded for later frame-by-frame analysis of shoe slips.

4   Results

4.1   Slips off Treads (Full Slips)

Full slips were recorded for 46% of arched outersoles (31% + 15%) at 82° strut angle, and 3% for continuous outersoles at 73° strut angle as indicated in Table 1 (full slips were not recorded for continuous outersoles at 82° or arched outersoles at 73°).

The average FMH/Nsg starting distance for all full slips was 27 mm (−20 mm, −32 mm and −30 mm for arched Neolite, arched leather and continuous Neolite outersoles respectively). No full slip occurred with the FMH at or in front of the nosing HTP. In other words, the full slips occurred as a continuation of a slips commencing with the MFH posteriorly to the nosing and midsole slope did not precipitate the slip: the slips were not overstep slips.

4.2   Partial Slips

As shown in Table 1, at the 82° strut angle, partial slips comprised 100% of continuous outersoles slips, and 69% and 85% of arched Neolite and arched leather outersoles slips respectively. At 73°, partial slips comprised 97% of continuous outersole slips (arched outersoles were not tested at 73°).

Table 1. Incidence (n.), length and maximum velocity of slips

	Continuous outersole			Arched outersole					
	Neolite			Neolite			Leather		
	n.	Length (mm)	Vel. (mm/s)	n.	Length (mm)	Vel. (mm/s)	n.	Length (mm)	Vel. (mm/s)
Initial strut angle: 82° *(n: 53)*									
Full slips	0% (0/6)	-	-	31% (4/13)	210	1500	15% (5/34)	210	1398
Partial slips	100% (6/6)	55	482	69% (9/13)	63	720	85% (29/34)	63	555
Initial strut angle: 73° *(n: 37)*									
Full slips	3% (1/37)	220	1220						
Partial slips	97% (36/37)	30	403						

4.3   Correlation of FMH/Nosing Distance and Slip Length

For the "Oxford" style shoes (Table 2), there was a positive correlation between FMH/Nsg distance and slip length, and between FMH/Nsg and velocity for the 82° strut angle (for all outersoles), but a negative correlation for 73° strut angle (for continuous Neolite outersoles). That is, for 82°, slip length and velocity increased as the FMH got closer to the nosing whereas, for 73°, slip length and velocity decreased as the FMH got closer. None of the correlations was very strong (from 0.5 to 0.8).

Table 2. Correlation of FMH/Nsg distance with slip length and velocity

Slip length	Cont. o'sole	Arched o'sole		Velocity	Cont. o'sole	Arched o'sole	
	Neolite	Neolite	Leather		Neolite	Neolite	Leather
Initial strut angle: 82° *(n: 53)*							
All slips	0.7	0.8	0.7	All slips	0.8	0.7	0.7
Partial slips	0.7	0.5	0.4	Partial slips	0.8	0.1	0.5
Initial strut angle: 73° *(n: 37)*							
All slips	−0.5	-	-	All slips	−0.6	-	-
Partial slips	−0.6	-	-	Partial slips	−0.7	-	-

Note: "Cont." = continuous; "o'sole" = outersole

4.4   High-Heeled Shoe

Only three tests were performed with the high-heels, at FMH/Nsg distances of from −3 mm to −90 mm (that is, 3 mm to 90 mm posteriorly to the nosing). Partial slips of approximately 13 mm average occurred, with a strong correlation between reducing FMH/Nsg distance and decreasing slip length (although the very small sample size precludes reliance upon this result). Reduced slip length with decreased FMH/Nsg distance is attributable to the vertical proximity of the heel to the tread; the absence of a full slip for these tests can also be attributable to this.

4.5   Maximum Velocity

At 82° initial strut angle for arched outersoles, for Neolite the average maximum velocity at the point of departure from the tread was 1,500 mm/sec for full slips and 720 mm/sec for partial slips and, for leather, it was 1,398 mm/sec for full slips and 555 mm/sec for partial slips. At 73°, for continuous outersoles, the average maximum velocity was 1220 mm/sec for full slips and 403 mm/sec for partial slips (there was no data for arched outersoles at 73°). The average maximum velocity of the high-heeled shoe (not shown) was 206 mm/sec.

It is evident that slip velocities for the continuous outersoles are considerably less than for the arched outersoles, especially for full slips, and less for the 73° strut angle than for the 82° strut angle.

5   Discussion

5.1   Arched Compared with Continuous Outersoles

The results here suggest that overstep slip risk is considerably greater for arched outersoles than for continuous outersoles, and that proximity of the FMH to the nosing, that is, a small FMH/Nsg distance, is much more instrumental in full slips for arched outersoles than for continuous outersoles.

Figure 4 indicates that full slips occurred when the FMH was from 20 mm to 50 mm posteriorly to the nosing. The full slips can be regarded as slips by the foresole that would have been partial slips if the FMH was further away from the nosing. This might also explain the full slips associated with the FMH/Nsg being anterior to the nosing; however, an alternative or contributing factor in this latter case might be the slope of the arched midsole.

The high correlation of slip velocity and slip distance can be expected because, as a slip progresses, the angle of the ground reaction force to the horizontal decreases and therefore also the force component parallel to the tread and the slip velocity increases.

The greater slip velocity of arched outersoles reinforces the foregoing conclusion of greater overstep slip risk of arched outersoles compared with continuous outersoles (except possibly for high-heeled shoes).

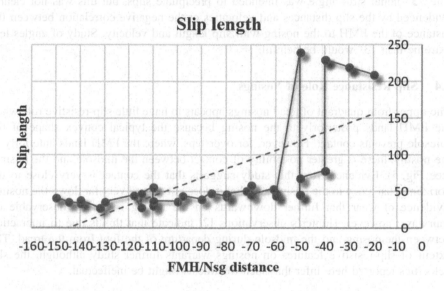

Fig. 4. Correlation of slip length and FMH distance from nosing FMH/Nsg)

5.2 Slip Length and Foot Declination

The role of the metatarsal heads and forefoot in partial slips is indicated by comparing slip length with the proportion of it represented by forefoot slip (corresponding with completion of foot declination—until which time body weight is borne by the forefoot and metatarsal heads).

Partial Slips

For continuous outersole shoes, the ratio of average slip length to average forefoot slip length is 87% for Neolite continuous outersoles, 75%, for Neolite arched outersoles, and 98% for leather arched outersoles. In other words, most of the slips occur before the heel engages the tread. It also indicates the heel's contribution to slip arrest (even if not protruberant).

Full Slips

For continuous outersole shoes, the ratio of average slip length to average forefoot slip length was 60% for Neolite continuous outersoles, 61%, for Neolite arched outersoles, and 48% for leather arched outersoles. In other words, most of the slips occur before the heel engages the tread. It also indicates the heel's contribution to slip arrest (even if they are not protruberant). The proportions are less than for partial slips. However, the motion graphs indicate that the heel plays a negligible slip retarding role as the foot slips off the tread.

5.3   Strut Angles

The 73° initial strut angle was intended to precipitate slips, but this was not clearly evidenced by the slip distances and velocities or the negative correlation between the distance of the FMH to the nosing with slip length and velocity. Study of angles less extreme than 73° would be helpful.

5.4   Slip Resistance Role of Nosings

The upper-front quadrant of tread nosings appears to have little slip-resistive role when the FMH lands posteriorly to the nosing because the typical convex shape of the foresole prevents contact. However, for oversteps, where the FMH lands anteriorly to the nosing, there *is* greater possibility of contact between the midsole and the nosing, (see Fig. 5). Evidence from this study suggests that the contact is very close to the horizontal tangency to the nosing and that it does not extend very far down the nosing. Evidence of wear that further downwards on the nosing is readily observable on stairways; however, Hunter's observations [2] indicate that this is due to interaction between the nosing and the midsole during departure of the foot from the tread. The extent of slip-resistive features on nosings warrants further study although, the slip velocities reported here infer that such treatments might be ineffectual.

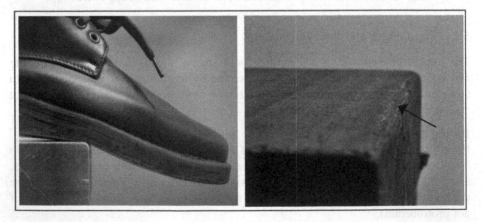

Fig. 5. Interaction of continuous outersole with upper-front quadrant of nosing (left photo); location of wear very near horizontal tangency of nosing (right photo)

5.5   Slip Arrest by Heels

Many tests for this study, resulted in slip arrest by the anterior edge of heels. The possible merit of this for fall avoidance might be obviated if the catching of the heel causes imbalance. There were also many instances of slips by continuous outersole shoes being arrested by the "perching" of the midsole on the tread nosing (see Fig. 5); this was a more gradual process and might be less likely to cause imbalance.

6   Conclusion

The results of this study suggest that, based on slip distance and velocity and the location of the first metatarsal heads with respect to the nosing, there is greater overstep slip risk for shoes with arched outersoles than for shoes with continuous (non-arched) outersoles. The findings also suggest that slip resistive attributes of the upper-front quadrant of nosings might help retard an overstep slip or the continuation of a slip beyond the nosing, but possibly only to a marginal extent, particularly in view of the high slip velocities of feet as they slip off the tread. The anticipated occurrence of overstep did not occur.

References

1. Hunter RA (2015) Loading profile of the landing foot on stair treads and their nosings during descent. In: Proceedings of the 19th triennial congress of the IEA, Melbourne, Australia
2. Hunter RA (2015) Foot and stair tread dynamics: a videographic study. In: Proceedings of the 19th triennial congress of the IEA, Melbourne, Australia
3. Hunter RA (2013) Nosing shape, nosing texture, tread texture and stairway slip resistance. In: Proceedings of the international conference of fall prevention and protection, 2013, National Institute of Occupational Health and Safety, Tokyo, Japan
4. Hunter RA (2011) Friction at tangential contact: in-situ slip testing of stair nosings. In: Proceedings of the international conference on slips, trips and falls, Health and safety Laboratory, Buxton, Derbyshire, UK

Increased Hand-Rung Force Is Associated with Increased Ladder Fall Risk

Erika M. Pliner and Kurt E. Beschorner

University of Pittsburgh, Pittsburgh, PA 15213, USA
beschorn@pitt.edu

Abstract. Falls from ladders are a frequent and severe injury. To better understand effective strategies to arrest a falling from ladder event, this study assessed 35 participants on their ability to arrest their body from falling after a ladder climbing perturbation. Peak hand-rung forces were compared with fall severity after an ascending climbing perturbation to assess if greater hand-rung force leads to reduced severity in the falling event. Severity of the falling event was measured by the weight supported in the safety harness (i.e. harness force). Higher harness forces were found to be associated with higher hand-rung forces. Hand-rung forces after a climbing perturbation do not reduce the severity in the falling event. Instead, this study suggests that hand-rung forces are caused by the severity in the falling event. This is critical knowledge because the hands are likely to decouple from the ladder if the severity of the falling event is greater than the grasping capability of the climber.

Keywords: Ladder falls · Hand-rung force · Hand decoupling

1 Introduction

Ladder falls are a global problem as indicated by multinational epidemiology reports [1–7]. In the occupational setting, ladders falls are the leading cause of fatal falls to lower levels [8]. Non-fatal ladder fall injuries are more severe than other Emergency Department treated injuries, resulting in 20 median days away from work [9]. A greater understanding on ladder falls is needed to create appropriate interventions that combat these frequent and severe injuries.

A climber can fall *with* a ladder or fall *from* a ladder. A fall *with* a ladder occurs when the ladder becomes unstable and the climber falls with the ladder. A fall *from* a ladder occurs when all limbs decouple from the ladder, causing the climber to fall from the ladder. Research has focused on preventing falls with ladders [10–12], but few studies have investigated fall from ladders. This is a critical gap in the literature because falls from ladders are just as frequent as falls with ladders [13].

After a ladder climbing perturbation (i.e. a foot decoupling event), the hands and feet attempt to reestablish contact with the ladder [14, 15]. These responses generate an upward force on the body to slow falling momentum and support the body [16].

© Springer Nature Switzerland AG 2019
S. Bagnara et al. (Eds.): IEA 2018, AISC 819, pp. 800–806, 2019.
https://doi.org/10.1007/978-3-319-96089-0_88

However, the relationship between the hand-rung force magnitude and fall severity is unknown. Higher hand-rung force may lower the severity of the fall. Alternatively, higher hand-rung force may be generated as a response to the severity of the falling event (i.e., higher hand-rung forces are generated in response to a more severe falling event). This study aims to investigate the relationship of hand-rung force with fall severity after an ascending ladder climbing perturbation.

2   Methods

2.1   Subjects

Thirty-five participants (24 ± 4 years of age; 75.5 ± 12.3 kg; 1.7 ± 0.1 m) were recruited for this study. Exclusion criteria consisted of musculoskeletal disorders, previous shoulder dislocations, osteoporosis/osteoarthritis, neurological/cognitive disorders, balance disorders or pregnancy. Approval was obtained by the Institutional Review Board at the University of Wisconsin-Milwaukee.

2.2   Experiment

Preparation. Participant anthropometric measurements were taken. Participants were equipped with athletic wear, standard shoes with a raised heel, shin guards and a safety harness. The safety harness was attached to a load cell (collection at 1 kHz) to measures the weight supported by the harness. Forty-seven reflective markers were placed on the participant's head (3 markers), torso (10 markers), arms (14 markers), legs (12 markers) and shoes (8 markers) and tracked by 13 motion capture cameras (collection at 100 Hz).

Custom-Built Ladder. A vertical, 12-foot ladder was custom-built in compliance with the US Occupational Safety and Health Administration (OSHA) standards. The rungs were 32 mm in diameter and spaced 205 mm apart [17]. The 8th and 9th rungs from the bottom of the ladder were each equipped with 4 uniaxial load cells (collecting at 2 kHz) to measure the applied horizontal and vertical forces (Fig. 1). The 4th rung (from the bottom) could be triggered to release when less than 5% of the participant's body weight remained on the previous rung. This simulated a ladder misstep and occurred at a point in time when a climber's foot is most likely to slip [14].

Procedures. Participants climb the ladder 30 times in both climbing directions (ascent,

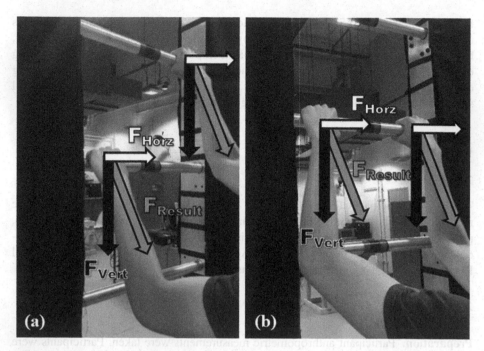

Fig. 1. Applied horizontal (F_{Horz}), vertical (F_{Vert}), and resultant (F_{Result}) force from the hands onto two (a) or one (b) rung(s).

descent), across three glove conditions (i.e. wearing no gloves, low friction gloves, high friction gloves). Participants were asked to climb at a comfortable but urgent pace to simulate the climbing speed of a regular-to-busy work day. Participants experienced 6 ladder climbing misstep perturbations (i.e. releasing of the fourth rung), one for each condition (2 climbing directions × 3 glove conditions). Order of perturbation was randomized. Three to six regular (unperturbed) climbs were performed prior to each perturbation to minimize perturbation anticipation. Only outcomes after ascending perturbations will be reported.

2.3   Analysis

Data Analysis

Fall Severity. Fall severity was quantified as the peak force supported by the safety harness between perturbation onset and end of perturbation (as determined by an algorithm that was previously described [18]). Harness force was normalized to body weight.

Hand-Rung Force. Hand-rung force was found for the moving hand (i.e. the hand that moved during/after the perturbation or was next to move in cases where the hand did not move), non-moving hand and combined hands. Classification of the moving and non-moving hand was kinematically determined from hand offsets and onsets of the 3rd metacarpal marker [15]. Individual hand-rung force (moving hand or non-moving hand) could only be found when one hand was in contact with the 8th or 9th rung (occurring 57 times for the moving hand and 71 times for the non-moving hand). Combined hand-rung force could only be found if both hands were grasping the 8th and/or 9th rung (occurring 63 times). Hand-rung force data was filtered using a second-order lowpass Butterworth filter with a cutoff frequency of 10 Hz [19]. The peak resultant (result) force was obtained after summing the forces from the horizontally (horz) and vertically (vert) mounted load cells per corresponding rung (Eq. 1). Peak resultant hand-rung force was normalized to body weight.

$$Result\ Force = \frac{\sqrt{\left(Horz_{left} + Horz_{right}\right)^2 + \left(Vert_{left} + Vert_{right}\right)^2}}{Body\ Weight} \tag{1}$$

Statistical Analysis
Two linear regression were performed with normalized harness force as the dependent variable. In the first linear regression, the normalized peak resultant hand-rung force of the moving hand and non-moving hand were entered into the model as the predictor variables. In the second linear regression, the normalized peak hand-rung force of the combined hands was entered into the model as the predictor variable. Square root transform was performed on normalized harness force to achieve normally distributed residuals. Normalized harness force and hand-rung forces were only assessed after an ascending perturbation. Effects of glove condition was not assessed because this has been previously reported [18, 20].

3 Results

The average (standard deviation) normalized harness force after an ascending perturbation was 0.17 (0.15). The average (standard deviation) normalized peak resultant hand-rung force of the moving, non-moving and combined hands after an ascending perturbation was 0.49 (0.23), 0.59 (0.16), and 1.07 (0.33), respectively. A higher normalized harness force was associated with a higher normalized peak hand-rung force of the non-moving ($p < 0.001$; $F = 14.580$) and combined ($p = 0.002$; $F = 10.414$) hands. Normalized peak hand-rung force of the moving hand increased with greater normalized harness force, but this was not significant ($p = 0.074$; $F = 3.321$) (Fig. 2).

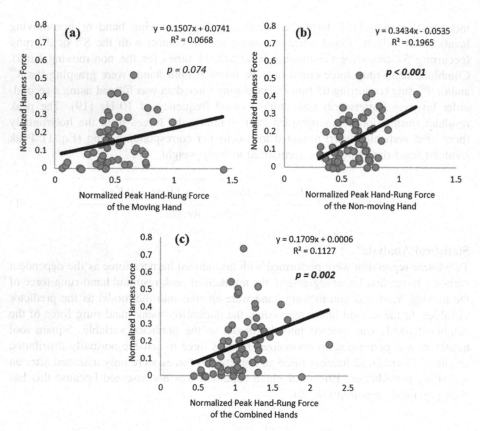

Fig. 2. Normalized harness force per normalized peak resultant hand-rung force for the moving (a), non-moving (b) and combined (c) hands after an ascending perturbation.

4 Discussion

This study investigated the relationship between peak resultant hand-rung forces and fall severity (i.e. harness force) after an ascending ladder climbing perturbation. We found increased hand-rung forces to not be associated with a reduction in fall severity. Instead, higher hand-rung forces were associated with a higher harness force. One explanation for this result is that the severity of the falling event may be influencing the observed hand-rung forces. Presumably, these hand-rung forces would increase until the hand decouples from the rung (referred to as the breakaway force [20–22]). The average hand-rung forces generated by each hand (49% to 59% of body weight) approach previously reported breakaway force values [20, 21], indicating that the hand is at risk of decoupling. Participants that do not reestablish their feet rely more on their upper body to arrest the falling event [16]. These participants will experience a ladder fall if their hand-rung forces surpass their grasping capabilities. Interventions that reduce the required upper body force to arrest a falling event is essential to prevent hand decoupling that leads to ladder fall outcomes. Easing the climber's ability to reestablish foot placement after a ladder climbing perturbation is a solution to reducing high hand-rung forces. Such interventions may be in the form of wider steps for lower rungs on the ladder to increase the climber's probability of foot-rung contact after a climbing perturbation.

There are limitations in this study. Hand-rung forces observed in this study may undermine the true hand-rung force experienced after a ladder climbing perturbation because participants were caught in a harness for safety reasons. In addition, only a horizontal, cylindrical ladder handhold was tested. Forces applied to other ladder handhold designs, shapes and orientations may vary with fall severity after a ladder climbing perturbation.

This study found higher normalized harness forces to be associated with higher hand-rung forces after a ladder climbing perturbation. The magnitude of hand-rung forces experienced after a ladder climbing perturbation is suggested to be caused by the severity of the falling event.

References

1. Ackland HM et al (2016) Danger at every rung: epidemiology and outcomes of ICU-admitted ladder-related trauma. Injury 47(5):1109–1117
2. Faergemann C, Larsen LB (2000) Non-occupational ladder and scaffold fall injuries. Accid Anal Prev 32(6):745–750
3. Häkkinen KK, Pesonen J, Rajamäki E (1988) Experiments on safety in the use of portable ladders. J Occup Accid 10(1):1–19
4. López MAC et al (2011) Occupational accidents with ladders in Spain: Risk factors. J Saf Res 42(5):391–398
5. Björnstig U, Johnsson J (1992) Ladder injuries: mechanisms, injuries and consequences. J Saf Res 23(1):9–18
6. Muir L, Kanwar S (1993) Ladder injuries. Injury 24(7):485–487
7. D'Souza AL, Smith GA, Trifiletti LB (2007) Ladder-related injuries treated in emergency departments in the United States, 1990–2005. Am J Prev Med 32(5):413–418
8. BLS (2012) Work-related fatal falls, by type of fall, 2010. R. D. census of fatal occupational injuries charts, U.S.D.o. Labor, Editor, Washington, DC
9. Socias CM et al (2014) Occupational ladder fall injuries-United States, 2011. MMWR Morb Mortal Wkly Rep 63(16):341–346
10. Chang C-C, Chang W-R, Matz S (2005) The effects of straight ladder setup and usage on ground reaction forces and friction requirements during ascending and descending. Saf Sci 43(7):469–483
11. Chang W-R, Chang C-C, Matz S (2004) Friction requirements for different climbing conditions in straight ladder ascending. Saf Sci 42(9):791–805
12. Hsiao H et al (2008) Extension-ladder safety: solutions and knowledge gaps. Int J Ind Ergon 38(11–12):959–965
13. Shepherd GW, Kahler RJ, Cross J (2006) Ergonomic design interventions–a case study involving portable ladders. Ergonomics 49(3):221–234
14. Schnorenberg AJ, Campbell-Kyureghyan NH, Beschorner KE (2015) Biomechanical response to ladder slipping events: effects of hand placement. J Biomech 48(14):3810–3815
15. Pliner EM et al (in review) Effects of upper body strength on ladder fall severity while controlling for hand and foot placement. Gait Posture

16. Pliner EM, Beschorner KE (2017) Applied hand-rung forces after a ladder perturbation. In: American Society of Biomechanics, Boulder, Colorado
17. OSHA, U.S.O.S.a.H.A. (2003) Stairways and ladders: a guide to OSHA rules, in US Department of Labor. Occupational Safety and Health Administration, Washington, DC
18. Pliner EM, Seo NJ, Beschorner KE (2017) Factors affecting fall severity from a ladder: Impact of climbing direction, gloves, gender and adaptation. Appl Ergon 60:163–170
19. Noé F, Quaine F, Martin L (2001) Influence of steep gradient supporting walls in rock climbing: biomechanical analysis. Gait Posture 13(2):86–94
20. Beschorner KE et al (2018) Effects of gloves and pulling task on achievable downward pull forces on a rung. Hum Factors 60(2):191–200
21. Young JG et al (2009) Hand-handhold coupling: effect of handle shape, orientation, and friction on breakaway strength. Hum Factors 51(5):705–717
22. Hur P, Motawar B, Seo NJ (2012) Hand breakaway strength model—effects of glove use and handle shapes on a person's hand strength to hold onto handles to prevent fall from elevation. J Biomech 45(6):958–964

Walkway Safety Evaluation and Hazards Investigation for Trips and Stumbles Prevention

Atena Roshan Fekr$^{(\boxtimes)}$, Gary Evans, and Geoff Fernie

Toronto Rehabilitation Institute—University Health Network,
University of Toronto, Toronto, Ontario, Canada
atena.roshanfekr@uhn.ca

Abstract. Falls are a major healthcare concern especially in the older population and tripping is a primary cause. Tripping is defined biomechanically as an event at which the lowest part of the foot makes unanticipated contact with either the walking surface or objects during the swing phase of the gait cycle. Identifying an obstacle or uneven surface as a tripping hazard is a subjective assessment, particularly when considering the vulnerabilities of people with knee/hip replacement, stroke, Parkinson disease and osteoarthritis. There is no universally accepted measurement tool or interpretation of the hazards. Walkways can pose a serious risk of tripping and falling because of different factors such as overloading, frost heave, and tree roots. The main goal of this study is to provide a new solution for extracting features of walkway tripping hazards that can be correlated with the Probability of Tripping (PoT). A new scanning device is proposed which uses an array of Time-of-Flight sensors. Before extracting the features we need to make sure the proposed device will provide us accurate measurements. Therefore, in this paper, we have addressed the main challenges associated with accurate sensor readout in a multi-sensory system. The results show that the presented scanning device will provide readouts with accuracy of ±2.5 mm.

Keywords: Falls · Trips · Stumbles · Time of Flight (ToF)

1 Introduction

Tripping during walking is the most common cause of falls in older adults [1], and the main reason of tripping is reduction in Minimum Foot Clearance (MFC) during walking. The MFC mostly occurs at mid-swing when the foot is at its maximum speed. MFC has been reported to be 1.29 ± 0.62 cm and 1.12 ± 0.5 cm for healthy young and old adults, respectively [2]. Therefore, even a very small unexpected change in the clearance between the feet and the walking surface will affect MFC and will increase the risk of tripping. The quality of sidewalks is critical in preventing trips and stumbles. The Americans with Disabilities Act [3] defined a trip hazard "as any vertical change of over 1/4 in. (6.4 mm) or more at any joint or crack". If the changes in level are between 1/4 and 1/2 in., then the hazard shall be beveled with a slope not steeper than 1:2. Changes in level exceeding 1/2 in. (13 mm) must comply with ramps or curb ramps

© Springer Nature Switzerland AG 2019
S. Bagnara et al. (Eds.): IEA 2018, AISC 819, pp. 807–815, 2019.
https://doi.org/10.1007/978-3-319-96089-0_89

guideline. In Canada, the Minimum Maintenance Standards (MMS) [4] were developed to provide municipalities with a defense against liability from actions arising with regard to levels of care on roads and bridges. In 2013, new sections are added which targeted the sidewalks in Ontario. Based on the amendment, a sidewalk needs to be repaired if the surface discontinuity (vertical discontinuity creating a step formation at joints or cracks) exceeds two centimeters" [4]. In addition, the minimum standard for the frequency of inspecting the sidewalks is once a calendar year. The sidewalk inspection is still a manual process which is performed by the inspectors who measure the height and depth of the hazards and determine if they exceed the MMS. According to the City of Hamilton report in 2016 [5], the inspection will cost about $60 K per year for the city and still the City's Risk Management department is concerned about the adequacy of inspection information. The main reason is that there exists no systematic way to monitor and scan the sidewalks. Cracked pavement, missed tiles, uneven and broken sidewalks are just a few common examples of outdoor walking surface hazards which cause injuries every year and still remain undetected. Walkway hazards are particularly dangerous since they are often difficult to see. The risk of not seeing the hazards can be increased because the people's attention might be distracted by diverse reasons such as texting, talking on the phone, holding a child's hand, or by simply looking at the merchandise in a store. In addition, the hazards may exist where there is not enough lighting at night, thus making the danger even more difficult to perceive.

Apparently, current sidewalk inspections are labor intensive and limited by human fallibility. Thus, automated detection of walkways hazards is required from both a safety and economic perspective. Automating the hazards detection however is not straightforward, whether in the form of an assistive or fully automated solution. In this paper, we have proposed a new cost effective portable scanning device for extracting different features from the hazards and eventually classify the hazards automatically.

Light Detection and Ranging (LiDAR) instruments have been used to extract the landscape characteristics [6–10], however based on a study in the US [8], the accuracy of LiDAR is typically 15 to 20 cm ($\sim$6 in.) and 1/3 to 1 m for vertical and horizontal measurements, respectively. Bathymetric LiDAR [9] which is used to measure water depth has a vertical accuracy of ±15 cm and horizontal accuracy of ±2.5 m. Lohani [9] also discussed the advantages of LiDAR technology compared to GPS, interferometry, and photogrammetry. The accuracy was reported as 5–15 cm for vertical and 30–50 cm for horizontal measurements. The error range of the laser scanning device was reported to be 5 to 200 cm in [10]. Albeit useful in topographic applications, we need more accurate devices for scanning sidewalks to be able to measure tripping hazards. Our idea is to use a 2-D array of Time of Flight (ToFs) sensors which is mounted on a portable base and will move and scan every 1000 mm $\times$ 200 mm area (with 20 ToFs). The ToFs provide accurate distance measurement with 1 mm resolution. After building the device, we will have a 3D imaging of the hazards to extract the important features such as average depth, maximum change, variation, length of the hazard, maximum slope in that particular hazard, etc. There are different challenges in using the ToFs next to each other to monitor the sidewalks. In this paper, we have addressed the main challenges and provide solutions to keep the accuracy high enough for extracting the relevant features of the hazards.

2 System Design

The VL53L0X is the smallest ToF laser-ranging module from STMicroelectronics which provides sufficiently accurate distance measurement. It can detect the "time of flight", or how long the light has taken to travel from the sensor to the pavement and back. This sensor is widely used in different applications such as robotics, area mapping and gesture recognition [12]. The sensor measures 50 mm to 1200 mm of range distance with Field Of View (FOV) 25°. The distance is estimated by measuring the phase shift φ between the emitted signal $s(t)$ and the reflected signal $r(t)$ presented in Fig. 1. The emitted and reflected signals are as follows:

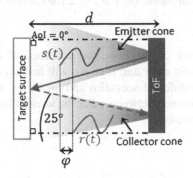

Fig. 1. ToF measurement based on calculating the phase shift φ between the emitted and reflected signals

$$s(t) = \sin(2\pi f_m t),$$

$$r(t) = R.\sin(2\pi f_m t - \varphi) = R.\sin\left[2\pi f_m \left(t - \tfrac{2d}{c}\right)\right], c = 3 \times 10^8 m/s$$

Where R is a reflection coefficient, c is the speed of light, f_m is the modulation frequency of $s(t)$ and $r(t)$ [6]. The distance d can be calculated from the following equation:

$$d = \frac{c}{2\pi f_m}.\varphi$$

In continuous measurement, the sampling rate is 50 Hz with ±5% error; however in the single mode the sensor data is sampled every 200 ms with less than ±3% error [11]. The ToFs communicate with the microcontroller via I²C. We have used MappyDot [12] to uniquely address 4 devices and to read the sensors either in single or continuous mode. The data of four ToFs is collected by an Arduino UNO as the I²C master shown in Fig. 2. In this paper, we will consider two main challenges in designing our sidewalk scanning device. First the accuracy of each individual sensor is determined and second the interference which occurs due to the small separation between the sensors in our design is measured.

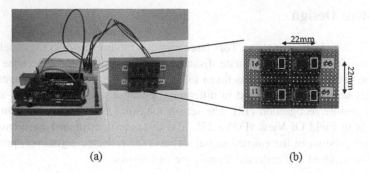

Fig. 2. (a) The hardware components, (b) The 2 × 2 ToFs sensors with unique addresses 0x08, 0x09, 0xA and 0xB

We will show how we can increase the accuracy of the individual sensor by compensating the systematic error and then we will investigate the interference effect on the sensor readouts in different scenarios and finally we will demonstrate that it is necessary to avoid interference between adjacent sensors by operating them sequentially rather than simultaneously.

3 Calibration

Calibration is a critical step in improving the accuracy of individual sensors in a multi-sensor system. Certain systems might require dedicated calibration procedures, which in general can be categorized under online or offline methods. Offline methods are mostly based on curve fitting, such as, the Least-Square (LS) method, to map raw sensor readings into corrected values [13] and compensate the systematic offset and gain. Note that when the gain and offset are both constant values and independent from the sensor measurements, then the calibration is translated into a linear curve-fitting function [14, 15]. In this paper, we have calibrated each sensor by considering the application of sidewalk scanning. We have used a box with a white object shown in Fig. 3. The box is put on a flat surface and the module is placed on a hole created on a white foam core material which restricts the ambient light inside the box. To minimize the impact of misalignment and keep the Angle of Incidence (AoI) close to 0°, a spirit level is used both inside the box and on the white foam core where the reading on the bubble tube is considered. Each sensor is calibrated using the least-square-method and a linear curve-fitting process with respect to a reference distance. Although the VL53L0X has a high immunity to ambient light [11], the experiment is performed with low ambient light at room temperature (23 °C). For our purpose, we assume that the maximum scanning height of the hazards is 300 mm. Therefore, a series of static acquisitions was performed by the ToFs using a white object varying the distance (d) in the range 50–300 mm with an increment of 50 mm. In each step (50 mm, 100 mm, 150 mm, 200 mm, 250 mm and 300 mm), 200 samples were collected from each sensor with a sampling rate of 5 Hz and the data was separated into two sets, called the training (70%) and the testing (30%) sets.

Fig. 3. The calibration setups

The linear curve fitting was done using the training set only. Then the model was used to calibrate the output values for the data in the testing set. The calibration equation for each sensor is presented in Fig. 4(a)–(d). It is worth noting that after calibration most of the data converges to the reference distances as shown in Fig. 4(e)–(h). Figure 4(i)–(l) show the box plots of the absolute error values before and after calibration. For each static acquisition, defined by specific values of d, the following quantities were computed:

$$Mean\,Absolute\,Error\,(MAE) = \frac{1}{N}\sum_{k=1}^{N}\left|d_{ref} - d_{ToF}\right|,\ MaxAE = \bigvee_{N}^{k=1}\left|d_{ref} - d_{ToF}\right|$$

$$Mean\,Error\,(ME) = \frac{1}{N}\sum_{k=1}^{N}(d_{ref} - d_{ToF}),\ SD = \sqrt{\frac{1}{N-1}\sum_{k=1}^{N}(d_{ref} - d_{ToF})^2}$$

As shown in Table 1, the maximum error after calibration decreases 81.70% on average while the average MAE% decreases from 6.88% to 1.37%. The standard deviations also show that the sensor readouts after calibration are more stable and reliable. Figure 5 demonstrates the distribution of the error values for all sensors and all 6 distances with/without calibration. It is observed that the errors are very close to fitting a normal distribution after calibration and therefore the mean error will properly represent the performance of the method which is almost zero (−0.02 mm). Another observation is that the error before calibration is kept increasing systematically by the distance shown in Fig. 6(e)–(h).

4 Discussion

The sensors are able to operate in both single and continuous modes. If multiple sensors operate simultaneously on the same frequency in the continuous mode, they interfere with each other. This interference depends on the distance of the object from the sensor and can cause large errors in depth measurement. In the proposed design the distance between the sensors is 22 mm, therefore considering the FOV of 25°; theoretically the system will suffer from interference if the detected object is further than

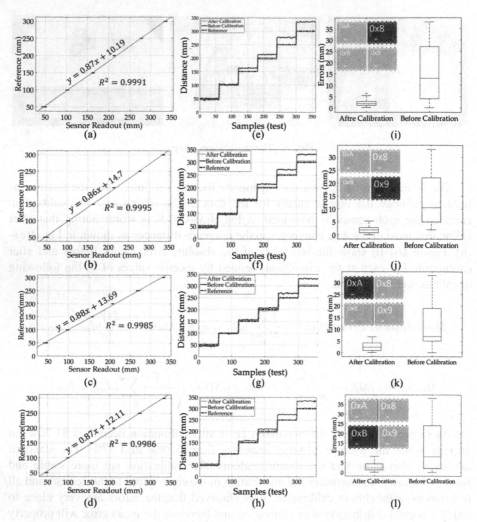

Fig. 4. (a)–(d) The training dataset versus the reference values and the linear curve fittings with the gain offset values, (e)–(h) the sensor readouts before and after calibration, (i)–(l) the box plot of the absolute errors before and after calibration

49.6 mm. Figure 7 shows the distance of the objects versus the overlap distance between two adjustment sensors.

In fact, we are able to reduce the interference by increasing the distance between the sensors; however the main objective is to scan the area and obtain the tripping hazards' features without sacrificing any resolution. In addition, we are not able to make any assumption about the height of the hazards, and different sensors in the array might read different parts of the hazards with various heights.

Figure 8(a) shows the effect of the interference on readouts of sensor 0x08. We have collected data from different combinations of the sensors listed in this figure.

Table 1. The errors before and after calibration for all 4 sensors

Sensor	MaxAE(mm)		MAE(mm)		MAE%		ME(mm) ± SD(mm)	
	Before calibration	After calibration	Before calibration	After calibration	Before calibration	After calibration	Before calibration	After calibration
0x08	38	6	16	2	8%	1.5%	−15 ± 13	−1 ± 3
0x09	33	5	14	2	7.5%	2%	−10 ± 14	1 ± 3
0xA	33	7	12	3	6%	2%	−9 ± 12	1 ± 2
0xB	38	8	13	3	6%	4%	−12 ± 13	0 ± 2

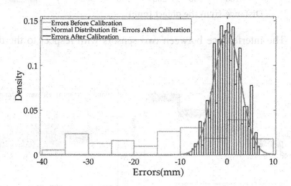

Fig. 5. The error distributions before and after calibration for all sensors

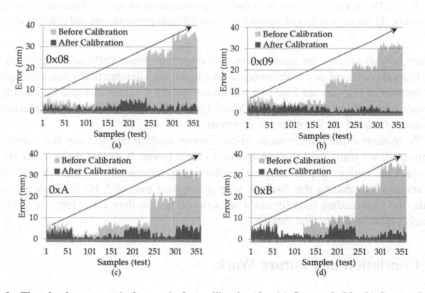

Fig. 6. The absolute errors before and after calibration for (a) Sensor 0x08, (b) Sensor 0x09, (c) Sensor 0xA, (d) Sensor 0xB

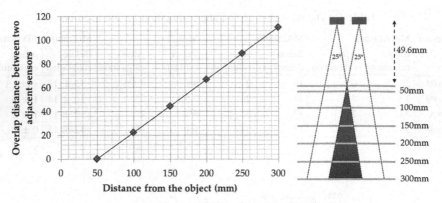

Fig. 7. The interference between two sensors with respect to the distance

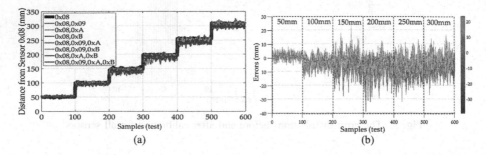

Fig. 8. (a) The distance readouts from 0x08 in continuous mode and in different combinations of sensors, (b) the error values of sensor 0x08 with different combinations and different distances

Figure 8(b) also shows the error values with respect to sensor 0x08 when operating in a single mode (without any interference). This figure also demonstrates that although the environmental impacts are kept to a minimum, the error values increase by moving the sensors further from the object. The Mean Error (*ME*) caused by the light interference increases from -1 ± 3 mm to -3 ± 6 mm.

To overcome the interferences in continuous reading and to keep the complexity minimized, we must read the sensors in a single mode. It means the microcontroller activates the sensors sequentially and performs the measurements, individually. Since the sensors are reading the distance with a sampling rate of 5 Hz in 'high accuracy' mode, the total reading time for our 4 sensors will be less than 1 s, which we think will be acceptable in a practical device.

5   Conclusion and Future Work

In this paper, we proposed a new device for measuring different features of sidewalks' tripping hazards. We discussed two important concerns for using ToF sensors in the design. The proposed calibration algorithm could provide results with an error of

±2.5 mm. The results also showed that parallel use of several ToFs with small separation may lead to an interference problem. Therefore, for this application, we must switch into single mode readout to prevent any interference between two adjacent sensors. The total reading time for our 4 sensors will be less than 1 s. In the future, we will extend our system into an array of 20 ToFs in which a scanning sequence will be used that avoids simultaneous measurements from sensors that are too closely spaced but results in an overall scanning time that is acceptable for a practical device. Choosing the best sequence of the sensors will be a function of the height of the device, as well. Our goal is to build a low-cost and portable device which can eventually be used by sidewalks inspectors for measuring the features of tripping hazards with high accuracy in a reasonable time.

References

1. Blake AJ, Morgan K, Bendall MJ, Dallosso H, Ebrahim SB, Arie TH et al (1988) Falls by elderly people at home prevalence and associated factors. Age Ageing 17:365–372
2. Winter DA (1992) Foot trajectory in human gait: a precise and multifactorial motor control task. Phys Ther 72:45–56
3. http://www.ada-compliance.com/ada-compliance/303-changes-in-level
4. Minimum Maintenance Standards for Municipal Highways Policy. Ontario Regulation 239/02.   http://www.selwyntownship.ca/en/townshiphall/resources/Council_Agenda_September_30_2014/4._a_Attch_-_Minimum_Maintenance_Policy.pdf
5. Digital Sidewalk Inspection - Municipal Engineers Association (MEA) Workshop (2016)
6. Bertuletti S, Cereatti A, Comotti D, Caldara M, Della Croce U (2017) Static and dynamic accuracy of an innovative miniaturized wearable platform for short range distance measurements for human movement applications. Sensors 17:1492
7. Glenn NF, Streutker DR, Chadwick DJ, Thackray GD, Dorsch SJ (2006) Analysis of LiDAR-derived topographic information for characterizing and differentiating landslide morphology and activity. Geomorphology 73:131–148
8. Light Detection and Ranging (LiDAR). http://web.pdx.edu/~jduh/courses/geog493f12/Week04.pdf
9. Lohani B (2008) Airborne altimetric LiDAR: principle, data collection, processing and applications. Note, IIT Kanpur, India
10. Huising EJ, Gomes Pereira LM (1998) Errors and accuracy estimates of laser data acquired by various laser scanning systems for topographic applications. ISPRS J Photogr Remote Sens 53(5):245–261
11. World's smallest Time-of-Flight ranging and gesture detection sensor. STMicrosystems
12. https://sensordots.org/mappydot
13. Feng J, Qu G, Potkonjak M (2004) Sensor calibration using non parametric statistical characterization of error models. In: Proceedings of IEEE Sensors, vol 3, pp 1456–1459
14. Roshan Fekr A, Janidarmian M, Radecka K, Zilic Z (2014) Multi-sensor blind recalibration in mHealth applications. In: 2014 IEEE Canada International Humanitarian Technology Conference - (IHTC), pp 1–4
15. Roshan Fekr A, Janidarmian M, Sarbishei O, Nahil B, Radecka K, Zilic Z (2012) MSE minimization and fault-tolerant data fusion for multi-sensor systems. In: IEEE 30th international conference on computer design (ICCD), pp 445–452

Author Index

Printed in the United States
By Bookmasters